Continuity and Discontinuity in the Peopling of Europe

Vertebrate Paleobiology and Paleoanthropology Series

Edited by

Eric Delson
Vertebrate Paleontology, American Museum of Natural History,
New York, NY 10024, USA
delson@amnh.org

Eric J. Sargis
Anthropology, Yale University
New Haven, CT 06520, USA
eric.sargis@yale.edu

Focal topics for volumes in the series will include systematic paleontology of all vertebrates (from agnathans to humans), phylogeny reconstruction, functional morphology, Paleolithic archaeology, taphonomy, geochronology, historical biogeography, and biostratigraphy. Other fields (e.g., paleoclimatology, paleoecology, ancient DNA, total organismal community structure) may be considered if the volume theme emphasizes paleobiology (or archaeology). Fields such as modeling of physical processes, genetic methodology, nonvertebrates or neontology are out of our scope.

Volumes in the series may either be monographic treatments (including unpublished but fully revised dissertations) or edited collections, especially those focusing on problem-oriented issues, with multidisciplinary coverage where possible.

For other titles published in this series, go to
www.springer.com/series/6978

Continuity and Discontinuity in the Peopling of Europe

One Hundred Fifty Years of Neanderthal Study

Proceedings of the international congress to commemorate "150 years of Neanderthal discoveries, 1856-2006", organized by Silvana Condemi, Wighart von Koenigswald, Thomas Litt and Friedemann Schrenk, held at Bonn, 2006, Volume I

Edited by

Silvana Condemi

UMR 6578 CNRS / Université de la Méditerranée/EFS, Laboratoire d'Anthropologie bioculturelle, Faculté de Médecine – Secteur Nord, Université de la Méditerranée, CS80011, Bd Pierre Dramard, 13344 Marseille Cedex 15, France

Gerd-Christian Weniger

Stiftung Neanderthal Museum, Talstrasse 300, D-40822 Mettmann, Germany

Springer

Editors
Silvana Condemi
Faculté de Médecine – Secteur Nord
UMR 6578
CNRS/Université de la Méditerranée/EFS
Laboratoire d'Anthropologie bioculturelle
CS80011, Bd Pierre Dramard
13344 Marseille Cedex 15
France
silvana.condemi@univmed.fr

Gerd-Christian Weniger
Stiftung Neanderthal Museum
Talstrasse 300, D-40822 Mettmann
Germany
weniger@neanderthal.de

ISSN 1877-9077 e-ISSN 1877-9085
ISBN 978-94-007-0491-6 e-ISBN 978-94-007-0492-3
DOI 10.1007/978-94-007-0492-3
Springer Dordrecht Heidelberg London New York

Library of Congress Control Number: 2011920687

Cover figure: The 16 fossil remains of the Neanderthal 1 fossil from 1856. Copyright holder Neanderthal Museum/S. Pietrek.

Printed on acid-free paper

Springer is part of Springer Science+Business Media (www.springer.com)

In memoriam Professor Francis Clark Howell

Conference Photos

Photos taken at the Congress: 150 Years of Neanderthal Discoveries: Early Europeans – Continuity & Discontinuity, held at the Rheinisches LandesMuseum in Bonn, Germany on July 21, 2006, taken by S. Condemi. Two photos from the 1956 Neanderthal Centenary conference, courtesy W. von Koenigswald (left, G.H.R. von Koenigswald with Indonesian crania, his daughter and nephew Wighart; right, S. Sergi G.H.R. and Wighart von Koenigswald).

Contents

Contributors

James C. M. Ahern
Department of Anthropology (3431), University of Wyoming,
1000 E. University Ave, Laramie, WY 82071, USA
jahern@uwyo.edu

Juan Luis Arsuaga
Departamento de Paleontología, Universidad Complutense de Madrid,
Avenida Complutense s/n, 28040 Madrid, Spain;
and
Centro UCM-ISCIII de Investigación sobre, Evolución y Comportamiento Humanos,
c/sinesio Delgado, 4, 28029 Madrid, Spain
jlarsuaga@isciii.es

Gal Avishai
Laboratory of Bio-Anthropology and Ancient-DNA, Institute of Dental Sciences,
Hebrew University-Hadassah, 91120 Jerusalem, Israel
avishai@cc.huji.ac.il

Shara Bailey
Department of Anthropology, New York University, 25 Waverly Place,
New York, NY 10003, USA
sbailey@nyu.edu

José María Bermúdez de Castro
Centro Nacional de Investigación sobre la Evolución Humana (CENIEH),
Burgos, Spain
josemaria.bermudezdecastro@cenieh.es

David Caramelli
Dipartimento di Biologia Evoluzionistica, Laboratori di Antropologia,
via del Proconsolo 12, 50122 Firenze, Italy
david.caramelli@unifi.it

José Miguel Carretero
Universidad de Burgos, Edificio I+D, Burgos, Spain
jmcarre@ubu.es

Rachel Caspari
Department of Sociology, Anthropology and Social Work, Central Michigan University,
Mt. Pleasant, MI 48859, USA
caspa1r@cmich.edu

Silvana Condemi
Laboratoire d'Anthropologie bioculturelle, Faculté de Médecine – Secteur Nord, Université de la Méditerranée, CS80011, Bd Pierre Dramard, Marseille Cedex 15 13344, France
silvana.condemi@univmed.fr

Christine Couture
Lab. d'Anthropologie des Populations du Passé, UMR 5199 CNRS, Université Bordeaux 1, 33405 Talence Cedex, France
c.couture@pacea.u-bordeaux1.fr

Love Dalén
Centro UCM-ISCIII de Investigación sobre, Evolución y Comportamiento Humanos, c/Sinesio Delgado, 4, 28029 Madrid, Spain
ldalen@isciii.es

Joan Daura
Grup de Recerca del Quaternari, SERP–Seminari d'Estudis i Recerques Prehistòriques, Dpto. de Prehistoria, Historia Antiga i Arqueologia, Universitat de Barcelona, c/Montalegre, 6, 08001 Barcelona, Spain
grupquaternari@hotmail.com

Brajković Dejana
Croatian Academy of Sciences and Arts, Institute for Quaternary Paleontology and Geology, Kovačića 5 10000, Zagreb, Croatia
dejbraj@hazu.hr

Yankel Gabet
Institute of Genetic Medicine, Kech School of Medicine, University of Southern California, 2250 Alcazar Street, IGM/CSC 240 L.A. 90033, USA
gabet@usc.edu

María Dolores Garralda
UMR 5199 Laboratoire d'Anthropologie des Populations du Passé, Université de Bordeaux, 33405 Talence, France;
U.D. de Antropología Física, Facultad de Biología, Universidad Complutense de Madrid, 28040 Madrid, Spain
mdgarral@bio.ucm.es

Mica Glantz
Department of Anthropology, Colorado State University, Ft. Collins, CO 80523, USA
mglantz@lamar.colostate.edu

Aida Gómez-Robles
Centro Nacional de Investigación sobre la Evolución Humana (CENIEH), Burgos, Spain
aida.gomez@cenieh.es

Naama Goren-Inbar
The Hebrew University of Jerusalem, Mount Scopus, Jerusalem 91905, Israel
goren@cc.huji.ac.il

Anders Götherström
Centro UCM-ISCIII de Investigación sobre, Evolución y Comportamiento Humanos, c/Sinesio Delgado, 4, 28029 Madrid, Spain;
Department of Evolutionary Biology, Uppsala University, Norbyvägen 18D, SE-75236 Uppsala, Sweden
Anders.Gotherstrom@ebc.uu.se

Philipp Gunz
Department of Human Evolution, Max Planck Institute for Evolutionary Anthropology,
Deutscher Platz 6, D-04103 Leipzig, Germany
gunz@eva.mpg.de

Catherine Hänni
Institut de Génomique Fonctionnelle de Lyon, IFR 128
Biosciences Lyon-Gerland, CNRS UMR 5242, INRA, UCB Lyon 1, Ecole Normale Supérieure de Lyon, Université de Lyon, 46 Allée d'Italie, 69364 Lyon Cedex 07, France
catherine.hanni@ens-lyon.fr

Thorolf Hardt
Institute of Anthropology (1050), Department of Biology (FB 10), Johannes Gutenberg-University Mainz, D-55099 Mainz, Germany
thorolf-hardt@gmx.de

Katerina Harvati
Department of Early Prehistory and Quaternary Ecology, and
Senckenberg Center for Human Evolution and Paleoecology,
Eberhard Karls Universität Tübingen, Rümelinstrasse 23, 72070 Tübingen, Germany
katerina.harvati@ifu.uni-tuebingen.de

Winfried Henke
Institute of Anthropology (1050), Department of Biology (FB 10),
Johannes Gutenberg-University Mainz, D-55099 Mainz, Germany
henkew@uni-mainz.de

Francis Clark Howell
University of California, Berkeley, USA

Jean-Jacques Hublin
Department of Human Evolution, Max Planck Institute for Evolutionary Anthropology,
Deutscher Platz 6, D-04103 Leipzig, Germany
hublin@eva.mpg.de

Ivor Janković
Institute for Anthropological Research, Amruševa 8, 10000 Zagreb, Croatia
ivor@inantro.hr

Olaf Jöris
Forschungsbereich Altsteinzeit des, Römisch-Germanisches Zentralmuseums,
Schloß Monrepos, D-56567 Neuwied, Germany
joeris@rgzm.de

Ivor Karavanić
Department of Archaeology, University of Zagreb,
I. Lučića 3, 10000 Zagreb, Croatia
ikaravan@ffzg.hr

Wighart von Koenigswald
Institut für Paläontologie der Universität Bonn, Nussallee 8, D-53113 Bonn, Germany
koenigswald@uni-bonn.de

Carles Lalueza Fox
Institute of Evolutionary Biology (CSIC-UPF), Barcelona (Spain),
Barcelona, Spain
carles.lalueza@upf.edu

Jadranka Mauch Lenardić
Croatian Academy of Sciences and Arts, Institute for Quaternary Paleontology and Geology, A. Kovačića 5, 10000 Zagreb, Croatia
jml@hazu.hr

Wu Liu
Institute of Vertebrate Paleontology and Paleoanthropology, Chinese Academy of Sciences, Beijing 100044, China
liuwu@ivpp.ac.cn

Roberto Macchiarelli
Dép. de Préhistoire, UMR 5198, MNHN, 1, rue René Panhard, 75013 Paris, France;
Dép. de Géosciences, Université de Poitiers, 40 av. du Recteur Pineau, 86022 Poitiers, France
roberto.macchiarelli@univ-poitiers.fr

Francesco Mallegni
Dipartimento di Biologia, Università degli Studi di Pisa, Via S. Maria 55, 56126, Pisa, Italy
fmallegni@biologia.unipi.it

Ann Margvelashvili
Centro Nacional de Investigación sobre la Evolución Humana (CENIEH), Burgos, Spain

Ignacio Martinez
Instituto de Salud Carlos III, Madrid, Spain
imartinezm@isciii.es

María Martinón-Torres
Centro Nacional de Investigación sobre la Evolución Humana (CENIEH), Burgos, Spain
maria.martinon.torres@gmail.com

Lucio Milani
Dipartimento di Biologia Evoluzionistica, Laboratori di Antropologia, via del Proconsolo 12, 50122 Firenze, Italy
lucio.milani@unifi.it

Ralph Müller
ETH Zurich, Institute for Biomechanics, HCI E 357.2, Wolfgang-Pauli-Str., 108093 Zürich, Switzerland
ram@ethz.ch

Elisabeth Nicholson López
Northwestern University, Chicago, IL 60611-3008, USA
lisaknlopez@gmail.com

Anthony J. Olejniczak
Department of Human Evolution, Max Planck Institute for Evolutionary Anthropology, Deutscher Platz 6, D-04103 Leipzig, Germany
Centro Nacional de Investigación sobre Evolucion Humana (CENIEH)
Paseo de la Sierra de Atapuerca, s/n Burgos, Spain
anthony.olejniczak@gmail.com

Ludovic Orlando
Centre for GeoGenetics, Natural History Museum of Denmark, University of Copenhagen, Oster Voldgade 5-7, Kobenhavn K 1350, Denmark
Lorlando@snm.ku.dk

Rolf Quam
Centro UCM-ISCIII de Investigación sobre Evolución y Comportamiento Humanos, c/Sinesio Delgado, 4, 28029 Madrid, Spain;
Division of Anthropology, American Museum of Natural History, Central Park West at 79th St., New York, NY 10024-5192, USA
rquam@amnh.org

Donald J. Reid
Department of Oral Biology School of Dental Sciences, University of Newcastle upon Tyne, Framlington Place NE2 4BW, Newcastle upon Tyne, UK
d.j.reid@ncl.ac.uk

Montserrat Sanz
Grup de Recerca del Quaternari, SERP–Seminari d'Estudis i Recerques Prehistòriques, Dpto. de Prehistoria, Historia Antiga i Arqueologia, Universitat de Barcelona, c/Montalegre, 6, 08001 Barcelona, Spain
grupquaternari@hotmail.com

Susana Sarmiento
Centro Nacional de Investigación sobrela Evolución Humana (CENIEH), Burgos, Spain
susanasarm@gmail.com

Nandini Singh
Department of Early Prehistory and Quaternary Ecology, and Senckenberg Center for Human Evolution and Paleoecology, Eberhard Karls Universität Tübingen, Rümelinstrasse 23, 72070 Tübingen, Germany
singhnandini@gmail.com

Fred H. Smith
Department of Sociology and Anthropology, Illinois State University, IL, USA
fsmith@ilstu.edu

Patricia Smith
Laboratory of Bio-Anthropology and Ancient-DNA, Institute of Dental Sciences, Hebrew University-Hadassah, 91120 Jerusalem, Israel
pat@cc.huji.ac.il

Tanya M. Smith
Department of Human Evolutionary Biology, Harvard University, 11 Divinity Ave, Cambridge, MA 02138, USA;
Department of Human Evolution, Max Planck Institute for Evolutionary Anthropology, Deutscher Platz 6, D-04103 Leipzig, Germany
tsmith@fas.harvard.edu

Roscoe Stanyon
Dipartimento di Biologia Evoluzionistica, Laboratori di Antropologia, via del Proconsolo 12, 50122 Firenze, Italy
roscoe.stanyon@unifi.it

Martin Street
Forschungsbereich Altsteinzeit des, Römisch-Germanisches Zentralmuseum, Schloß Monrepos, D-56567 Neuwied, Germany
street@rgzm.de

Maria Eulàlia Subira
Unitat d'Antropologia, Departament de Biologia Animal, de Biologia Vegetal i d'Ecologia, Edifici C, Universitat Autònoma de Barcelona, 08193, Bellaterra (Cerdanyola del Vallès) Barcelona, Spain
Eulalia.Subira@uab.es

Ian Tattersall
Division of Anthropology, American Museum of Natural History, Central Park West at 79th Street, New York, NY 10024, USA
iant@amnh.org

Thomas Terberger
Lehrstuhl für Ur- und Frühgeschichte, Historisches Institut, Ernst Moritz Arndt Universität Greifswald, Hans-Fallada-Strasse 1, D-17489 Greifswald, Germany
terberge@uni-greifswald.de

Anne-Marie Tillier
Laboratoire d'Anthropologie des Populations du Passé, UMR 5199-PACEA, Université Bordeaux 1, avenue des Facultés, 33405 Talence, France
am.tillier@pacea.u-bordeaux1.fr

Erik Trinkaus
Department of Anthropology, Washington University, Campus Box 1114, St. Louis, MO 63130, USA
trinkaus@wustl.edu

Bernard Vandermeersch
UMR 5199 Laboratoire d'Anthropologie des Populations du Passé, Université de Bordeaux, 33405 Talence, France
bvanderm@bio.ucm.es

Bence Viola
Department of Anthropology, University of Vienna, Althanstrasse 14, A-1091 Vienna, Austria
bence.viola@eva.mpg.de

Jean-Luc Voisin
USM 103, UMR 7194, Institut de Paléontologie Humaine, 1 rue René Panhard, 75013 Paris, France;
UMR 6578, Université de la Méditerranée Aix-Marseille 2, Faculté de Médecine, CS80011, Boulevard Pierre Dramard, 13344 Marseille Cedex 15, France
jeanlucv@mnhn.fr

Virginie Volpato
Abteilung Paläoanthropologie und Messelforschung, Forschungsinstitut und Naturmuseum, Senckenberganlage 25, 60325 Frankfurt am Main, Germany
vvolpato@senckenberg.de

Gerd-Christian Weniger
Stiftung Neanderthal Museum, Talstrasse 300, D-40822 Mettmann, Germany
weniger@neanderthal.de

Bernhard Weninger
Institut für Ur- und Frühgeschichte, Universität zu Köln, ^{14}C-Labor, Weyertal 125,

D-50923 Köln, Germany
b.weninger@uni-koeln.de

Ulrich Witzel
Research Group of Biomechanics, Ruhr-University Bochum, D 44780 Bochum, Germany
Ulrich.Witzel@ruhr-uni-bochum.de

Milford H. Wolpoff
Department of Anthropology, University of Michigan, Ann Arbor, MI 48109-1092, USA
wolpoff@umich.edu

Xiujie Wu
Institute of Vertebrate Paleontology and Paleoanthropology, Chinese Academy of Sciences, Beijing 100044, China
wuxiujie@ivpp.ac.cn

João Zilhão
ICREA Research Professor, University of Barcelona, Departament de Prehistòria, Història Antiga i Arqueologia – Seminari d'Estudis i Recerques Prehistòriques, C/ Montalegre 6, 08001 Barcelona, Spain
joao.zilhao@icrea.cat

Chapter 1
Neanderthals and Emergent Palaeoanthropology 50 Years Ago[1]

Opening address to the congress, 150 Years of Neanderthal Discoveries: Early Europeans – Continuity & Discontinuity, Rheinisches LandesMuseum, Bonn, Germany, July 21, 2006

F. Clark Howell

I know why you're here, the question is why am I here and I'm here because the organizers, let us say unanimously, said "would you please come to Bonn for the 150 years of Neanderthal?", and it took me a few minutes, maybe 6 or 7 seconds to say okay, and then they said what they would like me to do, and I said well, yeah, I really can do what I want, which is mostly what I do anyway, so in any case I came for several reasons. Not only because I happened to be here as Gabriele Uelsberg, the LandesMuseum Director, said for the first time in 1956, the first time I was in Europe was 1953, the same year that I received a doctorate, although I never stayed for the ceremony and they had to send it to my parents, because I was overseas, in London at Chris Stringer's museum, and Chris had just arrived I think recently, and you'll hear about that in some of the things I'm going to say to you. I'm going to talk to you about the 1950s, which is 50 years ago, so if you're under fifty there are some things that you might benefit from, from what I say. If you're over that, there are some things you might say, "well that's interesting, I really didn't know about that "or "that's not the way I heard it" or something like that. All that's possible. I really came because of something else. I was in Bonn in 1956. I was in Europe, all over with my new wife of 1 year, all over Western Europe, beginning in London and ending back in London, and all over Western Europe, Southern Europe, Central Europe and so on in a great big swoop over 3 months, looking at Neanderthal folks that I hadn't seen already in 1953. In 1953, they were all seen either in the British Museum (Natural History), now The Natural History Museum, let us not forget, and then in Paris.

And in Paris I was able to see things at the end of August; now you know if you go to Paris and you want to see something in a museum and it's not a public museum and it's the end of August, you can forget it. But I had connections, and these connections worked, I have had connections most of my life and mostly they work. I don't know why that is, serendipity, whatever it is, it's fantastic, so far so good, and in any case I was allowed to see things at the Institut de Paléontologie Humaine, including unpublished things, and including things that later were made into a fantastic dissertation and lots of other things by my colleague Bernard Vandermeersch, who's right here tonight, and that was because of Henri-Victor Vallois. I had connections to him, and he was extremely kind to me. I think one of the reasons is he never had any children, and I was a young man, you know, he thought young investigator, I don't think he had many students, frankly, anyway, and I was a young anatomist and he was an anatomist originally at Toulouse and so on and he thought, hmm why not. He was also the director of the Musée de l'Homme, at the same time as being the director of the IPH and the editor of *L'Anthropologie* along with Raoul Vaufrey, and he said "you can see anything you want in the Musée de l'Homme." Now Silvana Condemi is here and she can tell you, it's really tough to see anything, anywhere at the Musée de l'Homme sometimes, and I won't go into that any further, but it is, and everything was carte blanche for me. I cannot believe it, and I couldn't believe it then, I just took it for granted. Subsequently I learned that it wasn't so simple. But every time when I came to Europe later, and especially this big 1956 tour that I did, no obstacles were placed in my way. Now I didn't say anything about this in what I'm going to read to you, but I'm saying it sort of off the record, because it shouldn't be that way, it should not be that way. If a painting is done, dammit, you ought to be able to see it. And if you build an Aston Martin, they ought to be for sale if you have the money to buy one. And I feel that fossils are our heritage, if they're Hominidae, and everybody who qualifies should minimally be able to examine them, if they know what it means to examine somebody. You do not want an orthopedist examining you if he's an internist. I can tell you, it's not gonna work very well. You have to be equipped to do things and I insist that that's an important thing.

[1]F.C. Howell died of cancer on March 10, 2007, just 8 months after presenting this lecture. A recording made by W. von Koenigswald was transcribed by Ms. Evelyn Katz and edited by Eric Delson (both of the Department of Anthropology, Lehman College/CUNY, Bronx, NY 10468, USA), with editorial assistance from Susan C. Antón (Department of Anthropology, New York University).

F.C. Howell †
University of California, Berkeley, USA

S. Condemi and G.-C. Weniger (eds.), *Continuity and Discontinuity in the Peopling of Europe: One Hundred Fifty Years of Neanderthal Study*, Vertebrate Paleobiology and Paleoanthropology, DOI 10.1007/978-94-007-0492-3_1,

The real reason I'm here is because I happen to appreciate science in Germany. As a youngster after World War II and in college and so on the first language I learned, because I was illiterate in terms of my background and so on, was German, followed by a summer of French, followed by a year of Italian so I could cope with writings by Sergio Sergi and his father Giuseppe, and all these other people. Now they all write in English, so what use are these languages for me? Well they are useful, you know, when I want to order something or find the toilette, it works. Now I really came because of that, because I appreciate science in Germany, and what Germany did in science, in the history of science. You know the medical schools, those of you that don't know, in the United States, are based on the German system, and the teaching of anatomy and other things like that profited vastly from the German system. We didn't go anywhere else and many other parts of Europe, especially Central Europe, not to talk about Western Europe exactly, are based on systems that are essentially Germanic. So I'm here for that reason. I'm here for another reason, and I said this last night, so those who heard me say it could put their fingers in their ears, and that is because you have a Chancellor that I happen to admire greatly here. Now I haven't met her, and I haven't given her a backrub like somebody else in my country did the other day, he also does weird things all the time. But anyway she has the lead editorial page in Science that came out last Thursday, a week ago today. Now that's never been done before by any "politician". I call her a scientist because that's where she comes from, and she happens to be your Chancellor, so I would say chancellor. I would never use the word politician although I suppose to survive she has to occasionally be political. But I came because of that, and because of what she said in her editorial, which has to do with the people, and the future, and what is really important in life is education, and training of the young, and I would say training of the old. My God, they don't understand many things because they were poorly educated in my country, I insist it's my country, it's not my government, but it's my country, and I feel it's vital, and I feel that understanding of science is vital to everybody. You learn how to walk, you learn how to talk, you better learn some science, I don't care what you do with your life, you better know some science. And it's a lot more fun if you know science.

So I'm here for those reasons and also to tell you something about the 1950s. And I've written this because some people, including Eric Delson who's here tonight, will say Clark, why don't you ever write these things down and so on. And the answer is I write too much already, but I've written it down and there will be a book to come out of this and whatever I say tonight with alterations and emendations will probably be in there. So I'm going to tell you this is the way I thought when I wrote this.

We're gathered here tonight on a sesquicentennial occasion of a significant fossil discovery, the implications of which lie in a way at the very roots of human paleontology as a scientific endeavor. Due to serendipitous circumstances, I was present as well at that ceremonial occasion held in Düsseldorf and at the eponymous Neanderthal quarry locality which we had a gathering at, at a meeting memorable for its time no less than for its place with reference to the reinvigoration of basic science in the consequent decade after in excess of 6 years of Eurasian warfare. The invitation to speak here in an inaugural position affords me an opportunity to offer a personal perspective, hopefully insightful, on circumstances, events, situations, and personalities that reflect the emergence of paleoanthropolgical science within the sixth decade of the twentieth century. It's appropriate and worthy of a few reflections in respect to this first decade of the second half of the twentieth century. This was an interval of massive reconstruction in Europe via the Truman Doctrine and the Marshall Plan. Each of those individuals was absolutely unpolitical. Truman was utterly unpolitical, although he became a politician. He was a haberdasher. George Marshall, who was trained as a military man, he was really an intellectual. It was also a time of political reorientation and of attendant geographic and political polarization, of the progressive collapse or transformation of colonialism in Asia and in Africa, the development of nuclear power and weaponry, the initiation of space science and exploration, which was affected by rocketry and satellites, and it was the time of the very roots of computational science. Now all of those things are taken for granted, just ordinary, run of the mill, day to day things. Uprisings were common then as now, but in a different form in, for example, Algeria, Hungary, Kenya, Vietnam. There were local wars that continued in Korea, Palestine, Suez, Yugoslavia, and there was complete overturning of the government in China, by what became the PRC, and in Cuba, even the construction of a Berlin Wall in the case of that city, and the emergence of what was called an Iron Curtain and of the Warsaw Pact as a consequence of, a reaction to the establishment of NATO. Now all of you live with NATO although you probably don't see it or feel it, but NATO is all around you. You now live with something different and you also feel that and see it a little better and that's the European Union, and the one person who occasionally stands up and says hurrah is Winston Churchill, in that regard. Significant new representatives or heads of states then included Anthony Eden and Clement Atlee, Charles de Gaulle, Konrad Adenauer, the Shah of Iran for a moment, Nikita Khrushchev after Stalin's death in 1953, and he disowned Stalin in 1956, Mao Tse Tung, Ho Chi Minh, Dwight Eisenhower, and John Kennedy was entering the stage. We witnessed the first space flight in 1957. I was in Tanzania and watched the satellite go by. The first satellites were in 1957–1958, the first lunar rocket in 1959, the first commercial jet aircraft was in 1958, 4 years after I flew to Europe and to Africa on Boeing Stratocruisers and so on, four engine piston planes developed largely in World War II, by the British and by the Americans on the Allied side. The best part of that trip aside from Africa (my blood was infused with Africa forever after that), was that I flew first class because I had so much weight, and you could afford first class in those days, and when I got in my seat, I didn't pay attention to the person who was to the left of me. I took my gear and I put it up, I turned to the left and Marlene Dietrich was sitting next to me. Now I happened to look up at her, not down at her, but had I looked down I would have known it was Marlene Dietrich as well.

The first nuclear power plant was 1955, the first atomic submarine under the Arctic sea was 1957, the emergence of palaeo-oceanography occurred then, the first polio vaccine was in 1953, the first container shipping system appeared then, the first all purpose credit card appeared, and finally recognition of the helical structure of DNA was in 1953. Actually I always thought 1953 was the most important because that's when I got my doctorate, the first time anybody related to me anywhere, had gone to college much less got a doctorate. I was wrong. You know, I didn't know anything about Watson and Crick, I have to tell you, that spring I didn't know anything about anything like that, and I'd had courses that had to do with some of those things, way back, but I was off in another universe, getting out of here, studying real fossils, going to the field, those kinds of things. I missed the boat, often we miss the boat, we don't recognize what is in front of us, we perceive, but we do not understand what we perceive. Geochronology was enhanced then by the development of radiocarbon dating, so-called, and the applicability of the potassium argon methodology, known previously, to volcanic products of Cenozoic age, was affected through the development of mass spectrometer refinement. Now I was in universities in which each of those things occurred: Chicago, where I knew Williard Libby and people who worked with him, a man named Miller; and subsequently involved with people who were at University of California at Berkeley where I am now emeritus, because I knew Jack Evernden and Garniss Curtis. I went to one of their first major presentations in 1958 at an AAAS meeting in Philadelphia, the day after Christmas. I had been married 3 years, if you don't think my wife loved my leaving 1 day after Christmas, for a meeting in Philadelphia to hear somebody talk about dating volcanic rocks. It didn't go very well. Later she fell in love with Garniss Curtis, so it was all right, I guess, but it was hell. It'd been easier if we had had children, didn't have any children yet. Each occurred at institutions with which I was affiliated, and if I've not mentioned significant developments taken by all of you for granted, they probably were not yet envisioned or manifest, including those beepers that you have in your pocket, those cell phones, and all kinds of other things.

This perspective, however personal, reflects these manifold and varied experiences that I encountered and in which I participated within those years, at times noteworthy for the emergent formulation of our science and, of course, my incipient professional career began in those years. The initial 1950s constituted for me the completion of a University of Chicago education, a doctorate in anthropology and natural sciences in 1953, as I said. This was exactly 10 years after my secondary or high school (*Gymnasium*) graduation in 1943, after which, a few months later, I entered the U.S. Navy, trained, and they sent me overseas in the Pacific war theatre for nearly 3 years. And at the end of the war, when we were waiting to invade Japan and so on, we made a horrible mistake, but they still say it's a good idea, that we dropped two atom bombs on Japan just to tell them we were really there, and we didn't want them to stand up to us. I'm not happy about that, I'm not happy about war anymore. I was very warlike as a young person, I loved the military, I loved battles, most of what I remember in history, I wanted to be a civil war historian first of all, for the South, strange; that's cause I'm from a border state. Those 7 pre-professional university years afforded me a hitherto unperceived and hence unexpected and ever expanding intellectual, particularly scientific, immersion. My earliest Midwestern years were initially of a farm life in Kansas and subsequently in small cities, or in towns in Nebraska, then Indiana, then Wisconsin when I went into service. And I was scarcely prepared for advanced university life and as experienced by others of much more advantaged intellectual background and of course of economic status. I felt very much as a foreigner. But the amazing thing was I was much older than people of my same level in college, because we were all returning G.I.'s, and a number of us had never been to college before, we'd been in the military, and if it hadn't been for the G.I. bill we wouldn't have been in the military [sic – should be "in college"], we would have been doing whatever, working in hardware, I don't know what we would have done. With the financial support afforded to all military veterans, I was empowered to pursue an educational experience otherwise foreign to our family history. I'd already determined to seek a professional career in science, particularly in human evolutionary studies, as I thought about them then, even prior to entering military service, and these war years offered me some opportunity and time to read and to reflect and solidify such interests and potential goals. A postwar visit in August 1946, after my discharge, to New York City and its American Museum of Natural History, strongly reified my concerns when I personally met and visited with Franz Weidenreich and unexpectedly as well when we went to lunch with Ralph von Koenigswald, the latter had only arrived a few weeks previously from overseas wartime internment in Java. Weidenreich I'd first began to correspond with in my final year of high school, I was very audacious I just said, he's like anybody else, I wrote a letter to him. Unquestionably these interactions were of substantial significance to my future quest for a career in science. Luckily I made this visit as Weidenreich was to die only 2 years later at age 75. Weidenreich's own writings were very likely a major stimulus to my subsequent concerns with Neanderthals and their time and place in human evolution. Through attendance at professional meetings, travel to other universities, meeting those visiting scientists who often appeared at my own university, I soon became aware of and familiar with and eventually inducted within the sphere of scientists within evolutionary biology, in anatomy, in paleontology, biological anthropology, and prehistory and even Quaternary studies. So I did a degree that's anthropology but I did literally all my work after passing those horrendous day long exams in four anthropological fields, I did all my work in the natural sciences, aside from those languages that I had to pick up because otherwise I was a linguistic cripple. In what then was still [the] limited size and focus scope of a particular scientific endeavor, this was both possible and invaluable for a fledgling professional person like myself. I found encouragement and support not only within faculty, but also from scientists at other institutions, museums, foundations with whom I came in contact and interacted. So my own monitoring experience was then substantial, broad, and diverse. Such experiences greatly influenced my own attitudes, my own actions, subsequently, with reference to colleagues, associates, and students over the course of what became a long professional life, and ultimately globally, which I certainly never expected.

Consequent to the doctorate, I was very fortunate to be overseas for at least the summer months, or even longer on two occasions, for 6 out of 7 years of that decade. And I should mention that excludes 1955 when I married Betty Ann Tomsen of Danish descent and also resettled at the University of Chicago faculty from teaching anatomy previously at Washington University medical school in St. Louis. I never knew if St. Louis was sad or glad when I left. I have a feeling it was a mix because later the chairman made some remarks to me once in an elevator in New York. He said "my God Clark is it you?", I said yes. He said "it's too bad you left St. Louis"; I said where are you now, he said "I'm at Columbia". I said I guess we had reasons to leave, he said "I know we did". So, there you go. The years of 1953, 1956 and 1960 were years exclusively concentrated in Europe, 1959 I was in both Europe and the Levant and in Africa, and in 1954, and again in 1957, there were long stays. In Africa in 1954 I was there for over 7 months, and throughout most of sub-Saharan Africa and involved in both field and museum studies; and then subsequently in 1957, my wife and two students and I were in East Africa for over 8 months excavating at a large Paleolithic open air site in southern Tanganyika, now Tanzania. I have to tell you, you won't like it, it was better Tanganyika than Tanzania. Colonialism had something to say for itself, not everywhere but it did have something to say. You could walk down the street, you could leave the car unlocked, you could do all kinds of things. I know that people were in different straits but there was something to be said for it and I don't mind saying it.

Those four instances that involved Europe specifically, or even partially, are relevant to our concerns here today, as they mostly relate to issues of human evolution and prehistory within the Late Quaternary. Each afforded major opportunities to study hominid fossils, examine important artifact collections, often make visits to prehistoric sites widely over Western and Central Europe, not to mention the United Kingdom. And in 1953 there was an opportunity even to participate in the summer field season under the direction of Professor Hallam L. Movius at the Abri Pataud in the Dordogne, very close to the grotte or shelter Abri de Cro Magnon, and we stayed in the Cro Magnon Hotel. I took it for granted that one should stay in there if you were going to excavate, you know, just a moment away at Abri Pataud, which he had bought through the Peabody Museum in order to excavate. And I learned a lot then about doing Upper Paleolithic typology because that's what he said, "you're not gonna dig or anything, you're not even gonna see the site hardly, you're gonna stay in the basement of this farmhouse and you're gonna sort all the artifacts that come in and this is how you'll do that" and so on. And I said yes sir, yes sir, the way I'd been taught in the military; he was a colonel, of course, in the air force, but he was also my friend, and he was like a mentor and almost like a father to me, very stern but very understanding. And I did what I was told, and I learned a lot from him, and I also developed that summer a long and lasting friendship with François and Denise Bordes and many other people. I met Camille Arambourg that summer for the first time. He had a house in the Dordogne, and little did I know that later we'd work in the field together in Africa.

In several instances there were small conferences or symposia in which I was involved, and these included, in 1953, a gathering in London based at the Natural History Museum, but also held at a hotel nearby; at that meeting there was planning for future steps by the Wenner-Gren Foundation to support African paleoanthropology. And it was actually that moment when I, and others too, examined and experienced the demise of the Piltdown hoax at London's Natural History Museum at the hands of British colleagues. That means Weiner and LeGros Clark and so on, and Sherwood Washburn was there, and Charles Reed (a zoologist), myself, several other people. Augusto Azzaroli was there at the same time studying cervids from the [Cromer] Forest Bed. I didn't fully appreciate the moment as much as I might have, but I certainly knew, I never had any faith whatsoever for a minute about Piltdown ever. When I learned to read German, and I read Weidenreich's student Freidrich's long paper about that, then I read Gerritt Miller and some other things before I'd ever seen it, I said, there's something really weird about this,

this cannot possibly be, and so it wasn't. By the way, the person who was responsible for it was surely not Woodward, surely not Sir Arthur Keith, surely not ABCD, but somebody else. If you want to know, I'll tell you later, I'm almost sure who did it. And Chris Stringer will tell us whether I'm right or wrong. Sir Arthur Keith was still alive then and at the end of the summer, after I got back from France, I journeyed down to Down House in Kent to meet him and have tea with him. Later that summer, I'd sent him some of my papers and so on and he came down, he was a very tall man, very bent and so on, but I suppose about 6 ft 4 in. or so, very slender and long faced and so on, and he came forward like this – you could see all the veins and arteries. And he said, "Dr. Howell", very formal; I said, Clark Howell. He said, "I thought surely you were much older." I said, I will be. [Laughter] Sweet man, we talked for an hour and a half. He'd just finished his nap, and he was writing a book about Thomas Huxley.

In 1959–1961, I served as a participant, or a principal organizer, of Burg Wartenstein (Austria) Symposia of the Wenner-Gren Foundation. And on two occasions I attended larger and international congresses. In 1956 of course, the Neanderthal Centenary celebration in Düsseldorf, and in 1959 the Fourth Pan African Congress on Prehistory held in Leopoldville, now Kinshasa, and in the last instance prior to the Congress, I was among the first scientists to see, in Nairobi, Olduvai Hominid 5, or Zinj, after its discovery at Olduvai Gorge while I was in Ethiopia on a survey trip to the Lower Omo Basin. It took us 7 years to get permission to work in the Omo for 10 years. It was worth waiting for, it was painful to have to wait. Louis Leakey was wonderful, we had a nice dinner at their home in Karen, and Mary had a little kind of twinkle in her eye, and we never knew if it was the Scotch or whatever, and after dinner and so on they didn't always have dessert, but he said "maybe we'll have some cookies, you call them cookies don't you?" I said, you call them biscuits, he said that's right, so he brought out this metal box which had Danish biscuits or cookies in them, and he said "here open it up, it's a new one". And he opened it, and there was Zinj lying in the box, disassembled so the face was separate from most of the braincase. I couldn't believe it, I couldn't believe it. Later we all went to Kinshasa to the Pan African Congress, and at that time he offered Phillip Tobias (PVT) the chance of describing it, and he made the right choice, super guy to do it and the same with *Homo habilis*. Now some of these things we look at a little differently after they've been described, but the descriptions if they're very well done, the descriptions will always last, they'll stay there forever.

In 1950 the first major effort to bring human evolutionary studies into the framework of the modern evolutionary synthesis was a very major symposium called The Origin and Evolution of Man. It was held at Cold Spring Harbor, Long Island, New York with nearly 40 participants and over 100 registrants. I was fortunate to attend this as a student, and it played a very central role towards the redirection of the field and then crucial in my own future orientation towards studying human evolutionary biology. I'd met Ernst Mayr before, but this really, I mean you're together for a week, and you're walking around and drinking coffee, and everybody's equal, a graduate student is equal to the professor and so on, it was wonderful … and I got to know Dobzhansky very well. George Simpson was never easy to know by anybody, but he acknowledged "Mr. Howell", never "Clark". Anyway lots of people, very interesting meeting.

Another major and certainly a singular event was that of the Darwin Centennial, the largest gathering of which occurred at the University of Chicago in November 1959, with a plethora of participants and listeners, among a galaxy of scientists of international repute, mostly but not only in the Natural Sciences. It was fully and quickly published as *Evolution after Darwin* by the Chicago Press in three volumes in 1960 and then in 1962. Over nearly 3 months of that autumn, Sir Julian Huxley, I called him Julian, and I also called him Sir Julian, was an office neighbor of mine with whom regularly, morning, afternoon, lunch, whenever he felt like it, we discussed matters of common concern in regard to evolutionary biology. I'd read his volume Evolution, a Modern Synthesis, published in 1942, which got to the West Coast and I found it in 1944 and took it with me overseas, and so I'd read it, and I was very happy and honored to have a chance to get to know him. This was a man whose mind never went to sleep. He was always into something and so on. I believe that he was not given enough credit for the modern synthesis because, not only did he coin the word, but also he was too much of a synthesizer, and people often forgot all the basic science that he did, this man did a whopping amount of basic science, including unbelievable work in allometry. And he deserved everything that he ultimately got, including a very fine honorary doctorate at the University of Chicago on this very occasion, as did several other people, of course. But Julian Huxley, people thought he was sort of, you know, uppity British upper class and so on, I didn't think he was that way at all and sometimes people you know are frankly misread for whatever reason, and I won't go into details of that, but I'll just make the admonition, don't always believe what you see; I would say don't always believe what you think. Think twice and if it still goes that way, alright, maybe, but don't be so all fired sure, as my father used to say, don't be so all fired sure.

Many of the principal and major contributors to the modern evolutionary synthesis were present or represented depending on their health. Among many others of diverse fields among the 50 central participants at this centennial, this was the first such all inclusive symposium on evolution since the very seminal post war symposium in Princeton, 1947, (published in 1949) which was focused on genetics, paleontology, and evolution. That meeting built on major antecedent books by Ronald Fisher in 1929, J. B. S. Haldane in 1932, neither of whom I knew. Haldane, I would have love to have known, Fisher I think I could have done without probably. Theodosius Dobzhansky, 1937, who was a very dear man; Ernst Mayr, 1942, another mind never stopped running; and George Gaylord Simpson, 1944; all those books are customarily considered to reflect the consolidation and crystallization of the synthesis. I assume many, even most of you know this extensive reformulation within the life sciences, and perhaps even best through the volume edited by Ernst Mayr and William Provine called *The Evolutionary Synthesis, Perspectives on the Unification of Biology*, published by Harvard in 1980. Although that volume contains two informative though brief chapters devoted to embryology, among its broad ranging coverage of fields and research traditions, the developmental aspect was notably ignored or even absent within the traditional formulation of the modern synthesis. This was patently evident during the Darwin Centennial and was brought out there explicitly in that respect by Conrad (Hal) Waddington, with whom I managed to have several significant and long conversations on the subject. Waddington had a major influence on me, more than I knew until subsequently. Having studied embryology and development in the University, I considered this among other issues, including this significant role likely played by other natural scientists towards elaboration of the synthesis, as unfortunate and even unwarranted omissions. You can of course read about this and much else in Stephen Jay Gould's remarkable tome, the Structure of Evolutionary Theory, published shortly before his death.

In university, after acquiring foreign language capabilities, I'd actually read *Der Evolution der Organismes* edited by Gerhard Heberer, published in 1943, but which got to our university towards the end of the war, a volume that clearly revealed roots of the synthesis among some German and other natural scientists of the previous decades. And if, those of you who read German, useful insights into the role and participation of others are exemplified in the volume *Die Enstehung der Synthetische Theorie*, edited by Tom Junker and E. M. Engels, published in 1999 in this country, in Berlin. The critical role of development, now evo-devo, and developmental genetics is, of course, absolutely powerfully established within evolutionary biology now. It was not always so. I've always considered that developmental studies must constitute a central focus in human evolutionary biology as well, and I've said this over and over again. I consider, as have others, that the Neanderthal Centenary in 1956, published 1958, and the Burg Wartenstein1960 symposium, *Early Man and Pleistocene Stratigraphy in the Circum-Mediterranean Regions*, published 1962 in *Quaternaria*, constitute fundamental contributions to infrastructure of an emergent paleoanthropological science. This latter meeting, the one in *Quaternaria*, was organized by Alberto Carlo Blanc and myself, mostly during his second visit and residence at the University of Chicago in early 1959 as a visiting professor of paleoanthropology. His sudden death, which I learned of only 2 days before the symposium, when I visited Zurich and met Adolph Schultz, who said "I have very bad news for you", was a horrible and painful blow to me; however all the participants rose to the occasion and the symposium was an unbounded success. Blanc was another of the many people that I met in the course of my time in this world who was open, candid, translucent, generous beyond belief, etc. And there have been many such people. Each of these events facilitated, and the latter particularly emphasized, extensive discussion centered around precirculated drafts of papers – that was the Wenner-Gren plan.

An examination of each of these respective volumes is revelatory of focus and status of particular areas of interest, of fields of scientific endeavor, and of the nature and prevalence of theoretical frameworks, and it should be emphasized the extent to which congruence and even concilience was manifest as a consequence. I've broken down, and somebody can read it someday, the breakdown of people and their papers at the Neanderthal Centenary and at the workshop conference that I mentioned in parallel, which occurred later at Burg Wartenstein, and there were geologists at both but at the Burg Wartenstein Conference there were 14 instead of 4 geologists, we really had a slug of geologists because that's what we're trying to do. We had two paleontologists at Wartenstein, and there were four at the Neanderthal Congress. There were seven archaeologists at the Neanderthal Congress, and we had three archaeologists at Burg Wartenstein. And the paleoanthropologists, there were 12 at the Neanderthal Centenary and 1 at Burg Wartenstein. You can readily gain, even from a summary like this, something about the focus, the goals and the emphasis of these kinds of meetings. It should be mentioned that my passing participation (en route to Israel and East Africa) in an earlier Burg Wartenstein Conference in 1959 called *Social Life of Early Man* (published in 1961), enabled me to meet there Professor Francois Bourlière of Paris. This eventuated in our organization of a 1961 Burg Wartenstein Symposium called *African Ecology and Human Evolution*, published in 1963, that came to have a very major impact on naturalistic as well as paleoanthropological scientific studies in Africa. It's often thought to constitute a real turning point towards cross-disciplinary researches. I think that, and the preceding circum-Mediterranean one, certainly demonstrated that people of different disciplines can live and work and talk and associate together comfortably and freely with no incertitude or anything like that. Absolutely sure, and these were extremely important. The significance of this decade with particular regard to human evolutionary studies is exemplified by the very contributions that are assembled in the volume entitled *Ideas on*

Human Evolution: Selected Essays 1949–1961 edited by William Howells (with an s, no relation; [published by] Harvard, 1962), which some of you may be familiar with; if not, it's worthwhile reading those essays, if they were [originally] in German, they're translated into English. As a consequence of the aforesaid overseas travels, and museum and field researches coupled with participation in the aforementioned and other professional meetings, I gained an uncommon experience and a recognition of problems and the value to science of interdisciplinary researches. This occurred within a decade after my doctorate. So I considered myself not only unexpectedly fortunate but repeatedly and invaluably so. In these years, the most useful contribution to human paleontology were several editions of *Les Hommes Fossiles* by Boule and Vallois, including an English language edition in 1957. And volume 7 (on primate and human paleontology) of the *Traité de Paléontologie*, by Jean Piveteau; I eventually met him in Paris. Also the first such synthetic volume, *Les Néanderthaliens*, was published by Etienne Patte in 1955, a year before the Neanderthal Centennial, and he followed it 2 years later by a very useful monograph devoted to the Pech de l'Aze infant Neanderthal skull. I should add that some other major serials of interest to us had their genesis in the 1950s. We've done this [meeting] in association with DEUQUA; *Eiszeitalter und Gegenwart* saw its appearance in 1951, and I've been a DEUQUA member since 2 years before my doctorate, I thought it was a unique organization, and it was, but I've never been to a single meeting. What a shame. It's always at the wrong time of year. *Quaternaria* also started in 1954, *Vertebrata PalAsiatica* in 1956, *Radiocarbon* started in 1959, and *Current Anthropology* dawned in 1960, thanks to the hard efforts of Sol Tax. In the decade between 1951 and 1962, I contributed towards definition and critical evaluation of the so-called Neanderthal problem in ten published contributions. These were published variously in scientific journals, largely anthropological (five), and biological (one), a learned society, an encyclopedia, the International Geological Congress Commission, and in the Neanderthal Centenary Volume, *Hundert Jahre Neanderthaler*, which von Koenigswald edited. And I was to return repeatedly to the same or closely related topics in future decades as well. Overall these contributions reflect markedly the influence on me of the predominant framework of the modern evolutionary synthesis as I learned, experienced, and employed it in those times. Collectively their overall contents span much of the available and pertinent empirical data relevant towards efforts to evaluate and to seek to comprehend the role and relative place in hominid biological and behavioral evolution of those extinct antecedents of modern humankind. Many of the roots of numerous subsequent, more extensive and intensive scientific studies of specifically human paleontogical or even more broad-based paleoanthropological investigation may be similarly traced there. However, certainly there are major concerns now, unenvisioned half a century ago, especially in regards to technological developments and their elaboration in newly recognized or defined fields of research, concern and investigation; and development and applications of various methodological procedures that are innovative; and, last but not least, the enhancement of relevant theoretical frameworks including those based on hypothetico-deductive reasoning, going back of course to people involved with history of science. I think it's absolutely vital that if you do science, don't be a technician only. Have an appreciation of what you do in a historical perspective. It may not necessarily help you, it will not hurt you, it will make you happy, you will smile, it is fun, and it is important and it gets better all the time. History of science is a widely flourishing enterprise. An appreciation of past efforts and understandings of our scientific forerunners is an essential and requisite part of a scientific endeavor. Too often such history is ill known, it's ill appreciated, and it's poorly reported. In the coming days here in Bonn, we're assured of much that's hopefully new and even unexpected in our perpetual pursuit towards fuller understanding of the distant human past, and you've been very patient; thank you very much.

[Applause] Thank you. Thank you.

FCH: If you want to ask a question, please stand up, and state your name.

Q: [How about Piltdown?]

FCH: I'll tell you, if you want to know about Piltdown, you have a person who has been in charge of it for some years at the Natural History Museum, and if there is something specific you want to address, I am sure he would be happy to do that. That's my longtime friend, and I offer now my personal congratulations to a new FRS, [Chris Stringer]. What would you like him to answer?

Q: [Does he agree about who was the forger?]

FCH: ... You want me to say who I think it was ... There was, not now, there was a substantial *badinage* about this ... various people took it up in different ways. For example, the lamented Frank Spencer, a very fine historian of physical anthropology, G. A. Harrison, all sorts of people have written about this in different ways, and the Weiner book is still an outstanding book on the subject. Phillip Tobias is the one who really pursued the Arthur Keith association. He was unconvincing to me, in my opinion, you could ask why, but I won't go into it here – I felt that it was farfetched. My feeling was then, and for various reasons strongly, that it was a man named Martin Hinton; he was a worker in the museum and later sort of worked his way up. He's famous to people who know about murid rodents and so on, wrote a fantastic book for its time about arvicolids. I never met him, although he was still alive when I was there; he was retired by then. He was a great teaser, a taunter, a jokester; they found things associated in his *equipage* subsequent to his death and so on. I never talked to Chris about this. We talked about "business", and that's not "business". That's an event, in history. We can ask Chris, what he thinks about this … Are you there? …

CBS: I've actually written a [small] part of the story, because we had a 50th anniversary event

FCH: I remember you did, but I never saw it …

CBS: It was a 50th anniversary of the exposure, and so we had an exhibition, lectures, and I did a bit of extra research, and a new edition of Weiner's book was published. So I think you are right, that Hinton's behavior was certainly suspicious. My colleague Andy Currant unpacked a trunk from the attic above the old Keeper's office, with Hinton's initials on it, and in there were bones that had been cut and stained in a very similar manner to the Piltdown remains, and I think probably Kenneth Oakley secretly suspected that Hinton was involved. But my own work, and not just my own, Joe Weiner's work long ago, points to the fact that Dawson is still a very strong candidate. He found the first remains that we know of from the site, he found the last remains that we know of, at Piltdown 2, and I think it's a complex story. There are two sites, Piltdown 1 and 2, and Dawson is the only one who we can associate with the separate sites

FCH: with both of them …

CBS: and the fact is it's almost certain that the jawbone that was found at Piltdown 1, a tooth from that then turned up at Piltdown 2, and Dawson is the only link between those finds. So I think he's a very strong suspect; one can argue that he didn't have access to all the material or the knowledge to do it himself; with someone to help him, Hinton might come into the frame. But I think Hinton also comes into the frame because of the very weird object found in the last days of digging at Piltdown 1. They found, apparently under a hedge at the site, a large chunk of elephant bone that had been carved, and even at the time, some jokingly said "what's more appropriate for the earliest Englishman than something that looks like a cricket bat?"… [laughter] … It really *does* look like it, made of elephant bone, and of course that was faked too, it was carved on fossil bone with a steel knife, probably, and the fact is that I think one can look at the stuff that was found in Hinton's trunk and you can see that perhaps Hinton for whatever reason, maybe jokingly, maybe he just wanted to bring the thing to an end, he might well have planted the bat there, because it is so outrageous, and then to his horror, he saw it published as the oldest bone artifact in the world. [laughter] … And then strangely, straight after that of course, Dawson seems to start to lose interest in Piltdown 1, and he starts to go off and develop another site. So you can put 2 and 2 together and say, yes, Dawson did most of the stuff at Piltdown 1, Hinton planted that elephant bone, which is why he is then very evasive whenever Piltdown is mentioned. He really seems to be hiding something, but I think Dawson's still the main candidate.

FCH: There you go. OK, what else?

Q: Clark, you pointed to a lot of interesting developments in the '50s, you obviously were around for interesting developments in the 60's, 70's and subsequently. Did you experience similar things in subsequent decades? Or perhaps even more spectacular developments?

FCH: … A few years ago (in my first such appearance), Gerhard Bosinski (who like you was a student at the time of the last Neanderthal meeting), anyway at a certain point Bosinski asked me would I come and give a lecture to this group of people in Neuwied who were like friends, and I said OK. So I arrived and fortunately I had something written out, because often I just talk. He said well, we have to put it into German, and I almost fell over, and he said oh we'll translate it together. He said, is that [good], the way I rephrased. And I said well it's pretty much, but it takes so arduous, can't you shorten the distance between the verbs? So I gave the talk before this group of people, and Germans are pretty tolerant people, [with] very good manners, and they sat through this, gave me a hand, and I talked about some other things that would be of interest

Now, this is like a footnote, most people, many people more or less focus on one line of endeavor, they venture just at the edges, but they never sort of break the edges and go out. I haven't been like that. I was like that as a student, aside from trying to sort all kinds of things so I could find my way in the world, and so on but I never sort of said this what I'm gonna do for the next 125 years. I've done a lot of different things and some people here would say "I'm sure glad you did those things, but why don't you ever finish them?", and my answer is there'll be people behind me who will finish it, and they'll probably do a lot better than I did. But anyway I did the best I could, and I obviously have a short attention span in some regards. I mean, I'll pursue something like a birddog, you know, I just can't let loose of it, and then they'll be a point where it begins to relax, and then the next thing you know I'm over here, and that's because I really discovered there's something about something else that I didn't recognize before or I would have been more over there already. I don't know if you know exactly what I'm saying, but this really has to do with things that catch your interest, things that brute force you, pull you, twist you in a certain intellectual direction. I believe that composers have this kind of thing happen to them. I know nothing about music except adoration, but I believe that probably [they], maybe painters who shift their gears in the way they paint, like Van Gogh from the "Potato Eaters" to the [unintelligible]. I know in my own instance that I have been perpetually, everlastingly rejuvenated by this shift that transpires, and I think it will be judged in the future, not now, but it's certainly been important in terms of what little bit I've had to do with this thing that's become a science in my lifetime.

You might call it something else, but you certainly could not call it paleoanthropological sciences way back in the past, even in the 1930s by any means. You cannot call a Pre- and Protohistoric Science Congress a paleoanthropological congress, it's Pre- and Protohistoric Sciences, and the meetings of the American Association of Physical Anthropologists or the German Association of Human *Biologie* is not paleoanthropology. So when you talk about paleoanthropology it's something that is more inclusive, it's something that encircles a series of things within it, alright? And that didn't happen until subsequently, which is what I tried to indicate tonight without overstepping. And I believe that's true, and I believe you can see it. We still do not have the people who are doing history of [our] science and so on, the way there ought to be, I won't say should, the way there ought to be people, even yet. Richard Delisle did a recent book (he's a Canadian), and it's a good book, but it ain't the right book. And two people, for utterly different reasons, said so: one at the beginning, Milford Wolpoff, with whom I often disagree but sometimes think "that's an interesting idea," or Bernard Wood, with whom I equally disagree about certain things, but I have known him a long time. And they both were unhappy about this kind of thing. There was also another small book published that has to do with australopithecines, mostly *Australopithecus* and its coming into [favor] and so on. The best book that had to do with human evolutionary studies in the broad sense, but without enough of paleoanthropology as a whole in it, was Peter Bowler's book. He's a wonderful historian of science, there isn't anything he has written that is not worth reading.

I feel that people come into the world to do this. I think they could come from Germany, they could come from France, they could come from anywhere now. You have to have several languages, that's not a problem anymore and so on. There is plenty to say about all these things. And it's a happy, busy, creative world that we're all a part of, and I wish that Angela Merkel could talk to George W. Bush, and say "George there's really something important here, would you be willing to get off your bike and listen? ..." [laughter] It's not going to happen ... [applause].

Chapter 2
One Hundred Fifty Years of Neanderthal Discoveries: Continuity and Discontinuity

Silvana Condemi

The original Neanderthal fossil was discovered in Mettmann near Düsseldorf, Germany, in 1856 and the present volume is based on the conference that I had the honor of organizing in July 2006 with Professors Wighart von Koenigswald and Thomas Litt to commemorate the 150th anniversary of this discovery. It is in Bonn, at the Rheinisches Landesmuseum that the holotype of the Neanderthal taxon is kept, and it was at the Rheinische Friedrich-Wilhelms Universität in that same city that the conference was held.

The five symposia into which the conference was divided are the basis of two volumes. This first volume brings together the reworked versions of the papers and posters presented in three of these symposia (numbers 1, 4 and 5) on human paleontology (paleoanthropology in the European sense), which dealt with the origin, evolution and disappearance of the Neanderthal population.[1] The second volume, organized by N. Conard and J. Richter, assembles the papers presented in the two other symposia (numbers 3 and 4), concerning the way of life, the culture and the environment of this extinct population.

The purpose of this conference was to present the state of our knowledge concerning this fossil group, to examine questions that it raises in the present, and to evoke perspectives for future research.

If we compare the results of this conference with the centenary conference held 50 years ago in Germany, it becomes immediately apparent, not only that our understanding of the Neanderthal population has changed, but also that this change is closely interrelated with more general modifications in our conceptions of humanity since then. We also perceive that the geographical focus of research on this population has been greatly enlarged, corresponding to a shift in scientific paradigms. Europe was the focal point of interest during the first 100 years of study of this fossil, coinciding with the first century of paleoanthropology as a discipline, even when discoveries were made outside of the European continent. Investigations were carried out in relation to the European Neanderthals, since Europe was the pivotal axis around which paleoanthropological concerns in general turned. By contrast, the significance of Europe in paleoanthropology has clearly diminished over the past 50 years. Thanks to fossil discoveries throughout the ancient world and to developments in our methods of dating and of investigating fauna and industry, Europe has lost its primary place, and it is rather in light of discoveries in Africa and Asia that European fossils, in particular Neanderthals, are now interpreted.

In this introduction I will briefly summarize the principal moments that led to modifications in our view of Neanderthals and then, in the section that follows, present the different chapters included in this volume. These chapters clearly illustrate the changes in methods and techniques of contemporary research and the state of our knowledge regarding this fossil population.

An examination of studies published during the first 50 years following the discovery of Neanderthal brings to light an important change in our idea of the evolutionary history of humanity. After the initial discovery, the first researchers had to convince the scientific community that this specimen was not simply the remains of a pathological individual. They had to characterize its principal traits, while attributing to it a phylogenetic status and placing it, and Neanderthal discoveries that followed, in the theoretical framework of Darwinian evolution. In the scientific perspective of the second half of the nineteenth century, this *Homo neanderthalensis* (King, 1864) represented a chronological moment in the evo lutionary history of humanity, a ring in the larger chain which the discovery of Cro-Magnon in 1868 and *Pithecanthropus* in 1891 would further elaborate. From the standpoint of its anatomy,

[1] Symposium 1: "Outside Europe and Neanderthal Origins", conveners: S. Condemi and F. Schrenk; Symposium 2: "Neanderthal Palaeoenvironment", conveners: W.v. Koenigswald and T. Litt; Symposium 3: "Neanderthal Lifeways, Subsistence and Technology", conveners: N. Conard and J. Richter; Symposium 4: "Neanderthal Anatomy, Growth, Adaptation and Physical Variations", conveners: S. Condemi and W. Henke; Symposium 5: "Neanderthals and Modern Humans", conveners: G-C.Weniger and J. Orschiedt.

S. Condemi (✉)
Laboratoire d'Anthropologie bioculturelle, Faculté de Médecine – Secteur Nord, Université de la Méditerranée, CS80011, Bd Pierre Dramard, Marseille Cedex 15 13344, France
e-mail: silvana.condemi@univmed.fr

S. Condemi and G.-C. Weniger (eds.), *Continuity and Discontinuity in the Peopling of Europe: One Hundred Fifty Years of Neanderthal Study*, Vertebrate Paleobiology and Paleoanthropology, DOI 10.1007/978-94-007-0492-3_2, © Springer Science+Business Media B.V. 2011

its cognitive capacities and technical ability, Neanderthal was considered to be an "intermediary" being or "missing link". Interpreted in this manner, the geographical distribution of Neanderthals was taken to be quite vast since, as representative of the evolutionary history of humanity as a whole, it was considered to have been present throughout the ancient world.

Up until the Second World War, this approach to the Neanderthals and to their place in human evolution remained predominant (with the exception of the so-called theory of the "pre-sapiens"). Moreover, it is this perspective that we still encounter in the book *Hundert Jahre Neanderthaler* (von Koenigswald 1958), assembling the papers presented at the centennial Neanderthal conference in 1956. Indeed, in this book, papers dealing with European fossils which are still considered to be Neanderthal today (for example, Monte Circeo, Pech de l'Azé, or Spy) are presented alongside others from outside of Europe that we no longer take to be Neanderthals, such as the so-called Rhodesian fossils or the Solo fossils from Java – both defined in that conference as "tropical" Neanderthals – or the north African fossils (for example, Jebel Irhoud) – considered to be "Neanderthaloid".

At the same time, parallel to the 1956 centennial conference, a number of articles published in the 1950s already announced a more nuanced scientific perspective in which researchers entertained a less linear and more complex idea of human evolutionary history than that which had predominated since the initial discovery of the Neanderthal specimen. For example, the 1951 article of F. Clark Howell, "The Place of Neanderthal Man in Human Evolution", presents a synthetic approach to the Neanderthal fossils in which the attempt to understand their evolution includes analysis of their environmental and climatic context. Moreover, the multiplication of fossil discoveries, not only in Europe but throughout the ancient world (*Australopithecus* in Africa is a noteworthy example) portrays a much longer chronological range and an evolutionary history in which there are clear differences according to the continent in which the discovery was made.

It is during the 1960s, in a favorable political context, that a large number of excavations were undertaken in the Middle East, which were the source of numerous original studies. Here, too, we must remember the pioneering work of F. Clark Howell, notably in his article "Upper Pleistocene Stratigraphy and Early Man in the Levant" of 1959, which had a decisive impact on the scientific community. In his analysis, the author raised important questions concerning the Mousterian sites of the Levant, and regarding the phylogenetic status of certain fossils which had been found in these sites during the 1930s. In addition, a series of excavations undertaken in sites in the Middle East during the 1960s (beginning with those of R.S. Solecki at Shanidar, H. Suzuki at Amud, B. Vandermeersch at Qafzeh and A. Jelinek at Tabun) and a considerable number of studies on fossils, fauna, lithic industry, habitats, as well as dates from these sites have all been of particular significance for our present-day knowledge concerning Neanderthals. They have brought us to completely modify previous conceptions of Neanderthal as an intermediate fossil from the standpoint of anatomy, culture and chronology. Following these changes, new dating methods (TL, ESR) have led us to understand the need for a thorough revision of the idea of a succession of populations as it had been interpreted on the basis of the European model.

Beginning in the 1960s, these studies demonstrated that the geographic situation of Europe, a peninsula of Eurasia exhibiting particular environmental and climatic conditions, led to the differentiation of Neanderthals which, in overall evolutionary history, constituted a singular and marginal population. In addition, the multiplication throughout the 1960s and 1970s of fossil discoveries in Africa (for example, the different discoveries of *Australopithecus*, *Homo habilis*, etc.) brought to an end the Europocentric viewpoint in paleoanthropology stemming from the pre-war period. The whole series of these discoveries led to a still longer view of the evolutionary history of humanity and to a more differentiated perspective on human development according to the continents in which it occurred.

During the decades of the 1970s and 1980s, a number of paleoanthropological studies inaugurated an important shift in the approach to the Neanderthal fossil population. The quantification of anatomical traits, which had attained ever greater precision during the first century following the discovery of Neanderthal, was complemented during these decades with new kinds of approach to anatomical variability and to adaptive features among Neanderthals. The analysis of variation among modern human populations began to play a major role in Neanderthal studies, as the attempt was made to situate the data corresponding to Neanderthal specimens in a broader populational framework. It is in this scientific context that the Neanderthals were taken to represent a human fossil population, named *Homo sapiens neanderthalensis*, placed within the scale of variability of modern human populations.

Over the course of the 1980s, the multiplication of these anatomical studies of fossil populations, in particular of Neanderthals, accompanied by new methods of investigation such as cladistic analysis, and new tools such as the CT scan, led once again to a modification in approaches to the evolutionary history of the Neanderthal population. These anthropological studies, and those undertaken by prehistorians, aimed to better comprehend the way of life of this fossil population through investigation of the habitability of sites, the means and strategies of subsistence, or their paleo-environment. They have permitted us to clarify the question regarding not only the particularities of this population and its differences in regard to *Homo sapiens*, but also in relation to fossils that preceded them in Europe and throughout the ancient world. Thus, over the course of the 1980s the singularity of this population was placed ever more clearly in the forefront.

Moreover, from the very beginning of the 1980s the idea of the singularity of Europe concerned not only the Neanderthals, but the totality of European peopling since its origin. According to this viewpoint, *Homo erectus* had never been present in Europe although it existed in the other parts of the ancient world, or at least in Asia. This indicated a great difference in the entire evolutionary history of Europe in relation to that of Asia and Africa. Here again, at the "First Congress of Human Paleontology" held in Nice in 1982, it was F. Clark Howell who defended this position. In spite of the fact that part of the Congress was devoted to what the discoverers of the putative *Homo erectus* of Arago (including numerous comparative studies with "other" European *Homo erectus* specimens), Clark Howell (1986) supported the idea concerning the singularity of Europe from the most ancient periods onward!

At the end of the 1980s a number of researchers working on Neanderthals adopted the idea of the singularity of Europe since the most ancient available traces of its peopling, and they elaborated on the implications of this idea for the evolutionary schema. They focused on the evolutionary consequences of the marginal geographical position of Europe, on the particular evolutionary pattern of Neanderthals, and on their probable speciation. By the end of the 1980s researchers once again considered Neanderthals to represent a species different from their contemporaries in the rest of the ancient world. The name *Homo neanderthalensis* was reintroduced to characterize these fossils belonging to Europe. Although the point of view which attributes to Neanderthals the status of a species is not shared by all paleoanthropologists, certain chapters included in this volume support this position, which DNA studies since 1997 have tended to corroborate.

This brief historical survey permits us to appreciate the important modifications that have marked the study of Neanderthals. If, as we have noted, these changes have depended above all on fossil discoveries, they have also been due to new methods of study and investigation, not only in paleoanthropology, but also in disciplines which work in close collaboration with it.

It is in this broad context that the conference marking the 150th anniversary of the discovery of Neanderthal was held in 2006. In illustrating the contemporary state of our knowledge, we may also ask how long the view of Neanderthals presented here will retain its validity. We can anticipate certain technological and methodological developments in the future, such as those depending on the extraction of DNA which is well preserved in the crystal aggregates of fossil bones, and we can also expect new developments in research, above all in relation to the origin of the oldest European populations. It is nonetheless difficult to predict in any comprehensive way how Neanderthal studies will develop over the course of the next 50 years. Our brief historical investigation illustrates the primary importance of new fossil discoveries which may well overturn our current hypotheses. We may predict with high probability that research will develop extensively in areas which are not well known today, such as in certain regions of Africa and Asia. It is likely that the privileged and, indeed, nearly "exclusive" conception of relations and exchanges in peopling between Europe and Africa will be complemented, if not replaced, by the idea of relations and exchanges between Europe and Asia. If I have not dealt in any detail with the changes in our methods of study, it is clear that they have changed considerably over the past 50 years, which have witnessed the emergence of areas of investigation which could not previously have been imagined. Hence, although the helicoidal structure of DNA had just been determined, during the commemorative Neanderthal conference in 1956, it would have seemed fantastic to imagine the possibility of extracting DNA from the fossil bones of Neanderthals! There is an increasing tendency for studies of this population to gravitate from paleoanthropology to the field of paleobiology. It therefore seems certain that a different vision of the Neanderthals will emerge over the next 50 years.

F. Clark Howell presented the opening address to the conference in Bonn in 2006. This eminent researcher, who also participated in the centenary anniversary of the Neanderthal discovery in Düsseldorf, promised to provide us with a written text for this volume. His sudden death several months after the Bonn conference prevented him from completing this text. We therefore present in this volume a transcription of his oral talk.

In his opening address F. Clark Howell recalls the stages of his scientific development that we have briefly noted above. He places in relief the scientific context of the post-World War II period and the burgeoning of ideas that has been characterized as the "modern evolutionary synthesis" which fortified the Darwinian basis of natural selection in the process of evolution and led to a paradigm change in our general scientific outlook. At the same time, Howell indicates the change in the focus of scientific interest from Europe toward Africa. The narrative of his intellectual autobiography recalls changes that have marked our discipline, brought about not only through scientific influences but also through political developments over the past 50 years. It is fortunate that Clark Howell, animated as he was by an intense intellectual curiosity, was able to communicate his experience and his profound understanding to those around him. On a personal note, I remember with gratitude his kindness in permitting me, as a very young student, to use his large personal library, and his readiness to engage in discussions with me on the evolution of Neanderthals and on the relations between European Neanderthals and those of the Middle East. In recognition of his important contribution to our discipline, this book is dedicated to his memory.

Following this introduction, 24 other chapters make up this volume. It is not divided into sections as was the conference itself. The editors have preferred to present the chapters according to their thematic order in relation to the origin, evolution and disappearance of Neanderthals. In presenting contemporary responses to questions that have arisen in these areas, we have aimed to place in relief the debates these responses have fueled and the further questions they have raised. It is for this reason that more synthetically oriented chapters are found in this volume alongside other chapters which address a particular problem or topic. The broader interest of the latter type of chapter lies in its exposition of methods and techniques of analysis that are current in our disciplines.

At the beginning of the volume, six chapters focus on questions regarding classification, Neanderthal origins and the origin of the fossil populations that were contemporary to them, as well as on cultural and faunal changes that characterized their world. Thus, in Chap. 3, W. Henke and T. Hardt discuss the evolution and origin of 'early' *Homo* and they illustrate the difficulty of species recognition in paleoanthropological research. In the chapter which follows, Chap. 4, I. Tattersall likewise deals with the difficulty of species recognition, but in a perspective more directly centered on the European fossils. This author places in question the nomen *Homo heidelbergensis* as it is applied today in European Middle Pleistocene hominid systematics and also the use of the "accretion model" applied to Neanderthal evolution.

In Chap. 5, F. Mallegni also deals with the question of the origin of Neanderthals on the basis of a study of the fossil from Ceprano (Italy). Through cladistic analysis of the skull, he suggests that this fossil might represent an archaic species, different from *Homo antecessor* which he defines as *Homo cepranensis*. For Mallegni, the Italian specimen represents the holotype of this species, and Ceprano belonged to a population that gave rise to the European *Homo neanderthalensis*. This assumption regarding the oldest fossils of Europe is somewhat different from the interpretation of Bermudez de Castro et al. In Chap. 6, these authors, on the basis of a revision of features observed on the oldest Spanish fossils in Atapuerca, formulate a different hypothesis. For them, either a phylogenetic continuity exists between these archaic fossils, called *Homo antecessor*, and Neanderthals, or else both species shared a common ancestor.

Since researchers working on European specimens often refer to fossils found outside of this continent, we asked Wu Liu and Xiuje Wu to present the Chinese hominids. In Chap. 7, the authors summarize research advances made in recent years, including the discovery of new hominid fossil sites, and they examine studies related to these discoveries. Their data brings to light the significance of this recent work and the potential of China as a source of important fossils and insight into the relationship between Europe and Asia in years to come.

This first part of the volume concludes with two chapters which do not deal with particular fossil specimens, but with the dispersal of fossil populations, both toward and within Europe. In Chap. 8, N. Goren-Inbar focuses mainly on the geographical meeting point provided by the Levantine Corridor and places in relief the evolutionary and cultural background of Neanderthals. She argues that an understanding of the material culture and behavioral traits of this population requires the examination of earlier periods which form the background to Neanderthal abilities and productive capacity. Through her analysis of the Levantine culture, the author shows that particular aspects of the technologies characteristic of the Neanderthal era first emerged in the Lower Paleolithic.

In Chap. 9, W. von Koenigswald focuses on Pleistocene faunal exchange during the Middle and Late Pleistocene in Europe. He examines the impact of climate change and faunal turnover on human populations and concludes that migration as a result of such changes may account for morphological differences distinguishing various human fossils known from the Middle Pleistocene of Germany.

These last two chapters concerning the dispersal of fossil populations serve as a transition toward the second part of this volume, consisting of 13 chapters which deal with the anatomy and the diversity of the Neanderthal population. B. Vandermeersch and M. D. Garralda, in Chap. 10, summarize the data available for the principal human fossils in the European Middle and early Late Pleistocene. The authors illustrate the importance of Levantine fossils for the understanding of Neanderthals. Over the past 30 years, B. Vandermeersch has defended the thesis that there was a gradual evolution from the oldest European fossils to the Neanderthals (later called the accretion model). In this chapter, the authors do not exclude the possibility of geneflow from groups of Asian or African origin, above all in view of the breadth of their geographic dispersal and the diversity of environments to which they adapted.

This question of variability of the Neanderthal population is also the theme of the chapter by J.-L. Voisin. In Chap. 11, he examines the variability of Neanderthals by applying to them the concept of "speciation by distance". Since the differentiation of Neanderthals occurred in Western Europe, this hypothesis accounts for the morphological cline among Neanderthals, from the west (displaying more pronounced Neanderthal features) to the east. For Voisin, the shoulder girdle reflects this morphological cline, but only for architectural and not for functional features.

The particularities and variability of Neanderthals are also examined in this volume in terms of ontogenetic and growth patterns among this fossil population. Three studies deal with this theme. In the first, Chap. 12, A.-M. Tillier provides a synthetic view of difficulties we face in the effort to document distinct developmental stages among specimens originating from sites separated by vast distances or by large stretches in time. In spite of this limitation, the author illustrates the way in which comparative analyses may provide accurate information regarding the ontogenetic appearance of several diagnostic Neanderthal characteristics and reveal similarities and contrasts between Neanderthals and early modern children.

The two other studies on ontogenetic and growth patterns of Neanderthals are based on an examination of teeth. T. Smith et al., in Chap. 13, observe that studies of dental development have attained conflicting results regarding the similarity of Neanderthal growth and development to that of modern humans. Long-period lines on tooth crowns (perikymata) and roots (periradicular bands) are quantified, and crown formation, root development, and age at death are estimated. The authors show that Neanderthal dental development overlaps with the low end of modern human populations, and demonstrates a greater range of variation in Middle Paleolithic hominins than previously reported. On the basis of a different methodology, P. Smith et al. also examine Neanderthal growth patterns in relation to teeth. In Chap. 14, they consider tooth formation and the successive phases of dental development as a source of insight into ontogenetic processes. Using a three dimensional model to quantify developmental features, this study indicates that the characteristic features of Neanderthal teeth, expressed in intercusp distances and proportions, thin enamel and taurodont roots, represent a different partitioning of cell division and differentiation from that observed in *Homo sapiens sapiens*.

These studies are followed by further analyses of features found on Neanderthals which are open to debate. Their topics and methods permit us to gain a unique view of new approaches to the Neanderthal population.

The chapter by V. Volpato et al. (Chap. 15) employs synchrotron radiation microtomography (SR-μCT) to investigate the shaft length and the cross-sectional geometric properties of the humeri of the Regourdou 1 Neanderthal. The authors support the previous conclusion that Regourdou 1 was right-handed by only a modest degree of right dominance, which characterizes this individual with respect to the available Neanderthal data.

In Chap. 16, K. Harvati et al. explore mandibular traits that differentiate Neanderthals from modern humans (greater robusticity, a receding symphysis, a large retromolar space, a rounder gonial area, an asymmetric mandibular notch and a posteriorly positioned mental foramen). Using a morphological integration approach, their study assesses the effects of allometry and evaluates the influence of masticatory and paramasticatory activities on mandibular shape.

Following this chapter, Chap. 17 by P. Gunz and K. Harvati deals with the occipital bun, which is considered by some authors to be a derived Neanderthal trait and, therefore, it is often cited as evidence for admixture between Neanderthals and anatomically modern humans. On the basis of geometric-morphometric analysis, the authors suggest that the occurrence of "hemibuns" should not be used as evidence for admixture, since this morphology is not an independent trait, but a predictable correlate of the relative position of the temporal bone.

In Chap. 18, U. Witzel examines the relationship between skull function and skull shape by means of an application of Wolff's law through a deductive technique of structure synthesis. It aims to present a new method that can be used to test hypotheses regarding the relationship between structure and function during skull evolution.

D. Caramelli et al., in Chap. 20, and L. Orlando and C. Hänni, in Chap. 21, each give a review of current knowledge concerning Neanderthal DNA sequences and anticipate future challenges related to Neanderthal genomics. The authors illustrate that advances in ancient DNA technology have made possible a better understanding of the evolution of the Neanderthal gene pool and have permitted us to address the long-standing question of a possible genetic admixture with modern humans.

Chapter 19 presents the paleogeneticist's point of view according to which Neanderthals display a particular mtDNA sequence. In this new study of mtDNA, presented in this volume by J. L. Arsuaga et al., the sequence that the authors present is very short (52 bp), but it contains substitutions common to all Neandertals. There is also one substitution (16,243) that may to some degree vary with geological age. The authors conclude that the recovery of this short mtDNA fragment can be used as a diagnostic tool for taxonomic classification in European Late Pleistocene fossil human specimens.

The theme of the transition between Neanderthal and modern humans is analyzed in the third part of this volume comprising five chapters. Chapter 22 by O. Jöris et al. gives a critical review of the radiocarbon record available at the time of the demise of the last Neanderthals and the first appearance of anatomically modern humans in Europe. This paper provides a useful examination of the European radiocarbon-based chronometric record for the period between ca. 40.0 and 30.0 ka ^{14}C with reference to the stratigraphic evidence. From the point of view of dating, this chapter demonstrates how complex this period of transition is. It is here that the problem arises regarding the role of Neanderthals in the peopling of the Upper Paleolithic and the question of whether they were replaced by modern humans.

This question is developed in Chap. 23 by I. Jankovic et al. These authors place the genomic study of the fossil of Vindija in a larger context, which includes paleontological and archeological data for this site. The authors support the hypothesis of possible population interaction patterns between Neandertals and early modern humans in Europe.

A similar point of view is shared by E. Trinkaus in Chap. 24. On the basis of analysis of fossils originating in eastern Europe, the author argues that the later European Middle Upper Paleolithic (Gravettian) sample exhibits persistence of some of these and other Neanderthal/archaic features. He stresses the conclusion that simple models of an abrupt behavioral and phylogenetic transition for this period in Europe should be abandoned.

This conclusion is shared by J. Zilhao in Chap. 25. In opposition to the positions of a number of current researchers, the author supports the idea that, contrary to the "Human Revolution" paradigm that has predominated over the past 25 years, the Neanderthals cannot be considered to represent an evolutionary dead-end. He therefore argues that biogeography, demography and paleoethnography will permit us to reach different explanations to account for their differentiation and eventual demise.

Finally, in Chap. 26, M. Wolpoff and R. Caspari raise the question concerning the concept of modernity, or "humanness". In a discussion of an essay by Robert Proctor (2003), who suggested that a series of intellectual transitions gave rise to current understanding of the concept of "humanness", Wolpoff and Caspari explore transitions in our understanding of Neanderthal humanity.

At the conclusion of the arduous editorial process which was required for the publication of this volume, and also on behalf of my co-Editor, I would like to thank each of the authors for their presentations which shared with us their ideas concerning the state of research in our different fields. We regret the absence of articles by some participants in the Bonn conference, above all several paleoanthropologists who presented a different viewpoint than that found in this volume, namely that the Neanderthals did not contribute to the peopling of the Upper Paleolithic. It is understandable that those who had already written extensively on this theme did not feel that they had anything new to offer in print. I am deeply grateful also to all the reviewers who have helped to improve the papers. I would like to extend my gratitude to E. Delson for deciphering and editing the recording of the lecture by F. Clark Howell and for his invaluable assistance and advice during the different stages of preparation of this volume. I would also like to extend my heartfelt thanks to the Alexander von Humboldt Foundation, Bonn, for its generous support which made the Bonn conference possible.

References

Howell, F. C. (1951). The place of Neanderthal man in human evolution. *American Journal of Physical Anthropology, 9*, 379–416.

Howell, F. C. (1959). Upper Pleistocene stratigraphy and early man in the Levant. *Proceedings American Philosophical Society, 103*(1), 1–65.

Howell, F. C. (1986). Variabilité chez *Homo erectus*, et problème de la présence de cette espèce en Europe. *L'Anthropologie, t.90*(3), 447–481.

King, W. B. R. (1864). The reputed fossil man of the Neanderthal. *Quarterly Journal of Science, 1*, 88–97.

von Koenigswald, G. H. R. (Ed.). (1958). *Hundert Jahre Neanderthaler: 1856–1956*. Köln-Graz: Böhlau Verlag.

Proctor, R. N. (2003). Three roots of human recency. *Current Anthropology*, *44*(2), 213–239.

Chapter 3
The Genus *Homo*: Origin, Speciation and Dispersal

Winfried Henke and Thorolf Hardt

Abstract The taxonomical interpretation of the fossil record of our own genus *Homo* is still highly controversial. In spite of obvious major advances concerning the acquisition, analysis and interpretation of the fossil specimens there currently remain unbridgeable disagreements concerning the origin, speciation and dispersal of our genus. The polarized positions result foremost from different methodological approaches for species recognition (e.g., Tattersall 1986; Wolpoff and Caspari 1997; Schwartz 2000a; Wiesemüller et al. 2003; Jobling et al. 2004; Rothe and Henke 2006). Beside the Multiregional Evolutionary Model (MRE) there are different Recent African Origin Models (RAOMs) with varying numbers of speciation. This contribution discusses the evolution and origin of 'early' *Homo*, the assumed speciation events and the validity and credibility of traditional, newly described or reinterpreted *Homo*-species and the hypothesized patterns of dispersal. We ask for an explanation of the tremendous evolutionary success as well as the triggers for the incomparable career of *Homo*. Finally we will try to decipher the speciation process and dispersal patterns of our forerunners and to unravel the traces of our fossil ancestors.

Though paleoanthropologists, who seek the doorways to the past, have recently been very successful in developing new sets of keys (Henke and Tattersall 2007), the species recognition within the genus *Homo* remains a permanent challenge of paleoanthropological research. As the evolutionary process was a single real-historic-genetical event it is not a matter of taste how many species there have been. There is a need for sound taxonomic units strictly based on phylogenetic systematics as a prerequisite for progress in evolutionary biology (Wägele 2000; Wheeler and Meier 2000; Rothe and Henke 2001, 2006; Wiesemüller et al. 2003).

Keywords Hominin taxonomy • Species recognition • Speciation processes

W. Henke (✉) and T. Hardt (✉)
Institute of Anthropology (1050), Department of Biology (FB 10), Johannes Gutenberg-University Mainz, D-55099 Mainz, Germany
e-mail: henkew@uni-mainz.de; thorolf-hardt@gmx.de

Introduction

There are few problems in paleoanthropology that have been more continuously troublesome during the last century than defining the genus *Homo* because of its fragmentary hominid fossil record, the tendency to over-split hominid taxa and emphasize differences rather than similarities, and the lack of suitable methods to resolve the relative and absolute ages of fossil samples complained Howell (1978). Nearly a quarter of a century later Dunsworth and Walker (2002, p. 432) still bewail: "At the moment we are plagued with more questions than answers concerning the evolution of early *Homo*", and Smith's (2002, p. 456) conclusion in his review on the confusingly complicated patterns of migrations, radiations and continuity in Middle and Late Pleistocene humans is: "While these debates are not likely to end soon, there is clear evidence on several fronts that many participants practiced good science. [...] Along with an increasing amount of data pertinent to the patterns of modern human origins, these evolving models are certain to enhance greatly our understanding of the radiations and evolutionary history of our genus *Homo*."

Even though the methods of dating paleoanthropological and archaeological processes or events (see Wagner 2007) underwent a rapid progress during the last few decades, it is unrealistic to expect that the boundaries of the genus *Homo* are currently well defined and that the aspects of speciation and intrageneric relationships are solved. On the one hand there is a great improvement of the time resolution of the process of human evolution, and on the other hand there is to recognize a tremendous increase of the human fossil record by systematic explorations and meticulous excavations of the relevant fossiliferous layers (see Etter 1994; Henke 2005; Henke and Rothe 1994, 1999a; Johanson and Edgar 2007; Tattersall and Schwartz 2000), but in spite of both positive trends we have to notice the paradox consequence that we run the risk in paleoanthropology because of different genus and species concepts and contradictory phylogenetic theories to miss the wood for the trees (Rothe and Henke 2001; Wheeler and Meier 2000; Wägele 2000; Wiesemüller et al. 2003; Collard and Wood 2007).

S. Condemi and G.-C. Weniger (eds.), *Continuity and Discontinuity in the Peopling of Europe: One Hundred Fifty Years of Neanderthal Study*, Vertebrate Paleobiology and Paleoanthropology, DOI 10.1007/978-94-007-0492-3_3, © Springer Science+Business Media B.V. 2011

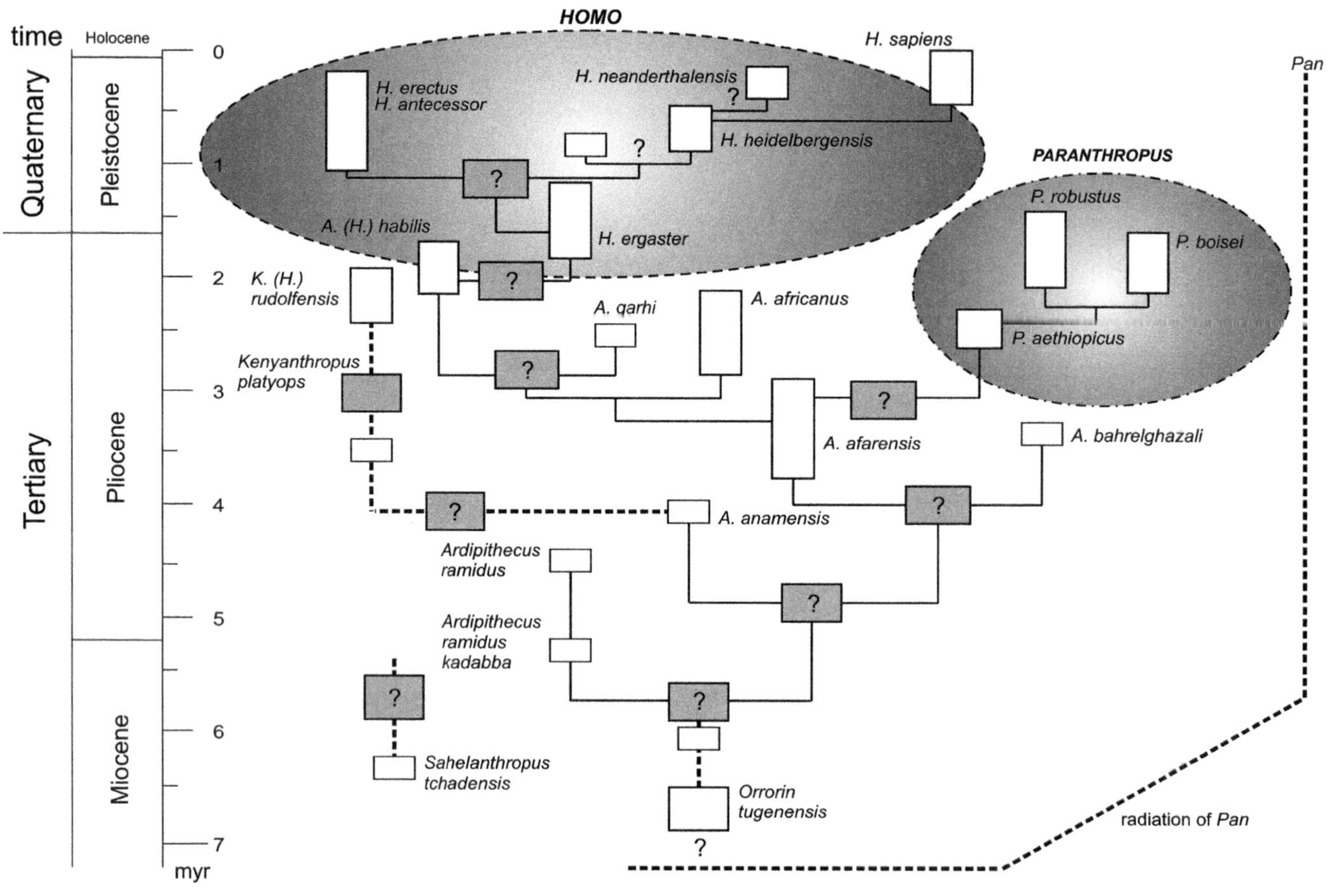

Fig. 3.1 Pedigree of hominin evolution following the extreme 'splitters' proposals that we do not favour – for further comments see text (Redrawn from Henke and Rothe 2003; Henke 2006a, modified)

Despite the rapidly increasing amounts of scientific data the controversies on human evolution are currently not minor than at that time when Clark Howell bewailed the over-split of hominid taxa, though the level has changed. While some commonly used junior synonyms like *Anthropopithecus* Dubois, 1892; *Pithecanthropus* Dubois, 1892; *Protanthropus* Haeckel, 1895; *Sinanthropus* Black, 1927, *Cyphanthropus* Pykraft, 1928; *Meganthropus* Weidenreich, 1945; *Telanthropus* Broom and Robinson, 1949 and *Atlanthropus* Arambourg, 1954, were dropped out and successively integrated in the genus *Homo* by a taxonomic revision during the sixties and seventies of the last century (e.g.. Groves and Mazák 1975), the number of *Homo*-species increased rapidly as well as the number of potential forerunners of the genus *Homo* (see pedigree in Fig. 3.1).

Due to the many different approaches ranging from evolutionary taxonomy (Simpson 1961; Mayr 1975) to numerical taxonomy or phenetics (Sneath and Sokal 1973) as well as phylogenetic systematics or cladistics (Hennig 1950; Wiley 1978; Ax 1984; Wägele 2000; Wiesemüller et al. 2003), we are actually far from a consensus concerning the taxonomically basic anthropological problem, how many fossil species of our genus have existed in the past and how they were related to one another. Despite the rapidly growing number of hominid fossil specimens, the diversification of disciplines participating in paleoanthropology and an obvious progress in methodology during the second half of the last century (Henke and Tattersall 2007; Hardt and Henke 2007; see Fig. 3.2), it is still highly controversial and open to discussion whether the process of human evolution was smooth or jumpy (see Wolpoff 1996–1997, 1999; Wolpoff and Caspari 1997; Henke and Rothe 1999a, b, 2001, 2005; Campbell and Loy 2000; Tattersall and Schwartz 2000; Tattersall 2007; Henke 2003a, b; Rothe and Henke 2006). While some scientists suggest that new hominin species originated through a slow accumulation of small-scale changes in existing lineages, i.e. a phyletic process called gradualism or phyletic transformation, others support a model which has been described as punctuated equilibrium (Gould and Eldredge 1977). The latter currently more widely accepted model likens the evolution of hominins to a series of starts and stops, i.e. periods of genetic and phenotypic stasis are periodically interrupted by rapid speciation events, which are the starts of new species (see Campbell and Loy 2000; Jobling et al. 2004; Futuyama 2007). Even if one chooses to accept that the most appropriate metaphor for the

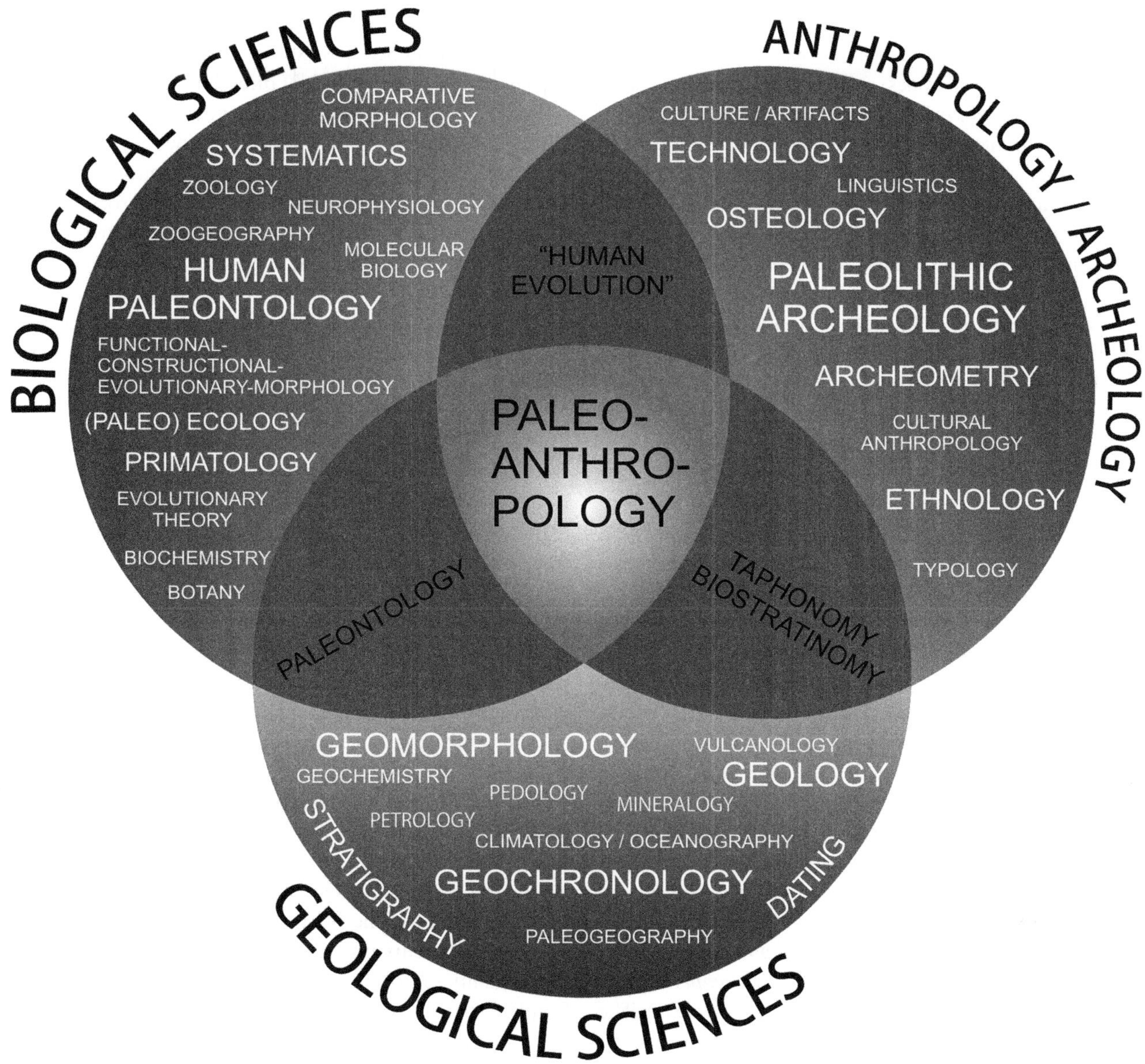

Fig. 3.2 VENN-Diagram presenting the participation of manifold scientific fields in palaeoanthropology (Modified after Delson 2000, from Henke and Tattersall 2007)

pattern of human evolution appears to be a branching bush, rather than a ladder to be climbed, one question still remains to be answered: How many hominin species were there?

This contribution aims to give an answer by reviewing the current alternative paleoanthropological ideas on the origin and diversification of the genus *Homo*. The review starts with a brief historical overview of the quests for centres of our generic origin, continues with the discussion on the earliest species of *Homo*, and asks for an explanation of the tremendous evolutionary success as well as the triggers for the incomparable career of our genus. Finally we will try to decipher the speciation process and dispersal patterns of our forerunners and to unravel the traces of our fossil ancestors.

Early Searches for the 'Cradle of Mankind'

The Neanderthal man from the Kleine Feldhofer Grotto in the Neander Valley near Düsseldorf was found by limestone workers in 1856 and described by the local teacher Johann Carl Fuhlrott (1803–1877). The fossil was the first early human specimen to be recognized as such. Discoveries like that from Engis (excavated in 1829) and another from Gibraltar (found in 1848) were made sooner, but their nature became evident much later. Fuhlrott's merit was that he had realized the significance of the fossils, which the limestone workers took for animal bones (Schott 1977, 1979; Schmitz and Thissen 2000). Fuhlrott fought, together with the anatomist Herrmann

Schaaffhausen (1816–1893), for their acceptance as ancient remains from the diluvial age (Zängl-Kumpf 1990). As the discovery of the fossil bones antedated the publication of Darwin's *Origin of Species*, this specimen has often been apostrophized as first proof for human evolution. (Bowler 1988; Desmond 1997; Corbey and Roebroeks 2001; Henke 2007). In his famous papers on *Evidences as to Man's Place in Nature*, Darwin's colleague Thomas Henry Huxley (1863) for the first time gave morphological arguments for our relationship with recent primates and pointed to the scarce fossil record known in his time. In spite of contrary statements by many historians, Huxley said virtually nothing about human origins but concentrated exclusively on demonstrating the physical resemblances of humans and apes. He concluded that humans were still sufficiently unique that they should be kept apart from the great apes, which he relegated to their own taxonomic group. Concerning the Neanderthal man from Germany, Huxley conducted a sophisticated comparison with anatomically modern skulls from Australian Aborigines and other aboriginal relicts, pioneering new ways of orienting and measuring skulls for easier comparison (Desmond 1997). He viewed the Neanderthals as a very "primitive race" of humans, "the most pithecoid of human crania yet discovered" (Huxley 1863, p. 205). The Huxley-biographer Desmond (1997) commented that Huxley was preparing the world for "ancient semihumans", which means that the idea of our early forerunners was at Darwin's time woolly and premature.

The Irish zoologist William King proposed in 1864 the species name *Homo neanderthalensis*, although his arguments for a separate species in the genus *Homo* were inadequate and imprecise. Since then, opinion has fluctuated as to whether the fossils should be considered as a separate species, *Homo neanderthalensis* or *Homo primigenius* or as a subspecies of *H. sapiens* (*H. sapiens neanderthalensis*). The "fate of the Neanderthals" (Brace 1962, 1964; Trinkaus and Shipman 1993) remained the trickiest controversy in paleoanthropology until now (Spencer 1984; Stringer and Gamble 1993; Tattersall 1995a, b; Krings et al. 1997, 2000; Henke and Rothe 1999a; Wolpoff 1999; Finlayson 2004; Henke 2006a, 2007; Harvati 2003, 2007).

What became very soon apparent in the second half of the penultimate century was the tremendous need for an extension of the fossil report – since that time colloquially termed as "missing links". Beyond it there was missing an improved comparative methodology to analyze and interpret the human fossils. As recently has been described by Henke (2006, 2007) and Henke and Rothe (2006) neither the Neanderthal fossils from the Feldhofer Grotto nor those from many other European sites had a path-breaking impact on the formation of paleoanthropology as biological science. Furthermore the discovery of the specimens from La Naulette (1866), Šipka (1880), Spy (1886), Taubach (1887) and the pivotal fossils and artefacts from Krapina (1899–1906) did not induce a paradigmatic shift in paleoanthropology though Dragutin Gorjanović-Kramberger's excavations had pushed the anthropological research in the right direction of detailed comparisons and hypothesis-testing (Gorjanović-Kramberger 1906; see Henke 2006b). Others asked whether Eugene Dubois' essential discovery of *Pithecanthropus erectus* in 1891 coincided with a paradigm shift in paleoanthropology (see Shipman and Storm 2002). The implication which could be derived from the fossils was that erect posture and bipedal walking antedated the process of encephalization, but even this finding was not convincing for the leading paleoanthropologists of that time. That paleoanthropology remained predominantly a highly narrative science in the early twentieth century became evident from the fatal Piltdown forgery (Spencer 1990a, b), named *Eoanthropus dawsoni*, and the fact that it took more than 40 years until John S. Weiner, Sir Kenneth Oakley, and Sir Wilfrid LeGros Clark jointly exposed the hoax, although we have to mention that there was much scepticism and rumour earlier (see Friederichs 1932; Weinert 1947). Whoever the players were in this black mark in science, they were aware of the attractiveness and fascination of fossils, and they obviously knew about the public appeal and the rare resources that help to decipher our place in nature (see Stringer and Gamble 1993; Foley 2001).

The embarrassing success of the Piltdown forgery unmasked the ethnocentrism and Euro-centrism of the European anthropologists of early twentieth century, in other words, hardly anybody of the protagonists of that time could image that our phylogenetic forerunners originated and lived in another continent than Europe the obvious centre of culture. The interpretation of the Piltdown fossil as a human precursor was partially responsible for the vehement dismissal of the first *Australopithecus* fossil from South Africa. Raymond Dart's interpretation of the Taung child as missing link between ape and man yielded a storm of controversy (Woodward 1925; Keith 1931). This is remarkable insofar as Dart's discovery matched the prophecy of Darwin (1871, p. 202): "It is, therefore, probable that Africa was formerly inhabited by extinct apes closely allied to the gorilla and chimpanzee: and, as these two species are now man's nearest allies, it is somewhat more probable that our progenitors lived on the African continent than elsewhere."

To the extent that these and other indications for an extra-European "cradle of humankind" were deliberately ignored, the Piltdown case is a telling example of cut and dried opinions (Spencer 1990a, b). Small wonder, that the scientific output of paleoanthropology during the first decades of the last century was far more redolent of stagnation than progress. The evolutionary biology at those times was characterized by Ernst Mayr as "chaotic" (Tattersall 2000a, p. 2). We can notice that even at the beginning of the twentieth century, Darwin's principles were widely misunderstood by anthropologists, who persisted in orthogenetic biological thinking or insisted on the theoretical split between natural sciences and humanities (see Henke and Rothe 2003, 2005, 2006). Paleoanthropological theory and methodology were nascent, and it took a total change to a population genetical thinking and a modern Synthesis by Theodosius Dobzhansky, Ernst Mayr and Julian Huxley in the first half of the last century and an integrated primatological approach during the second half, introduced by Sherwood Washburn (1953) in the fifties of the last century, to establish the theoretical frame of an innovative paleoanthropological research (see Foley 1987; Henke and Rothe 1994, 2003, 2006; Henke 2007; Henke and Tattersall 2007).

After the discovery of the Taung child Africa came more and more into the focus of paleoanthropology, firstly with surprising fossils from South Africa [*Australopithecus* (*Plesianthropus*) *transvaalensis*; *Paranthropus crassidens*; *Paranthropus robustus*] and secondly from East Africa [*Paranthropus* (*Zinjanthropus*) *boisei*, *Paranthropus* (*Paraustralopithecus*) *aethiopicus* (further newly excavated and described taxa see pedigree Fig. 3.1)]. A paleoanthropological sensation was the discovery of a human fossil from Olduvai (F.L.K.N.N. Site – Bed I) which was classified as *Homo habilis* by Louis S. B. Leakey et al. (1964). Phillip Tobias (1989a, b, 1991) retrospectively called this fossil a premature discovery after Stent who defines that "… a discovery is premature if its implications cannot be connected by a series of simple logical steps to canonical or generally accepted knowledge" (Stent 1972, p. 84).

Actually there is a highly controversial discussion on the phylogenetical status of *H. habilis* as the hypodigm of this taxon has been successively enlarged during the last 50 years and exhibits an extremely large variation (Alexeev 1986; Hartwig-Scherer and Martin 1991; Collard and Wood 2007; Susman 2008). The fact that there is no agreement about the classification of the earliest evidence of *Homo* is on the one hand an indication for a severe theoretical deficit of our taxonomical approaches and on the other hand strong evidence that we are at the roots of our genus. Currently there is nearly no doubt that Africa is the cradle of *Homo* and that our genus conquered the Old World soon after its emergence. What was the "Rubicon" *Homo* had crossed to finally reach this incomparable success of our own species – and how many species of this genus went extinct (Foley 1991)? These are essential paleoanthropological questions to be answered.

The Earliest Evidence of Homo

Within the theoretical concept of an adaptive evolutionary change from archaic hominids to human-like ones the traditional expectations concerning the genus *Homo* are full uprightness, successive reduction of the mastication apparatus as adaptation to changed food processing, improved encephalization, and evidence of cultural abilities. Whether the fossil species attributed to *Homo* had already developed a human-like society (Rothe and Henke 2006), a language, or art (Haidle 2007; Mithen 1998, 2007) can only be judged with great uncertainty, if at all. For example, allometrical effects may cover the real abilities, i.e., body sizes have to be taken into account to achieve an idea of the relative brain size (Hemmer 2007). Current research on cultural behaviors and underlying cognitive and linguistic competences in early hominins is a highly interdisciplinary enterprise (see Henke and Tattersall 2007), which gives only very rough information concerning the first appearance of the genus *Homo*.

We are of the opinion that there is a need for sound taxonomic units as a prerequisite for progress in evolutionary biology (Wägele 2000; Wheeler and Meier 2000; Rothe and Henke 2001, 2006; Wiesemüller et al. 2003), and do not agree with those who argue that discussions on species concepts are wasted effort. Though we are convinced that there is little hope that the discrepancies about the theory of speciation and the species concepts (e.g., biological species, chronological species, cladistic species, ecological species, evolutionary species, morphological species, phenetic species, phylogenetic species, reproductive species) will be eliminated in the near future and paleoanthropologists will reach a consensus on hominid systematics, we agree with those who regard taxonomy and systematics beside a thorough background in an increasingly broad spectrum of disciplines (see Fig. 3.2) as an essential prerequisite for phylogenetic discussions. Without an exact knowledge of the population genetical processes, a precise theory of speciation and an inter-subjective definition of taxonomic categories there is no chance to escape the narrative concepts of paleoanthropology and the image of an "Instant Science", as Kathy Chang labeled a paleoanthropology that knows the answers from the first glance (cited in White 2000).

Excurse

Here we will not step deeper into this discussion, but want to explain our viewpoint in brief. The most common definition of a species is the biospecies (Mayr 1969, 1975), i.e., a group of actually and potentially interbreeding natural populations, which is reproductively isolated from other species. Because this definition is applicable only to contemporaneous living organisms, alternative definitions have been proposed (Simpson 1961; Wiley 1978). Paleoanthropologists very often describe fossil species as morphospecies, based on morphological or anatomical similarity, or refer to temporally successive species in a single lineage, so-called paleospecies or chronospecies. We don't regard these species concepts as valid approaches. The most accepted definition of a species is given by Wiley (1978, p. 18) as "…a single lineage of ancestral descent populations of organisms which maintains its identity from other such lineages and which has its own evolutionary tendencies and historical fate" (see Wiesemüller et al. 2003; Henke 2005; Rothe and Henke 2006). The currently best method of choice to decipher phylogenies is the phylogenetic systematics or cladistics (see Hennig 1950; Wiley 1978; Ax 1984; Wägele 2000; Wiesemüller et al. 2003; Rothe and Henke 2006). A reconstruction of phylogenetic relationships is strictly based on the observation of synapomorphic traits. However we must be aware, paleoanthropologists are just modelling!

The Potential 'Candidates' for the 'Earliest' *Homo*

The discoverers of *H. habilis* (Leakey et al. 1964) described the *species nova* as being more advanced than *A. africanus* and not as humanized as *H. erectus*. But even 15 years after the species was labelled for the first time, the majority of competent scholars in the field did not accept *H. habilis* as a valid taxon. The reason was that there was a lack of 'morphological space' between *A. africanus* and *H. erectus*, as Stringer (1986) supposed in his article on "The Credibility of *Homo habilis*", i.e., the new species was strongly criticized in the beginning, for many experts argued that the Olduvai specimens could be classified within existing taxa. Meanwhile, there seems to be worldwide acceptance of the fact that specimens with smaller endocranial capacities need *not* be excluded from membership in *Homo* and that the fossils are not a normal variant of either the putative australopithecine ancestor or the putative descent *H. erectus*. Within the material, attributed to 'early *Homo*' (Wood 1992, 1996a, b), three species can be identified today, one of them resembles *Homo erectus* and is interpreted as either 'early African *Homo erectus*' or *Homo ergaster*, and the two other species are *Homo rudolfensis* and *Homo habilis* (*sensu stricto*).

Homo rudolfensis and Homo habilis (s. str.): Although Tobias (1991) made a strong case for only one species being represented among specimens attributed to *H. habilis*, other authors see evidence that there may be two species, one named *H. rudolfensis,* which was proposed in 1986 by the Russian anthropologist Valeri Alexeev for specimen KNM-ER 1470 (Alexeev 1986), and the other *H. habilis* [*s. str.*] (see e.g., Tobias 1989a, b; Wood 1992; Henke and Rothe 1994; Stringer 1996, 2001; Tattersall and Schwartz 2000; Collard and Wood 2007; Schrenk et al. 2007; Susman 2008), which are sister taxa within a monophyletic *Homo* clade. However, this phylogenetic interpretation is only a little more parsimonious than a polyphyletic one, explaining the features typifying each species as parallel developments (see also Bilsborough 1992, p. 132; Kennedy 1999; Rightmire 2007). Anyway, whatever may be the correct taxonomic solution, there are strong arguments against a *H. habilis* [*sensu lato*] concept while the hypodigm remains unsure. *H. rudolfensis* shows, apart from an increased average brain size of c. 750 cc, features of the face and masticatory apparatus that parallel those of *Paranthropus*, e.g., marked orthognathy, broader midface than upper face, and large palate, but *H. habilis s. str.* shows a moderate average brain size of 610 cc and progressive features of cranium, face, and jaws.

The postcranium of *H. rudolfensis* is evidently more derived, like that of later *Homo*, although the association with the skull fragments is not confirmed. In contrast, *H. habilis s. str.* shows a very plesiomorphic, australopithecine-like physique. The described combination of australopithecine-like cranio-dental features with a derived postcranium in *H. rudolfensis* and of progressive cranium and dentition with primitive body proportions in *H. habilis s. str.* (a highly putative association) allows the conclusion that neither of the two species represents a reliable ancestor of later *Homo* (see *H. ergaster*), because every interpretation has to take very unlikely evolutionary changes into account to explain these unusual morphological mosaics.

Fossils representing *H. habilis* [*s. l.*] were first described from Olduvai, where they were uncovered from layers that have also yielded australopithecine skeletal material (Bed I and Bed II), but the largest contribution to the hypodigm comes from another site in East Africa, Koobi Fora, on the northeast shore of Lake Turkana. Further remains of the species have been recovered from Members G and H of the Omo Shungura Formation. From the western shore of Lake Turkana, a cranial fragment from the Nachukui Formation has been described, and a fragmentary skull, Stw 53, was recovered from the South African cave of Sterkfontein in Member 5. The attribution of hominin material from Member 4 of Sterkfontein to *H. habilis s. l.* is uncertain, although material from Swartkrans Member I (Sk 847, Sk 27) was confirmed as belonging to *H. habilis s. l.* In addition to the fossils from sub-Saharan sites, there are reports on *H. habilis s. l.* fossils from sites beyond Africa, the Near East and Asia, which have only little reliability and acceptance (Fig. 3.3). The actual hypodigm concentrates especially on East Africa. A mandible from Uraha (U 501, Malawi), which was attributed to *H. rudolfensis* by Bromage et al. (1995) forms a link to the more northern sample (Schrenk et al. 1993, 2007).

The recognition of *H. habilis s. l.* and the dating of some specimens from Olduvai to about 2.0 Ma was the first compelling evidence for the existence of Tertiary man in the sense of a species of the genus *Homo* (Tobias 1989a, b). Due to the improvement of dating methods and the discovery of new remarkable fossils, it is currently accepted that the genus *Homo* emerged before the end of the Pliocene. Earliest well-flaked

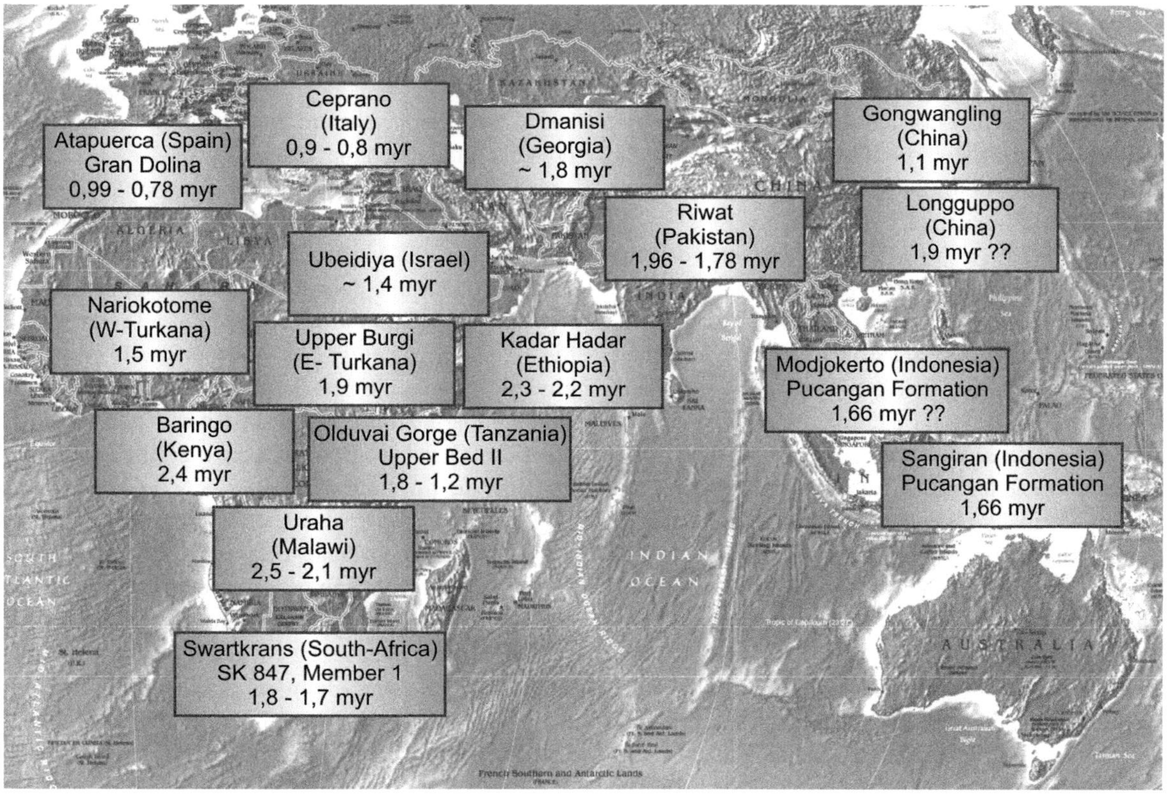

Fig. 3.3 Earliest traces of the genus *Homo* around the world with chronological setting (Henke 2003a, modified)

stone artefacts, identified as Oldowan-like in character, have been traced back to about 2.5–2.6 Ma (Semaw 2000); however, there is no proof that these implements testify to the presence of a particular hominin species. The definition of *H. habilis* is, essentially, an anatomical one, although ethological evidence may be added in support (Tobias 1989a, b, 1991). The most recent occurrences of *H. habilis* are from Olduvai Bed II, dating to about 1.6 Ma. The higher age of *H. rudolfensis* (2.5–1.8 Ma) in comparison to *H. habilis s. str.* (2.1–1.6 Ma) cannot be taken as evidence that this species is the better candidate for the direct *Homo* lineage, because the phylogenetic analysis has to be based on relevant diagnostic features (Table 3.1).

Tobias' (1991) review of the morphology of *H. habilis s. l.* lists the following critical morphological features of the first description, which have been strengthened and supplemented by subsequent studies.

The postcranium exhibits a very controversial morphological pattern: on the one hand there are distinct similarities with *H. sapiens* (e.g., clavicle, broad terminal phalanges, capitate metacarpophalangeal articulations, stout and adduced big toe, well marked foot arches) and on the other hand distinguished differences (e.g., scaphoid, trapezium, trochlea surface of the talus, robust metatarsal III). The partial skeleton OH 62, a *H. habilis s. str.*, which was found by Johanson et al. (1987),

Table 3.1 Diagnostic features of *Homo habilis s.l.*

- Absolute and estimated relative brain size (average 640 cc) with spectacular advance over australopithecines; exaggerated encephalization, brow ridges vertically thin
- Relatively open-angled external sagittal curvature to occipital
- Thin-walled braincase
- Light pneumatization of cranial bones
- Face moderately prognathous, but less marked than in *A. africanus*
- Retreating chin, with a slight or absent mental trigone
- Foramen magnum slightly in front of the basis cranii
- Large canines in comparison with australopithecines and *H. erectus*
- Canines large compared with premolars
- Petrous pyramid of the temporal bone lying in nearly transverse and coronal plane
- Cheek-teeth with reduced crown diameters and crown area in comparison to those of australopithecines
- Molar crowns small buccolingually and elongated mesiodistally
- Third molars tending to be smaller than second molars, especially P3, P4, M1 showing buccolingual narrowing of the crowns
- Lateral aspect of the frontal lobe exhibiting a pattern of sulci, typical of *Homo sapiens*
- Well developed bulges in Broca's area and in the inferior parietal lobule (part of Wernicke's area)
- Complex middle meningeal vascular pattern

Compiled by Tobias (1991)

has especially caused much discussion (Hartwig-Scherer and Martin 1991), because the estimated length and robustness of the humerus and forearm bones of OH 62 suggest that its proportions are remarkably ape-like, and the predicted weight/stature relationships are also more like that of Panini (Aiello and Dean 1990; Aiello 1996), but the phylogenetic status of 'Lucy's child' remains uncertain.

Although Tobias (1991) gave an extremely detailed description of *H. habilis s. l.*, listing 334 cranial, mandibular, and dental features, the question of what autapomorphic features define *H. habilis* remained controversial (Stringer 1986; Bilsborough 1992; Wood 1992; Henke and Rothe 1999; Grine 2001; Manzi 2001; Rightmire 2001a, b; Wood and Collard 2001, 2007). The main reason for the uncertainty in the interpretation of the diversity of early *Homo* is that the fossil hominin remains that are formally or informally allocated to *H. habilis* or declared to have affinities with this species vary from one author to another; in other words, there are multiple taxon solutions. The claimed heterogeneity of the *H. habilis s. l.* material from Olduvai (e.g., OH 7, 13, 16, 24), Koobi Fora (e.g., KNM-ER 1470, 1590, 1805, 1813, 3732), Omo (L894-1, Omo 75–14, Omo 222–2744), and Chemeron (KNM-BC1), as well as from Sterkfontein (e.g., Stw 53) caused different approaches to find a better supported classification, but the split into two species is taxonomically ambiguous (Wood 1996a, b; Henke and Rothe 1994, 1999). Cladistics has made little contribution to the search for distinctive features, or autapomorphies, of *H. habilis*. For example, Chamberlain and Wood (1987) concluded that when *H. habilis s. str.* and *H. rudolfensis* are separately included in a cladistic analysis, they are linked as sister taxa within a single clade defined by the feature states of elongated anterior basicranium, higher cranial vault, mesiodistally elongated M1 inf. and M2, and narrow mandibular fossa. Wood and Collard (2001, p. 141) concluded "…that if *Homo habilis sensu stricto* and *Homo rudolfensis* are included in *Homo* that genus fails both the 'grade' and the 'clade' tests."

The most complex cladistic analysis of early hominin relationships was conducted by Strait et al. (1997) and Strait and Grine (2004). Several different approaches agreed in indicating that the robust australopithecines form a clade, that *A. afarensis* is the sister taxon of all hominins, and that the genus *Australopithecus*, conventionally defined, is paraphyletic. Concerning *H. habilis*, the relationships of *A. africanus* and *H. habilis* were unstable in the sense that their positions varied in trees that were marginally less parsimonious than the favoured one.

The paleoecological scenario that explains the observed phylogenetic pattern states the possibility that bipedalism (and hence, the earliest hominins) evolved in response to changing ecological conditions in Africa during the late Miocene and early Pliocene. Vrba's (1988) faunal reconstructions indicate that hominin diversity between 2.5 and 1.5 Ma was possibly associated with environmental desiccation. After 2.5 Ma, hominin diversity is represented primarily by two distinct lineages, *Paranthropus* and *Homo*, which reacted to such desiccation by following different evolutionary trajectories (i.e., hypermastication vs. hypercephalization) (Strait et al. 1997, p. 56). The morphological changes demonstrate that the masticatory apparatus increased moderately in the early stages of human evolution. Subsequently it increased markedly in *Paranthropus* and decreased in *Homo*, a dichotomy that almost certainly represents a divergence in trophic adaptations (Fig. 3.4).

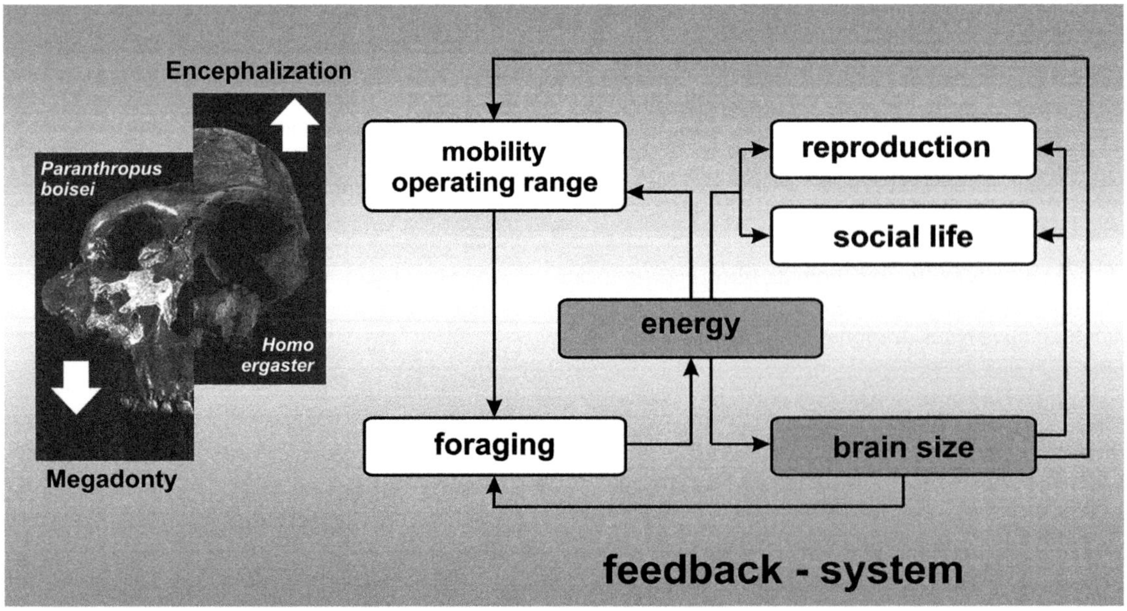

Fig. 3.4 Encephalization vs Megadonty: feedback-system, which describes the interdependencies of several components leading to the evolutionary 'success' of *Homo* (After Martin 1995; from Henke 2003a)

A cladistic analysis conducted by Wood and Collard (1999a, b) showed that neither *H. habilis* nor *H. rudolfensis* can be assumed with any degree of reliability to be more closely related to *H. sapiens* than they are to species allocated to other genera. In a recent contribution Collard and Wood (2007, p. 1575) confirmed their positions, concluding, that "…the available evidence still suggests that the adaptive strategies of *H. habilis* and *H. rudolfensis* were different from those operated by *H. erectus*, *H. ergaster*, *H. heidelbergensis*, *H. neanderthalensis*, and *H. sapiens*." Their proposal to remove *H. habilis* and *H. rudolfensis* from die genus *Homo* and to assign them to a different genus or pair of genera is not widely-accepted (e.g., Schrenk et al. 2007; Johanson and Edgar 2007; Susman 2008). However, the characterization of the *H. habilis s. l.* as "mysterious" hypodigm tells its own tale (Tattersall and Schwartz 2000).

Homo erectus (incl. Homo ergaster): At the end of the Basal Pleistocene (c. 1.5 Ma), *H. habilis* and *H. rudolfensis* disappeared from the fossil record, followed somewhat later by *P. robustus* and *P. boisei*. The exact reason for their extinction is not known, but there are indications from high faunal turnovers that climatic fluctuations may have caused dramatic ecological shifts. At the beginning of the Pleistocene, the first epoch of the Quaternary, which was characterized by a series of glacial and interglacial periods, a new hominin emerged, African *Homo erectus sive Homo ergaster*. There is a long-standing hypothesis that populations similar to this species were directly ancestral to the earliest members of the living species *H. sapiens*, whereas hypotheses concerning the link to hominin forerunners changed with the increase in the fossil record and are still under ongoing discussion (see e.g., Rightmire 1990; Henke and Rothe 1995, 1999; Henke 2003a, b, 2005, 2006a; Antón et al. 2007; Tattersall 2007; Hardt and Henke 2007).

The German zoologist Ernst Haeckel first linked humans to apes in a tree-like diagram in "General Morphology" from 1866 and suggested an Asian origin of humankind, based on the hypothesis of a near relationship to hylobatids (Haeckel 1866). The first fossil finds of *H. erectus* (formerly named *Pithecanthropus erectus*) were made in 1891 by Eugene Dubois in Central Java. The Dutch army doctor was convinced at that time that he had discovered the presumed 'missing link', the transitional form between apes and humans. The conviction that the new human taxon had been an erect bipedal creature resulted in the species name *erectus*. Because the Java man was the first non-European fossil in paleoanthropology, the discovery led to questioning of the European-centred world view that had so far been supported by the famous Neanderthal fossils. From now on, Asia was hesitantly accepted to be the place of human origin. South-East Asia and the Far East became in the first half of the twentieth century the centre of the search for the earliest human fossils, until it became evident that hominins of a *H. erectus* grade existed in the other continents of the Old World too.

During the last decades it became the favoured phylogenetic hypothesis that *H. erectus* (or a conspecific taxon named *H. ergaster*) originated in Africa from an earlier species of the same genus, *H. habilis s. l.*, forming an intermediate position in the human family tree between the ancient forerunner and 'archaic' *H. sapiens*. *H. erectus* was – sloppy verbalized – a sandwich-species. This species was said to be the conqueror of the Old World, the first hominin to emigrate out of Africa, successively reaching Asia and Europe. But its evolutionarily intermediate position in the human family tree has been questioned more and more during the last decades (e.g., Howell 1986, 1996; Franzen 1994a; Henke and Rothe 1995; Rightmire 1990; Tattersall 2000a, b, 2007; Schwartz and Tattersall 2002, 2003; Wood and Collard 2001; Henke 2003, Henke 2005, 2006; Henke and Rothe 2006; Antón et al. 2007; Harvati 2007).

Definitions of *H. erectus* (Dubois 1894) still rest on the Far Eastern fossils from Zhoukoudian and Java (Jacob 1975; Howells 1980; Weidenreich 1943). The species *H. erectus* came to replace a variety of contemporaneous geographically distinguished genera, including the genera *Pithecanthropus*, *Sinanthropus*, *Meganthropus*, and *Atlanthropus*. A taxonomic revision by Campbell (1965) deleted older genera and species, lumping these Early and Middle Pleistocene hominin taxa into a single species and separating them only on the subspecies level (*H. erectus erectus*, *H. e. modjokertensis*, *H. e. pekinensis*, *H. e. capensis*, *H. e. leakeyi*, *H. e. mauritanicus*, *H. e. heidelbergensis*, *H. e. ngandongensis*, *H. e. yuanmouensis*, *H. e. bilzingslebenensis*, *H. e. tautavelensis*, *H. e. georgicus*, and others). But this taxonomic approach was obviously not the complete solution to all problems, because the more fossils were found and the more precise the chronological dating became, the more complex the whole situation has become.

At the 100 years anniversary of "*Pithecanthropus*" Jens Lorenz Franzen (1994b, p. 9) asked among others the following burning questions: "What really is *Homo erectus*? Is it a good species? Behind which two or even more 'true' species may be hidden? – How can *Homo erectus* be defined, particularly if it is not the result of a speciation event but just a transitional phase of phylogenetic development on the way to modern man? – How can it be separated from 'archaic' *Homo sapiens*? – And then again, is *Homo erectus* the result of a cladogenetic event or is it the result of continuous transition? – Is it possible to distinguish between an Asian, and African and/or European branch of *Homo erectus*? – Should the African and European branches, if they really exist, be called species of their own?"

The answers were as diverse as the questions, by no means unexpected in paleoanthropology, especially in regard to the diversity and weaknesses in principal taxonomic approaches. Although some anthropologists regard *H. erectus* as a grade within a transitional phylogenetic model (e.g., Frayer et al. 1993; Thorne and Wolpoff 1981, 1992; Wolpoff 1980, 1992, 1996–1997, 1999; Wolpoff et al. 1994a, b; Wolpoff and Caspari 1997, 2000), other authorities hold the view that *H. erectus* originated from a hominin branch of African origin – possibly *H. ergaster* – in Asia and remained restricted to the Far East (e.g., Tattersall 1995a, b, 2007; Tattersall and Schwartz 2000). Some experts claim that the Asian and African wings of *H. erectus* exhibit no autapomorphic features and that the early Europeans originated from an African species other than *H. erectus*, named *Homo heidelbergensis* (Rightmire 1990, 1998). Is it a phantom, a chimera?

The recent discussion on the question of whether *H. erectus* was an ancestor of our own species or an evolutionary side branch received new stimuli from exciting fossils from Dmanisi (Georgia). From this site at the gates of Europe, in 1991 a fossil mandible and since 1999 more or less complete skulls and postcranial material of an unexpected high age (most reliable date 1.75 Ma) and surprising morphology indicate a much earlier Eurasian dispersal of *Homo* than was believed before (Bräuer et al. 1995; Henke 1995; Henke et al. 1999; Henke and Rothe 1999b; Gabunia et al. 2000a, b, 2001; Vekua et al. 2002; Rightmire et al. 2005).

Because early *Homo* fossil material from Africa that has been classified as *H. ergaster* (Groves and Mazák 1975; formerly attributed as 'African' *H. erectus*) has a maximum age of only c. 1.9 Ma, and the hominin fossils from Dmanisi (Gabunia et al. 2001; Vekua et al. 2002), Sangiran and Modjokerto (Java, Swisher III 1994; for discussion of the chronology see Langbroek and Roebroeks 2000), and Longgupo, China (Huang et al. 1995) may be of around the same age or a little younger, there is a severe problem of explaining the very early Eurasian dispersal and finding out which species was the pioneering emigrant (Fig. 3.5). The evidence from Longgupo Cave suggests by the given primitive nature of the premolar teeth, that the first hominid to occupy Asia may not have been *H. erectus*, but perhaps a variant of *H. ergaster* or even *H. habilis*. Etler et al. (2001) question this interpretation of the fossils from Longgupo. The mandibular fragment cannot be distinguished from penecontemporary fossil apes, especially the Late Micoene-Pliocene Chinese genus *Lufengpithecus*, while the incisor is indistinguishable

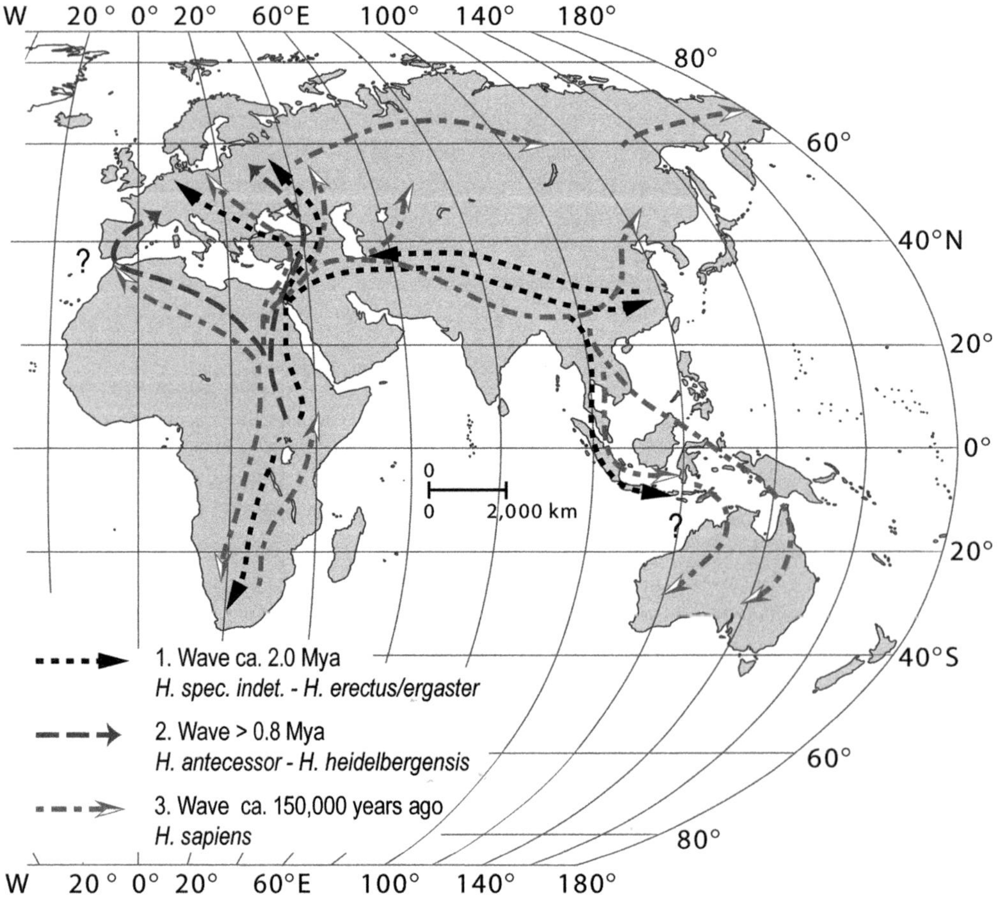

Fig. 3.5 Potential migration waves "Out of Africa" and periods of archaic *Homo* – populations supposed by several paleoanthropologists (After Hardt et al. 2007)

from those of recent land living east Asian people and may be intrusive in the deposit.

The Dmanisi mandible D 211 (Henke 1995; Bräuer et al. 1995) and the later excavated skulls with their small cranial capacities and plesiomorphic features (e.g., postorbital constriction) have been discussed as descendents of and "early" *H. ergaster* or an evolved *H. habilis* too (Gabunia et al. 2001). These specimens are actually the key fossils in the current discussion (Balter and Gibbons 2002) of the early *Homo*-dispersal from Africa. For a better understanding of the complex phylogenetic pattern and deciphering of the 'muddle in the middle', as Rightmire (1998) characterized the problems concerning Late Early and Middle Pleistocene human evolution, we have to analyze the regional and chronological hypodigm and discuss a possible African origin and early Asian dispersal (Henke and Rothe 1994, 1999a, b; Larick and Ciochon 1996; Wolpoff 1996–1997; Manzi 2001; Schwartz and Tattersall 2000, 2002). Only when species – in the sense of an evolutionary species (Wiley 1978) – have been adequately defined morphologically appropriate comparisons can be made and the distribution of character states across species is used to generate phylogenetic hypotheses. But until now, we do not have any consensus of the *H. erectus* hypodigm, which means that we have no agreement on the question of which fossils belong to the taxon that has been defined as *H. erectus* (Howells 1980, 1993; Rightmire 1990; Franzen 1994a, b; Henke and Rothe 1995; Howell 1996; Antón 2003; Antón et al. 2007; Tattersall 2007). Chronological information concerning the earliest traces of the genus *Homo* in the Old World is given in Fig. 3.6. *Homo* fossils which indicate the earliest appearance of *H. erectus* (or *H. ergaster*) or an earlier *Homo*-species out of Africa are described from Asia (incl. Dmanisi; see Fig. 3.7), while Europe has been invaded much later by hominins which have been classified as potential *H. erectus* or as *H. antecessor* followed by *H. heidelbergensis*, and *Homo* (*sapiens*) *neanderthalensis* and modern humans (Fig. 3.8) (Ullrich 1998).

Because those hominins that were found in Java since 1891 (formerly *Pithecanthropus*) and at Zhoukoudian, near

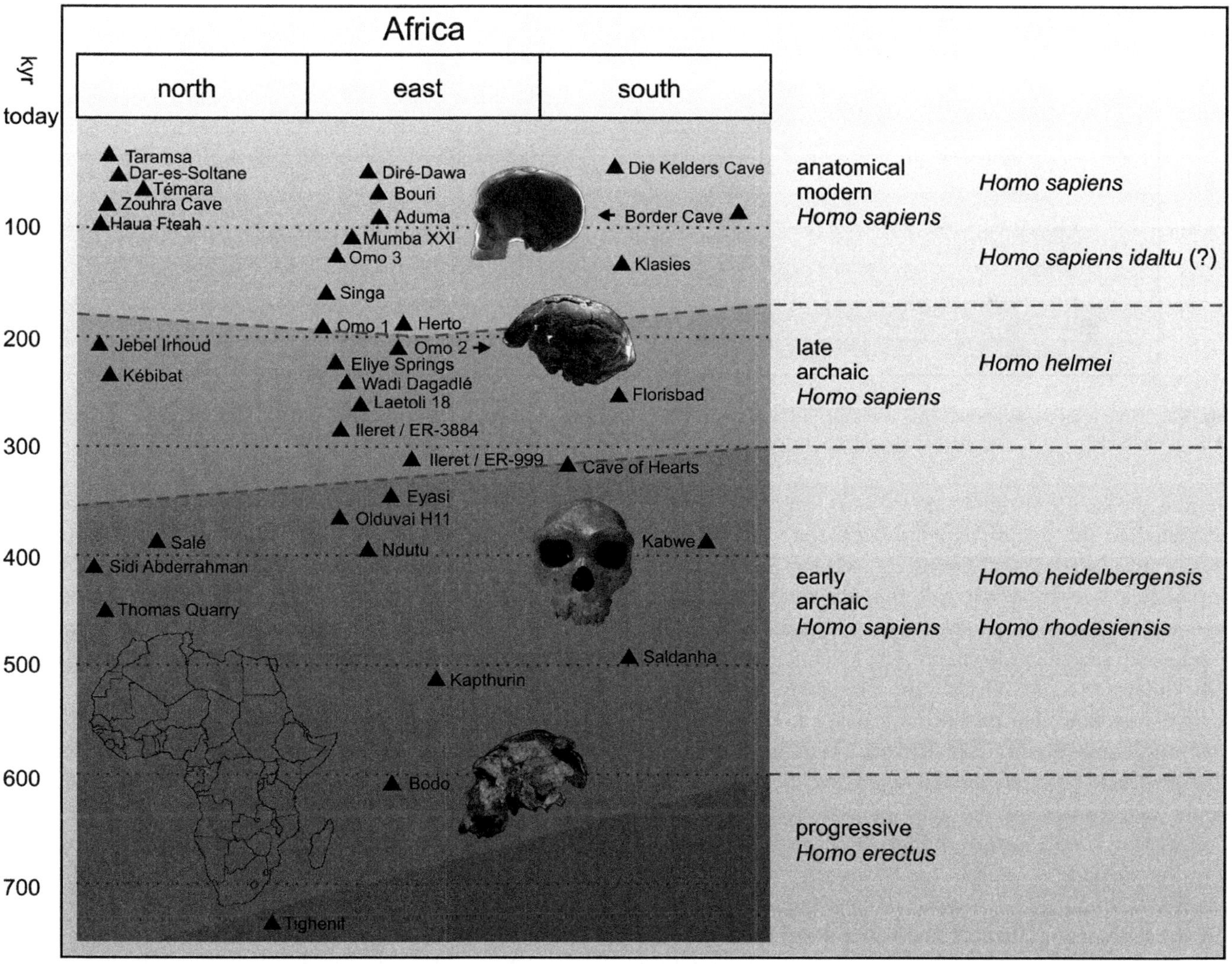

Fig. 3.6 Time scale of *Homo* fossils in Africa depicting taxa assignment and chronology (Hardt and Henke 2007)

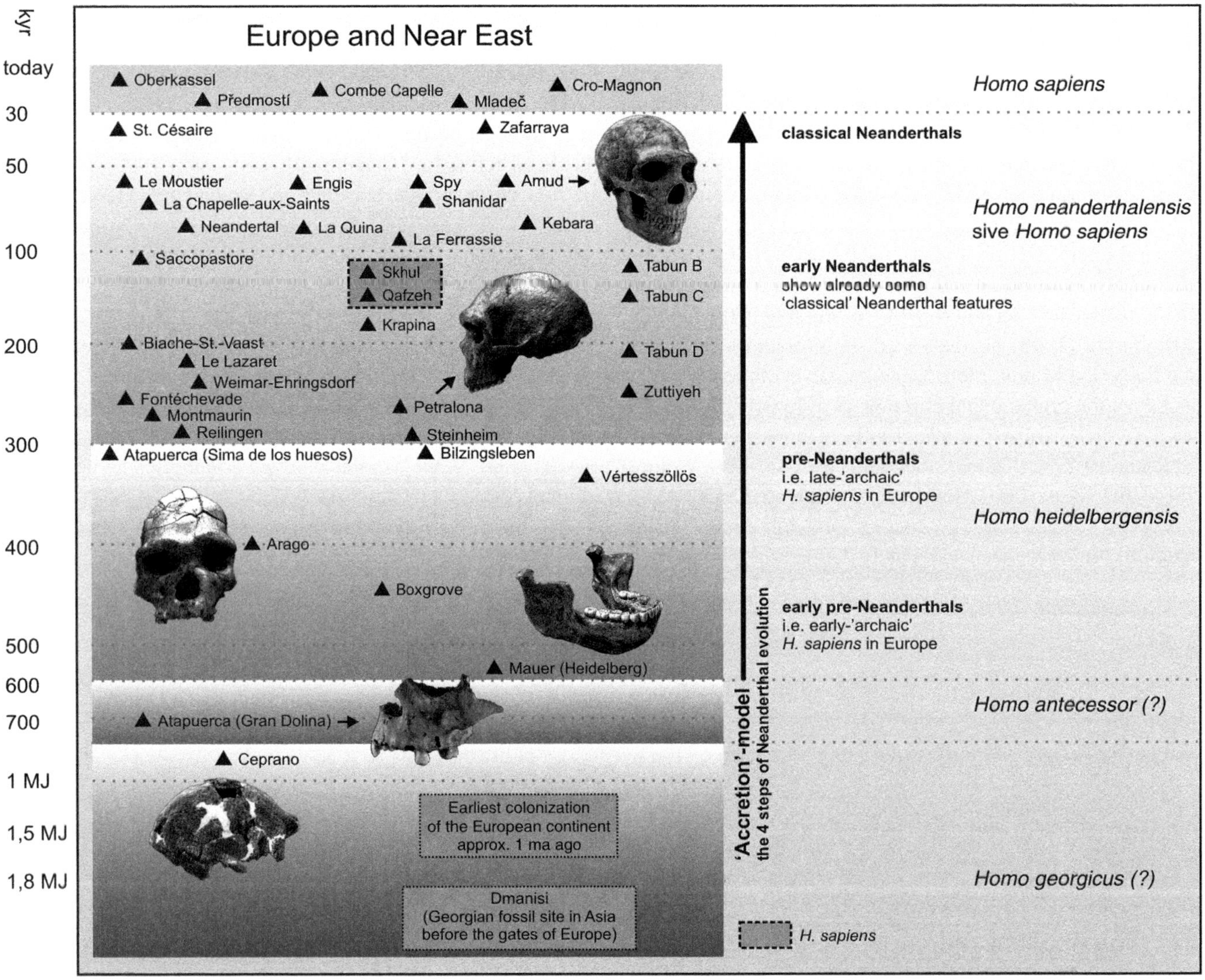

Fig. 3.7 Time scale of important *Homo* fossils in Europe and the Near East in chronological order and attribution to the steps of the gradual 'Accretion model' and the 'Archaic *Homo* model' respectively (Hardt and Henke 2007)

Beijing, in the 1930s, now known as *H. erectus* (formerly attributed to the genus *Sinanthropus*), were clearly more archaic than fossils from Europe (e.g., Mauer) and Northwest Africa (e.g., Ternifine), it was initially stated that *Homo* emerged in East Asia and dispersed westward. Since around 1960, when specimens from different localities in the eastern Rift Valley and South Africa were assigned to 'early *Homo*' for the first time, the picture has changed. Especially the hominin fossils around Lake Turkana (initially classified as *H. erectus* and currently termed *H. ergaster*) have proved to be the oldest ones and the more plesiomorphic forms too. Consequently, the actually preferred dispersal hypothesis sees the 'African *H. erectus*' or *H. ergaster* as the species which firstly immigrated to Asia and subsequently to Europe. Advances in dating methods and new finds from China and Indonesia indicate that early *Homo* may have arrived in East Asia by c. 2 Ma (Fig. 3.6). Alan Turner and O'Regan (2007) recently stated that the effects of tectonic and climatic changes on the Levantine route during the Plio–Pleistocene suggests that a late Pliocene dispersal should be given serious consideration, because the *Homo* migration can be seen as part of the pattern of dispersion by members of the terrestrial mammalian fauna (Torre et al. 1992; Hemmer 1999; Vrba 2007).

Judged from current archaeological and paleoanthropological evidences Europe was reached not earlier than c. 1.3 Ma (Arzarello et al. 2007; Carbonell et al. 2008), but we have to remember the rule that 'absence of evidence is no evidence of absence.'

From a palecological view, there has been several times admonished that those hypotheses should not be neglected which propose that *Homo* may have reached Europe from Far East Asia and not directly from Africa *via* the Levant (or even the Gulf of Sidra or the Street of Gibraltar). To prove

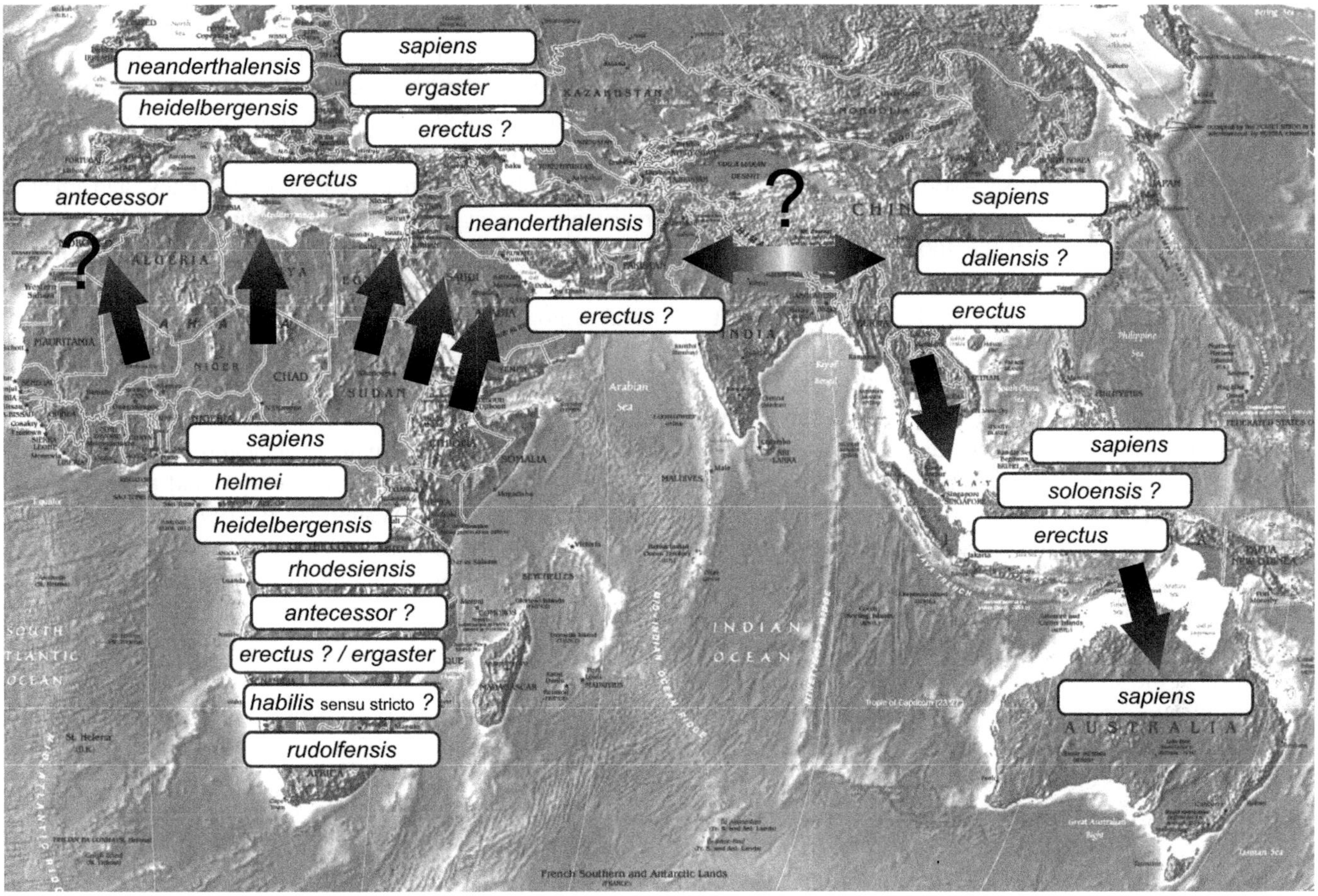

Fig. 3.8 Species of the Genus *Homo* ('splitters' view on dispersals out of Africa)

such dispersal scenarios, faunistic information should be taken much more into account, because of the co-evolution of hominin predators with carnivores (Felidae, Canidae). Further information can be gained from the dispersal pattern of the mammals that they scavenged or hunted (Henke et al. 1999; Torre et al. 1992; Turner and O'Regan 2007).

The regional fossil records of the Early and Middle Pleistocene hominins from Africa, Asia, and Europe demonstrate broadly similar morphological trends. There is – in the opinion of the gradualists – no convincing evidence to support a Middle Pleistocene speciation event leading to a distinct *H. sapiens* – quite the contrary, the proponents of the so-called multiregional theory of hominin evolution (see e.g., Wolpoff 1996–1997, 1999) point out that there is morphological continuity between *H. erectus* and *H. sapiens*. For example, the Ngandong skulls from Java, whose age may be no more than 34,000 years, have been described by some authors as *H. erectus* and by others as 'archaic' *H. sapiens* or a species of its own, *Homo soloensis*. Due to the repeated occurrence of fossil specimens exhibiting a morphologically intermediate pattern between *H. erectus* and *H. sapiens*, which is obviously incompatible with a punctuational interpretation of human evolution, there is cause for much debate on stability and change in *H. erectus* and need for a critical assessment of the inflation-like splitting of the genus *Homo* as practised by some protagonists (e.g., Schwartz 2000a, b; Schwartz and Tattersall 2002, 2003; Tattersall and Schwartz 2000).

Anthropologists who argue that a speciation event took place within the genus *Homo* during the Early and Middle Pleistocene describe many morphological characteristics of *H. erectus* (incl. *H. ergaster*) (Table 3.2), but character states that are autapomorphic are obviously rare and controversial (Andrews 1984; Bilsborough and Wood 1986; Bräuer and Mbua 1992; Henke and Rothe 1994; Howell 1986; Howells 1980, 1993; Hublin 1986; Kennedy 1991; Rightmire 1990, 1998, 2007; Wolpoff 1996–1997; Tattersall and Schwartz 2000). The splitting of the *Homo* hypodigm on the species level is highly controversial. If one does not take the extreme position of so-called 'lumpers' like Wolpoff et al. (1994a, b), who explain the variability within *Homo* by polymorphism and polytypism and merge *H. erectus* within the evolutionary species *H. sapiens*, there are diverse proposals by so-called 'splitters' ranging up to a maximum of a dozen *Homo*-species. As it is quite obvious that some of these decisions are much more gut feeling than brain generated, we should seriously ask for the reliability of the applied methodologies and the validity of the described taxa.

The following list describes some of the traits that set *H. erectus* apart from *H. habilis/H. rudolfensis* on the one hand and *H. sapiens* on the other hand. The schedule demonstrates the opinion of 'splitters', – so-called 'lumpers' do not accept these features as taxonomically valid (see also Henke and Rothe 1994, 1999a, b; Henke 2005; Table 3.2; Fig. 3.9).

Some anthropologists argue that *H. erectus* is an exclusively Asian taxon which possesses features not present in African specimens and in *H. sapiens* either. For this reason they see better arguments for an African than for an Asian origin of *H. sapiens*. Strong opposition comes from 'multiregionalists', because the supposed autapomorphic characters that seem to underline the uniqueness of the Asian sample are not truly independent characters, because they are correlated within functional units of adaptation. No anthropologist denies that there are differences between the Asian and African *Homo* samples under discussion, but the point is whether these differences are sufficient to warrant taxonomic recognition at the species level (see e.g., Bilsborough 1992; Henke and Rothe 1999a; Wiesemüller et al. 2003).

That *H. erectus* was fully upright and bipedal is still expressed by the species name, but until the virtually complete skeleton KNM-WT 15,000 (Walker and Leakey 1993) was discovered, we knew very little about the *H. erectus* postcranium. The skeleton of the c. 12-year-old boy from Nariokotome, West-Turkana, which has been dated to 1.6 Ma, enables us to reconstruct stature, limb proportions, locomotion, maturation, and gestation. In adulthood the young boy, who measured 1.68 m, would have been c. 1.80 m tall and weigh c. 47 kg. He was tall and thin, resembling present-day equatorial Africans. Rogers et al. (1996) discussed the behavioral implications of the archaeological and paleogeographical record and early *H. erectus* anatomy. The described physiological changes (e.g., secondary altriciality, longer period of maturation, increase in need for food, increase in long-distance locomotor efficiency, and greater resistance to heat stress), combined with the implied behavioral changes (e.g., greater parental investment, larger home ranges) and the archaeological evidence for a changed behavioral ecology (e.g., lithic standardization by the reduction of single platform cores, use of large flakes for cores) suggests that the 'early' *Homo* of the Early Pleistocene was less constrained than earlier hominins by the natural distribution of resources. This makes him an ideal candidate for emigrating pioneers (see Fig. 3.10). For this reason and from the total morphological pattern, Wolpoff et al. (1994, p. 341) see "no distinct boundary between *H. erectus* and *H. sapiens* in time and space." They regard the lineage as a single evolutionary species, but other authorities describe different earlier and further speciation processes, stating that the emergence of *Homo* has not been a single linear transformation of one species into another, but rather a "meandering, multifaceted evolution" (see Tattersall 2000b; and the splitter's pedigree version given in Fig. 3.1). The crucial paleoanthropological puzzle is to find a sound answer to the question: How many speciation processes took place, when and where? (see Hardt and Henke 2007).

Table 3.2 Diagnostic features of *Homo erectus*

- Cranial capacity ranging from 800 to 1,225 cc, thick brow ridge (torus supraorbitalis), especially in the later forms
- Special neurocranial proportions: wide cranial base, vault walls relatively vertical in their lower portions; long, flat, low braincase
- More arched than bell shaped contour of the braincase seen from the rear, occipital large and sharply angled
- Well marked nuchal plane bounded by a distinct nuchal ridge (occipital torus)
- Temporal lines distinct and slightly raised, especially anteriorly
- Dagittal keeling and parasagittal depression in Asian skulls only, occipital ankles
- The separation of the African sample from the Asian is highly questioned. Wood (1984) for example describes the following autapomorphies of the Asian *H. erectus*
- Occipital torus with sulcus above
- Angular torus and mastoid crest
- Sulcus on frontal behind torus
- Proportions and shape of occipital bone
- Relatively large occipital arc

Diagnostic features of *Homo erectus* as described by Groves and Mazák (1975) and others (see Henke and Rothe 1994)

Counting of Species – Purely a Matter of Taste?

How Many Homo-*Species Were There?*

Beside the lumper's model of gradualism there have been published many speciation models which interpret the hominin fossil record in quite different, obviously much more complex or even complicated ways ranging from moderate views of speciation to the view of extreme 'splitters'. As mentioned above, the currently preferred approach is the cladistic analysis. This method which has been developed by Hennig (1950) is based upon the presence and absence of larger numbers of traits or character states. Sister groups (adelphotaxa) are formed on the basis of sets of shared derived character states (synapomorphies), and species are defined on the basis of unique derived character states (autapomorphies). Similarities based on synapomorphic traits only are relevant for the reconstruction of a monophylum, a taxonomic group of organisms that has a single common ancestor (see also McHenry 1996; Wiesemüller et al. 2003; Rothe and Henke 2006). The cladistic principle sounds easy

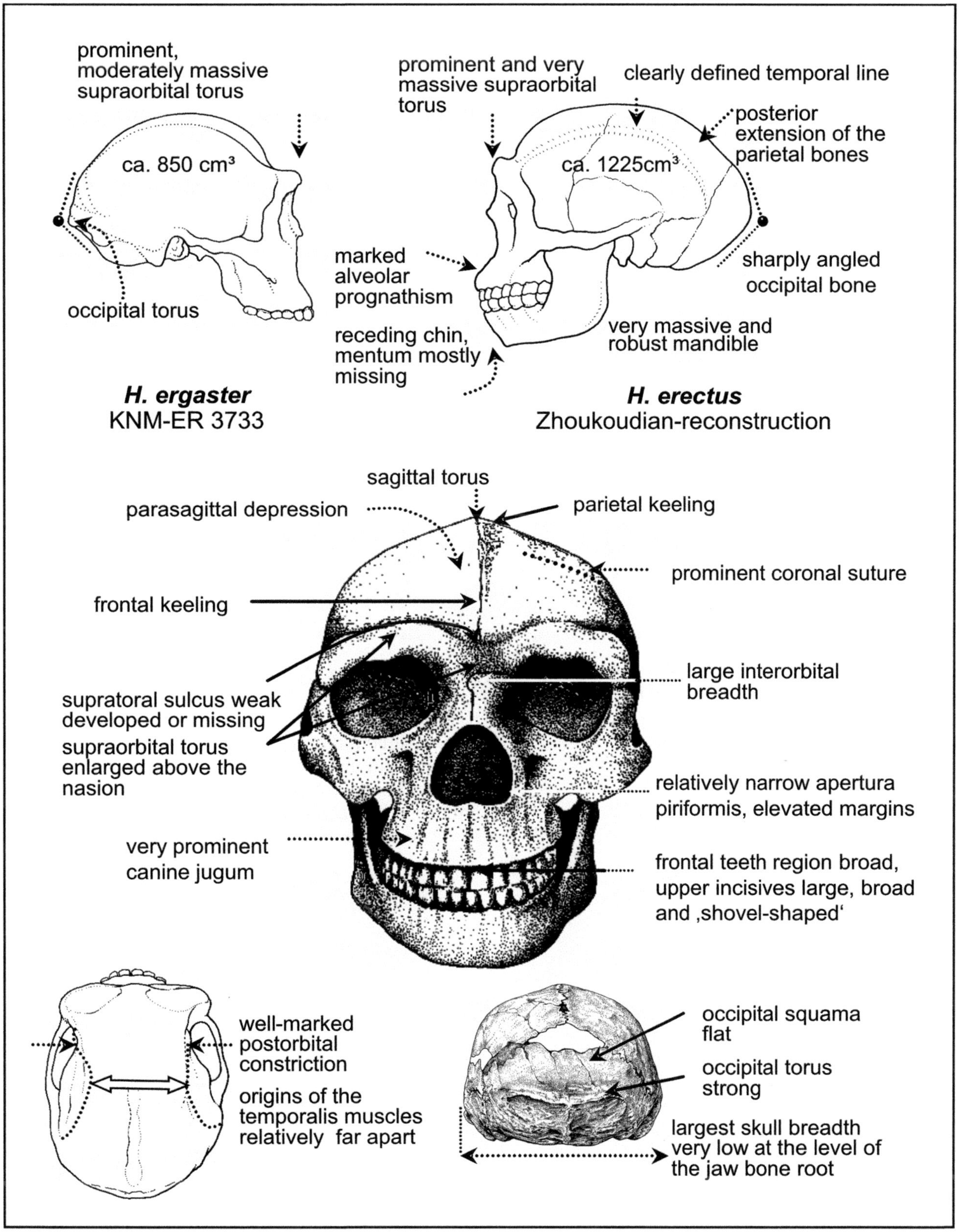

Fig. 3.9 Morphological features of the skull of *Homo ergaster* and *Homo erectus* demonstrated by the specimen KNM-ER 3733 and the reconstruction of the female *'Sinanthropus'* from Weidenreich (1943), Norma occipitalis of ZH XII (From Henke and Rothe 1999a, redrawn)

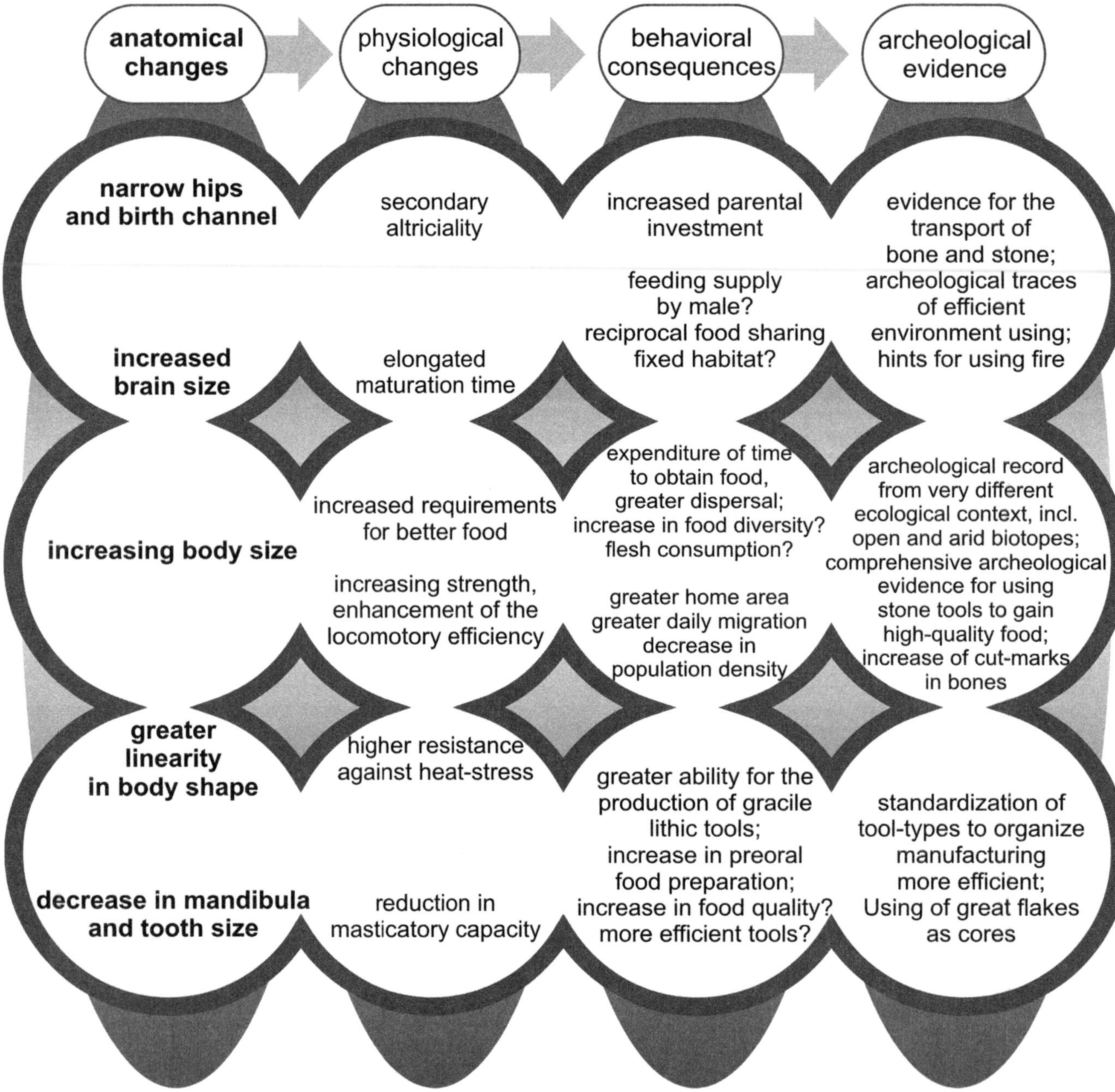

Fig. 3.10 Adaptational patterns of *H. erectus* (Rogers et al. 1996 modified)

but isn't, as the problems start with the choice and weighting of the characters, the scoring of character states and end with the arbitrariness of the principle of parsimony, the rule, that in the presence of two competing and otherwise equal explanations, the simpler of both should be considered the most likely. Ockham's razor, the theory regarding the simplest explanation as the best solution is highly controversial (Wiesemüller et al. 2003). Dilemmas are programmed and we have to solve them as we have to consider that the evolution of our own species has been a singular real-historic-genetical event and the paleoanthropological challenge is to reconstruct this unique phylogenetic incidence. Referring to the different taxonomical schools (see above) we have to state that we can not expect consensus due to the underlying principals and the applied methodological approaches. For this reason we have to discuss the opposed phylogenetic models against the methodological background (Rothe and Henke 2006).

Once again: lumpers' combine facts and evidences into simple, highly variable categories; taxonomically they create taxa that include variable morphs that might – to a 'splitter' – be better allotted to several taxa. The 'splitters' are convinced that where distinct morphs can readily be identified it would seem most productive to assume they represent

Table 3.3 Hominine taxonomy

Genus *Homo* Linnaeus, 1758 [including the following genera: *Anthropopithecus* Dubois, 1892; *Pithecanthropus* Dubois, 1892; *Protanthropus* Haeckel, 1895; *Sinanthropus* Black, 1927; *Cypanthropus* Pycraft, 1928; *Meganthropus* Weidenreich, 1945; *Atlanthropus* Arambourg, 1954; *Telanthropus* Broom and Robinson, 1949]; earliest appearance in the Pliocene, world-wide distribution

Species name and	First description	Age	Dispersal
Homo antecessor	(Bermudez de Castro et al., 1997)	EP	W-Europe
Homo cepranensis	(Mallegni et al., 2003)	EP	Italy
Homo erectus	(Dubois, 1892; Weidenreich, 1940)	P	Africa and Eurasia
Homo ergaster	(Groves and Mazák, 1975)	P	Africa and Eurasia
Homo floresiensis	(Brown et al., 2004)	LP–EHol	Indonesia
Homo georgicus	(Gabunia et al., 2002)	EP	W-Asia
Homo habilis	(Leakey et al., 1964)	Pli–P	Africa
Homo heidelbergensis	(Schoetensack, 1908)	MP	Africa and Europe
Homo helmei	(Dreyer, 1935)	MP	N-Africa
Homo mauretanicus	(Arambourg, 1963)	MP	N-Africa
Homo modjokertensis	(Koenigswald, 1950)	EP	Indonesia
Homo neanderthalensis	(King, 1864)	MP–LP	Eurasia
Homo palaeojavanicus	(Sartono 1981)	MP	SE-Asia
Homo rhodesiensis (cf. *heidelbergensis*)	(Woodward, 1921)	MP	Africa
Homo rudolfensis	(Alexeev, 1986; Wood, 1992)	Pli–P	E-Africa and Malawi
Homo sapiens	(Linnaeus, 1758)	P–today	Worldwide
Homo soloensis	(Dubois, 1940)	MP	SE-Asia

Hominine Taxonomy: Species designations of the genus *Homo*; temporal and geographic ranges. Except *Homo sapiens* all the other taxa went extinct (Adopted from; Mai et al. 2005; see also Henke 2003a, b, 2005)
Hol Holocene, *P* Pleistocene, *Pli* Pliocene, *E* early, *M* middle, *L* late

species unless there is compelling reason to believe otherwise, (see Campbell and Loy 2000; Mai et al. 2005). Is species recognition purely a matter of taste? Are species constructions – or do they really exist? (for different views see Wiley 1978; Ax 1984; Willmann 1985; Tattersall 1986; Wolpoff and Caspari 1997; Schwartz 2000a; Wiesemüller et al. 2003; Rothe and Henke 2006; Futuyama 2007). The list given in Table 3.3 demonstrates that more than a dozen species of the genus *Homo* have been described with increasing tendency during the last decades. For this reason we must ask how reliable the tremendous splitting is and how valid the definitions are.

Opposing Views of Lumpers and Splitters

Scenario I: Multiregional Evolution Model (MRE)

Jan Jelínek, a life-long proponent of regional continuity, influenced by the theories of Franz Weidenreich and effected by his own broad anthropological and archaeological research in the Old World continents and Australia asked: "Have we any solid scientific grounds on which to consider Middle Pleistocene European finds, with earlier morphological cranial changes, as *Homo sapiens* and the extra-European finds evolving in the same direction but in somewhat different degree and time sequence of adaptation into different conditions as *Homo erectus*?"

His answer was that the evidences from the global perspective should lead us sink *H. erectus* into *H. sapiens* (Jelinek 1978). He stated his view more precisely in the eighties: "If the differential diagnosis between *Homo erectus* and *Homo sapiens* cannot be other than by convention, and [...] this convention must be different for different geographical regions, then the value of such difference should be critically considered [...] It is time to replace the paleontological species with a biological one. [...] Paleontological taxonomy cannot be in contradiction with [...] biological facts" (Jelínek 1981, p. 88; see also Wolpoff and Caspari 1997, p. 253).

Together with Milford Wolpoff et al. (1994, 341) he proposed "to merge *Homo erectus* within the evolutionary species *Homo sapiens*". Wolpoff and Caspari (1997, p. 255) point to the fact that the evolutionary species "... retains the essence of the biological species – reproductive isolation (the reason it can maintain its identity and has unique evolutionary tendencies) – while avoiding many of its deficiencies: lack of time-depth, absence of morphological criteria for diagnosis, and, perhaps most important, emphasis on reproductive ties alone as a major cohesive force."

The 'multiregionalists' interpret the fossil evidences within the rationales of the evolutionary species concept as Jelínek proposed in the 1960s: "No species splits occurred when *H. sapiens* is said to originate from *H. erectus*; there was no division of one species into two, and therefore no species birthing..." (Wolpoff and Caspari 1997; p. 256).

The multiregional evolution model (MRE) states that there is no basis for distinguishing a species called *H. sapiens* from a species *H. erectus* (and other Middle and Late Pleistocene species); Fig. 3.11a reflects this gradualist view. The MRE traces all modern populations back to c. 2 Ma when humans migrated out of Africa for the first time. Since this phase there has been no speciation, but an interconnected web of ancient lineages existed in which the genetic contributions to all living peoples varied regionally and temporally (Thorne and Wolpoff 1981, 1992; Wolpoff 1999). We straighten out here that the MRE is in contrast to otherwise uttered opinions no candelabra-model. The MRE has its historic base in the polycentric evolution hypothesis of Franz Weidenreich, proposing "that the conditions associated with the initial migrations of humans from Africa … created the central and peripheral contrasts that affected the early establishment of regional features at the peripheries of the human range" (Wolpoff 1992, p. 26).

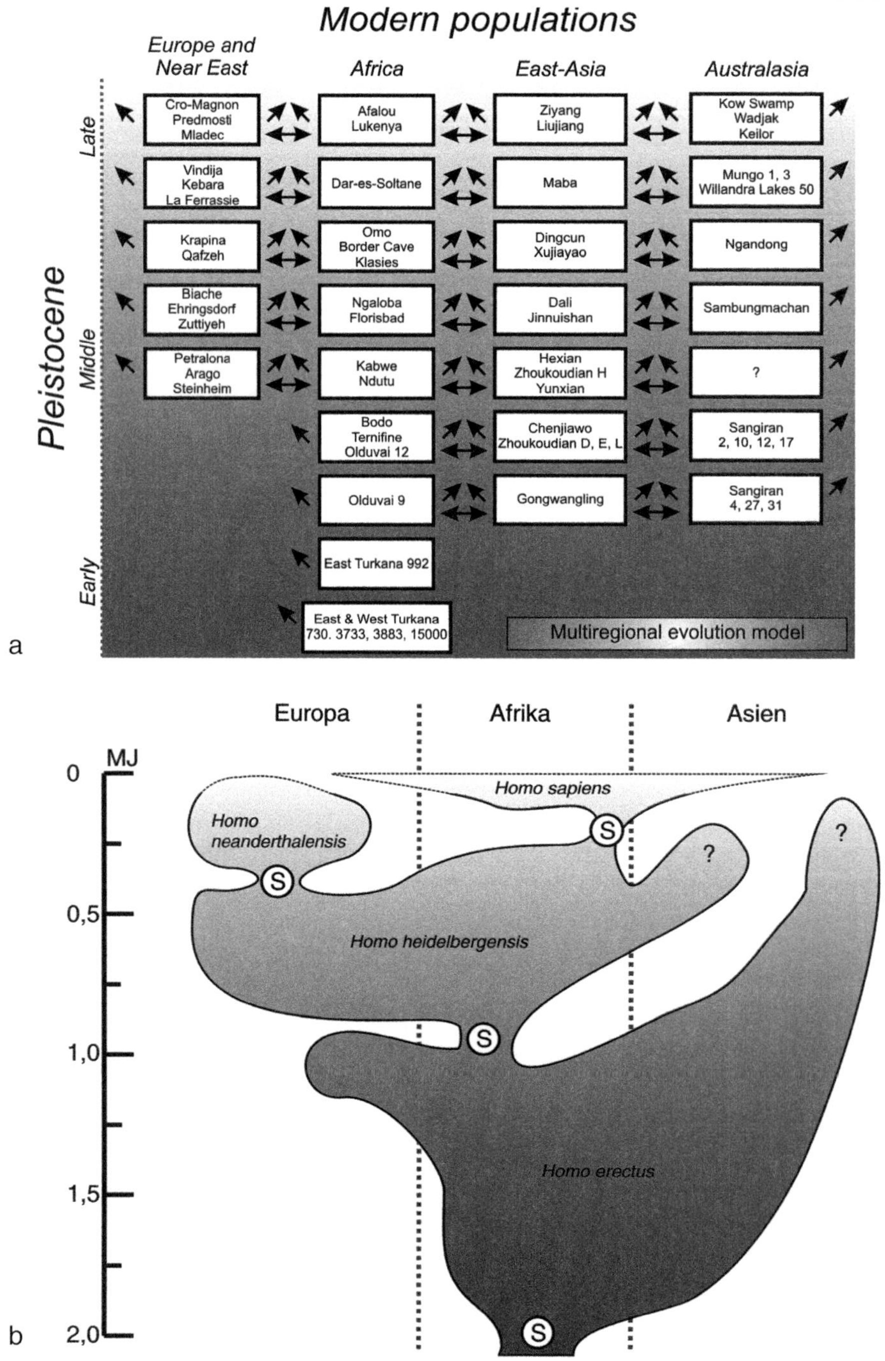

Fig. 3.11 (**a**) *MRE* (Multiregional evolution model) or the 'lumpers' view of human evolution (Frayer et al. 1993), (**b**) "Out of Africa"– model after Rightmire (2001a), (**c**) RAO-model after (Stringer 2002), (**d**) Out of Africa-model with hybridization after (Bräuer 2006, 2007)

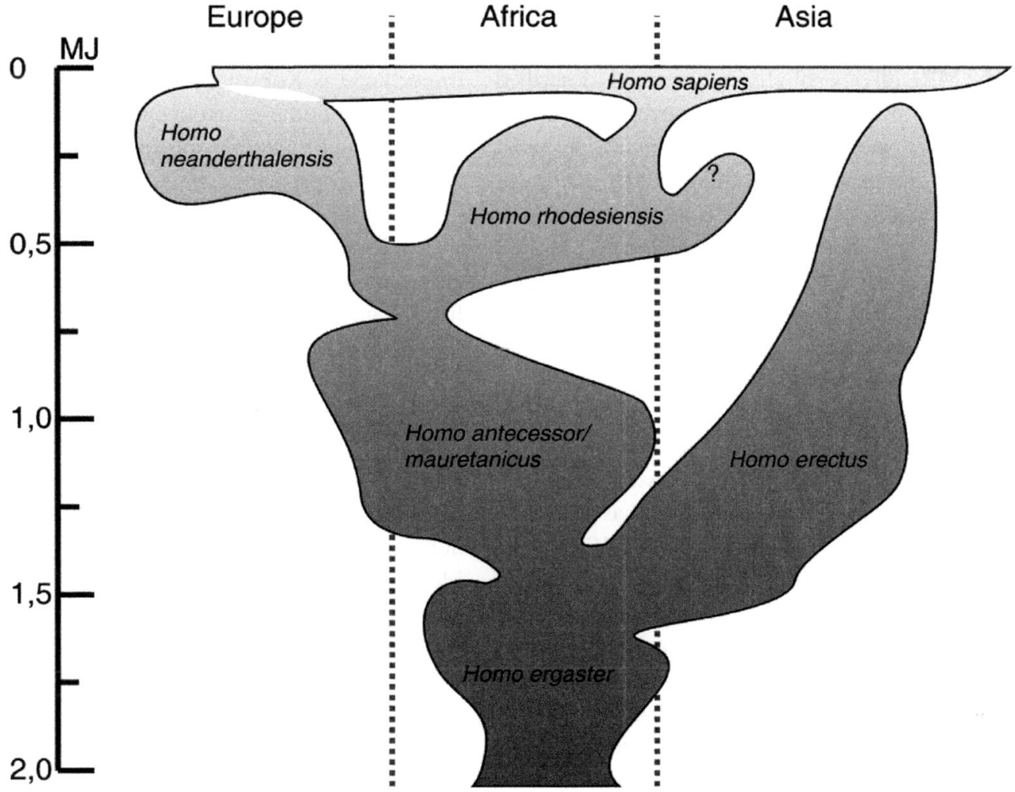

c

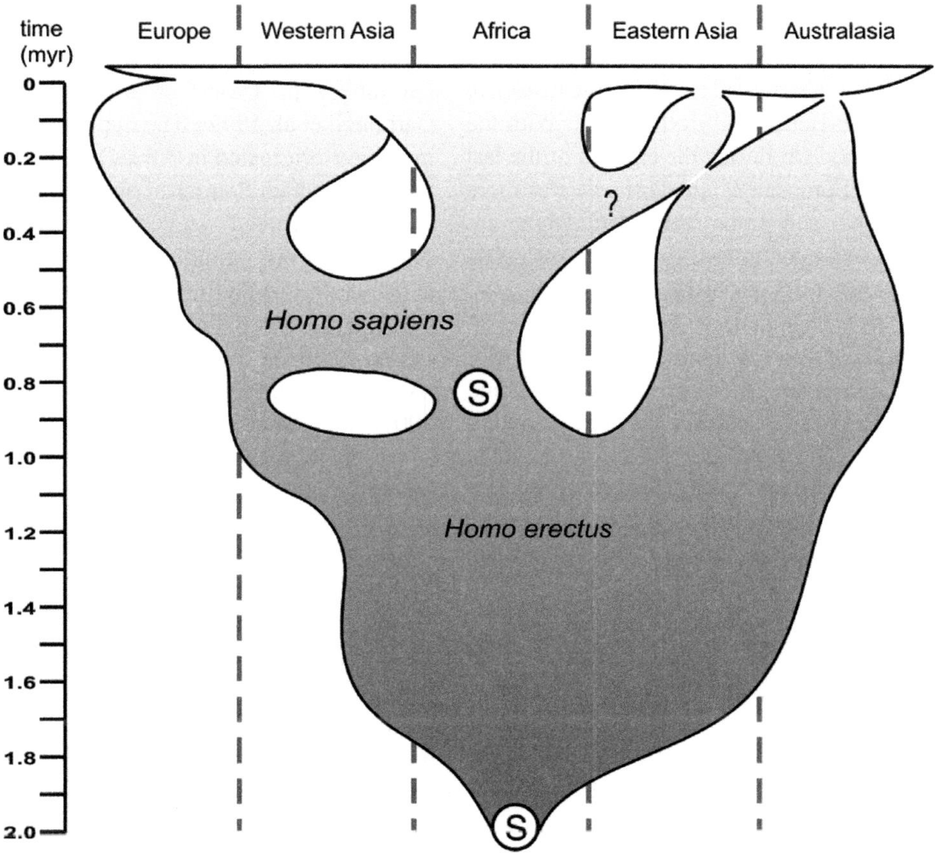

d

Fig. 3.11 (continued)

Scenario II: Recent African Origin Models (RAOM)

In opposite to the MRE all other phylogenetic models described here suppose at least one or more speciation events within the genus *Homo*. A widely accepted recent African origin model (RAOM) has been proposed by Rightmire (1998, 2001a, b; Fig. 3.11b). The pedigree describes three further speciations after the emergence of *Homo ergaster/erectus*. At around 0.8 Ma an evolutionary split in Africa is leading to *H. heidelbergensis*. While Rightmire leaves the phylogenetic role of *H. antecessor* in his model open (Rightmire 2001a), he recently discussed the hypothesis that this species could be a direct descent of *H. erectus* and the ancestor of *H. heidelbergensis* on the one side and *H. rhodesiensis* on the other side (Rightmire 2007; see also Stringer 2002, Fig. 3.11c, discussed below).

H. heidelbergensis, firstly described in 1907, has reached a reinterpretation during the last decade (Rightmire 1998). The hypothesis is that *H. heidelbergensis* has given on the one hand rise to *H. neanderthalensis*, who emerged in Europe, and on the other hand to *H. sapiens* or modern humans, who split in Africa and successively occupied all other continents. The traditional model of the sixties and seventies of the last century implied that *H. erectus* was a valid species in all the continents of the Old World. The small European hypodigm was divided in different subspecies such as *H. erectus heidelbergensis*, *H. erectus bilzingslebenensis*, *H. erectus tautavelensis* and some others. The subspecies-splitting was applied for the Asian and African branches of *H. erectus* as well (for scientific historical aspects see Groves 1989; Henke and Rothe 1994). With the rapid increase of the hypodigm during the eighties of the last century the validity of a European *H. erectus* taxon was more and more disputed (Henke and Rothe 1994, 1995; Rightmire 1998). However, meanwhile many paleoanthropologists believe that *H. erectus* never has reached Europe, but there are different causes for this change of view. While Wolpoff et al. (1994) – as just mentioned – and Wolpoff (1996–1997, 1999) do not believe that the taxon *H. erectus* is valid in any event, others regard those European specimens that were formerly discussed as belonging to *H. erectus* as a species of their own, which originated in Africa around 0.6 Ma. For taxonomic reasons it was named *H. heidelbergensis* (see Rightmire 1990, 1998; Lahr 1994; Groves and Lahr 1994; Condemi and von Koenigswald 1997; overview in Ullrich 1998; Wagner and Beinhauer 1997; Wagner et al. 1997; Hardt and Henke 2007). It is – as proponents of this classification claim – best described by specimens from Kabwe (formerly Broken Hill, Zambia), Bodo (Middle Awash, Ethiopia), Petralona (Greece), Arago (France) and Sima de los Huesos (Atapuerca, Spain) (more details see Figs. 3.6 and 3.7).

Concerning the validity of the derived features of *H. heidelbergensis* there is no consensus at all. Astonishingly enough, that the Kabwe/Broken Hill skull was formerly attributed to 'archaic' *H. sapiens* and some authorities adhere to this classification though the gradual split in 'early and late archaic' *H. sapiens* does not match with the rules of classification. However, this type of array is insofar of interesting importance as Kabwe 1 and Bodo are key fossils in all out-of-Africa hypotheses. Both specimens are recognised as representatives of the first appearance of people distinct from *H. erectus*. Rightmire (1998, p. 135) describes for *H. heidelbergensis* on the one hand a pattern of plesiomorphic *H. erectus*-like traits and on the other hand apomorphies like an evolved endocranial volume, more *sapiens*-like proportions of the frontal and temporal bone, as well as details of the nose and the palate as also present in the specimens from Elandsfontein (South Africa), Ndutu (Kenya) and European fossils from Petralona, Arago and others (see Figs. 3.6 and 3.7).

Newly unearthed fossils from Bouri (Middle Awash, Ethiopia) yield unique insights into unresolved spatial and temporal relationships of *H. erectus* and younger taxa. The hominin calvaria and postcranial remains from the Dakanihylo Member of the Bouri Formation are ~1.0 Ma; they are associated with abundant early Acheulean stone tools and a vertebrate fauna that indicates predominantly a savannah environment. Asfaw et al. (2002, p. 317) are convinced that the morphological attributes of the fossils "centre [them] firmly within *H. erectus*". They see strong indications "that African *H. erectus* was the ancestor of *Homo sapiens*." Furthermore, the new fossils from Bouri shed light on newly described c. 0.8 Ma fossils from the Gran Dolina of Atapuerca (Spain), which have been classified as *H. antecessor* (Bermudez de Castro et al. 1997; Arsuaga et al. 1999; Carbonell et al. 1999). The hypothesis that this *species nova* may have originated in Africa and given rise to *H. neanderthalensis*, which flourished between ca. 200 and 30 ka while *H. sapiens* evolved as adelphotaxon of *H. heidelbergensis* in Africa, is in our opinion a daring hypothesis. By new fossil evidence from Italy it is now even less reliable than before. Fresh arguments to drop the taxon *H. antecessor* come from the research of the 800–900 ka calvarium from Ceprano (Ascenzi et al. 2000a, b; Manzi 2001, 2004; Manzi et al. 2001, 2003; Bruner and Manzi 2007). The specimen exhibits a morphological pattern which separates it from archaic humans like *H. ergaster* and *H. erectus*. While Mallegni et al. (2003) consider the cranial features as indication for a separate species (*H. cepranensis*), others as Bruner and Manzi (2007) evaluate them as ancestral morphology of the *H. heidelbergensis/rhodesiensis* hypodigm. Both authors are pleasantly discrete when they state: "Nevertheless, a definite taxonomic interpretation of this fossil cannot be provided until more appropriate comparative fossil samples (both in terms of chronology and anatomical completeness) are recovered" (Bruner and Manzi 2007, p. 365).

As mentioned above, there are highly differing scenarios of the origin of 'anatomically' modern human beings. The irreconcilable standpoints of the proponents of the MRE and the RAOMs have been hardened by totally different species concepts. The reproach of Tattersall (cited by Flanagan, taken from Wolpoff 1999, p. 397) is that the 'multiregionalists' are "linking everything from *H. erectus* to *H. sapiens* into 'one big happy family' [...] Paleontologists do not give other animals such a special treatment." Is this really an appropriate argument if we consider the facts?

Wolpoff, the protagonist of the 'multiregionalists', comments on Tattersall's critique by quoting the evolutionary geneticist A. Templeton (2002): "We make far too much of our anatomical difference [...] Biologists who study, say, fruit flies know that each population can look quite distinct [...] and yet they are tempted to hastily split them into separate species. Why must we look at ourselves any differently?" (Wolpoff 1999, p. 397).

The crucial case of whether there was more than one *Homo*-species comes from the long-standing unsolved debate on the Neanderthal problem: in spite of contrary reports, especially comments on molecular biological results, there is in our opinion currently no solution to the problem by mtDNA (see Henke and Rothe 1999a, b) in spite of differing opinions and claims (Serre et al. 2004; Caramelli et al. 2006; Orlando et al. 2006; Krause et al. 2007), as the final decision can only be obtained by nDNA analyses. Recent conducted genomic research, studying the "phenotypic sweep" by Eswaran et al. (2005, p. 1) came to the conclusion "that as much as 80% of nuclear loci have assimilated genetic material from non-African archaic humans." These results coincide with those of Relethford (2001a) who regards a "Mostly out of Africa"-model, in which Africa contributes the most to accumulated ancestry in all regions, as the adequate MRE-version. In both cases there is no speciation accepted to explain the variability of Middle and Late Pleistocene *Homo*.

The different RAOMs suggests that the humankind descends from a very recent modern population that lived in Africa only around 200 ka and replaced early humans elsewhere. This hypothesis has been called by its opponents 'Garden of Eden model' because it pleads for a – very recent – single regional human origin. The main elements of this theory are derived from the chronological and morphological comparison of the African and non-African fossil specimens. The presence of regional continuity only in Africa and the first appearance of modern human features in this continent were taken as strong arguments for different replacement models (Afro–European *sapiens* hypothesis *with* hybridisation; Bräuer 1984; see further Bräuer 2006, 2007; Fig. 3.11d). Although Bräuer's hypothesis supported the idea of a replacement *with* admixture, other authorities advocated – in the beginning – for a *total replacement* from paleoanthropological (Stringer 1986; Stringer and Andrews 1988) or paleogenetic reasons (mtDNA research; Cann et al. 1987; Stoneking and Cann 1989; Wilson and Cann 1992). Especially those genetic studies, which revealed that an African woman from c. 200–150 ka– called Eve or 'Lucky Mother' – was our last common ancestor, gained much credit in public discussions and its populist summary 'we are all Africans'.

What we actually know about diversity and variability is that there is a great amount of *Homo* fossils from the late Middle and Late Pleistocene in Africa, Asia, and Europe whose morphology differs – in comparison to earlier specimens – only within a moderate range. The diachronic comparative analysis allows describing a trend to less massive faces and larger skullcaps, approaching step by step the pattern of [anatomically] modern humans in Africa. Because this gradual process is less convincingly verified in Asia and extremely controversial in Europe, so-called out-of-Africa hypotheses have reached a high rate of acceptance (Stringer 1982, 1992; Bräuer 1984; Cann 1992) compared to the gradualistic, anagenetic MRE (Thorne and Wolpoff 1992; Wolpoff 1996–1997; Wolpoff et al. 1984).

The debate has received a new dynamic by the results of evolutionary human genetic research. The results from mtDNA analyses of recent female samples (Cann et al. 1987; Cann 1992) were topped by sophisticated aDNA research on the classical Neanderthal specimens from Felderhofer Grotto (Krings et al. 1997), Mezmaiskaya Cave (Ovchinnikov et al. 2000), and other specimens (Pääbo et al. 2004; Serre et al. 2004). At least the popular science was convinced that the Neanderthal problem was solved by aDNA analyses; cover-headlines like "Neanderthals are not our ancestors" (see Krings et al. 1997) did not miss the target, the intended message that Neanderthal man belonged to a species of its own. But those who read the papers attentively will realize that these results do "not definitively resolve the question of a possible Neandertal contribution to the gene pool of modern humans since such a contribution might have been erased by genetic drift or by the continuous influx of modern human DNA into the Neandertal gene pool" (Serre et al. 2004, p. 0313).

In spite of contrasting assertions by the proponents of the different models and refining of the RAOMs by human demographic history (see Excoffier 2002), there is the conclusion that none of them is unequivocally supported by the available data, although the out-of-Africa model gained tremendous – in our judge highly undeserved – support by the aDNA results from Neanderthal fossils. The currently running Neanderthal Genome Project, a joint project of Germany's Max Planck Institute for Evolutionary Anthropology and Branford, Connecticut-based 454 Life Sciences Corp., to map the Neanderthal genome, or DNA code should be observed with critical scepticism. Of course can we expect very interesting genomic results; however, the prognosis is indicated that each taxonomic hypothesis will create new controversies. A cynic would say that those anthropologists who look for differences will find them; the crucial question will be how to evaluate them within taxonomical categories (Storch et al. 2001; Wägele 2000; Wiesemüller et al. 2003; Jobling et al. 2004; Rothe and Henke 2001, 2006; Futuyama 2007).

Clifford Jolly's (2001, p. 177) studies commemorate to circumspective interpretations. He looked for population genetical analogies from the papionin monkeys and their implications for human evolution and concluded: "An overall similarity in depth of genetic diversity suggests that papionin taxa such as *Papio* baboons, rather than extinct humans, may present the better analogy for human population structure of the "prereplacement" era. Neanderthals and Afro-Arabian "premodern" populations may have been analogous to extant baboons (and macaque) allotaxa: "phylogenetic" species, but "biological" subspecies."

Within the current version of his replacement model with hybridisation Bräuer (2007) sees good support from different lines of evidence that the European, African and possibly the Asian Middle Pleistocene lineages too belonged to one polytypic species *H. sapiens* (see Fig. 3.11d). We see in his statement the logical taxonomic consequence of the RAOM *with* hybridization and replacement.

In addition, concerning the evolution of European hominids Hublin (1998, p. 301) suggested that "...the development of the Neanderthal morphology results from an *accretion* phenomenon beginning in the middle of the Middle Pleistocene, around 450,000 BP..." due to drastic climatic changes in the Pleistocene. Caused by processes of geographical isolation through ice sheets, genetic drift, and intense selection as a result of the extremely cold environmental conditions the Neanderthal features run through a gradual development until they reached their terminal status (Couture and Hublin 2005; Harvati 2007; contra see Hawks and Wolpoff 2001). The proposed "Accretion model" is divided into four parts ranging from Stage 1 "early pre-Neanderthals" over Stage 2 "pre-Neanderthals" and Stage 3 "early Neanderthals" to Stage 4 "classic Neanderthals" with their typical features fully expressed (see Fig. 3.7). Consequentially, this model also offers two possibilities, namely to assign Neanderthals as an outstanding species *H. neanderthalensis* within a chrono- or morphospecies concept or as a subspecies of *H. sapiens* within an anagenetic concept. Whatever decision will be made, they will raise several new questions pertaining to African and European Middle Pleistocene lineages of *H. heidelbergensis* and *H. rhodesiensis* respectively.

There seems to be much more overlap with the MRE as former discussions indicated (Smith and Spencer 1984; Akazawa et al. 1992; Bräuer and Smith 1992; Wolpoff 1999; Relethford 2001a, b; Tobias et al. 2001). There is to recognize a gradual approximation to the MRE, which hypothesizes that modern humans evolved throughout the Old World as a single species roughly 2–1 Ma. But by far not all paleoanthropologists agree that speciation is a special case, an elusive phenomenon as e.g. Brace (1993), Frayer et al. (1993), Wolpoff and Caspari (1997, 2000), Wolpoff (1999), Grubb (1999) assert. Quite the contrary seems to be the case, not only in paleoanthropology (see Table 3.3) but in primatology too. The growing proliferation of specific names should be regarded with high scepticism, that's why Loring Brace (1993, p. 151) expressed in his paper with the hissing title "The creation of specific hominid names: Gloria in excelsis deo? or ego? or praxis?" the suspicion "that the vanity of the namer is frequently involved in the creation of new nomina". In any case, each inauguration of a new taxon must be based on agreed biological principles and a sound methodological concept, otherwise each further discussion is in vain, but as paleoanthropology is a fossil- and journalism-driven science there is little hope to escape this kind of discussions (White 2000, Henke 2006, 2007; for the general problem of popular sciences see Franck 1998). Unprofessional handling of the fossil specimens like LB 1 from Flores is a sad example how paleoanthropology is discredited to "paleopoesy" and looses its credibility. Neither such a fascinating discipline like paleoanthropology nor the rare specimens as witness of our evolutionary history merit this kind of para-scientific treatment in times where paleoanthropology has been grown up and consolidated as a serious discipline (see Henke and Rothe 1994, 1999, 2003; Henke and Tattersall 2007).

The "muddle in the middle" will continue as long as arbitrary species like *H. antecessor* are created by the analysis of very scanty fossil material. This kind of weak founded taxonomical proposals does not contribute to a consolidation of our diverse views on modern human origins The description of two lineages of species (*H. antecessor* – *H. heidelbergensis* – *H. neanderthalensis* and *H. antecessor* – *H. rhodesiensis* – *H. sapiens*) is extremely problematic as it describes no cladistic events but successive speciations. If one would exclude this kind of chronospecies and regard the changes as anagenetical process, it would be mere consistent to regard the European *H. heidelbergensis* as *H. neanderthalensis* (Stringer 1996; Arsuaga et al. 1997; Hardt and Henke 2007). Simultaneously *H. rhodesiensis* should be involved in the taxon *H. sapiens*.

Whether species like *H. helmei*, *H. soloensis* and *H. palaeojavanicus* which have been redefined by some authorities during the last years as well as the creation of the *species novae H. georgicus* (Gabunia et al. 2002), *H. cepranensis* (Mallegni et al. 2003) and especially *H. floresiensis* (Brown et al. 2004; see also Falk et al. 2005) are valid evolutionary species is to doubt, as a detailed description of their autapomorphies is missing.

Concluding Remarks

It is quite obvious that paleoanthropologists have, in spite of a tremendous increase of methodological approaches and facts, no clear picture of the speciation processes which took place in the genus *Homo*. Nevertheless most of them agree that our genus originated in Africa around 2 Ma. Soon after the first appearance of "early *Homo*" (most probably *Homo ergaster*) our forerunners immigrated into Asia and much later into Europe. After the first dispersal around c. 1.8 Ma there followed a second expansion out of Africa between 840 and 420 ka, followed by a third expansion wave around 150–100 ka as shown in Fig. 3.5. That an out-of-Africa dispersal took place again and again is beyond doubt, but whether this was a demic diffusion or a successive exodus of separate taxa with hybridization or replacement of the archaic populations is still highly controversial. The core problem is that it is not a matter of taste how many speciation events took place in this scenario. As this is a long-lasting problem we are afraid that paleoanthropologists don't have or don't apply the right methodological instruments to figure out how many species there have been. The way we see it is much more near the lumpers' view, than those of the splitters', as there are good arguments from different kinds of sources that our polymorphic and polytypic species *Homo sapiens* has relatively deep roots and that the dispersal was an early single event, followed by a permanent, uninterrupted but in the intensity regionally and chronologically highly varying gene flow with movements out of Africa to Asia and possibly back again (at least from Asia; Wolpoff 1999; Relethford 1999, 2001a; Zilhão 2006). Quite unexpected after 25 years of heated discussion Bräuer's RAOM approaches the MRE though there remain principal discrepancies. Bräuer (2008) states "that the African Middle Pleistocene lineage represents the species *Homo sapiens* and that the European Neandertal lineage, which derived from early African *Homo sapiens*, belongs to the same species except that there would be unequivocal evidence that Neandertals and modern humans were too different to be one species". Also Relethford's (2001a) "mostly out of Africa" model based on a population genetical approach point to a kind of compromise, as it only causes confusion to call in a splitter's manner any unit a species. However, this may be a wishful thinking. Gilbert et al. (2003, p. 259) put it this way: "We do not know where the gene flow barriers were among the sampled populations, nor do we know about unsampled populations/lineages." Discussing the taxonomic status of *H. rudolfensis* and *H. habilis* Wood (1996b, p. 112) supplementary concluded that "we are aware from the more recent fossil record of the major morphological "components" of *Homo*, but we remain ignorant about their functional interrelationships, the order in which they arose, and their relationships to the environmental and ecological pressures and constraints prevailing around 2 Ma." This view is still valid and remains a challenge for the next decade.

References

Aiello, L. (1996). Terrestriality, bipedalism and the origin of language. In W. G. Runciman, J. Maynard Smith, & R. I. M. Dunbar (Eds.), *Evolution and social patterns in primates and man* (pp. 269–289). Oxford: The British Academy, Oxford University Press.

Aiello, L., & Dean, C. (1990). *An introduction to human evolutionary anatomy*. San Diego: Academic.

Akazawa, T., Aoki, K., & Kimura, T. (Eds.). (1992). *The evolution and dispersal of modern humans in Asia*. Tokyo: Hokusen-Sha.

Alexeev, V. P. (1986). *The origin of the human race*. Moscow: Progress Publishers.

Andrews, P. (1984). *On the characters that define Homo erectus* (Courier Forschungsberichte Senckenberg, Vol. 69, pp. 167–198). Frankfurt am Main: Forschungsinstitut Senckenberg.

Antón, S. (2003). Natural history of *Homo erectus*. *Yearbook of Physical Anthropology, 46*, 126–170.

Antón, S., Spoor, F., Fellmann, C. D., & Swisher, C. C., III. (2007). Defining *Homo erectus*: Size considered. In W. Henke & I. Tattersall (Eds.), *Handbook of paleoanthropology* (Vol. 3, pp. 1655–1694). Berlin, Heidelberg, New York: Springer.

Arambourg, C. (1963). Le gisement de Ternifine. Part IC. *Le gisement de Ternifine. Part IC Mémoire, 32*, 37–190.

Arsuaga, J. L., Martínez, I., Gracia, A., & Lorenzo, C. (1997). The Sima de los Huesos crania (Sierra de Atapuerca, Spain). A comparative study. *Journal of Human Evolution, 33*, 219–281.

Arsuaga, J. L., Martínez, Lorenzo, C., Gracia, A., Muñoz, A., Alonso, O., & Gallego, J. (1999). The human cranial remains from Gran Dolina Lower Pleistocene site (Sierra de Atapuerca, Spain). *Journal of Human Evolution, 37*, 431–457.

Arzarello, M., Marcolinui, F., Pavia, G., Pavia, M., Petrinio, C., Petrucci, M., Rook, L., & Sardella, R. (2007). Evidence of earliest human occurrence in Europe: The site of Pirro Nord (Southern Italy). *Die Naturwissenschaften, 94*, 107–112.

Ascenzi, A., Biddittu, I., Cassoli, P. F., Segre, A. G., & Segre-Naldini, E. (2000a). A calvarium of late *Homo erectus* from Ceprano, Italy. *Journal of Human Evolution, 31*, 409–423.

Ascenzi, A., Mallegni, F., Manzi, G., Segre, A. G., & Segre Naldini, E. (2000b). A re-appraisal of Ceprano calvaria affinities with *Homo erectus*, after the new reconstruction. *Journal of Human Evolution, 39*, 443–450.

Asfaw, B., Gilbert, W. H., Beyene, Y., Hart, W. K., Renne, P. R., WoldeGabriel, G., Vrba, E. S., & White, T. D. (2002). Remains of *Homo erectus* from Bouri, Middle Awash, Ethiopia. *Nature, 416*, 317–320.

Ax, P. (1984). *Das phylogenetische System. Systematisierung der lebenden Natur aufgrund ihrer Phylogenese*. Stuttgart: Gustav Fischer Verlag.

Balter, M., & Gibbons, A. (2002). Were 'Little People' the first to venture out of Africa? *Science, 297*, 26–27.

Bermudez de Castro, J. M., Arsuaga, J. L., Carbonell, E., Rosas, A., Martínez, I., & Mosquera, M. (1997). A hominid from the Lower Pleistocene of Atapuerca, Spain: Possible ancestor to Neandertals and modern humans. *Science, 276*, 1392–1395.

Bilsborough, A. (1992). *Human evolution*. London: Blackie Academic & Professional.

Bilsborough, A., & Wood, B. A. (1986). The nature origin and fate of *Homo erectus*. In B. A. Wood, L. Martin, & P. Andrews (Eds.), *Major topics in primate and human evolution* (pp. 295–316). Cambridge: Cambridge University Press.

Bowler, P. J. (1988). *The non-Darwinian revolution: Reinterpreting a historical myth*. Baltimore: Johns Hopkins University Press.

Brace, C. L. (1962). Refocusing the Neanderthal problem. *American Anthropologist, 64*, 729–741.

Brace, C. L. (1964). The fate of the "classic" Neanderthals: A consideration of hominid catastrophism. *Current Anthropology, 5*, 3–66.

Brace, C. L. (1993). The creation of specific hominid names: Gloria in excelsis deo? or ego? or praxis? *Human Evolution, 8*, 151–166.

Bräuer, G. (1984). The "Afro-European *sapiens* hypothesis" and hominid evolution in East Asia during the late middle and upper Pleistocene. In P. Andrews & J. L. Franzen (Eds.), *The early evolution of man, with special emphasis on Southeast Asia and Africa*

(Courier Forschungsinstitut Senckenberg, Vol. 69, pp. 145–165). Frankfurt am Main: Forschungsinstitut Senckenberg.

Bräuer, G. (2006). Das Out-of-Africa-Modell und die Kontroverse um den Ursprung des modernen Menschen. In N. J. Conard (Ed.), *Woher kommt der Mensch?* (pp. 171–196). Tübingen: Attempto Verlag.

Bräuer, G. (2007). Origin of modern humans. In W. Henke & I. Tattersall (Eds.), *Handbook of paleoanthropology* (Vol. 3, pp. 1749–1780). Berlin, Heidelberg, New York: Springer-Verlag.

Bräuer, G., & Mbua, E. (1992). *Homo erectus* features used in cladistics and their variability in Asian and African hominids. *Journal of Human Evolution, 22*, 79–108.

Bräuer, G., & Smith, F. (Eds.). (1992). *Continuity or replacement: Controversies in Homo sapiens evolution*. Rotterdam: A.A. Balkema.

Bräuer, G., Henke, W., & Schultz, M. (1995). Der hominide Unterkiefer von Dmanisi: Morphologie, Pathologie und Analysen zur Klassifikation. *Jahrbuch des Römisch-Germanischen Zentralmuseums Mainz, 42*, 143–203. Tafel 57-64.

Bromage, T. G., Schrenk, F., & Zonneveld, F. W. (1995). Paleoanthropology of the Malawi Rift: An early hominid mandible from the Chiwondo Beds, northern Malawi. *Journal of Human Evolution, 28*, 71–108.

Brown, P., Sutikna, T., Morwood, M. J., Soejono, R. P., Jatmiko, Wayhu Saptomo, E., & Rokus Awe, Due. (2004). A new small-bodied hominin from the Late Pleistocene of Flores, Indonesia. *Nature, 431*, 1055–1061.

Bruner, E., & Manzi, G. (2007). Landmark-based shape analysis of the archaic *Homo* calvarium from Ceprano (Italy). *American Journal of Physical Anthropology, 132*, 355–366.

Campbell, B.G. (1965). The nomenclature of the Hominidae, including a definite list of hominid taxa. Occasional Papers Royal Anthropological Institute London 22.

Campbell, B. G., & Loy, J. L. (2000). *Humankind emerging* (8th ed.). Boston: Allyn and Bacon.

Cann, R. L. (1992). A mitochondrial perspective on replacement and continuity in human evolution. In G. Bräuer & F. Smith (Eds.), *Continuity or replacement: Controversies in Homo sapiens evolution* (pp. 65–73). Rotterdam: A.A. Balkema.

Cann, R. L., Stoneking, M., & Wilson, A. C. (1987). Mitochondrial DNA and human evolution. *Nature, 325*, 31–36.

Caramelli, D., Lalueza-Fox, C., Condemi, S., Longo, L., Milani, L., Manfredini, A., de Saint Pierre, M., Adoni, F., Lari, M., Giutini, P., et al. (2006). A highly divergent mtDNA sequence in a Neandertal individual from Italy. *Current Biology, 16*, R630–R632.

Carbonell, E., Esteban, M., Nájera, A. M., Mosquera, M., Rodríguez, X. P., Ollé, A., Sala, R., Vergès, J. M., Bermúdez de Castro, J. M., & Ortega, A. I. (1999). The Pleistocene site of Gran Dolina, Sierra de Atapuerca, Spain: A history of the archaeological investigations. *Journal of Human Evolution, 37*, 313–324.

Carbonell, E., Bermúdez de Castro, J. M., Parés, J. M., Pérez-González, A., Cuenca-Bescós, G., Ollé, A., Mosquera, M., van der Huguet, R., Made, J., et al. (2008). The first hominin of Europe. *Nature, 452*, 465–470.

Chamberlain, A. T., & Wood, B. A. (1987). Early hominid phylogeny. *Journal of Human Evolution, 16*, 119–133.

Collard, M., & Wood, B. A. (2007). Defining the genus *Homo*. In W. Henke & I. Tattersall (Eds.), *Handbook of paleoanthropology* (Vol. 1, pp. 1575–1610). Berlin, Heidelberg, New York: Springer-Verlag.

Condemi, S., & von Koenigswald, W. (1997). Der Unterkiefer von Mauer. In G. A. Wagner & K. W. Beinhauer (Eds.), *Homo heidelbergensis von Mauer. Das Auftreten des Menschen in Europa* (pp. 31–61). Heidelberg: Universitätsverlag C. Winter GmbH.

Corbey, R., & Roebroeks, W. (Eds.). (2001). *Studying human origins: Disciplinary history and epistemology*. Amsterdam: Amsterdam University Press.

Couture, C., & Hublin, J.-J. (2005). Les Néandertaliens. In O. Dutour, J.-J. Hublin, & B. Vandermeersch (Eds.), *Origine et Evolution des Populations Humaines* (pp. 129–150). Paris: Comité des Travaux Historiques et Scientifiques.

Darwin, C. (1871). *The descent of man and selection in relation to sex*. Reprinted 1981 by Princeton University Press ed. John Murray (1871), London.

Delson, E. (2000). Paleoanthropology. In E. Delson, I. Tattersall, J. A. Van Couvering, & A. S. Brooks (Eds.), *Encyclopedia of human evolution and prehistory* (2nd ed., pp. 504–505). New York: Garland.

Desmond, A. (1997). *Huxley*. London: Penguin Books.

Dreyer, T. F. (1935). A human skull from Florisbad, Orange Free State, with a note on the endocranial cast (by C.U. Ariëns-Kappers). *Proceedings of the Academy of Sciences, Amsterdam, 38*, 119–128.

Dubois, E. (1892). Paleontologische anderzoekingen op Java. *Versl Mijnw Batavia, 3*, 20–14.

Dubois, E. (1894). *Pithecanthropus erectus, eine menschenähnliche Übergangsform aus Java*. Batavia: Landesdruckerei.

Dubois, E. (1940). The fossil remains discovered by Dr. G. H. R. von Koenigswald and attributed by him to *Pithecanthropus erectus*, in reality remains of *Homo wadjakensis* (syn. *Homo soloensis*). *Proceedings of the Academy of Sciences, Amsterdam, 43*, 494–496, 842–851, 1268–1275.

Dunsworth, H., & Walker, A. (2002). Early genus *Homo*. In W. C. Hartwig (Ed.), *The primate fossil record* (pp. 419–435). Cambridge: Cambridge University Press.

Eswaran, V., Harpending, H., & Rogers, A. R. (2005). Genomics refutes an exclusively African origin of humans. *Journal of Human Evolution, 49*, 1–18.

Etler, D. A., Crummett, T. L., & Wolpoff, M. H. (2001). Longgupo: Early *Homo* colonizer or Late Pliocene *Lufengpithecus* survivor in South China. *Human Evolution, 16*, 1–12.

Etter, W. (1994). *Palökologie*. Basel, Boston, Berlin: Birkhäuser Verlag.

Excoffier, L. (2002). Human demographic history: Redefining the recent African origin model. *Current Opinion in Genetics & Development, 12*, 675–682.

Falk, D., Hildebolt, C., Smith, K., Morwood, M. J., Sutikna, T., Brown, P., Jatmiko, Saptomo, E. W., Brunsden, B., & Prior, F. (2005). The brain of LB1, *Homo floresiensis*. *Science, 308*, 242–245.

Finlayson, C. (2004). *Neanderthals and modern humans* (An Ecological and Evolutionary Perspective). Cambridge: Cambridge University Press.

Foley, R. A. (1987). *Another unique species: Patterns in human evolutionary ecology*. Harlow, New York: Longman Scientific & Technical.

Foley, R. A. (1991). How many hominid species should there be? *Journal of Human Evolution, 20*, 413–427.

Foley, R. A. (2001). In the shadow of the modern synthesis? Alternative perspectives on the last fifty years of paleoanthropology. *Evolutionary Anthropology, 10*, 5–14.

Franck, G. (1998). *Ökonomie der Aufmerksamkeit: Ein Entwurf*. München: Edition Hanser.

Franzen, J. L. (Ed.). (1994a). *100 years of Pithecanthropus the Homo erectus problem* (Courier Forschungsinstitut Senckenberg, Vol. 171). Frankfurt am Main: Forschungsinstitut Senckenberg.

Franzen, J. L. (1994b). The *Homo erectus* problem. In J. L. Franzen (Ed.), *100 years of Pithecanthropus the Homo erectus problem* (Courier Forschungsinstitut Senckenberg, Vol. 171, pp. 9–10). Frankfurt am Main: Forschungsinstitut Senckenberg.

Frayer, D. W., Wolpoff, M. H., Thorne, A. G., Smith, F. H., & Pope, G. (1993). Theories of modern human origins: The paleontological test. *American Anthropologist, 95*, 14–50.

Friederichs, H. F. (1932). Schädel und Unterkiefer von Piltdown ("*Eoanthropus dawsoni* Woodward") in neuer Untersuchung. *Zeitschrift für Anatomie und Entwicklungsgeschichte, 98*, 199–262.

Futuyama, D. J. (2007). *Evolution*. Heidelberg: Spektrum Akademischer Verlag.

Gabunia, L., Vekua, A., & Lordkipanidze, D. (2000a). Taxonomy of the Dmanisi crania. *Science, 289*, 55–56.

Gabunia, L., Vekua, A., Lordkipanidze, D., Swisher, C. C., III, Ferring, R., Justus, A., Nioradze, M., Tvalchrelidze, M., Antón, S. C., Bosinski, G., Jöris, O., de Lumley, M.-A., Masjuradze, G., & Mouskhelishvili, A. (2000b). Earliest Pleistocene hominid cranial remains from Dmanisi Republic of Georgia: Taxonomy, geological setting and age. *Science, 288*, 1019–1025.

Gabunia, L., Antón, S. C., Lordkipanidze, D., Vekua, A., Justus, A., & Swisher, C. C., III. (2001). Dmanisi and dispersal. *Evolutionary Anthropology, 10*, 158–170.

Gabunia, L., de Lumley, M.-A., Vekua, A., Lordkipanidze, D., & de Lumley, H. (2002). Discovery of a new hominid at Dmanisi (Transcaucasia, Georgia). *Comptes Rendus Palevolution, 1*, 243–253.

Gilbert, W. H., White, T. D., & Asfaw, B. (2003). *Homo erectus, Homo ergaster, Homo "cepranensis"*, and the Daka cranium. *Journal of Human Evolution, 45*, 255–259.

Gorjanović-Kramberger, K. (1906). *Der diluviale Mensch von Krapina in Kroatien* (Ein Beitrag zur Paläoanthropologie). Wiesbaden: C. W. Kreidel's Verlag.

Gould, S. J., & Eldredge, N. (1977). Punctuated equilibria: The tempo and mode of evolution reconsidered. *Paleobiology, 3*, 115–151.

Grine, F. E. (2001). Implications of morphological diversity in early *Homo* crania from Eastern and Southern Africa. In P. V. Tobias, M. A. Raath, J. Moggi-Cecchi, & G. A. Doyle (Eds.), *Humanity from African naissance to coming millennia* (pp. 107–116). Firenze/Johannesburg: Firenze University Press/Witwatersrand University Press.

Groves, C. P. (1989). A regional approach to the problem of the origin of modern humans in Australasia. In P. Mellars & C. Stringer (Eds.), *The human revolution: Behavioural and biological perspectives on the origins of modern humans* (pp. 274–285). Edinburgh: Edinburgh University Press.

Groves, C. P., & Lahr, M. M. (1994). A bush not a ladder: Speciation and replacement in Human evolution. *Perspectives in Human Biology, 4*, 1–11.

Groves, C. P., & Mazák, V. (1975). An approach to the taxonomy of the Hominidae: Gracile Villafranchian hominids in Africa. *Casopis pro Mineralogii a Geologii, 20*, 225–247.

Grubb, P. (1999). Evolutionary processes implicit in distribution patterns of modern African mammals. In T. G. Bromage & F. Schrenk (Eds.), *African biogeography, climate change, and human evolution* (pp. 150–164). New York: Oxford University Press.

Haeckel, E. (1866). *Generelle Morphologie der Organismen* (Vol. 2). Berlin: G. Reimer Verlag.

Haidle, M. (2007). Archeology. In W. Henke & I. Tattersall (Eds.), *Handbook of paleoanthropology* (Vol. 1, pp. 261–289). Berlin, Heidelberg, New York: Springer-Verlag.

Hardt, T., & Henke, W. (2007). Zur stammesgeschichtlichen Stellung des *Homo heidelbergensis*. In G. A. Wagner, H. Rieder, L. Zöller, & E. Mick (Eds.), *Homo heidelbergensis. Schlüsselfund der Menschheitsgeschichte* (pp. 184–202). Stuttgart, S: Konrad Theiss Verlag.

Hardt, T., Hardt, B., & Menke, P. (2007). Paleoecology: An adequate window on the past? In W. Henke & I. Tattersall (Eds.), *Handbook of paleoanthropology* (Vol. 1, pp. 503–554). Berlin, Heidelberg, New York: Springer.

Hartwig-Scherer, S., & Martin, R. D. (1991). Was "Lucy"more human than her "child"? Observation on early hominid postcranial skeletons. *Journal of Human Evolution, 21*, 439–449.

Harvati, K. (2003). Quantitative analysis of Neanderthal temporal bone morphology using 3-D geometric morphometrics. *American Journal of Physical Anthropology, 120*, 323–338.

Harvati, K. (2007). Neanderthals and their contemporaries. In W. Henke & I. Tattersall (Eds.), *Handbook of paleoanthropology* (Vol. 3, pp. 1717–1748). Berlin, Heidelberg, New York: Springer-Verlag.

Hawks, J. D., & Wolpoff, M. H. (2001). The accretion model of Neandertal evolution. *Evolution, 55*, 1474–1485.

Hemmer, H. (1999). Erstbesiedlung Europas nach Indizien der Großsäugergesellschaft. In W. Henke & H. Rothe (Eds.), *Stammesgeschichte des Menschen* (pp. 214–215). Berlin, Heidelberg, New York: Springer.

Hemmer, H. (2007). Estimation of basic life history data of fossil hominoids. In W. Henke & I. Tattersall (Eds.), *Handbook of paleoanthropology* (Vol. 1, pp. 587–620). Berlin, Heidelberg, New York: Springer.

Henke, W. (2003a). Evaluating human fossil finds. In G. Grupe & J. Peters (Eds.), *Decyphering ancient bones of bioarchaeological collections* (Documenta Archaeobiologicae, Vol. 1, pp. 59–76). Rahden/Westf: Verlag Marie Leidorf.

Henke, W. (2003b). Population dynamics during the European Middle and Late Pleistocene – smooth or jumpy? In H. Meller (Ed.), *Veröffentlichungen des Landesamtes für Archäologie Sachsen-Anhalt – Landesmuseum für Vorgeschichte* (Festschrift Dietrich Mania Band 57/I, pp. 247–258). Halle (Saale).

Henke, W. (2005). Human biological evolution. In F. M. Wuketits & F. J. Ayala (Eds.), *Handbook of evolution* (Vol. 2, pp. 117–222). Weinheim, New York: Wiley – VCH.

Henke, W. (2006a). Evolution und Verbreitung des Genus *Homo* – Aktuelle Befunde aus evolutionsökologischer Sicht. In N. J. Conard (Ed.), *Woher kommt der Mensch?* (pp. 104–142). Tübingen: Attempto Verlag.

Henke, W. (2006b). Gorjanović-Kramberger's research in Krapina – Its impact on paleoanthropology in Germany. *Periodicum Biologorum, 108*, 239–252.

Henke, W. (2007). Historical overview of paleoanthropological research. In W. Henke & I. Tattersall (Eds.), *Handbook of paleoanthropology* (Vol. 1, pp. 1–56). Berlin, Heidelberg, New York: Springer-Verlag.

Henke, W., & Rothe, H. (1994). *Paläoanthropologie*. Berlin, Heidelberg, New York: Springer-Verlag.

Henke, W., & Rothe, H. (1995). *Homo erectus* - valides Taxon der europäischen Hominiden? *Bulletin de Société Suisse d'Anthropologie, 1*, 15–26.

Henke, W., & Rothe, H. (1999a). *Stammesgeschichte des Menschen. Eine Einführung*. Berlin, Heidelberg, New York: Springer Verlag.

Henke, W., & Rothe, H. (1999b). Die phylogenetische Stellung des Neandertalers. *Biologie in unserer Zeit, 29*(6), 320–329.

Henke, W., & Rothe, H. (2001). *Entstehung des Genus Homo*. Ein Quantensprung in der Evolution des Menschen. Praxis der Naturwissenschaften. *Biologie in der Schule, 50*, 29–33.

Henke, W., & Rothe, H. (2003). *Menschwerdung*. Franfurt a. M: S. Fischer Verlag.

Henke, W., & Rothe, H. (2005). Ursprung, Adaptation und Verbreitung der Gattung *Homo*. Marginalien zur Evolution eines global player. In B. Kleeberg, T. Walter, & F. Crivellari (Eds.), *Urmensch und Wissenschaften. Eine Bestandsaufnahme* (pp. 89–124). Darmstadt: Wissenschaftliche Buchgesellschaft.

Henke, W., & Rothe, H. (2006). Zur Entwicklung der Paläoanthropologie im 20. Jahrhundert. Von der Narration zur hypothetiko-deduktiven Forschungsdisziplin. In D. Preuß, U. Hoßfeld, & O. Breidbach (Eds.), *Anthropologie nach Haeckel* (pp. 46–71). Stuttgart: Franz Steiner Verlag.

Henke, W., & Tattersall, I. (Eds.). (2007). *Handbook of paleoanthropology* (Vol. 1–3). Berlin, Heidelberg, New York: Springer.

Henke, W. in cooperation with Roth, H., & Simon, C. (1995). Qualitative and quantitative analysis of the Dmanisi mandible. In R. J. Radlanski, & H. Renz (Eds.), *Proceedings of the 10th international symposium on dental morphology (Berlin 1995)*(pp. 106–110). Berlin: "M" –Marketing Services.

Henke, W., Rothe, H., & Alt, K. W. (1999). Dmanisi and the early Eurasian dispersal of the genus *Homo*. In H. Ullrich (Ed.), *Hominid evolution: Lifestyles and survival strategies* (pp. 138–155). Gelsenkirchen, Schwelm: Edition Archaea.

Hennig, W. (1950). *Grundzüge einer Theorie der phylogenetischen Systematik*. Berlin: Deutscher Zentralverlag.

Howell, F. C. (1978). Hominidae. In V. J. Maglio & H. N. B. S. Cooke (Eds.), *Evolution of African mammals* (pp. 154–248). Cambridge: Harvard University Press.

Howell, F. C. (1986). Variabilité chez *Homo erectus* et problème de la presence de cette espèce en Europe. *L'Anthropologie, 90*, 447–481.

Howell, F. C. (1996). Thoughts on the study and interpretation of the human fossil record. In W. E. Meikle, F. C. Howell, & N. G. Jablonski (Eds.), *Contemporary issues in human evolution* (Wattis symposium series in anthropology, California Academy of Sciences Memoir 21, pp. 1–54). San Francisco.

Howells, W. W. (1980). *Homo erectus* – Who, when and where: A survey. *Yearbook of Physical Anthropology, 23*, 1–23.

Howells, W. W. (1993). *Getting here: The story of human evolution*. Washington: Compass Press.

Huang, W., Ciochon, R., Yumin, G., Larick, R., Qiren, F., Schwarcz, H., Yonge, C., de Vos, J., & Rink, W. (1995). Early *Homo* and associated artefacts from Asia. *Nature, 378*, 275–278.

Hublin, J. J. (1986). Some comments on the diagnostic features of *Homo erectus*. *Anthropos (Brno), 23*, 175–187.

Hublin, J. J. (1998). Climatic changes, paleogeography, and the evolution of the Neandertals. In T. Akazawa, K. Aoki, & O. Bar-Yosef (Eds.), *Neandertals and modern humans in Western Asia* (pp. 295–310). New York: Plenum.

Huxley, T. H. (1863). *Zeugnisse für die Stellung des Menschen in der Natur*. Braunschweig: Friedrich Vieweg und Sohn.

Jacob, T. (1975). Morphology and paleoecology of early man in Java. In R. H. Tuttle (Ed.), *Paleoanthropology, morphology and paleoecology* (pp. 311–325). The Hague: Mouton Publishers.

Jelinek, J. (1978). Comparison of Mid-Pleistocene evolutionary process in Europe and in South-East Asia. In *Proceedings of the symposium on natural selection* (pp. 251–257). Libice 1978, Praha.

Jelínek, J. (1981). Was *Homo erectus* always *Homo sapiens*? In D. Ferembach (Ed.), *Les Processus de l'Hominisation. L'Evolution Humaine. Les Faits. Les Modalités* (pp. 85–89). Paris: Editions CNRS.

Jobling, M. A., Hurles, M. E., & Tyler-Smith, C. (2004). *Human evolutionary genetics: Origins, peoples & disease*. New York: Garland Publishing.

Johanson, D., & Edgar, B. (2007). *Lucy und ihre Kinder* (2nd ed.). Heidelberg: Spektrum Akademischer Verlag.

Johanson, D. C., Masao, F. T., Eck, G. G., White, T. D., Walter, R. C., Kimbel, W. H., Asfaw, B., et al. (1987). New partial skeleton of *Homo habilis* from Olduvai Gorge, Tanzania. *Nature, 327*, 205–209.

Jolly, C. J. (2001). A proper study for mankind: Analogies from the Papionin monkeys and their implications for human evolution. *American Journal of Physical Anthropology, 116*, 177–204.

Keith, A. (1931). *New discoveries relating to the antiquity of man*. London: Williams and Norgate.

Kennedy, G. E. (1991). On the autapomorphic traits of *Homo erectus*. *Journal of Human Evolution, 20*, 355–371.

Kennedy, G. E. (1999). Is "*Homo rudolfensis*" a valid species? *Journal of Human Evolution, 36*, 119–121.

King, W. (1864). Reputed fossil man of the Neanderthal. *Quarterly Journal of Science, 1*, 88–97.

Krause, J., Lalueza-Fox, C., Orlando, L., Enard, W., Green, R. E., & Burbano, H. A., Hublin, J.-J., Hänni, C., Fortea, J., de la Rasilla, M., Bertranpetit, J., Rosas, A., Pääbo, S. (2007). The derived FOXP2 variant of modern humans was shared with Neandertals. *Current Biology, 17*, Report.

Krings, M., Stone, A., Schmitz, R. W., Krainitzki, H., Stoneking, M., & Pääbo, S. (1997). Neandertal DNA sequences and the origin of modern humans. *Cell, 90*, 1–20.

Krings, M., Capelli, C., Tschentscher, F., Geisert, H., Meyer, S., von Haeseler, A., Grosschmidt, K., Possnert, G., Paunovic, M., & Pääbo, S. (2000). A view of Neanderthal genetic diversity. *Nature Genetics, 26*, 144–146.

Lahr, M. M. (1994). The multiregional model of modern human origins: A reassessment of its morphological basis. *Journal of Human Evolution, 26*, 23–56.

Langbroek, M., & Roebroeks, W. (2000). Extraterrestrial evidence on the age of the hominids from Java. *Journal of Human Evolution, 38*, 595–600.

Larick, R., & Ciochon, R. L. (1996). The African emergence and early Asian dispersals of the genus *Homo*. *American Scientist, 84*, 538–552.

Leakey, L. S. B., Tobias, P. V., & Napier, J. R. (1964). A new species of the genus *Homo* from Olduvai Gorge. *Nature, 2002*, 7–9.

Linnaeus, C. (1758). *Systema naturae per regna tria naturae, secundum classes, ordines, genera, species cum characteribus, differentiis, synonymis, locis*. Holmiae: Laurentii Salvii.

Mai, L. L., Young Owl, M., & Kersting, M. P. (2005). *The Cambridge dictionary of human biology and evolution*. Cambridge: Cambridge University Press.

Mallegni, F., Carnieri, E., Bisconti, M., Tartarelli, G., Ricci, S., Biddittu, I., & Serge, A. (2003). *Homo cepranensis sp. nov.* and the evolution of African-European Middle Pleistocene hominids. *Comptes Rendus Palevolution, 2*, 153–159.

Manzi, G. (2001). The earliest diffusion of the genus *Homo* toward Asia and Europe: A brief overview. In P. V. Tobias, M. A. Raath, J. Moggi-Cecchi, & G. A. Doyle (Eds.), *Humanity from African naissance to coming millennia* (pp. 117–124). Firenze/Johannesburg: Firenze University Press/Witwatersrand University Press.

Manzi, G. (2004). Human evolution at the Matuyama-Brunhes boundary. *Evolutionary Anthropology, 13*, 11–24.

Manzi, G., Mallegni, F., & Ascenzi, A. (2001). A cranium for the earliest Europeans: Phylogenetic position of the hominid from Ceprano, Italy. *Proceedings of the National Academy of Sciences, 98*(17), 10011–10016.

Manzi, G., Bruner, E., & Passarello, P. (2003). The one-million-year-old *Homo* cranium from Bouri (Ethiopia): A reconsideration of its *H. erectus* affinities. *Journal of Human Evolution, 4*, 731–736.

Martin, R. D. (1995). Hirngröße und menschlche Evolution. *Spektrum der Wissenschaft, 9*, 58–57.

Mayr, E. (1969). *Principles of systematic zoology*. New York: McGraw Hill.

Mayr, E. (1975). *Grundlagen der zoologischen Systematik*. Hamburg: Paul Parey.

McHenry, H. M. (1996). Homoplasy, clades, and hominid phylogeny. In W. E. Meikle, F. C. Howell, & N. G. Jablonski (Eds.), *Contemporary issues in human evolution* (Wattis symposium series in anthropology, California Academy of Sciences Memoir 21, pp. 77–92). San Francisco.

Mithen, S. J. (1998). *The prehistory of the mind: A search for the origin of art, science and religion*. London: Thames & Hudson.

Mithen, S. J. (2007). The network of brain, language, and culture. In W. Henke & I. Tattersall (Eds.), *Handbook of paleoanthropology* (Vol. 3, pp. 1965–2000). Berlin, Heidelberg, New York: Springer.

Orlando, L., Darlu, P., Toussaint, M., Bonjean, D., Otte, M., & Hänni, C. (2006). Revisiting Neandertal diversity with a 100, 000 year old mtDNA sequence. *Current Biology, 16*, R400–R402.

Ovchinnikov, I. V., Götherström, A., Romanova, G. P., Kharitonov, V. M., Lidén, K., & Goodwin, W. (2000). Molecular analysis of Neanderthal DNA from the northern Caucasus. *Nature, 404*, 490–493.

Pääbo, S., Poinar, H., Serre, D., Jaenicke-Després, V., Hebler, J., Rohland, N., Kuch, M., Krause, J., Vigilant, L., & Hofreiter, M. (2004). Genetic analyses from ancient DNA. *Annual Review of Genetics, 38*, 645–679.

Relethford, J. H. (1999). Genetics of modern human origins and diversity. *Annual Review of Anthropology, 27*, 1–23.

Relethford, J. H. (2001a). *Genetics and the search for modern human origins*. New York, Weinheim: Wiley.

Relethford, J. H. (2001b). Ancient DNA and the origins of modern humans. *Proceedings of the National Academy of Sciences, 98*, 390–391.

Rightmire, G. P. (1990). *The evolution of Homo erectus: Comparative anatomical studies of an extinct human species*. Cambridge: Cambridge University Press.

Rightmire, G. P. (1998). Human evolution in the Middle Pleistocene: The role of *Homo heidelbergensis*. *Evolutionary Anthropology, 6*, 218–227.

Rightmire, G. P. (2001a). Morphological diversity in Middle Pleistocene *Homo*. In P. V. Tobias, R. Raath, J. Moggi-Cecchi, & G. Doyle (Eds.), *Humanity from African naissance to coming millenia* (pp. 135–140). Firenze/Johannesburg: Firenze University Press/ Witwatersrand University Press.

Rightmire, G. P. (2001b). Comparison of the Middle Pleistocene hominids from Africa and Asia. In L. Barham & K. Robson-Brown (Eds.), *Human roots: Africa and Asia in the Middle Pleistocene* (pp. 123–133). Bristol: Western Academic and Specialist Press.

Rightmire, G. P. (2007). Later Middle Pleistocene *Homo*. In W. Henke & I. Tattersall (Eds.), *Handbook of paleoanthropology* (Vol. 3, pp. 1695–1716). Berlin, Heidelberg, New York: Springer-Verlag.

Rightmire, G. P., Lordkipanidze, D., & Vekua, A. (2005). Anatomical descriptions, comparative studies and evolutionary significance of the hominin skulls from Dmanisi, Republic of Georgia. *Journal of Human Evolution, 50*, 115–141.

Rogers, M. J., Feibel, C. S., & Harris, J. W. K. (1996). Deciphering early hominid land use and behaviour: A multidisciplinary approach from the Lake Turkana basin. In C. C. Magori, C. B. Saanane, & F. Schrenk (Eds.), *Four million years of hominid evolution in Africa: Papers in Honour of Dr. Mary Douglas Leakey´s outstanding contribution in palaeoanthropology* (Kaupia, Vol. 6, pp. 9–19).

Rothe, H., & Henke, W. (2001). Methoden der Klassifikation und Systematisierung. Praxis der Naturwissenschaften – Biologie in der Schule. Köln/Leipzig: Aulis-Verlag Deubner, 50, 1–7

Rothe, H., & Henke, W. (2006). Stammbäume sind wie Blumensträuße – hübsch anzusehen, doch schnell verwelkt. In D. Preuß, U. Hoßfeld, & O. Breidbach (Eds.), *Anthropologie nach Haeckel* (pp. 149–183). Stuttgart: Franz Steiner Verlag.

Sartono, S. (1981). Pre-*Homo erectus* population in Java, Indonesia. In *X Congreso Union Internacional de Ciencias Prehistoricas y Protohistoricas* (pp. 47–86).

Schmitz, R. W., & Thissen, J. (2000). *Neandertal: Die Geschichte geht weiter*. Heidelberg: Spektrum Akademischer Verlag.

Schoetensack, O. (1908). *Der Unterkiefer des Homo heidelbergensis aus den Sanden von Mauer bei Heidelberg*. Engelmann, Leipzig: Ein Beitrag zur Paläontologie des Menschen.

Schott, L. (1977). Der Meinungsstreit um den Skelettfund aus dem Neandertal von 1856. Ausgrabungen und Funde. *Archäologische Berichte und Informationen, 22*, 235–238.

Schott, L. (1979). Der Skelettfund aus dem Neandertal im Urteil Rudolf Virchows. *Biologische Rundschau, 19*, 304–309.

Schrenk, F., Bromage, T. G., Betzler, C. G., Ring, U., & Juwayeyi, Y. (1993). Oldest *Homo* and Pliocene biogeography of the Malawi Rift. *Nature, 365*, 833–836.

Schrenk, F., Kullmer, O., & Bromage, T. (2007). The earliest putative *Homo* fossils. In W. Henke & I. Tattersall (Eds.), *Handbook of paleoanthropology* (Vol. 3, pp. 1611–1633). Berlin, Heidelberg, New York: Springer.

Schwartz, J. H. (2000a). The origin and identification of species. In M. Schultz, K. Christiansen, H. Greil, W. Henke, A. Kemkes-Grottenthaler, C. Niemitz, H. Rothe, W. Schiefenhövel, H. D. Schmidt, I. Schröder, H. Schutkowski, M. Teschler-Nicola, & U. Wittwer-Backofen (Eds.), *Proceedings: 3. Schittstelle Mensch – Umwelt in Vergangenheit, Gegenwart und Zukunft* (Kongress der Gesellschaft für Anthropologie, pp. 40–43). Göttingen: Cuvillier Verlag.

Schwartz, J. H. (2000b). Taxonomy of the Dmanisi crania. *Science, 28*, 55–56.

Schwartz, J. H., & Tattersall, I. (2000). What constitutes *Homo erectus*? *Acta Anthropologica Sinica, 19*(Suppl), 18–22.

Schwartz, J. H., & Tattersall, I. (2002). *The human fossil record. Volume 1: Terminology, craniodental morphology of genus Homo (Europe)*. New York: Wiley-Liss.

Schwartz, J. H., & Tattersall, I. (2003). *The human fossil record. Volume 2: Craniodental morphology of genus Homo (Africa and Asia)*. New York: Wiley-Liss.

Semaw, S. (2000). The world's oldest stone artefacts from Gona, Ethiopia: Their implications for understanding stone technology and patterns of human evolution between 2.6–1.5 Million Years ago. *Journal of Archaeological Science, 27*(12), 1197–1214.

Serre, D., Langaney, A., Chech, M., Teschler-Nicola, M., Paunovic, M., Mennecier, P., & Hofreiter, M. (2004). No evidence of Neandertal mtDNA contribution to early modern humans. *PLoS Biology, 2*(3), 0313–0317.

Shipman, P., & Storm, P. (2002). Missing links: Eugène Dubois and the origins of paleoanthropology. *Evolutionary Anthropology, 11*, 108–116.

Simpson, G. G. (1961). *Principles of animal taxonomy*. San Francisco: Freeman & Co.

Smith, F. (2002). Migrations, radiations and continuity: Patterns in the evolution of Middle and Late Pleistocene humans. In W. C. Hartwig (Ed.), *The primate fossil record* (pp. 437–456). Cambridge: Cambridge University Press.

Smith, F. H., & Spencer, F. (Eds.). (1984). *The origins of modern humans: A world survey of fossil evidence*. New York: Alan R. Liss.

Sneath, P. H. A., & Sokal, R. R. (1973). *Numerical taxonomy: The principles and practice of numerical classification*. San Francisco: Freeman & Co.

Spencer, F. (1984). The Neandertals and their evolutionary significance: A brief historical survey. In F. H. Smith & F. Spencer (Eds.), *The origins of modern humans: A world survey of fossil evidence* (pp. 1–49). New York: Alan R. Liss.

Spencer, F. (1990a). *The Piltdown papers 1908–1955*. Oxford: Natural History Museum, Oxford University Press.

Spencer, F. (1990b). *Piltdown – A scientific forgery*. Oxford: Natural History Museum, Oxford University Press.

Stent, G. S. (1972). Prematurity and uniqueness in scientific discovery. *Scientific American, 227*, 84–93.

Stoneking, M., & Cann, R. L. (1989). African origin of human mitochondrial DNA. In P. Mellars & C. B. Stringer (Eds.), *The human revolution: Behavioural and biological perspectives on the origins of modern humans* (pp. 17–30). Edinburgh: Edinburgh University Press.

Storch, V., Welsch, U., & Wink, M. (2001). *Evolutionsbiologie*. Berlin, Heidelberg, New York: Springer.

Strait, D. S., & Grine, F. E. (2004). Inferring hominoid and early hominid phylogeny using craniodental characters: The role of fossil taxa. *Journal of Human Evolution, 47*, 399–452.

Strait, D. S., Grine, F. E., & Moniz, M. A. (1997). A reappraisal of early hominid phylogeny. *Journal of Human Evolution, 32*, 1–82.

Stringer, C. B. (1982). Towards a solution of the Neanderthal problem. *Journal of Human Evolution, 2*, 431–438.

Stringer, C. B. (1986). The credibility of *Homo habilis*. In B. A. Wood, L. B. Martin, & P. Andrews (Eds.), *Major topics in primate and human evolution* (pp. 266–294). Cambridge: Cambridge University Press.

Stringer, C. B. (1992). Replacement, continuity and the origin of *Homo sapiens*. In G. Bräuer & F. G. Smith (Eds.), *Continuity or replacement: Controversies in Homo sapiens evolution* (pp. 9–24). Rotterdam: A.A. Balkema.

Stringer, C. B. (1996). Current issues in modern human origins. In W. Meikle, F. C. Howell, & N. Jablonski (Eds.), *Contemporary issues in human evolution* (Wattis symposium series in anthropology, California Academy of Sciences Memoir 21, pp. 115–134). San Fancisco.

Stringer, C. B. (2001). Modern human origins: Distinguishing the models. *African Archaeological Review, 18*, 67–75.

Stringer, C. B. (2002). Modern human origins: Progress and prospects. *Philosophical Transactions of the Royal Society of London, B, 357*, 563–579.

Stringer, C. B., & Andrews, P. (1988). Genetic and fossil evidence for the origin of modern humans. *Science, 239*, 1263–1268.

Stringer, C. B., & Gamble, C. (1993). *In search of the Neanderthals: Solving the puzzle of human origins*. London: Thames and Hudson.

Susman, R. L. (2008). Brief communication: Evidence bearing on the status of *Homo habilis* at Olduvai Gorge. *American Journal of Physical Anthropology, 139*, 356–361.

Tattersall, I. (1986). Species recognition in human paleontology. *Journal of Human Evolution*, 15, 165–175.

Tattersall, I. (1995a). *The fossil trail. How we know what we think we know about human evolution*. New York, San Diego, London: Hartcourt Brace & Company.

Tattersall, I. (1995b). *The last Neanderthal: The rise, success, and mysterious extinction of our closest human relatives*. New York: Nevraumont Publishing Company.

Tattersall, I. (2000a). Paleoanthropology: The last half-century. *Evolutionary Anthropology, 9*, 2–16.

Tattersall, I. (2000b). Once we were not alone. *Scientific American, 2*, 56–63.

Tattersall, I. (2007). *Homo ergaster* and its contemporaries. In W. Henke & I. Tattersall (Eds.), *Handbook of paleoanthropology* (Vol. 3, pp. 1633–1654). Berlin, Heidelberg, New York: Springer.

Tattersall, I., & Schwartz, J. H. (2000). *Extinct humans*. Boulder, Colorado: Westview Press.

Templeton, A. (2002). Out of Africa again and again. *Nature, 416*, 45–51.

Thorne, A. G., & Wolpoff, M. H. (1981). Regional continuity in Australasian Pleistocene hominid evolution. *American Journal of Physical Anthropology, 55*, 337–349.

Thorne, A. G., & Wolpoff, M. H. (1992). Multiregionaler Ursprung des modernen Menschen. *Spektrum der Wissenschaft H, 6*, 80–87.

Tobias, P.V. (1989a). The status of *Homo habilis* in 1987 and some outstanding problems. In G. Giacobini (Ed.), *Hominidae* (pp. 141–149). Proceedings of the 2nd International Congress on Human Paleontology, Turin, 1987. Mailand: Jaca Books.

Tobias, P.V. (1989b). A gradual appraisal of *Homo habilis*. In G. Giacobini (Ed.), *Hominidae* (pp. 151–154). Proceedings of the 2nd International Congress of Human Paleontology, Turin 1987. Mailand: Jaca Book.

Tobias, P. V. (1991). *Olduvai Gorge, vol. 4, Parts V-IX. The skulls, endocasts and teeth of Homo habilis*. Cambridge: Cambridge University Press.

Tobias, P. V., Raath, M. A., Moggi-Cecchi, J., & Doyle, G. A. (Eds.). (2001). *Humanity from African naissance to coming millennia*. Firenze/Johannesburg: Firenze University Press/Witwatersrand University Press.

Torre, D., Ficcarelli, G., Masini, F., Rook, L., & Sala, B. (1992). *Mammal dispersal events in the early Pleistocene of western Europe* (Courier Forschungsinstitut Senckenberg, Frankfurt 153, pp. 51–58). Frankfurt am Main: Forschungsinstitut Senckenberg.

Trinkaus, E., & Shipman, P. (1993). *The Neandertals: Changing the image of mankind*. New York: Alfred A Knopf.

Turner, A., & O'Regan, H. (2007). Zoogeography: Primate and early hominin distribution and migration patterns. In W. Henke & I. Tattersall (Eds.), *Handbook of paleoanthropology* (Vol. 1, pp. 421–440). Berlin, Heidelberg, New York: Springer-Verlag.

Ullrich, H. (1998). *Zur Diskussion um die Erstbesiedlung Europas und den ältesten Europäer* (Mitteilungen der Berliner Gesellschaft für Anthropologie,Ethnologie und Urgeschichte Bd. 19, pp. 57–74).

Vekua, A., Lordkipanidze, D., Rightmire, G. P., Agustí, J., Ferring, R., Maisuradze, G., Mouskhelishvili, A., Nioradze, M., Ponce de Leon, M., Tappen, M., Tvalchrelidze, M., & Zollikofer, C. (2002). A new skull of early *Homo* from Dmanisi, Georgia. *Science, 297*, 85–89.

von Koenigswald, G. H. R. (1950). Fossil hominids from the Lower Pleistocene of Java. *Proceeding of the Ninth International Geological Congress, London 1948, 9*, 59–61.

Vrba, E. (1988). Late Pliocene climatic events and hominid evolution. In F. E. Grine (Ed.), *Evolutionary history of the "robust" australopithecines* (pp. 405–426). New York: Aldine de Gruyter.

Vrba, E. (2007). Role of environmental stimuli in hominid origins. In W. Henke & I. Tattersall (Eds.), *Handbook of paleoanthropology* (Vol. 3, pp. 1441–1482). Berlin, Heidelberg, New York: Springer.

Wägele, J.-W. (2000). *Grundlagen der phylogenetischen Systematik*. München: Verlag Dr. Friedrich Pfeil.

Wagner, G. A. (2007). Chronometric methods in palaeoanthropology. In W. Henke & I. Tattersall (Eds.), *Handbook of paleoanthropology* (Vol. 1, pp. 311–338). Berlin, Heidelberg, New York: Springer.

Wagner, G. A., & Beinhauer, K. W. (Eds.). (1997). *Homo heidelbergensis von Mauer – Das Auftreten des Menschen in Europa*. Heidelberg: Universitätsverlag C. Winter Heidelberg GmbH.

Wagner, G. A., Fezer, U., Hambach, U., & von Koenigswald, W. (1997). Das Alter des *Homo heidelbergensis* von Mauer. In G. A. Wagner & K. W. Beinhauer (Eds.), *Homo heidelbergensis von Mauer – Das Auftreten des Menschen in Europa* (pp. 124–143). Heidelberg: Universitätsverlag C. Winter Heidelberg GmbH.

Walker, A. C., & Leakey, R. E. F. (Eds.). (1993). *The Nariokotome Homo erectus skeleton*. Cambridge: Harvard University Press.

Washburn, S. L. (1953). The strategy of physical anthropology. In A. L. Kroeber (Ed.), *Anthropology today: An encyclopedic inventory* (pp. 714–727). Chicago: University of Chicago Press.

Weidenreich, F. (1940). Some problems dealing with ancient man. *American Anthropologist, 42*, 375–385.

Weidenreich, F. (1943). The skull of *Sinanthropus pekinensis*: A comparative study on a primitive hominid skull. *Paleontologia Sinica, Neue Serie D, 10*, 1–485.

Weinert, H. (1947). *Menschen der Vorzeit*. Stuttgart: Ferdinand Enke Verlag.

Wheeler, Q. D., & Meier, R. (Eds.). (2000). *Species concept and phylogenetic theory: A debate*. New York: Columbia University Press.

White, T. D. (2000). A view on the science: Physical anthropology at the millennium. *American Journal of Physical Anthropology, 113*, 287–292.

Wiesemüller, B., Rothe, H., & Henke, W. (2003). *Einführung in die Phylogenetische Systematik*. Berlin, Heidelberg, New York: Springer.

Wiley, E. O. (1978). The evolutionary species concept reconsidered. *Systematic Zoology, 27*, 17–26.

Willmann, P. G. (1985). *Die Art in Zeit und Raum. Das Artkonzept in der Biologie und Paläontologie*. Hamburg: Paul Parey Verlag.

Wilson, A. C., & Cann, R. L. (1992). Afrikanischer Ursprung des modernen Menschen. *Spektrum der Wissenschaft, H. 6*, 72–79.

Wolpoff, M. H. (1980). *Paleoanthropology* (1st ed.). New York: Knopf.

Wolpoff, M. H. (1992). Theories of modern human origins. In G. Bräuer & F. Smith (Eds.), *Continuity or replacement: Controversies in Homo sapiens evolution* (pp. 25–63). Rotterdam: A.A. Balkema.

Wolpoff, M. H. (1996-1997). *Human evolution* (1996-1997th ed.). New York: The McGraw-Hill Companies, Inc., College Custom Series.

Wolpoff, M. H. (1999). *Paleoanthropology* (2nd ed.). Boston: McGraw-Hill.

Wolpoff, M. H., & Caspari, R. (1997). *Race and human evolution: A fatal attraction*. New York: Westview.

Wolpoff, M. H., & Caspari, R. (2000). The many species of humanity. *Prezglad Anthropologiczny - Anthropological Review, 63*, 2–17.

Wolpoff, M. H., Wu, X., & Thorne, A. G. (1984). Modern *Homo sapiens* origins: A general theory of hominid evolution involving the fossil evidence from East Asia. In F. H. Smith & F. Spencer (Eds.), *The origins of modern humans: A world survey of the fossil evidence* (pp. 411–483). New York: Alan R. Liss.

Wolpoff, M. H., Thorne, A. G., Jelinek, J., & Zhang, Yinyun. (1994a). The case of sinking *Homo erectus*. 100 years of *Pithecanthropus* is enough! In J. L. Franzen (Ed.), *100 years of Pithecanthropus: The Homo erectus problem* (Courier Forschungsinstitut Senckenberg, Vol. 171, pp. 341–361). Frankfurt am Main: Forschungsinstitut Senckenberg.

Wolpoff, M. H., Thorne, A. G., Smith, F. H., Frayer, D. W., & Pope, G. G. (1994b). Multiregional evolution: A world-wide source for modern human populations. In M. H. Nitecki & D. V. Nitecki (Eds.), *Origins of anatomically modern humans* (pp. 175–199). New York: Plenum Press.

Wood, B. A. (1992). The origin and evolution of the genus *Homo*. *Nature, 355*, 783–790.

Wood, B. A. (1996a). Human evolution. *BioEssays, 18*, 945–954.

Wood, B. A. (1996b). Origin and evolution of the Genus *Homo*. In W. E. Meikle, F. C. Howell, & N. G. Jablonski (Eds.), *Contemporrary issues in human evolution* (Wattis symposium series in anthropology, California Academy of Sciences Memoir 21, pp. 105–113). San Francisco.

Wood, B. A., & Collard, M. (1999a). The human genus. *Science, 284*, 65–71.

Wood, B. A., & Collard, M. (1999b). The changing face of genus *Homo*. *Evolutionary Anthropology, 8*, 195–207.

Wood, B. A., & Collard, M. (2001). Evolving interpretations of *Homo*. In P. V. Tobias, R. Raath, J. Moggi-Cecchi, & G. Doyle (Eds.), *Humanity from African naissance to coming millenia* (pp. 141–146). Firenze/Johannesburg: Firenze University Press/Witwatersrand University Press.

Woodward, A. S. (1921). A new cave man from Rhodesia, South Africa. *Nature, 108*, 371–372.

Woodward, A. S. (1925). The fossil anthropoid ape from Taung. *Nature, 155*, 235–236.

Zängl-Kumpf, U. (1990). *Hermann Schaaffhausen (1816-1893). Die Entwicklung einer neuen physischen Anthropologie im 19.* Frankfurt am Main: Jahrhundert. R.G. Fischer Verlag.

Zilhão, J. (2006). Neandertals and modern mixed, and it matters. *Evolutionary Anthropology, 15*, 183–195.

Chapter 4
Before the Neanderthals: Hominid Evolution in Middle Pleistocene Europe

Ian Tattersall

Abstract Hublin's (1998) influential "accretion model" essentially places all Middle Pleistocene European fossils in a single variable lineage culminating in *Homo neanderthalensis*. In this contribution I briefly examine the morphological justification (1) for regarding *Homo neanderthalensis* as a fully individuated Late Pleistocene entity, and (2) for the coexistence not of one but for two hominid clades (at least) in Europe during the Middle Pleistocene. One of those clades is entirely endemic to Europe and includes, along with the Neanderthals, hominids such as those from Steinheim, Reilingen and the Sima de los Huesos at Atapuerca. The other, broadly contemporaneous with it, shows none of the cranial synapomorphies of this "Neanderthal clade." Instead, it unites forms such as Mauer, Arago and Petralona with a cosmopolitan group of fossils that includes Kabwe and Bodo in Africa, and Dali and Jinniushan in China. It is to this group that the nomen *Homo heidelbergensis* applies; and as long as the Neanderthal-related Sima de los Huesos specimens continue to be misguidedly attributed to *Homo heidelbergensis*, major confusion will reign in European Middle Pleistocene hominid systematics.

Keywords Hominidae • Evolution • *Homo neanderthalensis* • *Homo heidelbergensis* • "Accretion model"

Introduction

As Harvati and Harrison (2006: 1) recently pointed out, "Neanderthals are the best represented and most studied group in the human fossil record." Yet while consensus seems slowly to be approaching on the status of these distinctive hominids as an independent species with a unique evolutionary history (e.g., Tattersall 1986; Rak et al. 1994; Arsuaga et al. 1997; Harvati et al. 2004; Lalueza-Fox et al. 2005 – though see Hawks and Wolpoff 2001; Ahern (2006) for opposing viewpoints), vigorous debate continues over the process of their emergence. Were the Neanderthals the product of an endemic linear evolutionary development in Europe, as Hublin's (1998) "accretion hypothesis" suggests, or were they a terminal species belonging to a wider European hominid radiation, as proposed by Schwartz and Tattersall (2005; see also Tattersall and Schwartz 2006)? And if the former, should the nomen *Homo neanderthalensis* be extended to all of the members of the evolving continuum, as mooted by Hublin (1998), or should the two ends of the putative lineage be recognized with different nomina, as favored by Rosas et al. (2006)?

Because two chapters in this book are devoted to the latest Early Pleistocene Atapuerca-Gran Dolina (Bermudez de Castro et al. 2011) and Ceprano fossils (Mallegni 2011), it does not seem necessary to preface the following remarks on European Middle Pleistocene hominid diversity with any detail about these important hominids. It may be worth noting, however, that links between the fragmentary earliest Atapuerca fossils (Gran Dolina, and now also Sima del Elefante) and later European hominids are hard to demonstrate. Certainly, Bermudez de Castro and colleagues (1997) cited features of the midface of their dentally primitive 780 Ka Gran Dolina form *Homo antecessor* that they believed might reflect an ancestral relationship both to the Neanderthal lineage in Europe, and to the one that ultimately gave rise to *Homo sapiens* in Africa. But to make this claim with any confidence requires demonstration that *Homo antecessor* constitutes the primitive sister of a *Homo neanderthalensis*-*Homo sapiens* clade that is clearly united by synapomorphies. And at the present stage of our knowledge this is virtually impossible to show (Schwartz and Tattersall 2005). What is more, the closest dental comparison of the Gran Dolina fossils, as both Hublin (2001) and Schwartz and T (2005) independently concluded, seems actually be with to the approximately contemporaneous Tighenif jaws from across the Mediterranean in Algeria.

I. Tattersall (✉)
Division of Anthropology, American Museum of Natural History, Central Park West at 79th Street, New York, NY 10024, USA
e-mail: iant@amnh.org

S. Condemi and G.-C. Weniger (eds.), *Continuity and Discontinuity in the Peopling of Europe: One Hundred Fifty Years of Neanderthal Study*, Vertebrate Paleobiology and Paleoanthropology, DOI 10.1007/978-94-007-0492-3_4,

As to the Ceprano calvaria, which is claimed be as old as 800–900 Ka, it does not seem firmly allied morphologically with the putatively somewhat younger material from the Gran Dolina (and, again, its closest comparison may be to another African specimen, this time the one million year-old cranium from Daka, in Ethiopia: Asfaw et al. 2002). The matter of resemblance is complicated by the fact that the adolescent Gran Dolina specimens are only very partially comparable with the Ceprano specimen in preserved anatomy, and not at all in developmental stage; but it seems highly dubious that equivalent adult Gran Dolina specimens, when found, will closely resemble Ceprano. The relationship, if any, between the earliest known hominid occupants of Europe and the much better documented Middle Pleistocene forms thus remains uncertain; and indeed there is a hiatus of around a quarter of a million years following Ceprano and Grand Dolina times before the hominid fossil record resumes in the region.

Homo neanderthalensis as an Individuated Entity

The strangely persistent notion that the Neanderthals are embraced in one way or another by our own species *Homo sapiens*, seems largely to be a pure artifact of history. The Feldhofer Neanderthal holotype was found in 1856, in a world in which the notion of evolution was yet to be articulated. It is thus hardly surprising that this unfamiliar but undeniably large-brained hominid was initially seen as representing a bizarre – either pathological, or wild and primitive – version of *Homo sapiens* (Schaaffhausen 1858; Virchow 1872). Much more remarkable is that even the most vociferous of Darwin's early advocates, Thomas Henry Huxley (Huxley 1863), opted to see the Neanderthaler as "the extreme term of a series leading gradually from it to the highest and best developed of human crania." This conclusion is perhaps easier to understand in light of the fact that, as Schwartz (2006) recently pointed out, Huxley's view was informed by a perception of racial hierarchy that allowed extremely broad bounds to human variation, as more and more resemblances were perceived and emphasized among specific human populations and apes. But, given its author, it is nonetheless surprising; and as Schwartz (2006) additionally noted, it initiated a ratchet effect: for, as ever more variants were admitted to *Homo sapiens*, the broader the perceived bounds of our species became. In any event, the staying power of those early conclusions can hardly be overemphasized; and after a period of frequent although far from exclusive marginalization in the early twentieth century, the Neanderthals became firmly re-established as a mere epiphenomenon of *Homo sapiens* after the linear thinking of the Evolutionary Synthesis took over the paleoanthropological mindset in the mid-twentieth century (Dobzhansky 1944; Mayr 1950).

However, as more hominid fossils of all kinds were subsequently discovered, it became increasingly difficult to justify the minimalist, unilinear, picture of human evolution painted by the acolytes of the Synthesis (Tattersall 1995). Today, it has become entirely untenable. The main implication for hominid systematics of this accumulating knowledge is that, while there are indeed several cases in which the specific separation of particular fossil hominids from *Homo sapiens* is debatable, *Homo neanderthalensis* is not one of them (Schwartz and Tattersall 2005). If morphology means anything at all, *Homo neanderthalensis* amply exceeds the minimal criteria necessary for recognizing a fully individuated species. It is merely icing on the cake that both the molecular (e.g. Lalueza-Fox et al. 2005) and developmental data (Ponce de León and Zollikofer 2001) now available strongly support this conclusion.

The upshot of all these recent advances is that the species *Homo neanderthalensis*, as recognized in Europe and western Asia over the past 200 Ka or so, is as distinctive an entity as can be found in the hominid fossil record. The multifarious apomorphies of the Neanderthal cranium have been listed too often to require repeating here in full (see Tattersall and Schwartz 2006 for a longer listing); but among them are highly characteristic double-arched and forwardly rounded brow ridges; inferomedially truncated orbits; narrow lower face and sharply retreating midface; medial projections within the nasal aperture; "en bombe" sagittal cranial profile, with a low lateral cranial profile behind a low frontal rise; a pitted suprainiac fossa; a long, narrow foramen magnum; a retromolar space; and low-set mandibular condyles. Despite some expected variety in their degree of expression, taken together these characteristics add up to a unique conformation of the cranium; and the recent composite reconstruction of a full Neanderthal skeleton by Sawyer and Maley (2005) serves as a powerful reminder of how distinctive (at least from *Homo sapiens*) these hominids were postcranially, too. This is especially notable in the wide pelvis that flares laterally, to match an inferiorly broad but upwardly tapering rib cage (Fig. 4.1). Indeed, the conformations of the thoracic and pelvic regions in particular are such as to suggest that in life substantial differences from our species existed not only in overall appearance but in gait (Rak 1990; Tattersall and Schwartz 2006). Distinctiveness of locomotor pattern and posture have also been suggested by analyses of inner ear morphology (Hublin et al. 1996); the sum of all this evidence makes it vanishingly unlikely that *Homo neanderthalensis* and *Homo sapiens*, two entities each with upwards of half a million years of independent evolutionary history since sharing a common ancestor (Green et al. 2006; Noonan et al. 2006; Tattersall and Schwartz 2006; Caramelli et al. 2011), would readily have recognized each other as potential breeding partners – even in the unlikely event of full genetic, developmental, or behavioral compatibility.

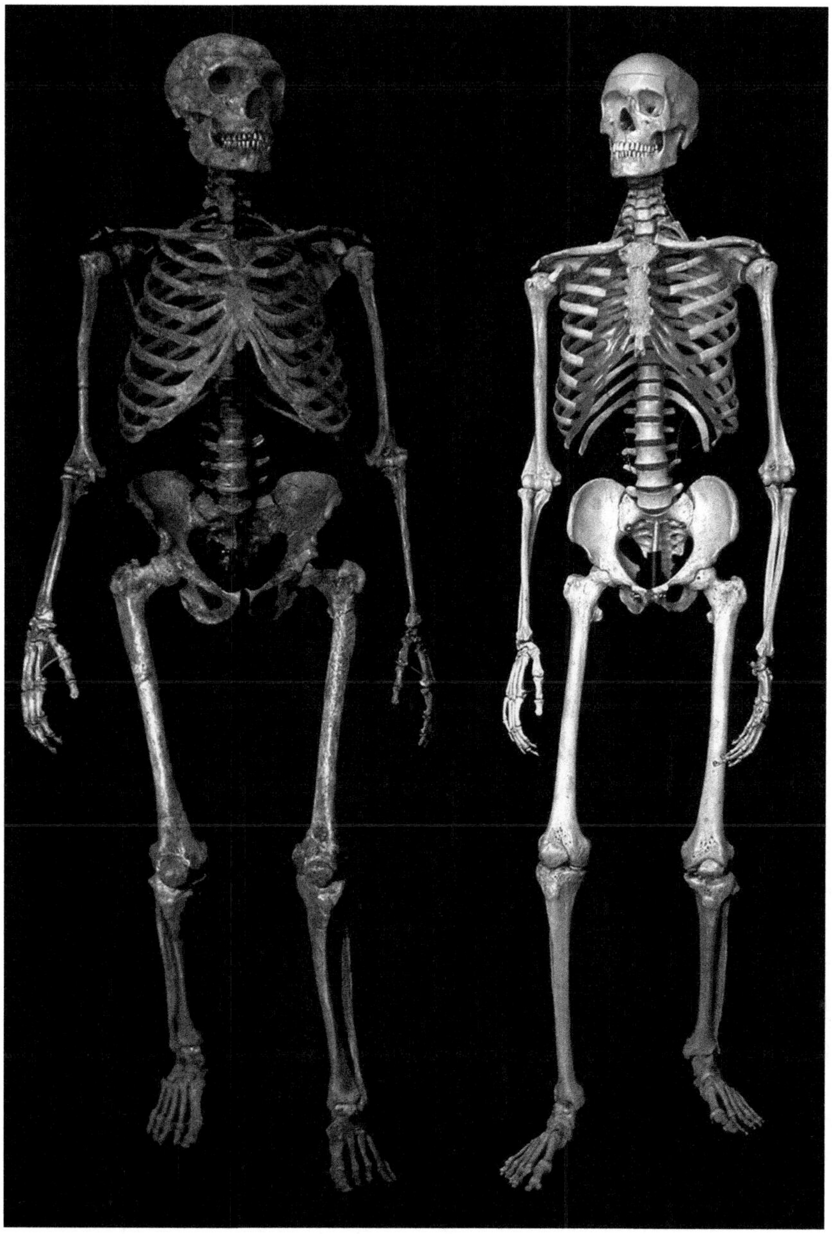

Fig. 4.1 *Front* and *side* views of a complete Neanderthal skeleton as reconstructed, using elements from five partial skeletons (principally La Ferrassie 1 and Kebara 2), by Sawyer and Maley (2005). Photo courtesy of Ken Mowbray

Whether or not the occasional act of mating may have occurred among members of the two species, there was clearly no biologically significant genetic interchange between them (e.g. Serre and Pääbo 2006).

The Middle Pleistocene Hominid Record in Europe

While there is no question that the Neanderthals of the Late (and latest Middle) Pleistocene constituted an entirely distinctive entity in themselves, a broader look at the range of morphologies exhibited by the full range of European Middle Pleistocene fossils nonetheless indicates that not all of the characteristics conventionally used to recognize Neanderthals are strictly speaking autapomorphies of *Homo neanderthalensis*. Certain of them are shared with other hominids that have not traditionally been identified as Neanderthals. Such hominids include the cranium from Steinheim in Germany, which is dated with less than total confidence to the first part of oxygen isotope Stage 7, around 225 Ka (Stringer and Gamble 1993), though it might be earlier. It is thus possibly penecontemporaneous with the earliest putative Neanderthals from Ehringsdorf (Blackwell and Schwarcz 1986; Grün et al. 1988), which present interpretive problems of their own (see Schwartz and Tattersall 2005); and it considerably antedates the earliest definitive Neanderthal specimen, from Biache-Saint-Vaast (late Stage 7, 175 Ka: Huxtable and Aitken 1988).

Few authorities have cared to allocate the Steinheim cranium to *Homo neanderthalensis*; but this specimen nonetheless shares a variety of features with Neanderthals, including very characteristic morphologies of the upper face such as orbital shape and supraorbital conformation. It also shows a large nasal fossa, with well defined anterior and posterior crests at its anterior margin and a hint of medial projections; an angulation along the anterior squamous suture; long and fairly straight paramastoid and anterior lambdoid sutures; a suprainiac fossa (albeit rather faint); and a rather rounded longitudinal profile of the posterior braincase. All of these are Neanderthal characters; but the Steinheim cranium is also decidedly non-Neanderthal-like in its posterior cranial profile and in the vertical side walls of its rather poorly-inflated braincase. It is, indeed, dissimilar to such an extent that almost all commentators have preferred to allocate it to "archaic *Homo sapiens*," rather than to characterize it as a Neanderthal.

Probably about as old as the Steinheim fossil (Ziegler and Dean 1998) is another German specimen, the posterior portion of a braincase from Reilingen. Early on attributed to *Homo erectus* (Czarnetzki 1989) or to "archaic *Homo sapiens*" (Adam 1989; Schott 1990), this fossil has more recently attracted comparison to the Neanderthals (Condemi 1996; Dean et al. 1998). Indeed, in its preserved morphology I can see nothing about this fossil to prevent its allocation to *Homo neanderthalensis*. The case for doing so is made especially compelling by the fact that it was apparently in the rear of the skull – which is all we have of Reilingen – that the full panoply of Neanderthal characteristics was last to become established. If this specimen is indeed a Neanderthal, and if its putative dating is correct, this would push back the stratigraphic range of *Homo neanderthalensis* back into the earlier part of Stage 7.

Farther back in time are the numerous fossils from the Sima de los Huesos at Atapuerca in Spain, now dated to >530 Ka (Bischoff et al. 2007). This amazing assemblage also shows numerous Neanderthal resemblances – postcranial as well as cranial, although the former may well simply represent the primitive condition for this subclade of *Homo* – but to a lesser degree than the Steinheim specimen does. The basic Neanderthal characters of the upper face are there, as well as the prenasal fossa and some projection of the frontal processes around the large nasal aperture; angulation along the anterior squamous suture; a long, straight parietomastoid suture; and a pitted suprainiac fossa. But in many ways the Sima hominids remain primitive relative to *Homo neanderthalensis*. They have a tent-shaped coronal cranial profile; they lack inflation of the infraorbital region; they have horizontal conchal crests in place of medial projections; there is no sharp retreat or downward narrowing of the midface; the zygomatic arches are deep and lie in front of a very short anterior lambdoid suture; and there is no clearly undercut occipital torus. Here is a hominid that clearly belongs in the clade to which the Neanderthals also belong; but which equally clearly cannot be assigned to the same species.

This last fact is something of which the describers of the Sima material were clearly aware, since they declined to refer their material to the species *Homo neanderthalensis* and instead placed it in *Homo heidelbergensis* (Arsuaga et al. 1997). This allocation implicitly rejects the reasoning of Hublin (1998), whose influential "accretion hypothesis" of Neanderthal evolution demands as a matter of logic that all "specimens involved in the [gradual] Neandertalization process" be included in *Homo neanderthalensis*, "even if they display only a few derived Neandertal features" (Hublin 1998: 302). In Hublin's view this "Neandertalization process" insensibly links older middle Pleistocene forms such as those from Mauer and Arago to the latest (Stages 5–3) "Classical" Neanderthals, via "Holstein-Hoxnian" forms such as Bilzingsleben and Sima, and "Saalian" ones such as Biache and Ehringdorf. On a practical level this tranquil progression has been somewhat upset by the recent redating of the Sima fossils to a far earlier period than previously thought; but it is actually much more disturbing from a procedural perspective. For while it might appeal to some vague notion of fairness, Hublin's "logic" actually amounts to a counsel of despair. This is because it makes nonsense of any attempt to recognize morphological entities – and, by extension, species – in this sector of the fossil record. Effectively, it demands species synonymy if any ancestor-descendant relationship is perceived between two distinguishable entities. Yet Nature clearly is, and always has been, "packaged" into the occasionally rather blurry entities we know as species; and since neither geographical distribution nor age (the only two other intrinsic qualities of fossils) has any direct bearing on systematic position, the only way we can hope to recognize those packages is via their morphology – as even advocates of species as extremely leaky vessels acknowledge (e.g. Holliday 2006). Just as with members of any lineage, the Sima fossils, the Steinheim cranium, and the whole Neanderthal assemblage can be represented as terminal entities on a cladogram, irrespective of what ancestor-descendant relationships may or may not exist among them. In other words, even an ancestor-descendant succession of species can be represented as a clade; and whatever one's viewpoint on the evolutionary pattern concerned, it is certainly most productive to view these hominids as members of such a clade – allowing each the opportunity to be analysed as a real entity.

Yet to reject the allocation of the the Sima fossils to *Homo neanderthalensis*, and instead to classify them in *Homo heidelbergensis*, as advocated for example by Arsuaga and colleagues (1997) and by Rosas et al. (2006), does not solve the implicit systematic problem. Indeed, it exacerbates that problem. This is not for any procedural reason. Instead, the difficulties result from the morphologies of the fossils themselves. *Homo heidelbergensis* is based on the Mauer jaw: hardly the most ideal of holotypes, but the holotype nonetheless. This specimen appears to be broadly contemporaneous (Wagner et al. 2010) with the Sima material as recently redated; and as Schwartz and I recently pointed out (Schwartz and Tattersall 2005), there is little in the preserved morphology of either form to suggest a close relationship between the two. Instead, the lower jaw from Mauer, distinctive as its robustness and very broad rami undoubtedly make it, shares a variety of significant features with the geologically rather younger mandibles from Arago. Similarities are especially visible in the quite large premolar and molar teeth of each, which resemble each other in both proportions and cuspal detail; but they are also evident in aspects of mandibular morphology (Schwartz and Tattersall 2005; Tattersall and Schwartz 2006). It seems perfectly reasonable to allocate the Mauer and Arago mandibles to the same species; and if this is done, we are entitled to view the associated Arago 21 partial cranium as emblematic of that species.

The face of Arago 21 is well preserved, and it shares a variety of features with the uncertainly dated Petralona cranium from Greece. Beyond Europe, it shares the same suite of characters with the hominids from Bodo, Kabwe and Saldanha in Africa, and perhaps also Dali and Jinniushan in China. Despite a certain heterogeneity in various details of morphology, all of these fossils make up a fairly cohesive group to which it is at least provisionally appropriate to give the name *Homo heidelbergensis*. And, as thus constituted, *Homo heidelbergensis* shows little in common with any member of the Neanderthal clade, the Sima form not excepted. The facial architecture of *Homo heidelbergensis* is characterized by a very broad and massive lower face, dominated above by superoinferiorly tall and well-developed supraorbital surfaces that peak in thickness around midorbit. The front surfaces above each eye twist up and back laterally below a blunt edge that demarcates each surface from the shallow posttoral sulcus behind. This is a far cry from the narrowing lower face of the Neanderthals, topped by gracefully arching supraorbitals whose surfaces roll smoothly out from the orbital roof and back on to the frontal. There is absolutely nothing here to suggest any specific relationship of *Homo heidelbergensis* to the Neanderthal clade – and nothing, either, to justify baptizing the Sima fossils *Homo heidelbergensis* simply because they are so evidently not *Homo neanderthalensis*. Sadly, this is not a message that has been widely grasped in paleoanthropology, and many authors seem to have passively accepted the notion that the very important and well-preserved Sima materials do indeed represent *Homo heidelbergensis*. But until the Sima materials have been properly conceded their own identity, major confusion will reign in studies of the Middle Pleistocene European hominid record. If you don't get the actors right, you'll never understand the play.

Given the essential contemporaneity of the Sima materials with bona fide *Homo heidelbergensis*, and the clear morphological gulf between them, it is clear that at least two hominid lineages coexisted in Europe during at least part of the Middle Pleistocene. Indeed, there may even have been more. A fair assortment of hominid fossils, mostly alas fragmentary, is known from Europe in the period between about 500 and 250 Ka. The Swanscombe partial cranium, about 400 Ka, has been most recently been viewed as a "primitive" Neanderthal (Hublin 1998; Stringer and Hublin 1999). On the face of it this specimen's claims to Neanderthal status are marginally weaker than those of Reilingen; but if it is indeed an early Neanderthal, this would place the origin of the species *Homo neanderthalensis* as early as Stage 11. An alternative interpretation would draw attention to features of the Swanscombe occipital that recall more closely the Sima fossils; but in either case, this English specimen is clearly a member of the larger "Neanderthal clade." More inscrutable is the Vérteszöllös cranium (350–250 Ka: Schwarcz and Latham 1990) from Hungary, whose affinities are hard to gauge, although no harder, perhaps, than those of the fragmentary Bilzingsleben crania (400–300 Ka: Schwarcz et al. 1988). Both forms have been rather mindlessly attributed to *Homo erectus*, a species for which there is no evidence whatever outside eastern Asia; but it is not clear that either fits easily into one of the two European groups already discussed.

The cosmopolitan *Homo heidelbergensis* is widely regarded as a possible progenitor, via an early African representative, of both the modern human and Neanderthal lineages. Whether or not this is a plausible scenario is not of concern here. However, what is clear is (1) that members of both the Neanderthal clade and of *Homo heidelbergensis* as broadly defined were present in Europe during the Middle Pleistocene; and (2) that European *Homo heidelbergensis* was not the progenitor of the Neanderthal clade. Vagaries of dating make it unclear from the fossil perspective exactly how long the cohabitation of these lineages lasted: but if Grün's (1996) dating of the Petralona cranium to as recent a date as 250–150 Ka is correct, it may be significant that this cranium is also perhaps the most derived of all known specimens of *Homo heidelbergensis*. This specialized aspect is most especially evident in the extraordinarily extensive cranial pneumatization shown by this specimen; and while this might conceivably be an individual feature, it might equally indicate that, in isolation, later representatives of European *Homo heidelbergensis* had become highly derived in intracranial sinus development – on a much more extensive scale than in any known member of the Neanderthal lineage. Only further fossil finds will test this notion; but meanwhile it does seem plausible to conclude that at least two hominid lineages were in some degree of competition in Europe over most of the Middle Pleistocene, with *Homo heidelbergensis* or its descendants becoming locally extinct at around the end of the Middle Pleistocene, leaving the field (temporarily) free for the newly evolved *Homo neanderthalensis*.

Conclusions

Numerous lines of evidence lead to the conclusion that the distinctive Late Pleistocene Neanderthal population of Europe and Western Asia merits recognition as the independent species *Homo neanderthalensis*. With the exception of some very late Middle Pleistocene forms such as Biache and probably also Reilingen, *Homo neanderthalensis* is restricted to the Late Pleistocene. However, there are several Middle Pleistocene forms that share some, but not all of their distinctive features with *Homo neanderthalensis*, and that are best regarded as members of the same larger clade. Such fossils most notably include the Steinheim cranium, and to a lesser extent the large assemblage of well preserved but mostly fragmentary fossils from the Sima de los Huesos at Atapuerca, now dated to >530 Ka. The latter fossils are approximately contemporaneous with the Mauer holotype of *Homo heidelbergensis*, which in turn bears significant resemblances to the mandibles included in the ~400 Ka hominid assemblage from Arago (Tautavel). Associated with these is the Arago 21 facial fossil, which in turn allows the allocation to *Homo heidelbergensis* of such relatively well preserved crania as those from Petralona in Greece; Bodo, Kabwe and Saldanha in Africa; and (probably) Dali and Jinniushan in China. None of these forms shows any of the morphological features that unite the "Neanderthal-clade," and there is little reason for regarding the European members of *Homo heidelbergensis* as early representatives of a lineage ultimately leading to the Neanderthals, as Hublin's (1998) "accretion hypothesis" suggests. Instead, it appears that two distinct hominid lineages coexisted in Europe during the Middle Pleistocene, the European branch of *Homo heidelbergensis* becoming extinct by the end of this period, while the lineage culminating in *Homo neanderthalensis* apparently monopolized this region throughout the Late Pleistocene prior to the incursion of *Homo sapiens*.

The researchers studying the Sima de los Huesos hominid materials have consistently allocated their fossils to the species *Homo heidelbergensis*, apparently for little reason other than that, as they correctly perceived, these fossils do not belong to *Homo neanderthalensis*. However, this effectively amounts to the usage of *Homo heidelbergensis* as a wastebasket taxon, since specimens reliably attributable to the latter species possess none of the apomophies that bind the Sima hominids to membership in the larger Neanderthal clade. As long as the Sima hominids are misattributed to *Homo heidelbergensis*, European hominid systematics will continue to be in a state of confusion. Since there is no adequate justification for attributing this remarkable Spanish hominid assemblage either to *Homo heidelbergensis* or to *Homo neanderthalensis*, it is clear that an alternative species nomen is required. To this writer's knowledge there is none currently available.

Acknowledgments I am most grateful to the organizers of the "150 Years of Neanderthal Discoveries" meeting for the privilege of attending this stimulating and historic event, and to Silvana Condemi and Gerd-C. Weniger for their invitation to contribute to this volume of proceedings. Thanks also to Eric Delson, and to Gary Sawyer and Ken Mowbray for the illustration.

References

Adam, K. D. (1989). Alte und neuer Urmenschenfunde I sudwest-Deutschland – Eine kritische Wardigung. *Quartär, 39/40*, 177–190.

Ahern, J. C. M. (2006). Nom-metric variation in recent humans as a model for understanding Neanderthal-early modern human differences: Just how "unique" are Neanderthal unique traits? In K. Harvati & T. Harrison (Eds.), *Neanderthals revisited: New approaches and perspectives* (pp. 255–268). Dordrecht: Springer.

Arsuaga, J. L., Bermudez de Castro, J. M., & Carbonell, E. (1997). Special issue on the Sima de los Huesos hominids and site. *Journal of Human Evolution, 33*(2/3), 105–421.

Asfaw, B., Gilbert, W. H., Bayene, Y., Hart, W. K., Renne, P. R., WoldeGabriel, G., Vrba, E. S., & White, T. D. (2002). Remains of *Homo erectus* from Bouri, Middle Awash, Ethiopia. *Nature, 416*, 317–320.

Bermudez de Castro, J. M., Arsuaga, J.-L., Carbonell, E., Rosas, A., Martinez, I., & Mosquera, M. (1997). A hominid from the lower Pleistocene of Atapuerca: Possible ancestors to Neandertals and modern humans. *Science, 276*, 1392–1395.

Bermudez de Castro, J. M., Martinon-Torres, M., Gomez-Robles, A., Margvelashvili, A., Arsuaga, J.L., Carretero, J.M., Martinez, I., & Sarmiento, S. (2011). The Gran Dolina-TD6 human fossil remains and the origin of Neanderthals. In: S. Condemi and G.-C. Weniger (Eds), *Continuity and discontinuity in the peopling of Europe* (pp. 67–76). Dordrecht: Springer.

Bischoff, J. L., Williams, R. W., Rosenbauer, R. J., Aramburu, A., Arsuaga, J Ll, Garcıa, N., & Cuenca-Besco, G. (2007). High-resolution U-series dates from the Sima de los Huesos hominids yields 600 N kyrs: Implications for the evolution of the early Neanderthal lineage. *Journal of Archaeological Science, 34*, 763–770.

Blackwell, B., & Schwarcz, H. (1986). U-series analysis of the Lower Travertine at Ehringsdorf, DDR. *Quaternary Research, 25*, 215–222.

Caramelli, D., Milani, L., Stanyon, R., & Fox, C. L. (2011). Towards Neanderthal paleogenomics. In: S. Condemi and G.-C. Weniger (Eds), *Continuity and discontinuity in the peopling of Europe* (pp. 219–222). Dordrecht: Springer.

Condemi, S. (1996). Does the human fossil specimen from Reilingen (Germany) belong to the *Homo erectus* or the Neanderthal lineage? *L'Anthropologie*, (Brno) *34*, 34/1-2, 69–77.

Czarnetzki, A. (1989). Ein archäischer Hominidencalvarium aus einer Kiesgrübe in Reilingen, Rhein-Neckar-Kreis. *Quartär, 39/40*, 191–201.

Dean, D., Hublin, J.-J., Holloway, R., & Ziegler, R. (1998). On the phylogenetic position of the pre-Neandertal specimen from Reilingen, Germany. *Journal of Human Evolution, 34*, 485–508.

Dobzhansky, T. (1944). On species and races of fossil and living man. *American Journal of Physical Anthropology, 2*, 251–265.

Green, R. E., Krause, J., Ptak, S. E., Briggs, A. W., Ronan, M. T., Simons, J. F., Du, L., Egholm, M., Rothberg, J. M., Paunovic, M., & Pääbo, S. (2006). Analysis of one million base pairs of Neanderthal DNA. *Nature, 444*, 330–336.

Grün, R. (1988). ERS dating of spring-deposited travertines. *Quaternary Science Reviews, 7*, 429–432.

Grün, R. (1996). A re-analysis of electron spin resonance dating results associated with the Petralona hominid. *Journal of Human Evolution, 30*, 227–241.

Grün, R., Schwarcz, H. P., Ford, D. C., & Hentzsch Grün, B. (1988). ESR dating of spring deposited travertines. *Quaternary Science Reviews 7*, 429–432.

Harvati, K., & Harrison, T. (2006). Neanderthals revisited. In K. Harvati & T. Harrison (Eds.), *Neanderthals revisited: New approaches and perspectives* (pp. 1–8). Dordrecht: Springer.

Harvati, K., Frost, S. R., & McNulty, K. P. (2004). Neanderthal taxonomy reconsidered: Implications of 3D primate models of inter- and intra-specific differences. *Proceedings of the National Academic Sciences USA, 101*, 1147–1152.

Hawks, J. D., & Wolpoff, M. H. (2001). The accretion model of Neandertal evolution. *Evolution, 55*, 1474–1485.

Holliday, T. W. (2006). Neanderthals and modern humans: An example of a mammalian syngameon? In K. Harvati & T. Harrison (Eds.), *Neanderthals revisited: New approaches and perspectives* (pp. 281–298). Dordrecht: Springer.

Hublin, J. J. (1998). Climatic changes, paleogeography, and the evolution of Neandertals. In T. Akazawa, K. Aoki, & O. Bar-Yosef (Eds.), *Neandertal and modern humans in Western Asia* (pp. 295–310). New York: Plenum Press.

Hublin, J.-J. (2001). Northwestern African Middle Pleistocene hominids and their bearingon the emergence of *Homo sapiens*. In L. Barham & K. Robson-Brown (Eds.), *Human roots: Africa and Asia in the Middle Pleistocene* (pp. 99–131). Bristol: Western Academic and Specialist Press.

Hublin, J.-J., Spoor, F., Braun, M., Zonneveld, F., & Condemi, S. (1996). A late Neanderthal from Arcy-sur-Cure associated with Upper Paleolithic artefacts. *Nature, 381*, 224–226.

Huxley, T. H. (1863). *Evidence as to Man's place in Nature*. London: Williams and Norgate.

Huxtable, J., & Aitken, M. (1988). Datation par thermoluminescence. In A. Tuffreau & J. Somme (Eds.), *Le gisement paleolithique moyen de Biache-Saint-Vaast* (Pas de Calais, Vol. 1). Mémoires de la Société Préhistorique Française 21,107–108.

Lalueza-Fox, C., Sampietro, M. L., Caramelli, D., Puder, Y., Lari, M., Calafell, F., Martinez Maza, C., Bastir, M., Fortea, J., Rasilla, M., Bertranpetit, J., & Rosas, A. (2005). Neanderthal evolutionary genetics: Mitochondrial DNA data from the Iberian Peninsula. *Molecular Biology and Evolution, 22*, 1077–1081.

Mallegni, F. (2011). The earliest European human peopling after the recent discoveries: Early Neanderthals or different lineages? In: S. Condemi & G.-C. Weniger (Eds), *Continuity and discontinuity in the peopling of Europe* (pp. 55–66). Dordrecht: Springer.

Mayr, E. (1950). Taxonomic categories in fossil hominids. *Cold Spring Harbor Symposia on Quantitative Biology, 15*, 109–118.

Noonan, J. P., Coop, G., Kudaravalli, S., Smith, D., Krause, J., Alessi, J., Chen, F., Platt, D., Pääbo, S., Pritchard, J. K., & Rubin, E. M. (2006). Sequencing and analysis of Neanderthal genomic DNA. *Science, 314*, 1113–1118.

Ponce de León, M. S., & Zollikofer, C. P. E. (2001). Neanderthal cranial ontogeny and its implications for late hominid diversity. *Nature, 412*, 534–538.

Rak, Y. (1990). On the differences between two pelvises of Mousterian context from the Qafzeh and Kebara caves, Israel. *American Journal of Physical Anthropology, 81*, 323–332.

Rak, Y., Ginsburg, A., & Geffen, E. (1994). Does *Homo neanderthalensis* play a role in modern human ancestry? The mandibular evidence. *American Journal of Physical Anthropology, 119*, 199–204.

Rosas, A., Bastir, M., Martinez-Maza, C., Garcia-Taberno, A., & Lalueza-Fox, C. (2006). Inquiries into Neanderthal craniofacial development and evolution: "accretion" versus "organismic" models. In K. Harvati & T. Harrison (Eds.), *Neanderthals revisited: New approaches and perspectives* (pp. 1–8). Dordrecht: Springer.

Sawyer, G. J., & Maley, B. (2005). Neanderthal reconstructed. *Anatomical Record New Anatomist, 283B*, 23–31.

Schaaffhausen, H. (1858). Zur Kentnis der ältesten Rassenschädel. *Bgl Verh Nat Bereins Preuss Rheinlandeu. Westphalens, 14*, 167–188.

Schott, L. (1990). *"Homo erectus reilingensis"* – Anspruch und Wirklichkeit eines Schädelfundes. *Biologische Rundschau, 28*, 231–235.

Schwarcz, H. P., & Latham, A. G. (1990). Absolute age determination of travertines from Vérteszöllös. In M. Kretzoi & V. Dobosi (Eds.), *Vérteszöllös: Site, man and culture* (pp. 549–555). Budapest: Akadémiai Kiado.

Schwarcz, H. P., Latham, A. G., Mania, D., & Brunnacker, K. (1988). The Bilzingsleben archaeological site: New dating evidence. *Archaeometry, 30*, 5–17.

Schwartz, J. (2006). "Race" and the odd history of paleoanthropology. *Anatomical Record New Anatomist, 289B*, 225–240.

Schwartz, J. H., & Tattersall, I. (2005). *The human fossil record, vol. 4: Craniodentalmorphologyofearlyhominids(GeneraAustralopithecus, Paranthropus, Orrorin) and overview*. New York: Wiley/Liss.

Serre, D., & Pääbo, S. (2006). The fate of European Neanderthals: Results and perspectives from ancient DNA analyses. In K. Harvati & T. Harrison (Eds.), *Neanderthals revisited: New approaches and perspectives* (pp. 211–220). Berlin: Springer.

Stringer, C. B., & Gamble, C. (1993). *In search of the Neanderthals*. London: Thames and Hudson.

Stringer, C. B., & Hublin, J. J. (1999). New age estimates for the Swanscombe hominid, and their significance for human evolution. *Journal of Human Evolution, 37*, 873–877.

Tattersall, I. (1986). Species recognition in the human fossil record. *Journal of Human Evolution, 15*, 165–175.

Tattersall, I. (1995). *The fossil trail: How we know what we think we know about human evolution*. New York: Oxford University Press.

Tattersall, I., & Schwartz, J. H. (2006). The distinctiveness and wider systematic context of *Homo neanderthalensis*. In K. Harvati & T. Harrison (Eds.), *Neanderthals revisited: New approaches and perspectives* (pp. 9–22). Berlin: Springer.

Virchow, R. (1872). Untersuchung der Neanderthal-Schädels. *Zeitschriftfar Ethnologie, 4*, 157–165.

Wagner, G. A., Krbetschek, M., Degering, D., Bahain, J.-J., Shao, Q., Falguères, C., Voinchet, P., Dolo, J.-M., Garcia, T., & Rightmire, G. P. (2010). Radiometric dating of the type-site for *Homo heidelbergensis* at Mauer, Germany. *Proceedings of the National Academy of Sciences USA*, doi/10.1073/pnas.1012722107.

Ziegler, B., & Dean, D. D. (1998). Mammalian fauna and biostratigraphy of the pre-Neanderthal site of Reilingen, Germany. *Journal of Human Evolution, 34*, 469–484.

Chapter 5
The Earliest European Human Peopling After the Recent Discoveries: Early Neanderthals or Different Lineages?

Francesco Mallegni

Abstract This report is a review of the European fossil record during the Pleistocene. We investigate the possibility of an anagenetic evolution of Neandertals starting from the most ancient hominins found at Ceprano, Gran Dolina, etc. It appears that the contribution of the Dmanisi population to later Europeans is very unlikely. We focus on the metric and morphological features of the oldest human fossil in Europe: the cranium from Ceprano. Among the characters observed in Ceprano, a few are also seen in more recent European *H. heidelbergensis* fossils (especially in Petralona) and many in African *H. heidelbergensis* fossils (especially in Bodo and Kabwe). We then consider the hypothesis that Ceprano could be ancestral to African *H. heidelbergensis* but not to European members of this taxon. A cladistic analysis seems to confirm this view. Lastly, in European *H. heidelbergensis*, we observe a continuity in characters that become more numerous approaching the beginning of the Late Pleistocene. These characters are typical of *Homo neanderthalensis* following the "Accretion-Model" hypothesis.

Keywords Ceprano • Dmanisi • Pleistocene • Phylogenetic analysis • Cladistic analysis

Extraordinary discoveries of remains belonging to hominids were made in the last decade of the twentieth century, as well as in the current century. These discoveries occurred in Italy at Ceprano (Ascenzi and Segre 1996, 2000; Clarke 2000; Gilbert et al. 2003; Mallegni et al. 2003), in Spain at Atapuerca (Arsuaga et al. 1993), and in southern Caucasus at Dmanisi (Gabunia and Vekua 1995a, b; Gabunia et al. 2000, 2002; de Lumley et al. 2006). The bones found at Dmanisi are considered the earliest representative remains of ancient humanity that reached the northern hemisphere of the Old World. After these discoveries, some paleoanthropologists developed new hypotheses about human evolution in these areas.

Humankind represented by the fossil record, which is more recent than the above-mentioned discoveries, is quite complex, as well as Humankind that preceded it in the African regions. Perhaps the skull found at Ceprano (Ascenzi and Segre 1996), in the middle of the Italian peninsula (Tyrrhenian area), and those discovered in La Gran Dolina at Atapuerca, in northern Spain (Arsuaga et al. 1993, 1997), could not be considered important enough to allow us to hypothesize a contribution to the later human lineages. In fact, at this moment, the skull of Ceprano is the only finding with a quite complex phenotype that seems to differentiate from that of other contemporaneous fossil records (for more details see Mallegni et al. 2003). The remains discovered in La Gran Dolina at Atapuerca are mostly represented by very fragmented bone remains, generally of young individuals(at least the bone fragments which allowed us to scientifically hypothesize their significance in the evolution of mankind).

We must proceed in order. We have already mentioned that in Southern Caucasus, and more precisely at Dmanisi, near Tiblisi in Georgia (Gabunia and Vekua 1995a, b; Gabunia et al. 2000, 2002), an exceptional series of remains belonging to hominids was exhumed from the deepest layers of the excavation site. At the beginning of the excavations, the site was believed to have contained only medieval remains. The most representative and prestigious of these bones are obviously skulls, but postcranial elements were found as well (de Lumley et al. 2006). All these remains belong to *Homo* and are datable between 1.8 and 1.75 Ka. This discovery represents, up until now, one of the indications of the earliest presence of hominids in the areas outside of Africa. It is slightly earlier than the discovery of the remains of Sangiran, in south-eastern Asia, which is about 1.6 Ka. To our knowledge, we can assume that these remains probably represent one of the most ancient migrations, perhaps the first, of hominids from the African cradle.

Since we must consider the evolution of human groups in the European territory, perhaps the remains of Dmanisi, which are so significant for the reasons mentioned above, are not so important in this ambit. In fact, up until now, we have not found bone remains of hominids in the European territory,

F. Mallegni (✉)
Dipartimento di Biologia, Università degli Studi di Pisa, Via S. Maria 55, 56126, Pisa Italy
e-mail: fmallegni@biologia.unipi.it

S. Condemi and G.-C. Weniger (eds.), *Continuity and Discontinuity in the Peopling of Europe: One Hundred Fifty Years of Neanderthal Study*, Vertebrate Paleobiology and Paleoanthropology, DOI 10.1007/978-94-007-0492-3_5,

nor any evidence of their culture, belonging to the chronologic period of at least 600 thousand years, which divides the remains of Dmanisi from the ones considered, for now, the most ancient in Europe. The Caucasian area, where Dmanisi is located, is separated from the Russian Sarmatian plain by a mountain range. These mountains may have represented an insurmountable barrier for these ancient groups, making it impossible to reach the Sarmatian plain, and the European territories. There were possible passages in the east and west of the mountain range due to the presence of small plains overlooking the Black Sea and the Caspian Sea, but they were probably unapproachable due to climatic factors (e.g., the intense cold). These groups probably were not adapted for living in these new and negative climatic factors. In particular, another insurmountable frontier was chiefly represented by the large Sarmatian rivers, the Don river, the Dnepr river, and the Volga river, which flow into the Azov Sea, the Black Sea, and the Caspian Sea.

After the period of the remains of Dmanisi discovered in the Southern Caucasus, there were new arrivals of hominids from the southern areas of the northern hemisphere (without considering the Oriental Asian areas). These arrivals were separated by at least two other remarkable chronological hiatuses. The first hiatus (which lasted 900–800 thousand years) is represented by the discovery of the skull of Ceprano and by the remains of La Gran Dolina of Atapuerca. The second hiatus could be associated with the arrival in the Middle East (approximately about 250 thousand years ago) of the human form which evolved in *H. sapiens* (remains of Zuttiyeh?). Otherwise, the second hiatus could be associated with the arrival in the Middle East of *H. sapiens* (approximately 100 thousand years ago), and it could be represented by the remains of Skull and Qafzeh (McCown and Keith 1939), which were defined by Vandermeersch (1981) as protocromagnoid. Their possible descendants appeared in Europe after 60 thousand years. In fact, the site of Pestera cu Oase, in Romania, represents for now the first presence of *sapiens* in the European territory, as it dates back to 40 thousand years (Trinkaus et al. 2003a, 2003b).

We can hypothesize that the same climatic and/or biogeographical factors (which in all probability occurred cyclically) impeded free access to the north-western areas of these *sapiens* whose cognitive and cultural apparatus was without doubt more effective than that of the hominids represented by the remains of Dmanisi.

It is a common belief that the most ancient Pleistocene humans, represented in the northern hemisphere by Dmanisi, would have found their future toward the east, traveling in the direction of the parallels and not of the meridians. At least, other human populations (except for those that arrived in Dmanisi) probably behaved in this way. They abandoned the African cradle and appeared in the Far East (in China, Java, and in the Indian subcontinent) in a period between 1.8 and 1.6 million years ago. However, with the ever-increasing number of excavations currently being carried out, it becomes increasingly likely that we will find ancient evidence of the same period (or immediately later) also to the west of Caucasus, in the European territory. Due to the argument proposed for this Congress, we must begin our dissertation from the remains of Ceprano and of La Gran Dolina of Atapuerca.

The skull of Ceprano was discovered by chance during road construction. Biddittu (aggregated to the Human Paleontology Institute of Rome) (Ascenzi and Segre 1996) recognized some bone remains belonging to the neurocranium in the soil removed by the scraper. The Human Paleontology Institute of Rome promptly organized a team that collected the human bone remains. In this way, we were able to restore part of the neural cranium, which allowed us to obtain some metrical measurements and morphological features (Fig. 5.1).

Unfortunately, the greater part of the remains of the upper side of the face was lost, the only bone elements found are so fragmented that any reconstruction of this side of the skull was impossible. We could only replace the frontal processes of the zygomatic bones. The right part of the neural cranium is in very good condition. This allows us to have a complete vision of the skull, by using CT images. The left part of the skull is deformed *ab antiquo*, probably due to the pressure of the soil on this side of the remains. The analysis of the earth inside the skull caused us to assume that the cranium dates back to between 900 and 800 thousand years (Ascenzi et al. 1996).

We took some reliable measurements following the Martin and Saller methodology (1956–1959) and the Wood method (1991). We also tested the presence, and the absence, of nonmetric characters (about 30) that Wood (1991) believes to be characteristic of *Homo erectus*. This kind of examination was also performed on other bones representing extinct species (Mallegni et al. 2003), both preceding and succeeding the skull of Ceprano. Some of these remains are African, belonging to *H. habilis*, *ergaster*, *rhodesiensis* (or to African *heidelbergensis*), others are Asian (*P. erectus*, *H. erectus*), and some others belong to the European area (*H. heidelbergensis*, and the so-called ante and pre-Neanderthals). Unfortunately, it was not possible to include the remains of La Gran Dolina of Atapuerca, which are called *H. antecessor* on the basis of the features of a juvenile mandible, because they were not suitable for this kind of observation due to the fact that the most well-preserved fossil records, as we have mentioned above, were juvenile (Arsuaga et al. 1997).

The features of these specimens were utilized for a cladistic analysis, following the method that was originally formulated by W. Hennig in 1996. This methodology codifies the features to formulate some hypotheses on the phylogenetic relationships among *taxa*. The results of the cladistic analysis were eight equally parsimonious trees (Mallegni et al. 2003)

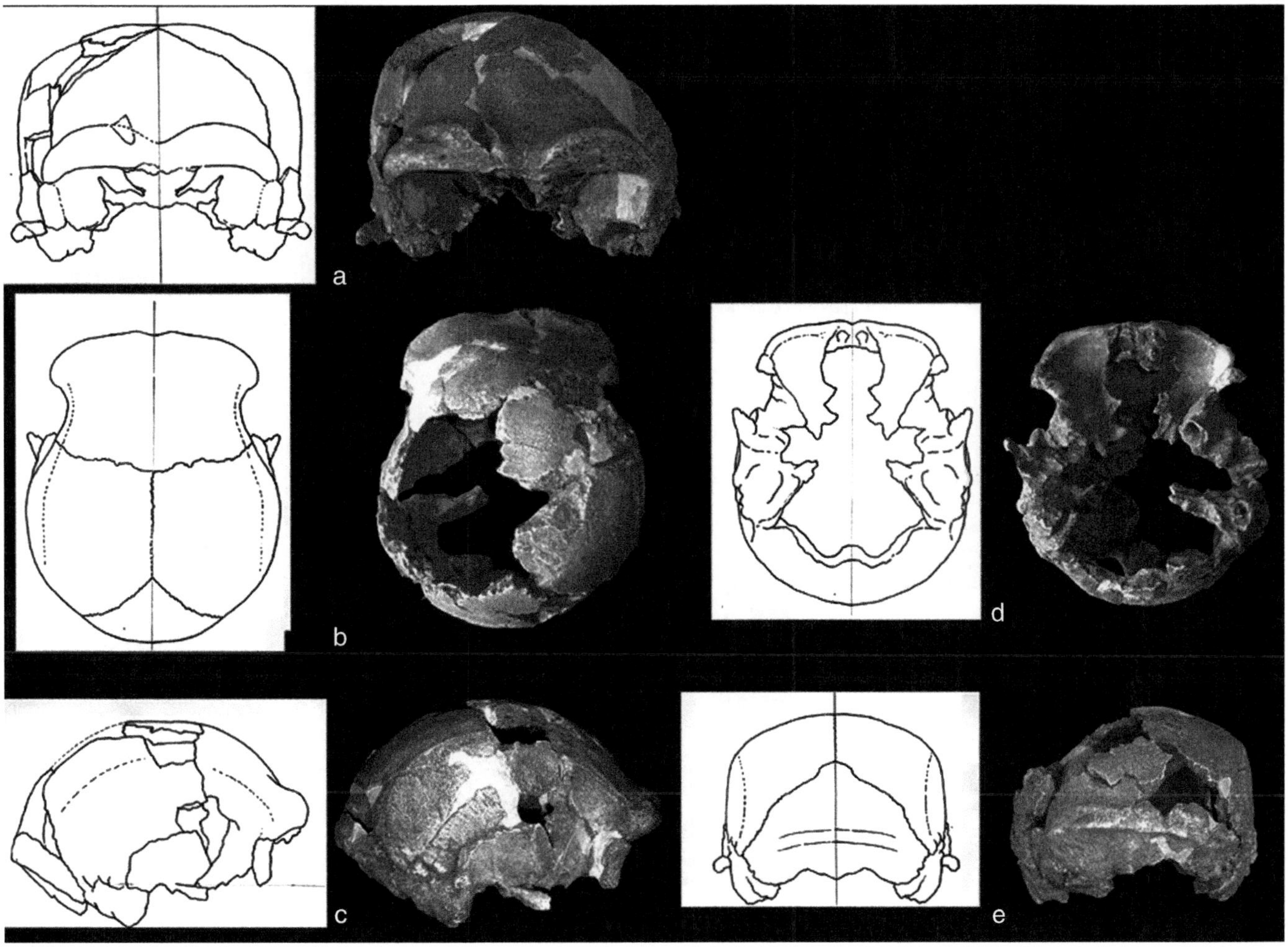

Fig. 5.1 The Ceprano calvarium, viewed in norma anterior (**a**), superior (**b**), lateralis (**c**), inferior (**d**), and posterior (**e**). In addition, some profiles obtained by means of the Mollison's dioptograph

which were synthesized in a single tree called "*strict consensus tree.*"

From the cladogram, it is possible to deduce that the Asian forms of *Homo erectus* (Zhoukhoudian and Sangiran) constitute a well-identifiable clade, separated from the rest of the taxonomic sample. The Asian *Homo erectus* is the "sister group" of a large clade including two main branches. The first branch includes Steinheim and the remains of Atapuerca and of Sima de los Huesos, while the second one includes Ceprano, Arago, Petralona, Kabwe, Saldaña, and Bodo. The samples coming from Dmanisi are definitely closer to the root of the whole cladogram (represented by OH-9), compared with these remains.

The bootstrap analysis provided a statistic support to these results, showing that the clade which includes Ceprano, Arago, Petralona, Kabwe, Saldaña, Bodo, Steinheim, and the remains of Atapuerca is clearly separated from *Homo erectus* Asiaticus, by the remains of Dmanisi and by OH-9.

On the basis of these results, we hypothesized the existence of a new species, *H. cepranensis*, due to the fact that the features which allowed us to identify this species appear for the first time in this specimen. In addition, they are also present in the later African specimens, and sometimes in the remains of the European ambit. On the other hand, these traits do not have anything in common with the fossil record of *H. erectus* coming from the Asian areas.

More detailed observations on the singular cranial elements of the remains of Ceprano emphasized some features which are not comprised in the list of Wood (1991) used for the cladistic analysis exposed above. These new characters are also different from those seen on the rare adult remains found at La Gran Dolina of Atapuerca. For example, the temporal bone (Fabbri and Mallegni 2005) presents an open digastric scissure, without bone bridges, and is not obliterated on the front side (Fig. 5.2), as we have noticed in numerous earlier specimens (KNM-ER 3733 and 3883). The Neanderthals and the so-called European pre-Neanderthals, as well as ATD 6–57 which was ascribed to *H. antecessor* (Arsuaga et al. 1993), almost always present a digastric scissure with the presence of a bone bridge. Therefore, we can

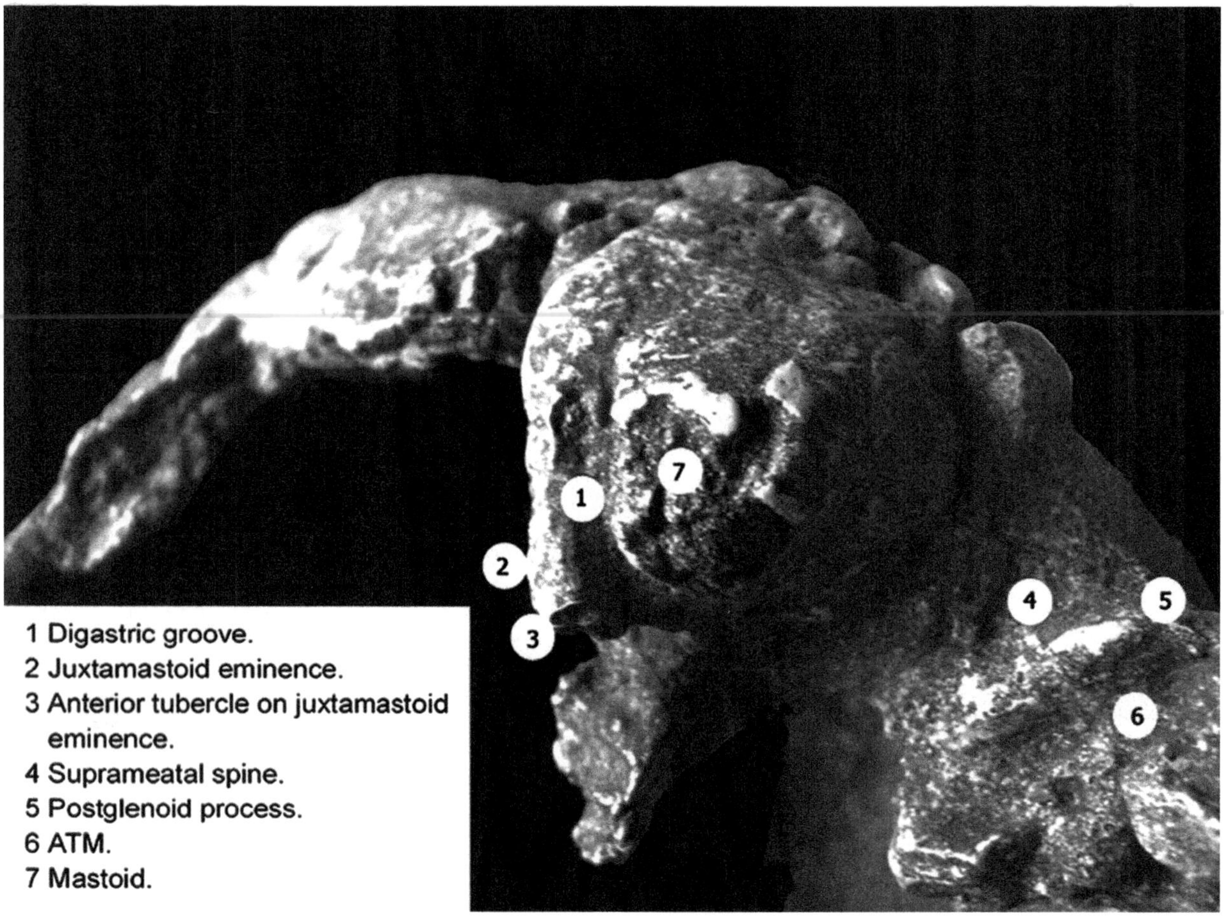

Fig. 5.2 The right mastoid region of Ceprano

only suppose that the future discoveries of adult temporal bones at Atapuerca could show the same features observed in the skull of Ceprano. It is interesting to note that some traits of Ceprano, especially those observable on the frontal and occipital bone, are present in the European specimens defined as *Homo heidelbergensis* (Petralona, Arago XXI and XLVII), as well as in the African specimens of the same species (Kabwe, Bodo, Saldaña) (Mallegni et al. 2003) (Figs. 5.3–5.7).

If the possibility of the phenomena of convergence is excluded, we must assume that the bones of Ceprano, which are more ancient than the other two European remains by at least 300–400 thousand years, are their ancestor, and of the African remains, as well (Kabwe, Bodo, Saldaña), following the cladistic results.

In addition to the cladistic analysis, we tried to carry out another examination based on the study of the physical dimensions of the skulls. The investigation was performed on the basis of six measurements which we compared with those of the 25 cranial remains belonging to modern humanity and to human fossil species. We employed the same data that Dean et al. (1998) utilized for the study of the cranial remains of Reilingen to which we added the data of *cepranensis*. The characters observed are: the maximum length of the cranium (this is a new datum, not utilized by Dean et al. (1998) due to the fact that the skull of Reilingen did not present the frontal bones), the maximum width of the skull, the biasteric width, the parietal arch, the parietal cord, and the lambda-inion cord. We analyzed these features simultaneously through the analysis of the principal components carried out by means of the program XLSTAT. The first two principal components are sufficient to interpret 77% of the variance. The first principal component is essentially determined by the parietal arch, by the parietal cord, and by the lambda-inion cord. The second principal component is determined by the maximum length of the cranium and by the biasteric width. The graphic with the two first main components shows a taxonomic distribution of the variables. In the remains of ER 3883, ER 3773, Zhoukhoudian X, XI, and XII, Sangiran 2 and 10, Sambungmachan 1, Solo I and Solo VI, the values of the first main component are lower than all the other values, approaching slightly to *Homo neanderthalensis* (La Quina 5). The Neanderthal remains and the remains of *Homo sapiens* usually have values which are lower compared with those of the second principal components.

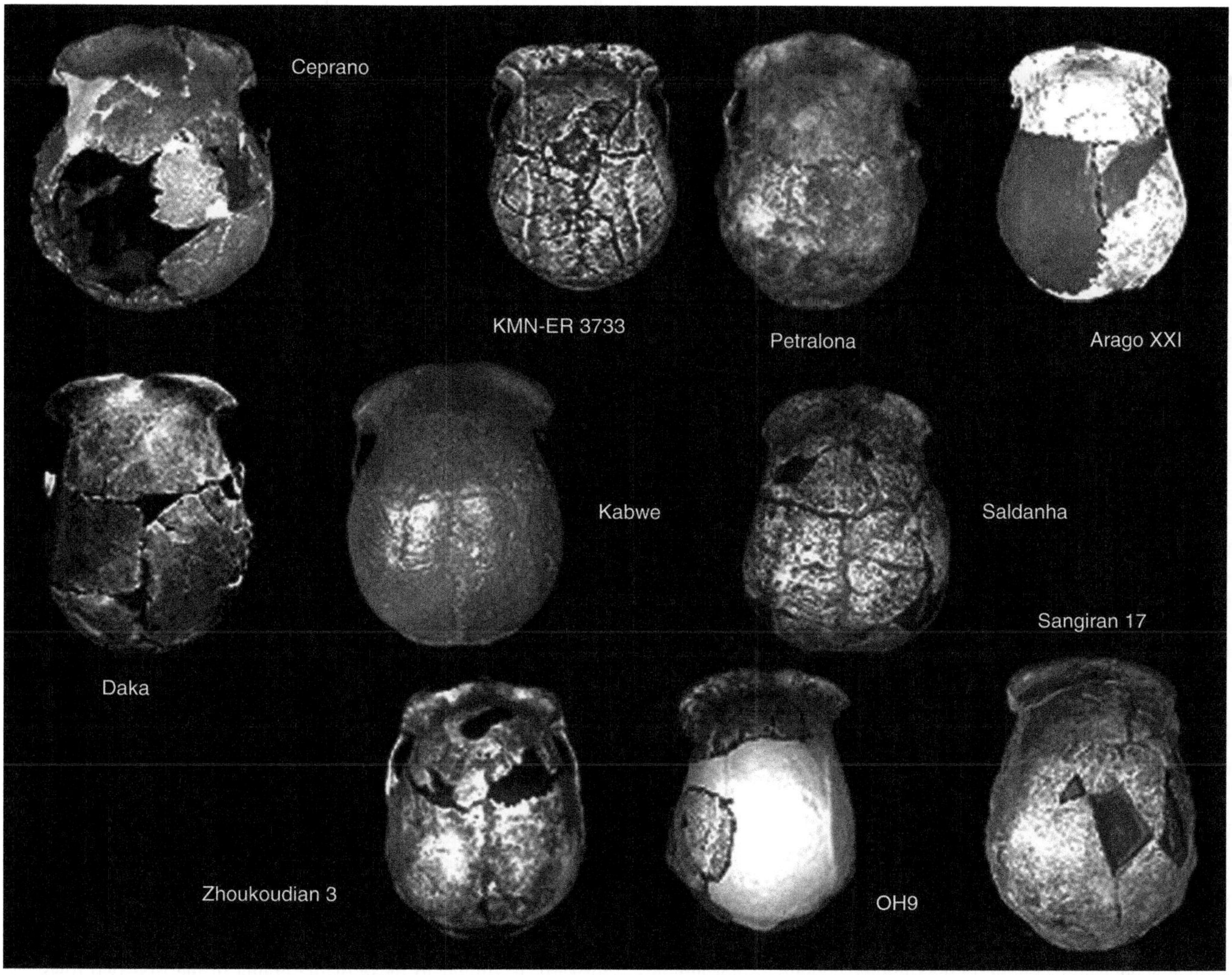

Fig. 5.3 Early and Middle Pleistocene crania in norma verticalis; it is possible to note the morphological affinities of the Ceprano cranium with those of Kabwe and Saldanha, and possibly with that of Petralona as well

The first principal component, in the remains of Solo VIII, IX, X, and XI, is similar to that of the Neanderthals and to that of *H. sapiens*, while the values of the second principal component remain higher. *Homo cepranensis* presents the highest ever observed values in the second principal component, while the value of the first component is positioned in the interval of the variation of the Neanderthal remains, of *H. sapiens* and of Solo VIII, IX, X, and XI. This fact indicates that the values of the parietal arch, parietal cord, and lambda-inion cord in *Homo cepranensis* and in the remains of Solo VIII, IX, X, and XI are homogeneous with those of *Homo neanderthalensis* and *Homo sapiens*. Anyway, the cranial maximum length and the bi-asterion width provide a different assessment to the morphometry of these species, rendering *H. cepranensis* and the remains of Solo well distinguishable from the remains of *H. neanderthalensis* and *H. sapiens*.

It is difficult to understand the significance of the morphometrical affinities between *Homo cepranensis* and the remains of Solo. The age of *H. cepranensis* has been estimated at about 800–900 thousand years, while that of the remains of Solo has been evaluated between 40 and 100 thousand years. The morphology of the calvarium does not indicate a close relationship between *H. cepranensis* and the remains of Solo. In light of these facts, we can surmise that the remains of Solo and *H. cepranensis* developed convergent dimensions of the bi-asterion width and of the maximum cranial length (Fig. 5.8). The results of this analysis are not different from those verified in the study carried out by Dean et al.(1998) on the skull of Reilingen, nor from those performed on the remains of Sambungmachan (Baba et al. 2003).

Obviously, it is not possible to make a comparison of the metrical data, nor of the morphological features, between the skull of Ceprano and the eponym of *H. heidelbergensis*

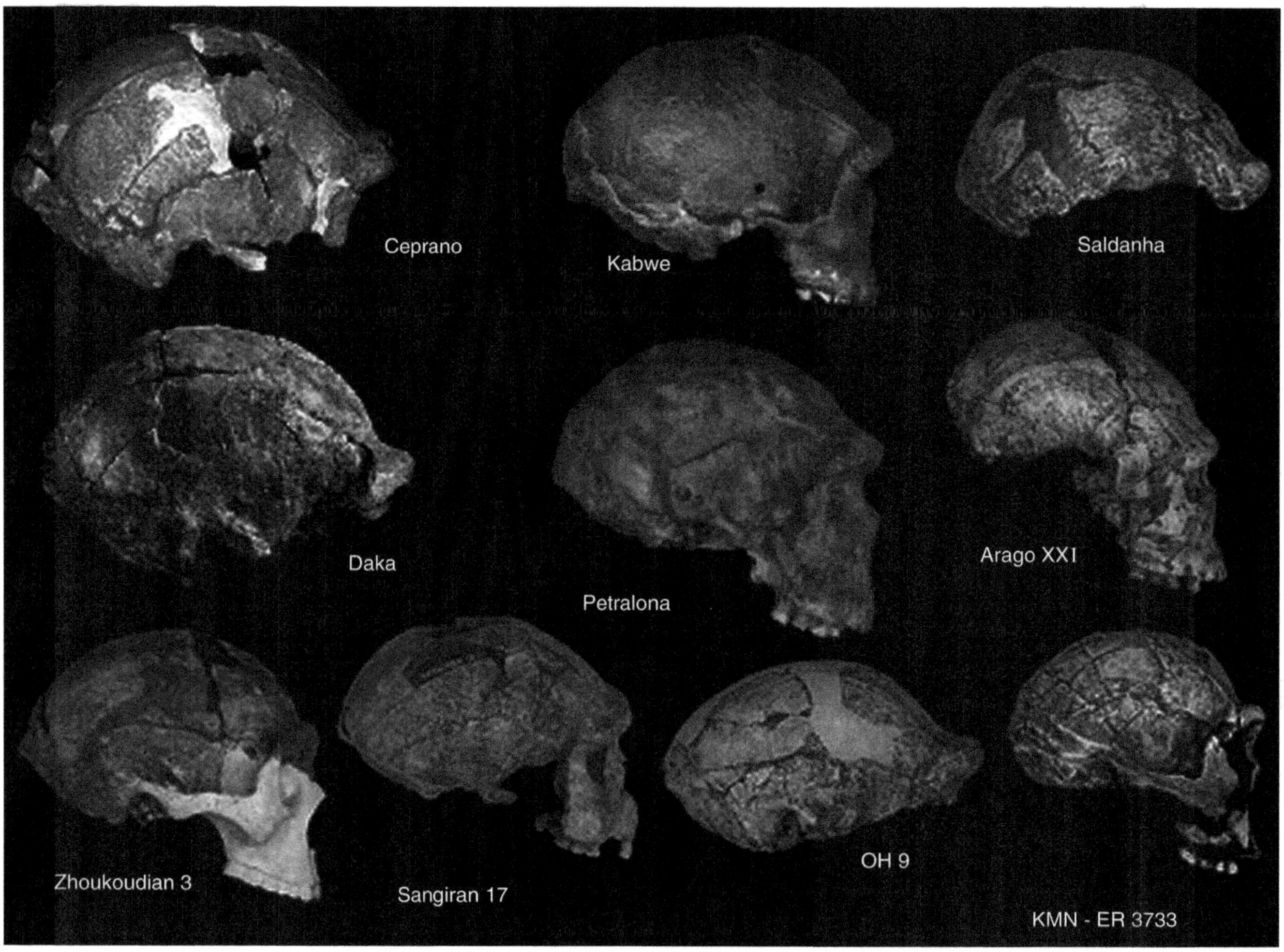

Fig. 5.4 Early and Middle Pleistocene crania in norma lateralis. Note how the lateral contour of the Ceprano cranium is morphologically close to those of Daka and Kabwe, while it differs (in the shape of the occipital profile) from Saldanha and partially from Petralona. The contour of the Ceprano frontal torus, which tends forward, is similar to those of the crania in the first two rows, as well as OH9, but differs from those of Asian *H. erectus* and of KNM-ER 3733

(that is the Mauer mandible), due to the fact that they are represented by different cranial parts. Perhaps the presence of the facial side of Ceprano could have permitted us at least to compare its dimension with the dimensions of the mandible (i.e., the distance between the two ATM), or the profile of the alveolar process and the denture of the two specimens (Fig. 5.9).

It is common knowledge that the species *heidelbergensis* was named and identified on the basis of the features of the mandible of Mauer (Schoetensack 1908), and this definition has been extended to the other two more recent European specimens as well (Petralona and Arago). The habit of naming and identifying new species on the basis of the mandible study is still widely diffused, even though we often have other parts of the cranium from the same sites at our disposal. The case of the remains of Mauer is different as they are the unique cranial remains from the site, and because they were the earliest European remains at the time of their discovery. In fact, the holotype of *H. antecessor* is a juvenile mandible, ATD6–5 and ATD6–96 (in spite of the presence of the facial bones and of the neurocranium of ATD6–69 and ATD6–15, which are both juvenile) (Arsuaga et al. 1997); the holotype of *H. georgicus* is the mandible D2600, despite the preceding discovery of the remains D2280, D2282, and the succeeding discovery of the remains D2700 and D3444 which are quite well preserved (de Lumley et al. 2006).

An excellent study carried out by Fabbri (2006) on some Pleistocene mandibles tends to demonstrate that this is not the most appropriate bone to use for the purpose of identifying a new species, because it does not present any unique and significant features. For example, *H. georgicus* D2600 differs mainly in two characteristics (its large dimensions and the protrusion of the inferior transversal torus) which are absent in the other lower jaw, D211, that is identified as the same species. However, these features are present in other mandibles assigned to *H. erectus* from South-Eastern Asia (i.e., Sangiran 6).

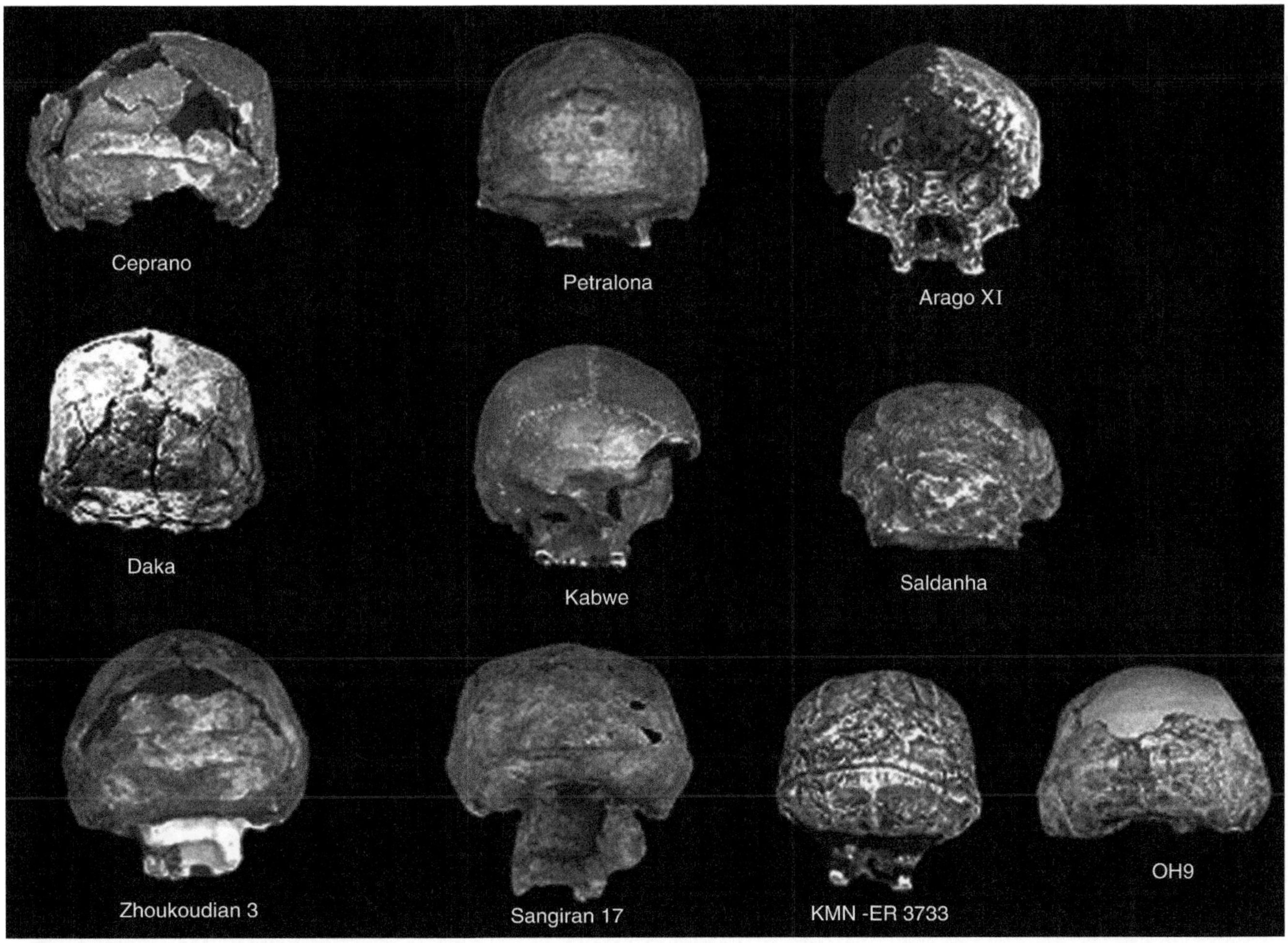

Fig. 5.5 Early and Middle Pleistocene crania in posterior view. The parietal bones begin to become vertical in the Ceprano and Daka crania, while in Petralona the *pars mastoidea* is still marked

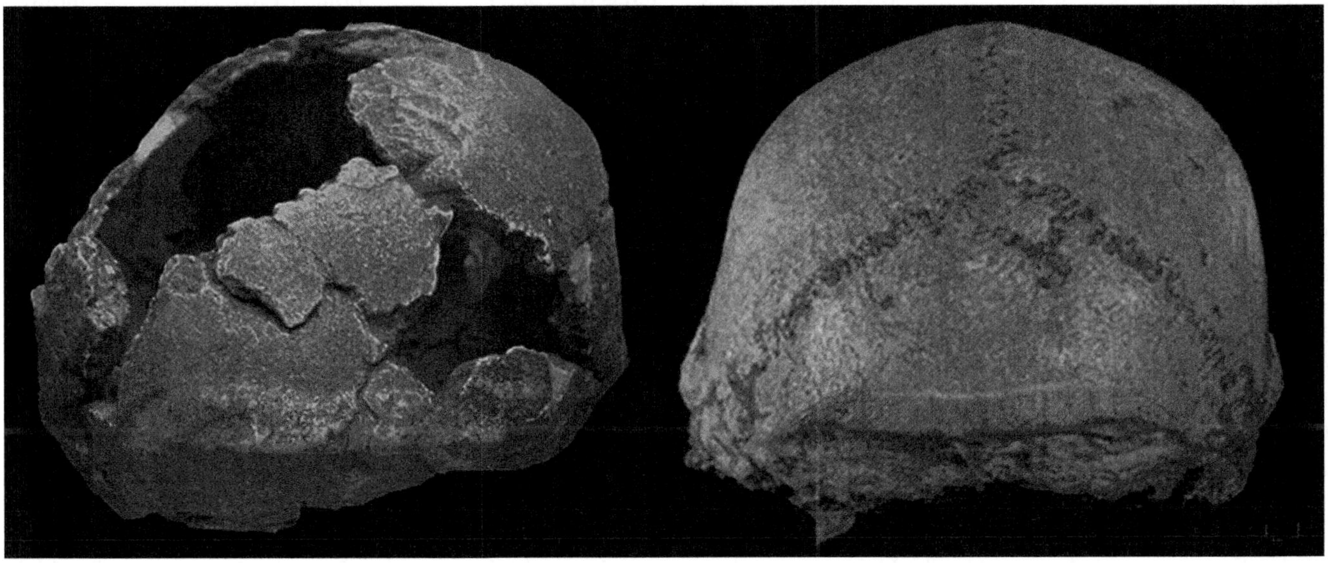

Fig. 5.6 Crania of Ceprano and Petralona in posterior view; the superior profiles of the occipital tori are similar in shape, even though in Ceprano the torus is more firm

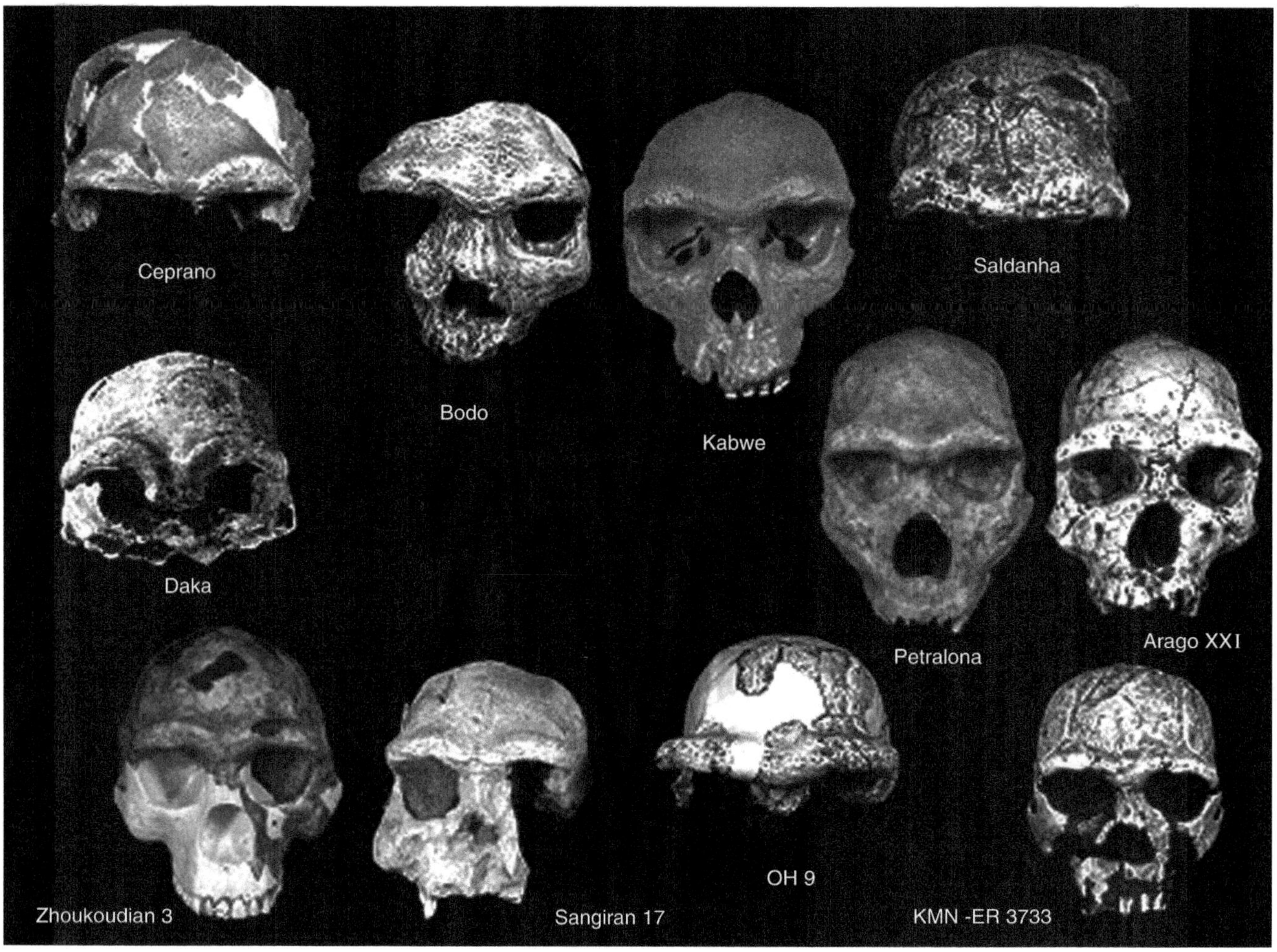

Fig. 5.7 Early and Middle Pleistocene crania in norma frontalis. The shapes of the frontal tori, the horizontal contour of the orbital superior margin, and the circumflex contour of the supraorbital margins are similar in Ceprano, Kabwe, Bodo, and Saldanha; there is also some similarity to Petralona

The distinctive traits of *H. antecessor* (mandible holotype ATD6–5 and ATD6–69) (Bermudez de Castro et al. 1997) are its small dimensions, a limited alveolar prominence, and the inclination of the mylohyoid sulcus. These features are present in a series of mandibles belonging to adolescent individuals from a wide geographical, taxonomic, and chronological range. We cannot exclude that among the remains found at Dmanisi and Atapuerca TD some species of the genus *Homo*, which have not yet been identified, are represented. The remains presently utilized to create these species do not present adequate characters. Recently Rightmire et al. (2006), as a result of their analysis on the cranial remains of D2280, D2282, and D2700, hypothesized that these bones belong to *Homo erectus* (subspecies *georgicus*), while in accordance with de Lumley et al. (2006), these remains still belong to *H. georgicus*.

The recent trend is to consider a certain number of remains of the European and African Middle Pleistocene, whose affinities with *erectus* and *neanderthalensis* are not clearly definable, as belonging to the *heidelbergensis* species. The latter has been revaluated in the last two decades (Rightmire 1985; Tattersall 1986). Inter alia, the species *heidelbergensis*, created on the basis of the Mauer mandible, is mostly formed by cranial remains which are not present in the findings of Mauer. The mandible, probably datable back to an early phase of the Middle Pleistocene (600 thousand years ago) (Schoetensack 1908), presents some features which are similar to those of the European fossils, which are considered by many scholars as the direct ancestors of the Neanderthals (i.e., Arago 2 and 3, Montmaurin, Bañolas). These features are: the *incisura submentalis*, the trace of the retromolar space, the tendency to uncover the tooth M3 in norma lateralis, horizontalization of the mental foramen, the value of the index of the distance mental foramen-M3 and of the distance mental foramen-incision.

The group of which this mandible is the holotype, i.e. the representatives (at least the European ones) of *H. heidelbergensis*, could be considered as the "stem group" which evolved into the European humanity of the Late Pleistocene. Dean et al. (1998) had already defined a similar evolution

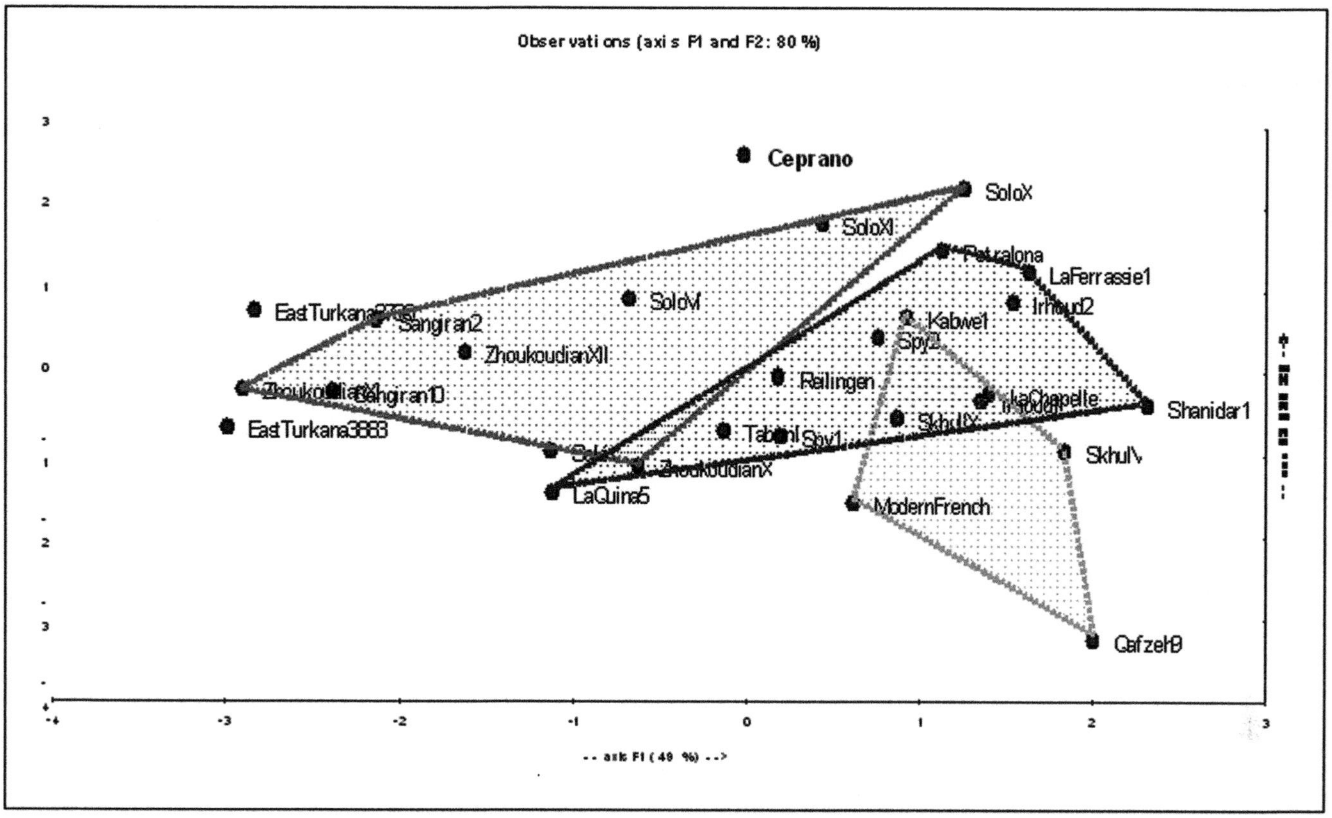

Fig. 5.8 Principal components analysis of 6 cranial variables on 25 Pleistocene and extant crania (following Dean et al. 1998) and Ceprano

model of the Neanderthals, which he called "Accretion Model." According to these authors, the mandible of Mauer could represent an "…early pre-Neanderthal," the very first phase of the evolutive history of this species. Fabbri assumes that considering the Mauer remains as a direct ancestor of *H. neanderthalensis*, would not provoke any confusion, such as considering the two different species, *H. heidelbergensis* and *H. neanderthalensis*, as pertaining to the same evolutive lineage. Therefore, we should consider *H. heidelbergensis* and *H. neanderthalensis* as belonging to a unique species. In Fabbri's opinion (2005) it is implausible to consider the possibility of the diffusion of the *heidelbergensis* species outside of Europe (in the African territories), because the majority of its remains cannot be compared to the holotype (the mandible of Mauer). The comparison is possible only between the European crania (Arago, Petralona) and the African skulls (Kabwe, Bodo, Saldaña) which are attributable to this species. The remains of Kabwe, Bodo, and Saldaña do not have many features in common with Arago and Petralona, especially in relation to the traits of the frontal bones, in particular the structure of the torus (which is more similar to that of Ceprano, particularly in Bodo, but in Saldaña and in Kabwe, as well). Also the structure of the face is different in the African remains of Kabwe and Bodo of which is possible to glimpse the canine fossa and the submalar incisure; however, the beginning of the extension of the maxillary sinus and the hint of the formation of a nasal bridge (due to the swelling of the frontal process of the upper jaw on both sides of the pyriform aperture) are absent. The nasal bridge begins to appear in the remains of Arago and Petralona and it is considered a characteristic of the Neanderthal forms.

There are some other considerations to keep in mind about the molecular analysis. Evidently, it is not possible to carry out the direct analysis between the DNA of the so-called European *heidelbergensis* and of the African homonyms. Nevertheless, quite recent data gave us some information about the analyses carried out on the remains of mt-DNA belonging to Neanderthal specimens (in the holotype of *Homo neanderthalensis*, in Mezmaiskaya and in Vindija – Krings et al. 1997; Ovchinnikov et al. 2000) and on other remains (the Gravettians of Paglicci 12 and 25 – Caramelli et al. 2003), which are slightly more recent than the first ones (but certainly belong to the *Homo sapiens* of the Italian Late Paleolithic). These data tend to demonstrate that there is a strong diversity between the two taxa; therefore, it is possible to hypothesize two different species, *neanderthalensis* and *sapiens*. The phenomenon of the differentiation probably ended about 200 Ka, but it could have started about 700–600 thousand years ago. We must ask ourselves how it is possible to hypothesize a specific affinity between the European remains of Arago and Petralona and the south-eastern African ones of Kabwe, Bodo, Saldaña, and Ndutu, if all of them

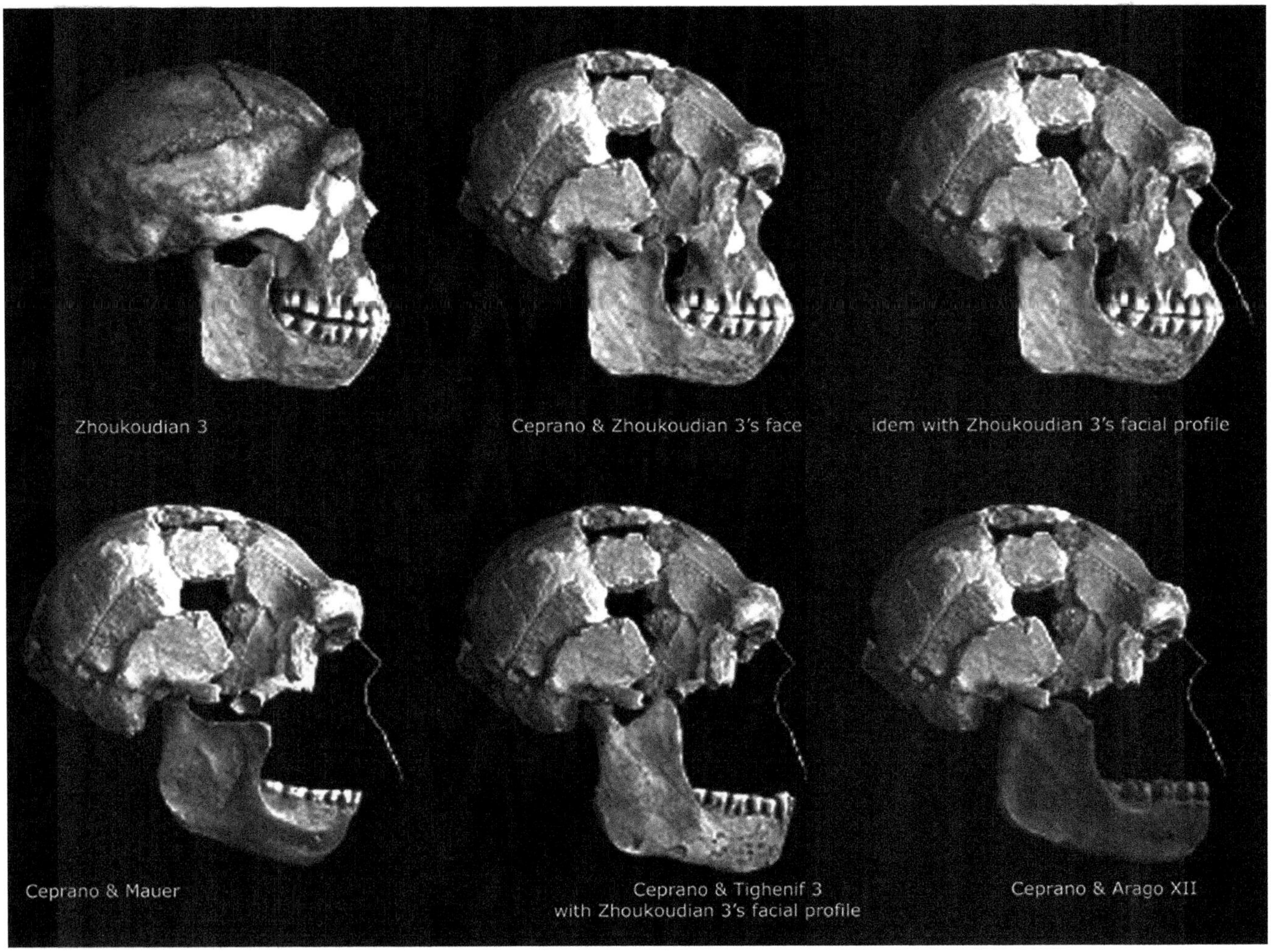

Fig. 5.9 The Ceprano calvarium, viewed in norma lateralis with a reconstructed face from Zhoukoudian 3 and/or mandibula from Zhoukoudian, Mauer, Tighenif III, and Arago XII

(except for Bodo) are datable at pre-600 thousand years. As already mentioned, it is in this period (700–600 thousand years ago) that we can theorize the beginning of the differentiation of the groups (occurred within 300–400 thousand years) in the two species, *neanderthalensis* and *sapiens*.

The noteworthy study carried out by Dean et al. (1998) on the fossil remains of Reilingen summarizes the state of the evolution of the remains which appeared before the complete manifestation of the classic Neanderthals. The research highlights the distinctive characters (autapomorphies) that emerge as time passes, through stages defined by the authors as "Neanderthal Stages" (from a climatic point of view "Isotope Stages"). These forms do not appear simultaneously, as in the case of the classic Neanderthals of the Late Pleistocene. These characteristics are expressed through the "Accretion Model," a theory of these authors. The phenotipy of some European Pleistocene remains is, without any doubt, quite complex. For example, in some parts of the remains of Steinheim, Bilzingsleben, Vertésszöllös, and Swanscombe, we can clearly note a variety of signs indicating an almost complete "neanderthalization." The contour of the skulls of Swanscombe and Steinheim tends, in norma posterior, to be roundish; this feature is similar to that of the Neanderthals. Both of them present an incipient suprainiac fossa and an increasing convexity of the occipital plain, as we can note in the remains of Vertésszöllös, and of Atapuerca Sima de los Huesos, as well. The remains of Steinheim present receding zygomatic bone, while the submalar incisure is scarcely delineated. The remains of Vertésszöllös lack the suprainiac fossa, while the cranium of Bilzingsleben presents a noticeable thickness of the occipital bone. A large part of these remains is deformed (i.e., Steinheim) due to the pressure of the soil, or more or less fragmented, or incomplete (i.e., Bilzingsleben, Vertésszöllös, Swanscombe), lacking in some parts that, if present, would have allowed us to define these remains more precisely.

Evidently, we have to keep in mind that there are other points of view such as that of Wolpoff who emphasizes that a mixture of the features defined by the "Accretion Model" is present in other Asian remains as well. According to Wolpoff, the hypothesis defined by this model would not be valid. However, we must consider that these features are also defined as apomorphies of the Neanderthals and not of other contemporary taxa. We have no notion whether some paleogenetic analysis on the remains of Ngandong (dating back to between 40 and 100 thousand years) and on the remains of *Homo floresiensis*, are currently underway. These remains are considered to be the last forms of the species *erectus*; such analyses could permit us to compare them to the data obtained on the mt-DNA of the Neanderthals, in order to determine the validity of the "Accretion Model." In any case, the final result of the evolution of the Neanderthals in the West and in the Middle East could suggest the theory that the environment influenced the beginning of different species (that is *erectus* in the eastern zones of Asia, and *sapiens* in Africa).

Returning to the European mindel-rissian specimens, their profile remains more complex. In fact, we could hypothesize that in the lacking parts of some of the remains, the same traits were probably present, which were also visible in the preserved parts of another remains, and so on. At the present stage of the researches, this phenomenon is interpreted by the hypothesis of the so-called mosaic features. The most significant explanation of these mosaic features is probably attributable to the possible segregation of the human groups in the territories which were geographically isolated. In fact, the Pyrenean range, the Alpine range, and the Balkan range, during the periods affected by the colder phases of the isotope stages (glacial acmes), formed real land pockets which were isolated due to the glacier expansions, the latter probably constituting a natural barrier for the human groups. In this way, the Iberian and the Italian pockets developed, as well as another wider pocket in the northern side of Europe (which included the areas of France, Germany, and of a large part of the Sarmatian plain), and the pocket in the south-east (which also included the Balkan peninsula). We can hypothesize that some partial genic drifts could develop in the human groups living in these land pockets. This phenomenon of isolation could have produced the beginning of new traits which (when the migration of human groups recommenced, at the end of the glacial acme) spread in the genic pool of the populations facilitated by crossbreeding. The period that elapsed was probably not long enough to enable the creation of new species. In this light, we could explain the presence of these mosaics of features in the more or less synchronous remains mentioned above.

The African fossil record, (Kabwe, Saldaña, and Bodo), which was formerly defined as Neanderthaloid, effectively does not present any Neanderthal features. It differentiates from its (more or less) contemporary remains (Arago and Petralona) as well, and from other mindel-rissian (*sensu lato*) European remains which are comprised in the stages 11–9. The African forms differ significantly from the northern ones in the cranial profiles, the structures of the frontal bones, and other facial bones (also in the details), even though some traits of the African remains recall, for example, Petralona. They probably originated from the same strain (perhaps the same strain to which the remains of Ceprano belonged). The ecological horizon and the problems of territory segregation worked out in different ways in the singular human groups.

In conclusion, this paper tends to identify, beginning from the earliest European hominids of the Middle Pleistocene, the presence of apomorphic Neanderthal features, at least in an early stage of development, or the existence of features that even though not exclusive, are very common in this species. These traits, at the current stage of our knowledge, seem to be present *in nuce* in the remains of Ceprano, while they are definitely absent in the remains of Dmanisi.

The presence of some traces of Neanderthal features in fossils dated approximately 700–600 thousand years ago (Mauer, Arago, and Petralona), and more certainly in fossils earlier of 350 thousand years ago (Atapuerca SH, Swanscombe, Bilzingsleben, Steinheim SH) (Bischoff et al. 1997) is in surprisingly perfect concordance with the paleogenetic data (Ingram et al. 2000). These data suggest a more ancient date for the identification of the Neanderthal evolutive lineage. Therefore, we assume that the numerous plesiomorphic traits, which are observable in the remains of Mauer, Petralona, and Arago, are not sufficient to allow us to hypothesize the existence of different species, nor of various evolutive lineages. On the other hand, the few and rudimental characters, which evoke the typical morphologies of the classic Neanderthals in some European Middle Pleistocene fossils, show the phylogenetic continuity of the European settlement from the beginning of the Middle Pleistocene until the arrival of *H. sapiens*.

Acknowledgments We are indebted to Ian Tattersall and Silvana Condemi and two anonymous reviewers for critical and helpful comments on the manuscript.

References

Arsuaga, J. L., Martinez, I., Gracia, A., Carretero, J. M., & Carbonell, E. (1993). Three new human skulls from the Sima de los Huesos in the Sierra de Atapuerca (Burgos, Spain). *Nature, 362*, 534–537.

Arsuaga, J. L., Bermúdes de Castro, J. M., & Carbonell, E. (Eds.). (1997). The Sima de los Huesos hominid site. *Journal of Human Evolution, 33*(Spec Issue 2–3), 105–421.

Ascenzi, A., & Segre, A. G. (1996). Artefacts and human teeth at the Fontana Ranuccio Middle Pleistocene site (Central Italy). *L'Anthropologie, XXXIV*(1–2), 39–46.

Ascenzi, A., & Segre, A. G. (2000). The fossil calvaria of *Homo erectus* from Ceprano (Central Italy): A new reconstruction. In M. Aloisi, B. Battaglia, E. Carafoli, & G. A. Danieli (Eds.), *The origin of humankind* (pp. 25–33). Venezia-Amsterdam: IVSLA-IOS.

Ascenzi, A., Biddittu, I., Cassoli, P. F., Segre, A. G., & Segre, N. E. (1996). A calvarium of late *Homo erectus* from Ceprano, Italy. *Journal of Human Evolution, 31*, 409–423.
Baba, H., Aziz, F., Kaifu, Y., Suwa, G., Kono, R. T., & Jacob, T. (2003). *Homo erectus* Calvarium from the Pleistocene of Java. *Science, 299*, 1384–1388.
Bermudez de Castro, J. M., Arsuaga, J., Carbonell, E., Rosas, A., Martinez, I., & Mosquera, M. (1997). A hominid from the lower Pleistocene of Atapuerca, Spain: Possible ancestor to Neandertals and modern humans. *Science, 276*, 1392–1395.
Bischoff, J. L., Fitzpatrick, J. A., Leo´n, L., Arsuaga, J. L., Falgue`res, C., Bahain, J. J. & Bullen, T. (1997). Geology and preliminary dating of the hominid-bearing sedimentary fill of the Sima de los Huesos Chamber, Cueva Mayor of the Sierra de Atapuerca, Burgos, Spain. *Journal of Human Evolution, 33*, 129–154.
Caramelli, D., Lalueza-Fox, C., Vernesi, C., Lari, M., Casoli, A., Mallegni, F., Chiarelli, B., Dupanloup, I., Bertranpetit, J., Barbujani, G., & Bertorelle, G. (2003). Evidence for a genetic discontinuity between Neandertals and 24,000-year-old anatomically modern Europeans. *Proceedings of the National Academy of Sciences of the United States of America, 100*, 6593–6597.
Clarke, R. J. (2000). A corrected reconstruction and interpretation of the *Homo erectus* skull from Ceprano, Italy. *Journal of Human Evolution, 39*, 433–442.
de Lumley, M. A., Gabunia, L., Vekua, A., & Lordkipanidze, D. (2006). Human remains from the Upper Pliocene – Early Pleistocene Dmanisi site, Georgia (1991–2000). Part I: The fossil skulls (D2280, D2282, and D2700). *L'Anthropologie, 110*, 1–110.
Dean, D., Hublin, J. J., Holloway, R., & Ziegler, R. (1998). On the phylogenetic position of the pre-Neanderthal specimen from Reilingen, Germany. *Journal of Human Evolution, 34*, 385–508.
Fabbri, P. F. (2006). Mandible and taxonomy of the Earliest European *Homo*. *Human Evolution, 21*(3–4), 289–300.
Fabbri, P. F., & Mallegni, F. (2005). *Il temporale del calvario di Campo Grande di Ceprano (Frosinone): Descrizione e confronti con reperti del Pleistocene inferiore e medio*. Atti XV Congresso degli Antropologi Italiani (pp. 219–231).
Gabunia, L., & Vekua, A. (1995a). A Plio-Pleistocene hominid from Dmanisi, East Georgia, Caucasus. *Nature, 373*, 509–512.
Gabunia, L., & Vekua, A. K. (1995b). La mandibule de l'homme fossile du Villafranchien superieur de Dmanisi (Georgie Orientale). *L'Anthropologie, 99*, 29–41.
Gabunia, L., Vekua, A., Lordkipanidze, D., Swisher, C. C., Ferring, R., Justus, A., Nioradze, M., Tvalchrelidze, M., Anton, S. C., Bosinski, G., Joris, O., de Lumley, M. A., Majsuradze, G., & Mouskhelishvili, A. (2000). Ealiest Pleistocene cranial remains from Dmanisi, Republic of Georgia: Taxonomy, geological setting, and age. *Science, 288*, 1019–1025.
Gabunia, L., de Lumley, M. A., Vekua, A., Lordkipanidze, D., & de Lumley, H. (2002). Découverte d'un nouvel hominide a Dmanisi (Transcaucasie, Georgie). *Comptes Rendus Palevol, 1*, 243–253.
Gilbert, W. H., White, T. D., & Asfaw, B. (2003). *Homo erectus, Homo ergaster, Homo "cepranensis,"* and the Daka cranium. *Journal of Human Evolution, 45*(3), 255–259.
Ingram, M., Kaessmann, H., Paabo, S., & Gyllensten, U. (2000). Mitochondrial genome variation and the origin of modern humans. *Nature, 408*, 708–713.
Krings, M., Stone, A., Schmitz, R. W., Krainitzki, H., Stoneking, M., & Paabo, S. (1997). Neandertal DNA sequences and the origin of modern humans. *Cell, 90*, 19–30.
Mallegni, F., Carnieri, E., Bisconti, M., Tartarelli, G., Ricci, S., Biddittu, I., & Serge, A. (2003). *Homo cepranensis* sp. nov. and the evolution of African–European Middle Pleistocene hominids. *Comptes Rendus Palevol, 2*(2), 153–159.
Martin, R., & Saller, K. (1956–1959). *Lehrbuch der Anthropologie*. Stuttgart: Fischer Verlag.
McCown, T. D., & Keith, A. (1939). *The stone age of Mount Carmel, Vol. II, the fossil human remains from the Levalloiso-Mousterian*. Oxford: Clarendon.
Ovchinnikov, I. V., Gotherstrom, A., Romanova, G. P., Kharitonov, V. M., Liden, K., & Goodwin, W. (2000). Molecular analysis of Neanderthal DNA from the northern Caucasus. *Nature, 404*, 490–493.
Rightmire, G. P. (1985). The tempo of change in the evolution of mid-Pleistocene Homo. In E. Delson (Ed.), *Ancestors: The hard evidence* (pp. 255–264). New York: Alan R. Liss.
Rightmire, G. P., Lordkipanidze, D., & Vekua, A. (2006). Anatomical descriptions, comparative studies and evolutionary significance of the hominin skulls from Dmanisi, Republic of Georgia. *Journal of Human Evolution, 50*, 115–141.
Schoetensack, O. (1908). *Der Unterkiefer des Homo heidelbergensis aus den Sanden von Mauer bei Heidelberg*. Leipzig: Wilhelm Engelmann.
Tattersall, I. (1986). Species recognition in palaeontology. *Journal of Human Evolution, 15*, 165–175.
Trinkaus, E., Milota, S., Rodrigo, R., Mircea, G., & Moldovan, O. (2003a). Early modern human cranial remains from the Petera cu Oase, Romania. *Journal of Human Evolution, 45*, 245–253.
Trinkaus, E., Moldovan, O., Milota, S., Bilgar, A., Sarcina, L., Athreya, S., Bailey, S., Rodrigo, R., Mircea, G., Higham, T., Ramsy, C., & van der Plicht, J. (2003b). An early modern human from the Petera cu Oase, Romania. *Proceedings of the National Academy of Sciences of the United States of America, 100*, 11231–11236.
Vandermeersch, B. (1981). *Les Hommes Fossiles de Qafzeh (Israel). Cahiers de Paléoanthropologie*. Paris: Editions du CNRS.
Wood, B. (1991). *Hominid cranial remains*. Koobi Fora Research project (Vol. 4). Oxford: Clarendon.

Chapter 6
The Gran Dolina-TD6 Human Fossil Remains and the Origin of Neanderthals

José María Bermúdez de Castro, María Martinón-Torres, Aida Gómez-Robles, Ann Margvelashvili, Juan Luis Arsuaga, José Miguel Carretero, Ignacio Martinez, and Susana Sarmiento

Abstract We present a revision of the main features with phylogenetic interest observed in the human fossil remains recovered from the Aurora Stratum of the TD6 level, Gran Dolina site (Sierra de Atapuerca, Spain) that have been assigned to *Homo antecessor*. Our aim is to test the hypothesis of a possible relationship between this species and the European Middle and early Late Pleistocene hominins, the so-called Neanderthal lineage. Some cranial, postcranial, and dental features are plesiomorphic for the genus *Homo* and thus, they are not useful for our purpose. Other morphologies are derived with regard to *H. ergaster/H. erectus*, and TD6 hominins share those traits with modern humans, with Neanderthals or with both lineages. In this context we hypothesize either that there exists a phylogenetic continuity between *Homo antecessor* and Neanderthals or that both species shared a common ancestor.

Keywords Human evolution • Pleistocene • *Homo antecessor* • Phylogenetic analysis.

J.M. Bermúdez de Castro (✉), M. Martinón-Torres (✉), A. Gómez-Robles, A. Margvelashvili, and S. Sarmiento
Centro Nacional de Investigación sobre la Evolución Humana (CENIEH), Burgos, Spain
e-mail: josemaria.bermudezdecastro@cenieh.es; maria.martinon.torres@gmail.com; aida.gomez@cenieh.es; susanasarm@gmail.com

J.L. Arsuaga
Departamento de Paleontología, Universidad Complutense de Madrid- Instituto de Salud Carlos III, Madrid, Spain
e-mail: jlarsuaga@isciii.es

J.M. Carretero
Universidad de Burgos, Edificio I+D, Burgos, Spain
e-mail: jmcarre@ubu.es

I. Martinez
Instituto de Salud Carlos III, Madrid, Spain
e-mail: imartinezm@isciii.es

Introduction

The origin of Neanderthals has been a matter of interest in Paleoanthropology, from the classical publications made by Boule (1911–1913), Howell (1951), Le Gros Clark (1955), or Boule and Vallois (1957), among others, to the most recent papers on ancient DNA recovered from fossil remains (Green et al. 2006; Noonan et al. 2006). There is a general agreement that the Neanderthals have deep roots in the Middle Pleistocene of Europe and they have been firmly related to *Homo heidelbergensis* (Stringer 1993a; Arsuaga et al. 1997). This species was named in 1908 by Otto Schoetensack to include the human fossil jaw found 1 year before at 25 m depth in the alluviate sand levels of the Neckar river, near the German village of Mauer, and 16 km SE from the city of Heidelberg. This jaw belonged to an individual who probably lived during the early Middle Pleistocene, between 640,000 and 735,000 years ago (Hambach 1996).

During the 1980s, some authors understood the necessity of grouping some fossil specimens with a morphology that was more evolved than that of the Early Pleistocene hominins (mainly *H. erectus* and *H. ergaster*) but still less derived than that of our own species. Thus, some Middle Pleistocene African and European specimens like those from Arago, Binzingsleben, Bodo, Kabwe, Petralona, or Swanscombe, which exhibited a combination of archaic and specialized traits not found in *H. erectus* and known as "archaic *H. sapiens*," were included in *H. heidelbergensis* taxon (Stringer 1985, 1993b; Rightmire 1988), with the Mauer mandible as the holotype of the species. Furthermore, this species was considered the Middle Pleistocene common ancestor for Neanderthals and modern humans (Stringer and McKie 1996; Tattersall 1996; Rightmire 1996, 1998a).

The study of the exceptional hypodigm of more than 5,600 human fossil remains (belonging to a minimum of 28 individuals) recovered so far from the Sima de los Huesos site of the Sierra de Atapuerca in northern Spain strongly suggests that Neanderthals appeared as a result of a local evolution of the Middle Pleistocene populations (e.g. Arsuaga et al. 1993; Arsuaga et al. 1997; Bermúdez de Castro 1993; Rosas 2001; Martinón-Torres 2006). The Sima

Vertebrate Paleobiology and Paleoanthropology, DOI 10.1007/978-94-007-0492-3_6,

de los Huesos hominins have been included in *H. heidelbergensis* by Arsuaga et al. (1997). The last radiometric studies (U-series) of a 14-cm thick in situ speleothem overlying the mud-breccia containing the human bones of the Sima de los Huesos site indicate an age of about 600,000 years for these hominins (Bischoff et al. 2006). These results confirm the deep roots of Neanderthals in the European Middle Pleistocene, and support the notion that *H. heidelbergensis* should be considered, together with *H. neanderthalensis* King, 1864, as a chrono species of the same "evolutionary" species (Arsuaga et al. 1997). In fact, except on rare occasions (e.g. the occipital from Vértesszöllös), most European Middle Pleistocene fossil specimens, such as those of Arago Petralona, Swanscombe, Steinheim and, of course, those from Sima de los Huesos site, exhibit one or more derived cranial and mandibular traits shared exclusively with Neanderthals (Neanderthal apomorphies). Furthermore, the dental proportions of the Mauer mandible, specially the buccolingual dimensions with respect to the total dental size, are similar to those of Neanderthals (Rosas and Bermúdez de Castro 1998; Bermúdez de Castro et al. 1999).

The fossil record suggests that at least from the middle of the Middle Pleistocene Europe was characterized by an endemism. This process was favored by the peculiar paleogeographical and paleoclimatological conditions of the European Peninsula during this long period (Hublin 1990). European hominins from this period would have evolved in isolation, probably without interbreeding with other non-European populations. Thus, the last common ancestor of Neanderthals and modern humans could not be represented in the European Middle Pleistocene fossil record (Arsuaga et al. 1997) currently available. The common ancestor of Neanderthals and modern humans should be more primitive and should also lack the specialized features characterizing each of these hominin lineages. Thus, we could track back in the fossil record to look for a more ancient hominin, who could be recognized by the presence of some features shared with the Neanderthal and modern human lineages.

Between 1994 and 1996, a rich Early Pleistocene assemblage of fossils and lithic industry was found at the so-called Aurora Stratum of the TD6 level of the Gran Dolina site, in the Railway Trench of the Sierra de Atapuerca. The assemblage, which initially included a total of 86 human fossil remains, has been significantly increasing during the 2003–2006 field seasons. These fossils exhibit a unique combination of a modern face and a primitive dentition that led to the Atapuerca research team to name a new species, *H. antecessor* (Bermúdez de Castro et al., 1997). We further concluded that this species might represent the last common ancestor to both the Neanderthal and modern human lineages, as an alternative to *H. heidelbergensis*.

In 2003, we made a comparative study of the TD6 and Sima de los Huesos human dental samples. The clear morphological differences found between both populations led us to propose a possible discontinuity between the Early Pleistocene European populations (represented by the TD6 hominins) and the European Middle Pleistocene populations (represented by the Sima de los Huesos hominins) (Bermúdez de Castro et al. 2003, and see also Carbonell et al. 2005). However, additional studies of the dentition (Martinón-Torres 2006; Martinón-Torres et al. 2007; Gómez-Robles et al. 2007) compelled us to reconsider this hypothesis. The aim of this report is to examine the present TD6 human sample in order to test the hypothesis of a possible phylogenetic relationship between *H. antecessor* and the *heidelbergensis/neanderthalensis* lineage.

The Age of theTD6 Level and the Human Fossil Sample

The first paleomagnetic investigation of the Gran Dolina site was performed by Parés and Pérez-González (1995). They found a paleomagnetic inversion at the TD7 level, 1 m above the Aurora Stratum, which they identified with the Matuyama–Brunhes boundary, based on the information furnished by the Gran Dolina fossil assemblages. The TD6 macromammals sample includes *Vulpes* sp., Canidae indet., Mustelidae indet., *Panthera* sp., *Felis silvestris*, *Ursus*, sp., Proboscidea indet., *Equus* sp., Stenoniano, *Stephanorhinus etruscus*, *Sus scrofa*, *Dama dama vallonetensis*, *Cervus elaphus cf. Acoronatus*, *Megalocerus cf. verticornis*, and *Capreolus capreolus* (van der Made 1999; García and Arsuaga 1999). This fossil assemblage is characteristic of the end of the Early Pleistocene and the beginning of the Middle Pleistocene. Among the microvertebrates, the presence in TD6 of *Mimomys savini*, also represented in TD7 and TD8, is noteworthy for the age determination of the Aurora Stratum (Cuenca-Bescós et al. 1999). Another study by Parés and Pérez-González (1999) confirmed that the Gran Dolina lower levels (TD1–TD6) displayed reversed polarity, whereas the upper levels (TD7–TD11) were normal. At the bottom of the TD section, these authors reported evidence of a short normal polarity event, which they interpreted as Jaramillo or Kamikatsura event. The electron spin resonance and U-series results obtained by Falguères and coworkers (1999) suggest an age range between 780,000 and 857,000 years for the Aurora Stratum.

The TD6 human hypodigm consists of more than 100 fragmented bones belonging to the cranial and postcranial skeleton. The sample includes more than 50 parts of clavicles, radii, femora, vertebrae, ribs, patellae, metacarpal and metatarsal bones, pedal, and manual phalanges. Fragments of frontal, parietal, temporal, occipital, maxillary, zygomatic, and sphenoid bones, as well as four mandibles, comprise the cranial sample. The dental sample consists of 5 deciduous and 37 permanent teeth. The human remains recovered until the 2005 season have been assigned to a minimum of nine individuals, identified by the maxillae, mandibles, and the teeth (Bermúdez de Castro et al. 2006). Some of the pieces recovered during 2003–2006 period are unpublished. Two mandibles found in 2006 are currently in the process of restoration.

Cranial and Mandibular Features

Arsuaga et al. (1999) made the study of the neurocranial and facial bones from TD6. We refer to these authors for a detailed anatomical description of the specimens and their particular features, and we will quote only those traits with taxonomical interest for the aim of this report.

The superior border of the temporal squama in the temporal bone fragment ATD6-20 is high and arced. This feature is a derived trait present in European and African Middle Pleistocene populations, modern humans, and Neanderthals, as well as in the Early Pleistocene Ceprano calvaria and later Asian Middle Pleistocene specimens. A high and arced squamosal suture is related to an increase in cranial capacity, which in TD6 hominins could be greater than 1,000 cc, according to the dimensions of the ATD6-15 frontal bone (Carbonell et al. 1995).

The facial skeleton is well represented in TD6 by five fragments, the most complete being ATD6-69 (Fig. 6.1). This specimen shows a fully modern pattern of midfacial morphology, therefore bearing no resemblance to the derived face of Neanderthals, who exhibit a characteristic midfacial prognathism (Rak 1986). This feature is also present in some of the European Middle Pleistocene specimens, such as Arago 21, Petralona, and those of Atapuerca-Sima de los Huesos. Although ATD6-69 belonged to an adolescent with incomplete facial growth, the specimens ATD6-19 (a small adult right zygomaxillary fragment), and ATD6-58 (an adult left large zygomaxillary fragment, lacking only the zygomatic process) also exhibit "modern" traits. ATD6-58 shows some expansion of the maxillary sinus that reduces the expression of the canine fossa.

The internal nasal cavity of ATD6-69 lacks the three Neanderthal apomorphies described by Schwartz and Tattersall (1996): development of an internal nasal margin bearing a well-developed and vertically oriented medial projection, swelling of the posterior-lateral wall of the nasal cavity as a result of a medially expanded maxillary sinus, and lack of an ossified roof over the lacrimal groove. Furthermore, nasal crests of ATD6-69 are similar to those of modern humans and lack the typical Neanderthal sharp lower margin formed by the lateral crest (Arsuaga et al. 1999). The lateral nasal crest is vertical and slightly concave, with its lower extremity behind the rhinion. The orientation of the lateral nasal walls in ATD6-69 clearly suggests that the nasal bones were elevated and forward sloping, a derived condition that *H. antecessor* shares with modern humans and Neanderthals.

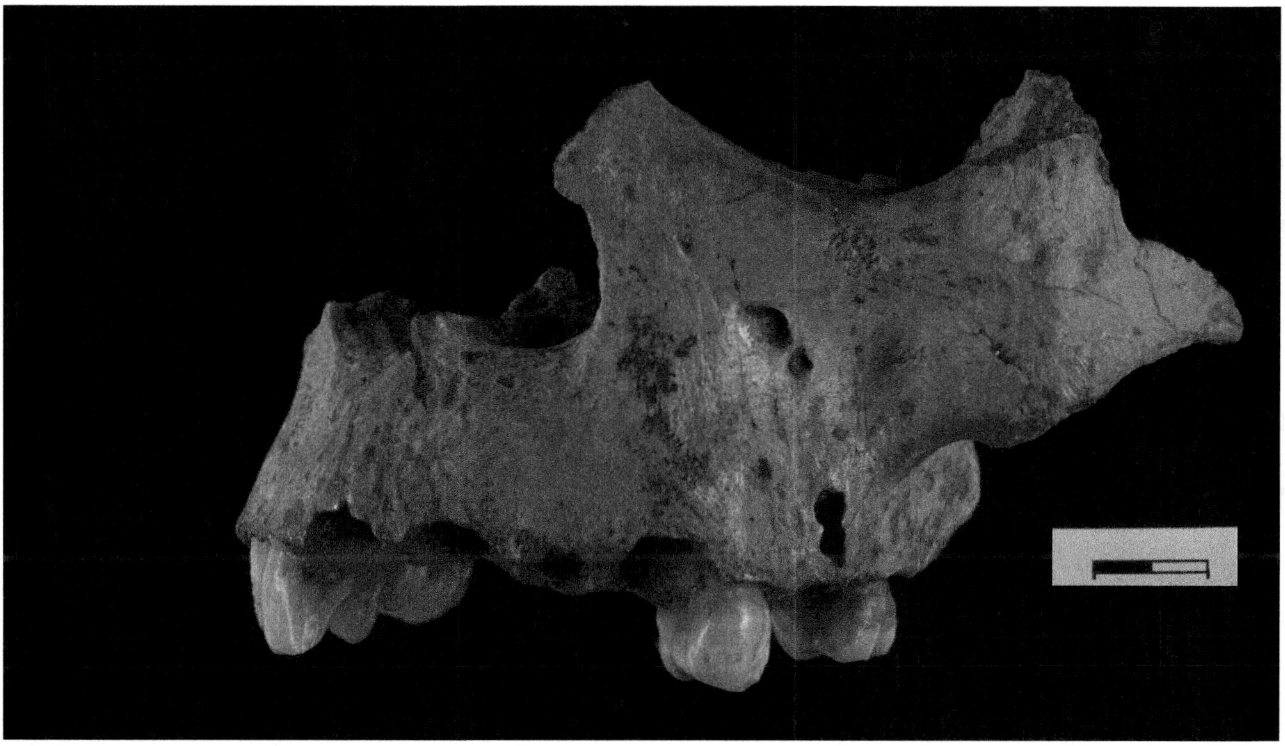

Fig. 6.1 ATD6-69 maxilla of *Homo antecessor*

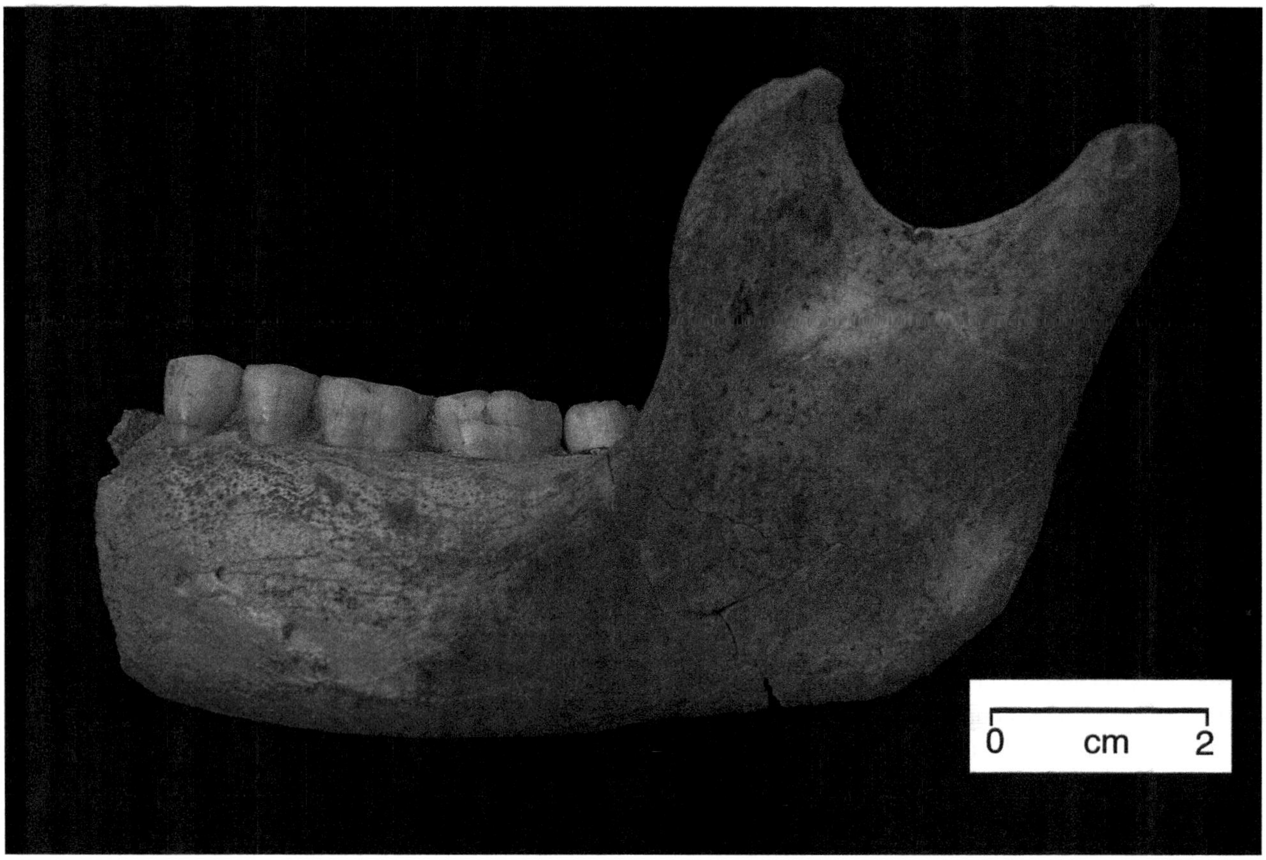

Fig. 6.2 ATD6-96 hemimandible of *Homo antecessor*

ATD6-69 also shows an anteriorly located incisive foramen, 5 mm behind the anterior alveolar margin. This incisive fossa opens up in the floor of the nasal cavity and it is also anteriorly placed. As a consequence, the incisive canal is nearly vertical. In *H. erectus/H. ergaster*, the incisive foramen is placed well behind the anterior alveolar margin and the incisive canal lies obliquely (Rightmire 1998b). ATD6-69 shares this trait with Neanderthals and modern humans, although we have also noticed a vertical incisive canal in the Buia cranium from the Early Pleistocene of Northern Danakil Depression in Eritrea.

The small mandibular fragment ATD6-5, which belongs to an adolescent, and the left half of a gracile adult mandible ATD6-96 (Fig. 6.2) exhibit a primitive structural pattern that is shared with all African and Asian Pleistocene specimens. Furthermore, none of the mandibular features considered apomorphic by Rosas (2001) in the European Middle and early Late Pleistocene hominin lineage are present in any of the two TD6 specimens. The position of multiple mental foramina at the level of P3–P4, the position of the lateral prominence at the level of M2, the low position of the mylohyoid line in relation to alveolar margin at the M3 level, the parallel trajectory of the mylohyoid line in relation to alveolar margin, the shallow relief of the pterigoyd fossa, and the lateral intersection between the mandibular notch and condyle are plesiomorphic traits observed in the TD6 specimens (Rosas and Bermúdez de Castro 1999; Carbonell et al. 2005). Some features of ATD6-96, such as the position of the M3 in relation to the ramus, the oblique inclination of the retromolar area, or the regular profile of the gonion are slightly derived with regard to the primitive status, ascertained in some *H. ergaster* specimens, as well as in some African (Tighenif) and Asian (Zhoukoudian) Middle Pleistocene mandibles. The absence of alveolar prominence in ATD6-5 and ATD6-96 is noteworthy, which contributes to their low corpus thickness and higher gracility in relation to the African Pleistocene specimens.

Dental Evidence

Most dental features of the TD6 hominins are plesiomorphic for the genus *Homo* and do not help to solve the question posed in this report. Thus, the upper lateral incisors are shovel shaped, the cingulum is present in the mandibular canines and premolars, the crown of the mandibular P3 is strongly asymmetrical with a well-developed talonid, the root morphology

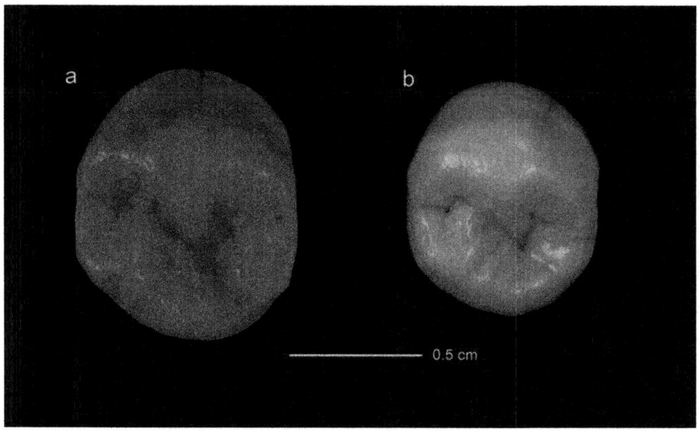

Fig. 6.3 Morphological comparison of the lower second premolars of Gran Dolina and Sima de los Huesos hominins. (**a**) ATD6-4 (Gran Dolina); (**b**) AT-221 (Sima de los Huesos)

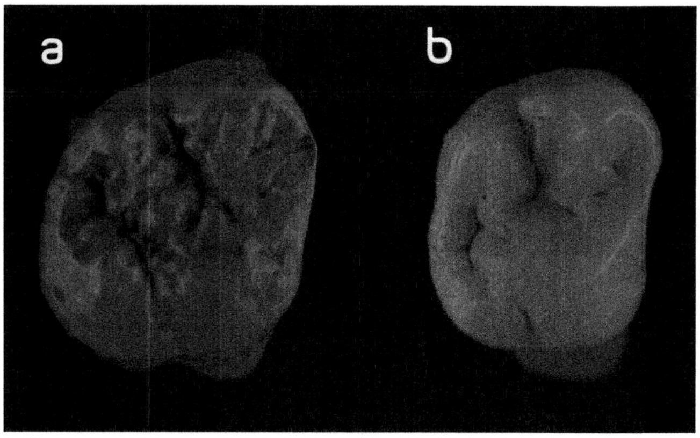

Fig. 6.4 Morphological comparison of the upper first molar of Gran Dolina and Sima de los Huesos hominins. (**a**) ATD6-11 (Gran Dolina); (**b**) AT-406 (Sima de los Huesos)

of the lower premolars is complex with two (MB+DL) roots, the hypoconulid is present and well developed in M1 and M2, the upper and lower first molars are, respectively, larger than the upper and lower second molars (M1>M2), and taurodontism is expressed.

Regarding upper incisors, *H. antecessor* shares with *H. erectus* and *H. neanderthalensis* high degrees of labial convexity (Martinón-Torres 2006). *H. erectus* and *H. antecessor* also present incipient forms of the "triangular shovel shape," typical of the Neanderthal populations (Martinón-Torres 2006). The thickened marginal ridges invade the lingual surface and define a deep and narrow longitudinal fossa, giving the occlusal surface a characteristic V-shape.

Lower canines show a derived morphology when compared to Plio-Pleistocene specimens like those assigned to *Homo habilis* and *Homo georgicus*, which resembles that of the Sima de los Huesos specimens. Lower canines in this species present an incisor-like conformation, smooth lingual surface, and long parallel marginal ridges. These traits, along with the P3>P4 sequence, might be an evidence of the relationship between the TD6 hominins and the European Middle Pleistocene populations. Still, TD6 lower canines still express cingulum, which is lost in later *Homo* populations.

Similarly, second lower premolars are more evolved in TD6 hominins than in their African counterparts (see Fig. 6.3). They are closer to the Neanderthal morphology displaying a particular combination of plesiomorphic traits (mesial metaconid, multiple lingual cusps, transverse crest, and assymetrical contour) in association with a reduced occlusal polygon (Martinón-Torres et al. 2006). Eventually, modern human morphology could originate from the TD6 conformation.

Finally, recent studies have shown that *H. antecessor* shares a derived conformation in their upper first molars (Gómez-Robles et al. 2007) with the European Middle Pleistocene populations and *H. neanderthalensis*. This shape consists of a rhomboidal occlusal polygon (consequence of the relative distal displacement of the lingual cusps), associated to a skewed outline with a protruding hypocone (Fig. 6.4). This conformation clearly differs from the *H. sapiens* shape, since this species keeps the primitive morphology observed in the African Pliocene species.

Postcranial Remains

Lorenzo et al. (1999) have made a comparative study of a sample comprising 22 hand and foot remains from TD6. They conclude that the morphology and dimensions of these remains are more similar to those of modern humans than to those of Middle and early Late Pleistocene hominins.

The adult clavicle ATD6-50 displays a set of quantitative and morphological traits shared with Neanderthals (Carretero et al. 1999). It is absolutely very long, has a low robusticity index, a pronounced shaft curvature, and relatively small epiphyses. These authors consider that ATD6-50 may present the primitive pattern (relatively longer clavicle) and, therefore, this bone is not useful for phylogenetic analysis. The same situation applies for the ATD6-43 radius. This specimen shows an absolutely and relatively long radial neck, a primitive feature that ATD6-43 shares with the Middle and early Late Pleistocene hominins, and differs from the derived condition (short neck length) observed in modern humans. In contrast, the diaphysis of ATD6-43 is long and straight, being more similar in these features to modern humans than to Neanderthals (Carretero et al. 1999). These authors conclude that the large absolute radial length of ATD6-43 suggests a high brachial index for *H. antecessor* and upper limb proportions more similar to *H. ergaster* and modern humans than to Neanderthals. Similarly, the right ATD6-22 and left ATD6-56 patellae are relatively narrow, with high patellar indices,

Table 6.1 Summary of some of the morphological traits that are derived in the TD6 fossil and their presence/absence in the Neanderthal (NEA) and modern humans (MH) lineage

	TD6	NEA	MH
Temporal squama high and arced	+	+	+
"Modern" midfacial morphology	+	–	+
Orientation of nasal lateral walls	+	+	+
Vertical incisive canal	+	+	+
Triangular shovel shape	+	+	–
Upper M1 morphology	+	+	–
Lower P4 morphology	+	+	–
General morphology of the hands	+	–	+
General morphology of the feet	+	–	+

similar to those of modern humans, and well above the Sima de los Huesos and Neanderthal values. Carretero et al. (1999) speculate that the Middle Pleistocene hominins, represented by the Sima de los Huesos sample, and the Neanderthals show the derived condition of an absolutely and relatively wider patellae (Table 6.1).

Discussion and Concluding Remarks

Bermúdez de Castro et al. (1997) hypothesized that *H. antecessor* could represent the last common ancestor to Neanderthals and modern humans. In order to test the hypothesis of a possible phylogenetic relationship between *H. antecessor* and the Neanderthals, we have described in this report some of the most important features of the available TD6 fossil evidence, including data of some new findings in the Aurora Stratum and new studies of the material recovered during the nineties of the last century. The TD6 mandibles are a good evidence of the evolutionary status of *H. antecessor*. They show a generalized morphology, with some plesiomorphic traits, and lack the robusticity that characterizes the African Early and Middle specimens, generally attributed to *H. ergaster*, the Javanese *H. erectus* mandibles, and some European specimens, such as Mauer and Arago 13. The TD6 mandibles lack the features considered apomorphic of the Neanderthal lineage, but there are no evidences against the possibility of an evolutionary continuity between the TD6 hominins and the European Middle Pleistocene populations.

Another important element for discussion is the presence of a modern midfacial topography in the TD6 hominins. Now we have proofs that the modern human face appeared in the Early Pleistocene, since it is present at least in the European fossil record (TD6). Obviously, this evidence points to a phylogenetic relationship between *H. antecessor* and the modern human lineage. In this context, it is necessary to know if the derived Neanderthal face can originate from morphologies similar to those observed in TD6 hominins. Since the specimen ATD6-58 exhibits a reduced expression of the canine fossa, Arsuaga et al. (1999) think that specimens such as AT-404 from Sima de los Huesos and Steinheim could present an intermediate morphological facial pattern between that of theTD6 hominins and that of Neanderthals.

Concerning teeth, most of the dental features of the TD6 hominins are plesiomorphic for the genus *Homo*. TD6 hominins share some sinapomorphies with *H. erectus* and *H. neanderthalensis*, particularly referred to the upper incisors morphology. The differences between the Early and Middle Pleistocene dental samples are strong, but the particular morphologies of the lower second premolars and upper first molars shared between these two groups could be pointing to a possible phylogenetic continuity between *H. antecessor* and *H. neanderthalensis*. Furthermore, Arago hominins show a suite of dental traits that deserves mention. Most Arago permanent teeth are large, especially those of the Arago 13 mandible. In this specimen the crown of the P3 is symmetrical and lacks cingulum and talonid. However, the apical fourth part of the root is divided in two components, MB and BL, each one with a single canal (seen by CT-scan). The crown of the P4 exhibits a well-developed talonid and the apical third of the root is also divided into two components, MB and DL, like in the TD6 sample. Also, the MB component has two root canals, and therefore, this tooth shows root morphology similar to that of Hominid 1 from TD6. On the other hand, Arago 13 and Arago 21 present a clear M1 < M2 size sequence, and the M2 and M3 of Arago 13 are hypo- and mesotaurodonts, respectively. In conclusion, Arago 13 shows a combination of the "Gran Dolina and Sima de los Huesos" dental traits. If the Arago hominins are related to the Sima de los Huesos hominins and both groups can be referred to *H. heidelbergensis* (in the sense of Arsuaga et al. 1999), then the dental evidence could support a phylogenetic link between *H. antecessor* and *H. heidelbergensis/H. neanderthalensis* (but see below).

Although the postcranial evidence from TD6 is limited, it seems that the preserved elements support better a relationship between *H. antecessor* and the modern human lineage than with the Neanderthal lineage.

In sum, from the dental and mandibular evidence, we realize that *H. antecessor* is a species that has preserved a certain number of primitive traits but, at the same time, is clearly derived with regard to *H. ergaster/H. erectus*. In the fragmentary evidence from the neurocranium, we can also perceive this evolution, probably due to a significant increase in the cranial capacity and associated features, such as the convexity of the superior border of the temporal squama. Concerning the middle and lower facial skeleton, it is noteworthy that the appearance of a "sapiens" pattern, which is clearly derived in relation to *H. ergaster/H. erectus* as well. Thus, it seems that *H. antecessor* could represent an event of speciation occurred in the Early Pleistocene very probably from *H. ergaster/H. erectus* or a related species.

An important additional evidence to this discussion is the adult calvaria recovered near Ceprano in Southern Latium, Italy (Ascenzi et al. 1996). Unfortunately, none of the human fossil remains recovered from TD6 is directly comparable with the Ceprano calvaria. Manzi et al. (2001) suggest that Ceprano specimen represents a unique morphological bridge between the clade *H. ergaster/H. erectus* and the African and European Middle Pleistocene hominins. In other words, Ceprano could be a representative of the last common ancestors of Neanderthals and modern humans as well. According to this idea, Manzi et al. (2001) consider the possible attribution of the Ceprano calvaria to *H. antecessor* or, alternatively, to another unnamed species. Mallegni et al. (2003) name the specimen from Ceprano as *Homo cepranensis sp. nov.*, a species related to the African Middle Pleistocene hominins, often referred as *H. rhodesiensis*.

At this point of the discussion it is important to mention that Hublin (2001) and Stringer (2003) considered that the human fossils recovered from Aurora Stratum of the level TD6 could be included in *H. mauritanicus* (considering the priority of the name assigned by C. Arambourg) together with those of Tighenif (Arambourg 1954), Rabat (Thomas and Vallois 1977), Thomas Quarry, and Sidi Abderrahman (Rightmire 1990). Hublin (2001) and Stringer (2003) have not presented a formal study to test their hypothesis which, however, has been tested by Bermúdez de Castro et al. (2007). These authors conclude that the Tighenif hominins, together with other contemporaneous (Thomas Quarry and Oulad Hamida 1), and perhaps later North African specimens (Sidi Abderrahman, Salé, and Rabat [Kebibat]) should be considered as a subspecies of the *H. ergaster* species, i.e. *H. ergaster mauritanicus*, and may be the result of an evolution in isolation in this African area. Thus, the TD6 hominins could belong to an exclusive Eurasian lineage, different from the North African group.

In 1996, Rightmire proposed that *H. heidelbergensis* was the result of an episode of speciation occurred in Africa or western Eurasia during the late Early or early Middle Pleistocene. The subsequent dispersal of *H. heidelbergensis* through Africa, Western Eurasia, and may be East Asia during the Middle Pleistocene gave rise to a wide-ranging species, which overlaps in time with late *H. erectus/H. ergaster*. We could agree with Rightmire's idea of a speciation during the Pleistocene, even though we think that this event may have probably occurred earlier, perhaps around 1 million years ago. As we stated above, *H. heidelbergensis* should be considered, together with *H. neanderthalensis*, as a chrono-species of the same "evolutionary" species. Most European Middle Pleistocene specimens exhibit clear evidences of their relationship with the Neanderthal lineage and, therefore, they cannot be ancestors of modern humans as well. As we have also stated above, the common ancestor of Neanderthals and modern humans should be more primitive and should also lack the specialized features characterizing each of these hominin lineages. Furthermore, the African Middle Pleistocene populations could be assigned to another species (i.e. *H. rhodesiensis*).

From the options presented by Rightmire for the geographic scenario of speciation event, we prefer the area of Western Eurasia. This area represents a true crossroads between Africa, Asia, and Europe, where we could expect to find a more generalized morphology instead of a specialized one. The most parsimonious hypothesis for the origin of the common ancestor to modern humans and Neanderthals would point to a region halfway to Africa (the origin of modern humans) and Europe (the origin of Neanderthals). From the morphological evidence observed in *H. antecessor*, this species would be related in some way to the speciation event occurred in this area during the Early Pleistocene. *H. antecessor* would represent either the true ancestor to the Neanderthal lineage or a dead evolutionary lineage replaced or genetically absorbed during the Middle Pleistocene by another population coming from the "mother area" (Carbonell et al. 2005, and see also Manzi et al. 2001). In this scenario, *Homo antecessor* and Neanderthals would have shared a common ancestor.

Acknowledgments The authors acknowledge the Gran Dolina field team, and in particular Eudald Carbonell, their dedication and effort made during the excavation of the TD6 level. This research was supported by funding from the Dirección General de Investigación of the Spanish Ministerio de Educación y Ciencia (MEC), Project N° CGL2006-13532-C03/BTE, and the Cátedra Atapuerca from the Fundación Atapuerca and Fundación Duques de Soria. Fieldwork at Atapuerca is supported by the Consejería de Cultura y Turismo of the Junta de Castilla y León. Aida Gómez-Robles has the benefit of a predoctoral FPU grant of the Spanish MEC.

References

Arambourg, C. (1954). L'hominien fossile de Ternifine (Algérie). *Comptes Rendus des Séances de l'Academie des Sciences, Paris, 239*, 893–895.

Arsuaga, J. L., Martínez, I., Gracia, A., Carretero, J. M., & Carbonell, E. (1993). Three new human skulls fromthe Sima de los Huesos site in Sierra de Atapuerca, Spain. *Nature, 362*, 534–537.

Arsuaga, J. L., Martínez, I., Gracia, A., & Lorenzo, C. (1997). The Sima de los Huesos crania (Sierra de Atapuerca, Spain). A comparative study. *Journal of Human Evolution, 33*, 219–281.

Arsuaga, J. L., Martínez, I., Lorenzo, C., Gracia, A., Muñoz, A., Alonso, O., & Gallego, J. (1999). The human cranial remains from Gran Dolina Lower Pleistocene site (Sierra de Atapuerca, Spain). *Journal of Human Evolution, 37*, 431–457.

Ascenzi, A., Biddittu, I., Cassoli, P. F., Segre, A. G., & Segre-Naldini, E. (1996). A calvarium of late *Homo erectus* from Ceprano, Italy. *Journal of Human Evolution, 31*, 409–423.

Bermúdez de Castro, J. M. (1993). The Atapuerca dental remains. New evidence (1987–1991 excavations) and interpretations. *Journal of Human Evolution, 24*, 339–371.

Bermúdez de Castro, J. M., Arsuaga, J. L., Carbonell, E., Rosas, A., Martínez, I., & Mosquera, M. (1997). A hominid from the Lower Pleistocene of Atapuerca, Spain: Possible ancestor to Neandertals and modern humans. *Science, 276*, 1392–1395.
Bermúdez de Castro, J. M., Rosas, A., & Nicolás, M. E. (1999). Dental remains from Atapuerca-TD6 (Gran Dolina site, Burgos, Spain). *Journal of Human Evolution, 37*, 523–566.
Bermúdez de Castro, J. M., Martinón-Torres, M., Sarmiento, S., & Lozano, M. (2003). Gran Dolina-TD6 *versus* Sima de los Huesos dental samples from Atapuerca: Evidence of discontinuity in the European Pleistocene population? *Journal of Arcaheological Science, 30*, 1421–1428.
Bermúdez de Castro, J. M., Carbonell, E., Gómez, A., Mateos, A., Martinón-Torres, M., Muela, A., Rodríguez, J., Sarmiento, S., & Varela, S. (2006). Paleodemografía del hipodigma de fósiles de homininos del nivel TD6 de Gran Dolina (Sierra de Atapuerca, Burgos): Estudio preliminar. *Estudios Geológicos, 62*, 145–154.
Bermúdez de Castro, J. M., Martinón-Torres, M., Gómez-Robles, A., Prado, L., & Sarmiento, S. (2007). *Comparative analysis of the Gran Dolina-TD6 (Spain) and Tighennif (Algerie) hominin mandibles*. Bull et Mém de la Soc d´Anthropologie, Paris.
Bischoff, J. L., Williams, R. W., Rosenbauer, R. J., Aramburu, A., Arsuaga, J. L., García, N., & Cuenca-Bescós, G. (2006). High-resolution U-series dates from the Sima de los Huesos hominids yields 600+ -66 kyrs: Implications for the evolution of the early Neanderthal lineage. *Journal of Arcaheological Science, 12*, 763–770.
Boule, M. (1911–1913). L'Homme fossile de La Chapelle-aux-Saints. *Annales de Paléontologie, 6*, 109–172; *7*, 105–192; *8*, 1–62.
Boule, M., & Vallois, H. (1957). *Fossil men*. London: Thames & Hudson.
Carbonell, E., Bermúdez de Castro, J. M., Arsuaga, J. L., Díez, J. C., Rosas, A., Cuenca-Bescós, G., Sala, R., Mosquera, M., & Rodríguez, X. P. (1995). Lower Pleistocene hominids and artifacts from Atapuerca-TD6 (Spain). *Science, 269*, 826–830.
Carbonell, E., Bermúdez de Castro, J. M., Arsuaga, J. L., Allue, E., Bastir, M., Benito, A., Cáceres, I., Canals, T., Díez, J. C., van der Made, J., Mosquera, M., Ollé, A., Pérez-González, A., Rodríguez, J., Rodríguez, X. P., Rosas, A., Rosell, J., Sala, R., Vallverdú, J., & Vergés, J. M. (2005). An Early Pleistocene hominin mandible from Atapuerca-TD6, Spain. *Proceedings of the National Academy of Sciences of the United States of America, 102*, 5674–5678.
Carretero, J. M., Lorenzo, C., & Arsuaga, J. L. (1999). Axial and appendicular skeleton of *Homo antecessor*. *Journal of Human Evolution, 37*, 459–499.
Cuenca-Bescós, G., Laplana-Conesa, C., & Canudo, J. I. (1999). Biochronological implications of the Arvicolidae (Rodentia, Mammalia) from the Lower Pleistocene hominid-bearing level of Trinchera Dolina 6 (TD6, Atapuerca, Spain). *Journal of Human Evolution, 37*, 353–373.
Falguères, C., Bahain, J.-J., Yokoyama, Y., Arsuaga, J. L., Bermúdez de Castro, J. M., Carbonell, E., Bischoff, J. L., & Dolo, J.-M. (1999). Earliest humans in Europe: The age of TD6 Gran Dolina, Atapuerca, Spain. *Journal of Human Evolution, 37*, 343–352.
García, N., & Arsuaga, J. L. (1999). Carnivores from the early Pleistocene hominid-bearing Trinchera Dolina 6 (Sierra de Atapuerca, Spain). *Journal of Human Evolution, 37*, 415–430.
Gómez-Robles, A., Martinón-Torres, M. B., de Castro, J. M., Margvelashvili, A., Bastir, M., Arsuaga, J. L., Pérez-Pérez, A., Estebaranz, F., & Martínez, L. (2007). A geometric morphometric analysis of hominin upper first molar shape. *Journal of Human Evolution, 53*, 272–285.
Green, R. E., Krause, J., Ptak, S. E., Briggs, A. W., Ronan, M. T., Simons, J. F., Du, L., Egholm, M., Rothberg, J. M., Paunovic, M., & Pääbo, S. (2006). Analysis of one million base pairs of Neanderthal DNA. *Nature, 444*, 330–336.
Hambach, U. (1996). Paläo-und gesteinsmagnetisch Untersuchungen im Quartär der Grube Grafenrain: Fundplatz des *Homo erectus heidelbergensis*. *Mannheimer Geschichtsblätter, NF, Beiheft 1*, 41–46.
Howell, F. C. (1951). The place of Neanderthal man in human evolution. *American Journal of Physical Anthropology, 9*, 379–415.
Hublin, J.-J. (1990). Les peuplements paléolitiques de l'Europe: Un point de vue paleobiogéographique. *Mémoires du Musée de Préhistoire D'Ile-de-France, 3*, 29–37.
Hublin, J.-J. (2001). Northwestern African Middle Pleistocene hominids and their bearing on the emergence of *Homo sapiens*. In L. Barham & K. Robson-Brown (Eds.), *Human roots: Africa and Asia in the Middle Pleistocene* (pp. 99–121). Bristol: Western Academic and Specialist Press.
Le Gros Clark, W. E. (1955). *The fossil evidence for human evolution (an introduction to the study of Paleoanthropology)*. Chicago: Chicago University Press.
Lorenzo, C., Arsuaga, J. L., & Carretero, J. M. (1999). Hand and foot remains from the Gran Dolina Early Pleistocene site (Sierra de Atapuerca, Spain). *Journal of Human Evolution, 37*, 501–522.
Mallegni, F., Carnieri, E., Bisconti, M., Tartarelli, G., Ricci, S., Biddittu, I., & Segre, A. (2003). *Homo cepranensis sp. nov.* and the evolution of African-European Middle Pleistocene hominids. *Comptes Rendus Palevol, 2*, 153–159.
Manzi, G., Mallegni, F., & Ascenzi, A. (2001). A cranium for the earliest Europeans: Phylogenetic position of the hominid from Ceprano Italy. *Proceedings of the National Academy of Sciences of the United States of America, 98*, 10011–10016.
Martinón-Torres, M. (2006). *Evolución del aparato dental en homínidos: estudio de los dientes humanos del Pleistoceno de la Sierra de Atapuerca.* Ph.D. dissertation, Universidad de Santiago de Compostela, Galicia.
Martinón-Torres, M., Bastir, M., Bermúdez de Castro, J. M., Gómez, A., Sarmiento, S., Muela, A., & Arsuaga, J. L. (2006). Hominin lower second premolar morphology: Evolutionary inferences through geometric morphometric analysis. *Journal of Human Evolution, 50*, 523–533.
Martinón-Torres, M., Bermúdez de Castro, J. M., Gómez, A., Bastir, M., Sarmiento, S., Muela, A., Arsuaga J. L. (2007). Gran Dolina-TD6 and Sima de los Huesos dental samples: Preliminary approach to some dental traits of interest for phylogenetic studies. In S. Bailey & J.-J. Hublin (Eds.), *Dental persepectives on human evolution* (pp. 65–79). Berlin: Springer.
Noonan, J. P., Coop, G., Kudaravalli, S., Smith, D., Krause, J., Alessi, J., Chen, F., Platt, D., Pääbo, S., Pritchard, J. K., & Rubin, E. M. (2006). Sequencing and analysis of Neanderthal genomic DNA. *Science, 314*, 1113–1118.
Parés, J. M., & Pérez-González, A. (1995). Paleomagnetic age for Hominid fossils at Atapuerca site, Spain. *Science, 269*, 830–832.
Parés, J. M., & Pérez-González, A. (1999). Magnetochronology and stratigraphy at Gran Dolina section, Atapuerca (Burgos, Spain). *Journal of Human Evolution, 37*, 325–342.
Rak, Y. (1986). The Neanderthal: A new look at an old face. *Journal of Human Evolution, 15*, 151–164.
Rightmire, G. P. (1988). *Homo erectus* and later Middle Pleistocene humans. *Annals Review Anthropology, 17*, 239–259.
Rightmire, G. P. (1990). *The evolution of Homo erectus, comparative anatomical studies of an extinct human species*. New York: Cambridge University Press.
Rightmire, G. P. (1996). The human cranium from Bodo, Ethiopia: Evidence for speciation in the Middle Pleistocene? *Journal of Human Evolution, 31*, 21–39.
Rightmire, G. P. (1998a). Human evolution in the Middle Pleistocene: The role of *Homo heidelbergensis*. *Evolutionary Anthropology, 6*, 218–227.

Rightmire, G. P. (1998b). Evidence from facial morphology for similarity of Asian and African representatives of *Homo erectus*. *American Journal of Physical Anthropology, 106*, 61–85.

Rosas, A. (2001). Occurrence of Neanderthal features in mandibles from the Atapuerca-SH site. *American Journal of Physical Anthropology, 114*, 74–91.

Rosas, A., & Bermúdez de Castro, J. M. (1998). The Mauer mandible and the evolutionary significance of *Homo heidelbergensis*. *Geobios, 31*, 687–697.

Rosas, A., & Bermúdez de Castro, J. M. (1999). The ATD6-5 mandibular specimen from Gran Dolina (Atapuerca, Spain). Morphological study and phylogenetic implications. *Journal of Human Evolution, 37*, 567–590.

Schoetensack, O. (1908). *Der Unterkiefer des Homo heidelbergensis aus des Sanden von Mauer bei Heidelberg. Ein Beitrag zur Paläontologie des menschen*. Leipzig: Engelmann.

Schwartz, J. H., & Tattersall, I. (1996). Significance of some previously unrecognized apomorphies in the nasal region of *Homo neanderthalensis*. *Proceedings of the National Academy of Sciences of the United States of America, 93*, 10852–10854.

Stringer, C. B. (1985). Middle Pleistocene hominid variability and the origin of Late Pleistocene humans. In E. Delson (Ed.), *Ancestors: The hard evidence* (pp. 289–295). New York: Alan R. Liss.

Stringer, C. B. (1993a). Secret of the pit of the bones. *Nature, 362*, 501–502.

Stringer, C. B. (1993b). New views on modern human origins. In D. T. Rasmussen (Ed.), *The origin and evolution of humans and humanness* (pp. 75–94). Boston: Jones and Bartlett.

Stringer, C. B. (2003). Out of Ethiopia. *Nature, 423*, 692–694.

Stringer, C. B., & McKie, R. (1996). *African exodus. The origin of modern humanity*. London: Jonathan Cape.

Tattersall, I. (1996). *The last Neanderthal*. New York: McMillan.

Thomas, A., & Vallois, H. V. (1977). Les dents de l'Homme de Rabat. *Bull et Mém de la Soc d'Anthropologie de Paris, 4*, 31–58.

van der Made, J. (1999). Ungulates from Atapuerca TD6. *Journal of Human Evolution, 37*, 389–413.

Chapter 7
The Hominid Fossils from China Contemporaneous with the Neanderthals and Some Related Studies

Wu Liu and Xiujie Wu

Abstract Among the more than 70 hominid fossil sites so far found in China, those of the Middle and Late Pleistocene are the most frequent. These Chinese hominids are contemporaneous to the Neanderthals. For the past decade, more attention has been paid to the field of Middle and Late Pleistocene human evolution in Chinese paleoanthropological studies, which resulted in a series of new hominid fossil finds and further understanding of human evolution in China. In this chapter, we briefly review the main hominid fossils found in China which are contemporaneous to the Neanderthals, and we report the research advances achieved in recent years including new hominid fossil sites and related studies.

Keywords Middle Pleistocene • Late Pleistocene • Human evolution • New sites

The Main Hominid Fossils Contemporaneous to the Neanderthals

Since the initial discovery of hominid fossils at the beginning of the twentieth century, more than 70 hominid fossil sites have been found in China. These fossils have been attributed to either *Homo erectus* or *Homo sapiens*. Most of these hominid fossils were found in Middle to Late Pleistocene deposits, and their chronological ages are approximately contemporaneous to those of the Neanderthals and their lineage. Figure 7.1 and Table 7.1 list some important Middle and Late Pleistocene hominid fossils found in China, respectively.

In this chapter, some of the hominid fossils listed in Table 7.1 are briefly described, either because of their well-preserved condition or because there have been very few studies on them since their discovery.

W. Liu (✉) and X. Wu
Institute of Vertebrate Paleontology and Paleoanthropology, Chinese Academy of Sciences, Beijing 100044, China
e-mail: liuwu@ivpp.ac.cn; wuxiujie@ivpp.ac.cn

Hexian

The hominid fossils found in Hexian, Anhui Province, include one skull cap, two cranial fragments, one mandible, and some isolated teeth (Fig. 7.2). These fossils were found in 1980 and 1981. Since then, a few papers were published giving simple descriptions of the fossils (Wu and Dong 1982; Wu 1983; Huang et al. 1982). These studies put the Hexian fossils into *Homo erectus*. But both the morphology and chronological age of the Hexian fossils pointed to a difference between Hexian and other *Homo erectus* fossils found in China. Recently, some new studies have been carried on Hexian fossils by Chinese colleagues, which address the morphological variation of *Homo erectus* in China and the endocast features of Hexian (Liu and Zhang 2004; Wu et al. 2006b).

Xujiayao

The Xujiayao site is located in Shanxi Province. The three excavations of Xujiayao in 1976, 1977, and 1979 respectively unearthed 20 hominid fossils including 12 pieces of parietal bones, 1 temporal bone, 2 pieces of occipital bone, 1 mandibular fragment, 1 child's maxilla, and 3 isolated teeth. Figure 7.3 displays some cranial fragments of the Xujiayao hominids. The chronological age of 125–104 ka makes these hominid fossils of great value to research on the origin of modern Chinese. Till now, only a few site reports with very simple descriptions of the hominid fossils have been made (Wu 1980, 1986).

Chaoxian

The hominid fossils found in Chaoxian, Anhui Province, include an occipital fragment, a maxillary fragment with both lateral incisors and right P^3-M^1, and three isolated teeth

S. Condemi and G.-C. Weniger (eds.), *Continuity and Discontinuity in the Peopling of Europe: One Hundred Fifty Years of Neanderthal Study*, Vertebrate Paleobiology and Paleoanthropology, DOI 10.1007/978-94-007-0492-3_7,

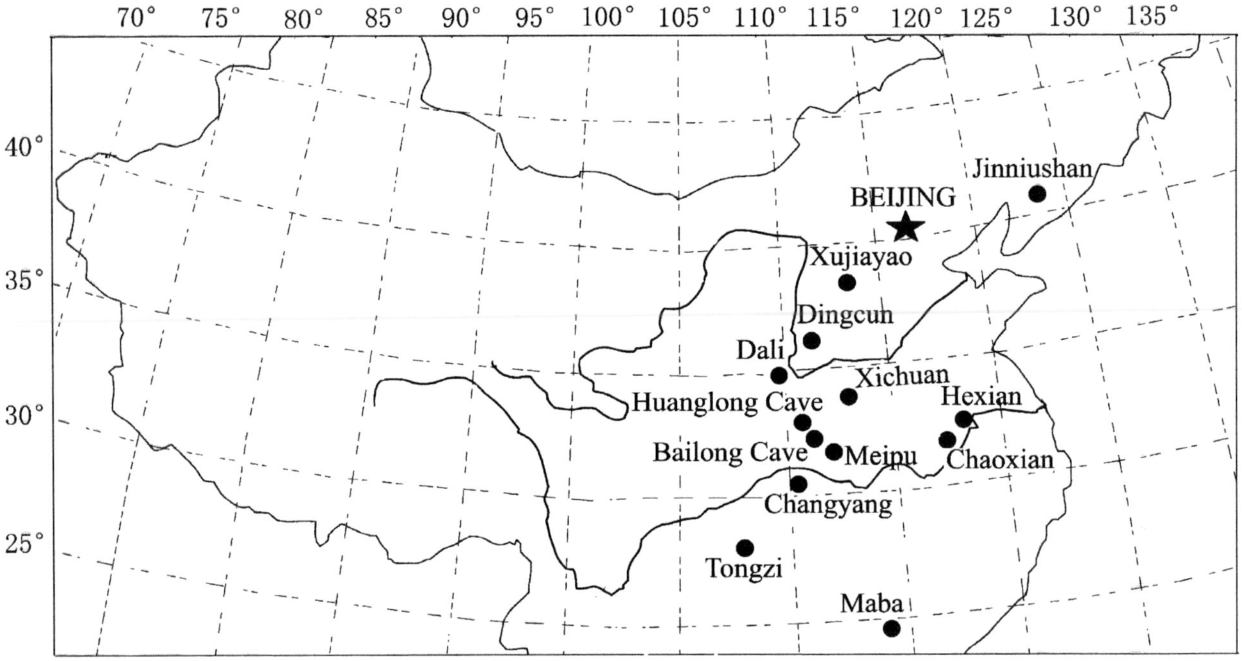

Fig. 7.1 The main hominid fossil sites in China contemporaneous to the Neanderthals

Table 7.1 The main human fossils of China contemporaneous with the Neanderthals

Sites	Main specimens	Geological epoch	Chronological dates
Hexian	1 skull cap, 2 cranial fragments, 1 mandible, and 9 isolated teeth	Middle Pleistocene	270–150 ka
Dali	1 cranium	Middle Pleistocene	209 ka
Jinniushan	1 cranium, 6 vertebrae, os coxae, 1 ulna	Middle Pleistocene	280 ka
Maba	1 skull-cap	Middle Pleistocene	135–129 ka
Xujiayao	15 cranial fragments, 2 jaw bones, and 3 isolated teeth	Late Pleistocene	125–104 ka
Dingcun	3 teeth; 1 parietal	Middle or Late Pleistocene	210–160 ka
Tongzi	6 teeth	Middle or Late Pleistocene	181–113 ka
Chaoxian	1 occipital; 1 maxilla with left I^2, P^2-M^2 and right I^2, P^1-M^1	Middle Pleistocene	200–160 ka
Changyang	Left maxilla with P^1 and M^1; isolated P_2	Middle Pleistocene	–
Meipu, Yunxian	4 isolated teeth	Middle Pleistocene	–
Xichuan	13 isolated teeth	Middle Pleistocene	–
Huanglong Cave	7 teeth	Late Pleistocene	103–94 ka
Loc 4, ZKD	1 premolar	Middle Pleistocene	250–110 ka
Bailong Cave	4 isolated teeth	Middle Pleistocene	–
Dadong, Panxian	3 isolated teeth	Middle Pleistocene	260-130 ka

(left P^4-M^2) (Fig. 7.4c). The Chaoxian site is only 50 km from Hexian, and the chronological age for the deposit yielding the hominid fossils is 200–160 ka, which have led some colleagues to propose that the overlapping of the time ranges between Hexian and Chaoxian suggest *Homo erectus* and archaic *Homo sapiens* may have coexisted in China (Chen and Zhang 1991). There was also a study on the tooth wear and tooth use of Chaoxian hominids (Zhang 1989).

Maba

The hominid fossil found at Maba in Guandong Province, south China, is only a skull cap composed of several fragments (Fig. 7.4a). After reconstruction, the Maba specimen contains the right orbital region and most of the skull cap including frontal, temporal, and occipital bones. Since Wu Rukang (Woo and Peng 1959) described the morphology of

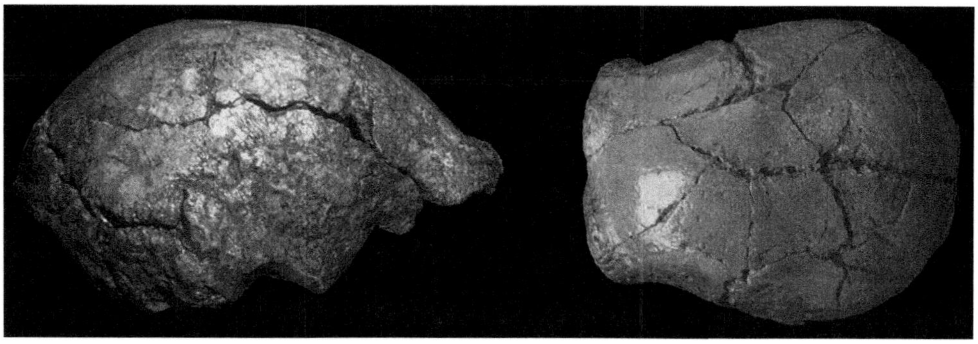

Fig. 7.2 Hominid cranium found in Hexian, Anhui Province of China

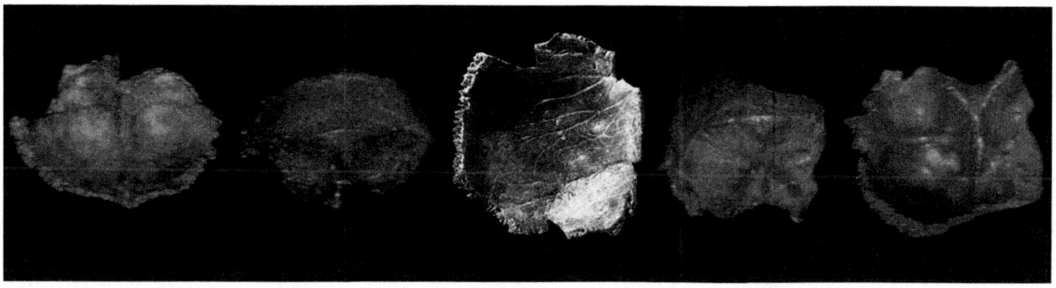

Fig. 7.3 Hominid fossils found in Xujiayao, Shanxi Province of China

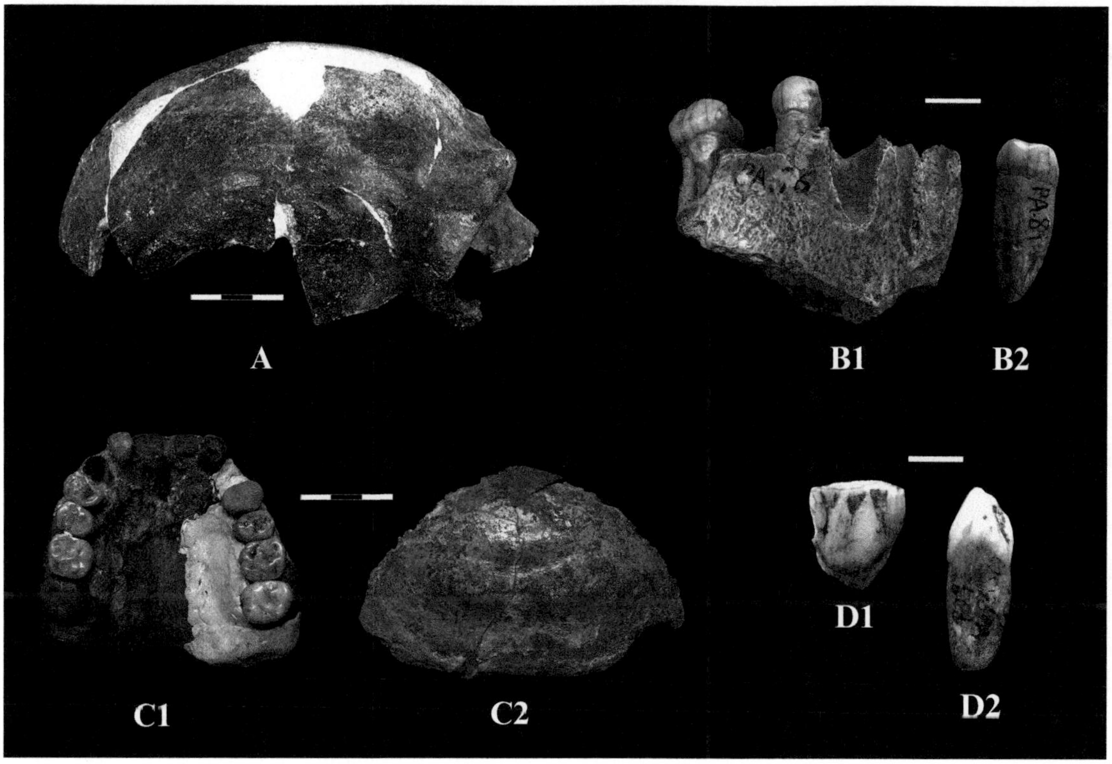

Fig. 7.4 Hominid fossils found in Maba (**a**), Changyang (**b**), Chaoxian (**c**), and Panxian (**d**)

Maba, no further specific study has been done on it. For the past decade, the main interest in Maba has been focused on its orbital shape. Wu Xinzhi (2004a, b) believes that the round-shaped orbit of Maba resembles that of European Neanderthals and differs from other Chinese hominids, suggesting gene flow between the two lineages (see the following section).

Xichuan

The 13 isolated hominid teeth (Fig. 7.5) collected from traditional Chinese medicine drug stores in Nanyang County and Xixia County, Henan Province, in 1973 supposedly came from Xichuan County, Henan Province. Because the teeth were not from excavations, there is no chronological age for them. According to Wu Rukang and Wu Xinzhi (1982), most of the teeth resemble those of *Homo erectus* morphologically. Because that was the only morphological description of these teeth, further studies are needed to clarify the morphological features of the Xichuan fossils and their relationship with other Middle and Late Pleistocene hominids in China.

Hominid Fossil Sites Recently Found in China

The research on Late Pleistocene human evolution in China has been playing an important role in the field of modern human origins not only for East Asia but around the world. Although many Late Pleistocene hominid fossils have been found in China, there are few dating between 100 and 50 ka, causing a big "fossil gap". For this reason, in recent years we have organized a series of field surveys and excavations in

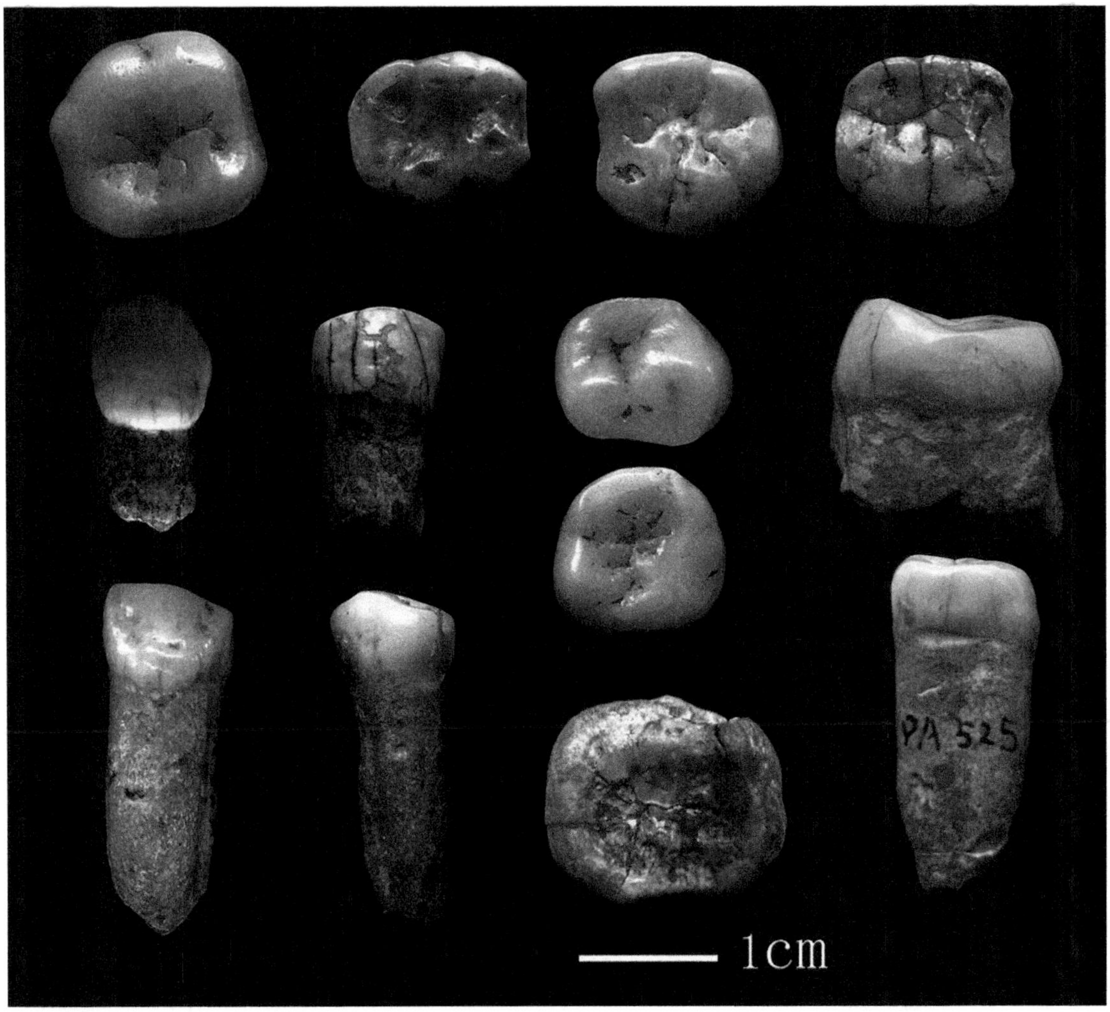

Fig. 7.5 Hominid teeth found in Xichuan, Henan Province

China trying to find more hominid fossils in this time period. In the past 5 years, several new Late Pleistocene human fossil sites have been found in China. Among them, one site near Zhoukoudian (ZKD) and four others in the West Hubei and Three Gorge region are the most important.

Tianyuan Cave

The Tianyuan Cave is located about 6 km southwest of the Zhoukoudian "Peking Man" site near Beijing. The cave was discovered in 2001, and excavations were carried out in 2003 and 2004, yielding both hominid and mammal fossils. The hominid fossils comprise 34 specimens including a mandible, teeth, and postcranial bones. The mammal fossils represent 29 species. No definite stone artifacts or other cultural remains have been found,but a great number of heavily fragmented bone flakes unearthed here suggest possible human activities. Preliminary analyses indicate that the fauna of the Tianyuan Cave is most similar to that of the ZKD Upper Cave and somewhat less close to the living fauna. The geological age should be Late Pleistocene. The results of absolute dating by several methods yielded an age of 42–39 ka, approximately the same age as that of Upper Cave (Tong et al. 2004; Shang et al. 2007). Detailed studies of the hominid fossils are underway.

West Hubei and Three Gorge Regions

The west Hubei and Three Gorge regions is a narrow area across the western part of Hubei Province with the Three Gorge area intermediate (Fig. 7.6). Since the first discovery of hominid fossils (a maxilla fragment and an isolated tooth) in Changyang County of this area in the 1950s (Chia 1957), 10 hominid fossil sites have been found (Liu et al. 2006d), including the Yunxian *Homo erectus* and Longgupo sites (Li and Etler 1992; Huang et al. 1995). In addition to these hominid fossil sites, more than 30 sites with stone artifacts and other evidence showing human activities have also been located. Since 2000, our field surveys have found four new Late Pleistocene human fossil sites, and some preliminary excavations have been carried out. All four sites have yielded hominid fossils, stone artifacts and other mammalian fossils.

Among the four new hominid fossil sites in the West Hubei and Three Gorge region, the Huanglong Cave is most important. The Huanglong Cave is located in Yunxi County, which is in the northwest of Hubei Province. From 2004 to 2006, three excavations were conducted at the Huanglong Cave. From these excavations, we found 7 hominid teeth (Fig. 7.7), some stone and bone tools, as well as nearly 3,000 mammal fossils. Preliminary dating analysis (U-series and ESR) indicates that the age of the human teeth is around 100 ka (Wu et al. 2006a).

Three other hominid sites, named Xinglong Cave, Leiping Cave, and Migong Cave, were found in the Three Gorge region within the Chongqing Municipality area (Liu et al. 2006d). Xinglong Cave is located in Fengjie County. In 2001, some mammal fossils were found in Xinglong Cave. The subsequent excavation found four human teeth, stone artifacts, ivory engravings, and other cultural remains. Based on biostratigraphic analysis and uranium series dating, the cave was dated about 100 ka (Gao et al. 2004). The second site recently found in the Three Gorge region is Leiping Cave, which is located in Wushan County. In 2004, while digging sand in the Leiping Cave, local farmers found a piece of human cranium and some mammal fossils. Then, our colleagues made a short excavation there unearthing more hominid fossils, stone tools, and mammal fossils. The hominid fossils found in the Leiping Cave include several cranial fragments (frontal, parietal, left and right temporal, and occipital bones), one upper incisor, and several limb bone fragments. According to the analysis of stratigraphy and faunal composition, we assume that the hominid probably lived in the early Late Pleistocene. The third new hominid fossil site, Migong Cave, is also located in Wushan County. Two hominid parietal fragments were found in 1999 and 2000, respectively. No formal excavation has been carried out in this site.

The field works in both past decades and recent years indicate that the West Hubei and Three Gorge regions are rich in hominid fossil and related materials. Some new sites from recent years' field works further make clear that hominids lived in these regions in the Late Pleistocene. For many years, it has been generally believed that the ages of most of the Late Pleistocene hominid fossils found in China are not earlier than 50 kyr BP. Although some of the them have been thought as early as more than 50 kyr BP, nearly all the dating is in debates because of either the unclear stratigraphic layer yielding the human fossils or methodology. Some colleagues even questioned the existence of the hominids around 100 ka in China. So, the hominid fossils and other related materials from Huanglong Cave and other sites will provide important information for research on the origin of modern Chinese.

Recent Studies on Middle and Late Pleistocene Human Evolution in China

For the past decade, with the advances in the research on modern human origins in East Asia, more attention has been paid to studies of Middle and Late Pleistocene human

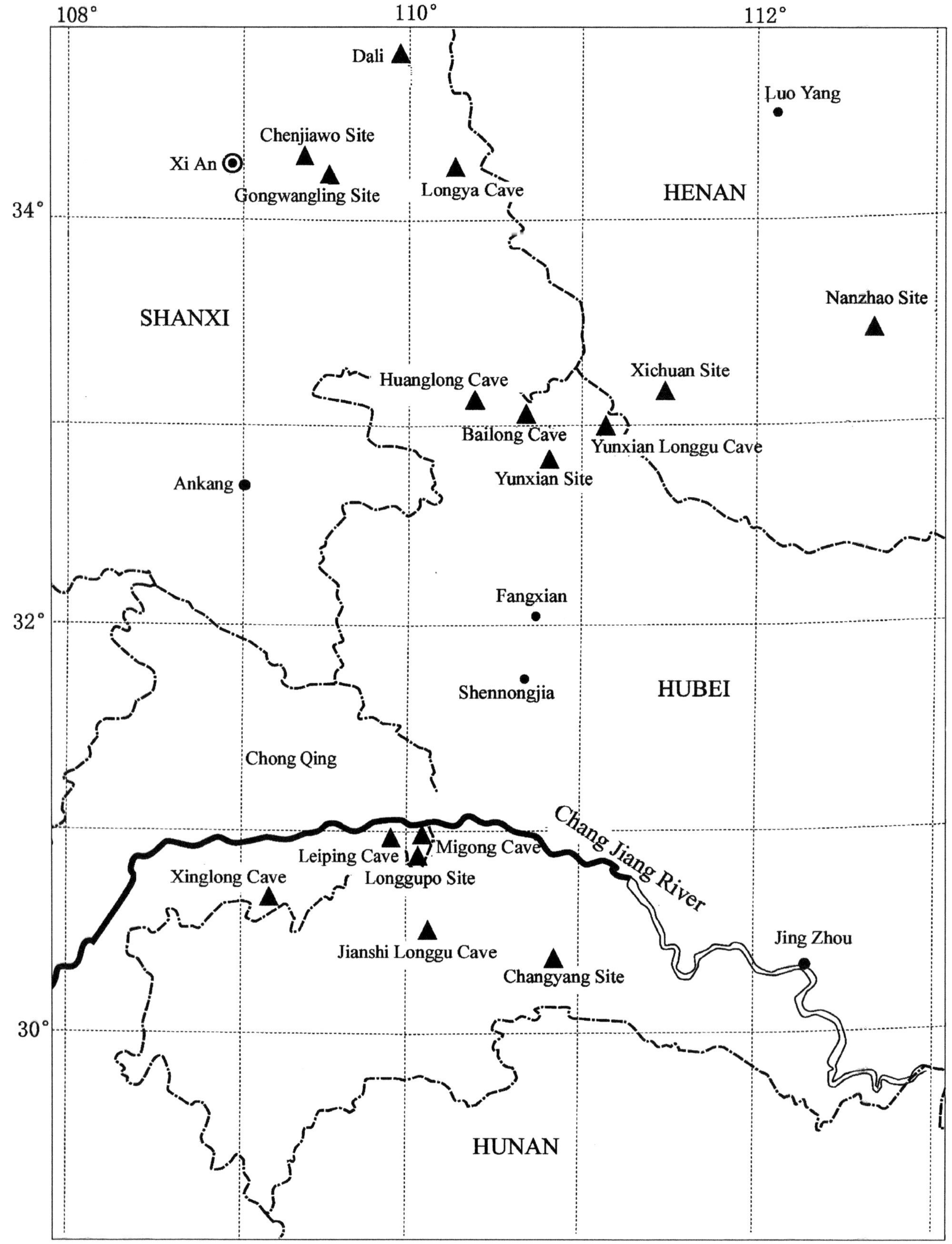

Fig. 7.6 The hominid fossil sites in the West Hubei and Three Gorge regions

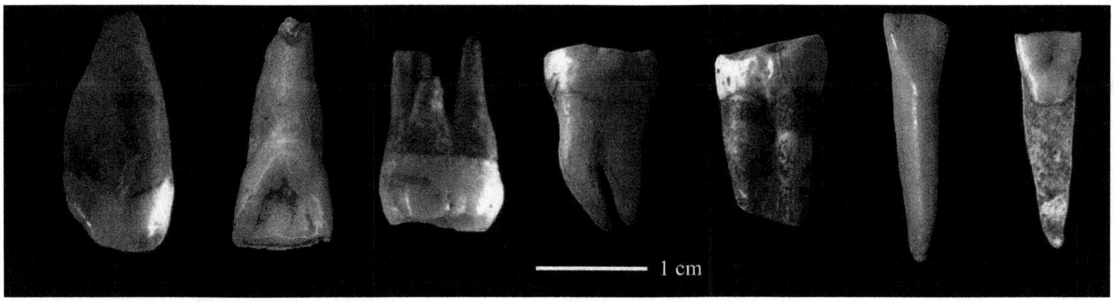

Fig. 7.7 The hominid teeth found in the Huanglong Cave in Hubei Province

evolution in China. A series of studies carried out by Chinese colleagues have examined different aspects of Middle and Late Pleistocene hominid fossils found in China, including their morphology, geographical variation, temporal change, and relationship with hominids from other parts of the world. Some of their results are described below.

Morphological Features on the Chinese Hominid Fossils Suggesting Gene Flow Between Middle-Late Pleistocene Hominids in China and Neanderthals

The origin of modern Chinese has been the key research question for the past two decades in China. As part of this topic, there have been some comparative studies on the cranial morphology between China and European Neanderthals. According to these studies, Wu Xinzhi (1998, 2004a, b) proposed the continuity of human evolution in China as evidenced by a group of common morphological features like shovel-shaped incisors and flatness of the face. There is also a morphological mosaic between *Homo erectus* and archaic *Homo sapiens* in China, adding further support for the regional continuity hypothesis. Besides, Wu Xinzhi's studies show that a few features commonly seen in the Neanderthal lineage can be identified on some Chinese hominid fossil skulls, probably suggesting gene flow between the Middle-Late Pleistocene hominids in China and the Neanderthals in Europe. Based on them, a so-called "continuity with hybridization" model for human evolution in China was proposed (Wu 1998).

The morphological features selected as evidence of gene flow between Chinese hominids and Neanderthals include: (1) protruding nasal saddle of cranium No. 2 from Yunxian and cranium No. 1 from Nanjing; (2) circular orbit and sharp inferolateral orbital margin of the Maba cranium; (3) the surface bulge between the piriform aperture and orbit in the Dali cranium and cranium No. 1 of Nanjing; (4) the chignonlike structure of the occipital region (bunning) on the crania of Ziyang, Liujiang, and Lijiang; and (5) more lateral orientation of the anterolateral surface of the frontosphenoidal process of the zygomatic bone in Upper Cave cranium No. 102 (see Fig. 7.8). Wu Xinzhi argues that in the Pleistocene these features are rare in China, but they are more frequent in Africa and Europe, especially in the Neanderthal lineage. The most reasonable explanation for their occurrence in Pleistocene China is that they are due to small amounts of intermittent gene flow from Europe.

However, these opinions and related studies are not widely accepted and even questioned by Chinese colleagues. Recently, some Chinese colleagues including the present authors conducted some comparative studies trying to investigate the morphological basis for gene flow between Chinese hominids and Neanderthals. Our preliminary results do not give support for a Neanderthal influence on Chinese hominids at least for some proposed morphological features (Liu et al. 2003; Zhang et al. 2004). Our studies indicate that the morphological features listed above are not strong enough to serve as evidence to support gene flow between Chinese Middle and Late Pleistocene hominids and European Neanderthals. The main problems for these five features is they are either atypical (like occipital bunning), or not exactly the same as Neanderthal features (Trinkaus 2006). Also, until now, there is no archaeological evidence to support human migration between east Asia and Europe in the Middle Pleistocene.

Studies on Nanjing No. 1 and No. 2 Hominid Fossils

In 1993 two fragmentary hominid crania were found in a cave site near Nanjing (Jiangsu Province) in east China and named Nanjing No. 1 and No. 2. The initial study (Wu and Li 2002) indicated morphological resemblance between the Nanjing and Zhoukoudian specimens, and both Nanjing No. 1 and No. 2 were put into *Homo erectus*. More recently, some further studies have been carried out on Nanjing hominid fossils. These studies involve different aspects of the Nanjing No. 1 hominid fossil including the nasal and facial

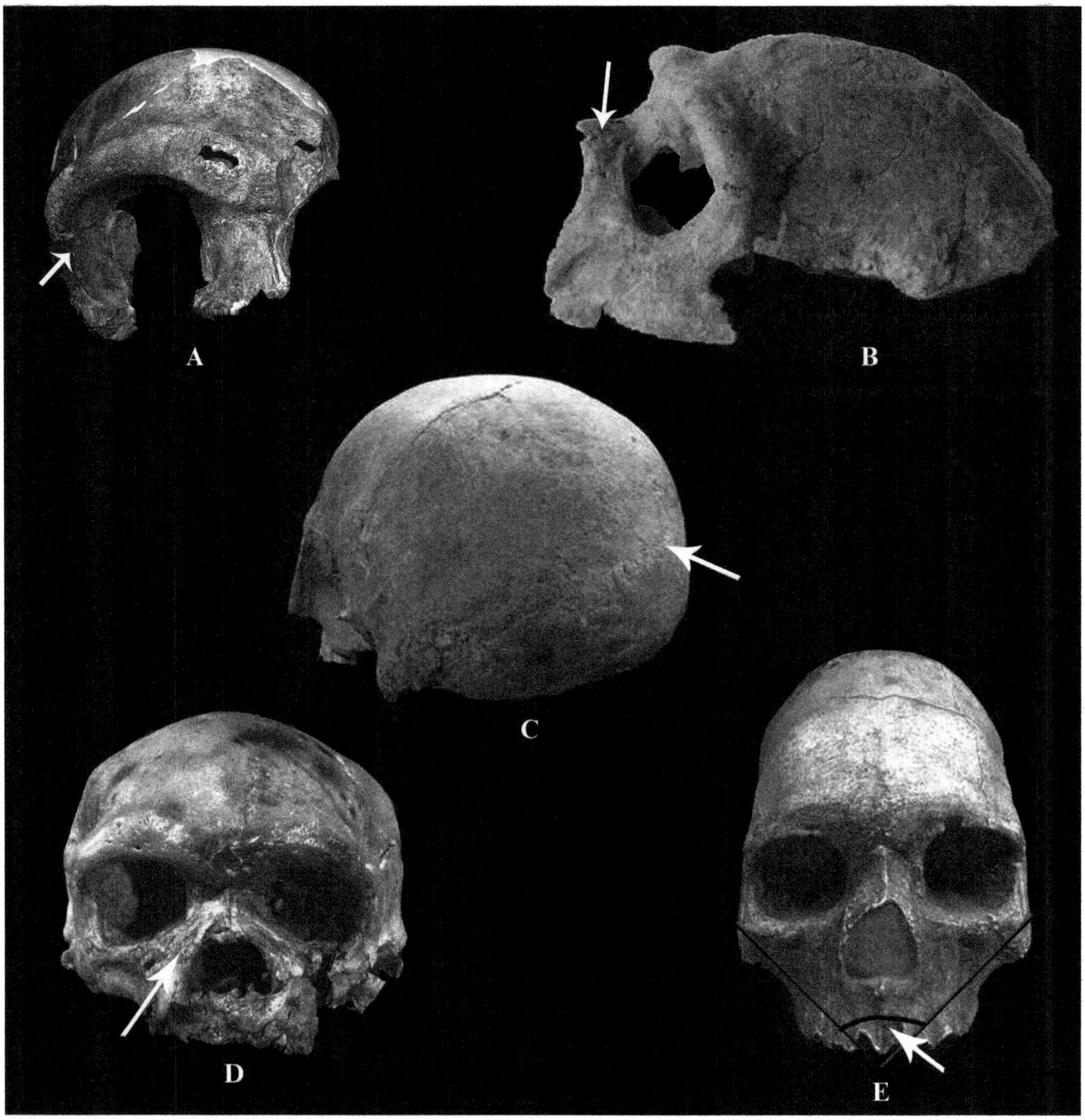

Fig. 7.8 Some Neanderthal features on the Chinese hominid fossils: (**a**) circular orbit and sharp inferolateral orbital margin of Maba; (**b**) protruding nasal saddle of Nanjing No. 1; (**c**) occipital bunning of Liujiang; (**d**) surface bulge between the piriform aperture and orbit of Dali; (**e**) more lateral orientation of anterolateral surface of the frontosphenoidal process of the zygomatic bone in UC 102

morphology (Zhang et al. 2004; Zhang and Liu 2005) and comparison of cranial morphology with *Homo erectus* from Eurasia and Africa (Liu et al. 2005). We also made a new reconstruction of the Nanjing No. 2 cranium, and its morphological features were studied (Zhang and Liu 2006).

As mentioned in the previous section, the highly projecting nasal bones of Nanjing No.1 have been believed to be evidence of gene flow from Europe. To further clarify this question, we studied the nasal morphology of Nanjing No. 1 and discussed the possibility of gene flow from European fossil hominids. Our observations show that highly projecting nasal bones have not been found in the crania from Africa, Europe, and West Asia during the time period of Nanjing *Homo erectus* or before. This feature appeared much later in Africa and Europe than in Asia. The available fossil evidence cannot support the western affinities of *Homo erectus* from Nanjing.

Nanjing No. 2 cranium has been believed to represent *Homo erectus*. However, its parietal and occipital bones were cracked into several fragments which were displaced from their normal positions. Recently, we made a new restoration and reconstruction of the Nanjing No. 2 cranium, which provides more anatomical details. Our study indicates that compared with *Homo erectus*, Nanjing No. 2 has larger parietal bones, relatively narrow upper squamous region of the occipital bone, and a probably larger cranial capacity. All these features suggest affinities with archaic *Homo sapiens*. Our statistical analyses of metric data further show the closer affinity of Nanjing No. 2 to archaic *Homo sapiens* than to *Homo erectus* from both China and Indonesia. The results of the study indicate that there must be some uncertainty in referring Nanjing No. 2 to *Homo erectus*; it is more likely that it belongs to *Homo sapiens* (*sensu lato*).

In addition to these studies of specific morphological features of Nanjing No. 1 and No. 2, further comparative analyses with hominids from both Africa and Eurasia were also carried out (Liu et al. 2005). Our morphological comparisons and metric analyses show that Nanjing No. 1 cranium shares typical *Homo erectus* features with African and European counterparts, demonstrating that *Homo erectus* is a widely distributed lineage that evolved during the million years after its Late Pliocene origins. The differences between Nanjing No.1 and Zhoukoudian suggest a certain level of regional variation in East Asian *Homo erectus*. Our detailed cranial morphological comparative study between Nanjing No. 1 and KNM-ER 3733 (Zhang and Liu 2006) indicate that even though Nanjing No. 1 is about 1 Myr later than KNM-ER 3733 in age, most of the cranial features of Nanjing No. 1 resemble those of KNM-ER 3733. The morphological resemblances between Nanjing No. 1 and KNM-ER 3733 are mainly in the calvaria region, whereas most features in the facial region differ between the two crania. We believe that the morphological similarity in the calvariae is diagnostic and supports the claim to refer KNM-ER 3733 to *Homo erectus*. In addition, the similarity suggests stability of some members of *Homo erectus* in the morphology of calvaria over a span of at least a million years. Unfortunately, the significance of the differences in facial skeleton of these crania is still uncertain.

Brain Evolution: Studies of Hominid Endocasts

For the past 5 years, we have conducted a series of studies on the brain evolution of Chinese hominids (Wu et al. 2004; Wu et al. 2006b). Our current studies mainly focus on endocasts of fossil hominids from Zhoukoudian and Hexian (Fig. 7.9). For the Zhoukoudian endocasts, we investigated the detailed history of the reconstruction and studies of the endocasts for this important site. Their gross morphology and metrics were also described (Wu et al. 2004). Based on these analyses, we further discussed future studies on Zhoukoudian endocasts. In addition. we reconstructed the endocasts of Hexian and Maba from their original fossil crania. The morphological features and metric data for the Hexian endocast have been studied (Wu et al. 2006b). We found that Hexian has many morphological features in common with ZKD. Metrical analyses show that the brain height, frontal breadth, cerebral height, frontal height, and parietal chord increased from *Homo erectus* to modern humans, while the length, breadth, frontal chord, and occipital breadth did not change substantially. We have made an endocast of Maba and plan to study it soon.

Some Recent Studies on Late Pleistocene Human Evolution in China

Among the Late Pleistocene hominid fossils so far found in China, the hominid crania and postcranial remains found from Upper Cave and Liujiang are the most complete and well preserved. The original studies (Weidenreich 1939;

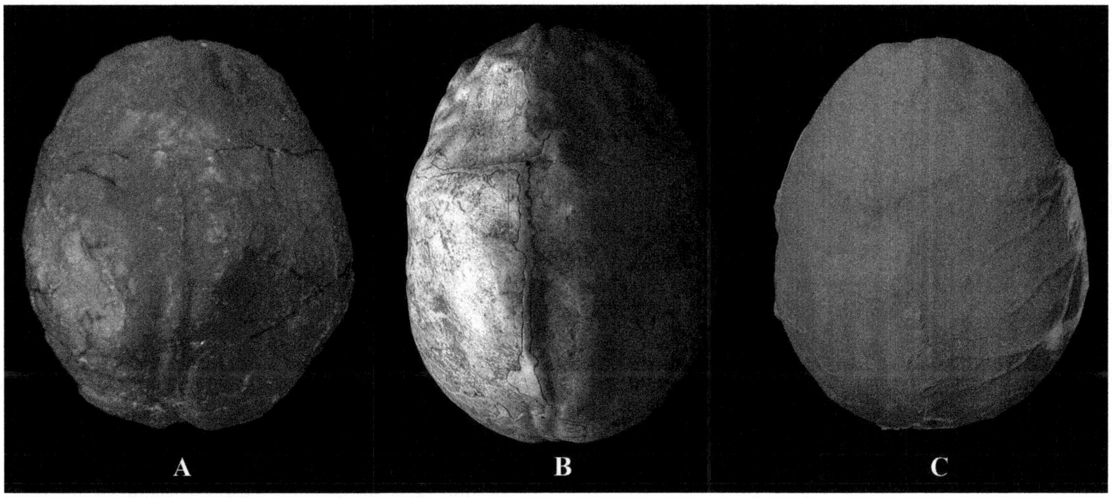

Fig. 7.9 Endocasts of hominids from Hexian (**a**), ZKD (**b**), and Maba (**c**)

Woo 1959; Wu 1960, 1961) indicated that both Upper Cave and Liujiang fossils preserve some primitive features of Late Pleistocene humans, but a group of modern Mongoloid features can also be identified. They have been regarded as an early type of ancestral Mongoloids.

Recently, great advances have been achieved in the study of Late Pleistocene human evolution which led to some new opinions on morphology and related evolutionary problems for Upper Cave and Liujiang. With such a background, we conducted new studies on these specimens (Liu et al. 2006a, b, c; Shang et al. 2006). Our studies cover several aspects related to the Upper Cave, Liujiang, and other problems of Late Pleistocene human evolution in China. Our comparative analyses of cranial morphology include the Upper Cave and Liujiang fossils, as well as modern Chinese specimens, including 1,180 skulls from different parts of China. Our results show: (1) The expression of most cranial features on the Liujiang cranium are within the variation ranges of modern Chinese, and only a few features of Liujiang have different expressions from modern Chinese; (2) A few primitive features such as a lower orbit can be observed on Liujiang cranium, indicating that it still preserves some Late Pleistocene human features, but compared with other Late Pleistocene humans, especially Upper Cave, the Liujiang cranium is more modern; (3) The differences between Liujiang and Upper Cave are mainly due to more primitive and robust features occurring on the Upper Cave crania. A few of these differences, such as the deeply depressed nasion of Upper Cave and broad nose of Liujiang, may be related to the climate or environmental adaptations. Based on these findings, we believe that the cranial morphology of Liujiang is very close to that of modern Chinese, and very few differences exist between them. Our studies do not support the opinion that the Liujiang cranium is more primitive than those of Upper Cave and Ziyang, and even put Liujiang as the earliest Late Pleistocene modern human in East Asia. The cranial differences between Liujiang and Upper Cave mainly reflect their evolution, and to a less extent the influences from their environments. Considering the similarity of cranial morphology between Liujiang and modern Chinese, and the uncertainty of the age of the fossiliferous layer of Liujiang, we believe that the current morphological evidence does not support the proposed earlier age for Liujiang.

The hominid fossils found in Liujiang include a complete cranium, the right os coxae, sacrum, two femur fragments and several vertebrae. Judged from the lack of duplication of elements, the comfortable articulation of the joint surfaces of adjacent bones and the similar texture of the bones, we are sure only a single individual is represented. This unusual discovery allows us to calculate body size, body proportions, and relative cranial capacity (encephalization quotient) for that individual rather reliably. Based on the measurements of the Liujiang cranium and reconstructed pelvis, we calculated the stature, body breadth, body weight, EQ index, and body proportion for the individual. Our result indicates that the Liujiang individual has body proportions (body height relative to body breadth) typical of warm adapted populations. Its encephalization quotient of 5.553 is greater than those of other Middle and Late Pleistocene humans such as Upper Cave and Jinniushan, and is closer to those of Minatogawa 2 and modern human populations. The body weight of 52.6 kg for Liujiang is also smaller than those of fossil humans living in higher latitude like Jinniushan, Upper Cave, and Neanderthals, and closer to those of Minatogawa, KNM-ER 3883, and KNM-ER 3733, which all lived in warmer climate region. We believe that the body size, body proportions, and relative cranial capacity (EQ) of the Liujiang individual suggest its resemblance to terminal Pleistocene and living humans.

Conclusion

The hominid fossils so far found in China indicate that Middle and Late Pleistocene was the time period when hominins were widespread living and evolving. More and wide morphological and behavioral variations occurred in these ancient humans. A series of newly found fossils and related studies provide evidence to support the existence of the early modern humans around 100 ka in China, which include the mandible and teeth found in Zhiren Cave of South China (Liu et al. 2010).

Acknowledgments The authors thank the organizers of the conference "150 Years of Neanderthal Discoveries: Early Europeans – Continuity & Discontinuity" for inviting Liu Wu to attend the conference. This work was supported by the Knowledge Innovation Program of the Chinese Academy of Sciences (KZCX2-YW-159), the National Natural Science Foundation of China (40972017) and the International Cooperation Program of MST of China (2007DFB20330 and 2009DFB20580).

References

Chen, T., & Zhang, Y. (1991). Palaeolithic chronology and possible coexistence of Homo erectus and Homo sapiens in China. *World Archaeology, 23*, 147–154.

Chia, L. (1957). Notes on the human and some other mammalian remains from Changyang, Hupei. *Vertebrate PalAsiatica, 1*, 247–257.

Gao, X., Huang, W., Xu, Z., et al. (2004). 120–150 ka human tooth and ivory engravings from Xinglongdong Cave, Three Gorges Region, South China. *Chinese Science Bulletin, 49*(2), 175–180.

Huang, W., Fang, D., & Ye, Y. (1982). Preliminary study on the fossil hominid skull and fauna of Hexian, Anhui. *Vertebrate PalAsiatica, 20*, 248–256.

Huang, W., Ciochon, R. L., Gu, Y., et al. (1995). Early *Homo* and associated artifacts from Asia. *Nature, 378*, 275–278.

Li, T., & Etler, D. (1992). New middle Pleistocene hominid crania from Yunnan in China. *Nature, 357*, 404–407.

Liu, W., & Zhang, Y. (2004). The cranial metric diversity of Chinese *Homo erectus*. *Acta Anthropologica Sinica, 22*, 89–104.

Liu, W., Mbua, E., Wu, X., et al. (2003). The comparisons of cranial features between Chinese and African Holocene humans, and their implications. *Acta Anthropologica Sinica, 22*, 89–104.

Liu, W., Zhang, Y., & Wu, X. (2005). A middle Pleistocene human cranium from Tangshan, Nanjing of southeast China: A comparisons with *Homo erectus* from Eurasia and Africa based on new reconstruction. *American Journal of Physical Anthropology, 25*, 253–262.

Liu, W., Gao, X., Pei, S., et al. (2006a). The paleoanthropological resources in West Hubei and three Gorges region, and some related studies. *Quaternary Sciences, 26*, 514–521.

Liu, W., He, J., Wu, X., et al. (2006b). The comparisons of cranial nonmetric features between upper cave skulls and modern north Chinese populations, and late Pleistocene human evolution in China. *Acta Anthropologica Sinica, 25*, 26–41.

Liu, W., Vialet, A., Wu, X., et al. (2006c). Comparaison de l'expression de certains caractères crâniens sur les hominidés chinois du Pléistocène récent et de l'Holocène (grotte supérieure de Zhoukoudian, sites de Longxian et de Yanqing). *L'Anthropologie, 110*, 258–276.

Liu, W., Wu, X., & Wang, S. (2006d). Some problems for the Late Pleistocene human cranium found in Liujiang of South China based on morphological analysis. *Acta Anthropologica Sinica, 25*, 177–194.

Liu, W., Jin, C., Zhang, Y., et al. (2010). Human remains from Zhirendong, South China, and modern human emergence in East Asia, *Proceedings of the National Academy of Sciences of the United States of America*, doi: 10.1073/pnas.1014386107.

Shang, H., Liu, W., Wu, X., et al. (2006). Upper Pleistocene human scapula from Salawusu, Inner Mongolia. *Chinese Science Bulletin, 51*, 2110–2115.

Shang H., Tong H., Zhang S., et al. (2007). An early modern human from Tianyuan Cave, Zhoukoudian, China. *PNAS, 104*, 6573–6578.

Tong, H., Shang, H., Zhang, S., et al. (2004). A preliminary report on the newly found Tianyuan Cave, a late Pleistocene human fossil site near Zhoukoudian. *Chinese Science Bulletin, 49*, 853–857.

Trinkaus, E. (2006). Modern human versus Neandertal evolutionary distinctiveness. *Current Anthropology, 47*, 597–618.

Weidenreich, F. (1939). On the earliest representatives of modern mankind recovered on the soil of East Asia. *Peking Natural History Bulletin, 13*, 161–174.

Woo, J. (1959). Human fossils found in Liukiang, Kwangsi, China. *Paleovertebrata & Paleoanthropology, 1*, 97–104.

Woo, J., & Peng, R. (1959). Fossil human skull of early paleoanthropic stage found at Mapa, Shaoquan, Kwangtung Province. *Vertebrata PalAsiatica, 3*, 176–182.

Wu, X. (1960). On the racial types of the upper cave man of Choukoudian. *Scientia Sinica, 10*, 998–1006.

Wu, X. (1961). Study on the upper cave man of Choukoudian. *Vertebrata PalAsiatica, 5*, 181–211.

Wu, M. (1980). Human fossils discovered at Xujiayao site in 1977. *Vertebrata PalAsiatica, 18*, 229–238.

Wu, M. (1983). *Homo erectus* from Hexian, Anhui found in 1981. *Acta Anthropologica Sinica, 2*, 109–115.

Wu, M. (1986). Study of temporal bone of Xujiayao Man. *Acta Anthropologica Sinica, 2*, 220–226.

Wu, X. (1998). Origin of modern humans of China viewed from craniodental characteristics of late *Homo sapiens*. *Acta Anthropologica Sinica, 17*, 276–282.

Wu, X. (2004a). On the origin of modern humans in China. *Quaternary International, 117*, 131–140.

Wu, X. (2004b). Fossil humankind and other Anthropoid primates of China. *International Journal of Primatology, 25*, 1093–1103.

Wu, R., & Dong, X. (1982). Preliminary study of *Homo erectus* remains from Hexian, Anhui. *Acta Anthropologica Sinica, 1*, 2–13.

Wu, R., & Li, X. (Eds.). (2002). *Homo erectus from Nanjing* (pp. 1–316). Nanjing: Jiangsu Science and Technology Publishing House.

Wu, R., & Wu, X. (1982). Human fossil teeth from Xichuan, Henan. *Vertebrata PalAsiatica, 20*, 1–9.

Wu, X., Liu, W., & Schepartz, L. (2004). The reconstruction and studies of the endocasts of Zhoukoudian *Homo erectus*. *Acta Anthropologica Sinica, 23(Supplement)*, 22–34.

Wu, X., Liu, W., Gao, X., et al. (2006a). Huanglong Cave, a new late Pleistocene hominid site in Yunxi of Hubei Province, China. *Chinese Science Bulletin, 51*, 2493–2499.

Wu, X., Schepartz, L., Falk, D., et al. (2006b). Endocast of Hexian *Homo erectus* from south China. *American Journal of Physical Anthropology, 26*, 445–454.

Zhang, Y. (1989). Tooth wear in early Homo sapiens from Chaohu and the hypothesis of use of anterior teeth as tools. *Acta Anthropologica Sinica, 8*, 314–319.

Zhang, Y., & Liu, W. (2005). Comparison of *Homo sapiens erectus* from Nanjing with those from Zhoukoudian and Sangian in facial morphology. *Acta Anthropologica Sinica, 24*, 171–177.

Zhang, Y., & Liu, W. (2006). Restoration of the Nanjing No. 2 fossil hominid calvarium: Morphology and taxonomic implications. *Acta Anthropologica Sinica, 25*, 267–276.

Zhang, Y., Liu, W., & Zhang, L. (2004). The morphology of nasal bones of *Homo erectus* from Nanjing and the possibility of gene flow from European fossil hominids. *Acta Anthropologica Sinica, 23*, 187–195.

Chapter 8
Behavioral and Cultural Origins of Neanderthals: A Levantine Perspective

Naama Goren-Inbar

Abstract The proceedings of the conference "150 years of Neanderthal discoveries – early Europeans: continuity and discontinuity" reflect the current state of the art as regards Neanderthals and their material culture in the Old World. The present contribution will focus on selected aspects of the world that predated the Neanderthals and their contemporaries. It draws mainly on data deriving from the meeting point of Africa and Eurasia (the Levantine Corridor) and focuses on the aspects that are most relevant for broadening our knowledge of the cultural background and evolution of the Neanderthals and early modern humans.

In order to better understand Neanderthal material culture and associated behavior, the archaeological remains should be viewed in conjunction and perspective with insights from an earlier period, namely the Lower Paleolithic. The issues addressed here include the first appearances of particular technological inventions pertaining to Mousterian/Middle Paleolithic technologies, the abilities of humans to learn, accumulate and share knowledge of their environment and its exploitation modes, as well as mobility patterns, migrations and colonization events.

Discoveries pertaining to Neanderthal populations in Europe have always been received with excitement and much scientific and lay interest. Over the years, many scholars have viewed these hominins as archaic and primitive creatures of limited abilities (and see discussion in Berman 1999; Speth 2004). Although recent opinion is subtler in its expression of this view of Neanderthal capabilities, the consensus on those of earlier hominins remains resolutely dismissive. Regrettably, this stance has resulted in the disregard of abundant data that suggest a strong correlation between ancient and modern behavioral patterns.

The Middle to Upper Paleolithic transition (with emphasis on the European record) and the disappearance and replacement by modern humans of the Neanderthals have been topics of extensive research. In contrast, although the transition from the Lower Paleolithic to the Middle Paleolithic occurred at ca. 250–300 ka across the whole of the Old World (e.g., Clark 1982a, b, 1988; Mercier et al. 2007; Tryon and McBrearty 2002, 2006; Tryon et al. 2005; Jaubert 2000–2001:157; Moncel 1995, 2005; but see Beaumont and Vogel 2006), it has been rather succinctly addressed and far less thoroughly investigated. This analytical bias towards the earlier period is no doubt partly due to its less direct involvement with our own species, but also a reflection of the absence of long uninterrupted sequences, taphonomic disturbances at the sites, lack of suitable dating methods, and meager publication in respect of the later period discussed here.

Yet despite all of the above, the available data indicate a continuity of hominin behavioral traits from the Lower to the Middle Paleolithic in diverse behavioral domains. Hominins of both periods share fundamental traits such as the ability to identify and occupy specific (favorable) landforms, the preference for specific ecological niches and habitats (e.g., Tuffreau et al. 1997; Roberts and Parfitt 1999; Pope 2002), successful exploitation of diverse resources and continuous survival in a given territory for a long period. Furthermore, hominin behavioral patterns that emerged during pre-Neanderthal times were later adopted, elaborated upon and widely distributed. These phenomena are evident in both the domains of planning and implementation. It will suffice to mention here the "domestication" and exploitation of fire (at Gesher Benot Ya'aqov (GBY), Alperson-Afil and Goren-Inbar 2006; Alperson-Afil et al. 2007; the complex modes of raw material acquisition and its transportation (e.g., at 'Ubeidiya, Bar-Yosef and Goren-Inbar 1993; at GBY, Madsen and Goren-Inbar 2004; at Tabun Cave, Verri et al. 2004, 2005); the emergence of species-specific targeting as a mode of game exploitation, indicating elaboration of hunting modes, weapons and efficient game processing (at Qesem Cave: Gopher et al. 2005; Lemorini et al. 2006 and at GBY: Rabinovich et al. 2008) and the presence, albeit rare, of non-utilitarian objects (a bead made of crinoid, GBY Goren-Inbar et al. 1991 and a figurine found at Berekhat Ram: Goren-Inbar 1986).

Of great interest, due to its high archaeological visibility, is the realm of stone tool production. It is in this domain that

N. Goren-Inbar (✉)
The Hebrew University of Jerusalem, Mount Scopus, Jerusalem 91905, Israel
e-mail: goren@cc.huji.ac.il

S. Condemi and G.-C. Weniger (eds.), *Continuity and Discontinuity in the Peopling of Europe: One Hundred Fifty Years of Neanderthal Study*, Vertebrate Paleobiology and Paleoanthropology, DOI 10.1007/978-94-007-0492-3_8, © Springer Science+Business Media B.V. 2011

particular aspects of the technologies characterizing the Neanderthal era are first observed in the Lower Paleolithic. Among these technologies are the Levallois flaking system, the soft hammer technique and the systematic production of blades. While all are widely represented in the Middle Paleolithic (MP) assemblages (and those of the Middle Stone Age [MSA]), and considered to some extent to be the hallmark of the MP, their origins are deeply rooted within the Lower Paleolithic and the Early Stone Age (ESA) material culture and technological sphere of knowledge.

Keywords Lower Paleolithic • Middle Paleolithic • Lithic technology

The Levallois Flaking System[1]

The entire record of the Eurasian MP and that of the MSA of Africa portrays technologies that have been first identified for earlier times – the Lower Paleolithic and in the ESA. Two technological modalities are particularly associated with these MP/MSA chrono-cultural units: those that exhibit Levallois characteristics and those that do not. While the MP Levantine record is characterized solely by the Levallois flaking system, classified as "Typical Mousterian" (after Bordes 1981), the European record is evidently much more diverse (Jaubert 2000–2001, with references for the last decade). However, particular technologies that are an integral part of these methods made their first appearance during much earlier times.

The Levallois flaking system has been mentioned in the context of the Levantine Acheulian by several researchers (Neuville 1951; Garrod and Bate 1937) and in syntheses of the Lower Paleolithic in the Levant and beyond (Bar-Yosef 1994; Clark 1975; Goren 1981). However, detailed accounts of the Levallois component of most of these LP assemblages have never been reported. Clearly, the dominance of surface sites as opposed to the small number of well-excavated sites hindered the establishment of a solid database as well as secure chronological assignment. The latest analyses of Acheulo-Yabrudian and Yabrudian assemblages from Tabun Cave (Gisis and Ronen 2006) as well as those deriving from Qesem Cave (Barkai et al. 2006; Gopher et al. 2005; Lemorini et al. 2006) did not yield Levallois products. This lack of evidence is inconclusive, as an absence of the Levallois flaking system from some of the sequences cannot be considered, as is frequently done (e.g., Gisis and Ronen 2006), an indication of a much later first appearance of the Levallois flaking system. Gopher et al. suggested: "Radial flake production at Qesem Cave appears to have been limited and opportunistic and we are quite confident that the Levallois concept and method are not represented in the Qesem Cave assemblages" (2005: 73). As the Acheulo-Yabrudian is the latest phase of the Lower Paleolithic, this view actually calls for the first appearance of the Levallois flaking system in the Middle Palaeolithic and its Levantine Mousterian occurrences.

Yet the Acheulian site of Berekhat Ram demonstrates a fully fledged Levallois component. Cores and flakes as well as flake tools recovered on site are typical products of the Levallois system (Goren-Inbar 1985). Given the frequent similarity in form of Levallois and handaxe manufacturing flakes, the sheer variety of Levallois products discovered within a clearly Acheulian assemblage, such as that of Berekhat Ram, rules out the possibility that the two have been confused.

As the Berekhat Ram site is older than 233 ka (Feraud et al. 1983), it seems that the first appearances of the Levallois technique are much earlier than generally assumed. Clearly, the Berekhat Ram assemblage demonstrates a fully developed Levallois flaking system with a large variety of the typical recurrent and preferential methods (Goren-Inbar 1985, figs. 5, 6, 14, 15).

When dealing with the question of Levallois antiquity, we have to integrate fragmentary data derived from different sites and distant geographical sources, despite obvious drawbacks, due primarily to the lack of dated Acheulian entities and sequences. Furthermore, there are preliminary indications for the development of the Levallois flaking system in Lower Paleolithic times, identifiable through the exploitation of specific morphologies (see below) that are prerequisites for the production of Levallois items, an extremely long process that required both skill (technology) and complex mental abilities.

A genuine Levallois flaking system does not appear in sub-Saharan Africa prior to 250 ka (e.g., McBrearty and Tryon 2005, 2006; Tryon 2003; Tryon and McBrearty 2002, 2006; but see Beaumont and Vogel 2006), which is the age of the early MSA (eMSA). However, if the production in the Final Acheulian of large predetermined flakes for the modification of bifacial tools is considered a variant of the Levallois method (and see explanations in the references above), then the first dated East African Levallois occurrence is assumed to have taken place between 285 and 509 ka in the Acheulian of the Kapthurin Formation in Kenya (ibid., and references therein). Despite the plethora of African Late and Final Acheulian sites and lithic assemblages, these cultural phases are generally poorly dated; the importance of the Kapthurin Formation data lies in their illustration of the potential of the African sequences.

Giant cores that exhibit centripetal scar patterning have been considered to represent the initial evolutionary phase of the Levallois method (i.e., Paddayya et al. 2006; Tryon and McBrearty 2006 and references therein). They are usually associated with the production of large flakes that were modified into handaxes and cleavers. While the technological

[1] In this paper I use Hovers's nomenclature of "Levallois flaking system" (in press) in order to avoid unnecessary confusion with regard to the Levallois terminology, defined as "method", "technology" and (frequently in English-language publications) "technique."

life-history of the handaxes is usually difficult to reconstruct due to the obliteration of earlier stages of modification by the sequential shaping of these tools, cleavers frequently bear scar patterns that could associate them with some of the observed large scars on the giant cores. If giant cores bearing other traits similar to the Middle Paleolithic Levallois flaking system (such as continuous acute-angle working edges and minimal preparation of the surface opposed to that of the débitage surface, which is observable in the small modifying scars) are accepted as the first appearance of this technique, then there is a case for suggesting that a more ancient origin should be attributed to the system.

The rich Acheulian horizons of GBY not only contain Levallois elements, but occur above the Brunhes-Matuyama boundary, thus dating the site to MIS 18–20 (Goren-Inbar et al. 2000; Feibel 2004). The presence of Levallois traits at such an early period may be an indication of the advanced mental and technological skills of the hominins responsible for the assemblage's manufacture (Madsen and Goren-Inbar 2004; Goren-Inbar and Saragusti 1996). This discovery is of particular significance since the age of the site is earlier than the early Levallois of Africa. Yet, it seems that the lack of archaic African evidence is most probably a result of poor dating rather than actual absence of lithic evidence. Further research is necessary in order to attempt to differentiate between scar patterning resulting from the application of the classical Levallois flaking system and that derived from various modification stages of bifacial tools. At GBY, as elsewhere, a possible key to identifying the existence of "genuine," fully intentional Levallois flaking lies within the component of small cores and blanks (Fig. 8.1). One should conclude that the Levallois flaking system is indeed a methodological and technological invention of a much greater antiquity than has previously been assumed.

It is quite obvious that archaeological data will continue to accumulate, changing dynamically the current low resolution as a consequence. Thus, the recent assignment of the Levallois flaking system in Europe to 300 ka (e.g. Tuffreau et al. 1997; for other areas in France, see Moncel 1999, 2003; Turq, personal communication, 2006; White et al. 2006) is but a single illustration of the fact that this system is much more archaic than previously perceived and that it predates the earliest recorded Neanderthal presence in Eurasia.

Cores on Flakes

As early as Pliocene times (2.34 ma ago), the splitting of individual stone nodules into several segments and subsequent utilization of the flaked material as cores were practiced (e.g., Delagne and Roche 2005). During the Middle Paleolithic the ability to exploit convex surfaces of "secondary products" (the ventral faces of flakes) as cores became a well-established trait of the Mousterian tool kit, a relatively common feature of assemblages of the Levant and across the Old World (e.g., Goren-Inbar 1988; Hovers 2007). In light of this, the observational skills and ability needed to exploit convex surfaces in order to produce predetermined flakes of different sizes are argued here to be traits of great antiquity. These were pivotal in the acquisition of the Levallois flaking system, the roots of which can be observed in the Acheulian. Acheulian assemblages from a wide geographical range have produced handaxes that were exploited as cores and from which large flakes were removed, thus destroying the symmetry of the item and most probably its original function. The exploitation of the convex surface of the handaxe necessitated a technological expertise that shares several characteristics with the Levallois flaking system.

The archaeological data indicate that the presence of cores on flakes was common in the Levantine Acheulian. They are known from sites as early as GBY but appear in larger quantities elsewhere, particularly towards the end of the Acheulian (e.g. the Amudian of Qesem Cave: Gopher et al. 2005).

Fig. 8.1 A small flint Levallois core from GBY (scale = 2 cm)

An increase in the use of convex surfaces is expressed in the exploitation and utilization of ventral faces of flakes on the one hand and the use of handaxes as cores on the other (DeBono and Goren-Inbar 2001). The latter has been reported recently from the excavations of the Acheulian open-air site of Revadim, where it was observed on 67% of the handaxes (Marder et al. 2006), and from the Acheulo Yabrudian assemblages of the Tabun and Misliya Caves (Gisis and Ronen 2006; Zaidner et al. 2006, respectively).

Soft Hammer Technique

The earliest Levantine evidence for the use of the soft hammer technique was discovered at GBY. It is expressed in the large quantity of *éclats de taille de bifaces*, small flakes resulting from the last stages of biface modification. Further evidence is available from typical features that are associated with the use of the soft hammer, which include particular crushing of the proximal area (dorsal face) adjacent to the striking platform as well as high frequencies of lipped striking platforms (Sharon and Goren-Inbar 1999; Goren-Inbar and Sharon 2006). The characteristic shallow scars resulting from the soft hammer technique are observed on many bifaces from the site. One should also mention a fragment of antler bearing damage signs typical of those created during experimental knapping (Fig. 8.2). These indications are further supported by even more conclusive finds from the younger Acheulian site of Boxgrove, dated to 0.5 ma, which supplied indications of the soft hammer technique in the form of microscopic flint flakes embedded in the antler surfaces that were used as a percussor (Pitts and Roberts 1997).

Evidence for the application of the soft hammer technique can therefore be demonstrated in the production of bifacial tools during the later phases of the Levantine Acheulian. Although it may have been used in a subsequent period for the production of specific types of scraper, there are no available studies clearly demonstrating its existence, despite the great potential of the Acheulo-Yabrudian and Yabrudian assemblages. Examination of the Middle Paleolithic European assemblages reveals ample evidence for the production of bifaces by the soft hammer technique (i.e., Soressi 2002). There is also a growing body of data indicating that this particular technique was applied during the modification of flakes (i.e., retouch) into Quina scrapers (Delagne and Jaubert, personal communications, 2006), and some indications from assemblages currently under study for the production of blanks, a mode that was to become widely common during the European Upper Paleolithic.

Systematic Blade Production in Lower Paleolithic Times

Systematic blade production was first recognized in the prehistoric record of the Levant in the 1920s, being identified by Garrod and by Rust (Garrod and Bate 1937; Rust 1950). Despite the advanced excavation methods, the secure stratigraphic contexts and the fact that the phenomenon recurred itself in different sites in this region (Syria, Lebanon, Israel), some scholars

Fig. 8.2 A damaged cervid (*Dama*?) antler from GBY (*left*) and an experimental antler percussor that was used to replicate basalt bifacial tools (scale divisions in cm)

Fig. 8.3 A sample of blades from the excavations of Qesem Cave

continuously doubted the integrity of the assemblages and refused to accept their assignment to the Lower Paleolithic (Bordes 1977). The systematic production of blades was apparently regarded as a marker of advanced abilities and thus outside the repertoire of non-modern hominins (Garrod 1962).

Recently acquired data demonstrate that blade production in the Lower Paleolithic was a systematic technological procedure, and much more common than previously thought. The latest discoveries in the Levant originate from excavations at Qesem Cave (coastal plain, Israel). These investigations have yielded unprecedented evidence of systematic blade production (Barkai et al. 2005; Gopher et al. 2005; Lemorini et al. 2006), representing a lengthy tradition documented in an extensively dated sedimentary sequence of some 7.5 m thickness (Barkai et al. 2003, 2006; Gopher et al. 2005) (Fig. 8.3).

It is of interest to note that the onset of the Middle Paleolithic in the Levant is characterized by the systematic production of blades. The Levantine record has furnished assemblages of similar technological as well as typological characteristics, although they were differently named (i.e., "Abu Zif", "Tabun D", "Humalian"; Meignen 1994 and references therein), all assigned to ca. 250–200 ka and stratigraphically underlying the long Mousterian sequences (Mercier et al. 2007, and references therein). Early dates for systematic blade production in Africa are known from the sequence at the Kapthurin Formation, Kenya (Tryon and McBrearty 2002, 2006). Beaumont and Vogel (2006) suggest much older age estimations for blade production of South African sites and consider its first appearance within the time range of the Late Acheulian. Evidently, the scarcity of other laminar assemblages in this continent is related to the poor resolution of the archaeological record rather than to an absence of know-how and technological abilities among the ESA hominins.

Adaptation, Mobility, Diffusion and Colonization

The processes involved in the colonization of the Old World are fundamental to understanding the behavior of Neanderthals and their contemporaries. In particular, the ability to survive, exploit and adapt in different ecological niches and under different environmental conditions provides some of the most attractive topics of research concerned with the Neanderthal world. Examination of the geographical distribution of Middle Paleolithic cultural entities in Eurasia indicates very widespread dispersion over large territories, evidenced through the spread of the Levallois flaking system and associated lithic traditions. This distribution is much more extensive (i.e., in the number of occurrences and territories) than any of its typo-technological predecessors.

The ways in which Neanderthals and other contemporaneous groups related to the Levantine environments are understood in part from repeated occupations of sites (mainly caves) and from their distribution in the landscape. Repeated occupation of the same sites in the Levant differs markedly from that in Europe, in that Levantine sites demonstrate greater similarity at the same site than between sites. The differences and similarities in tool kits and in the nature of the occupations at the sites are hypothesized by some to result from particular mechanisms of territorial behavior and demographic changes (e.g., Hovers 2009, 2001; Meignen et al. 2006; Wallace and Shea 2006), from functional behavior related to acquisition of raw material (e.g., Vermeersch 2002), and from social structure and behaviors that are assumed to reflect seasonal changes in the size of the paleo-community throughout the year cycles (Marks 1992). While task-specific sites existed during the Lower Paleolithic (i.e., quarries in the Levant: Barkai et al. 2006 and elsewhere:

Paddayya et al. 2006; Potts et al. 1999; Sampson 2006; kill sites: Goren-Inbar et al. 1994), the data on the geographic distribution of these earlier sites are very limited. However, they seem to reflect a large array of variability, most probably stemming from reasons similar to those mentioned above in relation to the behavior of the Neanderthals. These sites are located in different regions and ecological niches (from Western Europe to the Far East). The diffusion, migration or colonization mechanisms that enabled hominins to expand into these varied habitats and to behave in ways that resulted in these characteristic sites are under continuous debate. Several issues are central to this debate: (1) How many colonization episodes (waves) took place before the emergence of Neanderthals and modern humans? (2) What were the main dispersal routes? (3) What was the material culture possessed by hominins at the time of any particular sortie?

Hominin behavior in a given region is entirely dependent on recognition and knowledge of the potential subsistence resources. Examination of the Levantine record indicates that the Acheulian hominins had the ability to occupy remarkably diverse landscapes (Fig. 8.4). The distribution of sites clearly reflects the hominins' comprehensive knowledge of the terrain of the Levantine Corridor, as they are distributed throughout a variety of landforms, habitats and paleoclimatic zones. The Acheulian extensive occupation of the Levantine Corridor is but an initial chapter in the extremely prolonged duration of human exploitation of this area (Goren-Inbar and Speth 2004).

Central to this region is the close proximity of Lower Paleolithic sites to freshwater resources, including areas that are currently arid (i.e., the paleo-Lake Zihor: Ginat et al. 2003; Goren-Inbar et al. in preparation). It is evident that dependence on water sources (ancient lakes, riverbeds and springs: Por 2004) necessitated and dictated a particular mode of behavior (Bar-Yosef 1994; Gilead 1970), and that the distribution of sites in close proximity to these sources

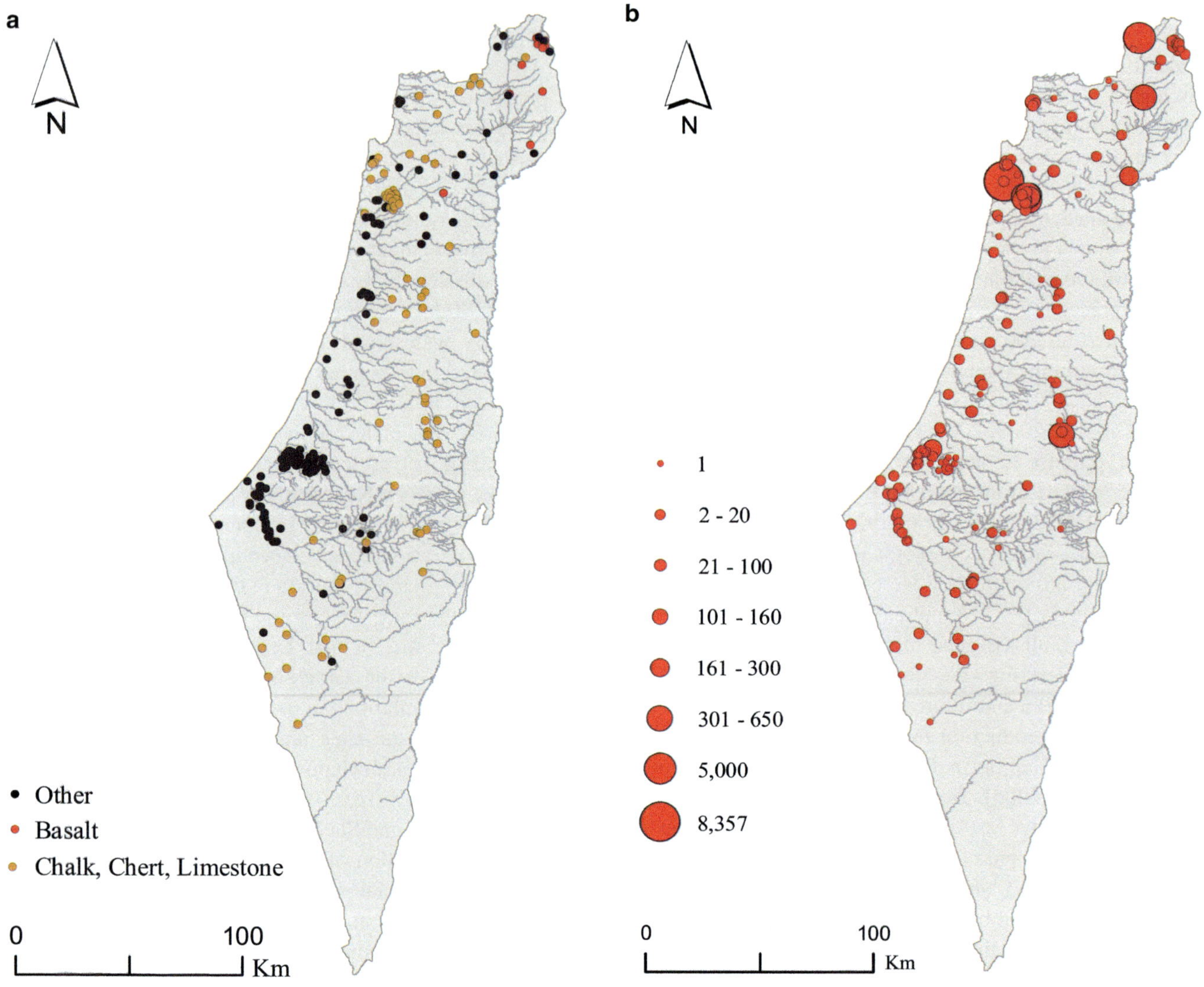

Fig. 8.4 Distribution map of Acheulian occurrences (sites and find spots); (**a**) on bedrock map; (**b**) according to the frequency of handaxes

is indicative of a thorough knowledge of the landscape and its optimal exploitation. This aspect of behavior underwent a major change during the Middle Paleolithic in Eurasia, when extensive occupations are less dependent upon proximity to water bodies and hence much more extensive than previously recorded.

In-depth knowledge of the environment in antiquity is also expressed through its exploitation for subsistence purposes. The paleobotany study at GBY, for example, yielded a unique opportunity on the one hand to examine the plants that existed during the paleoclimate oscillations of MIS 18–20, and on the other hand to gain an insight into the edible plant components of the lake-edge depositional sequence. The analysis of the wood fragments, fruits and seeds provided direct evidence of the paleo-vegetation of the early Lower Paleolithic Levant. Analyses of the organic material revealed a typical Mediterranean vegetation, which mirrors that of present-day natural habitats in the area, with an additional rich assemblage of submerged and lakeshore vegetation (Goren-Inbar et al. 2001, 2002, 2004; Melamed 1997, 2003). These finds indicate a wealth of edible species (fruits, grains, rhizomes, etc.), which permit greater resolution of the modeling of Acheulian subsistence, diet and behavior in this region. It further provides additional insight into the diversity of the exploitable sources and the depth of knowledge that enabled Acheulian hominins to effectively exploit these seasonal landscapes.

The behavioral system that concerns stone artifacts involves an intricate network of acquisition, selection, multistepped production and transportation of various items to focal points on the landscape. Although little is known of the functional aspects of the artifacts, patterns similar to those described above, though more limited, have been reported for the Acheulian of Europe (Lhomme et al. 2000) and are widely documented for the Middle Paleolithic in the context of Neanderthal behavior.

Some newly acquired Levantine data allow a better understanding of additional aspects of mobility, primarily those concerning transportation of raw material and artifacts. Previously published data on Lower Paleolithic hominin behavior were mainly concerned with raw material acquisition of particular rock types, sizes and shapes, transportation into the sites and a variety of other properties. These are only some aspects of the much wider realm of hominin mobility strategies. Mobility patterns seem to have become more extensive and complex along the Pliocene and Pleistocene time trajectory and may indicate quite a high degree of sophistication expressed in foresight and behavioral complexity.

We were recently able to demonstrate that extensive mobility to and from the sites took place at the Acheulian occupations of GBY. This mobility is expressed in the transportation of different raw materials into the archaeological horizons and involves the introduction of particular clast sizes, including basalt, flint and limestone that are not an integral component of the local sedimentary sequence (Goren-Inbar et al. 2000). In addition to this type of import, it was further demonstrated that there was active selection of raw material such as that expressed in the basalt archive at the site (basalt slabs of particular morphology and density and their various end products). This enables traits such as shape, size and quality of the imported objects to be more accurately identified (Madsen and Goren-Inbar 2004). The study of the basalt items further demonstrated that some of the basalt blocks (the giant cores) were introduced to the site, while others were exploited elsewhere and only the derived roughouts or nearly finished tools (handaxes and cleavers) were imported to the site.

It appears that each of the Acheulian sites in the GBY sequence reveals a different scenario of lithic inventories, although typologically and technologically they all belong to the same Acheulian tradition (ibid.). Despite these differences, there is a common denominator: the behavioral complexity that has emerged from analyses of mobility patterns is a recurring characteristic. The pattern discussed above is clearly more complex than the one identified at the earlier site of 'Ubeidiya, where the basalt bifaces (handaxes, trihedrals, quadrihedrals and picks) were introduced to the site as nearly finished tools, with only minimal additional knapping taking place on site (Bar-Yosef and Goren-Inbar 1993).

A similar pattern can be observed in the flint component of the Acheulian assemblages from GBY (Goren-Inbar and Sharon 2006; Sharon and Goren-Inbar 1999). Clearly, what is documented is only a small fraction of a complex web of movement of raw materials and artifacts at various stages of manufacture in and out of each site (Goren-Inbar and Sharon 2006). The identification of particular mobility patterns is extremely informative, as they suggest repeated movements within a given territory and extensive/in-depth knowledge of the area.

Recent studies of Acheulian sites in Europe, both on the continent and in England, have resulted in the identification of complex mobility patterns similar to those described above for GBY. These were identified, among others, through detailed refitting analyses (Hallos 2005; Pope 2002). The results contribute much to the understanding of behavioral traits and confirm that the high mobility identified at earlier times in the Levant prevailed in Europe at later times.

Acquiring a better understanding of mobility patterning is fundamental in any attempt to enlarge the scope of our knowledge of diffusion and colonization. Thus, when attempting to reach this objective, one should aim to utilize the most trustworthy evidence and avoid data that have not been subjected to rigorous investigation, particularly in terms of geology, paleontology and stratigraphical integrity.

A wealth of knowledge on archaeological sites and the geographical distribution of prehistoric cultures in the Old World has been accumulated over the last 150 years, since the discovery of the first Neanderthal. It is our scholarly goal to formulate additional hypotheses and examine previous postulated ones based on the currently available data, preferably pristine data. It is invalid at the present state of knowledge to consider fragmentary and selected topics (e.g., skeletal remains, paleontological assemblages, and scanty or unsound archaeological data) in order to gain further understanding of diffusion and colonization processes, as has recently been done (e.g., the lithics of 'Erq el-Ahmar or the age of the 'Ubeidiya Formation as in Dennell and Roebroeks 2005). Recent discussion of the impressive hominin skeletal finds from Dmanisi has raised the question as to whether this represents a permanent or transient colonization. Attempting to describe hominin behavioral abilities at Dmanisi, the authors suggested: "but there is no certainty that hominins managed to colonize this region on a long-term basis. Indeed it seems likely that many of the earliest dispersals from Africa into Eurasia resulted in occupations that were ephemeral, and the Early Pleistocene record does not document any continuity of populations through southern Asia to the Far East (Dennell 2003). The distribution of the first colonists to cross this landscape (the dynamics of such populations in respect to environmental change, and the extent of gene exchange among parapatric groups are entirely unknown." (Rightmire et al. 2006:139). This view clearly overlooks and contradicts extensive archaeological evidence from two geological formations in Israel, which demonstrate the abilities of hominins to colonize territories in the Early Pleistocene and occupy them for a **very long duration**. While only a segment of these formations are exposed, each of them reveals a scenario of repeated hominin occupations in lake-margin environments. The older 'Ubeidiya Formation is of Early Pleistocene age (Eisenman et al. 1983; Tchernov 1986), while the younger Benot Ya'akov Formation (BYF) is dated to the Early and Middle Pleistocene and assigned to MIS 18–20 (Feibel 2001, 2004; Goren-Inbar et al. 2000). While each of these two depositional sequences is located in a different segment of the Dead Sea Rift, **the estimated duration of each is considered to be some 100 kyr** (Feibel 2004). At both formations a very long sequence of occupations is documented: the 'Ubeidiya Formation has so far yielded over 70 different sites (Bar-Yosef and Goren-Inbar 1993; Shea and Bar-Yosef 1998), while at GBY over 13 sites were excavated (Goren-Inbar 2004; Goren-Inbar et al. 2000). Due to the fragmentation of both formations by prolonged tectonic activity of the Dead Sea Transform and the minimal and highly fragmentary size of the outcrops, it is impossible to demonstrate that the two archaeological localities (formations), situated in two adjacent Dead Sea Rift basins, were **continuously** or periodically occupied beyond the 100 kyr of the known data. Indeed, a series of core drillings to a depth of ca. 80 m that were carried out at the GBY site failed to reach the base of the BY Formation (Goren-Inbar 2004). These drillings furnished geological and archaeological data indicating that the sequence is much longer than previously assumed. Similarly, while 'Ubeidiya is indeed younger than the Dmanisi record, it should be noted that the 1.4 ma Member Fi (Eisenmann et al. 1983), which yielded most of the archaeological horizons, is only one segment of the sedimentary record of the 'Ubeidiya Formation (Bar-Yosef and Goren-Inbar 1993). The Levantine data reflect an entirely different scenario than the one suggested above. Early hominins could and did colonize new territories and these occupations were prolonged, as indicated below.

While the entire extent of the Benot Ya'akov and 'Ubeidiya Formations is unknown, it seems most likely that the Levantine Corridor was continuously occupied by hominins with different cultural traditions. The particular origin of each of the traditions can be traced back to Africa, as they differ morpho-technologically from one another (Goren-Inbar and Saragusti 1996; Saragusti and Goren-Inbar 2001). The presence of two distinctly different traditions is viewed as reflecting different African dispersals. The generally low resolution of the very early archaeological record in the Levantine Corridor may mask additional evidence testifying to additional sorties.

The mobility pattern is crucial for understanding processes of diffusion and colonization. The available data illustrate the problematic nature of archaeological resolution, a difficulty that is pertinent to all archaeological periods but is of greater amplitude where Lower Paleolithic sites are concerned, due to the impact of natural processes affecting them over a much longer time span. Yet, the archaeological records of both Acheulian sites clearly illustrate the colonization of two distinct Acheulian episodes within the Levantine Corridor. 'Ubeidiya is definitely not an outlier due to its great age, as was recently suggested by Foley and Lahr (2003: 114). On the contrary, it is in full accord with many other assemblages that are classified in Africa as "Developed Oldowan" and share the same cultural inventories, some of the faunal components and the overall age. Furthermore, it contributes extensively to a better understanding of the mechanism of diffusion, as it is currently viewed. When repeated visits to the same locality are considered together with high levels of mobility, and such a trend is repeated at archaeological sites in different geographical regions, there is reason to suggest an incremental pattern of diffusion (see terminology in Bar-Yosef and Belfer-Cohen 2001). Such findings support the argument for a slow dispersal rate as suggested by Anton and Swisher (2004).

The control of fire is considered to be one of the most important cultural innovations in hominin lifestyle (Goren-Inbar et al. 2004; Alperson-Afil et al. 2009). Evidence from GBY provides strong evidence for the presence of hearths in the archaeological layers, thus attesting to the ability of hominins to control fire as early as MIS 18. Furthermore, additional evidence currently under investigation indicates that this ability may have existed at a much earlier date at the site. Needless to say, such a cultural tool may have been fundamental in facilitating hominin dispersal and furthering their ability to cope in increasingly seasonal environments. This evidence, the oldest in Eurasia, is chronologically followed by evidence from various Acheulian sites in the Levant (i.e., Tabun Cave: Garrod and Bate 1937) and most recently by the newly recovered burned bone data from Qesem Cave (Lemorini et al. 2006). Clearly, the extensive and rich evidence for the use of fire at these Middle Paleolithic sites testifies to its importance amongst both Neanderthals and modern humans. The roots of this important aspect of paleo-behavior are embedded in the Lower Paleoloithic and supported by the presence of hearths at the site of Beeches Pit, England (Gowlett 2006; Gowlett et al. 2005; Hallos 2005).

In summary, the Lower Paleolithic Levantine record, though segmented, indicates that predetermination expressed in the use of the Levallois flaking system and systematic blade production existed long before the appearance of the Neanderthals. These findings, and the large volume of the evidence, clearly refute some current terminologies that attempt to characterize pre-Neanderthal cultural entities by lumping the diverse cultural occurrences into a few classes – the "Modes" (Carbonell et al. 1999; Clark 1961; Foley 1996; Foley and Lahr 1997, 2003; Lahr and Foley 1998). We have demonstrated above that particular technological inventions characterizing the Middle Paleolithic and Middle Stone Age originated and achieved their full forms in earlier times – during the Lower Paleolithic and Early Stone Age. Furthermore, their coexistence in earlier times rules out the attempts to use each of these traits as an independent indicative marker for particular time segments in human cultural history. Taking the above into consideration, it is evident that attempting to differentiate the various dispersals Out of Africa on the basis of these "markers" and their possible cultural significance is of questionable value.

The review of the pre-Neanderthal archaeological data presented above demonstrates that some of the current schemes of dispersal modes are misleading, as they do not consider the wealth of newly acquired data and thus mask some major prehistoric inventions on the one hand and blur their diffusions on the other. Indeed, there is some awareness of the effect that the suggested schemes are faulty, as expressed by Foley and Lahr (2003: 113) "there are continuities between them". But this awareness is insufficient and the continuous use of Modes I to IV, and hence the lumping together of cultural innovations and apparent behaviors that are clearly of archaic nature, is extremely misleading.

We have described in this article the great similarity in abilities – both cognitive and cultural – between hominins predating modern humans and Neanderthals, based mainly on newly acquired data concerning selected aspects that existed during the Lower Paleolithic and ESA. Clearly, the similarities are more extensive during the later phases of the Lower Paleolithic than those observed in the Pliocene African sites. This similarity is revealed through a wide array of multidisciplinary studies, adding to the growing scope of our understanding of the Neanderthals and their contemporaries. It seems that the current state of research requires the formulation of additional questions that will attempt to achieve a better and more precise definition of the domains in which the abilities of Lower Paleolithic hominins differ from those of the Neanderthals.

In a recently published article Tryon and McBrearty (2006: 492) described their findings of the transition between the Early Stone Age to the Middle Stone Age: "Combined evidence from the tools and flake production methods suggest an incremental and mosaic pattern of change in hominin adaptive strategies during the Acheulian–MSA transition." In this study, we have similarly demonstrated that in the Levantine Corridor, and at earlier times, such phenomena are discernible. It seems that modern human abilities, both mental and technological, evolved quite early in time and changed at different rates. In order to gain additional and thorough understanding of modern behavior, it is mandatory to enlarge our knowledge of the cultural entities that are assigned to the Early and Middle Pleistocene times. Comparative study of behavioral traits of archaic hominins and the data pertaining to the Neanderthal era will enable us to explore aspects of cultural and non-cultural evolutionary traits. It will further contribute to the abundant attempts to decipher the evolutionary tempo of the cultural, behavioral and mental abilities of different hominin taxa, and hence enable us to understand ourselves better.

Acknowledgements I would like to thank the organizers of this conference, and in particular S. Condemi, for inviting me to take part in this important celebration of 150 years since the first Neanderthal discovery. This paper was written during a sabbatical year in Cambridge. Clare Hall College and the Department of Biological Anthropology, University of Cambridge, provided excellent facilities. M. Lahr and R. Foley excelled in their hospitality at the new Leverhulme Centre for Human Evolutionary Studies (LCHES), Cambridge.

Thanks are due to N. Alperson-Afil for the production of Fig. 8.4, and to G. Laron for the photographs of Figs. 8.1 and 8.2; Fig. 8.3 was photographed by P. Shrago and permission to use it was generously given by R. Barkai and A. Gopher. E. Hovers commented on and corrected the text, and A. Belfer-Cohen contributed thoroughly in both extended comments and many editorial aspects to the present version. V. Ling's talent made this contribution significantly easier to read.

References

Alperson-Afil, N., & Goren-Inbar N. (2006). Out of Africa and into Eurasia through controlled use of fire: Evidence from Gesher Benot Ya'aqov, Israel. *Archaeology Anthropology & Ethnology of Eurasia, 28*, 63–78.

Alperson-Afil, N., Richter, D., & Goren-Inbar, N. (2007). Phantom hearths and controlled use of fire at Gesher Benot Ya'aqov, Israel. *PaleoAnthropology, 2007*, 1–15.

Alperson-Afil, N., Sharon, G., Zohar, I., Biton, R., Melamed, Y., Kislev, M. E., Ashkenazi, S., Rabinovich, R., Werker, E.,

Hartman, G., & Goren-Inbar, N. (2009). Spatial orgnaization of hominins activities at the Acheulian Gesher Benot Ya'aqov, Israel. *Science, 326*, 1677–1680.
Antón, S. A., & Swisher, I. C. C. (2004). Early dispersals of Homo from Africa. *Annual Review of Anthropology, 33*, 271–296.
Barkai, R., Gopher, A., Lauritzen, S. E., & Frumkin, A. (2003). Uranium series dates from Qesem Cave, Israel, and the end of the lower Palaeolithic. *Nature, 423*, 977–979.
Barkai, R., Gopher, A., & Shimelmitz, R. (2005). Middle pleistocene blade production in the Levant: An Amudian assemblage from Qesem Cave, Israel. *Eurasian Prehistory, 3*(2), 39–74.
Barkai, R., Gopher, A., & LaPorta, P. C. (2006). Middle Pleistocene landscape of extraction: Quarry and workshop complexes in northern Israel. In N. Goren-Inbar & G. Sharon (Eds.), *Axe age: Acheulian tool-making from quarry to discard* (pp. 7–44). London: Equinox.
Bar-Yosef, O. (1994). The lower paleolithic of the near East. *Journal of World Prehistory, 8*(3), 211–265.
Bar-Yosef, O., & Belfer-Cohen, A. (2001). From Africa to Eurasia. *Quaternary International, 75*, 19–28.
Bar-Yosef, O., & Goren-Inbar, N. (1993). *The lithic assemblages of 'Ubeidiya*. Jerusalem: Institute of Archaeology, Hebrew University.
Beaumont, P. B., & Vogel, J. C. (2006). On a timescale for the past million years of human history in central South Africa. *South African Journal of Science, 102*, 217–228.
Berman, J. C. (1999). Bad hair days in the Paleolithic: Modern (re)construction of cave man. *American Anthropologist, 10*(2), 288–304.
Bordes, F. (1977). Que sont le Pré-Aurignacien et le Iabroudien? In B. Arensburg & O. Bar-Yosef (Eds.), *Moshe stekelis memorial volume* (pp. 49–55). Jerusalem: Israel Exploration Society.
Bordes, F. (1981). Vignt-cinq ans après: le complexe mousterién revisité. *Bulletin de la Société Préhistorique Française, 78*, 77–87.
Carbonell, E., Mosquera, M., Rodríguez, X. P., Sala, R., & Made, J. V. D. (1999). Out of Africa: The dispersal of the earliest technical systems reconsidered. *Journal of Anthropological Archaeology, 18*, 119–136.
Clark, G. (1961). *World prehistory: In new perspectives*. Cambridge: Cambridge University Press.
Clark, J. D. (1975). A comparison of the Late Acheulian industries of Africa and the Middle East. In K. W. Butzer & G. L. Isaac (Eds.), *After the Australopithecines: Stratigraphy, ecology and culture change in the middle Pleistocene* (pp. 605–659). Chicago: Aldine.
Clark, J. D. (1982a). The transition from Lower to Middle Palaeolithic in the African continent. In A. Ronen (Ed.), *The transition from lower to middle Palaeolithic and the origin of modern man* (pp. 235–255). BAR International Series 151, Oxford.
Clark, J. D. (1982b). The cultures of the Middle Palaeolithic/Middle Stone Age. In J. D. Clark (Ed.), *The Cambridge history of Africa, from the earliest times to c. 500 BC* (pp. 248–341). Cambridge: Cambridge University Press.
Clark, J. D. (1988). The Middle stone age of East Africa and the beginnings of regional identity. *Journal of World Prehistory, 2*(3), 235–305.
deBono, H., & Goren-Inbar, N. (2001). Note on a link between Acheulian and Levallois technologies. *Journal of the Israel Prehistoric Society, 31*, 9–23.
Delagnes, A., & Roche, H. (2005). Late Pliocene hominid knapping skills: The case of Lokalalei 2 C, West Turkana, Kenya. *Journal of Human Evolution, 48*, 435–472.
Delagnes, A., & Ropars, A. (Eds.). (1996). *Paléolithique Moyen en Pays de Caux (Haute-Normandie): Le Pucheuil, Etoutteville – deux gisements de plein air en milieu loessique*. Paris: Éditions de la Maison des Sciences de l'Homme.
Dennell, R. (2003). Dispersal and colonisation, long and short chronologies: How continuous is the Early Pleistocene record for hominids outside East Africa? *Journal of Human Evolution, 45*, 421–440.
Dennell, R., & Roebroeks, W. (2005). An Asian perspective on early human dispersal from Africa. *Nature, 438*, 1099–1104.
Eisenmann, V., Ballesio, R., Beden, M., Faure, M., Geraads, D., Guerin, C., & Heintz, E. (1983). Nouvelle interpretation biochronologique des grands mammifères d'Ubeidiya, Israël. *Geobios, 16*(5), 629–633.
Feibel, C. S. (2001). Archaeological sediments in lake margin environments. In J. K. Stein & W. R. Farrand (Eds.), *Sediments in archaeological contexts* (pp. 127–148). Salt Lake City: University of Utah Press.
Feibel, C. S. (2004). Quaternary lake margins of the Levant Rift Valley. In N. Goren-Inbar & J. D. Speth (Eds.), *Human paleoecology in the Levantine corridor* (pp. 21–36). Oxford: Oxbow Books.
Feraud, G., York, D., Hall, C. M., Goren, N., & Schwarcz, H. P. (1983). $^{40}Ar/^{39}Ar$ age limit for an Acheulian site in Israel (sample site: Berekhat Ram). *Nature, 304*(5923), 263–265.
Foley, R. (1996). The adaptive legacy of human evolution: A search for the environment of evolutionary adaptedness. *Evolutionary Anthropology, 4*(2), 194–203.
Foley, R., & Lahr, M. M. (1997). Mode 3 technologies and the evolution of modern humans. *Cambridge Archaeological Journal, 7*(1), 3–36.
Foley, R., & Lahr, M. M. (2003). On stony ground: Lithic technology, human evolution, and the emergence of culture. *Evolutionary Anthropology, 12*, 109–122.
Garrod, D. (1962). The Middle Palaeolithic of the Near East and the problem of Mount Carmel Man. *The Journal of the Royal Anthropological Institute of Great Britain and Ireland, 92*(2), 232–259.
Garrod, D. A. E., & Bate, D. M. A. (1937). *The stone age of Mount Carmel: Excavation at Wadi Mughara*. Oxford: Oxford University Press.
Gilead, D. (1970). *Early Paleolithic cultures in Israel and the Near East*. Ph.D. thesis, Hebrew University, Jerusalem.
Ginat, H., Zilberman, E., & Saragusti, I. (2003). Early Pleistocene lake deposits and lower Paleolithic finds in Nahal (Wadi) Zihor, southern Negev desert, Israel. *Quaternary Research, 59*, 445–458.
Gisis, I., & Ronen, A. (2006). Bifaces from the Acheulian and Yabrudian layers of Tabun Cave, Israel. In N. Goren-Inbar & G. Sharon (Eds.), *Axe age: Acheulian tool-making from quarry to discard* (pp. 137–154). London: Equinox.
Gopher, A., Barkai, R., Shimelmitz, R., Khalaily, M., Lemorini, C., Hershkovitz, I., & Stiner, M. (2005). Qesem Cave: An Amudian site in central Israel. *Journal of the Israel Prehistoric Society, 35*, 69–92.
Goren, N. (1981). The Lower Palaeolithic in Israel and adjacent countries. In P. Sanlaville, & J. Cauvin (Eds.), *Préhistoire du Levant* (pp. 193–205). C.N.R.S. Colloque 598, Lyon.
Goren-Inbar, N. (1985). The lithic assemblages of Berekhat Ram Acheulian site, Golan Heights. *Paleorient, 11*(1), 7–28.
Goren-Inbar, N. (1986). A figurine from the Acheulian site of Berekhat Ram. *Mitekufat Haeven, 19*, 7–12.
Goren-Inbar, N. (1988). Too small to be true? A reevaluation of cores on flakes in Levantine Mousterian assemblages. *Lithic Technology, 17*(1), 37–44.
Goren-Inbar, N. (2004). The Gesher Benot Ya'aqov Acheulian Site – The environment and behavior of early/middle Pleistocene Settlers in the Hula Valley. In A. P. Derevianko & T. I. Nokhrina (Eds.), *Arkheologiya i paleoekologiya Evrasii [Archaeology and Paleoecology of Eurasia]* (pp. 267–278). Novosibirsk: Institute of Archaeology and Ethnography SB RAS Press.
Goren-Inbar, N., & Saragusti, I. (1996). An Acheulian biface assemblage from the site of Gesher Benot Ya'aqov, Israel: Indications of African affinities. *Journal of Field Archaeology, 23*, 15–30.
Goren-Inbar, N., & Sharon, G. (2006). Invisible handaxes and visible Acheulian biface technology at Gesher Benot Ya'aqov, Israel. In N. Goren-Inbar & G. Sharon (Eds.), *Axe age: Acheulian tool-making from quarry to discard* (pp. 111–135). London: Equinox.
Goren-Inbar, N., & Speth, J. D. (Eds.). (2004). *Human paleoecology in the Levantine Corridor*. Oxford: Oxbow Books.

Goren-Inbar, N., Lewy, Z., & Kislev, M. E. (1991). The taphonomy of a Jurassic bead-like fossil from an Acheulian occupation at Gesher Benot Ya'aqov. *Rock Art Research, 8*(2), 133–136.

Goren-Inbar, N., Lister, A., Werker, E., & Chech, M. (1994). A butchered elephant skull and associated artifacts from the Acheulian site of Gesher Benot Ya'aqov, Israel. *Paleorient, 20*(1), 99–112.

Goren-Inbar, N., Feibel, C. S., Verosub, K. L., Melamed, Y., Kislev, M. E., Tchernov, E., & Saragusti, I. (2000). Pleistocene milestones on the Out-of-Africa corridor at Gesher Benot Ya'aqov, Israel. *Science, 289*, 944–974.

Goren-Inbar, N., Sharon, G., Melamed, Y., & Kislev, M. (2001). Nuts, nut cracking, and pitted stones at Gesher Benot Ya'aqov, Israel. *PNAS, 99*(4), 2455–2460.

Goren-Inbar, N., Werker, E., & Feibel, C. S. (2002). *The acheulian site of Gesher Benot Ya'aqov: The wood assemblage*. Oxford: Oxbow Books.

Goren-Inbar, N., Alperson, N., Kislev, M. E., Simchoni, O., Melamed, Y., Ben-Nun, A., & Werker, E. (2004). Evidence of hominin control of fire at Gesher Benot Ya'aqov, Israel. *Science, 304*, 725–727.

Gowlett, J. A. J. (2006). The early settlement of northern Europe: Fire history in the context of climate change and the social brain. *Comptes Rendus Palevolution, 5*, 299–310.

Gowlett, J. A. J., Hallos, J., Hounsell, S., Brant, V., & Debenham, N. (2005). Beeches Pit: Archaeology, assemblage dynamics and early fire history of Middle Pleistocene site in East Anglia, UK. *Eurasian Prehistory, 3*(2), 3–38.

Hallos, J. (2005). "15 minutes of fame": Exploring the temporal dimension of Middle Pleistocene lithic technology. *Journal of Human Evolution, 49*, 155–179.

Hovers, E. (2001). Territorial behavior in the middle Paleolithic of the Southern Levant. In N. J. Conard (Ed.), *Settlement dynamics of the middle Paleolithic and middle stone age* (pp. 123–152). Tübingen: Kerns Verlag.

Hovers, E. (2007). The many faces of cores-on-flakes: A perspective from the Lavantine Mousterian. In S. McPherron (Ed.), *The identification and study of alternative core technologies*. Cambridge Scholarly Press, Cambridge.

Hovers, E. (2009). *The organization of Mousterian Lithic technology*. Oxford University Press, Oxford.

Jaubert, J. (2000–2001). Les sociétés du Paléolithique Moyen en France: Principaux de ces dix dernières années. *Zephyrus, 53–54*, 153–175.

Lahr, M. M., & Foley, R. (1998). Towards a theory of modern human origins: Geography, demography, and diversity in recent human evolution. *Yearbook of Physical Anthropology, 41*, 137–176.

Lemorini, C., Stiner, M. C., Gopher, A., Shimelmitz, R., & Barkai, R. (2006). Use-wear analysis of an Amudian laminar assemblage from the Acheuleo-Yabrudian of Qesem Cave, Israel. *Jounal of Archaeological Science, 33*, 921–934.

Lhomme, V., Connet, N., & Chaussé Chaussé, C. (2000). Essai d'interprétation du site Paléolithique Inférieur de Spoucy 1 (Yonn). *Gallia Préhistoire, 42*, 1–44.

Madsen, B., & Goren-Inbar, N. (2004). Acheulian giant core technology and beyond: An archaeological and experimental case study. *Eurasian Prehistory, 2*(1), 3–52.

Marder, O., Milevski, I., & Matskevich, Z. (2006). The handaxes of Revadim Quarry: Typo-technological considerations and aspects of intra-site variability. In N. Goren-Inbar & G. Sharon (Eds.), *Axe age: Acheulian tool-making from quarry to discard* (pp. 223–242). London: Equinox.

Marks, A. E. (1992). Typological variability in the Levantine Middle Paleolithic. In H. L. Dibble & P. A. Mellars (Eds.), *The middle Paleolithic: Adaptation, behavior and variability* (pp. 127–141). Philadelphia: University of Pennsylvania Museum Press.

McBrearty, A., & Tryon, C. (2006). From Acheulian to Middle Stone Age in the Kapthurin Formation, Kenya. In E. Hovers & S. L. Kuhn (Eds.), *Transitions before the transition: Evolution and stability in the middle Paleolithic and middle Stone Age* (pp. 257–277). New York: Springer.

Meignen, L. (1994). Paléolithique moyen au Proche-Orient: le phénomène laminaire. In S. Révillion & A. Tuffreau (Eds.), *Les industries laminaires au Paléolithique moyen* (pp. 125–159). Paris: Éditions CNRS.

Meignen, L., Bar-Yosef, O., Speth, J. D., & Stiner, M. C. (2006). Middle Paleolithic settlement patterns in the Levant. In E. Hovers & S. L. Kuhn (Eds.), *Transitions before the transitions: Evolution and stability in the middle Paleolithic and middle Stone age* (pp. 149–169). New York: Springer Science+Business Media, Inc.

Melamed, Y. (1997). *Reconstruction of the landscape and the vegetarian diet at Gesher Benot Ya'aqov archaeological site in the lower paleolithic period*. M.Sc. thesis, Bar-Ilan University, Ramat Gan.

Melamed, Y. (2003). *Reconstruction of the Hula valley vegetation and the Hominid vegetarian diet by the lower Palaeolithic botanical remains from Gesher Benot Ya'aqov*. Ph.D. thesis, Bar-Ilan University, Ramat Gan.

Mercier, N., Valladas, H., Froget, L., Joron, J.-L., Reyss, J.-L., Weiner, S., Goldberg, P., Meignen, L., Bar-Yosef, O., Belfer-Cohen, A., Chech, M., Kuhn, S.L., Stiner, M.C., Tillier, A.-M., Arensburg, B., Vandermeersch, B., (2007) Hayonim Cave: A TL-based chronology for this Levantine Mousterian sequence. *Journal of Archaeological Science, 34*, 1064–1077.

Moncel, M.-H. (1995). Contribution à la connaissance du Paléolithique moyen ancien (antérieur au stade isotopique 4): l'example de la Vallée de l'Ardèche et de la moyenne Vallée du Rhône (France). *Préhistoire Européenne, 7*, 81–109.

Moncel, M.-H. (1999). Les assemblages lithiques du site Pleistocène moyen d'Orgnac 3 (Ardèche, Vallée du Rhône, France). Université de Liège, ERAUL, Liège.

Moncel, M.-H. (2003). L'exploitation de l'espace et la mobilité des groupes humains au travers des assemblages lithiques à la fin du Pléistocène moyen et au début du Pléistocène supérieur. La Moyenne Vallée du Rhône entre Drôme et Ardèche. BAR International Series 1184, Oxford.

Moncel, M.-H., Moigne, A.-M., & Combier, J. (2005). Pre-Neandertal behaviour during isotopic stage 9 and the beginning of stage 8. New data concerning fauna and lithics in the different occupation levels of Orgnac 3 (Ardèche, South-East France): Occupation types. *Journal of Archaeological Science, 32*, 1283–1301.

Neuville, R. (1951). *Le Paléolithique et Mésolithique du désert de Judée*. Paris: Archives de l'Inst. de Paléont.

Paddayya, K., Jhaldiyal, R., & Petraglia, M. D. (2006). The Acheulian quarry at Isampur, Lower Deccan, India. In N. Goren-Inbar & G. Sharon (Eds.), *Axe age: Acheulian tool-making from quarry to discard* (pp. 45–73). London: Equinox.

Pitts, M., & Roberts, M. (1997). *Fairweather Eden: Life in Britain half a million years ago as revealed by the excavations at Boxgrove*. London: Century.

Pope, M. I. (2002). *The significance of Biface-Rich assemblages: An examination of behavioural controls on Lithic assemblage formation in the lower Paleolithic*. Ph.D. Thesis. University of Southampton, Southampton.

Por, D. F. (2004). The Levantine waterway, riparian archaeology, paleolimnology, and conservation. In N. Goren-Inbar & J. D. Speth (Eds.), *Human Paleoecology in the Levantine Corridor* (pp. 5–20). Oxford: Oxbow Books.

Potts, R., Behrensmeter, A. K., & Ditchfield, P. (1999). Paleolandscape variation and early Pleistocene hominid activities: Members 1 and 7, Olorgesailie formation, Kenya. *Journal of Human Evolution, 37*, 747–788.

Rabinovich, R., Gaudzinski, S., & Goren-Inbar, N. (2008). Systematic butchering of fallow deer (*Dama*) at the early Middle Pleistocene Acheulian site of Gesher Benot Ya'aqov (Israel). *Journal of Human Evolution, 54*, 134–149.

Rightmire, G. P., Lordkipanidze, D., & Vekua, A. (2006). Anatomical descriptions, comparative studies and evolutionary significance of the hominin skulls from Dmanisi, Republic of Georgia. *Journal of Human Evolution, 50*, 115–141.

Roberts, M. B., & Parfitt, S. A. (1999). *Boxgrove: A middle Pleistocene Hominid site at Eartham Quarry, Boxgrove,West Sussex*. London: English Heritage.

Rust, A. (1950). *Die Höhlenfunde von Jabrud (Syrien)*. Neumünster: Karl Wachholtz.

Sampson, C. G. (2006). Acheulian quarries at hornfels outcrops in the upper Karoo region of South Africa. In N. Goren-Inbar & G. Sharon (Eds.), *Axe age: Acheulian tool making from quarry to discard* (pp. 76–107). London: Equinox.

Saragusti, I., & Goren-Inbar, N. (2001). The biface assemblage from Gesher Benot Ya'aqov, Israel: Illuminating patterns in "Out of Africa" dispersal. *Quaternary International, 75*, 85–89.

Sharon, G., & Goren-Inbar, N. (1999). Soft percussor use at the Gesher Benot Ya'aqov Acheulian site? *Mitekufat Haeven, 28*, 55–79.

Shea, J. J., & Bar-Yosef, O. (1998). Lithic assemblages from new (1988–1994) excavations at Ubeidiya: A preliminary report. *Mitekufat Haeven, 28*, 5–19.

Soressi, M. (2002). Le Moustérien de tradition acheuléenne du sud-ouest de la France. Discussion sur la signification du faciès à partir de l'étude comparée de quatre sites: Pech-de-l'Azé I, Le Moustier, La Rochette et la Grotte XVI. Université de Bordeaux I, Bordeaux.

Speth, J. D. (2004). News flash: Negative evidence convicts Neanderthals of gross mental incompetence. *World Archaeology, 36*(4), 519–526.

Tchernov, E. (1986). *Les mammifères du Pleistocène inférieur de la Vallée du Jordain à Oubeidiyeh*. Paris: Association Paléorient.

Tryon, C. A. (2003). *The Acheulian to Middle Stone Age transition: Tephrostratigraphic context for archaeological change in the Kapthurin Formation, Kenya*. Ph.D. dissertation, University of Connecticut, Storrs.

Tryon, C. A., & McBrearty, S. (2002). Tephrostratigraphy and the Acheulian to Middle Stone Age transition in the Kapthurin Formation, Baringo, Kenya. *Journal of Human Evolution, 42*, 211–235.

Tryon, C. A., & McBrearty, S. (2006). Tephrostratigraphy of the Bedded Tuff Member (Kapthurin Formation, Kenya) and the nature of archaeological change in the later Middle Pleistocene. *Quaternary Research, 65*, 492–507.

Tryon, C. A., McBrearty, S., & Texier, P.-J. (2005). Levallois lithic technology from the Kapthurin Formation, Kenya: Acheulian origin and Middle Stone Age diversity. *African Archaeological Review, 22*(4), 199–229.

Tuffreau, A., Marcy, K.-L., & Tuffreau, A. (1997). Land-use and site function in Acheulean complexes of the Somme Valley. *World Archaeology, 29*(2), 225–241.

Vermeersch, P. M. (Ed.). (2002). *Palaeolithic quarrying sites in upper and middle Egypt*. Belgium: Leuven University Press.

Verri, G., Barkai, R., Bordeanu, C., Gopher, A., Hass, M., Kaufman, A., Kubik, P., Montanari, E., Paul, M., Ronen, A., Weiner, S., & Boaretto, E. (2004). Flint mining in prehistory recorded by *in situ* produced cosmogenic 10Be. *Proceedings of the National Academy of Sciences, 101*(21), 7880–7884.

Verri, G., Barkai, R., Gopher, A., Hass, M., Kubik, P. W., Paul, M., Ronen, A., Weiner, S., & Boaretto, E. (2005). Flint procurement strategies in the late lower Palaeolithic recorded by *in situ* produced cosmogenic ^{10}Be in Tabun and Qesem Caves (Israel). *Jounal of Archaeological Science, 32*, 207–213.

Wallace, I. J., & Shea, J. J. (2006). Mobility patterns and core technologies in the middle Paleolithic of the Levant. *Journal of Archaeological Science, 33*(9), 1–17.

White, M., Scott, B., & Ashton, N. (2006). The Early middle Palaeolithic in Britain: Archaeology, settlement history and human behaviour. *Journal of Quaternary Science, 21*(5), 525–541.

Zaidner, Y., Druck, D., & Weinstein-Evron, M. (2006). Acheulo-Yabrudian handaxes from Misliya Cave, Mount Carmel, Israel. In N. Goren-Inbar & G. Sharon (Eds.), *Axe age: Acheulian tool-making from quarry to discard* (pp. 243–266). London: Equinox.

Chapter 9
Discontinuities in the Faunal Assemblages and Early Human Populations of Central and Western Europe During the Middle and Late Pleistocene

Wighart von Koenigswald

Abstract The middle and late Pleistocene history of Central and Western Europe includes several intervals of faunal change involving both local extinction and immigration of new species from elsewhere. Substantial faunal turnovers correspond to times of climate change. For many species, Central and Western Europe was a peripheral part of their geographic range and thus an area of temporal occurrence. The evolution of these taxa can be traced to core areas elsewhere. An important question concerns the extent to which human populations were similarly affected by climate change and faunal turnover. The successive groups of humans that populated Central and Western Europe did not necessarily originate in the same core area, and different areas of origin may explain morphological differences distinguishing various human fossils known from the middle Pleistocene of Germany.

Keywords Pleistocene faunal exchange • Central Europe • Mauer • Bilzingsleben • Steinheim • Neanderthal

Introduction

Central Europe north of the Alps is fairly rich in human fossils of middle and late Pleistocene age due to intensive research during the twentieth century. Most of the human remains do not come from proper archaeological excavations and their age is often uncertain. Nevertheless they have played an important role in the history of paleoanthropology. Several finds were formally named and thus represent the types for species or subspecies. Fossils from Central Europe represent different stages of human evolution, and the finds from Mauer, Bilzingsleben, Steinheim, Reilingen, and Weimar-Ehringsdorf have been interpreted as a more or less consistent evolutionary lineage from archaic humans to pre-Neanderthals, and proto-Neanderthals leading to classical Neanderthals (Hublin 1990; Condemi 1998).

The mammalian fauna of the middle and late Pleistocene in Central Europe shows repeated alternation of an interglacial *Elephas* assemblage and a glacial *Mammuthus* assemblage. These faunal assemblages immigrated, but became extinct locally when ecological conditions were unfavorable. Multiple climatic oscillations mean that the same faunas reoccur and disappear repeatedly (Koenigswald 2006).

This paper raises the question of whether human populations in Central Europe followed a similar pattern of repeated immigration and disappearance. It is assumed that humans expanded their range from southern France or Spain repeatedly (Gamble et al. 2004). But the occurrence of Asian species in the European mammalian fauna during specific interglacial periods suggests the possibility that human populations migrated from other regions to Central Europe during the middle Pleistocene as well.

Detailed faunal lists were given in Koenigswald and Heinrich (1999). For the hominid fossils mostly the original names are used to ensure the correlation of the specimens to the type sites. The stratigraphic scheme given in Fig. 9.1 shall not camouflage the often neglected uncertainties and discrepancies still existing.

Specific Geographic Conditions in Central and Western Europe

Central and Western Europe was pivotal in the elucidation of Pleistocene history. It was in this region, that the first traces of the great expansion of glaciers during the Pleistocene were recognized (Agassiz 1840; Torell 1875) and, later, that the alternation of glacial and interglacial periods was deduced from successive generations of moraines and boulder clays. In no other part of the world did this very complex sequence of events become as apparent (Geikie 1894; Keilhack 1899; Penck and Brückner 1901–1909).

Similarly, the fossil occurrences of mammals that live today in the Arctic, such as *Rangifer, Gulo,* and lemmings

W. von Koenigswald (✉)
Institut für Paläontologie der, Universität Bonn, Nussallee 8, D-53113 Bonn, Germany
e-mail: koenigswald@uni-bonn.de

S. Condemi and G.-C. Weniger (eds.), *Continuity and Discontinuity in the Peopling of Europe: One Hundred Fifty Years of Neanderthal Study*, Vertebrate Paleobiology and Paleoanthropology, DOI 10.1007/978-94-007-0492-3_9,

Geology		small mammal stratigraphy	significant immigrants	important faunal localities	Interglacial conditions	Supposed OIS
Holocene		*Arvicola terrestris*-faunas				1
Late Pleistocene	Weichselian					
	Eemian	*Arvicola cantianus-terrestris* faunas	*Elephas antiquus* *Hippopotamus* *Bubalus*	Lehringen Taubach		5e
Middle Pleistocene	Warthe Drente	late *Arvicola cantianus* faunas	*Elephas antiquus* *Bubalus*	W. Ehringsdorf Steinheim/Murr		?7
	Dömnitz Hosteinian		*Elephas antiquus*	Schöningen Bilzingsleben		?9 ?11
	Elsterian					
	Cromerian Complex	early *Arvicola cantianus* faunas *Arvicola*	*Elephas antiquus* *Hippopotamus* *Arvicola*	Kärlich G Mosbach Mauer		
	Brunhes	*Mimomys* + *Mimomys savini*-faunas		Süssenbron Voigtstedt		
Lower Pleistocene	Matuyama	*Mimomys savini*-faunas with *M. pusillus*	*Hippopotamus*	Untermaßfeld		>19

Fig. 9.1 Stratigraphic scheme, indicating the supposed position of faunal localities and the various interglacial phases in Central Europe (From Koenigswald 2006)

(*Lemmus* and *Dicrostonyx*), was recognized in the fossil record (Nehring 1880, 1890). Koken (1912), for example, when describing the Pleistocene faunas of Paleolithic sites in Germany, differentiated between glacial faunas characterized by *Mammuthus primigenius* and interglacial faunas characterized by *Elephas antiquus*.

In Western and Central Europe, faunas and floras can often be more readily attributed to a glacial or interglacial environment than to a specific age, since very similar faunal and floral assemblages appeared in all the glacial or all the interglacial faunal periods, respectively (Fig. 9.2). In the Pleistocene flora, studied in pollen profiles from lake deposits, the sequence of immigration of the various plant taxa is very characteristic for each interglacial (Litt 1994). In contrast, Pleistocene faunal localities reflect neither the sequence of immigration nor do they always enable successive interglacials to be distinguished.

The geography of Europe, specifically the orientation of the mountain ranges, is responsible for the great difference between glacial and interglacial floras and faunas. The Pyrenees and the Alps are both oriented from west to east and thus they form an ecological barrier between Central Europe and the Mediterranean region. They also buffer minor climatic oscillations. Any influence from the Arctic is ameliorated in the Mediterranean region, and during times of glaciation minor climatic oscillations did not affect the glacial fauna in the north. Faunal turnovers occurred in Western and Central Europe only between glacial and interglacial periods. Glacials and interglacials differed not only in their mean annual temperatures, but even more so in the magnitude of annual temperature fluctuation and relative humidity. Glacial periods had a strong continental climate, while interglacials were characterized by a strong maritime influence. Changes in humidity were particularly important in catalyzing the drastic faunal changes between *Mammuthus* assemblages and *Elephas* assemblages.

The Pleistocene fauna of the Mediterranean is characterized by successive waves of faunal immigration (Azzaroli et al. 1982; Kotsakis 2006), but not by comparable faunal change between glacial and interglacial periods. In this respect Central and Western Europe differ very much from North America or most other regions of the world. The orientation of the mountain belts is most significant and may be the explanation for the difference. In North America the Rockies and Appalachians are oriented north-south and, in

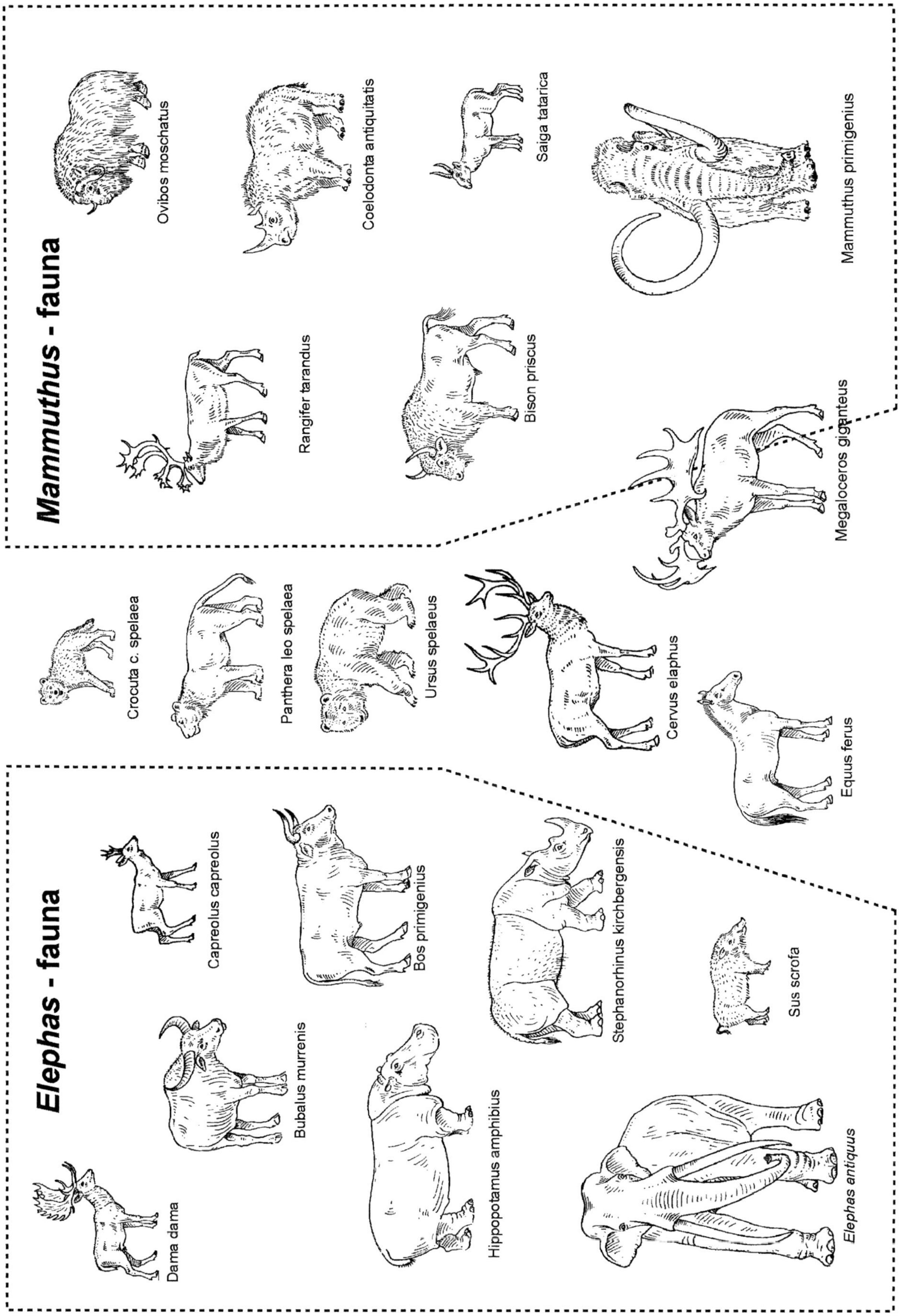

Fig. 9.2 The main components of the interglacial *Elephas* fauna and the glacial *Mammuthus* fauna. They occurred alternatively in Central and Western Europe

contrast to Europe, cold winds from the north may reach much farther south. The Great Plains between the two mountain ranges maintained a continental climate. In contrast to the climatic alternation in Europe between continental and maritime conditions, Siberia has always maintained a very dry and continental climate. In Siberia interglacial faunas hardly differ from glacial assemblages in the same area (Sher 2006). The vast area north of the Himalayan belt was continuously continental even as the Pleistocene climate changed. Hence, it is not surprising that faunas of the glacial and interglacial periods in Siberia and North America did not change to the same extent that they did in Western or Central Europe.

Immigration and Local Extinction are the Pattern of Faunal Exchange in Western and Central Europe

Mammuthus assemblages characterizing glacial periods in Western and Central Europe immigrated from the northeast. In contrast, *Elephas* assemblages expanded their range from Mediterranean regions using the major river valleys of the Rhine and the Danube as immigration routes. The faunal exchange was very intensive, since only a few herbivores occurred in both assemblages. Carnivores were generally not as much affected. Following the onset of unfavorable climate conditions in Western and Central Europe, species that could not tolerate the climate disappeared. The term "emigration" is often used for this process, deduced from the idea that clever humans knew how to escape an unfavorable situation. However, for the fauna, such disappearances should more precisely be called a "local extinction" (Koenigswald 2003). The biological reaction to an unfavorable climate is reduction in the number of offspring. If unfavorable conditions last for several generations, this leads to extinction in a local area. Due to this repeating pattern of immigration and local extinction, Western and Central Europe was an area of "temporal occurrence", or temporary occurrence, for most mammalian species (Koenigswald 2003). There are only a few genera among herbivores that occur in both glacial and interglacial faunas, e.g., *Ursus*, *Cervus*, and *Arvicola*. Nevertheless even in these genera it cannot be excluded that there was some replacement of populations, since subspecies are generally not recognizable in the fossil record. Central Europe and the northern part of Western Europe was a typical area of "temporal occurrence" (Fig. 9.3). Although some taxa like rhinos and elephants are represented by more derived forms in subsequent periods, their evolution seemingly took place in areas where the taxa were present continuously. Such areas are defined as "core areas," and these are the areas from which the species migrated into Central Europe again. Only in the core areas was there any evolutionary continuity. Nevertheless, even in the regions of temporal occurrence these newly evolved forms may have reappeared in each subsequent phase.

The extent to which human populations were involved in a similar pattern of repeated migration, like the mammalian fauna, remains an open question. Certainly human populations expanded their territories when conditions were favorable, but whether human populations suffered local extinctions, or rather were able to escape to more favorable areas can be questioned. In Central Europe a simple southward migration might have been stopped by the Alps. The disappearance of Vikings in Greenland is an historical example of local extinction of a once flourishing human population due to climatic changes.

In the early part of the middle Pleistocene, humans occurred in Central Europe only during interglacial periods, and thus with each warming a new immigration is probable. Only during the (late?) Saalian Complex did pre-Neanderthals coexist in Central Europe in the biome with *Mammuthus fauna*. During the Eemian, humans hunted game of the *Elephas* assemblage, while during the Weichselian humans, this time in the form of the classical Neanderthals, again lived alongside the *Mammuthus* assemblage. The fossil record shows the presence of humans during this interval of glacial-interglacial-glacial conditions, but not enough is known to indicate whether these populations were continuously present in areas north of the Alps, or whether different populations alternated in the region.

Biostratigraphy of the Middle and Late Pleistocene in Central Europe

The multiple faunal exchanges in Western and Central Europe are characterized by a repeated immigration of taxa. Thus it is difficult to do any stratigraphy using first and last occurrence records or datums (FAD, LAD). The presence of taxa was not continuous, and some taxa were irregular in their reoccurrence. The time span of the middle and late Pleistocene was too short, compared to the Tertiary, for major evolutionary changes, and continuous evolution of certain forms can be expected only in the core areas.

The rodent *Arvicola*, a vole, is one of the few taxa present in Western and Central Europe throughout the middle and late Pleistocene, and it shows evolutionary changes that can be used as stratigraphic markers (Koenigswald and Heinrich 1999). A biostratigraphic framework based on *Mimomys* and *Arvicola* voles is summarized here. One species in particular, *Mimomys savini,* characterizes mammalian faunas at the beginning of the middle Pleistocene, which are found at Voigtstedt and Süssenborn in Thuringia, and the Upper

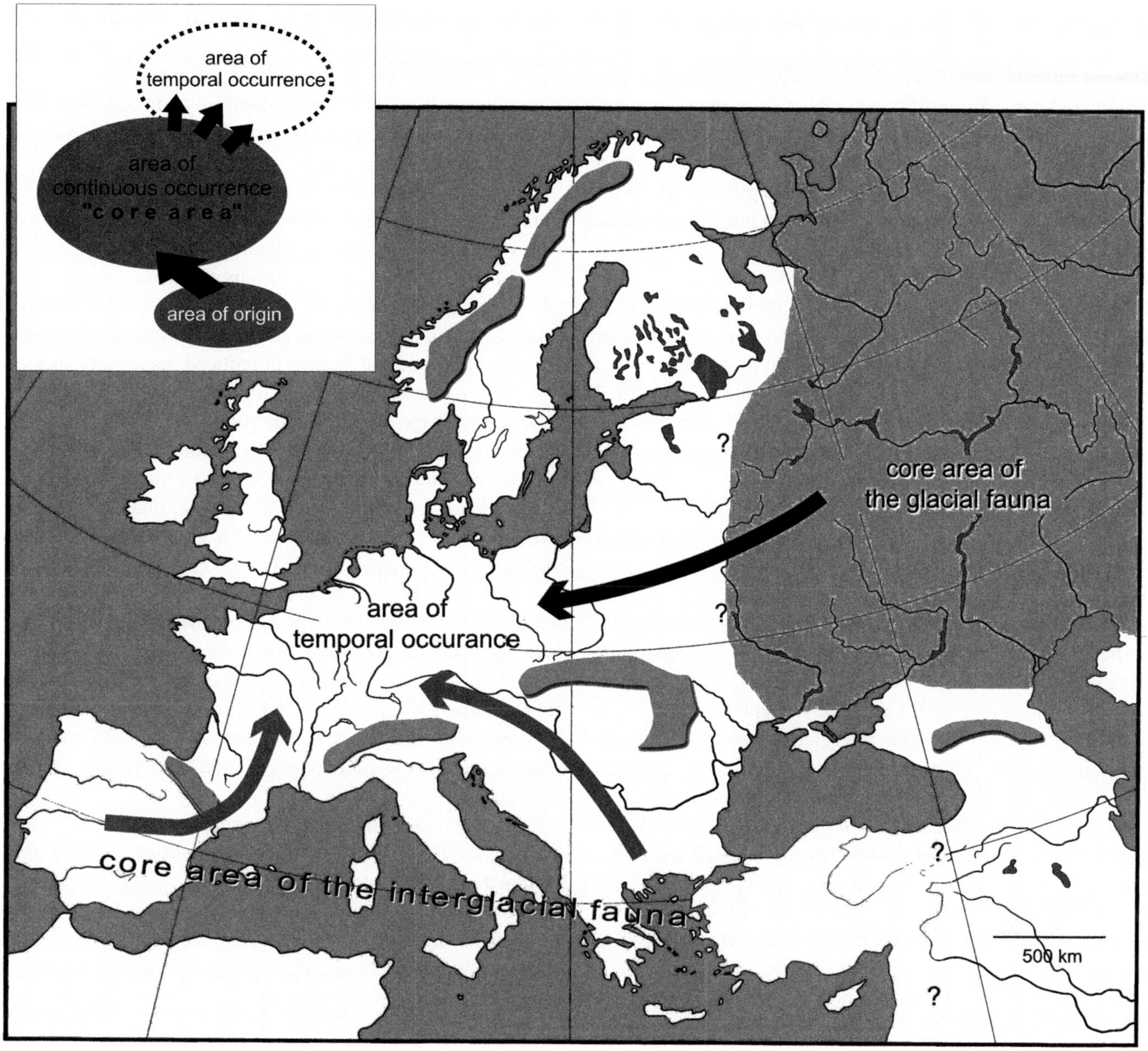

Fig. 9.3 Central and Western Europe as an area of temporal occurrence for both, the glacial *Mammuthus* fauna and the interglacial *Elephas* fauna

Freshwater Bed of the Cromer Forest Bed Series in East Anglia. In this way, the *Mimomys savini* faunas correlate with part of the Cromerian complex.

During the Cromerian, *Mimomys savini* was replaced by *Arvicola cantianus* (Koenigswald 1973). This species was recently cited as *Arvicola mosbachensis*, a younger synonym (Maul et al. 2000). *Mimomys* differs in having rooted molars, while those of *Arvicola* are rootless. *Arvicola* most likely evolved from *Mimomys*, however, not in Central Europe but in the south, where it might have been present somewhat earlier. Stratigraphically, *Arvicola cantianus* occurs in Central Europe in the late Cromerian. Although *Arvicola* first appears with interglacial *Elephas antiquus*, this vole remains in Central Europe even after the onset of cold climate conditions, and after the *Mammuthus* fauna immigrated. These faunas, known as "early *Arvicola cantianus* faunas," are thought to have antedated the Elsterian. They show a greater diversity of insectivores and rodents than faunas after the Elsterian. Typical sites with early *Arvicola cantianus* faunas are Mauer, Mosbach II, and Kärlich G.

After the Elsterian, *Arvicola cantianus* remained an index fossil, but the diversity of other small mammals became reduced. "Late *Arvicola cantianus* faunas," which range from the Elsterian to the Holsteinian and the Saalian complexes, occur in both glacial and interglacial phases. Within this time period, a change can be observed in the enamel thickness of molar teeth of *Arvicola,* which is quantified as SDQ (for *Schmelzband-Differenzierungs-Quotient*; Heinrich 1982). This evolutionary trend can be used to some degree as a stratigraphic indicator. It is still not possible to correlate all faunal levels with the OIS [Oxygen isotope stages] because the position of the Elsterian is still unclear. According to Sarnthein et al. (1986) and Parfitt et al. (2005), the Anglian (=Elsterian) represents OIS 12 and Schreve (2001) correlates the Hoxnian with OIS 11 based on the supposed sequence of faunas and terraces in England. Stringer and Hublin (1999) discuss the Hoxnian as OIS 9 and the post-Anglian "Swancombe stage" as OIS 11. In Central Europe the Holsteinian is defined as the first post-Elsterian interglacial. The type locality of the Holsteinian was dated carefully and correlated with OIS 9 (Geyh and Müller 2005, 2006 but Scourse 2006; Nitychoruka et al. 2006). The possibility that two post-Elsterian interglacials show the same pollen signal was rejected by the geological evidences from the type region.

The stratigraphic positions of interglacials before or within the Saalian complex is problematic due to differing interpretations of the geology in the Elbe-Saale region. Mania and Thomae (2006) postulate four interglacial phases in the Holstein Complex before the Drenthe (first Saalian ice advance), and two additional ones before the Warthe (second Saalian ice advance). Litt et al. (2005) accept one interglacial phase, or at the most two interglacial phases, after the Holsteinian and before the Drenthe, but none between Drenthe and Warthe. Thus he correlates Drenthe and Warthe with OIS 6 (Litt 2006). The *Elephas* assemblages at the relevant sites are nearly identical. It is possible that the occurrence of *Bos primigenius* in Steinheim/Murr and at Schöningen indicates an age younger than Bilzingsleben II.

Faunas of the Eemian (OIS 5e) and the early Weichselian reflect a transition from *Arvicola cantianus* to *Arvicola terrestris. Arvicola terrestris* then continues from the Weichselian into the Holocene (OIS 1). The large mammals of the *Mammuthus* assemblage are thus very similar to those of the Saalian.

Ecology and Biostratigraphy of German Localities with Human Remains

Middle Pleistocene

Most sites of the middle and late Pleistocene in Germany that have yielded remains of archaic humans, pre-Neanderthals or classical Neanderthals, have also produced mammalian faunas that enable a solid assessment of the ecology and also facilitate biostratigraphic correlation.

The site of Mauer near Heidelberg, where the famous mandible, the holotype of *Homo heidelbergensis* was collected, yielded a typical interglacial fauna (Schötensack 1908). The *Elephas* assemblage includes *Hippopotamus antiquus*, indicating a strong maritime influence with mild winters. Presence of both early *Arvicola cantianus* and *Stephanorhinus hundsheimensis* in this fauna shows that it preceded the Elsterian biostratigraphically.

The paleoecology of Bilzingsleben II was similar. Here remains of the robust "*Homo erectus bilzingslebenensis*" were found. The vegetation indicates full interglacial conditions, and the fauna represents a typical *Elephas* assemblage (Mania et al. 1997). However, compared with the early *Arvicola cantianus* faunas preceding the Elsterian, the diversity of small mammal faunas is reduced, especially that of insectivores and rodents. Thus Bilzingsleben II is regarded as an early stage of late *Arvicola cantianus* faunal zone (Koenigswald and Heinrich 1999). Geologically the site is located on Elsterian till and thus it is most likely Holsteinian in age. Expansion of the Elsterian ice sheet makes it unlikely that human populations survived in Central Europe. The small corridor between the Scandinavian ice shield and the mountain glaciers of the Alps was only a few hundred kilometers wide. It was most probably not favorable for big game, and it was definitely not favorable for the *Elephas* fauna. The interglacial fauna totally disappeared from Central Europe during the Esterian, and then immigrated from the Mediterranean again when the climate ameliorated. Thus, most probably, the human population also re-immigrated with the interglacial fauna.

Two other sites bearing human remains and related to late *Arvicola cantianus* faunas are Steinheim/Murr and Weimar-Ehringsdorf, which represent interglacial conditions preceding the Eemian. The river deposits of Steinheim/Murr, where the delicate cranium of "*Homo steinheimensis*" was excavated, produced a very diverse interglacial *Elephas antiquus* assemblage. The interglacial fauna from Steinheim/Murr was traditionally referred to the Holsteinian (Adam 1954 a, b, 1966, 2003), but this was based on a now-rejected stratigraphic scheme that included only two interglacial periods (Holsteinian and Eemian) after the Elsterian. Steinheim/Murr is definitively older than Eemian and might belong to OIS 7. Thus between Bilzingsleben II and Steinheim, a cold phase is most probable, during which the *Elephas* faunal assemblage disappeared from Central Europe and was replaced by a *Mammuthus* fauna, which is present at a lower level at Steinheim/Murr.

The interglacial layer at Steinheim/Murr is of special interest because it seems that two bovids, *Bos primigenius* and *Bubalus murrensis* (Berckhemer, 1927), occur at this level for the first time in Central Europe (Fig. 9.4). Ecologically *Bubalus* indicates a warm climate with mild winters, i.e. a maritime influence. The origin of *Bubalus* is of great significance. It is an Asian genus and the closest relative of *Bubalus murrensis* in the extant fauna is *Bubalus arnee* from India. However, in

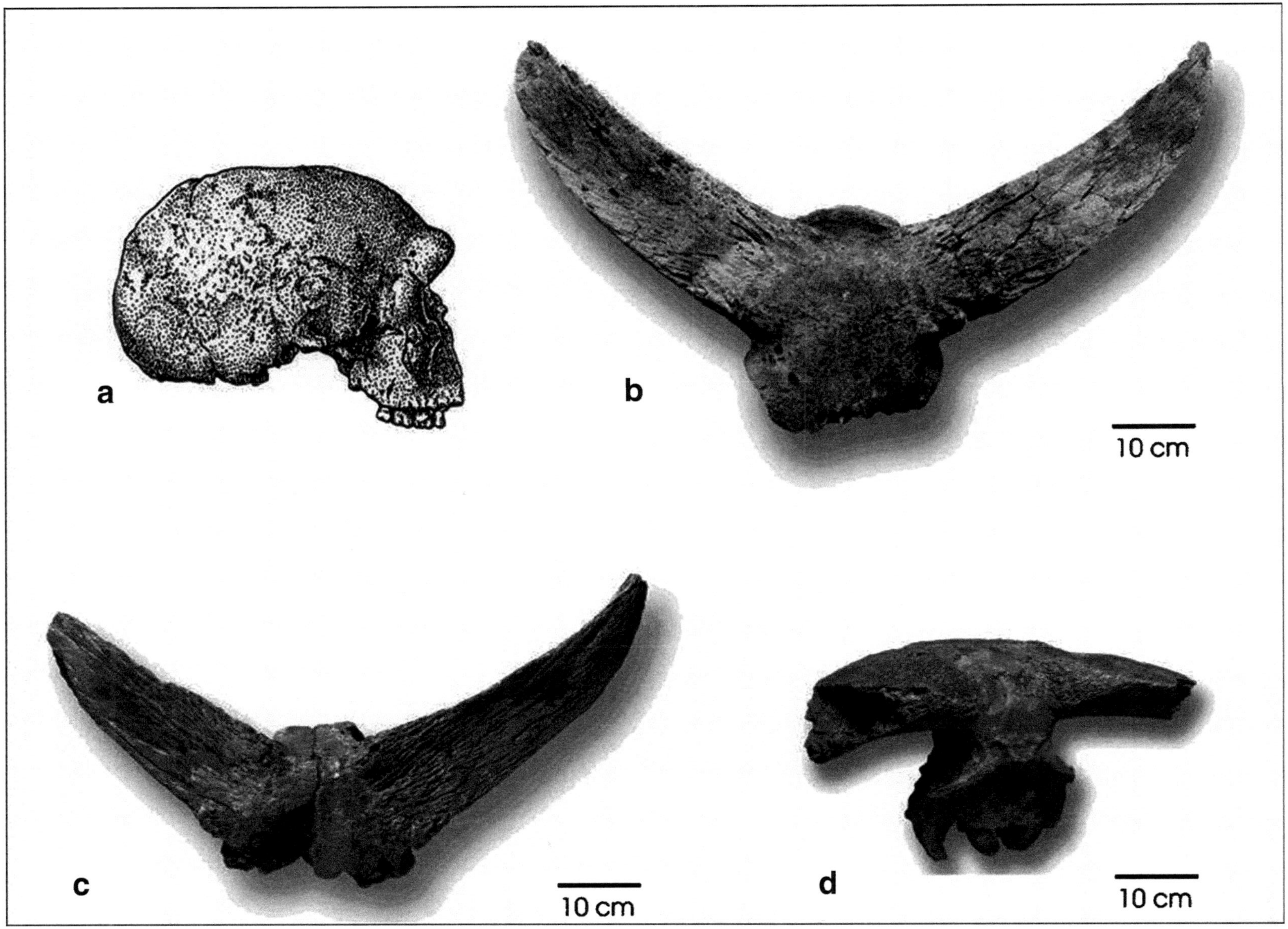

Fig. 9.4 Fossils of the *Elephas* fauna. *Bubalus murrensis*, the Indian water buffalo, is an exotique immigrant from Asia. (**a**): Human cranium from the middle Pleistocene interglacial of Steinhein/Murr. (**b**): Partial cranium of *Bubalus murrensis* (the water buffalo, an exotic immigrant from Asia) from the same level as the carnium of "*Homo steinheimensis*". (**c**) and (**d**): Two cranial parts of *Bubalus murrensis* from a later reoccurrence during the Eemian, found in sand deposits in the northern Rheinebene near Darmstadt. The triangular shape of the horn cores and the flattened front side are characteristic (**c**: Wolfskehlen [Felsbergmuseum, Beedenkirchen] and **d**: Geinsheim [Coll. F. Menger, Groß-Rohrheim]). (**a**: From Adam et al. 1995; **b**: photo by Hans Lumpe Stuttgart, **c** and **d**: photos by Georg Oleschinski, Bonn)

terms of morphology, Chinese species of *Bubalus* from the late middle Pleistocene are very close to *Bubalus murrensis* (Young 1936; Koenigswald 1986). Unfortunately the fossil record of *Bubalus* is very limited, and the full history of its dispersal is still unknown, but the presence of *Bubalus* shows that Asian immigrants are definitively present in the *Elephas* fauna of Steinheim II.

The *Homo* cranium from Steinheim is much more gracile than that of Bilzingsleben. Besides the discussed assumption that the cranium represents a female from same population as Bilzingsleben, it cannot be excluded that "*Homo steinheimensis*" immigrated at a younger interglacial together with Asian mammals. This might open the question of whether middle Pleistocene hominids always immigrated from the same core area, most probably from the southwest, or if morphological differences may indicate different areas of origin, including possibly an independent immigration from the east. Thus it seems problematic and premature to include "*Homo steinheimensis*" in *Homo heidelbergensis* (Johanson 1998). Such a synonymy may obscure obvious differences that have been known for a long time (Gieseler 1974; Ziegler 2006).

Weimar-Ehringsdorf is a travertine deposit of predominately interglacial character (Kahlke 1974, 1975). Human remains and a rich *Elephas antiquus* fauna were excavated from the lower travertine, which is of particular interest here. This site was traditionally regarded as Eemian, but several lines of evidence indicate that the Weimar-Ehringsdorf fauna represents an older interglacial period. Some authors assume it represents an interglacial period within the Saalian complex, but according to Litt et al. (2005) there is no evidence for a full interglacial within the Saalian. Thus, most probably, Weimar-Ehringsdorf antedates the Saalian ice advances (Drenthe and Warthe). *Arvicola* is at an evolutionary stage that is intermediate between *Arvicola* from Bilzingsleben and *Arvicola* from typical Eemian sites such as Taubach and Burgtonna. Biostratigraphic correlation between these different interglacial sites is difficult because the typical *Elephas antiquus* assemblages of the late middle Pleistocene and Eemian are very similar. In the lower travertine, which produced the human remains, two Asian immigrants (*Cyrnaonyx antiqua*, an otter, and *Ursus thibetanus*, a small bear) are remarkable (Heinrich and Fejfar 1988; Koenigswald and Heinrich 1999).

The Saalian

Prior to the Saalian Complex in Central Europe, human remains or artifacts were found in interglacial environments only. But during Saalian time (OIS 6–8), the first indication was found that humans lived in Central Europe during cold periods as well.

Acheulean hand axes from Markleeberg are often cited as the oldest evidence that humans lived under glacial conditions in Central Europe (Baumann et al. 1983). The gravels at this site are thought to represent glacial conditions in the early Saalian complex. However, no faunal remains have been found directly associated with the tools. The Körbisdorf gravels at Markleeberg, which represent the main terrace of the early Saalian Complex, yielded the famous skeleton of *Mammuthus primigenius* from Pfännerhall (Töpfer 1957). But at other sites, the gravels produce significant amounts of wood remains, suggesting that they may represent other climatic conditions and not just typical glacial conditions. Thus, there is no clear-cut evidence regarding the ecology of these deposits.

The loess from Ochtendung has produced very good evidence that pre-Neanderthals were contemporary with a typical *Mammuthus primigenius* fauna. Human artifacts were excavated in association with *Rangifer*, *Coelodonta*, and *Mammuthus* remains (Bosinski et al. 1995). Nearby, at a site with similar conditions, a fragmentary human calvarium was discovered (Berg et al. 2000). Age dating is very good, since the sites are situated on top of small volcanic cones that erupted about 200,000 years ago. These sites are regarded as being about 170 ka and thus correspond to a late phase of the Saalian complex, most probably OIS 6. The Ochtendung calvarium represents the earliest well-dated evidence of humans coexisting with a *Mammuthus* assemblage in Central Europe.

The Eemian

During the interglacial conditions of the Eemian, humans hunted the various mammals that made up the *Elephas* assemblage. Human remains are rare. The travertine of Taubach near Weimar (Kahlke 1977) is definitively younger than Weimar-Ehringsdorf and correlates with the Eemian. It produced a human molar. Two other sites, Gröbern near Leipzig (Mania et al. 1990), and Lehringen near Verden/Aller (Thieme and Veil 1985; Houben 2003) are to be mentioned here, although they did not yield human remains. Both sites are lake deposits with an extensive pollen record, and thus stratigraphic correlation to the Eemian is secure. These sites are significant because of the discovery of skeletons of *Elephas antiquus* together with artifacts. The elephant carcasses were butchered near the shore, but this does not necessarily imply that humans killed these animals (Koenigswald 2002).

In ecological reconstructions, late Pleistocene deposits of the Rhine River are of great interest, as they have produced both a *Mammuthus* fauna and the typical *Elephas antiquus* assemblage. Years of observations have shown that the glacial fauna comes from the upper part of the section, while thick trunks of black oak (*Quercus* sp.) characterize the lower section with faunal remains from the last interglacial. In addition to typical taxa of the interglacial fauna, *Bubalus murrensis* and *Hippopotamus amphibius* occur in several Rhine sand pits (Koenigswald 1988, 2006). Their excellent preservation and frequency preclude the possibility of redeposition from older sediments. Correlation to the last interglacial is plausible, first because of the geological situation, and also because *Hippopotamus* occurs frequently on the British Isles during the Ispwichian, but not during the preceding interglacial period.

The occurrence of *Bubalus* and *Hippopotamus* in Rhine sand pits is very significant for reconstruction of the paleoecology (Fig. 9.4). According to the fossil flora, the mean annual temperature was only 2° or 3° higher than today. Thus, *Hippopotamus* and *Bubalus,* which live in subtropical regions today, do not indicate a paleoclimate that was much warmer than that of the present day. Extant animals can tolerate lower temperatures but stay in the water to escape from cold winds. This means mild winters but cooler summers. Thus a strong maritime influence on the climate can be postulated for the Rhine area for at least part of the Eemian. This maritime influence certainly tapered off towards the east; at least *Hippopotamus* did not occur farther to the east during the Eemian.

In one of the sandpits of the Rhine River at Reilingen, a human calvarium was discovered during commercial quarrying. Morphologically the calvarium has been attributed to very different evolutionary stages (Czarnetzki 1989; Adam 1989; Condemi 1996; Ziegler & Dean 1998). The Reilingen sandpit has produced both glacial and interglacial faunal elements including *Hippopotamus*. Since high groundwater levels continuously obscure the stratigraphic section, precise stratigraphic attribution of faunal remains is difficult. Löscher (1989) argued for an Eemian age, while Ziegler and Dean (1998) consider an older age.

The Weichselian

It is assumed that the *Mammuthus primigenius* fauna immigrated into Central Europe at the beginning of the Weichselian and replaced the interglacial *Elephas antiquus* fauna. However, the palynological record shows several climatic oscillations at the beginning of the Weichselian. Unfortunately, the faunal record is not complete enough to identify exactly when the *Mammuthus primigenius* fauna did appear, nor does it reveal the sequence in which various species appeared. Different mammals undoubtedly expanded their ranges at slightly different times, according to their specific ecological requirements, but this has not been detected yet in the fossil record. The Weichselian *Mammuthus primigenius* fauna is very similar to that of the late Saalian, and while large mammals cannot be used to differentiate between the two time intervals, the evolutionary stage of *Arvicola* may be helpful since most Weichselian sites show the typical *Arvicola terrestris.*

The *Mammuthus* faunal assemblage has no ecological equivalent in the extant mammalian fauna. Thus ecological reasons are most probably responsible for the extinction of many of the large mammals. The surviving species occupy quite different biotopes. One group lives in an arctic environment, mostly in the open tundra with lemmings, caribou, and muskoxen. Another group lives in Central Asia under very continental conditions: the Saiga antelope and various rodents are included here. However, neither the tundra nor the Central Asian mountain regions can be used as a model for glacial conditions in Central Europe.

The classical Neanderthal of the Weichselian in Central Europe coexisted with a *Mammuthus* assemblage and thus lived under glacial conditions. However, on this point, one has to realize that the *Mammuthus* assemblage represents a cold steppe environment and not necessarily the extreme conditions present during the maximal extent of the glaciers. During the last glacial maximum, it is known that the number of large mammals and the number of humans were greatly reduced, or both were possibly even absent. Thus the presence of human remains or artifacts preserved with remains of a *Mammuthus* assemblage does not prove that people occupied an area throughout an entire glacial period.

In the early and middle Weichselian, sites with Mousterian cultural remains, the culture related to the Neanderthal population, are well represented in Central Europe. Many of these sites produced a rich contemporaneous *Mammuthus primigenius* fauna. However, human remains are not very common.

In northern Germany, the fragmentary human carnium from Salzgitter-Lebenstedt was found together with a typical *Mammuthus primigenius* assemblage including *Rangifer tarandus* (Kleinschmidt 1953; Gaudzinski 1998). From the lake deposits of Königsaue a complex stratigraphy of the early Weichselian with Mousterian artifacts was described (Mania and Toepfer 1973).

In southern Germany, several caves contain occupation sites with a Mousterian culture, which show the full diversity of the *Mammuthus primigenius* fauna. Typical sites include various caves near Blaubeuren, such as Sirgenstein (Koken 1912), or in the valley of the Lone near Ulm, such as Bockstein (Lehmann 1969), and Hohlenstein-Stadel (Gamble 1999). The latter produced a shaft of a human femur (Kunter and Wahl 1992). In the Weinberghöhlen near Neuburg/Donau, Mousterian levels (I-G'), as well as the later ones, provide the *Mammuthus primigenius* fauna (Koenigswald et al. 1974). Very important is the Sesselfelsgrotte in Franconia, which produced the fetus of a Neanderthal baby and two milk teeth (Rathgeber 2006). The fauna is not yet fully studied, but it is again a typical *Mammuthus* fauna. *Mammuthus* faunas lived under glacial conditions, most likely with some permafrost, but most probably not during phases when Scandinavian and Alpine glaciers were at their maximal extent.

The cave site Hunas in Franconia yielded a human tooth attributed to *Homo neanderthalensis* (Alt et al. 2005). According to recent dates, the relevant level of cave sediments is from the early Weichselian (Rosendahl et al. 2006) and not older as supposed by previous authors (Heller 1983).

Minor climatic changes, such as interstadials, do not seem to change the *Mammuthus primigenius* assemblage very much. The occurrence of *Ovibos moschatus* might indicate restricted periods of especially dry and cold conditions, but such finds are not known from Mousterian layers. The lack of exact dating of fossils older than 50,000 years precludes any meaningful consideration of whether Neanderthals lived in the area continuously, during the cold stages of OIS 4, or not.

The fauna does not indicate any significant ecological change at the time when modern humans arrived in Central Europe. The *Mammuthus primigenius* fauna remained unaltered until the beginning of the last glacial maximum (LGM), as shown at the various Aurignacian and Gravettian sites. After the LGM (last glacial maximum), the glacial fauna is characterized once more by a highly continental climate, but the diversity of the fauna is significantly reduced. Thus, changes in the fauna happened distinctly later than the arrival of modern humans.

Conclusions

In Central and Western Europe, the major climatic changes that took place during the middle and late Pleistocene led to repeated immigration of a *Mammuthus* assemblage and an *Elephas* assemblage. The immigration of these faunas occurred in alternations, and transitional stages are virtually unknown. It is more likely that each faunal assemblage underwent local extinction when a changed environment favored the other.

During most of the middle Pleistocene, humans occurred in Central Europe only during interglacial periods, and they seem to have immigrated, along with the interglacial *Elephas antiquus* fauna, from Mediterranean regions. Thus, based on the faunal record and ecological analysis, the series of human remains from Mauer, Bilzingsleben, Steinheim, Weimar-Ehringsdorf, and Neanderthal do not represent genetic continuity through continuous settlement but multiple re-immigrations.

Human immigrants did not necessarily originate from the same area each time, e.g., southern France or Spain, where continuous human occupation was probable. Immigration of humans from the southeast cannot be excluded during some interglacials, especially when eastern faunal elements occur too, as at Steinheim/Murr.

Even when humans were able to live under glacial conditions with a *Mammuthus* fauna, since the Saalian, it is not certain whether areas north of the Alps were continuously occupied. During the time of maximal extension of the Scandinavian and Alpine ice sheets, life north of the Alps was probably difficult, for fauna and for humans.

Despite its fairly rich fossil record, Central Europe probably played only a marginal role in the evolution of the Neanderthal lineage during the middle Pleistocene. The faunal history is very complex in this area, due to major changes in a sometimes harsh environment, and this led to repeated faunal replacements.

Acknowledgements My studies on Pleistocene faunal communities were supported by the Deutsche Forschungsgemeinschaft in several projects throughout several decades. I am indebted to many colleagues for inspiring discussions on Pleistocene faunas of Central Europe and human interaction. I want to thank especially Silvana Condemi, Marseilles; Wolf-Dieter Heinrich, Berlin; Thijs van Kolfschoten, Leiden; Thomas Litt, Bonn; Elaine Turner, Neuwied; and Reinhard Ziegler, Stuttgart and to the two reviewers Mikael Fortelius, Helsinki and Eric Delson, New York. English is not my mother tongue and the text was kindly revised by Carole Gee and Philip Gingerich. Many thanks to D. Kranz and G. Oleschinski, Bonn, who provided the artwork and photos, and to Frank Menger, Groß-Rohrheim, and the Felsbergmuseum Beedenkirchen for the loan of the *Bubalus* skulls.

References

Adam, K. D. (1954a). Quartärforschung am staatlichen Museum für Naturkunde Stuttgart. *Beiträge zur Naturkunde, 167*, 1–14.

Adam, K. D. (1954b). Die mittel-pleistozänen Faunen von Steinheim an der Murr (Württemberg). *Quaternaria, 1*, 131–144.

Adam, K. D. (1989). Alte und neue Urmenschen-Funde in Südwestdeutschland – eine kritische Würdigung. *Quartär, 39*, 177–190.

Adam, K. D. (2003). Der *Homo steinheimensis* im Spannungsfeld von Alt- und Neumensch. *Veröffentlichungen des Landesamtes für Archäologie Sachsen-Anhalt Landesmuseum für Vorgeschichte, 57*, 29–42.

Adam, K. D., Bloos, G., & Ziegler, R. (1995). Steinheim/Murr, N of Stuttgart – locality of *Homo steinheimensis*. In W. Schirmer (Ed.), *Quaternary field trips in Central Europe* (Vol. 2, pp. 727–728). München: Pfeil.

Agassiz, L. (1840). *Études sur les glaciers*. Switzerland (Jentand Gassman): Neuchâtel.

Alt, K. W., Kaulich, B., Reisch, L., Vogel, H., & Rosendahl, W. (2005). The Neanderthalian molar from Hunas, Germany. *HOMO Journal of Comparative Human Biology, 12*, 187–200.

Azzaroli, A., De Giuli, C., Ficarelli, G., & Torre, D. (1982). Late Pliocene to early mid-Pleistocene mammals in Eurasia: Fauna succession and dispersal events. *Paleogeography, Paleoclimatology, Palaeoecology, 66*, 77–100.

Baumann, W., Mania, D., Toepfer, V., & Eissmann, L. (1983). Die Paläolithischen Neufunde von Markkleeberg bei Leipzig. *Veröffentlichungen des Landesmuseums für Vorgeschichte Dresden, 16*, 1–280.

Berckhemer, F. (1927). *Buffelus murrensis* n. sp. – Ein diluvialer Büffelschädel von Steinheim a. d. Murr. *Jahreshefte des Vereins für vaterländische Naturkunde in Württemberg, 83*, 146–158.

Berg, A.v, Condemi, S., & Frechen, F. (2000). Die Schädelkalotte des Neanderthalers von Ochtendung/Osteifel: Archäologie, Paläoanthropologie und Geologie. *Eiszeitalter und Gegenwart, 50*, 56–68.

Bosinski, G., Street, M., & Baales, M. (1995). The palaeolithic and mesolithic of the Rhineland. In W. Schirmer (Ed.), *Quaternary fieldtrips in Central Europe* (pp. 829–999). München: Pfeil.

Condemi, S. (1996). Does the human fossil specimen from Reilingen (Germany) belong to the *Homo erectus* or to the Neanderthal lineage? *L'Anthropologie, 34*, 69–77.

Condemi, S. (1998). The Neanderthals: A cold adapted European middle Pleistocene population? *L'Anthropologie, 36*, 35–42.

Czarnetzki, A. (1989). Ein archaischer Hominidencalvariarest aus einer Kiesgrube in Reilingen, Rhein-Nekar-Kreis. *Quartär, 39*(40), 191–201.

Gamble, C. (1999). The Hohlenstein-Stadel revisited. In The role of humans in the accumulation of European lower and middle palaeolithic bone assemblages. Monographien des Römisch-Germanischen Zentralmuseums Mainz 42 (pp. 305–324).

Gamble, C., William, D., Pettitt, P., & Richards, M. (2004). Climate change and evolving human diversity in Europe during the last glacial.– *Philosophical Transactions Royal Society London B*, 359:243–254.

Gaudzinski, S. (1998). Knochen und Knochengeräte der mittelpaläolithischen Fundstelle Salzgitter-Lebenstedt (Deutschland). *Jahrbuch des Römisch-Germanischen Zentralmuseums Mainz, 45*, 163–220.

Geikie, J. (1894). *The great ice age, and its relation to the antiquity of man* (p. 850). London: Stanfort.

Geyh, M. A., & Müller, H. (2005). Numerical ^{230}TMJ dating and a palynological review of the Holsteinian/Hoxnian interglacial. *Quaternary Science Reviews, 24*, 1861–1872.

Geyh, M. A., & Müller, H. (2006). Missing evidence for two Holstein-like Interglacials. Reply to the comments by J.D. Scourse on: Numerical 230Th/U dating and a palynological review of the Holsteinian/ Hoxnian Interglacial. *Quaternary Science Reviews, 25*, 3072–3073.

Gieseler, W. (1974). *Die Fossilgeschichte des Menschen* (pp. 1–247). Stuttgart: Fischer.

Heinrich, W. D. (1982). Zur Evolution und Biostratigraphie von *Arvicola* (Rodentia, Mammalia), im Pleistozän Europas. – Zeitschrift geologischer Wississenschaften 10(6): 683–735.

Heinrich, W. D., & Fejfar, O. (1988). Fund eines Lutrinen (Mammalia: Carnivora, Mustelidae) aus dem Unteren Travertin von Weimar-Ehringsdorf in Thüringen. *Zeitschrift geologischer Wissenschaften, 16*(6), 515–529.

Heller, F. (Ed.) (1983). Die Höhlenruine Hunas bei Hartmannshof (Landkreis Nürnberger Land) – Eine paläontologische und urgeschichtliche Fundstelle aus dem Spät-Riß. Quartär-Bibliothek 4 (pp. 1–407).

Houben, C. (2003). Die Wirbeltierfauna aus dem letzten Interglazial von Lehringen (Niedersachsen, Deutschland). *Eiszeitalter und Gegenwart, 52*, 25–39.

Hublin, J. J. (1990). Les peuplements paléolithiques de l'Europe: un point de vue paléobiogéographique. *Mémoire de Musèe de Préhistoire d'Ile de France, 3*, 29–37.

Johanson, D. (1998). *Lucy und ihre Kinder*. In: *Paléolithique Moyen Récent et Paléolithique Supérieur Ancien en Europe*, ed. C Farizy. Nemours (p. 272). Heidelberg: Spektrum.

Kahlke, H. D. (Ed.) (1974). Das Pleistozän von Weimar Ehringsdorf, Teil 1. – Abhandlungen Zentrales Geologisches Institut, Paläontologische Abhandlung 21 (pp. 1–351).

Kahlke, H. D. (Ed.) (1975). Das Pleistozän von Weimar-Ehringsdorf, Teil II. – Abhandlungen. Zentrales Geologisches Institut, Paläontologische Abhandlung 22 (pp. 1–594).

Kahlke, H. D. (Ed.) (1977). Das Pleistozän von Taubach bei Weimar. Quartärpaläontologie 2 (pp. 1–509).

Keilhack, K. (1899). Die Stillstandslagen des letzten Inlandeises und die hydrographische Entwicklung des pommerschen Küstengebietes. *Jahrbuch der Preussischen Geologischen Landesanstalt, 19*, 90–152.

Kleinschmidt, A. (1953). Die zoologischen Funde der Grabung Salzgitter-Lebenstedt. *Eiszeitalter und Gegenwart, 3*, 166–188.

Koenigswald, W. v. (1973). Veränderungen in der Kleinsäugerfauna von Mitteleuropa zwischen Cromer und Eem (Pleistozän). *Eiszeitalter und Gegenwart, 23/24*, 159–167.

Koenigswald, W. v. (1986). Beziehungen des pleistozänen Wasserbüffels (*Bubalus murrensis*) aus Europa zu den asiatischen Wasserbüffeln. *Zeitschrift für Säugetierkunde, 51*, 312–323.

Koenigswald, W. v. (Ed.) (1988). Zur Paläoklimatologie des letzten Interglazials im Nordteil der Oberrheinebene. Paläoklimaforschung 4. Stuttgart: Fischer.

Koenigswald, W. v. (2003). Mode and causes of the Pleistocene turnovers in the mammalian fauna of Central Europe. *Deinsia, 10*, 305–312.

Koenigswald, W. v. (2006). Mammalian faunas from the interglacial periods in Central Europe and their stratigraphic correlation. In F. Sirocko, M. Claussen, M. F. Sanchez Goñi, & T. Litt (Eds.), *The climate of past interglacials* (pp. 445–454). Amsterdam: Elsevier.

Koenigswald, W. v. & Heinrich, W. D. (1999). Mittelpleistozäne Säugetierfaunen aus Mitteleuropa – der Versuch einer biostratigraphischen Zuordnung. *Kaupia, 9*, 53–112.

Koenigswald, W. v. (2002). Lebendige Eiszeit – Klima und Tierwelt im Wandel. Darmstadt (Wissenschaftliche Buchgesellschaft) und Stuttgart (Thesis), 1–190.

Koenigswald, W. v., Müller-Beck, H., & Pressmar, E. (1974). Archäologie und Paläontologie in den Weinberghöhlen von Mauern (Bayern) Grabungen 1937-1967. Archäologica Venatoria 3 (pp. 1–152).

Koken, E. v. (1912). Geologie und Tierwelt der paläolithischen Fundstellen Deutschlands. In R. R. Schmidt (Ed.), *Die Diluviale Vorzeit Deutschlands*. Stuttgart (Schweitzerbart) (pp. 159–226).

Kotsakis, T. (2006). Evolution of the mammalian Pleistocene faunas in the Mediterranean area. Terra Nostra, 2006/2 (pp. 30–35).

Kunter, M., & Wahl, J. (1992). Das Femurfragment eines Neandertalers aus der Stadelhöhle des Hohlensteins im Lonetal. Fundberichte aus Baden-Württemberg 17/1 (pp. 111–124).

Lehmann, U. (1969). Die Fauna. In R. Wetzel, & G. Bosinski (Ed.), *Die Bocksteinschmiede im Lonetal*. Veröffentlichungen des staatlichen Amtes für Denkmalpflege Stuttgart A 15 (pp. 133–167).

Litt, T. (1994). Paläoökologie, Paläobotanik und Stratigraphie des Jungquartärs im nordmitteleuropäischen Tiefland. *Dissertationes Botanicae, 227*, 1–185.

Litt, T. (2006). Climate, vegetation and mammalian faunas in Europe during the middle Pleistocene Interglacials (MIS 7, 9, 11). In F. Sirocko, M. Claussen, M. F. Sanchez Goñi, & T. Litt (Eds.), *The climate of past interglacials* (pp. 351–357). Amsterdam: Elsevier.

Litt, T., Ellwanger, D., Villinger, E., & Wansa, S. (2005). Das Quartär in der Stratigraphischen Tabelle von Deutschland 2002. *Newsletter of Stratigraphy, 41*, 385–399.

Löscher, M. (1989). Das Alter des Reilinger Schädels aus geologischer Sicht. *Quartär, 39/ 40*, 203–208.

Mania, D., & Thomae, M. (2006). Pleistozänstratigraphie und Paläolithikum im mittleren Elbe-Saale-Gebiet. 73. Tagung der Arbeitsgemeinschaft Norddeutscher Geologen 6.-9. Juli 2006, Halle.

Mania, D., & Toepfer, V. (1973). Königsaue. *Veröffentlichungen des Landesmuseums für Vorgeschichte in Halle, 26*, 1–164.

Mania, D., Mania, U., Heinrich, W. -D., Fischer, K., Böhme, G., Turner, A., Erd, K., & Mai, D. -H. (1997), Bilzingsleben V: *Homo erectus* – seine Kultur und Umwelt. 1-265. Bad Homburg-Leipzig (Ausbildung und Wissen).

Mania, D., Thomae, M., Litt,.T, & Weber, T. (eds.) (1990). Neumark-Gröbern: Beiträge zur Jagd des mittelpaläolithischen Menschen. (p. 320) Berlin: Deutscher Verlag der Wissenschaften.

Maul, C., Rekovets, L., Heinrich, W. D., Keller, Th., & Storch, G. (2000). *Arvicola mosbachensis* (Schmidtgen 1911) of Mosbach 2: A basic sample for the early evolution of the genus and a reference for further biostratigraphical studies. *Senckenbergiana Lethaea, 80*(1), 129–147.

Nitychoruka, J., Binka, K., Ruppert, H., & Schneider, J. (2006). Holsteinian interglacial = Marine isotope stage 11? *Quaternary Science Reviews, 25*, 2678–2681.

Nehring, A. (1880). Uebersicht über vierundzwanzig mitteleuropäische Quartär-Faunen. – Zeitschrift der Deutschen geologischen Gesellschaft 32: 468–505.

Nehring, A. (1990): Ueber Tundren und Steppen der jetzt- und Vorzeit, mit besonderer Berücksichtigung ihrer Faunen Berlin (p. 257).

Parfitt, S. A., et al. (2005). The earliest record of human activity in northern Europe. *Nature, 438*, 1008–1012.

Penck, A., & Brückner, E. (1901–1909). Die Alpen im Eiszeitalter: 1–3. Leipzig (Tauchnitz).

Rathgeber, T. (2006). Neandertaler-Kinder aus der Sesselfelsgrotte im Altmühltal, Bayern Deutschland). In G. Ülsberg(Ed.), Roots/ Wurzeln der Menschheit (Rhein. Landesmuseum Bonn) (p. 326).

Rosendahl, W., Kaulich, B., Alt, K. W., Hambach, U., & Reisch, L. (2006). The Neandertalian site Hunas: 50ky (OIS 5b-OIS 3) climate and environment in Southern Germany. Terra Nostra 2006/2 (pp. 132–133).

Sarnthein, M., Stremme, H. E., & Mangini, A. (1986). The Holstein interglaciation: Time-stratigraphic position and correlation to stable-isotope stratigraphy of deep-sea sediments. *Quaternary Research, 26*, 283–298.

Schötensack, O. (1908). Der Unterkiefer des *Homo heidelbergensis* aus den Sanden von Mauer bei Heidelberg. Leipzig.

Schreve, D. C. (2001). Differentiation of the British late middle Pleistocene interglacials: The evidence from mammalian biostratigraphy. *Quaternary Science Reviews, 20*, 1693–1705.

Scourse, J. (2006). Comment on: Numerical 230Th/U dating and a palynological review of the Holsteinian/Hoxnian interglacial by Geyh and Müller. *Quaternary Science Reviews, 25*(2006), 3070–3071.

Sher, A. (2006). Pleistocene faunal and environmental evolution: Adding a north eastern dimension to the European story. Terra Nostra 2006/2 (pp. 40–45).

Stringer, C. B., & Hublin, J. J. (1999). New age estimates for the Swancscombe hominid, and their significance for human evolution. *Journal of Human Evolution, 37*, 873–877.

Thieme, H., & Veil, S. (1985). Neue Untersuchungen zum eemzeitlichen Elefanten-Jagdplatz Lehringen, Ldkr. Verden. Die Kunde N.F. 36 (pp. 11–58).

Toepfer, V. (1957). Die Mammutfunde von Pfännerhall im Geiseltal. *Veröffentlichungen des Landesmuseums für Vorgeschichte Halle, 16*, 1–58.

Torell, O. (1875). Über einen gemeinschaftlich mit Herrn Berendt und Orth nach den Rüdersdorf Kalkbergen gemachten Ausflug. *Zeitschrift der deutschen geologischen Gesellschaft, Berlin, 27*, 961–962.

Young, C. C. (1936). New finds of fossil *Bubalaus* in China. *Bulletin Geological Society China, 15*, 505–519.

Ziegler, R. (2006). Der Urmensch von Steinheim an der Murr, Baden-Württemberg (Deutschland). In G. Uelsberg (Ed.), Roots/Wurzeln der Menschheit. Bonn (Rheinisches Landesmuseum) (p. 314).

Ziegler, R., & Dean, D. (1998). Mammalian fauna and biostratigraphy of the pre-Neandertal site of Reilingen, Germany. *Journal of Human Evolution, 34*, 469–484.

Chapter 10
Neanderthal Geographical and Chronological Variation

Bernard Vandermeersch and María Dolores Garralda

Abstract There is now a reasonable sample of human fossils from the European Middle and beginning Upper Pleistocene. However, our ability to fully understand their evolutionary relationships and the part they played in the ancestry of the Neanderthals remains uncertain. Part of the reason for this is the fragmentary nature of many of the finds, with fossils preserving different anatomical features, making detailed anatomical comparisons difficult or impossible. An equally important obstacle to our knowledge of this part of human biological history are the often difficult to interpret and conflicting dates that have been obtained for many of these finds. Nevertheless, a number of fossils, including the sizable sample from the Sima de los Huesos, testify to the European ancestry of the Neanderthals, although the possible presence and gen-flow from groups of Asian or maybe African origins cannot be excluded.

At present, our knowledge on the Neanderthals must consider the huge geographic area where they have been identified, the chronological span throughout more than 140 ky, and the very diverse environments to which they adapted. The second part of this chapter is the summary of the main points about the Neanderthal variability and biodynamics.

Keywords Neanderthal variability • Middle Pleistocene hominins • Human evolution • Europe • Western Asia • Dating

B. Vandermeersch (✉)
UMR 5199 Laboratoire d'Anthropologie des Populations du Passé, Université de Bordeaux, 33405 Talence, France
e-mail: bvanderm@bio.ucm.es

M.D. Garralda (✉)
UMR 5199 Laboratoire d'Anthropologie des Populations du Passé, Université de Bordeaux, 33405 Talence, France
and
U.D. de Antropología Física. Facultad de Biología, Universidad Complutense de Madrid, 28040 Madrid, Spain
e-mail: mdgarral@bio.ucm.es

Introduction

Neanderthal variability can be studied from two points of view, chronological and geographical. The first one consists not only of reconstructing the origin and evolution of the Neanderthal morphology, but also in determining if prior to their evolution there were several species coexisting in Europe, or only one. Further, considering the latter possibility there are two current interpretations: either the Neanderthals have a long history as an evolutionary species, *Homo neanderthalensis*, or they derived from *Homo heidelbergensis*.

Geographical variability can only be analyzed at the end of the Neanderthal lineage, during OIS 4, when the maximum territorial expansion of this group seems to be documented, and the human remains are more numerous.

We will deal shortly with these two questions. Our aim is not to provide answers, but to focus on some current problems in the study of Neanderthal evolutionary biology.

Chronological Variability

To analyze this problem it is necessary to examine the chronology of some European fossils, in relation to their morphology, in order to consider the possibility of the presence of one or two species in Europe at the beginning of the Middle Palaeolithic (~at about 250 ky). An additional aspect of this question is the identification of *Homo heidelbergensis* in Europe.

Mauer

The Mauer fossil presents two important and unsolved problems: its unknown stratigraphical position (thus, its age is uncertain) and the fact that it is an isolated mandible (Fig. 10.1). Both uncertainties make it very difficult to interpret this specimen from a phylogenetic perspective.

Accidentally discovered in 1907, at the bottom of a loess quarry, the exact level from where the fossil originated is

S. Condemi and G.-C. Weniger (eds.), *Continuity and Discontinuity in the Peopling of Europe: One Hundred Fifty Years of Neanderthal Study*, Vertebrate Paleobiology and Paleoanthropology, DOI 10.1007/978-94-007-0492-3_10, © Springer Science+Business Media B.V. 2011

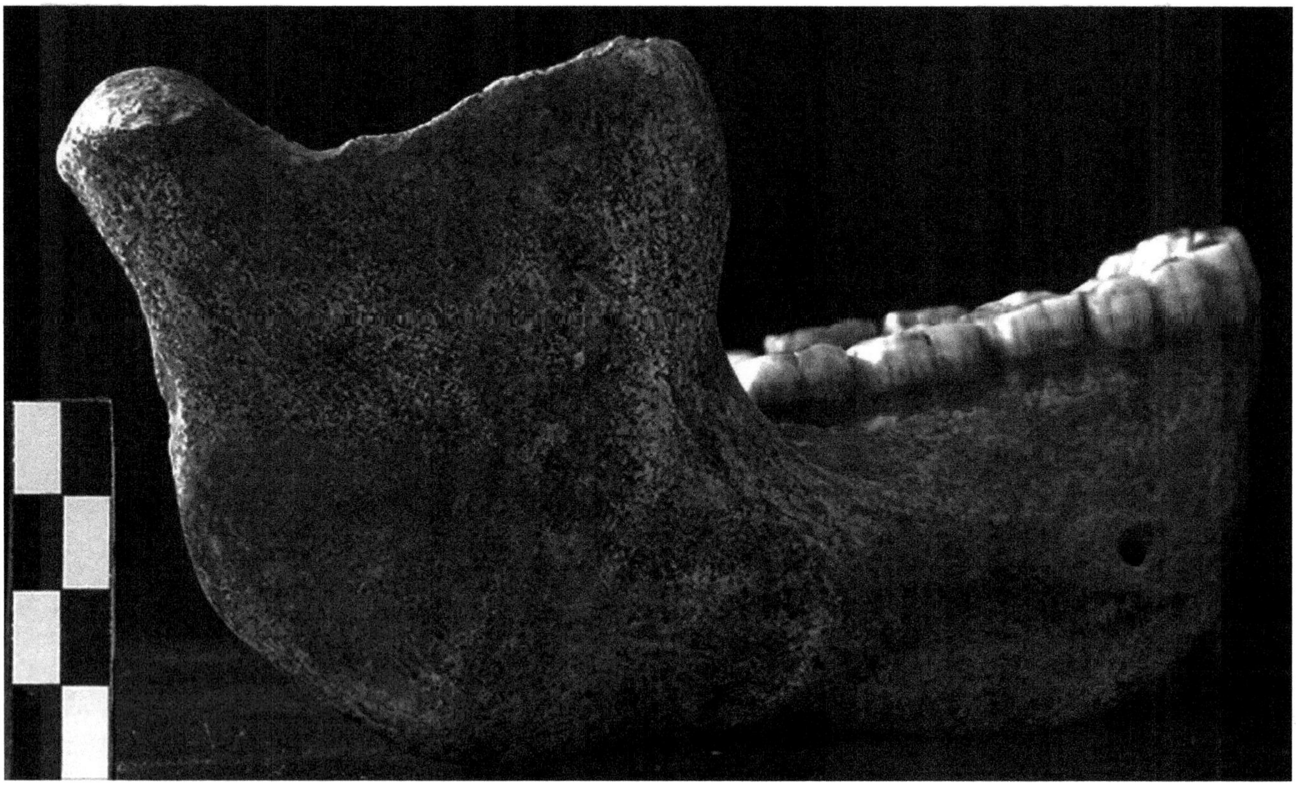

Fig. 10.1 Mandible of Mauer (Germany), *right lateral view* (Photo courtesy of A. Mounier)

unknown, and the quarry has since been altered. Consequently, it is impossible to replace it in its correct chronostratigraphical position. However, Wagner and Beinhauer (1997) have produced a remarkable work combining the analysis of ancient documents with a new study of the site furnishing more precise information about this discovery. According to their data, the mandible comes from the inferior sands of Mauer (Fundschicht) and can be assigned to isotopic stages 13–15. Thus, age estimation for this fossil is between 474 and 621 ky (Wagner and Beinhauer 1997).

In the monograph published in 1908, Schoetensack focused on characteristics which he called "primitive," creating the species *Homo heildelbergensis*. But at that time, the age of the fossil was totally unknown and the interpretations later proposed depended on the antiquity attributed to the fossil. For Piveteau, for instance, the mandible dated from the "Final Villafranchian"; however, in the "Traité de Paléontologie" (1957), he examined it together with the Neanderthals, but in a separate chapter. During the symposium celebrating the centenary of the Neanderthal discovery, von Koenigswald (1958) included Mauer in a list of German Neanderthals, but without comment because of the fossil's primitive morphology. More recently, the title of the book commemorating the 85th anniversary of the Mauer discovery is: "Schichten – 85 Jahre *Homo erectus heidelbergensis*" (Beinhauer and Wagner 1992), and still more recently, another publication reattributed to the fossil the name of the species created by Schoetensack (Wagner and Beinhauer 1997). These remarks illustrate the difficulties in the interpretation of this mandible. It possesses unquestionable archaic characteristics, but resembles neither Asiatic or African *Homo erectus*, nor the Neanderthals. It does not present typical Neanderthal features such as the backward position of the mental foramen, the retromolar space, the relation of the extremity of the sigmoid notch and the condyle, or the development of the anterior teeth (Condemi and Koenigswald 1997).

For some researchers, Mauer and some other European fossils, such as Arago and Petralona, represent one European species, *Homo heidelbergensis*, which could be a "grade" on the Neanderthal lineage (Rosas and Bermúdez de Castro 1998).

Arago

The "Caune de l'Arago" (Tautavel, France) is one of the major sites of the ancient Palaeolithic of southern France. The excavations, directed by H. de Lumley, uncovered numerous human remains, including the anterior part of a skull and two mandibles (H. de Lumley and M.-A. de Lumley 1971; de Lumley 1982). Most of them, especially the skull N° 21, come from layer III, with an age of around 450 ky on the basis of the associated mammals (Iacumin et al. 1996). Direct dating of cranium Arago 21 by uranium series and gamma spectrometry gave a similar result, but with a high degree of uncertainty (Yokoyama and Nguyen 1981).

If the age of 450 ky is correct, this fossil is possibly contemporaneous with, or slightly older than those from the Sima de los Huesos at Atapuerca (Spain), which is in accordance with the morphological data. The Arago skull presents archaic characteristics, such as a very receding frontal, the morphology of the *torus supra-orbitalis,* and the shape of the palate. It also possesses Neanderthal features such as the development of the facial region and the obliquely backward malar bone. These characteristics are more developed on Neanderthal skulls, but there is no doubt that Arago is on the lineage which evolved to the later Neanderthals.

However, the mandibles Arago 13 and Arago 2 do not present derived Neanderthal characteristics and they are very different from each other. Their dimensions and robustness are different, but both have some similarities with the Mauer mandible, demonstrating the difficulty of interpreting an isolated jaw. Considering that the Arago mandibles lack real Neanderthal characteristics, while the upper face displays some of them, it is possible that Mauer, a little more ancient, could correspond to a human group displaying similar "mosaic" morphology.

Boxgrove

An incomplete tibia and two isolated teeth were uncovered at the Boxgrove site (Sussex, England), and dated from the OIS 13, around 500 ky, confirmed by geological, macro- and microfaunal data (Stringer et al. 1998; Stringer 2006). Thus, these remains are penecontemporaneous with the Mauer mandible. Associated with Acheulean artifacts, the tibia is exceptionally robust. The study of diaphyseal sections shows proportions comparable with those of the Neanderthals and seem to reflect cold adaptation (Trinkaus et al.1999).

However, because it is an adaptation known in various species of homeotherm vertebrates and appears independently in various human populations, it remains uncertain if this resemblance to the Neanderthals has any phylogenetic significance. Nevertheless, this fossil may represent, with Mauer, Arago, and Swanscombe, "early members of a western European lineage that culminated in the last glacial Neanderthals" (Trinkaus et al. 1999). It has been attributed to *Homo* cf. *heidelbergensis* by its discoverers who, however, remarked that the morphological characteristics of the tibia largely overlap various populations, making it difficult to propose a precise taxonomic status for Boxgrove.

Atapuerca – Sima de los Huesos

At least 27 individuals, represented by skulls, mandibles, and postcranial elements, have been recovered from the site of La Sima de los Huesos at Atapuerca Hill (Burgos, Spain). Although there are important chronological problems, the site has become the most important European Middle Pleistocene site (Arsuaga et al. 1997).

Combined U-series and ESR, dating of both speleothem and human bones, appeared to provide a minimum age of about 200 ky and suggestive evidence of possible entry prior to 320 ky (Bischoff et al. 2003). More recently, however, reanalysis of the speleothem produced dates of around 530/600 ky (Bischoff et al. 2007). The presence of *Panthera leo* suggests a maximum age of 600 ky because this is the lower limit of its presence at the Iltalian site of Isernia. The rodent *Mimomis savini* is absent in the Sima de los Huesos, while it is present in layer TD8 of the Gran Dolina, dated to about 596/615 ky (ESR and U-series on mammal teeth, Falguères et al. 1999). Thus, the Sima de los Huesos sample is probably less than 600 ky, with a reasonable estimate of their age between 350 and 450 ky.

Morphologically, the Sima de los Huesos fossils appears younger than Mauer and Boxgrove, and possibly also Arago, but they are much more complete, with remarkably preserved skulls.

In their study of the crania, Arsuaga et al. (1997) emphasized the presence of archaic characteristics which are absent on the Neanderthals (the sagittal keel, for instance), but they also documented the presence of derived features, such as a protruding middle face. But the Neanderthal features do not have the same development as on the typical Neanderthals, and the authors cited above considered that the skulls of Saccopastore (Italy; Fig. 10.2) were more similar to the Sima de los Huesos specimens than the typical Neanderthals, some of them of very recent chronology.

There is no doubt that the Sima de los Huesos series represents a stage in the process of "Neanderthalization," which took place in Europe throughout the Middle Pleistocene and the beginning of the Upper Pleistocene. These series were on the Neanderthal lineage, even if all the apomorphies are not yet present. Skull N° 5, for instance, presents a face with many of the traits associated with Neanderthal midfacial prognatism, but the shape and morphology of the vault are quite different from those of the Neanderthals.

Vérteszöllös

The site of Vérteszöllös is situated on a terrace of the Atalér river, 50 km to the west of Budapest. It was a quarry where, in 1965, an occipital and a deciduous molar were found associated with fauna and a Clactonian industry (Vértes 1965). Four layers were identified at the site, the human remains coming from the lowest.

The fauna is abundant and has been attributed to an interstadial of the Mindel glaciation (Kretzoi and Vértes 1965). Initial results by the U-series technique gave an age

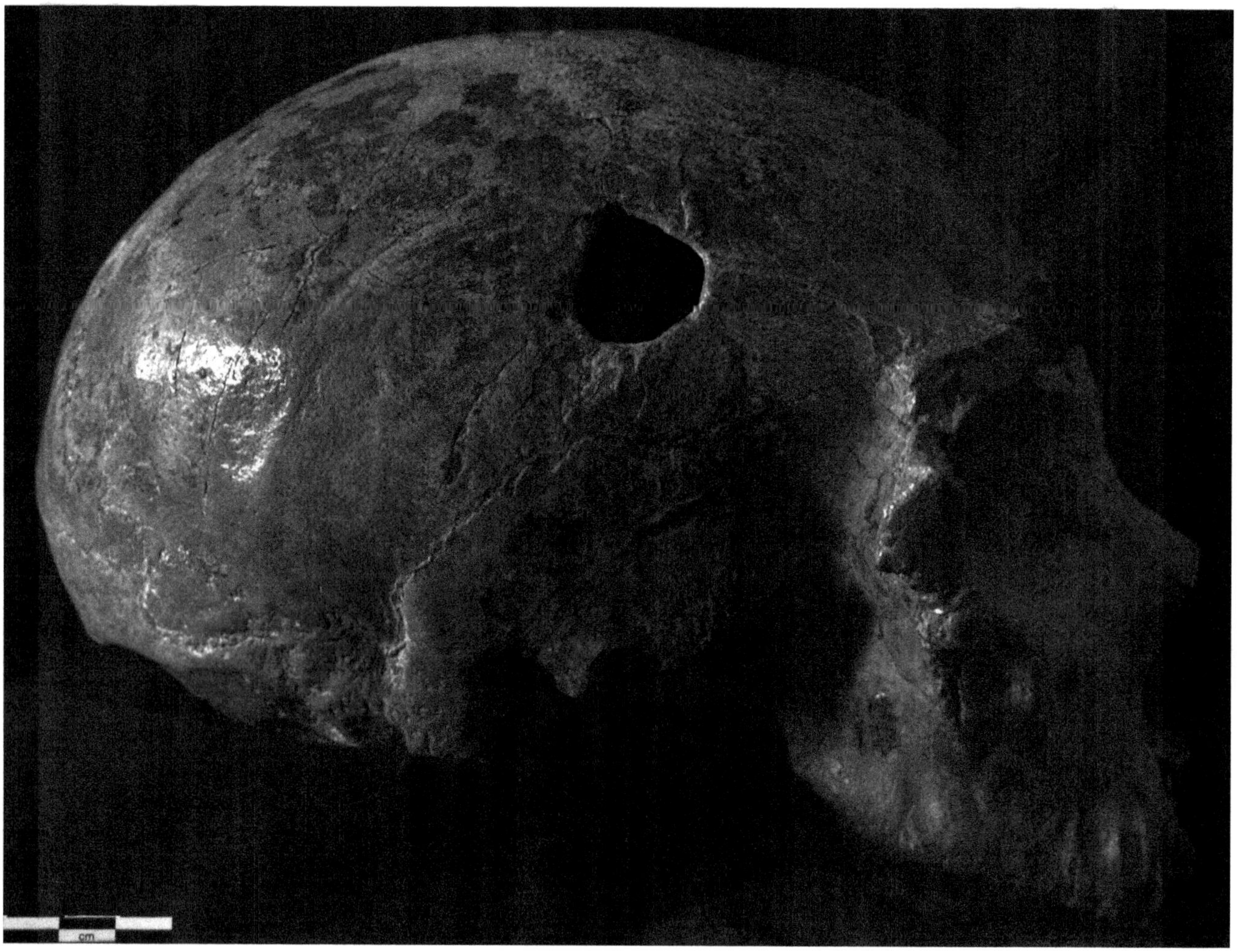

Fig. 10.2 Skull of Saccopastore 1 skull (Italy), *right lateral view* (Photo courtesy of A. Mounier)

of between ~250 and 475 ky (Cherdyntsev et al. 1965); later determinations were between ~250 and 350 ky (Cherdyntsev 1971). More recently, other samples indicated ~185/210 ky (Schwarcz and Lathan 1990). These radiochronological dates are very imprecise, but the abundant macro- and microfauna indicate a Mindel interstadial, and an age of ~350 ky. Vérteszöllös could be more or less contemporaneous with la Sima de los Huesos, or slightly younger, but in order to know the chronological position of the Vérteszollos remains, it would be necessary to obtain more precise dates.

The occipital, described by Thoma (1966), does not possess the general morphology of the Neanderthal occipital in the *torus*, or the suprainiac fossa, but it has a *torus* directed upwards toward the sagittal region, which resembles that of ancestral populations, although it has a tubercle at the end of the right side. Hublin (1988) suggested the possibility that this morphology could have evolved to the Neanderthalian supra-iniac fossa if the area progressively became more and more concave, simultaneously decreasing in height and developing the tubercles at both extremities.

But, strictly speaking, Vérteszöllös occipital lacks Neanderthal apomorphies, and it must be remembered that the fossil corresponds to a period when the human remains reflect a wide diversity. We ignore the morphology of other regions of the Vérteszöllös skull.

Bilzingsleben

Bilzingsleben is an open air site, in the Wipper Valley (Thuringia, Germany), 35 km north of Erfurt. The archaeological deposits are in the lowest part of a travertine sequence, and excavations directed by D. Mania discovered numerous human fragments representing at least three individuals. Two adult skulls have been partially reconstructed.

The industry is a variety of micro-Clactonian and the associated fauna is abundant (Mania et al. 1980). Paleontological and paleobotanical data indicate a moist climate which could correspond to the Holstein interglacial,

~280–300 ky (Mania and Thomae 2006). But U-series analysis gave an older age, around 350 ky (Schwarcz et al. 1988).

Cranial vaults are thick. The frontal fragments possess projecting *torus* and very receding squamas. There are numerous occipital fragments, angulated and with a very robust *torus*. Those fossils lack Neanderthal traits, and with such a morphology it is difficult to integrate them into the Neanderthal lineage. Vlček (1989) and Mania et al. (1994) assigned them *Homo erectus*.

In conclusion, the Bilzingsleben human remains appear to be different from the Pre-Neanderthals. They also seem more recent than Arago, and penecontemporaneous to Vérteszöllös and La Sima de los Huesos. This interpretation means that *two lineages* were present in Europe during the Middle Pleistocene, but several cautions must be observed. The first is that the Neandertal lineage probably evolved by accretion and fragmentary fossils such as those of Bilzingsleben may not present apomorphies that are present, at the same time, on others. Additionally, the inaccuracy of many dates and, consequently, that the chronological position of the European fossils is still unclear.

Swanscombe

The site of Swanscombe (Kent, England) is a stratified deposit of gravels and clays. At the lowest part of the deposits were found, in 1935, 1936, and 1955, the two parietals and the occipital from the same skull, with a rich fauna and an Acheulean industry.

The fossil is assigned to the Holstein interglacial on the basis of the fauna, and could be ~300 ky old, although geomorphological investigations of the terrace estimate an age of about 400 ka (Stringer and Hublin 1999). The incomplete skull is very different from the Bilzingsleben human remains, because the occipital is rounded, without a centrally strong *torus*. On the contrary, the weak occipital *torus* displays a bilateral projection and is surmounted by a central suprainiac fossa and the occipital plane is strongly convex.

Steinheim

The skull of Steinheim (Germany) was discovered in 1933, in a river deposit of gravels and clays with a diversified fauna assigned to the OIS 7, ~225 ky (Adam 1954a, b, 1985). There was no lithic industry.

The skull is crushed and deformed but many characteristics can be observed. Hublin (1988) showed clearly that the occipital area can be integrated into the Neanderthal evolutionary line. The face is short with a concave area below the orbits and an angulated malar bone. This aspect could be accentuated by the *post mortem* distortion, but the face was not typically Neanderthal, and the fossil can represent another example of the mosaic evolution.

Petralona

The Petralona skull (Greece) was found accidentally (in 1959), in a cave 37 km from Thessaloniki. Because the sediments were covered by a stalagmitic floor that partially covered the skull, the stratigraphical position of the fossil is unknown. Liritzis (1980) identified two layers, one brown-red and another paler. The same brown-red layer was identified at the top of the stalagmitic deposit of the cave. Unfortunately, the U-series dates are very inaccurate, ranging from 150 ky to more than 350 ky. ESR determinations show a similar range, from 127±37 to 340 ky. More recently, Grün (1996) reanalyzed the ESR dates and concluded that the age was ~150–250 ky, which is in accordance with most of U-series results.

Morphologically, the skull (Figs. 10.3 and 10.4) presents, like Arago 21, a Neanderthal-like protruding mid-face, associated with some archaic features, such as a prominent and angular *torus occipitalis*.

Biache-Saint-Vaast

Two partial skulls were exhumed from Biache-Saint-Vaast (France) in a terrace of the river Scarpe. Biache-Saint-Vaast 1 was found in place in 1976, while Biache-Saint-Vaast 2 was found later, fragmented and mixed with faunal remains.

Biache 1 was just above the layer IIA. The pollen, the microfaune, and the molluscs correspond to a temperate climate event in the Saale glaciation (Tuffreau et al. 1978). The layer is dated by thermoluminescence at ~175 ky (Huxtable and Aitken 1988). The industry is an abundant Mousterian of La Ferrassie type (Tuffreau 1988).

On Biache 1, only the half posterior part of the skull is preserved, exhibiting characteristic Neanderthal morphology, with suprainiac fossa, small mastoid apophysis, and protruding occipitomastoid crest. The general pattern of the vault is also typical, with a transverse profile "*en bombe*" and an occipital curvature very similar to those of La Chapelle-aux-Saints or La Ferrassie (Rougier 2003).

The problems of estimating populational morphology by taking into account only one individual are well demonstrated with the Biache sample: the unpublished and fragmentary male adult Biache 2 corresponds to a very robust individual with a thick and protruding *torus supraorbitalis*, unexpected when considering Biache 1. The differences in what parts of each fossil are preserved make attemps at morphological comparisons extremely difficult (Biache 2,

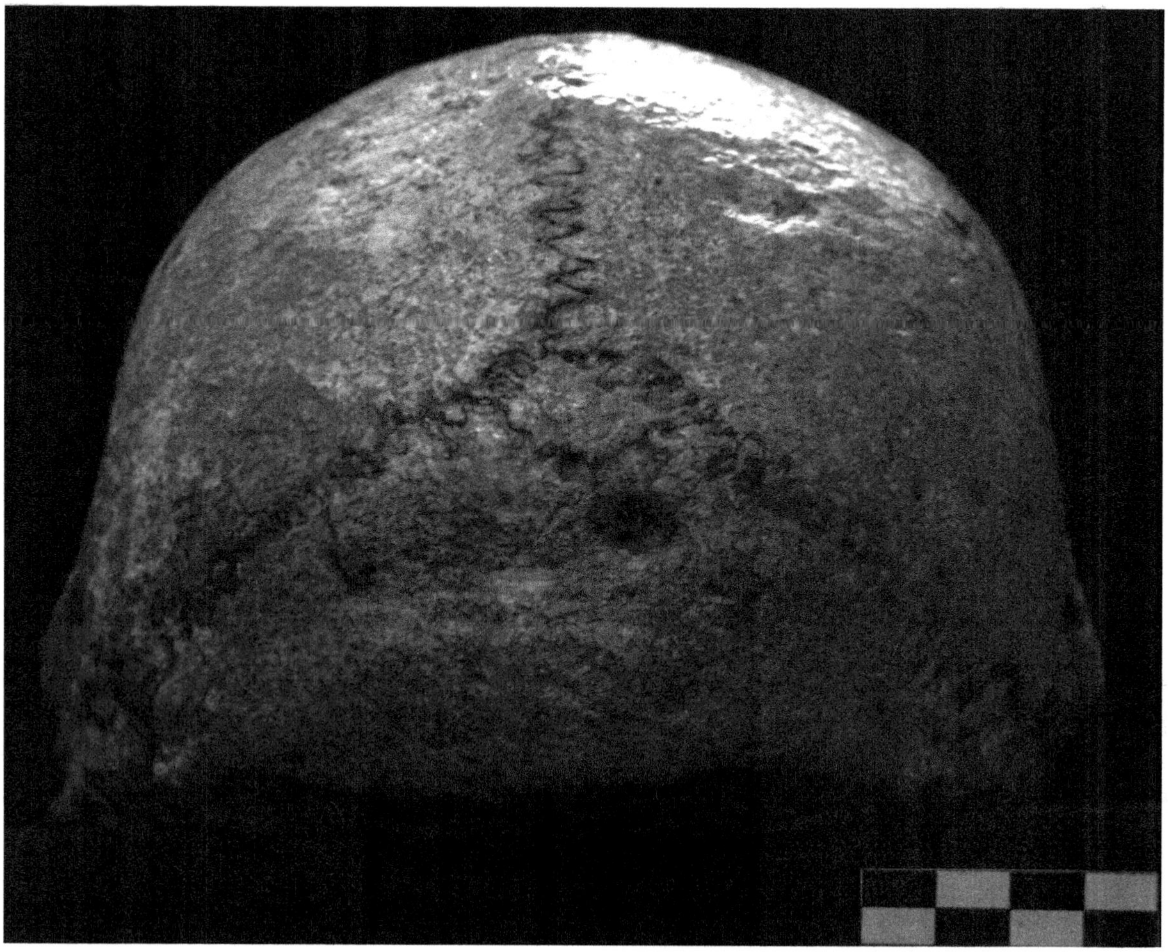

Fig. 10.3 Skull of Petralona skull (Greece), *occipital view* (Photo courtesy of A. Mounier)

for example, lacks the posterior portion of the vault present in the presumed young female, Biache 1).

After the time of the Biache-Saint-Vaast fossils, ~175 ky (OIS 6), all European human remains can be assigned to the Neanderthal group.

Conclusions

There are some other European fossils between 500 and 150 ky, such as Fontana Ranuccio or Castel di Guido (Italia) which are possibly older than 400 ky, and those from Apidima (Grecia), Reilingen (Germany), and Montmaurin (France), but all of them are difficult to place chronologically. Although they were not discussed in the present paper, they do not alter our conclusions.

Taking into account their morphological characteristics, the Bilzingsleben fossils – probably a little more recent than those from Arago and La Sima de los Huesos – diverge from the Preneanderthals from the two latter sites. This observation may perhaps indicate that until about 350 ky ago, or a little earlier, two evolutionary lineages existed in Europe, one of late *Homo erectus*, another of Preneanderthals. Nevertheless, caution must be exercised, not only because of the difficulties associated with comparisons of fragmentary fossil specimens, often preserving different parts of their anatomy, but also due to the tentative nature of the chronological placement of many of them. Thus, the presence in Europe of two populations throughout the Middle Pleistocene can be only a hypothesis.

This raises several questions about the concept of *Homo heidelbergensis*. As usually presented for Europe, it is possible to distinguish two periods in the Neanderthal lineage, two chronospecies, *Homo heidelbergensis* (Preneanderthals) and *Homo neanderthalensis*. However, this results in the lumping together into the same taxon, *H. heidelbergensis*, isolated or fragmentary bones, such as the Mauer mandible or the Boxgrove tibia, too incomplete to be rigorously interpreted, with other more complete fossils, such as Arago, Swanscombe, Petralona, or La Sima de los Huesos, which present a mixture of archaic and modern features. The problem

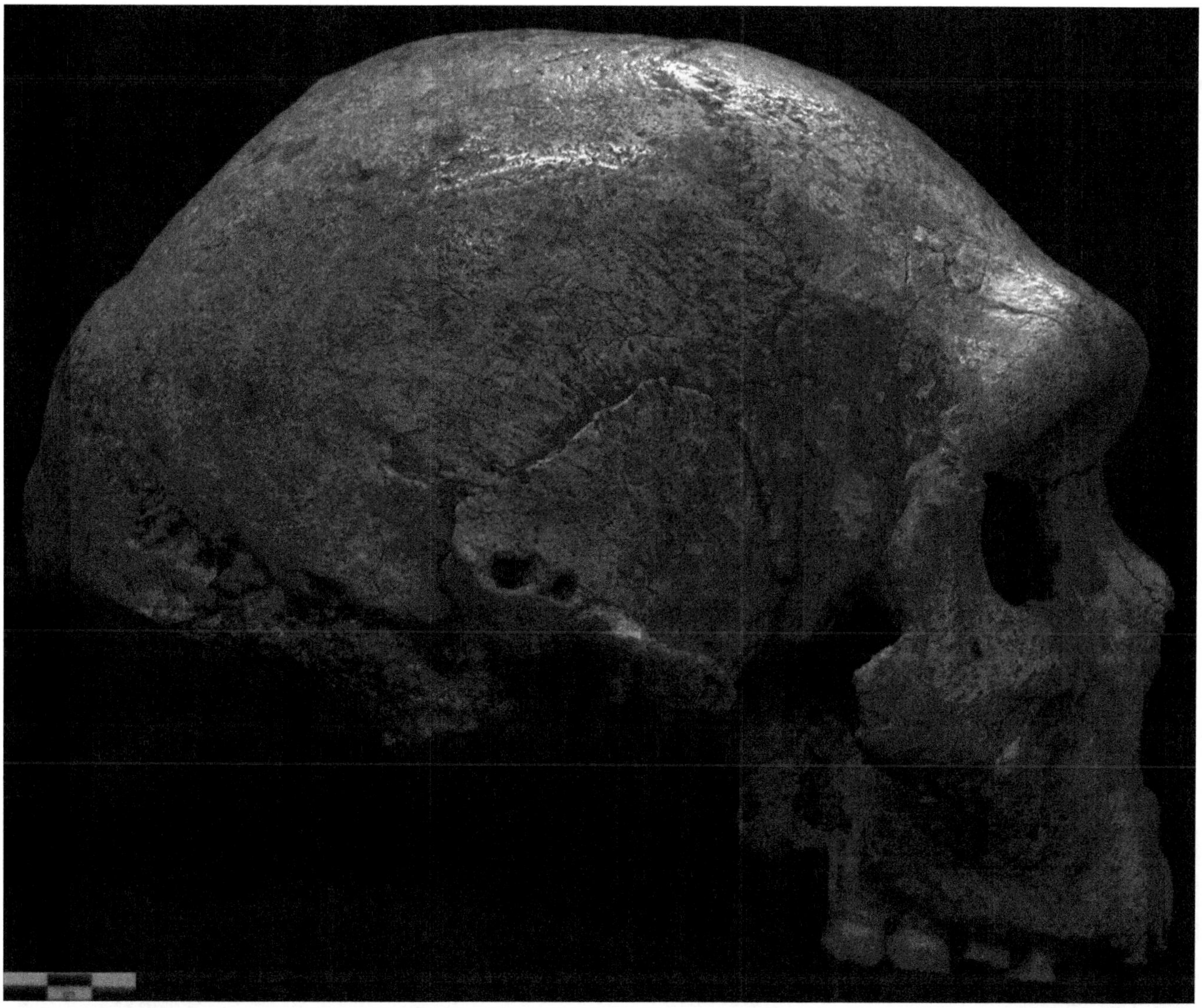

Fig. 10.4 Skull of Petralona skull (Greece), *right lateral view* (Photo courtesy of A. Mounier)

is that several of these modern features are generally considered as Neanderthal apomorphies. In any event, if the taxon *H. heidelbergensis* is to be preserved, it will be necessary to establish a functional diagnosis, which, in our opinion, does not exist, and this would inevitably entail a revision of the apomorphies of the species *H. neanderthalensis.*

Faunal studies (Koenigswald, present volume) demonstrate that during most of the Middle Pleistocene, humans are predominantly documented in Central Europe during interglacial periods and seem to have migrated with elements from the Mediterranean and Southeast regions. So, very probably, human fossils document not a "genetic continuity" but multiple migrations from different origins and in small groups.

Consequently, the biological history of Europe between ~500 and ~300 ky was probably more complex than usually considered. The possibility that gene flow existed, among the diverse groups moving throughout Eurasia, cannot be excluded, although, if such was the case, it was necessarily moderated because of the scant number of individuals. This complexity is also probably reflected in much older fossils such as Ceprano or Gran Dolina, especially the ATD6-96 mandible (Carbonell et al. 2005).

The Diversity of the Neanderthals

Known Neanderthal remains extend over a huge geographic area across all European regions not covered by ice, and a great part of Asia, from the north of the Black Sea, Turkey, Near and Middle East (Syria, Israel, Iraq, Iran) to Central Asia, where, in Uzbekistan, was found the Teshik Tash site. At present, the eastern and northernmost fossils are the isolated teeth from Denisova and Okladnikov in the Altai Mountains, in the south central region of Siberia.

Such a vast geographical area, covering more than 11,000,000 km^2, deserves consideration of the very diverse environments, many of them periglacial, where these Neanderthal populations lived and to which they adapted. Accordingly, living conditions, the flora and the fauna varied during the warm and cold periods, depending also on the different latitude or altitude of each region. These different paleoenvironments must have had important consequences on the biodynamics of the human groups, but are difficult to appreciate because of the fragmentation and dispersal of the human remains.

Chronologically, Neanderthal remains have been identified between 170/160 ky and perhaps 30 ky. There is thus a partially documented evolutionary history of more of ~130 ky, more than 5,200 generations. This is important to take into consideration because throughout this period, numerous macro- and microevolution factors must have acted with different intensity, according to the circumstances of each population. Geochronological studies indicate that the oldest typical Neanderthal remains appear during a warm and humid period, the Riss-Würm (or Eemien) interglacial, which corresponds to OIS 5. Nevertheless, most of the fossils are attributed to the two cold periods of the Early Würm (OIS 4), whereas the most recent findings corresponded to the Hengelo interstadial (OIS 3) with mild climatic conditions.

There are ~400 Neanderthal individuals discovered throughout this vast territory, although most of them are isolated bones and teeth, frequently fragmentary and incomplete. To estimate the biological diversity of these human groups, and therefore to identify regional populations (representing clinal groups), it would be necessary to have skeletal series, or at least several complete individuals, which is not the case. In reality, the whole sample is poor and incomplete, and the example of the Spy crania (Figs. 10.5 and 10.6) shows that individual variation can be significant. Genetics demonstrate that intrapopulational variation is much more important than that existing between two different populations.

In the Spy case, the two craniums (Figs. 10.5 and 10.6) present relatively significant differences. Spy II is shorter, with higher forehead and less prominent *torus*; its sagittal profile shows higher cranial vault and the occipital less extended towards the rear. It also has been shown that Spy I is slightly careened, while Spy II is not.

Those differences attracted the attention of several researchers, for example, in 1930 Hrdlička approximated Spy II to modern morphology. Later, Thoma (1975) rejected

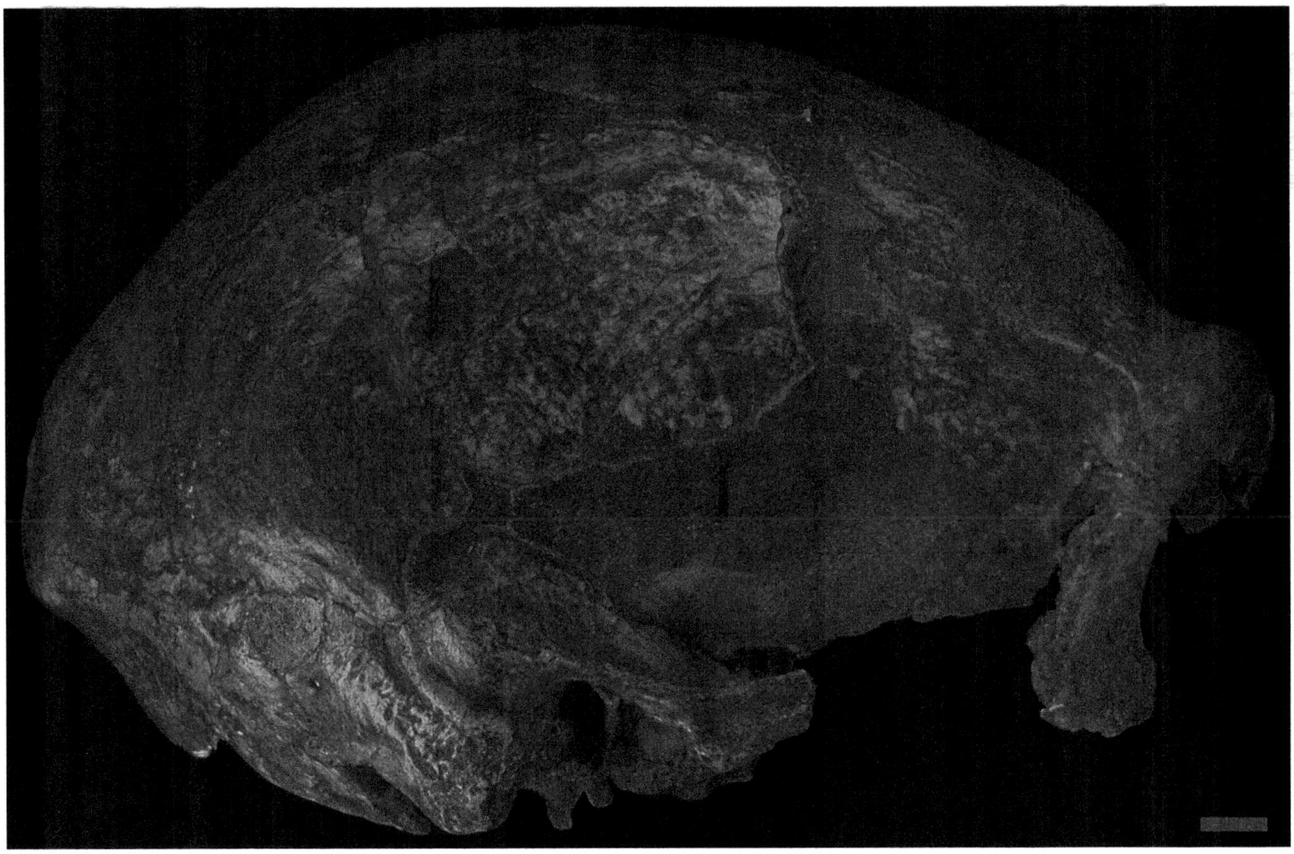

Fig. 10.5 Skull of Spy 1 (Belgium), *right lateral view* (Photo courtesy of the Institut Royal des Sciences Naturelles de Belgique, Bruxelles)

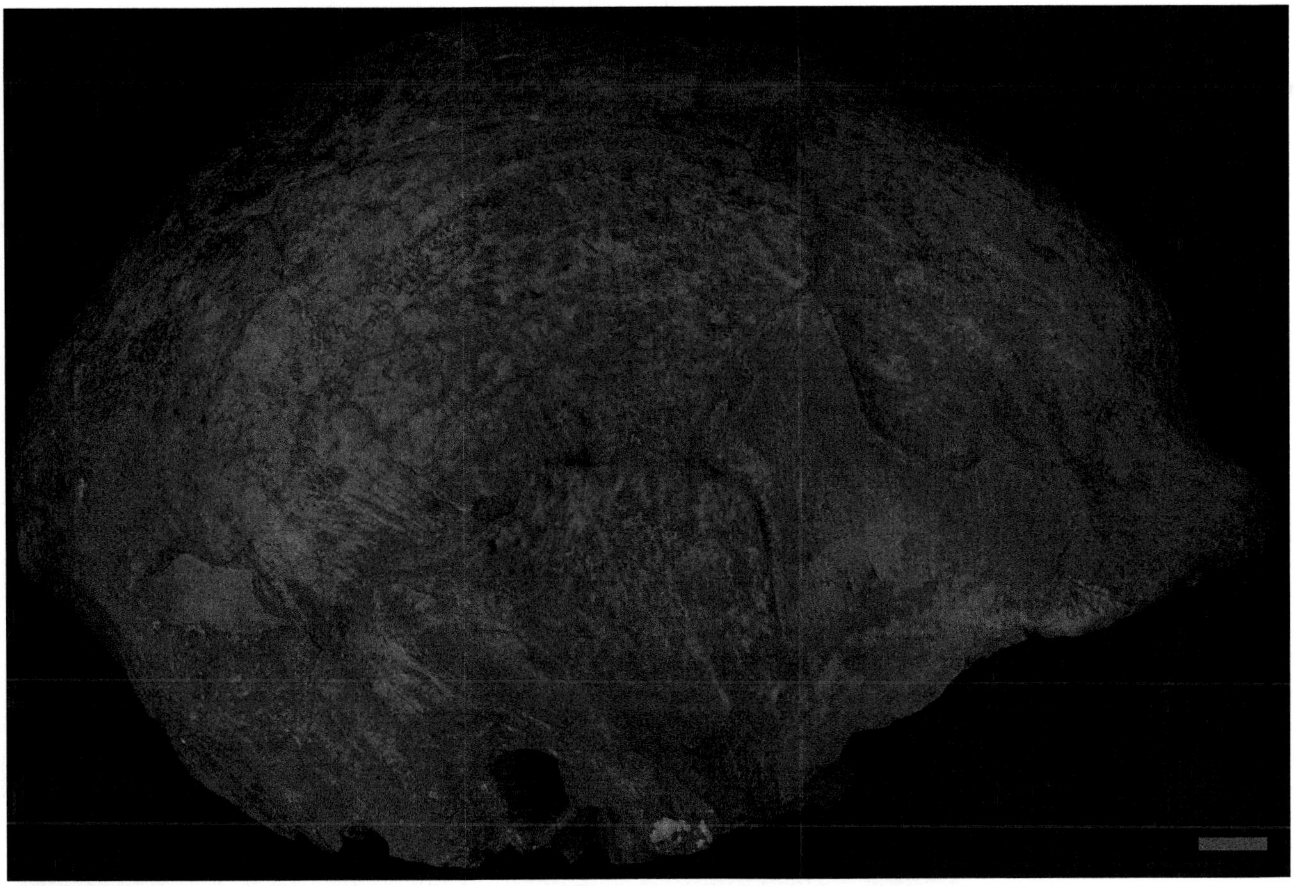

Fig. 10.6 Skull of Spy 2 (Belgium), *right lateral view* (Photo courtesy of the Institut Royal des Sciences Naturelles de Belgique, Bruxelles)

this interpretation and argued that both were Neanderthals, in spite of the evident individual differences. They were discovered very close together and at the same level, but could this variation be the reflection of their recent chronology, or of the presence of new genes in the population they represent?

At Krapina (Croatia), one of the earliest and more interesting Neanderthal samples, the presence of numerous individuals provides an estimate of sexual dimorphism as well as individual variability, well illustrated by the large collection of teeth, mandibles, postcranial skeletons, and incomplete crania. For example, the Krapina 5 skull, recently published by Caspari and Radovcic (2006), possesses great robusticity in contrast with other adult crania from the same site.

Unfortunately, the European fossil data set is so limited that it is not possible to use modern analytical criteria to examine problems of chrono-spacial, individual, sexual, intra- and interpopulational variability. Considering the available data, the morphology of the Mediterranean Neanderthals appears to be somewhat different from that of Neanderthals inhabiting higher latitudes. Biological differences also may have existed between groups living in Western and Central Europe, but, at present, this idea will remain tentative until new fossil finds fill the numerous gaps in the record. In any case, there is sufficient variability amongst the known Neanderthal sample to initiate a preliminary investigation of these differences.

The Saint-Césaire individual (Fig. 10.7) is a good example. The cranial gracility and the small dimensions of the teeth are remarkable compared to other Neanderthals. Taking into account its chronology and cultural context, the small size of the Saint-Césaire dentition might be related to the end of the Neanderthal lineage, individual variation, or to sexual dimorphism provided that the Charentian fossil is a female.

In the study of the L'Hortus human remains, de Lumley (1972, 1973) wrote that those coming from the upper levels were more slender, with smaller teeth, than the Neanderthals found in the lower layers, suggesting the existence of a "Mediterranean population" which differed from other groups. But, to date, there has been no additional data to confirm this hypothesis.

Undoubtedly, the biodynamics of different Neanderthal populations was influenced by many evolutionary factors including normal variation, natural selection for diverse environments, and small population size. However, because of the enormous area occupied by the Neanderthals and the time range of their existence, an understanding of the influence of each of these factors will be difficult to calculate.

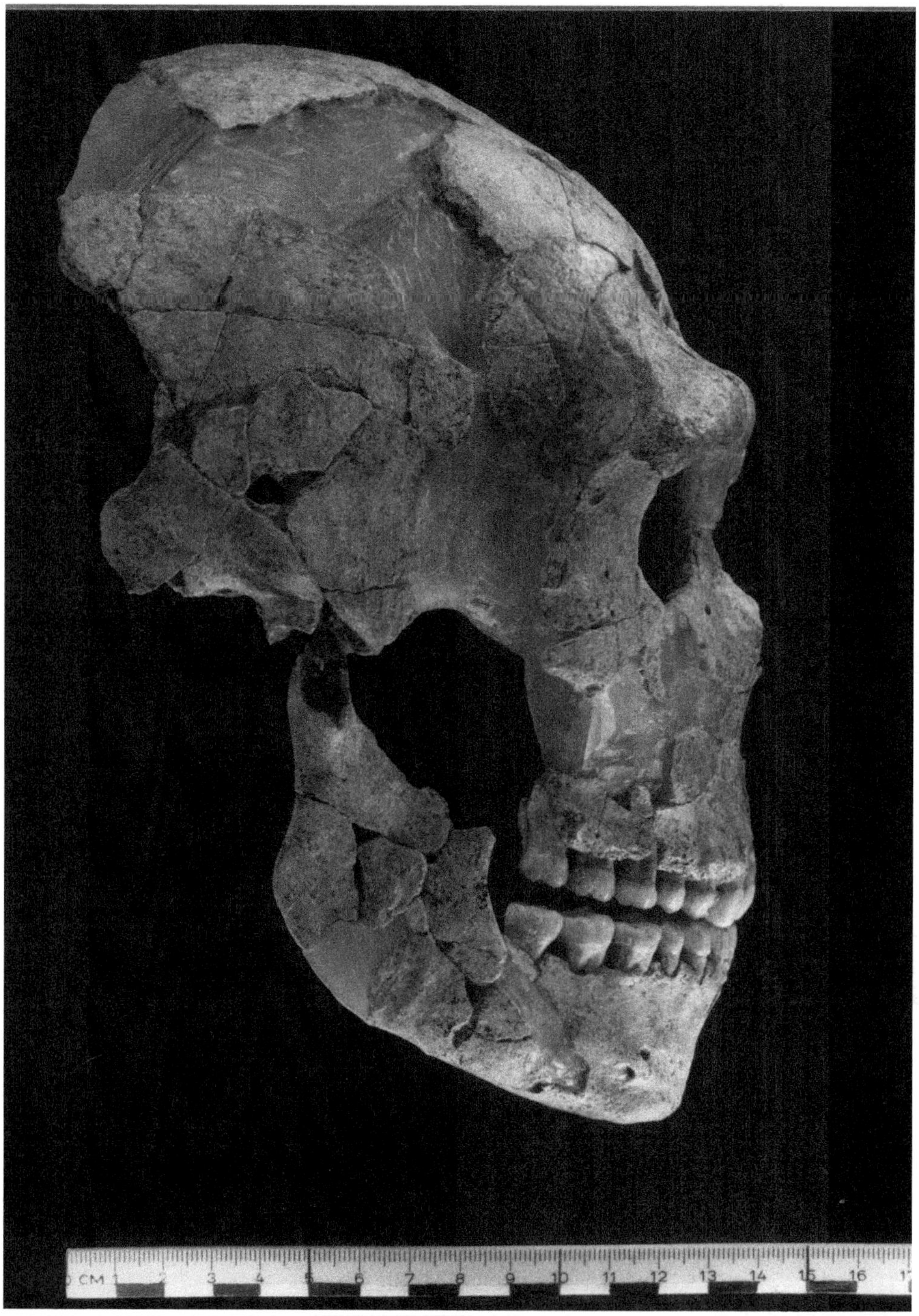

Fig. 10.7 Skull of Saint-Césaire (France) (Photo B. Vandermeersch)

It is also noteworthy that the data coming from the analyses of the mtDNA of ~16 Neanderthal remains has revealed the existence of polymorphisms, especially between that of the earliest, Scladina, and samples from later-in time fossils. However, in our opinion the differences cannot be exclusively attributed to the influence of genetic drift or natural selection, since genetic processes usually have a more complex background.

From the morphological point of view, Neanderthal diversity appears more evident when the European fossils are compared with those of the Near East. In this comparison, two vast regions are separated by the Mediterranean, the Black Sea, and the foothills of the Caucasus. Demographic movements and contacts between human groups in South-Western Asia and Western Europe were probably infrequent at the beginning of the Upper Pleistocene, so the presence of variation between them is not surprising.

Indeed, grouping all the fossils from South Western Asia can underestimate the distance separating those sites (for example, Amud and Shanidar are about 1,000 km apart), as well as their different chronologies and environmental conditions.

A major problem is the uncertain antiquity of several of these fossils, especially the skeleton of the presumed female Tabun C1, considered amongst the oldest fossil specimens from the Levant. Tabun C1 was uncovered by Garrod in 1932 (Garrod and Bate 1937) in circumstances that make it impossible to establish if it was deposited during the formation of level C, or if it appeared in a *fossa* excavated from the most recent level B (Bar-Yosef and Callander 1999).

This uncertainty of its precise stratigraphic placement is crucial to the reconstruction of Neanderthal evolution in the Levant. Depending on the attribution of the fossil to level B or C, the earlier placement would document the presence of Near Eastern Neanderthals at about 170 ky (Jelinek1992), more or less comtemporary with the European Neanderthals, or, with a placement in level B, to a more recent time, around 90 ky. This situation is even more complicated when the isolated mandible (Tabun C2) from level C is considered. The morphology of this mandible has been variously interpreted as similar to that of other Levantine Neanderthals or as that of an "An atomically Modern Human."

The other sites have been dated to around 60 ky, for example Kebara (Valladas et al. 1998), and the somewhat younger Amud (Schwarcz and Rink 1998). From Iraqui Kurdistan, the Shanidar series seems to be between 60 and 46 ky (Trinkaus 1983). In comparison with the European Neanderthals (Fig. 10.8), those from the South West Asia have a more vertical forehead, a more elevated neurocranium, and less prominent occipitals. Because of the higher vault, the transversal contour is not "en bombe" as is common in the European fossils.

The central region of the face, although well developed, does not present the same backward and outward obliquity as the European Neanderthals. In the Shanidar sample, a small concavity corresponding to the *fossa canina* is present, and the malar is slightly angled below the infero-external angle of the orbit. The zygomatic arch is thicker than in the European Neanderthals, and its root is situated a bit higher in relation with the auditory meatus. The mastoid apophyses are more prominent.

Since these traits are more or less pronounced, depending on the specimen, it is reasonable to suggest that the most complete crania from Shanidar and the incomplete Amud 1 show peculiarities (autapomorphies?) specific to the South Western Asia Neanderthals. That leads to the question as to

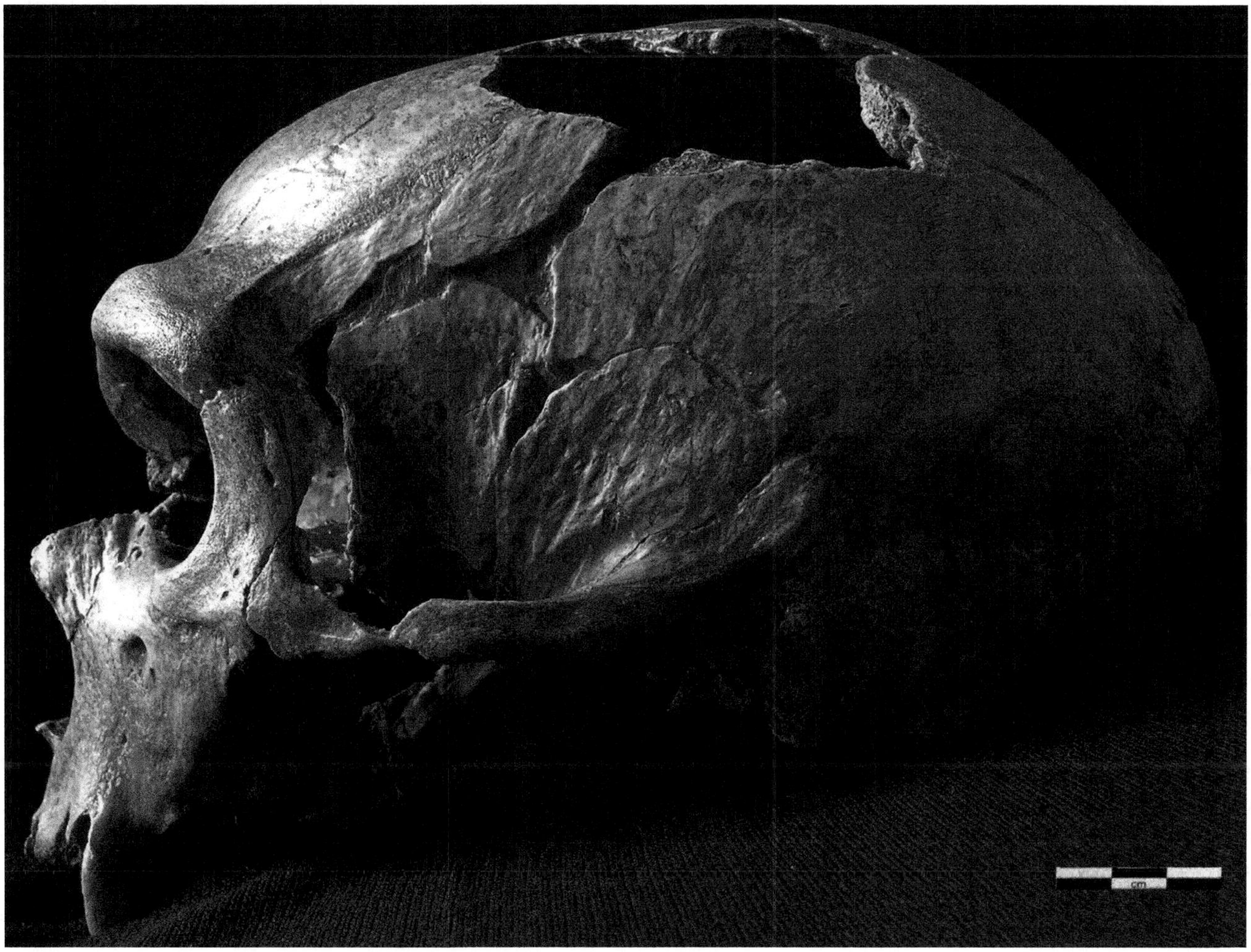

Fig. 10.8 Skull of La Chapelle aux Saints (France), *left lateral view* (Photo courtesy of A. Mounier)

when the Asiatic populations diverged from the European, assuming that both had the same ancestral origin.

At the moment, there are no Near or Middle Eastern fossils with Neanderthal apomorphies as ancient as the European Pre-Neanderthals. It is possible that the Neanderthal lineage originated and evolved in the region of today's Europe, and that from there some groups moved, perhaps under environmental pressures, to warmer regions. If the date of around 170 ky for Tabun C1 is correct, it suggests the presence of the Neanderthals in Israel prior to OIS4. Condemi's (1991) study of the Saccopastore crania pointed out several traits in common with Tabun C1, Shanidar, and Amud 1, in comparison with the so-called "classic Neanderthals." If the general characteristics of the latter correspond to cold adaptations, as generally suggested, it is possible that these features had not developed in the groups already inhabiting less rigorous climates. Though a reasonable interpretation for the moment, it can only be considered a working hypothesis. The archaeological record, particularly well known in Israel, clearly demonstrates the presence of ancestral populations whose morphology is still unknown.

Moreover, what about the Neanderthals from Central Asia? Geographic continuity between Eastern Europe and Asia over a considerable time period has been well documented, probably resulting in significant biocultural interactions. The Teshik-Tash child's skeleton in Uzbekistan (dated between 50 and 30 ky) documents the presence of Neanderthals in that region; the cranial features, however, reveal a morphology somewhat different from that observed in European Neanderthal children, perhaps reflecting variability or gene flow. There are many Mousterian sites in Central Asia, but this vast region, with the exception of the few human fossils found in the Altai caves (Denisova end Okladnikov), is at present almost totally unknown.

Conclusions

The diverse data just summarized about the Neanderthals provide a complex, but still incomplete view of their morphological characteristics, as well as their long evolutionary history, which appears to be linked to the late Pre-Neanderthals and, through them, to still earlier Eurasian origins. Nevertheless, many problems remain unsolved.

There are many gaps in the study of human biological history and this is especially the case when the Neanderthals are considered. After 150 years of polemic, compelling anthropological, genetic, and cultural data has now been presented to reject the often cited image of the Neanderthals as a morphologically uniform population, the result of an almost linear evolution in Europe. The peopling of Europe was probably varied until the Holstein period and later the remains of the Neanderthals, through their long evolutionary history, and the vast geographic territory in which they appear, reflect chrono- and geographically diversified populations, which we have just begin to glimpse.

Acknowledgments The authors would like to give special thanks to Prof. Alan Mann for the careful revision of the text and his valuable comments.

References

Adam, K. D. (1954a). Die zeitliche Stellung des Urmenschen Fundsicht von Steinhein an des Murr Innerhalb des Pleistozans. *Eiszeialter u. gegenwart, 4/5*, 18–21.

Adam, K. D. (1954b). Die mittelpleistozänen Faunen von Steinheim an der Murr (Württemberg). *Cuaternaria, 1*, 131–144.

Adam, K. D. (1985). The chronological and systematic position of the Steinheim skull. In E. Delson (Ed.), *Ancestors: The hard evidence* (pp. 272–276). New York: Alan Liss.

Arsuaga, J. L., Martínez, I., Gracia, A., Lorenzo, C., & García, N. (1997). The Sima de los Huesos crania (Sierra de Atapuerca, Spain), a comparative study. *Journal of Human Evolution, 33*, 219–281.

Bar-Yosef, O., & Callander, J. (1999). The woman from Tabun: Garrod's doubts in historical perspective. *Journal of Human Evolution, 37*(6), 879–885.

Beinhauer, K. W., & Wagner, G. A. (1992). *Schichten – 85 Jahre Homo erectus heidelbergensis von Mauer*. Mannheim: Braus. Reiss-Museum des Stadt.

Bischoff, J. L., Shamp, D. D., Aramburu, A., Arsuaga, J. L., Carbonell, E., & Bermúdez de Castro, J. M. (2003). The Sima de los Huesos Hominids date to beyond U/Th equilibrium (>350 kyr) and perhaps to 400–500 kyr: New radiometric dates. *Journal of Archaeological Science, 30*, 275–280.

Bischoff, J. L., Williams, R. W., Rosenbauer, R. J., Aramburu, A., Arsuaga, J., García, N., & Cuenca-Bescós, G. (2007). High-resolution dates from the Sima de los Huesos hominids yields 600 ± 66 kyrs: Implications for the evolution of the early Neanderthal lineage. *Journal of Archaeological Science, 34*, 763–770.

Carbonell, E., de Bermúdez Castro, J. M., Arsuaga, Jl, Allue, E., Bastir, M., Benito, A., Cáceres, I., Canals, T., Díez, J. C., van der Made, J. C., Mosquera, J., Ollé, A., Pérez-González, A., Rodríguez, J., Rodríguez, X. P., Rosas, A., Rosell, J., Sala, R., Vallverdú, J., & Vergés, J. M. (2005). An Early Pleistocene hominin mandible from Atapuerca-TD6 Spain. *Proceedings of the National Academy of Sciences of the United States of America, 102*(16), 5674–5678.

Caspari, R., & Radovcic, J. (2006). New reconstruction of Krapina 5, a male Neanderthal cranial vault from Krapina Croatia. *American Journal of Physical Anthropology, 130*(3), 294–307.

Cherdynstsev, V. V. (1971). *Uranium 234*. Tel Aviv: Israel Program for Scientific Translation.

Cherdynstsev, V. V., Kazachevsky, I. V., & Kuzmina, E. A. (1965). Age of Pleistocene carbonate formation according to thorium and uranium isotopes. *Geokhimiya, 9*, 1085–1092.

Condemi, S. (1991). *Les Hommes fossiles de Saccopastore et leurs relations phylogénétiques* (Cahiers de Paléoanthropologie). Paris: CNRS.

Condemi, S., & Koenigswald, W. (1997). Der Unterkiefer von Mauer. In G. A. Wagner & K. W. Beinhauer (Eds.), *Homo heidelbergensis von Mauer. Das Auftreten des Menschen in Europa* (pp. 200–214). Heidelberg: HVA.

de Lumley, M.-A. (1972). Les Néandertaliens de la grotte de l'Hortus (Valflaunes, Hérault). In H. de Lumley (Ed.), *La Grotte de l'Hortus* (pp. 375–385). Marseille: Université de Provence.

de Lumley, M.-A. (1973). *Anténéandertaliens et Néandertaliens du Bassin Méditerranéen occidental Européen*. Paris: CNRS.
de Lumley, H. (Ed.). (1982). *L'Homo erectus et la place de l'homme de Tautavel parmi les hominidés fossiles*. Nice: Colloque. Prétirage Cong. Intern. Pal. Hum.
de Lumley, H., & de Lumley, M. A. (1971). Découverte de restes anté-néandertaliens datés du début du Riss à la Caune de l'Arago (Tautavel, Pyrénnées-Orientales. *Comptes Rendus de l'Académie des Sciences Paris, 272*, 1729–1742.
Falguères, C., Bahain, J. J., Yokoyama, Y., Arsuaga, J. L., Bermúdez de Castro, J. M., Carbonell, E., Bishoff, J. L., & Dodo, J. M. (1999). Earliest humans in Europe. The age of TD6 Gran Dolina, Atapuerca, Spain. *Journal of Human Evolution, 33*, 343–352.
Garrod, D. A., & Bate, M. A. (1937). *The stone age of Mount Carmel: Excavation at the Wadi-el-Mughara*. Oxford: Clarendon.
Grün, R. (1996). A re-analysis of electron Spin resonance dating results associated with the Petralona hominid. *Journal of Human Evolution, 30*, 227–241.
Hrdlička, A. (1930). The skeletal remains of early man. Smithsonian miscellaneous collections. *American Ethnology, 83*, 267–270.
Hublin, J. J. (1988). Les plus anciens représentants de la lignée néandertalienne. In E. Trinkaus (Ed.), *L'Homme de Néandertal 3 – L'anatomie* (pp. 81–94). Liège: ERAUL.
Huxtable, J., & Aitken, M. J. (1988). Datation par thermoluminescence. In A. Tuffreau et J. Sommé (Eds.), *Le gisement paléolithique moyen de Biache-Saint-vaast (Pas-de-Calais, Volume 1- Stratigraphie, Environnement, Etudes archéologiques (première partie)*. Mémoires de la Société Préhistorique Française T. 21, pp. 107–108.
Iacumin, P., Cominotto, D., & Longonelli, A. (1996). A stable isotope study of mammal skeletal remains of mid-Pleistocene age, Arago cave, eastern Pyrennees, France. Evidence of taphonomic and diagenetic effects. *Palaeogeography, Palaeoclimatology, Palaeoecology, 126*, 151–260.
Jelinek, A. (1992). The Middle Paleolithic in Southern Levant, with comments on the appearance of modern *Homo sapiens*. In A. Ronen (Ed.), *The transition from Lower to Middle Paleolithic and the origin of modern man* (Vol. 151, pp. 57–104). Oxford: BAR Intern Serie.
Koenigswald, G.H.R. Von (1958). Liste der deutschen Neanderthal-funde. In G. H. R. von Koenigswald (Ed.), *Hundert Jahre Neanderthaler* (pp. 21–26). Köln-Graz: Böhlau-Verlaf.
Koenigswald, W. V. (2011). Climatic changes, faunal diversity and environment of the Neanderthals in central and Western Europe during the Middle and Upper Pleistocene. In: S. Condemi and G.-C. Weniger (Eds), *Continuity and discontinuity in the peopling of Europe* (pp. 101–112). Dordrecht: Springer.
Kretzoi, M., & Vertés, L. (1965). Lower Palaeolithic hominid and pebble industry in Hungary. *Nature, 208*, 205.
Liritzis, Y. (1980). 230Th/234U dating of speleothems in Petralona. *Anthropos Athens, 7*, 215–241.
Mania, D., & Thomae, M. (2006). Pleistozänstratigraphie und Paläolithikum im mittleren Elbe-Saale-Gebiet. 73. *Tagung der Arbeitsgemeinschaft Norddeutscher Geologen 6–9 Juli 2006*. Halle.
Mania, D., Toepfer, V., & Vlcěk, E. (1980). Bilzingsleben I. *Homo erectus*: Seine Kultur und seine Umwelt. *Veröffentlichungen des Landesmuseuns für Vorgeschichte in Halle* Bd 32.
Mania, D., Mania, U., & Vlcěk, E. (1994). Latest finds of skull remains of Homo erectus from Bilzingsleben (Thuringia). *Naturwissenschaften, 81*, 123–127.
Piveteau, J. (1957). *Traité de paléontologie. Tome VII. Primates. Paléontologie humaine*. Paris: Masson.
Rosas, A., & Bermúdez de Castro, J.-M. (1998). The Mauer Mandible and the evolutionary significance of *Homo heidelbergensis*. *Geobios, 31*, 687–697.
Rougier, H. (2003). L'Homme de Biache-Saint-Vaast dans son contexte Européen. In R. Debrosse & A. Thévenin (Eds.), *Préhistoire de l'Europe: Des origines à l'Âge du Bronze* (pp. 21–26). Paris: CTHS.
Schoetensack, O. (1908). *Der Unterkiefer des Homo heidelbergensis aus den Sanden von Mauer bei Heidelberg*. Leipzig: Wilhem Engelmann.
Schwarcz, H., & Lathan, A. (1990). Absolute age determination of travertines from Vérteszöllös. In M. Kretzoi & V. Dobosi (Eds.), *Vérteszöllös: Site. Man and culture*. Budapest: Akademia Kiado.
Schwarcz, H., & Rink, J. W. (1998). Progress in ESR and U-series chronology of the Levantine Paleolithic. In T. Akazawa, K. Aoki, & O. Bar-Yosef (Eds.), *Neandertals and modern humans in Western Asia*. New York: Plenum.
Schwarcz, H., Grün, R., Latham, A. G., Mania, D., & Brunnacker, K. (1988). The Bilzinsgleben archaeological site: New dating evidence. *Archaeometry, 30*, 5–17.
Stringer, C. B. (2006). *Homo britanicus. The incredible story of human life in Britain*. London: Penguin, Allan Lade.
Stringer, C. B., & Hublin, J. J. (1999). New age estimates for the Swanscombe hominid and their significance for human evolution. *Journal of Human Evolution, 37*, 873–877.
Stringer, C. B., Trinkaus, E., Roberts, E., Parfitt, M. B., & Macpahil, R. I. (1998). The Middle Pleistocene human tibia from Boxgrove. *Journal of Human Evolution, 37*, 873–877.
Thoma, A. (1966). L'occipital mindélien de Vérteszöllös. *L'Anthropologie, 70*, 495–534.
Thoma, A. (1975). Were the Spy fossils evolutionary intermediates between Classic Neandertal and Modern Man? *Journal of Human Evolution, 4–5*, 387–410.
Trinkaus, E. (1983). *The Shanidar Neandertals*. New York: Academic.
Trinkaus, E., Stringer, C. B., Ruff, C. B., Hennessy, R. J., Roberts, M. B., & Parfitt, S. A. (1999). Diaphyseal cross-sectional geometry to the Boxgrove Middle Pleistocene human tibia. *Journal of Human Evolution, 37*, 1–25.
Tuffreau, A. (1988). L'industrie lithique du niveau IIA. In A. Tuffreau, J. Sommé (Eds.), *Le gisement paléolithique moyen de Biache-Saint-Vaast (Pas-de-Calais). Volume 1 – Stratigraphie, Environnement, Etudes archaéologiques (première partie)*. Mémoires de la Société Préhistorique Française, T. 21, pp. 171–183.
Tuffreau, S., Chaline, J., Munaut, A., Piningre, J.-F., Poplin, F., Puissegur, J.-J., Sommé, J., & Vandermeersch, B. (1978). Premiers résultats de l'étude du gisement paléolithique de Biache-Saint-Vaast (Pas-de-Calais). *Comptes Rendus de l'Académie des Sciences, Paris, sér. D, 286*, 457–459.
Valladas, H., Mercier, N., Joron, J. L., & Reyss, J. L. (1998). GIF laboratory dates for Middle Paleolithic levant. In T. Akazawa, K. Aoki, & O. Bar-Yosef (Eds.), *Neandertals and modern humans in Western Asia*. New York: Plenum.
Vertés, L. (1965). Discovery of *Homo erectus* in Hungary. *Antiquity, 39*, 303.
Vlcěk, E. (1989). *Homo erectus* in Europa. *Ethnographische-Archäologische Zeifschrift, 30*, 287–305.
Wagner, G. A., & Beinhauer, W. (Eds.). (1997). *Homo heidelbergensis von Mauer. Das Auf treten des Menschen in Europe*. Heidelberg: HVA.
Yokoyama, Y., & Nguyen, H.-V. (1981). Datation directe de l'homme de Tautavel par spectrométrie gamma, non-destructive, du crâne humain fossile Arago XXI. *Comptes Rendus de l'Académie des Sciences.Paris, Série II, 292*, 927–930.

Chapter 11
A Preliminary Approach to the Neanderthal Speciation by Distance Hypothesis: A View from the Shoulder Complex

Jean-Luc Voisin

Abstract Neanderthal extinction is still under debate and there are two main schools of thought on this topic: (1) Neanderthals and modern humans are two distinct species and (2) Neanderthals and modern humans are a single species, with or without two subspecies. Recently, a new hypothesis has risen up, which takes into account arguments from both schools: the Neanderthal speciation by distance (i.e. Voisin 2006c). This hypothesis is based on a morphological cline from East to West in Neanderthal populations. In other words, the farther those populations lived to the west, the more they displayed pronounced Neanderthal characters. The aim of this study is to test the speciation by distance hypothesis in Neanderthal in regard to the shoulder complex. The shoulder girdle displays a morphological cline from East to West, but only for architectural characters and not for functional ones. This cline could be better explained by a result of a speciation by distance induced by genetic drift than by a different response to any physical activities. This study tends to confirm the speciation by distance model for Neanderthal, even if more studies are needed to confirm it firmly.

Keywords Geographical cline • Europe • Near East • Shoulder girdle • Clavicle • Scapula

Introduction

Neanderthal extinction is still under debate and there are two main schools of thought on this topic: (1) Neanderthals and modern humans are two distinct species (i.e., Rak 1993; Hublin et al. 1996; Stringer and McKie 1996; Bermùdez de Castro et al. 1997; Krings et al. 1997; Stringer 1998, 2002; Bocquet-Appel and Demars 2000; Hublin 2000; Arsuaga et al. 2001; Bräuer 2001; Schillaci and Froehlich 2001; Rak et al. 2002; Harvati 2003; Harvati et al. 2004); and (2) Neanderthals and modern humans are a single species, with or without two subspecies (i.e. Thoma 1965; Trinkaus 1983, 1991; Smith et al. 1989a, 2005; Smith 1991; Smith and Trinkaus 1991; Frayer 1992; Wolpoff et al. 2000; Relethford 2001, 2003; Ahern et al. 2002; Curnoe and Thorne 2003). But recently, a new hypothesis has risen up, which takes into account arguments from both schools: the Neanderthal speciation by distance (Moncel and Voisin 2006; Voisin 2006c).

Extreme examples of speciation by distance are "ring species" or speciation by circular overlap. "Ring species provide dramatic evidence that normal genetic divergence within one species can build up to a sufficient level to generate two species" (Ridley 2004: 388). Among vertebrates, fully convincing examples of ring species are few and include the Californian salamander *Ensatina eschscholtzii* (Ridley 2004), the herring gull *Larus argentatus* and lesser black-backed gull *Larus fuscus* (Mayr 1974), and the greenish warbler *Phylloscopus trochiloides* (Irwin et al. 2001a, 2005). In central Siberia, two distinct forms, *P. trochiloides viridanus* and *P. trochiloides plumbeitarsus,* are sympatric without interbreeding (Fig. 11.1), and therefore may be considered two species. These two forms are nevertheless connected by a chain of interbreeding populations encircling the Tibetan plateau to the south (*P. trochiloides ludlowi, P.t. trochiloides, P.t. obscuratus*), and traits change gradually in consecutive populations (Irwin et al. 2001a). There is no obvious species boundary along this chain, and the two terminal "species" *viridanus* and *plumbeitarsus* are connected by gene flow (Irwin et al. 2001b).

Between the two forms living in central Siberia, morphological traits change gradually in consecutive populations encircling the Tibetan plateau, in the same manner as those of western to eastern Neanderthals. Thus, just before the spread of modern humans into Europe about 40,000 years ago, there was a chain of Neanderthal populations throughout Europe and the Near East, more or less connected by

J.-L. Voisin (✉)
USM 103, UMR 7194, Institut de Paléontologie Humaine,
1 rue René Panhard, 75013 Paris, France
and
UMR 6578, Université de la Méditerranée Aix-Marseille 2,
Faculté de Médecine, CS80011, Boulevard Pierre Dramard,
13344 Marseille Cedex 15, France
e-mail: jeanlucv@mnhn.fr

S. Condemi and G.-C. Weniger (eds.), *Continuity and Discontinuity in the Peopling of Europe: One Hundred Fifty Years of Neanderthal Study*,
Vertebrate Paleobiology and Paleoanthropology, DOI 10.1007/978-94-007-0492-3_11,

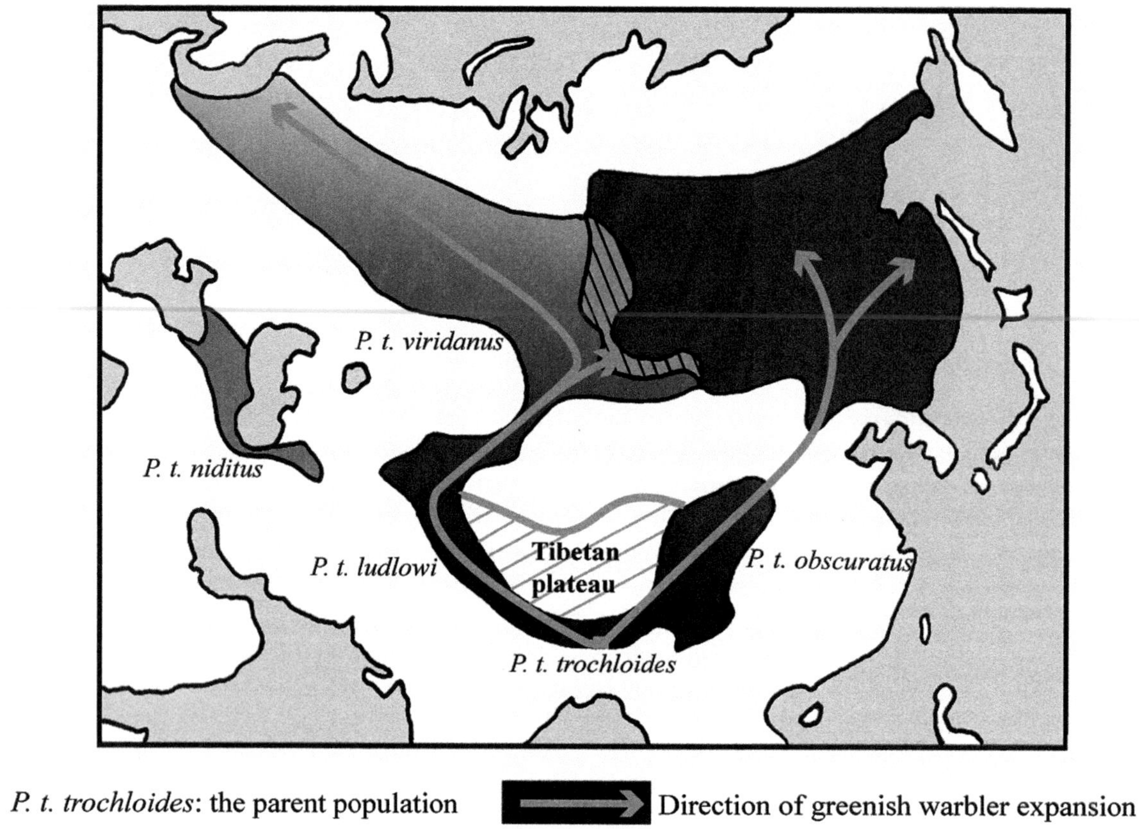

Fig. 11.1 The greenish Warblers (*Phylloscopus trochiloides*) ring species. The break in the population in west China is inferred to be recent and caused by deforestation (After Irwin et al. 2001a and Voisin 2006c)

Table 11.1 Summary of Neanderthal characteristics in West Europe and Near-East

West Europe	Near-East
Pronounced Neanderthal characters: *"Hyper Neanderthal" morphology*	Slightly pronounced Neanderthal characters: *"Hypo Neanderthal" morphology*
No Neanderthal characters within first post-Neanderthal populations.	Neanderthal characters within first post-Neanderthal populations.

gene flow (the gene flow rate would have varied as the ice sheets expanded and receded) and displaying a morphological cline from East to West. In other words, in the west part of the Neanderthal distribution area, human groups are characterized by pronounced Neanderthal characters and in the East part, populations are characterized by slightly pronounced Neanderthal characters (Table 11.1). Moreover, Neanderthal features seem to subsist in Central Europe and Near East post-Neanderthal populations (i.e. Smith et al. 1989b, 2005; Frayer 1992; Wolpoff et al. 2001, 2004; Trinkaus et al. 2003a, b; Janković et al. 2006; Ahern 2006; Hawks 2006).

The aim of this study is to test the speciation by distance hypothesis in Neanderthal in regard to the shoulder complex. I used shoulder girdle bones because they are the most characteristic of Neanderthal postcranial ones (Heim 1974, 1982; Vandermeersch 1981; Voisin 2004, 2006a) and also because the upper limb capacities depend on the shoulder complex. More extensive studies are currently in progress both on cranial and postcranial evidence and also on teeth (Voisin and Condemi, in press). The present work is more a preliminary report than a full conclusive paper and its aim is to show that a morphological cline clearly existed in Neanderthal populations and affected special features.

Materials and Methods

Materials

We studied 18 Neanderthal scapula and 17 clavicles (Table 11.2), which were completed for some characters by additional individual data from the literature. The sample is completed by 33 clavicles and 29 scapulas of modern humans (MH) (*Homo sapiens sapiens*) from several parts of the world (Europe 10, Africa 7, North America 5, Asia 7, unknown 4 (the last four, only for clavicles)). These

Table 11.2 Fossil remains used in this study

Clavicle	Scapula
Régourdou (L) and (R)	/
La Ferrassie I (L)[a] and (R)[a]	La Ferrassie I (L)[a] and (R)[a]
Neanderthal (R)	Neanderthal (R)
/	Spy (L) and (R)
La Chapelle-aux-Saints (L)[a]	/
Krapina 142 (R)[a]	Krapina 125 (L)[a]
Krapina 143 (R)[a]	Krapina 127 (R)[a]
Krapina 144 (R)[a]	Krapina 128 (R)[a]
Krapina 145 (R)[a]	Krapina 129 (R)[a]
Krapina 149 (R)[a]	Krapina 130 (L)[a]
Krapina 153 (L)[a]	Krapina 131 (L)[a]
Krapina 154 (L)[a]	Krapina 132 (R)[a]
Krapina 155 (L)[a]	Krapina 134 (R)[a]
Krapina 156 (L)[a]	Krapina 135 (R)[a]
/	Vindija 209 (L)
Kebara (L) and (R)	Kebara (L) and (R)
/	Tabun I (L)[a]

(L) and (R) mean respectively left and right
[a]Original remains

specimens are housed in the Département "*Hommes, Natures, Sociétés*" *du Musée de l'Homme*, Paris (France), the *Institut de Paléontologie Humaine*, Paris (France), the Croatian Natural History Museum, Zagreb (Croatia), and the *Institut Royal des Sciences Naturelles de Belgique*, Bruxelles (Belgium).

Methods

Study of the Clavicles

Due to its complexity, the morphology of the clavicle will be approached in regard to its curvatures. When projected on two perpendicular planes, one cranial and one dorsal, the clavicle morphology can be decomposed in elementary curvatures, as shown in Fig. 11.2.

The middle arc of curvature is estimated according to Olivier's method (1951a) as the proportion between the length of the chord and the height of the curvature (Fig. 11.2):

Cranial plane:
The acromial curvature (external one): e/h.100
The sternal curvature (internal one): f/g.100.

Dorsal plane:
The acromial curvature (inferior one): e'/h'.100
The sternal curvature (superior one): f'/g'.100

Three other characters will also be used: the total length (measured with a calliper as the greatest length of the bone), the presence or absence of a costo-clavicular ligament insertion, or costal tuberosity, and of a subclavius sulcus.

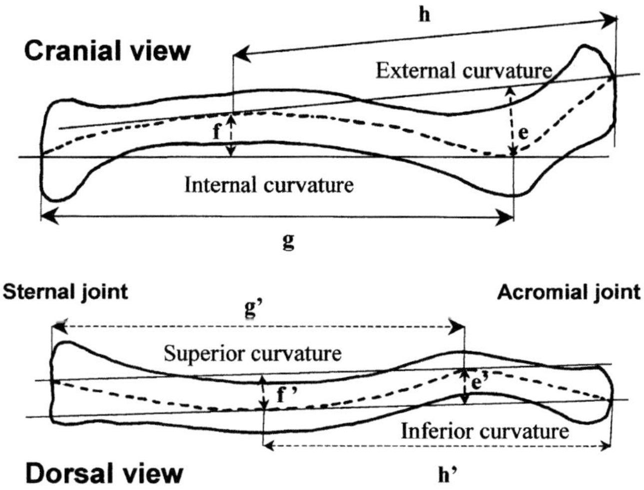

Fig. 11.2 Determination of curvatures on a right clavicle of *Pan troglodytes* (Olivier 1951a)

Study of the Scapulas

As the majority of the scapula remains are damaged, only three measurements could be used on most fossils:

The glenoid fossa index:
(Breadth of the glenoid fossa/height of the glenoid fossa) × 100.

The breadth and the height of the glenoid fossa are measured as suggested by Martin (1928). This method is less precise than that proposed by Vallois (1928–1946), but as most works on Neanderthal used the Martin technique, I will also use it for obtaining data comparable with those of other authors.

The scapula neck index (Larson 1995):
Neck length/root square (breadth of the glenoid fossa × height of the glenoid fossa) × 100.

This index is associated to arm movement, especially to the abduction function of the infraspinatus (Larson 1995). Neck length is the minimum width of the infraspinous fossa (between A and B) at the neck of the scapula (Fig. 11.3), measured with a caliper.

The angle A:
It is the angle between glenoid great axis and the ventral bar (the prominent buttress just medial to the axillary border on the ventral surface of the scapula). This angle is correlated to the axillo-glenoid angle (Stern and Susman 1983) and its measure does not need a complete scapula, which is very rare in fossil records. The higher the angle values, the more the glenoid fossa is oriented cranially.

The morphology of the axillary border and the associated sulcus:
Ventral, bisculate, or dorsal. This character is associated to arm movements and could be the result of a high muscle activity or could have a more phylogenetic relationship

(Trinkaus 1977; Heim 1982; Frayer 1992; Odwak 2006; Trinkaus 2006, 2008).

The measurements and distribution of the variables have been computed with ViStat 6.4 ® (Young 2001). Graphics showing the range of variation of each variable are represented by the mean and +/− two times standard deviation.

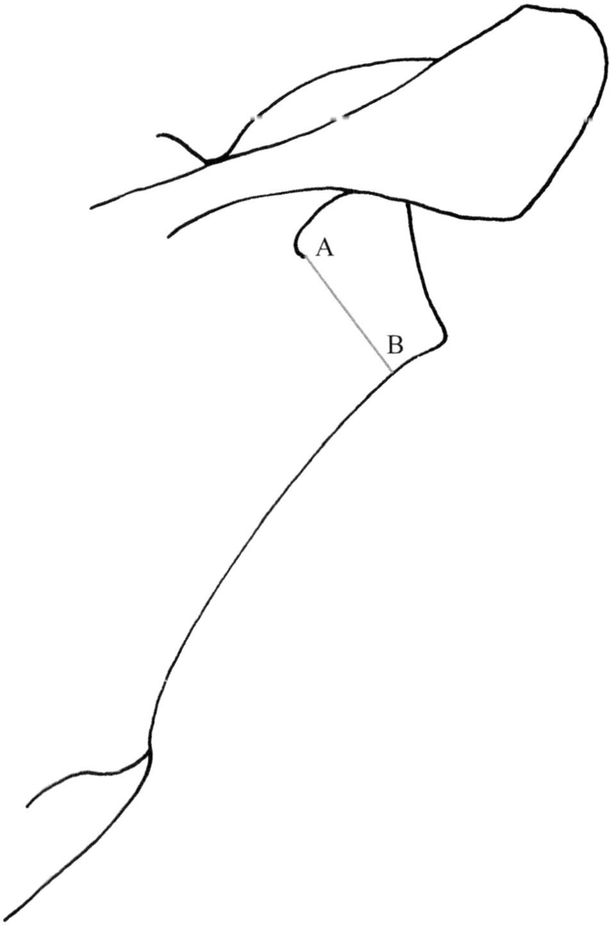

Fig. 11.3 Neck length or the minimum width of the infraspinous fossa (From A to B) (After Larson 1995)

Results

The Clavicle

Curvatures in Cranial View

In superior view, all fossil clavicles studied here are distributed within the range of variation of modern human ones (Fig. 11.4, Table 11.3) and there is no evidence for a cline for this feature. This result shows that Neanderthal clavicles, in superior view, are less S-shaped than classically described (Boule 1911–1913; Patte 1955; Heim 1974, 1982; Vandermeersch and Trinkaus 1995) and display no differences with modern human ones in cranial view. This result is confirmed by other recent works (Voisin 2000, 2001, 2004, 2006a).

Curvatures in Dorsal View

In dorsal view, modern human clavicles can be classified into three morphological groups (Fig. 11.5), or types, according to Olivier's studies on more than 800 clavicles from Europe, Africa, America, and Australia (Olivier 1951b, 1954, 1955; Olivier et al. 1954). Type I possesses only an inferior curvature, and is the most frequent. Type II clavicles are far less common, and display two curvatures in dorsal view, a superior one at the sternal end, and an inferior one at the acromial end. Type III clavicles show a superior curvature at the acromial end, and none at the sternal part. Type III clavicles are by far the least frequent form (Olivier 1951b, 1954, 1955; Olivier et al. 1954).

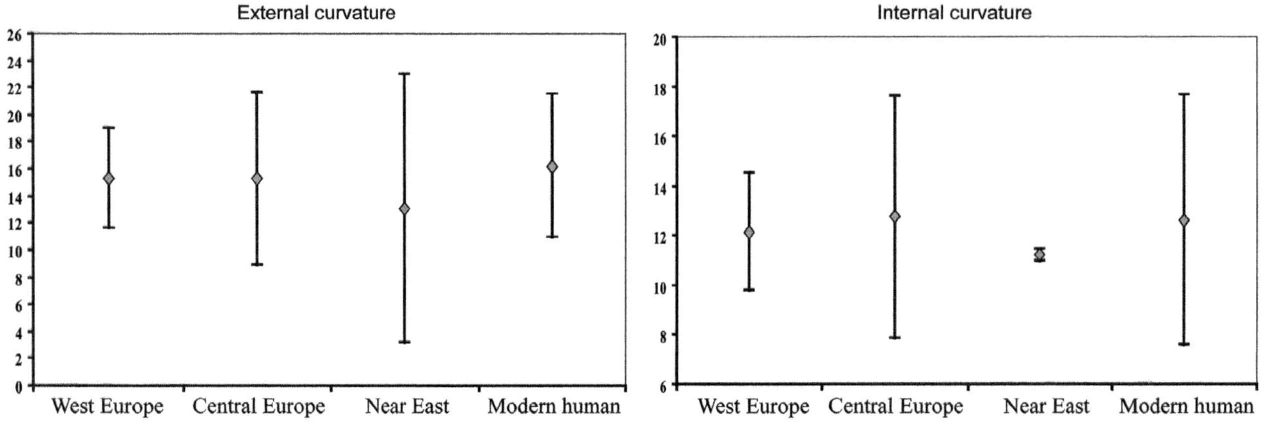

Fig. 11.4 Mean and range of variation of clavicle curvatures in cranial view in West Europe, Central Europe, and Near-East Neanderthals

Table 11.3 Values of clavicle curvatures in superior view

Individual	External curvature	Internal curvature	Region
Régourdou (R)	14.46	10.00	West Europe
Régourdou (L)	13.16	11.86	
La Ferrassie 1 (R)	14.22	13.24	
La Ferrassie 1 (L)	17.65	12.32	
Neanderthal (R)	16.67	13.16	
La Chapelle-aux-Saints (L)	/	12.24	
Mean	**15.23**	**12.14**	
Krapina 142 (R)	17.40	16.70	Central Europe
Krapina 143 (R)	13.90	12.10	
Krapina 154 (L)	14.10	11.40	
Krapina 149 (R)	20.00	/	
Krapina 144 (R)	12.00	/	
Krapina 155 (L)	18.10	/	
Krapina 153 (L)	10.80	10.30	
Krapina 145 (L)	/	11.30	
Krapina 156 (L)	/	14.70	
Mean	**15.19**	**12.75**	
Kebara (L)	16.51	11.11	Near-East
Kebara (D)	9.52	11.29	
Mean	**13.02**	**11.20**	
Modern Human Standard deviation/Mean	2.65/16.12	2.52/12.62	

Table 11.4 Values of clavicle curvatures in dorsal view

Individual	Inferior curvature	Superior curvature	Region
Régourdou (R)	7.41	8.04	West Europe
Régourdou (L)	3.03	3.21	
La Ferrassie 1 (R)	5.41	5.83	
La Ferrassie 1 (L)	8.23	2.61	
Neanderthal (R)	7.41	6.18	
La Chapelle-aux-Saints (L)	7.37	/	
Mean	**6.48**	**5.17**	
Krapina 142 (R)	6.9	5.9	Central Europe
Krapina 143 (R)	6.3	0	
Krapina 154 (L)	6.3	6.6	
Krapina 149 (R)	13.3	/	
Krapina 144 (R)	9.3	/	
Krapina 155 (L)	6.8	/	
Krapina 153 (L)	3.8	3.9	
Krapina 145 (L)	/	7.2	
Krapina 156 (L)	2.2	/	
Mean	**6.86**	**4.72**	
Kebara (L)	4.94	0	Near East
Kebara (R)	3.38	0	
Mean	**4.16**	**0**	
Modern human standard deviation/mean	2.37/4.98	1.70/1.15	

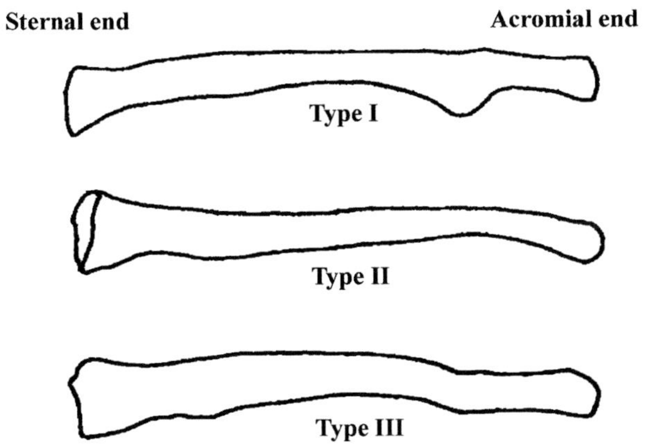

Fig. 11.5 The three types of modern human clavicles (For a definition, see text and Voisin 2006a)

Nearly all clavicles that do not belong to modern or Upper Paleolithic humans display two curvatures in dorsal view: an inferior one at their acromial extremity and a superior curvature at their sternal extremity (Voisin 2004, 2006a, 2008). However, some modern human clavicles display two curvatures in dorsal view, but their morphology is different from that of Neanderthal (Voisin 2004, 2006a, 2008). The superior curvature, when present, is less pronounced and less frequent in modern humans (Table 11.4). Moreover, some Neanderthal clavicles (Régourdou left and right, La Ferrassie I right, Krapina 153 and 154) display a superior curvature that is even more pronounced than the inferior one (Table 11.4). However, Kebara (right and left) and Krapina 143 display a modern morphology, showing only the inferior curvature. As a whole, 78% of Neanderthal clavicles show two curvatures in dorsal view (Voisin 2004, 2006a, 2008), and none with only one curvature are from West Europe.

The Length

Neanderthal clavicles have an average length similar to that of Upper Paleolithic and modern human ones (Voisin 2004, 2006a) (Table 11.5). However, Neanderthal populations are heterogeneous for this character (Fig. 11.6, Table 11.5) and clavicles from Western Europe are longer than those from Krapina and those from the Near East. This observation, confirmed by other authors (Trinkaus 1983; Nara 1994), may show that Neanderthal clavicles display a trend toward a reduction in size from West to East, even if Kebara clavicles stand close to western Neanderthal values.

Table 11.5 Clavicle total length values

Individual	Total length	Region
Régourdou (R)	152.0	West Europe
Régourdou (L)	146.0	
La Ferrassie 1 (R)	173.5	
La Ferrassie 1 (L)	178.0	
Neanderthal (R)	140.0	
Mean	**157.9**	
Krapina 153 (L)	145.0	Central Europe
Krapina 142 (R)	149.0	
Krapina 143 (R)	130.9	
Krapina 154 (L)	118.0	
Mean	**135.7**	
Kebara (L)	162.0	Near East
Kebara (R)	150.5	
Shanidar 3 (R)[a]	149.0	
Shanidar 1 (L)[b]	150.0	
Tabun 1 (R)[c]	135.0	
Tabun 1 (L)[c]	135.0	
Mean	**146.9**	

[a, b] From Trinkaus (1981, 1982)
[c] From Heim (1982)

Costal Tuberosity and Subclavius Sulcus

All Neanderthal clavicles studied here display both a costal tuberosity and a subclavius sulcus, except Krapina 153 and 149, which do not show respectively any costal tuberosity or subclavius sulcus. These two characters display no clinal variation in Neanderthal populations, and their frequencies are close to that of modern humans (Table 11.6).

Table 11.6 The presence or the absence of the costal tuberosity and the subclavius muscle in Neanderthals

Individual	Costal tuberosity	Subclavius sulcus
Shanidar 1 (L)[a]	Present	Present
Shanidar 1 (R)[a]	Present	?
Shanidar 3 (R)[a]	Present	?
Kebara (L)	Present	Present
Kebara (R)	Present	Present
Tabun 1 (L)	Present	/
Tabun 1 (L)	Present	/
Krapina 153	Absent	Present
Krapina 143 (R)	Present	?
Krapina 142 (R)	Present	Present
Krapina 156 (L)	Present	Present
Krapina 155 (L)	/	Present
Krapina 144 (R)	/	/
Krapina 145 (R)	Present	Present
Krapina 149 (R)	/	Absent
Krapina 154 (L)	Present	Present
Krapina 157 (L)	Present	/
Régourdou (R)	Present	Present
Régourdou (L)	Present	Present
La Ferrassie I (R)	Present	Present
La Ferrassie I (L)	Present	Present
La Chapelle-aux-Saints (L)	Present	/
Neanderthal (R)	/	Present
Neanderthal frequency (in%)	**94.73**	**92.86**
Modern human frequency (%)	**93.9**	**81.8**

/ Character not observable because part of the bone is missing
? No information available from literature
[a]From Trinkaus (1982, 1983)

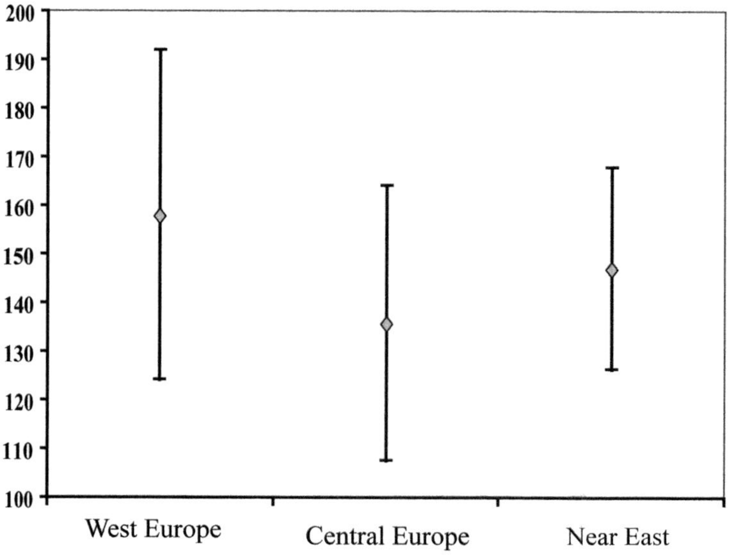

Fig. 11.6 Mean and range of variation of clavicle length in West Europe, Central Europe, and Near-East Neanderthals

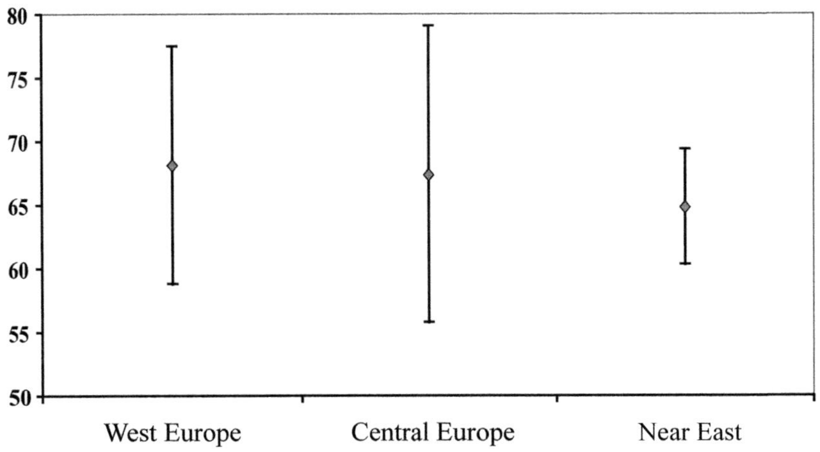

Fig. 11.7 Mean and range of variation of the glenoid index in West Europe, Central Europe, and Near-East Neanderthals

However, the two characters, taken together, are more frequent than on modern clavicles. In other words, it is more frequent to find the costal tuberosity and the subclavius sulcus on the same clavicle in Neanderthals (91.7%) than in modern humans (75.8%).

The Scapula

The Glenoid Index

The glenoid index, which is narrower in Neanderthals than in modern humans (i.e. Stewart 1962; Heim 1974, 1982; Voisin 2000; Trinkaus 2006), does not show any clear variation between Neanderthals in general and any clinal variation in particular (Fig. 11.7, Table 11.7).

The Scapula Neck Index

The scapula neck index seems to show slight differences from East to West (Table 11.7, Fig. 11.8). The average index is higher in Near East than in central or West Europe. In the Near East, the values are closer to those of modern humans (Voisin 2000). But, for this index, the data comes from only two scapulas from the same individual, Kebara, so that it is still impossible to conclude that the arm movement capacities are different between European and Near-East Neanderthals.

Angle *A*

The angle between glenoid axis and the ventral bar, or angle *A*, seems to be higher in Central European populations than in those of the Near East and West Europe (Table 11.7, Fig. 11.9). Nevertheless, Neanderthal populations are not homogenous for this character, which, at the same time, shows no evidence of an East to West morphological cline. However, the Kebara left value does not seem accurate, this scapula having possibly suffered during the time it spent underground (Vandermeersch 1991).

Table 11.7 Data for the glenoid and scapula neck index as well as for angle *A* (angle between glenoid axis and ventral bar). Some glenoid index values are taken from Vandermeersch (1981, 1991) and Trinkaus (2006)

Individual	Glenoid index	Scapula neck index	Angle *A*
Kebara (R)	63.9	91.6	148.0
Kebara (L)	61.5	93.5	133.0
Shanidar 1 (L)	65.8		
Tabun 1 (L)	65.5	89.5	141.5
Amud 1	67.5		
Mean	**64.8**	**91.5**	**140.8**
Krapina 127 (R)	65.4	75.2	155.0
Krapina 129 (R)	61.5	67.1	147.0
Krapina 125 (L)	68.0	81.4	148.0
Krapina 132 (R)	67.1	73.7	146.5
Krapina 130 (L)	77.3	86.2	147.0
Krapina 131 (L)	63.3	91.3	149.0
Krapina 133 (R)	69.1	90.2	
Vindija 209 (L)	77.0		138.5
Mean	**68.6**	**80.7**	**147.3**
Ferrassie I (R)	68.1	77.1	132.5
Ferrassie I (L)	65.2	80.8	141.0
Spy (L)	75.0	93.8	
Spy (R)	72.3	83.2	135.0
Neanderthal (R)	63.2	81.1	145.0
Mean	**68.8**	**83.2**	**138.4**
Modern human (standard deviation/mean)	**4.0/80.3**	**6.8/83.6**	**5.3/146.4**

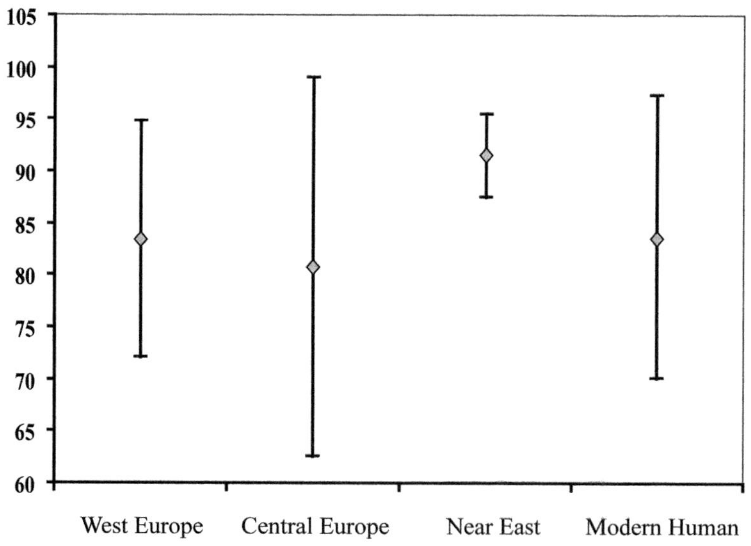

Fig. 11.8 Mean and range of variation of the scapula neck index in West Europe, Central Europe, and Near-East Neanderthals

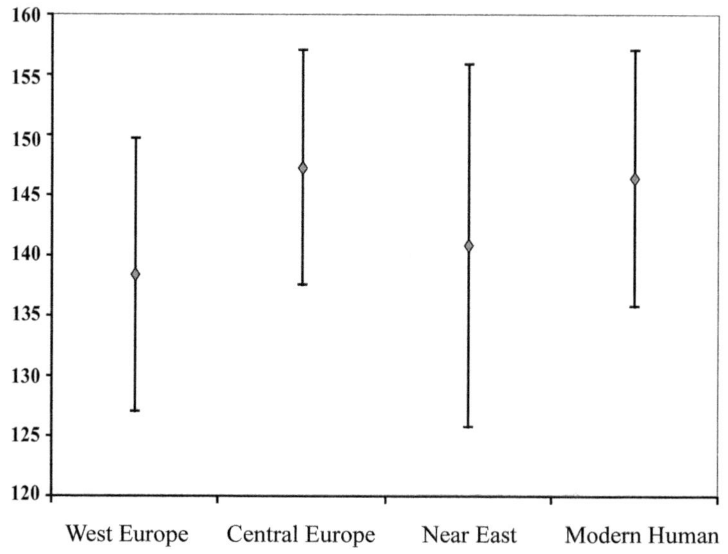

Fig. 11.9 Mean and range of variation of the angle between glenoid axis and the ventral bar, or angle A in West Europe, Central Europe, and Near-East Neanderthals and in modern humans

The Axillary Border

The morphology of the axillary border, and especially the position of the axillary sulcus (ventral, bisulcate, or dorsal), of the Neanderthal scapula has been well described and studied because of its possible distinctiveness between Neanderthals and Modern Humans and/or its relations to arm movements (i.e. Boule 1911–1913; Vallois 1928, 1932, 1946; Stewart 1962, 1964; Heim 1974, 1982; Trinkaus 1977, 1982, 1983; Vandermeersch 1981; Voisin 2000; Busby 2006; Odwak 2006; Trinkaus 2006). For a long time, a sulcus in dorsal position was considered as characteristic of Neanderthals (Boule 1911–1913; Vallois 1928, 1932, 1946; Stewart 1962, 1964) in comparison to modern humans, which were considered to possess only a ventral one, or sometimes a dorsal as well as a ventral one (bisulcate scapula). More recent works show that, in some modern human populations, it is possible to find a dorsal sulcus with a high frequency (Heim 1974, 1982; Trinkaus 1977, 1982, 1983, 2006; Vandermeersch 1981; Odwak 2006). Moreover, some Neanderthal scapulas also display the bisulcate type, and some also only possess a ventral sulcus. The bisulcate type is more frequent in Near East Neanderthal populations than in others (Table 11.8). Moreover, only a small proportion of scapulas from Central Europe have no dorsal sulcus, and none are known from West Europe. In other words, the distribution of the axillary border morphology is not homogenous within Neanderthal populations.

Table 11.8 The position of the axillary sulcus on Neanderthal scapulae. Some morphological data are taken from Boule (1911–1913), Fraipont (1927), Trinkaus (1982, 1983, 2006), Vandermeersch (1991), Frayer (1992) and Odwak (2006)

Ventral	Bisulcate	Dorsal	Percent (%)[a]
	Shanidar 3 (R)		37.5
		Shanidar 1 (L)	
		Shanidar 2 (L)	
		Shanidar 4 (R)	
		Tabun 1 (L)	
		Amud 1	
	Kebara (L)		
	Kebara (R)		
	Krapina 125 (L)		20
Krapina 127 (R)			
		Krapina 128 (R)	
		Krapina 129 (R)	
		Krapina 130 (L)	
		Krapina 131 (L)	
		Krapina 132 (R)	
		Krapina 134 (R)	
		Krapina 135 (R)	
		Vindija 209 (L)	
		La Ferrassie 1 (R)	0
		La Ferrassie 1 (L)	
		La Ferrassie 2 (R)	
		Neanderthal (R)	
		Spy (R)	
		Spy (L)	

[a] Frequencies, in percentage of non dorsal axillary sulcus

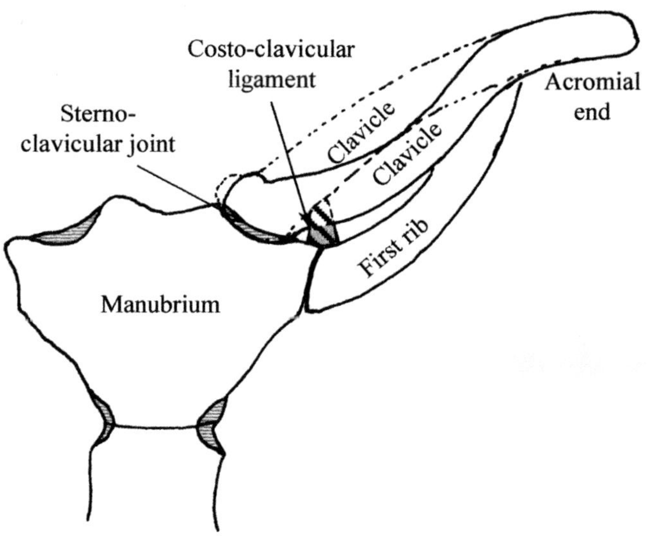

Fig. 11.10 Clavicles associated with a high scapula in regard to the thorax. *Dotted line*, human clavicle (with a unique inferior curvature); *full line*, great ape clavicle (with two curvatures in *dorsal view*). Note the costo-clavicular length difference with the two types of clavicle (After Voisin 2006b)

Discussion

The Clavicle

In Neanderthal, clavicle length and curvatures in dorsal view show a clinal variation from East to West. Clavicles may display only one curvature in dorsal view in the Near-East, contrary to those of western Neanderthals, which always display two. At the same time, clavicles become longer from East to West (with the exception of Kebara). Curvatures in dorsal view are associated to shoulder architecture (Voisin 2006a, b). As the costoclavicular ligament limits horizontal and vertical clavicle movements at the sternoclavicular joint, an elongation of this latter ligament increases the mobility and weakness of the joint and involves greater muscular control, exerted by the subclavius muscle. An important superior curvature permits to associate a high scapula with respect to the thorax and a clavicle with its medial end nearly parallel to the manubrium (Fig. 11.10). This condition avoids the elongation of the costoclavicular ligament. In other words, two pronounced curvatures in dorsal view are associated to scapulas located higher on the thorax than scapulas associated to clavicles with only an inferior curvature (Voisin 2004, 2006a, b). Hence, the curvatures cline within Neanderthals, in dorsal view, allows concluding the scapula position becomes progressively higher on the thorax from East to West within this population.

In the Near East, Neanderthal shoulder architecture is very similar to that of modern humans, whereas it becomes clearly different from ours in West Europe. Moreover, for a same shoulder breadth, the more the scapula sits high on the thorax, the longer is the clavicle. Thus, the geographic cline in clavicle length observed in this study is consistent with the dorsal curvatures cline.

At the same time, some clavicle characters do not display any geographical cline, like the curvatures in superior view and the subclavius sulcus and costal tuberosity frequencies. As demonstrated previously (Voisin 2006a, b), curvatures in superior view are related to arm movements, especially arm elevation, which are needed for both carrying and throwing objects like spears. There does not seem to be any difference in arm movement capacity from East to West in Neanderthal populations. The subclavius sulcus and costal tuberosity are present on most Neanderthal clavicles, and the frequency of these two characters is higher than in modern humans (Voisin 2000). As the scapula glenoid index, this character could be considered as a Neanderthal character, and may be due to a higher level of physical activities than in modern humans (Ray 1959; Jit and Kaur 1986).

Clavicle characters showing an East to West morphological cline are architectural, and, on the contrary, characters not displaying any morphological cline are functional.

The Scapula

Among scapula characters, the narrower morphology of the Neanderthal glenoid fossa is well known (i.e. Vallois 1928, 1932, 1946; Heim 1974, 1982; Trinkaus 2006). According to Churchill and Trinkaus (1990) and Trinkaus (2006), this narrower morphology "is related to habitual degrees of loading in medial and lateral hyperrotation of the glenohumeral articulation" (Trinkaus 2006, p. 344). This morphology is associated to peculiar arm movements and characteristic of Neanderthals and, as several other clavicle functional characters, the glenoid fossa morphology does not display any East to West cline.

The minimum width of the infraspinous fossa at the neck of the scapula, or scapula neck, determine the extension of the subscapularis fossa (Larson 1995) and also the importance of the infraspinous fossa, even on fragmentary remains (Voisin 2000). This index shows no clinal variation from East to West, and subsequently there is no variation of the subscapularis and infraspinous fossa along the Neanderthal distribution area. Furthermore, variation in Neanderthals does not differ from that of modern humans (Voisin 2000). In other words, the infraspinous and subscapularis muscles display no variation from West to East in Neanderthal populations, and are similar to those of modern humans. As other shoulder girdle functional characters, the minimum width of the infraspinous fossa at the neck of the scapula does not display any clinal variation.

The angle A, between the glenoid fossa and ventral bar, is correlated to the axillo-glenoid angle (Stern and Susman, 1983) and give a good overview of the orientation of the glenoid fossa. The orientation of the glenoid fossa is associated to arm movements for locomotion and/or manipulation. Central Europe and Near-East Neanderthals seem to have very similar values for this angle. On the contrary, these values tend to be lower in Western Europe. This difference could be explained by the fact that the scapula is situated higher on the thorax in Western than in Near-East or Central Europe Neanderthals. With lesser values, the glenoid fossa would have been oriented less cranially than in modern humans or in Eastern Neanderthals, and compensate for the elevation of the scapula in Western Neanderthals.

The morphology of the axillary border displays a morphological cline from West to East. According to several authors like Smith (1976), Trinkaus (1977), and Odwak (2006), it is an acquired character because the three types of sulci exist within Neanderthal, even on the same site like Krapina. Trinkaus (1977) suggest that the dorsal sulcus increases the attachment of the area for *Teres minor*. This latter muscle, along with infraspinatus, is a primary lateral rotator of the humerus and helps retaining the humeral head in the glenoid fossa. As Odwak (2006) wrote, this morphology may reflect overall osseous changes related to muscular hypertrophy and muscle re-orientation as well as to robusticity. The elevation of the shoulder in Neanderthals, as their clavicles show it, could be responsible for a reorientation of the shoulder muscles and especially the *Teres minor*. Thus, the morphological cline observed for the axillary sulcus may reflect the elevation of the shoulder girdle and the muscle hypertrophy that characterize Neanderthal populations. Even if this variation of the axillary sulcus is not genetically determinated, it reflects the Neanderthal shoulder architecture and it is more a by-product of the shoulder elevation than a real adaptation.

Thus, like clavicle curvatures, scapula traits associated to arm movement do not present any clinal variation from East to West and are homogenous within Neanderthal population. On the contrary, scapula characters in relation to shoulder architecture display an East to West cline.

Conclusion

A geographical cline in some characters of the shoulder girdle architecture seems to be evident from East to West in Neanderthal populations. Because of the scarcity of the remains at our disposal, it is not possible to draw any firm conclusion yet, and more studies, some of them in progress now, are needed on the subject. However, characters which seem to display a geographical gradient from East to West in Neanderthal populations are architectural, like clavicle morphology in dorsal view. On the contrary, functional characters, like clavicles curvatures in superior view, do not show any geographical cline.

Functional characters could be the result of various activities, and thus differences could be explained by behavior changes. On the contrary, architectural characters are mostly inherited and thus seem to be less a response to any physical activities than a result of a speciation by distance induced by genetic drift.

More studies are needed to conclude about the geographical cline from East to West in Neanderthal populations, but this first work shows that a geographical cline probably exists and is more a matter of architecture than a matter of function.

Acknowledgments I would first like to thank Silvana Condemi for her proposal to participate in the "150 years of Neanderthal discoveries – Early Europeans – continuity & discontinuity." I would also thank Professors Dominique Grimaud-Hervé, Jakov Radovcić, and Philippe Menecier, who allowed me to work, respectively, on the collections of the *Institut de Paléontologie Humaine* (Paris, France), the Croatian Natural History Museum (Zagreb, Croatia), and the *Département des collections du Musée de l'Homme* (Paris, France).

References

Ahern, J. C. M. (2006). Non-metric variation in recent humans as a model for understanding Neanderthal-early modern human differences: Just how "unique" are Neanderthal unique traits. In K. Harvati & T. Harrison (Eds.), *Neanderthals revisited – new approaches and perspectives* (pp. 255–268). Dordrecht: Springer.

Ahern, J. C. M., Lee, S. H., & Hawks, J. D. (2002). The late Neandertal supraorbital fossils from Vindija cave, Croatian: A biased sample? *Journal of Human Evolution, 43*, 419–432.

Arsuaga, J. L., Martinez, I., & Gracia, A. (2001). Analyse phylogénétique des hominidés de la Sierra de Atapuerca (Sima de los Huesos et Gran Dolina TD 6): l'évidence crânienne. *L'Anthropologie (Paris), 105*, 161–178.

Bermùdez de Castro, J. M., Arsuaga, J. L., Carbonell, E., Rosas, A., Martinez, I., & Mosquera, M. (1997). A hominid from lower Pleistocene of Atapuerca, Spain: Possible ancestor to Neandertals and modern humans. *Science, 276*, 1392–1395.

Bocquet-Appel, J. P., & Demars, P. Y. (2000). Neanderthal contraction and modern human colonization of Europe. *Antiquity, 74*, 544–552.

Boule, M. (1911–1913). L'Homme fossile de la Chapelle-aux-Saints. *Annales de Paléontologie, 6, 7 & 8*, 111–172, 121–192, 111–170.

Bräuer, G. (2001). The "Out-of-Africa" model and the question of regional continuity. In P. V. Tobias, M. A. Raath, J. Moggi-Checchi, & G. A. Doyle (Eds.), *Humanity from African naissance to coming millennia* (pp. 183–189). Firenze & Witwatersrand: Firenze University Press, Witwatersrand University Press.

Busby, A. M. (2006). A multivariate analysis of the ontogeny of the scapular axillary border. *Periodicum Biologorum, 108*, 364–371.

Churchill, S. E., & Trinkaus, E. (1990). Neandertal scapular glenoid morphology. *American Journal of Physical Anthropology, 83*, 147–160.

Curnoe, D., & Thorne, A. (2003). Number of ancestral human species: A molecular perspective. *Homo, 53*, 201–224.

Fraipont, C. (1927). Sur l'omoplate et le sacrum de l'homme de Spy. *Revue Anthropologique (Paris), 37*, 189–195.

Frayer, D. W. (1992). The persistence of Neanderthal features in post-Neanderthal Europeans. In G. Bräuer & F. H. Smith (Eds.), *Continuity or replacement: Controversies in Homo sapiens evolution* (pp. 179–188). Rotterdam: A.A. Balkema.

Harvati, K. (2003). The Neanderthal taxonomic position: Models of intra- and inter-specific craniofacial variation. *Journal of Human Evolution, 44*, 107–132.

Harvati, K., Frost, S. R., & McNulty, P. (2004). Neanderthal taxonomy reconsidered: Implication of 3D primate models of intra- and inter specific differences. *Proceedings of the National Academy of Sciences of the United States of America, 101*, 1147–1152.

Hawks, J. (2006). Selection on mitochondrial DNA and the Neanderthal problem. In K. Harvati & T. Harrison (Eds.), *Neanderthals revisited – new approaches and perspectives* (pp. 221–238). Dordrecht: Springer.

Heim, J. L. (1974). Les Hommes fossiles de la Ferrassie (Dordogne) et le problème de la définition des Néandertaliens classiques. *L'Anthropologie (Paris), 78*, 81–112.

Heim, J. L. (1982). Les hommes fossiles de la Ferrassie (II). *Archive de l'Institut de Paléontologie Humaine, 38*, 1–272.

Hublin, J. J. (2000). Modern non-modern hominid interactions: A mediterranean perspective. In O. Bar-Yosef & D. Pilbeam (Eds.), *The geography of Neandertals and modern humans in Europe and the greater Mediterranean* (pp. 157–182). Haward: Peabody Museum of Archaeology and Ethnology.

Hublin, J. J., Spoor, F., Braun, M., Zonneveld, F., & Condemi, S. (1996). A late Neanderthal associated with upper Palaeolithic artefacts. *Nature, 381*, 224 226.

Irwin, D. E., Bensch, S., & Price, T. D. (2001a). Speciation in a ring. *Nature, 409*, 333–337.

Irwin, D. E., Irwin, J. H., & Price, T. D. (2001b). Ring species as bridges between microevolution and speciation. *Genetica, 112–113*, 223–243.

Irwin, D. E., Bensch, S., Irwin, J. H., & Price, T. D. (2005). Speciation by distance in a ring species. *Science, 307*, 414–415.

Janković, I., Karavanić, I., Ahern, J. C. M., Brajković, D., Lenardić, J. M., & Smith, F. H. (2006). Vindija cave and the modern human peopling of Europe. *Collegium Antropologicum, 30*, 315–319.

Jit, I., & Kaur, H. (1986). Rhomboïd fossa in the clavicles of North Indians. *American Journal of Physical Anthropology, 70*, 97–103.

Krings, M., Stone, A., Schmitz, R. W., Krainitzki, H., Stoneking, M., & Pääbo, S. (1997). Neandertal DNA sequences and the origin of modern humans. *Cell, 90*, 19–30.

Larson, S. G. (1995). New characters for the functional interpretation of primate scapulae and proximal humeri. *American Journal of Physical Anthropology, 98*, 13–35.

Martin, R. (1928). *Lehrbuch der anthropologie* (2nd ed.). Jena: Verlag von Gustav Fischer.

Mayr, E. (1974). *Populations, Espèces et Evolution*. Paris: Hermann.

Moncel, M. H., & Voisin, J. L. (2006). Les "industries de transition" et le mode de spéciation des groupes néandertaliens en Europe entre 40-30 ka. *Comptes Rendus Palevol, 5*, 183–192.

Nara, T. (1994). Etude de la variabilité de certains caractères métriques et morphologiques des néandertaliens. Ph.D. dissertation, Université de Bordeaux I.

Odwak, H. (2006). Axillary border morphology in modern human and Neanderthals. *Periodicum Biologorum, 108*, 353–364.

Olivier, G. (1951a). Technique de mesure des courbures de la clavicule. *Comptes Rendus de l'Association des Anatomistes. XXXIXe Réunion (Nancy), 69*, 753–764.

Olivier, G. (1951b). Anthropologie de la clavicule. *Bulletins et Mémoires de la Société d'Anthropologie de Paris 10ème série, 2*, 67–99.

Olivier, G. (1954). Anthropologie de la clavicule. *Bulletins et Mémoires de la Société d'Anthropologie de Paris 10ème série, 5*, 144–153.

Olivier, G. (1955). Anthropologie de la clavicule. *Bulletins et Mémoires de la Société d'Anthropologie de Paris 10ème série, 6*, 282–302.

Olivier, G., Chabeuf, M., & Laluque, P. (1954). Anthropologie de la clavicule. *Bulletins et Mémoires de la Société d'Anthropologie de Paris 10ème série, 5*, 35–46.

Patte, E. (1955). *Les Néanderthaliens*. Paris: Masson.

Rak, Y. (1993). Morphological variation in *Homo neanderthalensis* and *Homo sapiens* in the Levant; a biogeographic model. In W. H. Kimbel & L. B. Martin (Eds.), *Species, species concepts, and primate evolution* (pp. 523–536). New York: Plenum.

Rak, Y., Ginzburg, A., & Geffen, E. (2002). Does *Homo Neandertalensis* play a role in modern human ancestery? The mandibular evidence. *American Journal of Physical Anthropology, 119*, 199–204.

Ray, L. J. (1959). Metrical and non-metrical features of the clavicle of the Australian Aboriginal. *American Journal of Physical Anthropology, 17*, 217–226.

Relethford, J. H. (2001). Absence of regional affinities of Neandertal DNA with living humans does not reject multiregional evolution. *American Journal of Physical Anthropology, 115*, 95–98.

Relethford, J. H. (2003). *Reflections of our past*. Boulder: Westview.

Ridley, M. (2004). *Evolution* (3rd ed.). Oxford: Blackwell.

Schillaci, M. A., & Froehlich, J. W. (2001). Nonhuman primate hybridization and the taxonomic status of Neanderthals. *American Journal of Physical Anthropology, 115*, 157–166.

Smith, F. H. (1976). The Neanderthal remains from Krapina, Northern Yugoslavia: An inventory of the upper limb remains. *Zeitschrift für Morphologie und Anthropologie, 67*, 275–290.

Smith, F. H. (1991). The Neandertals: Evolutionary dead ends or ancestors of modern people? *Journal of Anthropological Research, 47*, 219–238.

Smith, F. H., & Trinkaus, E. (1991). Les origines de l'homme moderne en Europe centrale: un cas de continuité. In J. J. Hublin & A. M.

Tillier (Eds.), *Aux origines d'Homo sapiens* (pp. 251–290). Paris: Presse Universitaire de France.
Smith, F. H., Falsetti, A. B., & Donnelly, S. M. (1989a). Modern human origins. *American Journal of Physical Anthropology, 32*, 35–68.
Smith, F. H., Simek, J. F., & Harrill, M. S. (1989b). Geographic variation in supraorbital torus reduction during the later Pleistocene (c. 80 000–15 000 BP). In P. Mellars & B. Stringer (Eds.), *The human revolution* (pp. 172–193). Edinburgh: Edinburgh University Press.
Smith, F. H., Jankovic, I., & Karavanic, I. (2005). The assimilation model, modern human origins in Europe, and the extinction of Neandertals. *Quaternary International, 137*, 7–19.
Stern, J. T., & Susman, R. L. (1983). The locomotor anatomy of *Australopithecus afarensis. American Journal of Physical Anthropology, 60*, 279–317.
Stewart, T. D. (1962). Neanderthal scapulae with special attention to the Shanidar Neanderthals from Iraq. *Anthropos, 57*, 779–800.
Stewart, T. D. (1964). The scapula of the first recognized Neanderthal skeleton. *Bonner Jahrbucher, 164*, 1–14.
Stringer, C. B. (1998). Chronological and biogeographic perspectives on later human evolution. In T. Akazawa, K. Aoki, & O. Bar-Yosef (Eds.), *Neandertals and modern humans in western Asia* (pp. 29–37). New York: Plenum.
Stringer, C. B. (2002). Modern human origins: Progress and prospects. *Philosophical Transaction of the Royal Society, London, B Biological Science, 357*, 563–579.
Stringer, C. B., & McKie, R. (1996). *African exodus – the origins of modern humanity*. London: Pimlico.
Thoma, A. (1965). La définition des Néanderthaliens et la position des hommes fossiles de Palestine. *L'Anthropologie (Paris), 69*, 519–533.
Trinkaus, E. (1977). A functional interpretation of the axillary border of the Neandertal scapula. *Journal of Human Evolution, 6*, 231–234.
Trinkaus, E. (1981). Neanderthal limb proportions. In C. B. Stringer (Ed.), *Aspect of human evolution* (pp. 187–224). London: Taylors & Francis.
Trinkaus, E. (1982). The Shanidar 3 Neandertal. *American Journal of Physical Anthropology, 57*, 37–60.
Trinkaus, E. (1983). *The Shanidar Neandertals*. New York: Academic.
Trinkaus, E. (1991). Les hommes fossiles de la grotte de Shanidar, Irak: évolution et continuité parmi les hommes archaïques tardifs du Proche-Orient. *L'Anthropologie (Paris), 95*, 535–572.
Trinkaus, E. (2006). The Krapina scapulae. *Periodicum Biologorum, 108*, 341–351.
Trinkaus, E. (2008). Kiik-Koba 2 and Neandertal axillary border ontogeny. *Anthropological Science, 116*, 231–236.
Trinkaus, E., Marks, A. E., Brugal, J. P., Bailey, S. E., Rink, W. J., & Richter, D. (2003a). Later Middle Pleistocene human remains from the Almonda Karstic system, Torres Novas, Portugal. *Journal of Human Evolution, 45*, 219–226.
Trinkaus, E., Milota, S., Rodrigo, R., Mircea, G., & Moldovan, O. (2003b). Early modern human cranial remains from the Pestera cu Oase, Romania. *Journal of Human Evolution, 45*, 245–253.
Vallois, H. V. (1928). L'omoplate humaine. Etude anatomique et anthropologique. *Bulletins et Mémoires de la Société d'Anthropologie de Paris série 7, 9*, 129–168.
Vallois, H. V. (1932). L'omoplate humaine. Etude anatomique et anthropologique. *Bulletins et Mémoires de la Société d'Anthropologie de Paris série 8, 3*, 3–153.
Vallois, H. V. (1946). L'omoplate humaine. Etude anatomique et anthropologique. *Bulletins et Mémoires de la Société d'Anthropologie de Paris série 9, 7*, 16–99.
Vandermeersch, B. (1981). *Les Hommes fossiles de Qafzeh (Israël)*. Paris: Cahiers de Paléoanthropologie C.N.R.S édition.
Vandermeersch, B. (1991). La ceinture scapulaire et les membres supérieurs. In O. Bar-Yosef & B. Vandermeersch (Eds.), *Le squelette Moustérien de Kebara 2* (pp. 157–178). Paris: CNRS édition.
Vandermeersch, B., & Trinkaus, E. (1995). The postcranial remains of the Regourdou 1 Neandertal: The shoulder and arm remains. *Journal of Human Evolution, 28*, 439–476.
Voisin, J. L. (2000). L'épaule des hominidés. Aspects architecturaux et fonctionnels, références particulières à la clavicule. Ph.D. dissertation, Museum National d'Histoire Naturelle (Paris).
Voisin, J. L. (2001). Evolution de la morphologie claviculaire au sein du genre *Homo*, conséquence architecturale et fonctionnelle sur la ceinture scapulaire. *L'Anthropologie (Paris), 105*, 449–468.
Voisin, J. L. (2004). Clavicule: approche architecturale de l'épaule et réflexions sur le statut systématique des néandertaliens. *Comptes Rendus Palevol, 3*, 133–142.
Voisin, J. L. (2006a). Krapina and other Neanderthal clavicles: A peculiar morphology? *Periodicum Biologorum, 108*, 331–339.
Voisin, J. L. (2006b). The clavicle, a neglected bone; morphology and relation to arm movements and shoulder architecture in Primates. *The Anatomical Record. Part A, 288A*, 944–953.
Voisin, J. L. (2006c). Speciation by distance and temporal overlap: A new approach to understanding Neanderthal evolution. In K. Harvati & T. Harrison (Eds.), *Neanderthals revisited: New approaches and perspectives*. Berlin: Springer, Kluwer.
Voisin, J. L. (2008). Omo I Kibish clavicle: Archaic or modern? *Journal of Human Evolution, 55*, 438–443.
Wolpoff, M. H., Hawks, J., & Caspari, M. (2000). Multiregional, not multiple origins. *American Journal of Physical Anthropology, 112*, 129–136.
Wolpoff, M., Hawks, J., Frayer, D., & Hunley, K. (2001). Modern human ancestry at the peripheries: A test of the replacement theory. *Science, 291*, 293–297.
Wolpoff, M., Mannheim, B., Mann, A., Hawks, J., Caspari, R., Rosenberg, K. R., Frayer, D. W., Gill, G. W., & Clark, G. (2004). Why not the Neanderthals? *World Archaeology, 36*, 527–546.
Young, F. (2001). ViSat 6.4 on www.visualstats.org.

Chapter 12
Facts and Ideas in Paleolithic Growth Studies (Paleoauxology)

Evidence from Neanderthals in Europe

Anne-Marie Tillier

Abstract In the study of prehistoric populations, the identification of taxonomic markers derives from the study of preserved adult individuals. A complementary approach for understanding morphological differences between populations involves an investigation of ontogenetic and growth patterns. Within the Neanderthal population, the specimens employed to document distinct developmental stages originated from sites often separated by hundreds (or thousands) of kilometers and thousands of years in time. In the reconstruction of maturation patterns in Neanderthals, there are difficulties related to the availability of the fossil record itself, the evaluation of geographical and temporal variations, and the choice of appropriate comparative reference samples. Among Neanderthals, the importance of sexual dimorphism in the manifestation of morphological variation and age-related changes during the growth period cannot be evaluated. However, comparative analyses provide accurate information regarding the ontogenetic appearance of several diagnostic Neanderthal characteristics and reveal similarities and contrasts between Neanderthal and early modern children. This contribution attempts to address few aspects of the ongoing debate regarding maturational events and life history in nonadult Neanderthals from western and central Europe.

Keywords Children • Diagnostic traits • Health status • Middle Paleolithic • Ontogeny • Skeletal variation • Temporal and/or regional changes

Introduction

Neanderthals have been found in sites geographically dispersed in Western Eurasia and over long time duration, probably more than 250,000 years. Most of the human remains were associated with a Middle Paleolithic (i.e. Mousterian *sensu largo*) archaeological context; very few have been found with Lower Upper Paleolithic industries. The Neanderthal fossil record has played a major role in the development of studies from an auxological perspective that addressed the question of origin of modern human-like pattern of growth.

In the last two decades, paleoanthropologists have attempted to improve the methods of analyzing immature skeletons, and the impact of methodological advancements in dental and skeletal analyses has stimulated the emergence of paleoauxology. Hypothetical scenarios have emerged concerning life-history stages in Neanderthals and contradictory hypotheses have been produced focusing on growth and development. The specimens employed to document distinct developmental stages within the entire Neanderthal population originated from sites often separated by hundreds (or thousands) of kilometers and thousands of years in time. However, a general lack of appreciation for individual variation within and between groups (geographically and chronologically dispersed) characterizes studies comparing nonadult fossil specimens.

Comparative growth studies in extant populations have demonstrated that environmental differences and selective pressures have effects on the growing child. Accordingly, there is no guarantee that human groups, geographically distant in western Eurasia and evolving in different sites and habitats during several tens of thousands of years, shared similar growth patterns. In the present context, we think that there are limitations to the meaning of large-scale comparative analyses of immature fossil specimens in terms of paleoauxological reconstruction. This is the reason why this contribution will focus on individuals originating in western and central Europe.

Sampling Nonadult Neanderthals in Europe

Besides the geographical and chronological dispersals, reconstruction of Neanderthal childhood from skeletal studies has had to face practical difficulties, such as bone preservation

A.-M. Tillier (✉)
Laboratoire d'Anthropologie des Populations du Passé, UMR 5199-PACEA, Université Bordeaux 1, avenue des Facultés, 33405 Talence, France
e-mail: am.tillier@pacea.u-bordeaux1.fr

S. Condemi and G.-C. Weniger (eds.), *Continuity and Discontinuity in the Peopling of Europe: One Hundred Fifty Years of Neanderthal Study*, Vertebrate Paleobiology and Paleoanthropology, DOI 10.1007/978-94-007-0492-3_12,

and sampling limitations. While nonadult individuals represent no more than 25% of the overall Neanderthal sample, given the preservation of the fossil record, some skeletal growth stages are less well documented than others. Until the recent rediscovery of the Le Moustier 2 skeleton from South-western France (Maureille 2002), fetus and infant remains (i.e. individuals less than 1 year old) were really fragmentary. Age distribution (estimated age at death based on tooth development) of non-adult Neanderthal individuals from European sites shows a prevalence of the second age-class (1–4 year), followed by the class 10–14 year. At the present time, late adolescence is represented by a unique specimen, Le Moustier 1 (Ullrich 2005).

Within the fossil record, specimens with both skull and infracranial bones preserved are rare (e.g. Roc de Marsal, Le Moustier 1 and 2 in Southwestern France). There are a very small number of individuals represented by almost complete skulls or crania (e.g. Pech de l'Azé1, La Quina H18 in France, Engis 2 in Belgium, Subalyuk 2 in Hungary). A majority of Neanderthal immature finds are isolated bones, especially vault elements (e.g. Krapina 1 and 2 in Croatia) or mandibles (e.g. Scladina in Belgium). In a few cases, immature Neanderthals preserve only infracranial elements (e.g. La Ferrassie 6 in France, Kiik-Koba 2 in Crimea, few bones from Cova Negra in Spain) and consideration of the individual's growth profile is rather difficult to assess in the lack of close association with accurate age estimation.

Only a few sites restricted to western and central Europe have a large enough sample size to evaluate individual variation or to permit a within-site comparison between nonadult and adult skeletal morphologies. Yet, even within a single site, the question of contemporaneity arises, as illustrated by the Krapina and Le Moustier sites. At Le Moustier, the history of site excavations and circumstances of discoveries of the specimens led to confusion over the stratigraphical provenance and chronological attribution of the two nonadult individuals (Maureille 2002; Maureille and Turcq 2005). At the Croatian site, two partial immature skulls, Krapina1 and 2, have been unearthed: they originated respectively from level 8 in the upper part of the stratigraphic sequence and level 3 down in the sequence (e.g. Smith 1976; Radovčic et al. 1988).[1] Krapina1, geologically more recent, represents a single specimen (of unknown dental age estimates) within the large Krapina hominid sample, and the evaluation of several aspects of its peculiar morphology in terms of affinities to its predecessors remains questionable (see Minugh-Purvis et al. 2000; Sansilbano-Collilieux and Tillier 2006).

In this respect, intrapopulation variation might have occurred throughout time that cannot be neglected in the reconstruction of maturation patterns. Potential evolutionary trends in a regional Neanderthal sample are supported by the Pech de l'Azé and Roc de Marsal children found in sites from the Dordogne separated by less than 25 km (Tillier 1996; Soressi et al. 2007). Evidence of ontogenetic variation in skull morphology between the two individuals of the same age class (differences in facial size and in robustness of the symphyseal region of the mandible) was emphasized by Tillier and radiometric dates recently obtained for both sites provides the basis for a new level of understanding of this individual variation. An early age around 70,000 years BP appears appropriate for the Roc de Marsal deposit associated to with a Mousterian assemblage (Sandgathe et al. 2005), while the Pech de l'Azé specimen (the only fossil hominid found in association with a Mousterian of Acheulean tradition type B) is more recent (Soressi et al. 2007), dated to the early part of oxygen isotope stage 3.

[1] Radiometric techniques applied to animal enamel tooth placed the section (Rink et al. 1999) between 110,000 and 137,000 years BP.

Becoming a Neanderthal

Neanderthals are commonly seen as a rather homogeneous group morphologically, and skull features traditionally play a major role in their anatomical definition (e.g. Stringer et al. 1984; Hublin and Tillier 1991; Hublin 1998; Condemi 2006). The taxonomic affiliation of immature remains is inferred from the identification of distinguishing anatomical features that have been proposed to describe adult Neanderthal skeletal morphology. However, it must be said that taxonomic identification of immature skeletal remains cannot avoid the significance of morphological changes directly related to growth and maturation processes.

Fetus, Neonates, and Infants

Apart from Le Moustier 2 (still being studied), the first steps of the ontogenetic trajectory in Neanderthals are documented by fragmentary remains, the taxonomic assignment of which are rather difficult to assess. Based on the examination of a few Eurasian Upper Pleistocene specimens (La Ferrassie 5 and 4, L'Hortus I and Ib, Arcy-sur-Cure in France; Shanidar 7 and 9 in Iraq; Kebara 1 and Qafzeh 13 in Israel), scholars have not been able to identify in early infancy the specific population features that characterize adult cranial morphology (Vlček 1970; de Lumley 1973; Trinkaus 1983; Hublin et al. 1996; Tillier 1986, 1995, 1999; Tillier et al. 2003). On the contrary, Heim (1982: 16–17) described on the two La Ferrassie fragmentary frontal bones a faint thickening of the supraorbital region that was interpreted as an early development of a Neanderthal supraorbital torus. Heim argued that such a morphology had been previously identified on the Staroselye child from Crimea. Yet, this specimen from Crimea is fully modern in its cranial morphology (Alekseev 1976; Tillier, in Ronen 1982: 315), and moreover its Mousterian attribution is no longer supported (Marks et al. 1997).

Apart from the Le Moustier 2 skeleton, infracranial evidence remains fragmentary and limits the inference of the early appearance of characteristics thought to be diagnostic of Neanderthals. Detailed descriptions of upper and lower limb remains belonging to Neanderthals under 1 year old exist in the literature: they mainly refer to French and Crimean specimens (Vlček 1973, 1975; Heim 1982). However, morphological features of the infant infracranial skeleton in modern humans are far from being well known; comparative studies between fossil and recent populations focus on a few infants who derive from sedentarized populations (i.e. regional osteological collections used in forensic anthropological studies): this can explain the general emphasis on diaphyseal robustness, curvature of long bones (ex: radius), or early appearance of muscle markings (see below, the example of the first metacarpal), in Neanderthal infant appendicular skeletons. In practice, some degree of morphological overlap between Neanderthal and Upper Paleolithic infant specimens cannot be avoided. Finally, the state of preservation of the postcranial skeleton limits body size and limb proportion reconstructions in Neanderthal babies.

Age-Related Changes in Individual Distinguishing Features

At birth, only ear ossicles and the bony labyrinth within the temporal bone are fully grown and have their mature shape and size. There is clearly considerable variability in ear ossicle measurements among extant modern humans (Arensburg et al. 1981; Arensburg and Tillier 1983). The Neanderthal ossicles are preserved in three specimens (Biache St Vaast 1, La Ferrassie 3, and Le Moustier 1; Heim 1982; Ponce de Leon and Zollikofer 1999; Rougier 2003) and are described as different in size and shape from those of modern humans. But, as noted by Spoor (2002: 296), "a larger sample (…) would be required to come to definitive conclusions."

Information concerning selected traits that can be used in the distinctiveness of Neanderthal bony labyrinth morphology can be obtained from recent publications (e.g., Hublin et al. 1996; Spoor and Zonneveld 1998; Thompson and Illerhaus 1998; Ponce de Leon and Zollikofer 1999; Spoor et al. 2003). The data collected from CT analyses reveal a certain degree of individual variation within the Neanderthal sample as illustrated by the Le Moustier 1 labyrinth. This specimen does not exhibit a "fully typical Neanderthal-like" morphology and it is suggested that such a morphology can be viewed, either as "(..) a sign of gene flow between Neanderthals and modern migrants," or as a "more extreme form of (*Neanderthal*) normal range of variation" (Spoor et al. 2003 :162). Related to the former proposal is the assumption that Le Moustier 1 might be considered as a late Neanderthal in the sample, a condition which cannot be established with regard to the historical context of the discovery, as mentioned before. As emphasized by Ponce de Leon and Zollikofer (2005: 256), the Le Moustier 1 labyrinth morphology questions the apomorphies of otic characters.

With regard to the external skull development, Vlček (1970) was the first to suggest that individual distinguishing features can be recognized in immature Neanderthals of at least 2 years of age. Hublin (1980), Heim (1982), Tillier (1986), and Minugh-Purvis (1988) have come to a similar conclusion. For instance, the occipital morphologies of young specimens from La Chaise Abri-Suard and La Ferrassie sites (Fig. 12.1) exhibit coexistence of a fully developed suprainiac fossa and bilaterally protruding occipital torus, a low occipital height, a highly convex occipital squama. Similarly, compared to that of modern children, the size of the Neanderthal mastoid process relative to that of the juxtamastoid eminence changed less during ontogeny; in other words, adult Neanderthals accentuated a juvenile trait in this part of the temporal bone (Tillier 1983, 1988).

Few researchers argued that similarities and contrasts between Neanderthal and modern human skull morphology evolved as postnatal bone maturation progressed (e.g. Tillier 1983, 1986; Coqueugniot 1999; Williams 2001). Neanderthal juveniles have an absent or reduced supraorbital torus, a more vertical forehead, a more rounded cranial vault (Fig. 12.2).

Similarly, the shape of the foramen magnum in juvenile Neanderthals shows some degree of morphological overlap with modern humans. The Engis 2 child skull from Belgium (ca. 4 years at death), exhibits an elongated foramen magnum, as shown by the value of the breadth/length index (Tillier 1983). On this young specimen, neither the anterior nor the posterior intra-occipital synchondroses are fused. Older Neanderthals from France with a well-preserved basicranium, such as the two adults, i.e. La Ferrassie 1 and La Chapelle-aux-Saints, exhibit higher values of the index than that of the Belgian child (Creed-Miles et al. 1996; Tillier 1998, 2000).

Interestingly, Arsuaga et al. (1997) documented within-site age-related individual variation of the foramen magnum shape at Sima de Los Huesos (Spain) that supported similar conclusions about age-related changes among other archaic members of the genus *Homo*. Furthermore, comparative data deriving from analyses of modern children demonstrate that the occurrence of an elongated foramen magnum can be, indeed, interpreted as an age-related trait (Coqueugniot and Le Minor 2002). The changing proportions of the foramen magnum throughout life are in fact partly related to complete closure of both intra-occipital and spheno-occipital

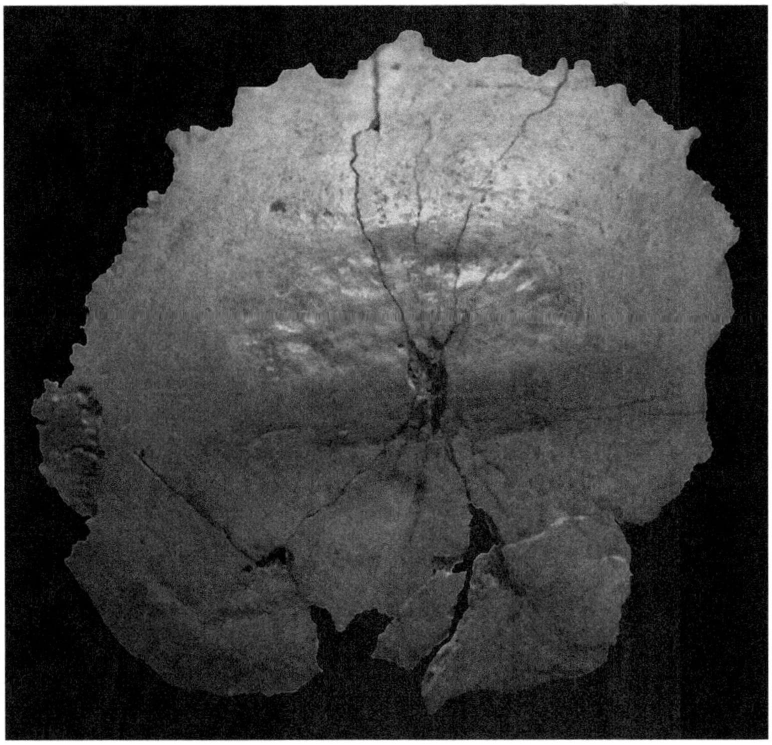

Fig. 12.1 External view of the occipital bone of the La Ferrassie 8 child showing the development of the suprainiac fossa and bilateral protruding torus on a young child, ca. 2 years old at death (Photo J.-L. Heim)

synchrondroses. On the basis of this constancy through the juvenile years, the presence of an elongated foramen magnum on a very young fossil specimen does not necessarily align it with Neanderthals (Creed-Miles et al. 1996; Tillier 1998; Coqueugniot 1999; versus Rak et al. 1994).

Anatomical traits of the Neanderthal facial skeleton (e.g. midfacial projection, anterior maxillary inflation and straightness of the maxillo-zygomatic profile in norma facialis, double-arched browridge with large pneumatization, posterior positioning of the mental foramen) appear later in ontogeny. A posterior localization of the mental foramen can be observed before the eruption of the first permanent molar in young Neanderthals (Fig. 12.3, Coqueugniot 2000). By contrast, other facial features seem to be fully developed in late childhood (Tillier 1983, 1986, 1987; Minugh-Purvis 1988; Williams 2001) and these changes in developmental patterns occur when children have anterior and posterior permanent teeth (Fig. 12.4a, b).

In conclusion, it is clear that the phylogenetic assignment of immature Neanderthal cranial remains is easier for some skeletal parts than for others, as age-related changes affect the skull morphology (e.g. cranio-facial shape). Only few diagnostic features of the cranial vault are known to be present at early childhood (e.g. Heim 1982; Hublin 1980; Tillier 1983).

Infracranial Skeletal Morphology and Neanderthal Traits Revisited

The appendicular skeleton and the pelvic girdle are thought to document morphological differences between human groups within the Late Pleistocene hominid sample (e.g. Churchill 1994; Trinkaus 1992; Trinkaus et al. 1998). Most of the features described on the Neanderthal upper and lower limbs can be seen as reflections of the high level of muscularity of these hominids. Consequently, it is not surprising that their differentiation on subadult upper and lower limbs cannot be identified in the early stages of development. Each new discovery of specimens contributes to the knowledge of the range of variation in postcranial skeletal morphology among nonadult Neanderthals and ontogenetic appearance of features currently found in adults (Arsuaga et al. 2007).

As a matter of fact, comparative analysis of Neanderthal and Upper Paleolithic adult skeletal remains provide evidence that some of the diagnostic features described on the Neanderthal appendicular skeleton (i.e. morphology of the scapular axillary border, muscular markings on long bones, claviculo-humeral index, etc.) have also been identified in Lower or Middle Upper Paleolithic skeletons from Europe (e.g. Jelinek 1992; Frayer 1992; Trinkaus et al. 2006).

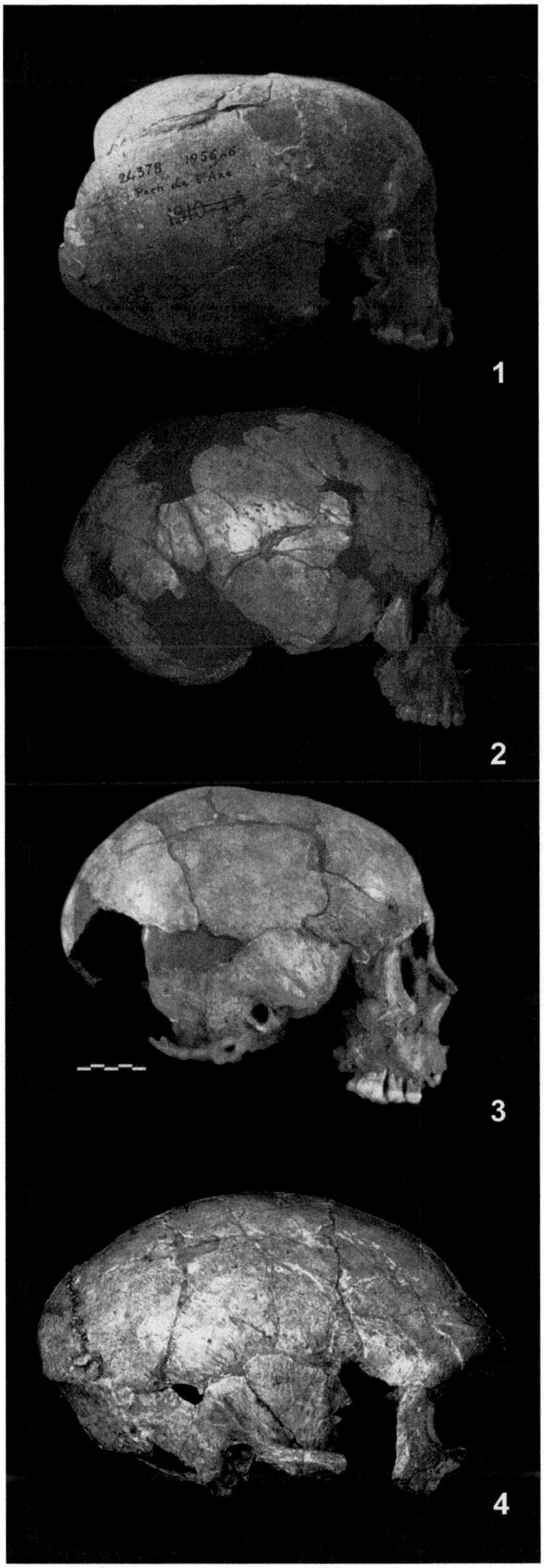

Fig. 12.2 Skull development in Neanderthals from Southwestern France. *1* Pech de l'Azé I child (ca. 2 years old at death), *2* Roc de Marsal (ca. 3 years old), *3* La Quina H18 (ca. 7 years old), and *4* the adult La Quina 5. Besides age-related changes in skull morphology (supraorbital torus, frontal curvature, height/length skull ratio, occipital bun), it should be reminded that Roc de Marsal and the two La Quina specimens are earlier (attributed to oxygen isotope stage 5 and 4 respectively) than Pech de l'Azé I (Photos A.-M. Tillier)

Until recently, the comparison between Neanderthal and modern children excluded Upper Paleolithic specimens. Furthermore, given the scarcity of available comparative material of earlier European archaic humans, it remains to be seen what features were already present in earlier, pre-Neanderthal populations in order to clarify the polarities of the traits.

The Superior Pubic Ramus

Of particular interest here, a few scholars (e.g. McCown and Keith 1939; Stewart 1960; Trinkaus 1976; Rosenberg 1988; Rak 1990) have identified diagnostic features of the Neanderthal hipbone. Special emphasis was given to the superior pubic ramus which, according to the authors, was described as very long and vertically flat among Neanderthals, while short and thick in modern humans. Consequently, the superior pubic ramus was thought to have a shape which had a potential in assessing the Neanderthal affinities of a specimen. Additional data and new discoveries led to a revision of the phylogenetic status of pubic characteristics and an elongated pubic ramus should be regarded as plesiomorphic: it was already present on earlier fossil remains than Neanderthals, as documented by australopithecines, the Sima de Los Huesos specimens in Spain, and was described in early Levantine hominids from Tabun and Skhul (e.g. Arensburg and Belfer-Cohen 1998; Arsuaga et al. 1999; Bonmati and Arsuaga 2005; Tillier 2005a).

Within the immature Neanderthal sample, the number and state of preservation of the hip bone remains limit consistent metric analysis (Majó 2000; Majo and Tillier 2003). Unfortunately, the pubic skeletal part is lacking or poorly preserved among four specimens (Roc de Marsal, Le Moustier 1 and 2, Kiik-Koba 2 in Crimea). From the La Ferrassie rockshelter, two of the immature individuals provided information on the development of juvenile pubic bones (Heim 1982). La Ferrassie 8, a child ca. 2 years old, preserves a large portion of the two superior pubic rami, besides the two nearly complete ilia and one fragment of the left ischium. La Ferrassie 6 hipbone consists of a right pubis and ischium (minus the superior part of the acetabulum), two thirds of the left ilium, as well as fragments of the left ischium and right ilium. No teeth were preserved for this specimen and the age at death of the child was estimated between 3 (Heim 1982) and 4–5 years (Tompkins and Trinkaus 1987; Majo and Tillier 2003).

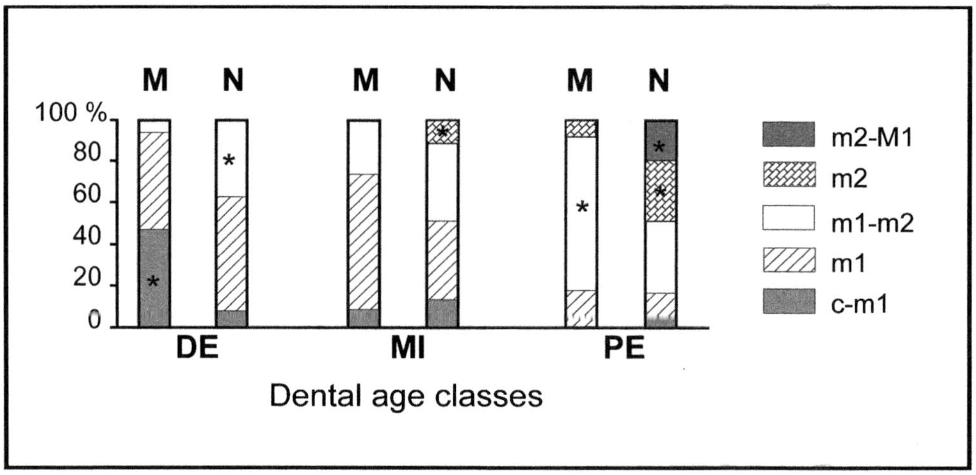

Fig. 12.3 Position of the mental foramen and evolution during growth in Neanderthal (*N*) and Modern (*M*) mandibles. Classes in dental development: DE = deciduous dentition (c, m1, m2), MI = deciduous dentition + first permanent molar (M1), PE = Permanent dentition. A posterior positioning of the mental foramen in Neanderthal mandibles is not dependent on the first permanent molar (*) statistically significative difference (After Coqueugniot 2000)

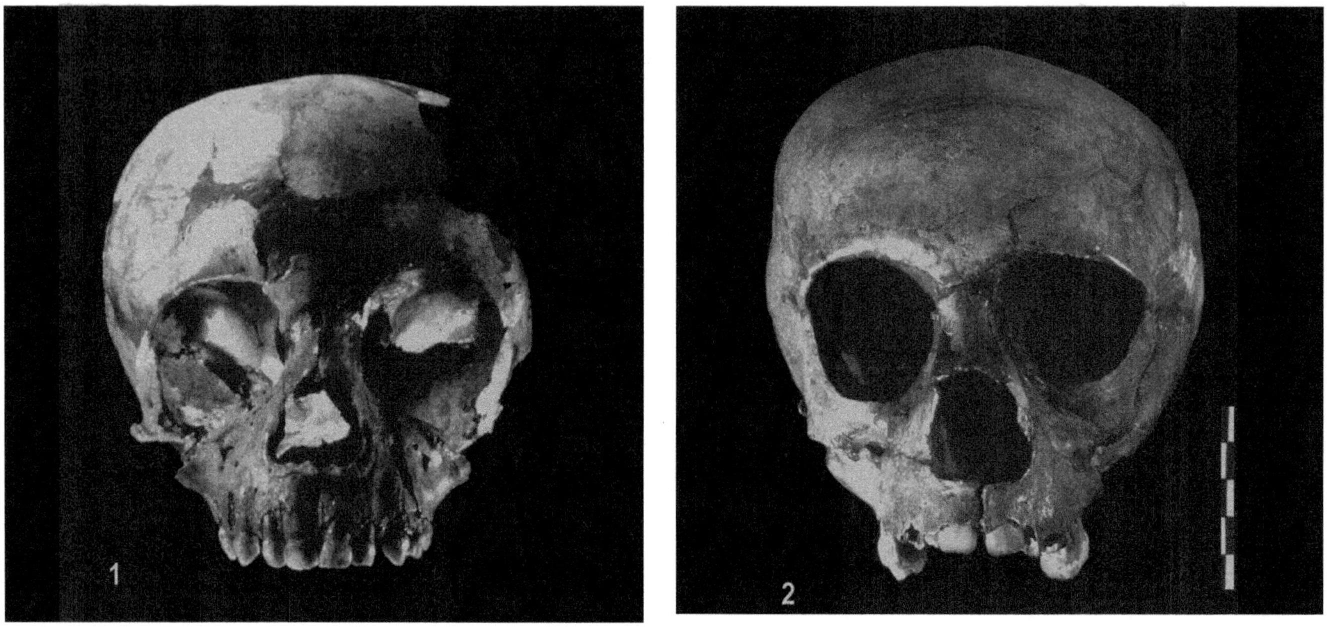

Fig. 12.4 Roc de Marsal (*1*) and La Quina H18 (*2*) skulls in frontal view. The skull of the oldest child (La Quina18, ca. 7 years old at death) is narrower but exhibits supraorbital torus, supraglabellar depression, and straight maxillo-zygomatic inferior profile, three features absent on Roc de Marsal (Photo A.-M. Tillier)

On the basis of a comparative analysis between the La Ferrassie specimens and modern children, the assumption of age-related changes in the pubic shape among Neanderthals cannot be rejected.[2] The elongation of the superior pubic ramus is present only on La Ferrassie 6, obviously older in developmental age than La Ferrassie 8 (considering bone maturation and diaphyseal lengths). But, in contrast with the adult shape, the immature La Ferrassie 6 pubis exhibits no variation in height between the ventral and dorsal margins of the superior ramus (Tompkins and Trinkaus 1987; Majó 2000; Majo and Tillier 2003). The age at which this last pubic change occurs in Neanderthals before adulthood is unknown, due to the lack of bones from older individuals.

Interestingly, comparative analysis of hipbone growth conducted between the La Ferrassie individuals and Levantine early modern humans from Qafzeh (Majo and Tillier 2003) revealed a certain degree of morphological

[2] It is clear that drawing definitive conclusions on the basis of two specimens is premature.

overlap in ramus pubic elongation between the two groups within similar age-classes. Such results lead one to reconsider the use of the pubic morphology in the assessment of Neanderthal affinities in Eurasian Middle Paleolithic children. Hipbone remains of Upper Paleolithic specimens belonging to similar age-classes (Henry-Gambier 2001; Bruzek and Trinkaus 2002) are not complete enough to permit metrical comparisons. The incomplete pubic bone of Lagar Velho (age 4.5–5 years) from Portugal is described as "relatively short compared to those of both Late Pleistocene and recent children of similar age" (Bruzek and Trinkaus 2002: 431), while the Late Upper Paleolithic specimen from Grotte des Enfants in Italy (GE2, age ca. 2 years, Henry-Gambier in Henry-Gambier (Ed) 2001: 78) possesses a pubic length that falls within the range of variation of modern children of similar ages (Tillier 1999).

Lower Limb Segment Proportions

Adult Neanderthals exhibit (to maintain the same tense) relatively short tibial length relative to femoral length and such segmental proportions of the lower limb are commonly seen as a possible reflection of cold-climatic adaptation (e.g. Trinkaus 1981; Holliday 1995). Measurements defined on adult long bones are not applicable for immature specimens, due to patterns of bone maturation (Tillier 2005b). Furthermore, secondary ossification centers are mainly lacking (i.e. proximal and distal epiphyses) and diaphyseal lengths incompletely preserved in a majority of Neanderthal children (e.g. Kiik-Koba 2, Roc de Marsal, la Ferrassie 3, etc.).

A single child, La Ferrassie 6 (Fig. 12.5) allows an accurate evaluation of segment proportions, illustrated by the value of the crural index (tibio-femoral intermetaphyseal length ratio). Despite the lack of reliable age estimation for La Ferrassie 6 (no teeth are preserved), comparison between fossil and recent children provides evidence of individual and interpopulation variations (Table 12.1). The crural index, lower in La Ferrassie 6 than in the two Levantine early modern immature skeletons, is closer to few values in European Upper Paleolithic individuals. Furthermore, data collected from recent populations document large variability. Given the range of variation reported in crural index values, it would be fruitful to increase the sample of data collected for Neanderthals with new discoveries.

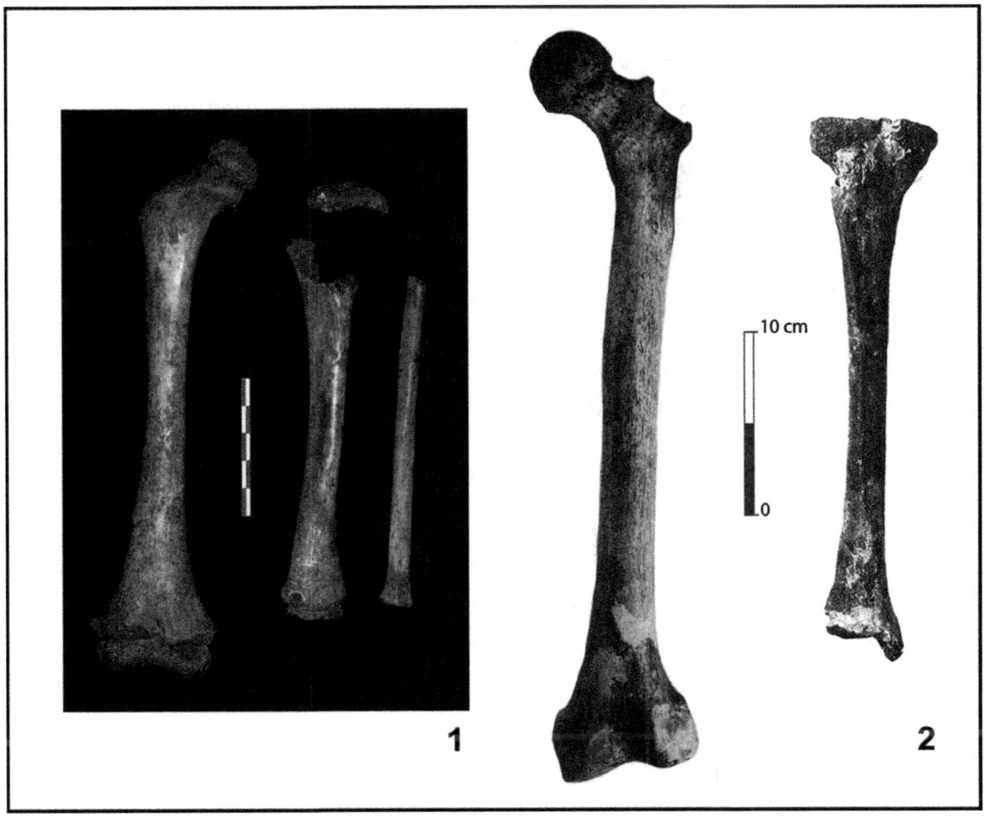

Fig. 12.5 Within-site pattern of lower limb growth in Neanderthals from Southwestern France. Both La Ferrassie 6 child ca. 4 years old (*1*) and La Ferrassie 2 adult (*2*) exhibit short tibial length relative to femoral length; however, the crural index (using maximum lengths including epiphyses) is lower on the adult (74.6) than on the child (77.6 minus cartilage thickness). The percentage of adult size attained is 45% for the juvenile femur (maximum length including epiphyses) and 47% for the tibia (Photo J.-L. Heim and A.-M. Tillier)

Table 12.1 Variation in crural index values (tibio-femoral intermetaphyseal length ratio) among children. Comparison between the Neanderthal La Ferrassie 6 child, other Paleolithic specimens, and recent children

Children and their geographical provenance	Ages at death	Crural index
Europe: Middle (MP) and Upper Paleolithic (UP)		
La Ferrassie 6 (MP)	ca. 4 years?	76.5 (1)
Grotte des Enfants 2 (UP)	ca. 2 years	82.0 (2)
Grotte des Enfants 1 (UP)	ca. 3 years	79.0 (2)
Arene Candide 11 (UP)	ca. 3 years	83.0 (2)
Lagar Velho (UP)	ca. 4 years	79.0 (3)
Arene Candide 8 (UP)	ca. 7 years	78.4 (2)
Southwestern Asian: Middle Paleolithic		
Dederyeh 1	ca. 2 years	80.5 (4)
Skhul 1	ca. 3 years	85.7 (1)
Recent samples geographically dispersed		
Coimbra (Portugal, $N=20$)	7–16 years	77.0 ± 4.2 (1)
Japanese children ($N=46$)	Unknown	79.2 ± 2.8 (4)
Spitalfields (Great Britain, $N=26$)	0–5 years	83.2 ± 3.0 (1)
South African black children ($N=61$)	Unknown	83.5 ± 2.4 (4)

Bone length measurements of fossil specimens employed in the comparison derive from the following sources: (1) Tillier 2000, (2) sources in Henry-Gambier 2001, (3) Ruff et al. in Zilhao and Trinkaus 2002, (4) Kondo and Dodo *in* Akazawa and Muhesen 2002). Individual data from cross-sectional reference samples were collected either by the author (Spitalfieds and Coimbra collections) or by her Japanese colleagues

Thumb Proportions and Morphology

On the first metacarpal, adult Neanderthal handbones exhibit hypertrophied insertion of the opposing muscle of the thumb (*opponens pollicis musculus*), and a long distal phalange relative to the proximal phalange of the pollex. It was suggested that such features influenced the effectiveness of manipulative movements and precision grip (Vlček 1973; Trinkaus 1983; Villemeur 1994). Preservation of handbones in immature specimens is quite exceptional, especially complete thumbs, as shown by Roc de Marsal (ca. 3 years old at death) and La Ferrassie 3 (skeletal age presumed to be close to 10 years, according to Heim 1982) specimens. Young individuals such as Kiik-Koba 2 (less than 1 year old) and Roc de Marsal do not manifest specific development of attachment for the opposing muscle of thumb in the first metacarpals (Madre-Dupouy 1992 and author's observations versus Vlček 1973), in contrast with the oldest child, La Ferrassie 3. This specimen also exhibits an elongated distal pollical phalange ($I=91.3$, versus 69.8 for Roc de Marsal) similar to that of adult Neanderthals (between 85.2 and 89.3 for La Ferrassie 1; 84.3 La Ferrassie 2). However, as already noted in the case of pubic growth, additional data are required to confirm an age-related change for the appearance of specific pollical features during childhood and to precisely determine the age trends in timing of their appearance.

Yet, this short review suggests that variation in the appearance of traits during infracranial skeleton growth in Neanderthals cannot be ignored. To a certain degree, the skeletal remains of very young Neanderthals are generally similar to those of modern humans of the same developmental age. Additional insight into the variation of ontogenetic pattern and age-related differences in skeletal morphology among young Neanderthals is necessary to draw definitive conclusions. Most of the bone features related to the requirements of manipulation and locomotion are associated with late development of interosseous ligamentous areas and muscle attachments. The reconstruction of skeletal maturational events would benefit from the accumulation of data with new discoveries and further studies.

Is There a Neanderthal Growth Pattern?

Central to the argument in favor of a distinct Neanderthal growth pattern is the idea of a unique modern-like pattern denied by the existent data bringing evidence of interpopulation diversity. Differences in population patterns of growth and development are well documented in modern living population samples (e.g. Eveleth and Tanner 1990; Bogin 1991). Nutritional, social, and environmental factors contribute to the differences in growth rate and development. It would not be surprising that Neanderthals living over a period of tens of thousands of years should manifest differences in growth patterns.

Cranial Size

The last two decades have seen an increasing concern with distinct interpretations of ontogenetic patterns in Neanderthals that have employed brain growth and cranial size. Few scholars (e.g. Trinkaus 1984; Dean et al. 1986) have asserted that a nonmodern human-like pattern in Neanderthal fetal growth resulted in an enlarged brain size at birth, while others (e.g. Tillier 1986, 1995; Minugh-Purvis 1988) have argued that such an assertion could not be supported by the available fossil record.

The Devil's Tower child from Gibraltar has also served as a model to assume larger braincases in juvenile Neanderthals than in recent children, and to suggest (e.g. Dean et al. 1986; Stringer et al. 1990; Trinkaus and Tompkins 1990) a distinct growth rate of the skull in Neanderthals. Other scholars (e.g. Tillier 1988; Coqueugniot 1994; Minugh-Purvis 1988) have argued from a larger comparative sample that individual variation in skull size existed among Neanderthal subadults, within the same age-class as Devil's Tower (as shown by

Pech de l'Azé, Subalyuk 2, Engis 2) or in older specimens (e.g. La Quina H18). Cranial size assessments of Neanderthal were based on direct measurements of sufficiently preserved specimens. Moreover, the specificity of the modern reference sample selected in the comparison to Neanderthals influenced the results of the analysis, as all recent children did not show similarly-sized braincases (Coqueugniot 1994, 1999).

New endocranial volume estimations were proposed for three Neanderthal children (e.g. Roc de Marsal, Devil's Tower, and Engis 2), using computer-assisted analysis and reconstruction of missing skeletal parts (Zollikofer et al. 1995; Coqueugniot and Hublin 2007). Based on their results, the authors suggested that brain growth could be faster during early childhood in Neanderthals than in early modern humans. However, one should keep in mind that there is a discrepancy between authors's results and cranial size for Engis 2, and that the data employed in the comparative analysis were collected by different techniques. Furthermore, the influence of sexual dimorphism in the manifestation of cranial size variation during the growth period cannot be evaluated among young Neanderthals.

Dental Development

During the last two decades, dental studies have been employed to suggest a non–modern-like pattern of growth in Neanderthals. Scholars have explored tooth enamel microstructures and timing of permanent crown formation (incisors and/or molars) to support a shorter period of dental growth in Neanderthals (e.g. Dean et al. 1986; Stringer et al. 1990; Ramirez-Rozzi and Bermudez de Castro 2004). Others studies have revealed similarities in perikymata counts between Neanderthal and modern human teeth (e.g. Mann et al. 1990, 1991; Tillier et al. 1995; Guatelli-Sternberg et al. 2005; Ramirez-Rozzi 2005; Monge et al., 2005). The resulting consequence is a clearer picture of the Neanderthal position within this modern human range.

Furthermore, better knowledge of modern human diversity, including both extant and fossil samples (e.g. Liversidge 2003; Liversidge and Molleson 2004; Monge et al. 2006; Reid and Dean 2006) reinforces the idea that in modern populations, significant variation exists in enamel incremental markings and perikymata packing patterns.

Following Reid and Dean (2006), more useful information can be accrued using molar microstructures that appear to be less variable than those of incisors. Using high-resolution micro–computed tomography analysis of two molars from La Chaise-de-Vouthon in France (one deciduous and one permanent), Macchiarelli et al. (2006) suggested that distinctiveness of tooth formation and dental growth in Neanderthal children is questionable. Recently, Smith et al. (2007), from their analysis of the Belgian Neanderthal specimen from the Scladyna Cave, concluded that postcanine teeth had probably different formation times in Neanderthals, in contrast with anterior teeth. Following these authors, the Scladyna Neanderthal child presents molars that were "(...) characterized by shorter periods of overall crown formation than modern humans (...)" (Smith et al. 2007: 20221).

Discussion

As we already mentioned (Tillier 2000), patterns of growth in Neanderthals are difficult to assess without reference to larger comparative samples. It is clear, from an auxological perspective, that investigations of Neanderthal growth trajectories can benefit from the use of enlarged samples that document the diversity of subadult modern dental and skeletal morphologies within past (i.e. Middle and Upper Paleolithic specimens) and recent populations geographically dispersed. Recent modern human samples used in comparative analyses mostly originate from regional collections, i.e. mainly European ones. The interpretation of observed differences (cranial size, number and packing pattern of perikymata on teeth, development of muscle markings, limb proportions, etc.) between Neanderthals and modern children must consider the significance of inter- and intrapopulation variation within distinct samples, following in this sense a methodological approach generalized in the study of adult individuals. Such data will markedly contribute to our ability to better assess the phylogenetic status of the features associated with Neanderthals and to more fully understand their biological implications.

Various lines of evidence dealing with variation of dental maturation in living humans and fossil hominids (e.g. Liversidge 2003; Liversidge and Molleson 2004; Monge et al. 2006; Guatelli-Sternberg et al. 2005; Reid and Dean 2006; Macchiarelli et al. 2006) challenge the use of data collected from studies of individual cases (Ramirez-Rozzi and Bermúdez de Castro 2004; Smith et al. 2007) in the identification of species or lineages, and the reconstruction of life history in past populations. Scholars professionally concerned with odontology studies know that there are ethnic differences in the eruption schedules for deciduous and permanent dentition; there is a wealth of literature on interpopulation variations. Huge differences in tooth emergence between captive and wild chimpanzees recently documented (Zilhman et al. 2004) also questions the use of tooth emergence times as population biological markers.

Finally, it is clear that the paucity of well-preserved skeletons (a common problem for growth-related researches) limits (1) the evaluation of individual variation among Neanderthals, and (2) the reconstruction of Neanderthal trajectories of limb bone lengths related to dental age (versus Thompson and Nelson 2000). In addition, the influence of sexual dimorphism in the manifestation of morphological variation during the growth period cannot be evaluated. It is widely accepted that there are no accurate morphological criteria and reliable methods for a sex estimation of immature skeletons. For Paleolithic specimens, including Neanderthals, investigation in genetic markers on X/Y chromosomes remains rather problematic.

In most cases, the relationship between individual variation and anomalies in developmental patterns is rather difficult to assess. From limb bone lengths and body size reconstruction of the Neanderthal Le Moustier 1 adolescent, Thompson and Nelson (2000: 489) suggested that this individual had experienced a very late growth spurt. Besides the fact that "(…) all measurements included herein must be considered to be estimates" (Thompson and Nelson 2005:280), it is of concern to consider that the delay in skeletal growth achievement might reflect either prolonged growth related to sex or individual developmental disturbances in this individual (dental age estimates of 15.5 ± 1.25 years, according to Thompson and Nelson 2005) rather than delayed infracranial growth in Neanderthals relative to modern humans.

Life History of Neanderthal Children

The Neanderthal immature sample under examination includes infants and children who were the nonsurvivors in the population, a common situation in the study of archaeological samples (Saunders and Hoppa 1993). Although they serve to reconstruct growth patterns and maturational events during childhood within the Neanderthal lineage, it should be kept in mind that they might illustrate cases of abnormal developmental processes. The European Neanderthal sample represents a relative abundance of immature individuals to whom an age less than 10 years can be assigned, but the real level of infant mortality cannot be established from the small number of individuals (less than 1 year) uncovered.

Indicators of Nonspecific Stress

Evidences of growth difficulties, such as tooth enamel hypoplasia (Ogilvie et al. 1989; Brennan 1991; Garcia-Sanchez et al. 1994; Skinner 1996; Guatelli-Sternberg et al. 2004), have been examined among Neanderthals. But authors have employed different scoring techniques (i.e. all types of enamel hypoplasia checked or examination limited to linear enamel hypoplasia, number of teeth scored, etc.) and this could easily explain the varying results in the analysis of permanent teeth. A crucial test in the discussion of nonspecific stress markers lies in the developmental degree of enamel defects and the number of teeth affected.

Neanderthal children exhibit a low frequency of hypoplastic enamel defects on deciduous teeth (Skinner 1996; author's observations). The Devil's Tower child is quite unique in the number of teeth affected, three of the four deciduous teeth preserved and two of the four permanent molars, a condition that indicates, according to Skinner (1996: 844), two stress episodes, at birth and around 3.5 years. Guatelli-Sternberg et al. (2004) focused on linear enamel hypoplasia affecting permanent teeth (from Southern France individuals and Krapina sample); they reported a low percent (less than 40%) of individuals bearing at least two (or more) teeth affected, a result that did not support higher levels of environmental stress episodes among Neanderthals than in the comparative sample of Inuit foragers.

Anomalies, Bone Lesions

If associated skull and infracranial remains are taken into consideration, the fossil evidence in support of developmental defects of bones comes from a small number of young individuals. This evidence, however, is not complete enough at the moment to allow us to draw definitive conclusions about the mortality and stress patterns in the Neanderthal immature sample from Europe. Furthermore, due to their structure, immature cranial bones tend to be less resistant to postmortem soil pressure than adult ones and postmortem skull deformations may occur. Such postmortem alterations cannot be rejected in the case of postbregmatic depression observed on the Krapina 1 and Subalyuk 2 specimens (Pap et al. 1995; Sansilbano-Collilieux and Tillier 2006).

The number of reported cases of antemortem cranial alterations is small. The Devil's Tower cranium is represented by three bones, and among them the frontal is well preserved with the exception of the orbital roofs. The inner table of this bone exhibits a rugged surface with digital impressions clearly visible on a frontal X-ray (Fig. 12.6). Such impressions affecting the bone are known in recent children to be associated with increased intracranial pressure; they usually vanish around the age of 4 years (Maroteaux 1982). An examination of Neanderthal children indicates that similar alteration of the inner frontal surface is present on (at least) two other specimens, Krapina 1 and La Chaise-Abri Suard 17 (Sansilbano-Collilieux and Tillier 2006). In both cases, individual age at death of the children is unknown,[3] while the

[3] We personally feel that, in the absence of the dentition, the reliability of age estimates in children older than perinates, based on single cranial measurements, is questionable.

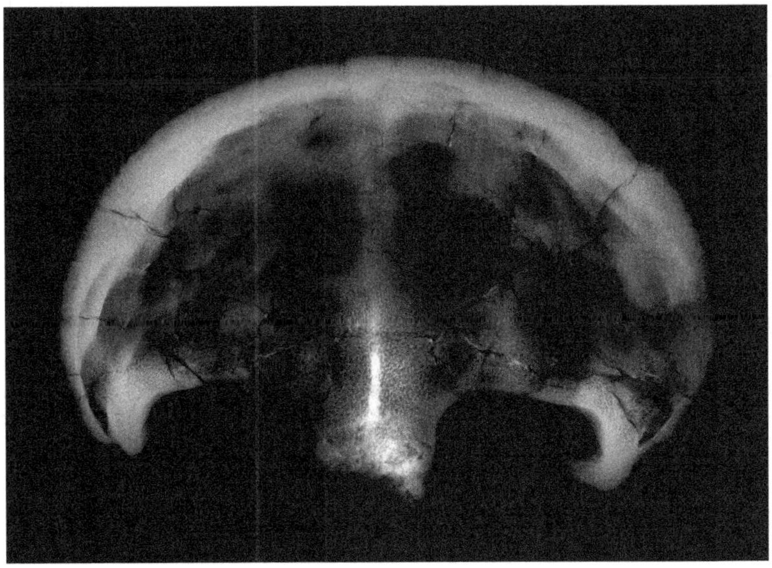

Fig. 12.6 X-ray picture of the Devil's Tower frontal bone (by courtesy of T. Molleson, BMNH, London) showing the distribution of the digital impressions

Devil's Tower child died during the fifth year postnatal (dental age estimates). Interestingly, the developmental pattern observed on Devil's Tower and Krapina 1 frontal bones is associated with enlarged cranial size relative to other specimens (e.g. Dean et al. 1986; Tillier 1988; Minugh-Purvis et al. 2000), and, to dental indication of a stress episode around 3 years for Devil's Tower, while the Krapina 1 temporal exhibits peculiarities in the internal sinus patterns (Minugh-Purvis et al. 2000; Sansilbano-Collilieux and Tillier 2006).

A second immature individual recovered from the Krapina rockshelter, Krapina 2, is represented by the posterior region of the cranium and possesses internal features that deviate substantially from the condition found in adults from the site. In this specimen, the occipital bone exhibits displacement of endocranial structures with dissociation between external and internal axes of the bone, absence of the occipital crest and large vermian fossa (Sansilbano-Collilieux and Tillier 2006).

On the basis of a new computerized reconstruction of the Le Moustier 1 skull, Ponce de Leon and Zollikofer (2005) state that this adolescent shows evidence of mild plagiocephaly and healed condylar neck fracture of the left side of the mandible.

The relationships between the aforementioned various bone abnormalities and the death of the individuals is rather difficult to assess, as most of them are not lethal in extant human populations. The only available evidence of bone lesion related to the cause of death might come from the Lazaret specimen in Southeastern France. The preserved portion of the cranium, a right parietal bone, showed evidence of a localized bone lesion on both external and internal surfaces which has been interpreted as a meningomia-induced tumor (de Lumley 1973; Dastugue and de Lumley 1976), a condition quite exceptional for a child supposed to be less than 10 years old at death and dated to oxygen isotope stage 6.

Concluding Remarks

Comparative analyses provide accurate information regarding the ontogenetic appearance of some diagnostic Neanderthal characteristics in the skull and infracranial skeleton. Collected data reveal similarities and contrasts between Neanderthal and modern children, but additional information is needed to confirm an age-related change for the appearance of specific features during childhood and to precisely determine the timing of their appearance. The interpretation of morphological differences as reflections of temporal and/or regional changes, as well as an investigation of within-site variation in growth-related features, would benefit from the accumulation of data with further studies and new discoveries.

In view of the aforementioned problems, we feel that the definition of a Neanderthal growth pattern and the characterization of within-population variation are far to be reached. Future research that incorporates juvenile skeletons from geographically distinct populations including hunter-gatherers will undoubtedly lead to a greater understanding of the evolutionary significance of juvenile Neanderthal skeletal remains.

Comparative growth studies in extant populations have demonstrated that environmental differences and selective pressures have effects on the growing child. Accordingly, there is no guarantee that Neanderthals, evolving in different sites and habitats during several tens of thousands of years, shared similar growth patterns with living humans. There is no need to assert that possible variability in developmental pattern and somatic growth in Neanderthals was due to limitations in social and cognitive faculties. In 2007 we are still far from a consensus on the nature of the biological patterns that contributed to the development of Neanderthal skeletal morphology.

N. B. Since the acceptance of the manuscript, a new study of neurocranial size growth has been performed, using virtual reconstruction of incomplete infant and juvenile skulls (Ponce de Leon et al. 2008). Based on their reconstruction of the neonate specimen from the Mezmaiskaya Cave from Russia, considered as a Neanderthal (a questionable attribution, see Barriel and Tillier 2002), the authors established that "Neanderthal brain size at birth was similar to that in recent *Homo sapiens* (…)". Data from older specimens lead the authors to conclude that postnatal brain growth rate was different between the two samples (i.e. Neanderthals and recent children) used in the comparative analysis. In this context, these authors suggest that additional evidence for brain size growth in early anatomically modern humans is still requested, joining us in this point (Tillier 1995: 65).

Acknowledgements I would like to deeply thank Silvana Condemi (UMR 6578, Marseille) and Wighart von Koenigswald (Institut für Paläontologie, Bonn) for inviting me to participate in the International Conference held in Bonn "*150 years of Neanderthal Discoveries. Early Europeans – Continuity & Discontinuity*" and to contribute to the Symposium *Neanderthal Anatomy, Adaptation, Physical and Cultural Variations*. I wish to express my gratitude to directors and colleagues from several Institutes for their kind permission to access original fossils under their care, in Belgium (J-M. Cordy), Croatia (J. Radovčic), England (C.B. Stringer, R. Kruszinski, and T. Molleson), Hungary (I. Pap), Russia (E. Godina and N. Miklaschevskaia), and France (J-J. Cleyet-Merle, Y. Coppens, J-L. Heim, A. Langaney, and J-P. Mohen). I am grateful to M. Seurin, H. Coqueugniot (LAPP-UMR 5199/PACEA, Talence), J-L. Heim (Musée de l'Homme, Paris) for providing assistance in the illustration; to G. Sarma for improving the English text. The research was financially supported by the *Centre National de la Recherche Scientifique* (LAPP-UMR 5199/PACEA, University Bordeaux 1). Thanks are also due to the anonymous reviewers for their constructive and input comments.

References

Akazawa, T., & Muhesen, S. (Eds.). (2002). *Neanderthal Burials. Excavations of the Dederiyeh Cave, Afrin, Syria*. Kyoto: International research Center for Japanese Studies.

Alekseev, V. P. (1976). Position of the Staroselye find in the Hominid system. *Journal of Human Evolution, 5*, 413–421.

Arensburg, B., & Belfer-Cohen, A. (1998). Sapiens and Neandertals: Rethinking the Levantine Middle Paleolithic Hominids. In T. Akazawa, K. Aoki, & O. Bar Yosef (Eds.), *Neandertals and modern humans in Western Asia* (pp. 311–322). New York/London: Plenum.

Arensburg, B., & Tillier, A-M. (1983). A New Mousterian child from Qafzeh (Israel): Qafzeh 4a. *Bulletins et Mémoires de la Société d'Anthropologie de Paris, 10*, 61–69.

Arensburg, B., Harell, M., & Nathan, H. (1981). The human middle ear ossicles: Morphometry and taxonomic implications. *Journal of Human Evolution, 10*, 199–205.

Arsuaga, J. L., Martinez, I., Gracia, A., & Lorenzo, C. (1997). The Sima de Los Huesos crania (Sierra de Atapuerca, Spain). A comparative study. *Journal of Human Evolution, 33*(2/3), 219–281.

Arsuaga, J.-L., Lorenzo, C., Caretero, J.-M., Gracia, A., Martinez, I., Garcia, N., Bermudez de Castro, J. M., & Carbonell, E. (1999). A complete human pelvis from the Middle Pleistocene of Spain. *Nature, 399*, 255–258.

Arsuaga, J.-L., Villaverde, V., Quam, R., Martinez, I., Carreterro, J.-M., Lorenzo, C., & Gracia, A. (2007). New Neandertal remains from Cova Negra (Valencia, Spain). *Journal of Human Evolution, 52*, 31–58.

Barriel, V., & Tillier, A-M. (2002). L'enfant de Mezmaiskaya (Caucase) examiné dans une double perspective paléogénétique et paléoanthropologique. *Bulletins et Mémoires de la Société d'Anthropologie de Paris,Tome, 14*(1-2), 163–191.

Bogin, B. (1991). *Patterns of human growth* (Cambridge Studies in Biological Anthropology, Vol. 3). Cambridge: Cambridge University Press.

Bonmati, A., & Arsuaga, J.-L. (2005). Inventory and preliminary description of Middle Pleistocene pelvis remains from the site of the Sima de Los Huesos, Atapuerca, Spain. *American Journal of Physical Anthropoogy, 126*(S40), 76.

Brennan, M. (1991). *Health and disease in the middle and upper Paleolithic of Southwestern France*. Ph.D. dissertation, New York University, New York.

Bruzek, J., & Trinkaus, E. (2002). The pelvic morphology. In J. Zilhao & E. Trinkaus (Eds.), *Portrait of the artist as a child. The Gravettian human skeleton from the Abrigo do Lagar Velho and its Archaeological context* (Trablhos de Arqueologia, Vol. 22, pp. 427–434). Lisboa: Insituto Portugues de Arqueologia.

Churchill, S. E. (1994). *Human upper body evolution in the Eurasian Later Pleistocene*. Ph.D. dissertation, University of New Mexico, Albuquerque.

Condemi, S. (2006). 150 years of Neanderthal studies: Old questions, new answers. In E. B. Bodzsár & C. Susanne (Eds.), *Human evolution: Facts and factors* (Biennal Books of EAA, Vol. 4, pp. 33–48).

Coqueugniot, H. (1994). Equations d'estimation de la capacité crânienne chez l'enfant: application paléoanthropologique. *Anthropologie, Brno, XXXII*(3), 243–250.

Coqueugniot, H. (1999). Le crâne d'Homo sapiens en Eurasie : croissance et variation depuis100000 ans. BAR International Series 822, Oxford.

Coqueugniot, H. (2000). La position du foramen mentonnier chez l'enfant: révision ontogénétique et phylogénétique. *Bulletins et Mémoires de la Société d'Anthropologie de Paris, 12*(3–4), 227–246.

Coqueugniot, H., & Hublin, J.-J. (2007). Endocranial volume and brain growth in immature Neandertals. *Periodicum Biologorum, 109*(4), 379–385.

Coqueugniot, H., & Le Minor, J.-M. (2002). Fermeture des synchondroses intra-occipitales: Implication dans la configuration du foramen magnum. *Comptes Rendus Palevolution, 1*, 35–42.

Creed-Miles, M., Rosas, A., & Kruszynski, R. (1996). Issues in the identification of Neandertal derivative traits at early post-natal stages. *Journal of Human Evolution, 30*, 147–153.

Dastugue, J., & de Lumley, M.-A. (1976). Les maladies des hommes préhistoriques. In H. de Lumley (Ed.), *La Préhistoire Française* (pp. 612–622). Paris: Editions duCNRS.

de Lumley, M-A. (1973). Anté-néandertaliens et Néandertaliens du Bassin Méditerranéen occidental européen. Etudes Quaternaires, Mémoire2, Université de Provence, Marseille.

Dean, C., Stringer, C. B., & Bromage, T. G. (1986). Age at death of the Neanderthal child from Devil's Tower, Gibraltar and the implications for studies of general growth and development in Neanderthals. *American Journal of Physical Anthropology, 40*, 301–309.

Eveleth, P. B., & Tanner, J. M. (1990). *Worldwide variation in human growth* (2nd ed.). Cambridge: Cambridge University Press.

Frayer, D. (1992). The persistence of Neanderthal features in post-Neanderthal Europeans. In G. Brauer & F. H. Smith (Eds.), *Continuity or replacement? Controversies in Homo sapiens evolution* (pp. 179–188). Rotterdam: A.A. Balkama.

Garcia-Sanchez, M., Tillier, A-M, Garralda, M.-D., & Vega-Toscano, G. (1994). Les dents d'enfant des niveaux moustériens de la grotte de Carihuela (Grenade, Espagne). *Paleo, 6*, 79–86.

Guatelli-Sternberg, D., Larsen, C. S., & Hutchinson, D. L. (2004). Prevalence of the duration of Linear enamel hypoplasia: A comparative study of Neandertals and Inuit foragers. *Journal of Human Evolution, 47*, 65–84.

Guatelli-Sternberg, D., Reid, D. J., Bishop, T. A., & Larson, C. S. (2005). Anterior tooth growth periods in Neandertals were comparable to those of modern humans. *Proceedings of the National Academy of Sciences of the United States of America, 102*, 14197–14202.

Heim, J.-L. (1982). *Les enfants néandertaliens de La Ferrassie*. Paris, Masson: Fondation Singer Polignac.

Henry-Gambier, D. (Ed.). (2001). *La sépulture des enfants de Grimaldi (Baoussé-Roussé, Italie). Anthropologie et palethnologie funéraire des populations de la fin du Palélithique supérieur.* Paris: CTHS.

Holliday, T. W. (1995). *Body size and proportions in the Late Pleistocene Western olds world and the origins of modern humans*. Ph.D. dissertation, University of New Mexico, New Mexico.

Hublin, J.-J. (1980). La Chaise (Abri Suard), Engis 2 et La Quina H18: développement de la morphologie occipitale externe chez l'enfant prénéandertalien et néandertalien. *Comptes Rendus de l'Académie des Sciences,D, Paris, 291*, 669–672.

Hublin, J.-J. (1998). Climatic changes, paleogeography, and the evolution of the Neandertals. In T. Akazawa, K. Aoki, & O. Bar-Yosef (Eds.), *Neandertals and modern humans in Western Asia* (pp. 295–310). New York: Plenum.

Hublin, J.-J., & Tillier, A.-M. (1991). L' «*Homo sapiens*» en Europe occidentale: gradualisme et rupture. In J.-J. Hublin & A.-M. Tillier (Eds.), *Aux Origines d'Homo sapiens. Nouvelle Encyclopédie Diderot* (pp. 291–327). Paris: Presses Universitaires de France.

Hublin, J.-J., Spoor, F., Braun, M., Zonneveld, F., & Condemi, S. (1996). A late neandertal from Arcy-sur-Cure associated with Upper Paleolithic artefacts. *Nature, 381*, 224–226.

Jelinek, J. (1992). Gravettian shoulderblades; their morphological variability and other interesting features. *Anthropologie, Brno, XXX*(3), 45–50.

Liversidge, H. M. (2003). Variation in modern dental development. In J. L. Thompson, G. E. Krovitz, & A. J. Nelson (Eds.), *Patterns of growth and development in the genus Homo. Cambridge studies in biological and evolutionary anthropology* (pp. 73–113). Cambridge: Cambridge University Press.

Liversidge, H. M., & Molleson, T. (2004). Variation in crown and root formation and eruption of human deciduous teeth. *American Journal of Physical Anthropology, 123*, 172–180.

Macchiarelli, R., Bondioli, L., Debenath, A., Mazurier, A., Tournepiche, J.-F., Birch, W., & Dean, C. (2006). How Neanderthal molar teeth grew. *Nature, 444*, 748–751.

Madre-Dupouy, M. (1992). *L'enfant néandertalien du Roc de Marsal. Etude analytique et comparative. Cahiers de Paléoanthropologie*. Paris: Editions duCNRS.

Majó, T. (2000). L'os coxal non-adulte : approche méthodologique de la croissance et de la diagnose sexuelle. Application aux enfants du Paléolithique moyen. Thèse en Sciences Biologiques et Médicales, Université Bordeaux 1, option Anthropologie. (Unpublished)

Majó, T., & Tillier, A-M. (2003). A new approach to the quantitative analysis of postcranial growth in Neandertals and modern humans. Evidence from the hipbone. In J. L. Thompson, G. E. Krovitz, & A. J. Nelson (Eds.), *Patterns of Growth and Development in the Genus Homo. Cambridge studies in biological and evolutionary anthropology* (pp. 361–385). Cambridge: Cambridge University Press.

Mann, A. E., Lampl, M., & Monge, J. (1990). Décompte de périkymaties chez les enfants néandertaliens de Krapina. *Bulletins et Mémoires de la Société d'Anthropologie de Paris, 2*, 213–224.

Mann, A. E., Lampl, M., & Monge, J. (1991). The uses of scanning electron microscopy in the reconstruction of Neanderthal lifeways. *American Journal of Physical Anthropology, 12*, 124.

Marks, A. E., Demidenko, Yu E, Monigal, V. I., Ferring, C. R., Burke, A., Rink, J., & McKinney, C. (1997). Staroselye and the Staroselye child: New excavations, new results. *Current Anthropology, 38*(1), 112–123.

Maroteaux, P. (1982). *Maladies osseuses de l'enfant*. Paris: Flammarion, coll. Médecine Sciences.

Maureille, B. (2002). La redécouverte du nouveau-né néanderthalien Le Moustier 2. *Paléo, 14*, 221–238.

Maureille, B., & Turcq, A. (2005). Excavations at the Le Moustier sites and their importance in French Archaeology. In H. Ullrich (Ed.), *The Neandertal adolescent Le Moustier1. New aspects, new results* (pp. 21–31). Berlin: Staatliche Museen zuBerlin – Preussisicher Kulturbesitz.

McCown, T. D., & Keith, A. (1939). *The Stone Age of Mount Carmel* (Vol. II). Oxford: Clarendon University Press.

Minugh-Purvis, N. (1988). *Patterns of craniofacial growth and development in Upper Pleistocene Hominids*. Ph.D. dissertation, University of Pennsylvania, Philadelphia.

Minugh-Purvis, N., Radovčic, J., & Smith, F. H. (2000). Krapina1: A juvenile Neandertal from the Early Late Pleistocene of Croatian. *American Journal of Physical Anthropology, 111*, 393–424.

Monge, J., Maureille, B., & Mann, A. E. (2005). La dentition humaine. In O. Dutour, J.-J. Hublin, & B. Vandermeersch (Eds.), *Objets et méthodes en Paléoanthropologie* (pp. 103–135). Paris: CTHS.

Monge, J. M., Tillier, A-M, & Mann, A. E. (2006). Perikymata number and spacing on Early Modern Teeth: Evidence from Qafzeh Cave, Israel. *Bulletins et Mémoires de la Société d'Anthropologie de Paris, 18*(1–2), 25–33.

Ogilvie, M. D., Curran, B. K., & Trinkaus, E. (1989). The incidence and patterning of dental enamel hypoplasias among the Neandertals. *American Journal of Physical Anthropology, 79*, 25–41.

Pap, I., Tillier, A.-M., Arensburg, B., & Chech, M. (1995). The Subalyuk Neanderthal remains (Hungary): A re-examination. *Annales Historico-Naturales, Musei Nationalis Hungarici, 88*, 233–270.

Ponce de Leon, M., & Zollikofer, C. (1999). New evidence from Le Moustier1: computer-assisted reconstruction and morphometry of the skull. *The Anatomical Record, 254*, 474–489.

Ponce de Leon, M., & Zollikofer, C. (2005). Looking through stones and bones: The new computer-assisted reconstruction of the Le Moustier1cranium and its implication for Neanderthal ontogeny. In H. Ullrich (Ed.), *The Neandertal adolescent Le Moustier1. New aspects, new results* (pp. 244–264). Berlin: Staatliche Museen zuBerlin – Preussisicher Kulturbesitz.

Ponce de Leon, M., Golonova, L., Doronichev, V., Romanova, G., Akazawa, T., Kondo, O., Ishida, H., & Zollikoffer, C. (2008). Neanderthal brain size at birth provides insights into the evolution of human life history. *PNAS, 105*(37), 13764–13768.

Radovčic, J., Smith, F. E., Trinkaus, E., & Wolpoff, M. H. (1988). *The Krapina Hominids. An illustrated catalog of skeletal collection*. Zagreb: Mladost.

Rak, Y. (1990). On the differences between two pelvises of Mousterian context from Qafzeh and Kebara Caves, Israel. *American Journal of Physical Anthropology, 81*, 323–332.

Rak, Y., Kimbel, B., & Hovers, E. (1994). A Neandertal infant from Amud Cave, Israel. *Journal of Human Evolution, 26*, 313–324.

Ramirez-Rozzi, F. (2005). Âge au décès de l'enfant néandertalien de l'Hortus. *Bulletins et Mémoires de la Société d'Anthropologie de Paris, 17*(1–2), 47–55.

Ramirez-Rozzi, F., & Bermúdez de Castro, J. M. (2004). Surprisingly rapid growth in Neanderthals. *Nature, 428*, 936–939.

Reid, D. J., & Dean, M. C. (2006). Variation in modern human enamel formation times. *Journal of Human Evolution, 50*, 329–346.

Rink, W. J., Schwarcz, H. P., Radovčic, J., Rukavina, D., Karavanic, I., & Richter, D. (1999). *ESR Dating at the Croatian Palaeolithic sites of Krapina, Vindija and Mujina pecina*. Paper presented at the Krapina Neandertals and Human Evolution in Central Europe, Zagreb 24 Aug

Ronen, A. (Ed.) (1982). The Transition from Lower to Middle Palaeolithic and the Origin of Modern Human. Oxford: BAR International Series 151.

Rosenberg, K. R. (1988). The functional significance of the Neandertal pubic length. *Current Anthropology, 29*, 595–617.

Rougier, H. (2003). Etude descriptive et comparative de Biache-Saint-Vaast1(Biache-Saint-Vaast, Pas-de-Calais, France). Thèse Université Bordeaux1, Spécialité Anthropologie Biologique (unpublished).

Sandgathe, D., Dibble, H. L., McPherron, S. J. P., Turcq, A. (2005). Roc-de-Marsal (Campagne duBugue,Dordogne). Rapport d'opération pour l'année2005. http://www.oldstoneage.com.

Sansilbano-Collilieux, M., & Tillier, A-M. (2006). Variation in intracranial morphology within the Krapina Hominid sample. Evidence from Krapina1and2. *Periodicum Biologorum, 108*(4), 443–447.

Saunders, S. R., & Hoppa, R. D. (1993). Growth deficit in survivors and non survivors: Biological mortality bias in subadult skeletal samples. *Yearbook of Physical Anthropology, 36*, 127–151.

Skinner, M. (1996). Developmental stress in immature Hominines from Late Pleistocene Eurasia: Evidence from enamel hypoplasia. *Journal of Archaeological Science, 23*, 833–852.

Smith, F.H. (1976). The Neanderthal Remains from Krapina. A descriptive and Comparative Study. University of Tennessee, Knoxville, Report of Investigations.

Smith, T. M., Toussaint, M., Reid, D. J., Olejniczak, A. J., & Hublin, J.-J. (2007). Rapid dental development in a Middle Paleolithic Belgian Neandertal. *PNAS, 104*(51), 20220–20225.

Soressi, M., Jones, H. L., Rink, W. J., Maureille, B., & Tillier, A-M. (2007). The Pech-de-l'Azé I Neandertal child: ESR,uranium-series and AMS C14 dating of its context. *Journal of Human Evolution, 52*, 455–466.

Spoor, F. H. (2002). The Auditory Ossicles. In J. Zilhao & E. Trinkaus (Eds.), *Portrait of the artist as a child. The Gravettian human skeleton from the Abrigo do Lagar Velho and its archaeological context* (pp. 293–296). Lisboa: Trabalhos de Arqueologia 22.

Spoor, F., & Zonneveld, F. (1998). Comparative review of the human bony labyrinth. *Yearbook of Physical Anthropology, 41*, 211–251.

Spoor, F., Hublin, J.-J., Braun, M., & Zonneveld, F. (2003). The bony labyrinth of Neanderthals. *Journal of Human Evolution, 44*, 141–165.

Stewart, T. D. (1960). Form of the pubic bone in Neanderthal man. *Science, 131*, 1437–1438.

Stringer, C. B., Hublin, J.-J., & Vandermeersch, B. (1984). The origin of anatomically modern humans in Western Europe. In F. H. Smith & F. Spencer (Eds.), *The origin of modern humans* (pp. 51–135). New York: Alan R. Liss.

Stringer, C. B., Dean, M. C., & Martin, R. D. (1990). A comparative study of cranial and dental development within recent British sample and among Neandertals. In C. E. J. Rousseau (Ed.), *Primate life history and evolution* (pp. 15–152). New York: Wiley.

Thompson, J. L., & Illerhaus, B. (1998). A new reconstruction of the Le Moustier1skull and investigation of internal structuresusing 3-D-mCT data. *Journal of Human Evolution, 35*, 647–665.

Thompson, J. L., & Nelson, A. J. (2000). The place of Neandertals in the evolution of hominid patterns of growth and development. *Journal of Human Evolution, 38*, 475–495.

Thompson, J. L., & Nelson, A. J. (2005). The postcranial skeleton of Le Moustier 1. In H. Ullrich (Ed.), *The Neandertal Adolescent Le Moustier 1. New Aspects, new results* (pp. 265–281). Berlin: Staatliche Museen zuBerlin-Preuβischer Kulturbesitz.

Tillier, A-M. (1983). Le crâne d'enfant d'Engis: un exemple de distribution des caractères juvéniles, primitifs et néanderthaliens. *Bulletins de la Société Royale Belge d'Anthropologie et de Préhistoire., 94*, 51–75.

Tillier, A-M. (1986). Quelques aspects de l'ontogenèse du squelette crânien des Néanderthaliens. In V. V. Novotny & A. Mizerova (Eds.), *Fossil man – New facts, new ideas* (pp. 207–216). Brno: Anthropos. 23.

Tillier, A-M. (1987). L'enfant de La Quina H18 et l'ontogénie des Néanderthaliens. In B. Vandermeersch (Ed.), *Préhistoire de Poitou-Charentes. Problèmes actuels* (pp. 201–206). Paris: CTHS.

Tillier, A-M. (1988). La place des restes de Devil's Tower (Gibraltar) dans l'ontogenèse des Néanderthaliens. *Bulletins et Mémoires de la Société d'Anthropologie de Paris, 4*, 257–266.

Tillier, A-M. (1995). Neanderthal ontogeny: A new source for critical analysis. *L'Anthropologie, 33*(1–2), 63–68.

Tillier, A-M. (1996). The Pech-de-l'Azé and Roc-de-Marsal Children (Middle Paleolithic, France): Skeletal evidence for variation in Neanderthal Ontogeny. *Human Evolution, 11*(2), 113–119.

Tillier, A-M. (1998). Ontogenetic Variation in Late Pleistocene Homo Sapiens from the Near East: Implications for Methodological Bias in Reconstructing Evolutionary Biology. In T. Akazawa, K. Aoki, & O. Bar-Yosef (Eds.), *Neandertals and modern humans in Western Asia* (pp. 381–389). New York: Plenum.

Tillier, A-M. (1999). *Les enfants moustériens de Qafzeh. Interprétation phylogénétique et paléoauxologique. Cahiers de Paléoanthropologie*. Paris: CNRS.

Tillier, A-M. (2000). Palaeoauxology applied to Neanderthals. Similarities and contrasts between Neanderthal and Modern Children. In A-M. Tillier (Ed.), *Children in the past. Paleoauxology, demographic anomalies, taphonomy and mortuary practises* (pp. 109–120). Brno: Anthropologie. XXXVIII/1.

Tillier, A-M. (2005). La croissance des enfants dans les populations du passé. Questions méthodologiques et perspectives en paléoanthropologie. In O. Dutour, J.-J. Hublin, & B. Vandermeersch (Eds.), *Objets et méthodes en Paléoanthropologie* (pp. 247–270). Paris: CTHS.

Tillier, A-M. (2005a). The Tabun C1 skeleton: A Levantine Neanderthal? In D. Kaufman, M. Evron-Weinstein (Eds.), The Prehistory of the Eastern Mediterranean: Studies presented to Professor A. Ronen. *Journal of the Israel Prehistoric Society*, 35, 439–450.

Tillier, A-M., Mann, A. E., Monge, J., & Lampl, M. (1995). L'ontogenèse, la croissance de l'émail dentaire et l'origine de l'homme moderne: l'exemple des Néanderthaliens. *Bulletins de la Société Royale Belge d'Anthropologie et de Préhistoire, 106*, 97–104.

Tillier, A-M., Arensburg, B., Vandermeersch, B., & Chech, M. (2003). New human remains from Kebara Cave (Mount Carmel). The place of the Kebara hominids in the Levantine Mousterian Fossil Record. *Paleorient, 29*(2), 35–62.

Tompkins, R., & Trinkaus, E. (1987). La Ferrassie 6 and the development of Neandertal pubic Morphology. *American Journal of Physical Anthropology, 73*, 223–239.

Trinkaus, E. (1976). The morphology of European and Southwest Asian Neandertal pubic bones. *American Journal of Physical Anthropology, 44*, 95–104.

Trinkaus, E. (1981). Neanderthal limb proportions and cold adaptation. In C. B. Stringer (Ed.), *Aspects of human evolution* (pp. 187–224). London: Taylor & Francis.

Trinkaus, E. (1983). *The Shanidar Neandertals*. New York: Academic.

Trinkaus, E. (1984). Neanderthal pubic morphology and gestation length. *Current Anthropology, 25*(4), 509–514.

Trinkaus, E. (1992). Morphological contrasts between the Near Eastern Qafzeh-Skhul and Late Archaic human samples: Grounds for a behavioral difference? In T. Akazawa, K. Aoki, & T. Kimura (Eds.), *The evolution and dispersal of modern humans in Asia* (pp. 277–294). Tokyo: Hokusen-sha.

Trinkaus, E., & Tompkins, R. L. (1990). The Neandertal life cycle: The possibility, probability and perceptibility of contraste with recent humans. In C. E. J. Rousseau (Ed.), *Primate life history and evolution* (pp. 153–180). New York: Wiley.

Trinkaus, E., Ruff, C. B., & Churchill, S. E. (1998). Upper limb versus lower limb loading patterns among Near Eastern Middle Paleolithic Hominids. In T. Akazawa, K. Aoki, & O. Bar Yosef (Eds.), *Neandertals and modern humans in western Asia* (pp. 391–404). New York: Plenum.

Trinkaus, E., Smith, F. H., Stockton, T. C., & Shackelford, L. L. (2006). The human postcranial remains from Mladeč. In M. Teschler-Nicola (Ed.), *Early Modern Humans at the Moravian Gate. The Mladeč Caves and their remains* (pp. 385–445). Wien, New York: Springer.

Ullrich, H. (2005) (Ed.). The Neandertal adolescent Le Moustier1. New aspects, new results. Staatliche Museen zuBerlin – Preussisicher Kulturbesitz, Berlin.

Villemeur, I. (1994). *La Main des Néandertaliens. Comparaison avec la main des hommes de type moderne. Morphologie et mécanique Cahiers de Paléoanthropologie*. Paris: CNRS Editions.

Vlček, E. (1970). Etude comparative onto-phylogénétique de l'enfant du Pech de l'Azé par rapport à d'autres enfants néanderthaliens. In D. Ferembach, P. Legoux, R. Fenart, R. Empereur-Buisson, & E. Vlček (Eds.), *L'enfant du Pech de l'Azé. Archives de l'Institut de Paléontologie Humaine* (pp. 149–180). Paris: Mémoire 33.

Vlček, E. (1973). Post-cranial skeleton of a Neandertal child from Kiik Koba, URSS. *Journal of Human Evolution, 2*, 537–546.

Vlček, E. (1975). Morphology of the first metacarpal of Neanderthal individuals from the Crimea. *Bulletins et Mémoires de la Société d'Anthropologie de Paris, 2*(XIII), 257-–276.

Williams, F. L'Engle (2001). *Heterochronic perturbations in the craniofacial evolution* of *Homo* (Neandertals and modern humans) and *Pan* (*P. troglodytes* and *P. paniscus*). Ph.D. dissertation, University of Massachusetts, UMI Number 3000356.

Zilhao, J., & Trinkaus, E. (Eds.). (2002). *Portrait of the artist as a child. The Gravettian human skeleton from the Abrigo do Lagar Velho and its Archaeological Context* (Trablhos de Arqueologia, Vol. 22). Lisboa: Intituto Portugues de Arqueologia.

Zilhman, A., Bolter, D., & Boesch, C. (2004). Wild chimpanzee dentition and its implications for assessing life history in immature hominin fossils. *PNAS, 101*(29), 10541–10543.

Zollikofer, M. S., Ponce de León & R.D. Martin, (1995) Neanderthal computer skulls. *Nature, 375*, 283–285.

Chapter 13
Dental Development and Age at Death of a Middle Paleolithic Juvenile Hominin from Obi-Rakhmat Grotto, Uzbekistan

Tanya M. Smith, Donald J. Reid, Anthony J. Olejniczak, Shara Bailey, Mica Glantz, Bence Viola, and Jean-Jacques Hublin

Abstract Studies of dental development have reported conflicting results regarding whether Neanderthal growth and development was similar to that of modern humans. The discovery of a partial permanent maxillary juvenile dentition (OR-1) from the Obi-Rakhmat Grotto, Uzbekistan, provides the opportunity to assess dental development and age at death in a Paleolithic hominin with strong Neanderthal similarities using incremental dental features. Long-period lines on tooth crowns (perikymata) and roots (periradicular bands) were quantified, and crown formation, root development, and age at death were estimated. An anomalous upper molar was determined to be a left M^2 with a rare developmental condition (gemination). Perikymata numbers for OR-1 were similar to modern southern African population means, but were less than modern northern European and Neanderthal means. Root extension rates were estimated to be similar to (or slightly higher than) modern human values, although few modern comparative data are available. Assuming the long-period line periodicity of this individual fell within a Neanderthal distribution (6–9 days), the maximum age at death of OR-1 is estimated at 8.1 years, but is more likely to have been 6.7–7.4 years (7 or 8 day periodicity). Modern European human developmental standards would suggest an age at death of approximately 8–9 years. These results are consistent with other studies suggesting that Neanderthal dental development overlaps with the low end of modern human populations, and demonstrates a greater range of variation in Middle Paleolithic hominins than previously reported. It is clear that perikymata number alone does not distinguish these taxa; data on long-period line periodicity and molar eruption would yield additional insight into Neanderthal life history.

T.M. Smith (✉)
Department of Human Evolutionary Biology, Harvard University, 11 Divinity Ave, Cambridge, MA 02138, USA
and
Department of Human Evolution, Max Planck Institute for Evolutionary Anthropology, Deutscher Platz 6, D-04103 Leipzig, Germany
e-mail: tsmith@fas.harvard.edu

D.J. Reid
Department of Oral Biology School of Dental Sciences, University of Newcastle upon Tyne, Framlington Place NE2 4BW, Newcastle upon Tyne, UK
e-mail: d.j.reid@ncl.ac.uk

A.J. Olejniczak and J.-J. Hublin
Department of Human Evolution, Max Planck Institute for Evolutionary Anthropology, Deutscher Platz 6, D-04103 Leipzig, Germany
e-mail: anthony.olejniczak@gmail.com; hublin@eva.mpg.de

S. Bailey
Department of Anthropology, New York University, 25 Waverly Place, New York, NY 10003, USA
e-mail: sbailey@nyu.edu

M. Glantz
Department of Anthropology, Colorado State University, Ft. Collins, CO 80523, USA
e-mail: mglantz@lamar.colostate.edu

B. Viola
Department of Anthropology, University of Vienna, Althanstrasse 14, A-1091 Vienna, Austria
e-mail: bence.viola@eva.mpg.de

Keywords Crown formation • Root formation • Perikymata • Periradicular band • Gemination • Neanderthal • Extension rate • Life history • Incremental feature

Introduction

Recent studies of hominin dental tissues have utilized incremental features to infer patterns of life history (the scheduling of development and the timing of reproductive events) (e.g., Bromage and Dean 1985; Dean et al. 2001; Ramirez Rozzi and Bermudez de Castro 2004; Guatelli-Steinberg et al. 2005; Smith et al. 2007a, b; reviewed in Smith 2008). Previous histological work on juvenile Neanderthal dentitions has been limited to studies of individuals from Devil's Tower, Gibraltar (Dean et al. 1986; Stringer et al. 1990; Stringer and Dean 1997), Montgaudier Cave, France (Mann and Vandermeersch 1997); Hortus, France (Ramirez Rozzi 2005); Scladina, Belgium (Smith et al. 2007b); and Dederiyeh, Syria (Sasaki et al. 2002). These studies were primarily focused on determining age at death from counts

S. Condemi and G.-C. Weniger (eds.), *Continuity and Discontinuity in the Peopling of Europe: One Hundred Fifty Years of Neanderthal Study*, Vertebrate Paleobiology and Paleoanthropology, DOI 10.1007/978-94-007-0492-3_13,

of temporal lines in tooth enamel. The number and spacing of external long-period growth lines has also been studied in considerable samples of Neanderthals and Upper Paleolithic modern humans (Mann et al. 1991; Ramirez-Rozzi 1993a; Ramirez Rozzi and Bermudez de Castro 2004; Guatelli-Steinberg et al. 2005, 2007; Guatelli-Steinberg and Reid 2008; Reid et al. 2008). Histological studies of internal enamel development have been conducted on four permanent teeth from Tabun, Israel; La Chaise-de-Vouthon, France; Scladina, Belgium; and Lakonis, Greece (Dean et al. 2001; Macchiarelli et al. 2006; Smith et al. 2007b, 2009). Several of these studies have reported that the Neanderthal dentition developed in a shorter time than that of modern humans, although in some cases Neanderthals appear to overlap with the low or 'rapid' end of the human range.

The discovery of a juvenile Middle Paleolithic hominin from the Obi-Rakhmat Grotto in Uzbekistan (Glantz et al. 2004, 2008) possessing several isolated, associated developing teeth presents the rare opportunity to assess dental development in a central Asian Paleolithic hominin. Metric and morphological analyses of the dentition suggest that this individual most closely resembles a Neanderthal (Glantz et al. 2004, 2008; Bailey et al. 2008). This study aims to assess whether the duration of crown formation and developmental stage at death in this individual supports the proposed 'rapid developmental profile' based on dental evidence from other Neanderthals. Surface manifestations of long-period incremental features on the tooth crowns and roots were quantified, the degree of root formation was assessed, and crown formation and root development prior to death were estimated using Neanderthal cuspal formation times and a range of likely long-period increment periodicity values. Age at death was estimated using Neanderthal initiation ages, which were added to the time of crown and root formation. These data were compared with data on incremental development in modern humans from northern England and southern Africa, as well as a large sample of Neanderthals (Macchiarelli et al. 2006; Reid and Dean 2006; Smith et al. 2007b, c; Guatelli-Steinberg and Reid 2008; Reid et al. 2008; Smith et al. 2010).

Material and Methods

Six isolated teeth (left upper I2, C, P3, P4, M1, M2) were recovered from the site of Obi-Rakhmat Grotto (Fig. 13.1), which are considered to belong to a single individual due to

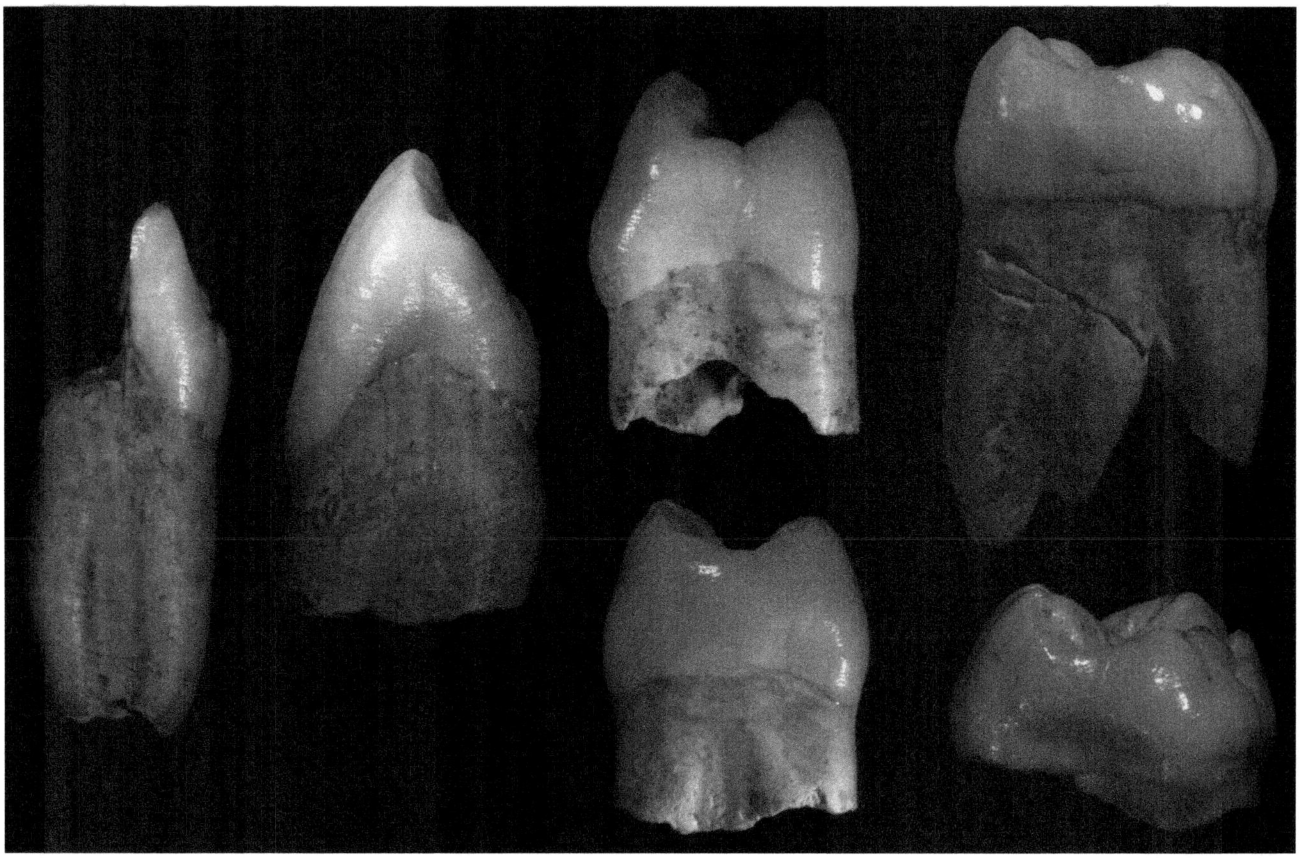

Fig. 13.1 The left maxillary dentition of the Obi-Rakhmat hominin. From *left* to *right*: lateral incisor, canine, third and fourth premolars (upper and lower teeth, respectively), first and second molars (upper and lower teeth, respectively)

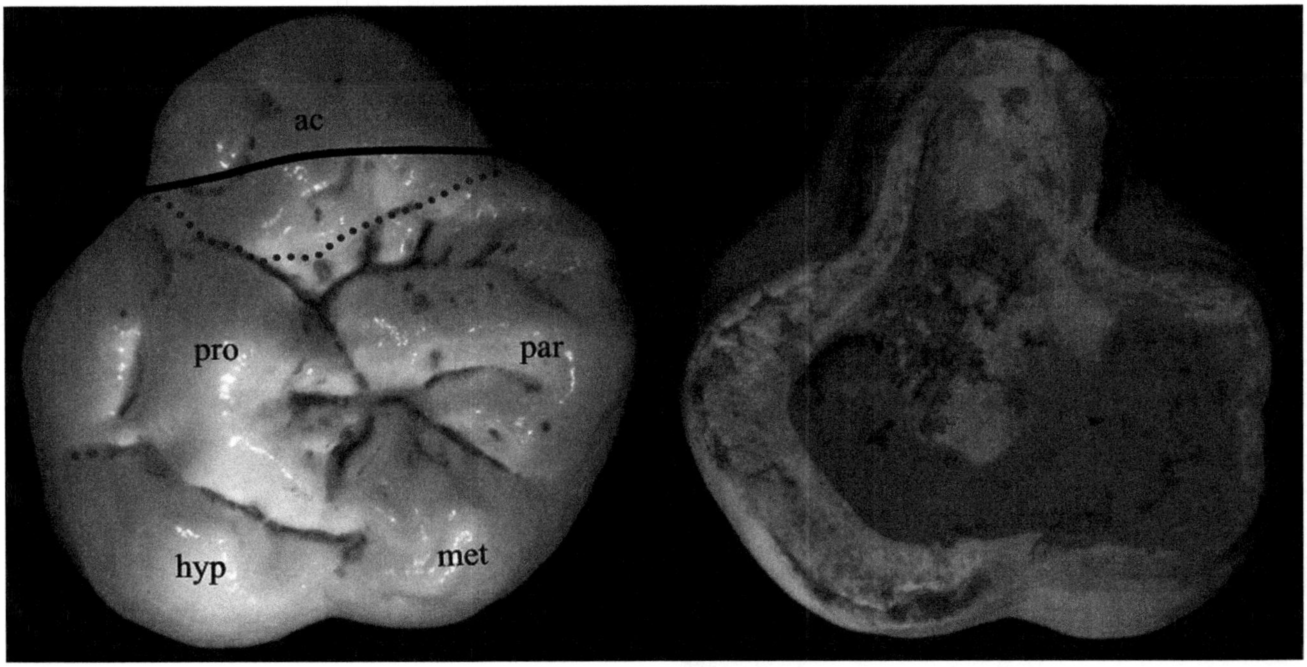

Fig. 13.2 Interpretation of anomalous molar tooth as an upper left second molar. *Left*- maxillary cusps are indicated: *par*- paracone, *pro*- protocone, *met*- metacone, *hyp*- hypocone, *ac*- accessory cusp believed to represent molar gemination (fusion of a supernumary tooth - defined by *dotted* line). The *solid line* is the altered mesial border shown in Fig. 13.3. *Right*- underside of the developing anomalous molar, showing difference in root thickness (and root length by implication) likely due to pathology

their preservation, stage of development, and physical proximity (Glantz et al. 2004, 2008). The teeth have been described in several sources, which have noted that the most posterior molar tooth displays an anomalous morphology that has been interpreted differently by different scholars (Glantz et al. 2004; Bailey et al. 2008). During the course of this study, the anomalous posterior molar was judged to be a left M^2 based on the orientation of the cusps and ridges, as well as root morphology. It is suggested that the major accessory cusp (and supporting root) represent an accessory cusp fused to the mesial aspect of the second molar (Fig. 13.2). The smaller accessory cusp on the lingual aspect of the protocone is interpreted as Carabelli's cusp. The interpretation of this tooth as a left M^2 is supported by comparison with the left M^1 (Fig. 13.3), and when taken together, this row shows a consistent orientation of the *crista obliqua* as well as typical hypocone reduction from M^1 to M^2. This interpretation is similar to an illustration in Glantz et al. (2004: Figure 5, p. 87).

The teeth were originally molded and cast using high-resolution impression materials (3 M Espe Imprint II, Vantico Araldite 2020), and computed tomographic (CT) scans of the original material were made at the Medical University of Innsbruck, Austria. However, the slice thickness of the resultant CT scans, as well as the image quality, were not adequate to yield accurate linear measurements of enamel thickness or quantification of tissues volumes (Olejniczak et al. 2007). Developmental times for postcanine teeth were estimated for individual cusps, which do not necessarily initiate and complete formation simultaneously (Ramirez-Rozzi 1993b; Reid et al. 1998). Crown formation time was determined as the sum of cuspal and imbricational formation time (methods reviewed in detail in Smith 2008). Cuspal formation times were taken from a histological study of several Neanderthals (Smith et al. 2010). Imbricational formation times were assessed from repeated counts of long-period lines (perikymata) on the surface of each crown/cusp (Fig. 13.4), which were made by two individuals (T.S. and D.R.) using stereomicroscopy at 50× magnification. A slight estimate was made for light wear on the M^1 mesiolingual cusp. Perikymata number was multiplied by a range of probable periodicity values (discussed below).

Root length was assessed from casts and photographs of the original teeth, and corrections were made for minor amounts of missing root. Long-period lines known as periradicular bands, which are equal to internal long-period Andresen lines (Smith et al. 2007b; Smith and Reid, 2009), were counted from casts at 50× magnification (Fig. 13.4). Counts of perikymata and periradicular bands were multiplied by a range of Neanderthal periodicity values (6–9 days: Smith et al. 2010) to yield imbricational formation and root formation times, respectively. Root extension rate ranges

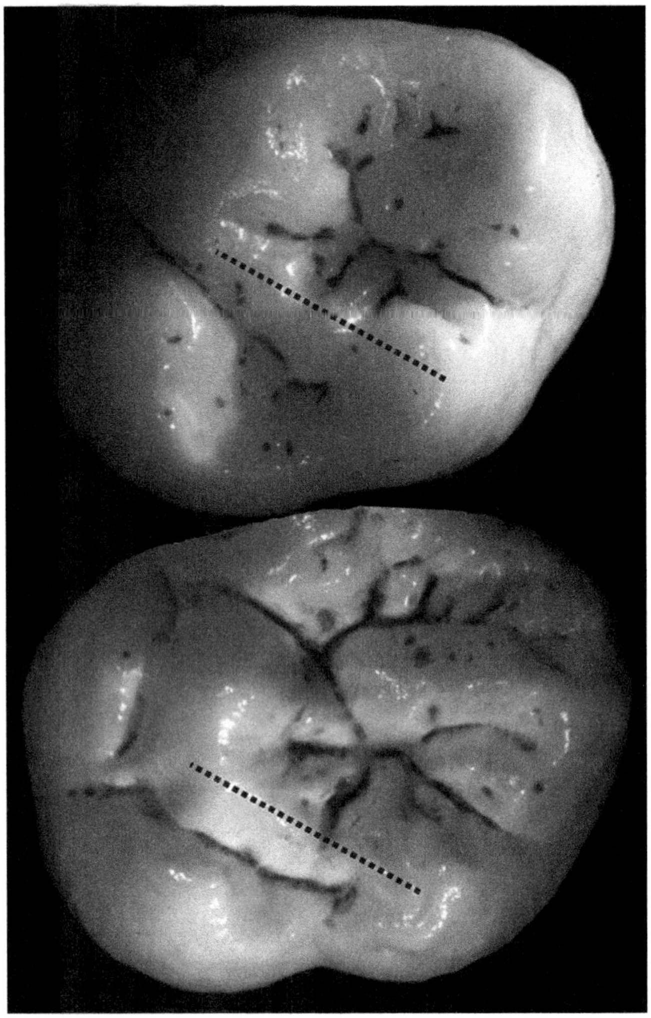

Fig. 13.3 Reconstruction of LM^{1-2} with the geminated cusp from the LM^2 removed (see Fig. 13.2). The M^1 is shown above the M^2, and the *crista obliqua* is indicated with a *dotted line* on both teeth

were estimated for each intact root by dividing the total root length by the product of the respective number of long period lines multiplied by the minimum and maximum estimated periodicity values (6 and 9 days). Finally, age at death was estimated from each cusp by adding histologically derived initiation ages from the Scladina Neanderthal (Smith et al. 2007b) to the range of estimated crown and root formation times. Alternative models for initiation ages are considered in the discussion.

Results

Developmental variables, crown formation time, and age at death are presented in Table 13.1. Estimated extension rates for intact roots were estimated as follows: 14.0 mm of I2 distal root formed at 8.0–12.1 μm/day (assuming a 9 or 6 day periodicity, respectively), 5.0 mm of mesial canine root formed at 4.9–7.4 μm/day, 4.6 mm of P3 buccal root formed at 4.2–6.3 μm/day, 5.4 mm of P3 lingual root formed at 4.4–6.7 μm/day, 5.0 mm of P4 buccal root formed at 4.7–7.1 μm/day, 4.8 mm of P4 lingual root formed at 4.6–6.9 μm/day, 15.4 mm of M1 mesiobuccal root formed at 6.0–9.1 μm/day, and 10.6 mm of M1 lingual root formed at 4.4–6.7 μm/day. The duration of M1 root formation was estimated to be between 4.6 and 7.0 years for the mesiobuccal root.

The mean maximum likely age at death was 8.1 years using a periodicity of 9 days. However, using the modal periodicity of 11 Neanderthals, 7 or 8 days, yields an age at death of 6.7–4 years.

Discussion

Anomalous Molar Morphology

As noted above, the morphology of the unusual posterior molar has been interpreted in a number of ways, resulting in different classifications as an upper right or upper left second or third molar. We believe this tooth shows a rare condition where either a supernumerary tooth has fused with the second molar during crown development, or the developing second molar underwent additional division during formation, resulting in additional cusps in a process known as gemination (e.g., Kronfeld 1939; Tsesis et al. 2003). Several clinical case studies describe these conditions in mandibular molars (e.g., Turell and Zmener 1999; Nunes et al. 2002; Tsesis et al. 2003), noting that the clinical distinction between tooth fusion and gemination is subtle. No evidence of separate pulp chambers was found in CT slices, which suggests that this represents an instance of gemination, although had the tooth completed formation it may have been easier to eliminate the possibility of fusion of a supernumerary tooth. Other aspects of this molar are also pathological, only a few millimeters of mesial and lingual roots are present, while the buccal root was estimated to be almost twice as long based on its thickness at the cervical margin (Fig. 13.2). It is quite possible that the presence of an accessory cusp/root affected the position of the tooth in the crypt, and may have caused the lingual root to develop later or slower than the buccal roots.

We also note that although it is uncommon to find a Carabelli's cusp on an upper second molar that is larger than that of the first molar, the frequency of Carabelli's cusp (grade 3 or larger) on the UM2 in Neanderthals is 58% (Bailey, unpublished data). Other Neanderthal upper second molars have been observed to show large Carabelli's cusps (e.g., Ehringsdorf, Arago, Krapina DP#3).

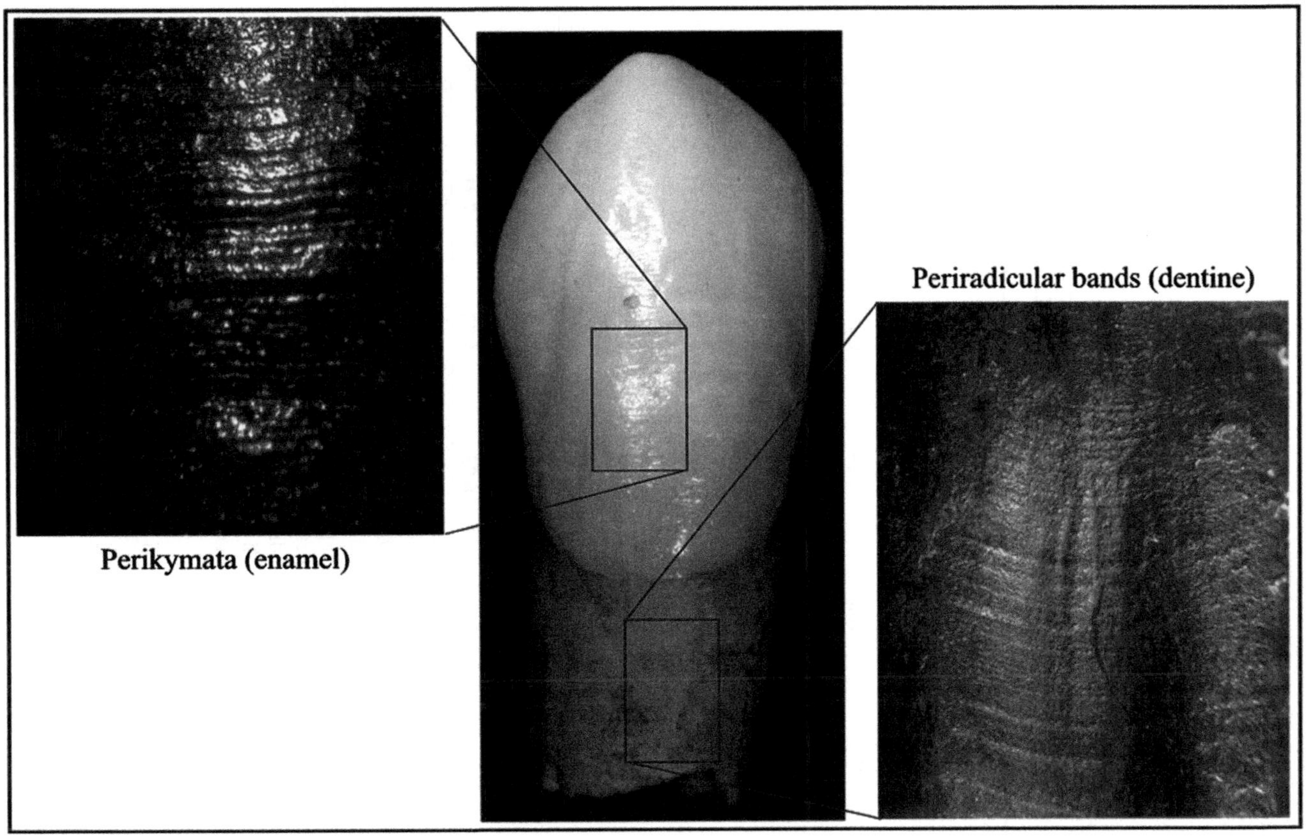

Fig. 13.4 Perikymata (*left*) and periradicular bands (*right*): long-period growth lines preserved on the surface of the developing canine (*center*) from the Obi-Rakhmat Grotto hominin

Developmental Implications

Long-period line (perikymata) numbers for the Obi-Rakhmat tooth crowns are generally lower than mean values for modern humans from northern Europe, but are similar to southern African modern human values (Table 13.2). Values from OR-1 are also lower than Neanderthal mean perikymata numbers, but are similar to the juvenile Neanderthal from Hortus, with the exception of the canine (Fig. 13.5) (Ramirez Rozzi 2005; Guatelli-Steinberg and Reid 2008; Reid et al. 2008). It appears that perikymata number is variable in modern humans, and ranges encompass most Neanderthal values (Mann et al. 1991; Guatelli-Steinberg et al. 2005; Guatelli-Steinberg and Reid 2008; Reid et al. 2008; Smith et al. 2010). The individuals from Obi-Rakhmat Grotto and Hortus expand Neanderthal perikymata ranges for maxillary teeth reported by Guatelli-Steinberg and Reid (2008).

If the periodicity of OR-1 fell at the lower end of the Neanderthal (or modern human) range (6–8 days), crown and root long-period line counts would be consistent with other studies that suggest that Neanderthals show a slightly more rapid period of dental development than some modern human populations. However, because an inverse relationship exists between Retzius line number and periodicity in modern humans (Reid and Ferrell 2006), it may not be the case that imbricational formation time is lower than in modern humans. If this individual had a periodicity of 10 or more days (seen in 31 of 365 modern humans: Smith et al. 2007c) the time represented by perikymata would be equal to or greater than modern human means. While we consider this quite unlikely, we cannot exclude this possibility.

Relatively little data exist regarding root extension rates in living or fossil hominins, particularly for human maxillary teeth (reviewed in Dean 2006; Smith 2008). The estimated rates of M^1 root extension calculated for OR-1 (6.0–9.1 μm/day) are fairly similar to the overall extension rate (6.3 μm/day) reported for the mandibular first molar from La Chaise (Macchiarelli et al. 2006) and the maxillary first molar from Scladina (min rate 6.6 μm/day: Smith et al. 2007b) as well as longitudinal data from modern human first molars (Dean 2006; Macchiarelli et al. 2006). This result represents additional (albeit indirect) evidence that periradicular bands are equivalent to other long-period lines, and may therefore be used to assess the rate and duration of root development (also see Dean 1995; Smith et al. 2007b; Smith and Reid, 2009).

Histological analysis of the Obi-Rakhmat juvenile is dependent on several parameters that must be estimated,

Table 13.1 Estimation of crown formation time, root formation time, and age at death in the Obi-Rakhmat Grotto hominin

Max Teeth	Init Age	Cusp Time	Pkg	CFT (7)	CFT (8)	Prd	RT (7)	RT (8)	Death 6 Day	Death 7 Day	Death 8 Day	Death 9 Day
I2	205	224	130	1134	1264	194	1358	1552	6.50	7.39	8.28	9.16
C	102	210	137	1169	1306	113	791	904	4.96	5.65	6.33	7.02
P3												
Buccal	617	225	99	918	1017	123	861	984	5.96	6.56	7.17	7.78
Lingual	617	322	82	896	978	135	945	1080	6.14	6.73	7.33	7.92
P4												
Buccal	750	295	85	890	975	117	819	936	6.18	6.74	7.29	7.84
Lingual	750	298	78	844	922	116	812	928	6.06	6.59	7.12	7.65
M1												
mb	−13	211	74	729	803	283	1981	2264	6.14	7.39	8.37	9.35
ml		287	~68	763	831							
M2												
mb		351	83	932	1015							
ml		382	81	949	1030	28+	196+	224+				
db		364	75	889	964							
dl		342	85	937	1022							
								Ave	**6.03**	**6.72**	**7.41**	**8.10**

Max Teeth = Left maxillary teeth available: second incisor (I2), canine (C), third and fourth premolars (P3 & P4), and first and second molars (M1 & M2). For molar cusps: mb = mesiobuccal cusp (paracone), ml = mesiolingual cusp (protocone), db = distobuccal cusp (metacone), dl = distolingual cusp (hypocone). Init Age = Initiation ages taken from the Scladina Neanderthal (Smith et al. 2007b) save for the P4, which was estimated to be intermediate between Neanderthal P3s and recent human P4 data (Dean et al. (1993). Cusp = Cuspal formation times taken from Neanderthal values in Smith et al. (2010). Cuspal thickness was assumed to be equal between P3 and P4. Pkg = Perikymata, the number of long-period lines on the enamel surface counted from casts of the original teeth. Slight estimations were made for light wear on the ml cusp of the M1, as indicated by '~'. CFT = cuspal enamel formation plus the total number of perikymata multiplied by 7 and 8 (model Neanderthal periodicity values). Prd = Periradicular bands, the number of long-period lines on the root surface counted from casts of the original teeth. Slight estimations were made for broken root apices. RT = Root formation time calculated as the number of periradicular bands multiplied by model Neanderthal periodicity values, 7 and 8. Death = age at death calculated as the age at initiation plus the time of crown and root formation using a range of Neanderthal periodicity values of 6-9 (Smith et al. 2010). All times and ages are in days except for the last four columns, which are in years. Age at death was not calculated for the M2 as the mb and db root was missing, and the ml and dl root length appeared to be foreshortened/pathological

Table 13.2 Perikymata (long-period lines) counts in the Obi-Rakhmat juvenile compared to values from two modern human populations and a sample of Neanderthals

Tooth	Cusp	OR-1	African	European	Neanderthals
C		137	135 ± 31	148 ± 24	141 ± 10
P3	Buccal	99	80 ± 19	122 ± 17	115
	Lingual	83	83 ± 15	106 ± 14	–
P4	Buccal	85	87 ± 10	107 ± 20	108
	Lingual	78	76 ± 15	84 ± 16	–
M1	mb	86	80 ± 13	92 ± 14	86 ± 8
	ml	68	87 ± 16	89 ± 14	79
	db	76	81	–	96 ± 9
	dl	69	84	67	–
M2	mb	83	93 ± 15	85 ± 14	94 ± 6
	ml	81	87 ± 17	95 ± 13	77
	db	75	–	78	–
	dl	85	–	78	–

Max Teeth = Left maxillary teeth: canine (C), third and fourth premolars (P3 & P4), and first and second molars (M1 & M2). For molar cusps: mb = mesiobuccal cusp (paracone), ml = mesiolingual cusp (protocone), db = distobuccal cusp (metacone), and dl = distolingual cusp (hypocone). Mean values for southern African, northern European, and Neanderthal populations are followed by standard deviations when possible. Data are from Reid et al. (2008) and Smith et al. (2010)

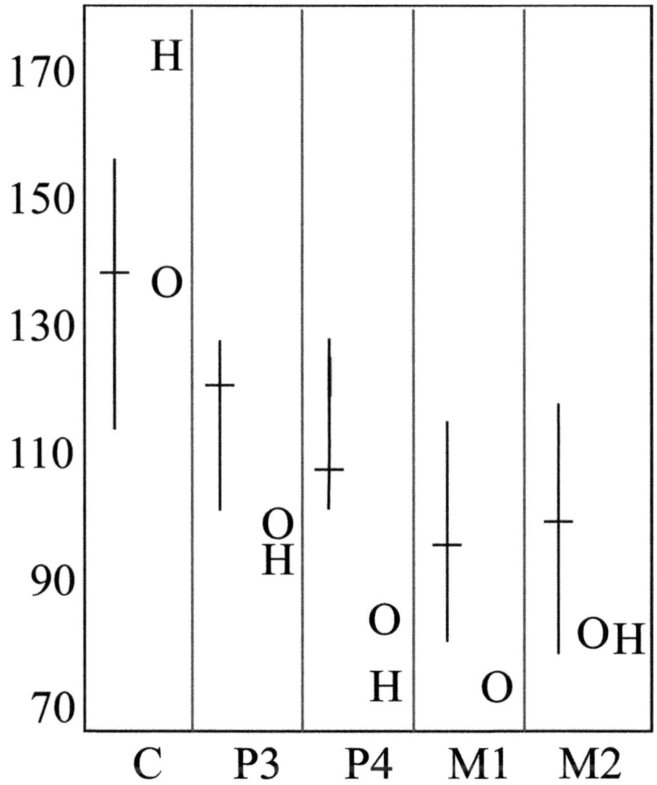

Fig. 13.5 Long-period line (perikymata) numbers in the Obi-Rakhmat individual (O), Hortus II-III individual (H), and a sample of Neanderthals (vertical range bars with mean values indicated by horizontal bars). Data for post-canine teeth are from buccal/mesiobuccal cusps. Tooth types and data are from Table 13.2, Guatelli-Steinberg and Reid (2008), and Ramirez Rozzi (2005)

barring physical or virtual sectioning of the dentition, which may lead to some degree of uncertainty in the final age at death. These estimates include the cuspal formation time, long-period line periodicity, and initiation age. Because Neanderthal molars possess thinner cuspal enamel than modern humans (Smith et al. 2007b), which is a consistent pattern across the dentition (Smith et al. 2010), it is likely that estimated times from the Scladina Neanderthal are more accurate than those derived from modern humans. Similarly, modal long-period line values from other Neanderthals (7 or 8 days) are likely to be more accurate than values derived from other taxa. It has been demonstrated that the mean estimated age at death changes by 0.7 years in this individual when the periodicity is increased or decreased by 1 day. Finally, estimates of initiation age may represent an additional source of error. Very few histological estimates are available for the maxillary dentition, which include an individual of African origin (Dean et al. 1993), four or less European individuals (Reid et al. 1998), and several teeth from the Scladina Neanderthal (Smith et al. 2007b). The Scladina juvenile's initiation ages differed by less than 2 months from the African individual for the tooth types available in this study (UI2, UC, UM1, UM2). Using European initiation ages from Reid et al. (1998) would increase the mean age at death by approximately 3–4 months, which is rather unlikely given numerous studies that have reported early tooth initiation in Neanderthals (reviewed above). In short, a range of possible ages is reported in this study that reflect the most accurate picture of Neanderthal dental development currently available, and the error associated with these ages is likely to be on the order of months rather than years.

Relatively little is known about maxillary dental development in modern human populations due to limitations in radiographic techniques and the time-consuming nature of histological studies. Published standards for modern humans of European origin with a mandibular developmental stage equivalent to the Obi-Rakhmat juvenile suggest an age at death of approximately 8–9 years (Smith 1991). The histological approach in this study yields a most likely age between 6.7 and 7.4 years, which is near the low end of modern European ranges. Global variation in crown formation times and eruption ages is still poorly understood. It is clear that African populations show more rapid anterior and premolar dental development than European populations (Dean et al. 1993; Reid and Dean 2006; Reid et al. 2008), younger initiation ages (Dean et al. 1993; Reid et al. 1998; Liversidge 2008), and younger ages at dental eruption for certain tooth positions (Liversidge 2003). Given the range of modern human variation documented to date, it appears that traditional assessments of age at death in juvenile Paleolithic hominins (e.g., Tillier 2000) should not be based on comparisons with modern European juveniles (Smith et al. 2007b, 2010). It is possible that an African developmental model is more accurate; should this be the case, the 'rapid developmental profile' reported for Neanderthals may be due, in part, to limited comparative samples. In conclusion, while the Obi-Rakhmat hominin shows slightly more rapid development than northern European modern humans, additional data are needed regarding initiation ages, long-period line periodicity, and eruption ages in order to resolve debates over life history differences between modern humans and Neanderthals.

Acknowledgements The authors acknowledge the excavators of Obi-Rakhmat: Andrei Krivoshapkin, Patrick Wrinn, Anatoly Derevianko, and the rest of the Obi-Rakhmat team. We appreciate the comments of two reviewers, as well as the invitation to contribute to this volume by Silvana Condemi. Debbie Guatelli-Steinberg also provided helpful assistance by making comparative data available. Funding was provided by the Max Planck Society, the EVAN Marie Curie Research Training Network MRTN-CT-019564, and Harvard University.

References

Bailey, S. E., Glantz, M., Weaver, T., & Viola, B. (2008). The affinity of the dental remains from Obi-Rakhmat Grotto, Uzbekistan. *Journal of Human Evolution, 55*, 238–248.

Bromage, T. G., & Dean, M. C. (1985). Re-evaluation of the age at death of immature fossil hominids. *Nature, 317*, 525–527.

Dean, M. C. (1995). The nature and periodicity of incremental lines in primate dentine and their relationship to periradicular bands in OH 16 (*Homo habilis*). In J. Moggi-Cecchi (Ed.), *Aspects of dental biology: Paleontology, anthropology and evolution* (pp. 239–265). Florence: International Institute for the Study of Man.

Dean, M. C. (2006). Tooth microstructure tracks the pace of human life-history evolution. *Proceedings of the Royal Society B, 273*, 2799–2808.

Dean, M. C., Stringer, C. B., & Bromage, T. G. (1986). Age at death of the Neanderthal child from Devil's Tower, Gibraltar and the implications for studies of general growth and development in Neanderthals. *American Journal Physical Anthropology, 70*, 301–309.

Dean, M. C., Beynon, A. D., Reid, D. J., & Whittaker, D. K. (1993). A longitudinal study of tooth growth in a single individual based on long- and short-period incremental markings in dentine and enamel. *International Journal of Osteoarchaeology, 3*, 249–264.

Dean, C., Leakey, M. G., Reid, D., Schrenk, F., Schwartz, G. T., Stringer, C., & Walker, A. (2001). Growth processes in teeth distinguish modern humans from *Homo erectus* and earlier hominins. *Nature, 414*, 628–631.

Glantz, M. M., Viola, T. B., & Chikisheva, T. (2004). New hominid remains from Obi-Rakhmat grotto. In A. P. Derevianko (Ed.), *Obi-Rakhmat grotto* (pp. 77–99). Novosibirsk: Institute of Archaeology and Ethnography SB RAS Press.

Glantz, M., Viola, B., Wrinn, P. J., Chikisheva, T., Derevianko, A., Krivoshapkin, A. I., Islamov, U., Suleimanov, R. H., & Ritzman, T. (2008). New hominin remains from Uzbekistan. *Journal of Human Evolution, 55*, 223–237.

Guatelli-Steinberg, D., & Reid, D. J. (2008). What molars contribute to an emerging understanding of lateral enamel formation in Neandertals vs. modern humans. *Journal of Human Evolution, 54*, 236–250.

Guatelli-Steinberg, D., Reid, D. J., Bishop, T. A., & Larsen, C. S. (2005). Anterior tooth growth periods in Neanderthals were comparable to those of modern humans. *Proceedings of the National Academy of Sciences of the United States of America, 102*, 14197–14202.

Guatelli-Steinberg, D., Reid, D. J., & Bishop, T. A. (2007). Did the lateral enamel of Neandertal anterior teeth grow differently from that of modern humans? *Journal of Human Evolution, 52*, 72–84.

Kronfeld, R. (1939). *Histopathology of the teeth and their surrounding structures*. Philadelphia: Lea & Febiger.

Liversidge, H. (2003). Variation in modern human development. In J. L. Thompson, G. E. Krovitz, & A. J. Nelson (Eds.), *Patterns of growth and development in the Genus Homo* (pp. 73–113). Cambridge: Cambridge University Press.

Liversidge, H. (2008). Timing of human mandibular third molar formation. *Annals of Human Biology, 35*, 294–321.

Macchiarelli, R., Bondioli, L., Debénath, A., Mazurier, A., Tournepiche, J.-F., Birch, W., & Dean, C. (2006). How Neanderthal molar teeth grew. *Nature, 444*, 748–751.

Mann, A., & Vandermeersch, B. (1997). An adolescent female Neandertal mandible from Montgaudier Cave, Charente, France. *American Journal of Physical Anthropology, 103*, 507–527.

Mann, A. E., Monge, J. M., & Lampl, M. (1991). Investigation into the relationship between perikymata counts and crown formation times. *American Journal of Physical Anthropology, 86*, 175–188.

Nunes, E., de Moraes, I., Novaes, P., & de Sousa, S. (2002). Bilateral fusion of mandibular second molars with supernumerary teeth: Case report. *Brazilian Dental Journal, 13*, 137–141.

Olejniczak, A. J., Grine, F. E., & Martin, L. B. (2007). Micro-computed tomography of primate molars: Methodological aspects of three-dimensional data collection. In S. E. Bailey & J.-J. Hublin (Eds.), *Dental perspectives on human evolution: State of the art research in dental paleoanthropology* (pp. 103–116). Dordrecht: Springer.

Ramirez Rozzi, F. (2005). Age at death of the Neanderthal child from Hortus. *Bulletins et mémoires de la Société d'anthropologie de Paris, 17*, 47–55.

Ramirez-Rozzi, F. V. (1993a). Microstructure and development of the enamel tooth of the Neanderthal from Zafarraya, Spain. *Comptes Rendus de l'Academie des Sciences, 316*, 1635–1642.

Ramirez-Rozzi, F. V. (1993b). Tooth development in East African *Paranthropus*. *Journal of Human Evolution, 24*, 429–454.

Ramirez Rozzi, F. V., & Bermudez de Castro, J. M. (2004). Surprisingly rapid growth in Neanderthals. *Nature, 428*, 936–939.

Reid, D., & Dean, M. C. (2006). Variation in modern human enamel formation times. *Journal of Human Evolution, 50*, 329–346.

Reid, D. J., & Ferrell, R. (2006). The relationship between number of striae of Retzius and their periodicity in imbricational enamel formation. *Journal of Human Evolution, 50*, 195–202.

Reid, D. J., Beynon, A. D., & Ramirez-Rozzi, F. V. (1998). Histological reconstruction of dental development in four individuals from a medieval site in Picardie, France. *Journal of Human Evolution, 35*, 463–477.

Reid, D. J., Guatelli-Steinberg, D., & Walton, P. (2008). Variation in modern human premolar enamel formation times: Implications for Neanderthals. *Journal of Human Evolution, 54*, 225–235.

Sasaki, C., Suzuki, K., Mishima, H., & Kozawa, Y. (2002). Age determination of the Dederiyeh 1 Neanderthal child using enamel cross-striations. In T. Akazawa & S. Muhesen (Eds.), *Neanderthal burials: Excavations of the Dederiyeh Cave Afrin, Syria* (pp. 263–267). Kyoto: International Research Center for Japanese Studies.

Smith, B. H. (1991). Standards of human tooth formation and dental age assessment. In M. A. Kelley & C. S. Larsen (Eds.), *Advances in dental anthropology* (pp. 143–168). New York: Wiley-Liss.

Smith, T. M. (2008). Incremental dental development: Methods and applications in hominoid evolutionary studies. *Journal of Human Evolution, 54*, 205–224.

Smith, T. M., & Reid, D. J. (2009). Temporal nature of periradicular bands ("striae periradicales") on mammalian tooth roots. In: T. Koppe, G. Meyer, & K.W. Alt (Eds.), *Comparative Dental Morphology* (pp. 86–92). Basel: Karger.

Smith, T. M., Tafforeau, P. T., Reid, D. J., Grün, R., Eggins, S., Boutakiout, M., & Hublin, J.-J. (2007a). Earliest evidence of modern human life history in North African early *Homo sapiens*. *Proceedings of the National Academy of Sciences of the United States of America, 104*, 6128–6133.

Smith, T. M., Toussaint, M., Reid, D. J., Olejniczak, A. J., & Hublin, J.-J. (2007b). Rapid dental development in a Middle Paleolithic Belgian Neanderthal. *Proceedings of the National Academy of Sciences of the United States of America, 104*, 20220–20225.

Smith, T. M., Reid, D. J., Dean, M. C., Olejniczak, A. J., Ferrell, R. J., & Martin, L. B. (2007c). New perspectives on chimpanzee and human dental development. In S. E. Bailey & J.-J. Hublin (Eds.), *Dental perspectives on human evolution: State of the art research in dental paleoanthropology* (pp. 177–192). Dordrecht: Springer.

Smith, T. M., Tafforeau, P., Reid, D. J., Pouech, J., Lazzari, V., Zermeno, J. P., Guatelli-Steinberg, D., Olejniczak, A. J., Hoffman, A., Radovcic, J., Masrour, M., Toussaint, M., Stringer, C. & Hublin, J-J. (2010). Dental evidence for ontogenetic differences between modern humans and Neanderthals. *Proceedings of the National Academy Sciences of the United States of America 107*, 20923–20928.

Smith, T. M., Harvati, K., Olejniczak, A. J., Reid, D. J., Hublin, J.-J., & Panagopoulou, E. (2009). Brief communication: Dental development and enamel thickness in the Lakonis Neanderthal molar. *American Journal of Physical Anthropology, 138*, 112–118.

Stringer, C. B., & Dean, M. C. (1997). Age at death of Gibraltar 2 – a reply. *Journal of Human Evolution, 32*, 471–472.

Stringer, C. B., Dean, M. C., & Martin, R. D. (1990). A comparative study of cranial and dental development within a recent British sample and among Neandertals. In C. J. De Rousseau (Ed.), *Primatelife history and evolution* (pp. 115–152). New York: Wiley-Liss.

Tillier, A. M. (2000). Neanderthal ontogeny: A new source for critical analysis. *Anthropologie, XXXVIII*(1), 109–120.

Tsesis, I., Steinbock, N., Rosenberg, E., & Kaufman, A. Y. (2003). Endodontic treatment of developmental anomalies in posterior teeth: Treatment of geminated/fused teeth- report of two cases. *International Endodontic Journal, 36*, 372–379.

Turell, I., & Zmener, O. (1999). Endodontic management of a mandibular third molar fused with a fourth molar. *International Endodontic Journal, 32*, 229–231.

Chapter 14
Computerized Reconstruction of Prenatal Growth Trajectories in the Dentition: Implications for the Taxonomic Status of Neandertals

Patricia Smith, Gal Avishai, Ralph Müller, and Yankel Gabet

Abstract The hierarchical pattern of tooth formation means that successive phases of development can be identified in fully formed teeth offering a unique insight into ontogenetic processes. The spatial geometry of the cusps expressed in the topography of the dentin-enamel junction (DEJ) records the partitioning of cell proliferation and differentiation as well as the timing of these events. The final stage of development is expressed in the topography of the fully formed crown (OES). Here, the overlying shell of enamel increases crown volume, while modifying cusp relations seen at the DEJ reflecting local variations in enamel thickness.

Using serial scans taken with a micro-CT at 16 μm we have developed a three-dimensional model that enables us to identify, and more importantly quantify, all these developmental features. We have applied this model to reconstruct growth trajectories and their impact on tooth size and cusp relationships in teeth of varying size and tooth classes. The results are used to interpret the extent of developmental variation expressed in Neandertal molars. They indicate that the characteristic features of Neandertal teeth, expressed in intercusp distances and proportions, thin enamel and taurodont roots represent a different partitioning of cell division and differentiation from that observed in *Homo sapiens sapiens*.

Our findings indicate the existence of differences between Neandertals and other hominins in genes controlling the switch-on/switch-off mechanism that regulates the timing, rate and spatial organization of cell proliferation and differentiation of epithelial derived tissues. The thin Neandertal enamel results in teeth that wear down fast, resulting in loss of dental function. We propose that the changes observed in Neandertal teeth, may be secondary to those occurring in other organs developing from epithelial-mesenchymal interaction, with greater adaptive significance.

Keywords Tooth Formation • Ontogenesis • Neandertal - *Homo sapiens sapiens* • micro-CT

P. Smith (✉) and G. Avishai (✉)
Laboratory of Bio-Anthropology and Ancient-DNA,
Institute of Dental Sciences, Hebrew University-Hadassah,
91120 Jerusalem, Israel
e-mail: pat@cc.huji.ac.il; avishai@cc.huji.ac.il

R. Müller
ETH Zurich, Institute for Biomechanics, HCI E 357.2,
Wolfgang-Pauli-Str., 108093 Zürich, Switzerland
e-mail: ram@ethz.ch

Y. Gabet
Institute of Genetic Medicine, Kech School of Medicine,
University of Southern California, 2250 Alcazar Street,
IGM/CSC 240 L.A. 90033, USA
e-mail: gabet@usc.edu

Introduction

Over 150 years after their first discovery, there is still no consensus concerning the taxonomic status of the Neandertals. Hrdlička (1930: pp. 319), complained that Neandertal skeletal remains were too few and too poorly preserved to provide adequate samples for the morphometric analyses necessary to resolve this issue. Today, the number of fossils available for analysis has greatly increased, demonstrating that the temporal and geographic distribution of Neandertals was greater than originally perceived and was accompanied by a correspondingly larger range of morphometric variation (Bailey and Hublin 2006; Harvati 2003; Hublin 1998, 2011; Rosas et al. 2006; Smith et al. 2005; Stringer 2002; Stringer et al. 2004; Wolpoff et al. 2001). Moreover, some features originally considered as having arisen *de novo* in Neandertals have been reported in other fossil hominins including Upper Paleolithic *Homo sapiens sapiens* (Soficaru et al. 2006; Trinkaus et al. 2003), renewing speculation as to the possibility of gene flow between them. Even aDNA techniques (Gutiérrez et al. 2002; Hebsgaard et al. 2007; Pääbo et al. 2004; Weiss and Smith 2007) have not yet provided a conclusive answer to the question of Neandertal admixture. The application of theories that relate the timing and pattern of developmental variation to evolutionary diversity provides an alternative starting point for investigating the phylogenetic status of Neandertals.

S. Condemi and G.-C. Weniger (eds.), *Continuity and Discontinuity in the Peopling of Europe: One Hundred Fifty Years of Neanderthal Study*, Vertebrate Paleobiology and Paleoanthropology, DOI 10.1007/978-94-007-0492-3_14,

While an 'Evo-Devo' approach cannot be directly applied to the study of fossil remains, the concepts can be utilized to reconstruct growth trajectories from examination of the fully formed teeth. In contrast to the bony skeleton, tooth crown form and size once finalized does not undergo remodeling or repair so that the dental phenotype is not affected by environmental factors related to function (Dahlberg 1961, 1985). Most importantly, the hierarchical pattern of tooth formation means that the partitioning of cell division and differentiation, expressed during morphogenesis, is recorded in the topography of the dentin-enamel junction (DEJ) and outer enamel surface (OES) (Korenhof 1979; Kraus 1952; Sasaki and Kanazawa 2000; Smith et al. 1997, 2000).

Both oral ectoderm and neural crest ecto-mesenchyme participate in tooth formation. This process is governed by numerous growth factors some of which act specifically on epithelial derivatives and others primarily on mesenchymal tissues (Jernvall and Thesleff 2000; Butler 1956, 1967, 2000; McCollum and Sharpe 2001; Zaho et al. 2000). Each tooth germ is initiated by down-growth of a fold in the oral epithelium. This forms a cap-shaped structure, the dental organ, that partially encloses a local condensation of mesodermal cells. The cusps are initiated one after another in response to growth factors secreted by the primary and secondary enamel knots (Jernvall et al. 2000). Biomineralization begins when cells of the inner enamel epithelium and underlying layer of cells of the dental papilla differentiate into ameloblasts and odontoblasts that form enamel and dentin respectively. The process starts at the cusp tips and proceeds apically (Fig. 14.1).

Cusp initiation in all hominids follows a constant sequence of protoconid, metaconid, hypoconid, entoconid and hypoconulid (Butler 1956, 1967, 2000; Kraus 1952; Kraus and Jordan 1965; Swindler and McCoy 1965; Swindler et al. 1968; Swindler and Meekins 1991; Winkler et al. 1996), but the amount of growth between the initiation of successive cusps varies as does the duration and rate of further growth (Butler 1967, 2000; Jernvall et al. 2000). Differences in growth rates along the inner and outer slopes of the cusps, as well as those between adjacent cusps, are reflected in their height, shape (that is thin or rounded), as well as the order of coalescence with adjacent cusps. These features are preserved in the topography of the DEJ, on which is superimposed a layer of enamel of varying thickness, that increases the volume of the tooth while modifying the outline defined at the DEJ. Thus, crown size and cusp pattern seen on the outer enamel surface of the tooth reflect not only the order of cusp initiation but also the spatial organization of cell division and differentiation as reflected in cusp size, proportions and location defined at the DEJ, modified by the overlying thickness of enamel.

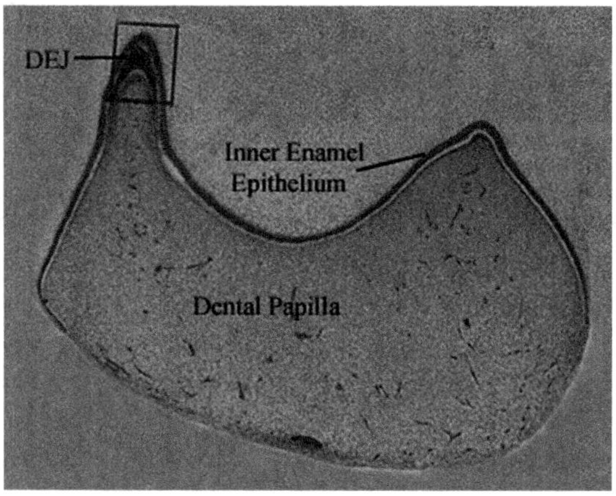

Fig. 14.1 Developing lower 2nd deciduous molar showing initial phase of biomineralization. Note that enamel and dentin formation have begun on the protoconid, enclosed in the *rectangle*, but not yet on the metaconid (Modified from Kraus and Jordan 1965)

Recent advances in imaging techniques have now made it possible to obtain extremely accurate high resolution serial scans that differentiate between dentin and enamel and can be compiled to create accurate 3-D reconstructions of both tissues. Using this approach we have developed a model that facilitates quantification of individual cusps and their relationship to one another at the DEJ and OES. The accuracy of the model has been tested using developing tooth germs (Avishai et al. 2004) and found to provide a reliable estimate of the partitioning of cell proliferation and biomineralization. These processes are reflected in cusp volume and distances between cusp tips and coalescence points. The model is used here to examine the effect of variation in growth trajectories on crown size and cusp pattern of contemporary molars and applied to interpret the significance of developmental variation that results in the characteristic morphology of Neandertal molars.

The Model

Our model is derived from serial micro-CT scans taken at 16 μm intervals. We used a standard protocol for scanning and subsequent reconstructions and analyses that was developed by R.M. The landmarks were located by G.A. directly from the micro-CT workstation. They included XYZ coordinates of cusp tips and points of coalescence between cusps, the cemento-enamel junction defining the base of the crown and enamel and dentin contours of individual cusps and crown circumference. The total measurement error calculated from reconstructions derived from repeated scans was 1.2% (Avishai et al. 2004).

We have now scanned 30 lower second deciduous and first permanent molars of recent humans using the method detailed in Avishai et al. (2004). We oriented all specimens along a horizontal plane defined by the cusp tips of the protoconid, metaconid and hypoconid, which are the first cusps

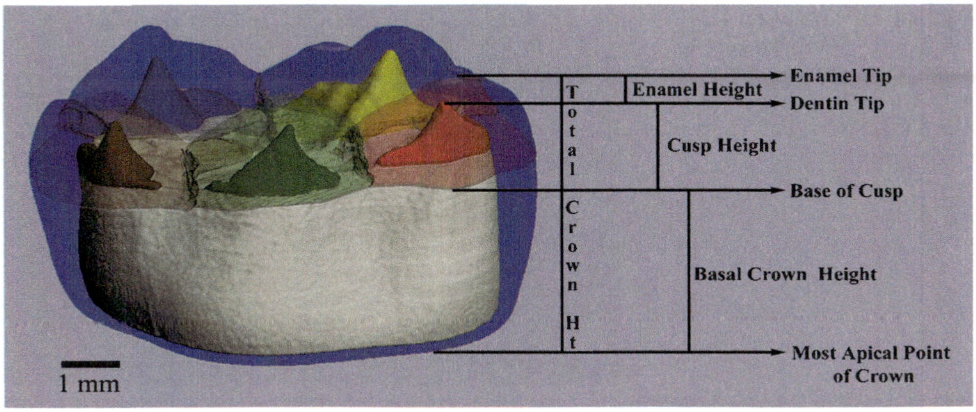

Fig. 14.2 Buccal view of micro-CT reconstruction of lower second deciduous molar showing major reference points measured

initiated (Kraus and Jordan 1965), and used this as our reference plane for evaluating subsequent tooth development. Computerized reconstructions of each tooth were compiled using an algorithm developed by R.M. The vertical distance between cusp tips was used to estimate the order of cusp initiation and the distance between the most occlusal and most apical points of coalescence of the cusps, defined as Maximal Fusion Height (MFH). This was used to assess the amount of growth between them. A line joining the first two coalescence points of each cusp was used to construct a virtual plane defining the base of each cusp and used in subsequent calculations. The area of cusp base was defined by morphing to compensate for height differences between coalescence points. Analyses included calculation of: cusp height and volume at the DEJ and OES, intercusp distances, enamel thickness and volume, cross-sectional area and thickness of enamel at standard points along the cusps, crown volume and maximum perimeter and area of the crown.

For ease of interpretation we partitioned Euclidean distances between points of measurements into their vertical and horizontal components, with apical growth reflected in dentin cusp height (DCH) and maximum coalescence height difference (MFH), while horizontal growth is reflected in distances' between cusp tips. The ratio of horizontal to vertical cusp distance provides an estimate of the amount and direction of growth between successive cusps (Fig. 14.2).

In modern deciduous and permanent molars, cusps first coalesce on their outer surfaces, while still continuing to grow independently on their inner slopes (Fig. 14.3). The protoconid first coalesces with the metaconid and hypoconid, the next coalescence takes place between the hypoconulid and entoconid. This is followed by the hypoconid-hypoconulid. The metaconid-entoconid coalescence is the last to occur. Once this occurs, the cusps are united together by a continuous ring of dentin fixing their relations to one another, even though they continue to grow on their inner slopes. The amount of further growth on their inner aspects before coalescence is reflected in the depth of grooves and fovea.

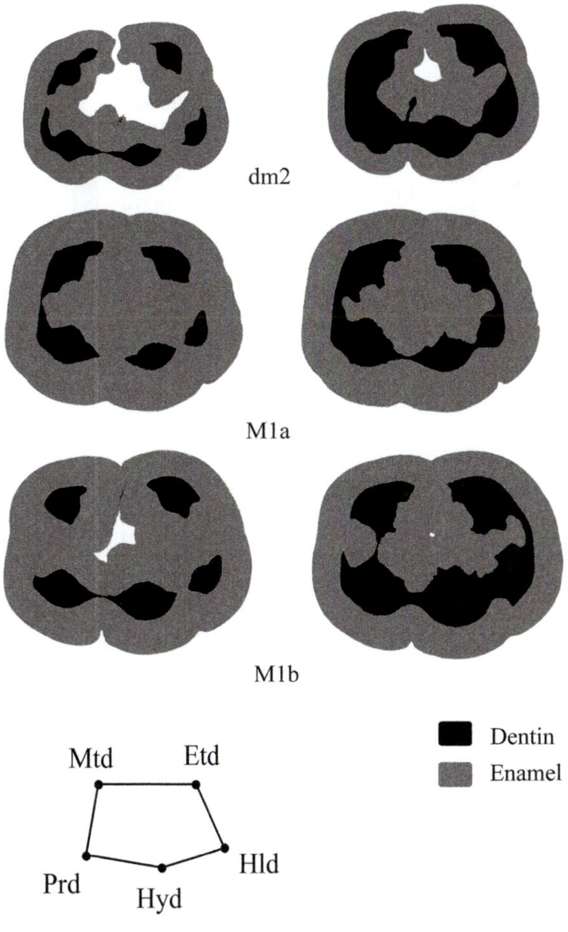

Fig. 14.3 Reconstructed axial slices of three lower molars at first point of coalescence (*at left*) and last coalescence prior to metaconid-entoconid coalescence (*at right*). All slices oriented as outlined in diagram at *bottom*. *Top* – second deciduous molar, note: thin enamel. *Middle* – small first permanent molar. *Bottom* – large first permanent molar

Since cusps expand in basal area as they grow, their height before coalescence is reflected in intercusp distances and overall dimensions of the crown. Occasionally additional folds appear in the inner enamel epithelium, in which independent

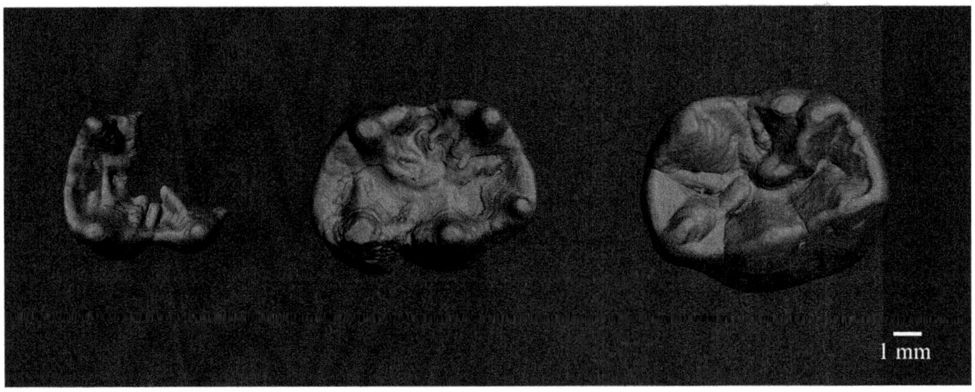

Fig. 14.4 Occlusal representation of three lower 2nd deciduous molars, showing changes in surface topography in different stages of development

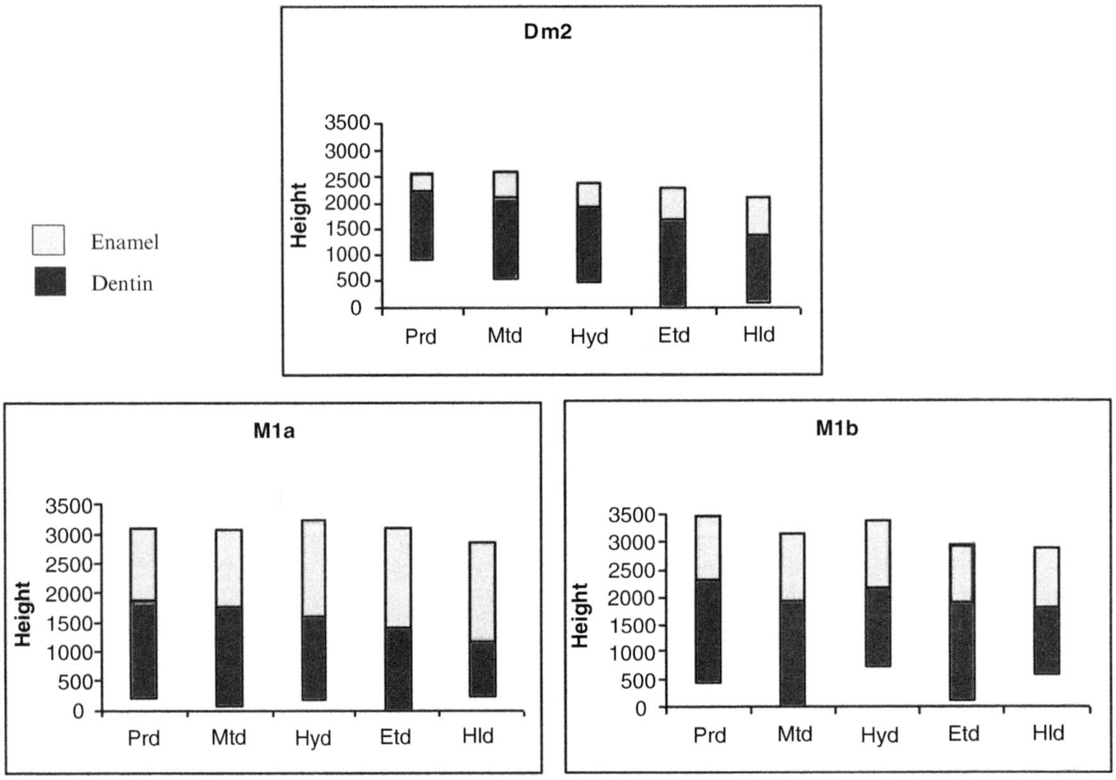

Fig. 14.5 Vertical location of enamel tip (*top of bar*) and cusp base (*bottom of bar*). Height measured in μm. Note differences in enamel and dentin height between cusps and between teeth. The metaconid is tallest in the permanent molars and the entoconid tallest in the deciduous molar. Enamel is thicker in the permanent teeth than in the deciduous tooth and is thickest on the hypoconid and hypoconulid

centers of bio-mineralization develop, producing extra ridges and cuspules (Fig. 14.4). Measurements taken at the DEJ show that the amount of vertical growth expressed by MFH is correlated with horizontal growth expressed by intercusp distances (Smith et al. 2007).

The spatial components of growth expressed by distances between cusp tips and coalescence points at the DEJ indicate major differences between tooth classes as well as between large and small teeth of the same class (Fig. 14.5 and Table 14.1). In the deciduous molars, the entoconid, which is the fourth cusp initiated, is consistently the tallest followed by the metaconid, protoconid and hypoconulid (Fig. 14.5). The hypoconid, the third cusp initiated, is consistently the shortest. At the DEJ of the first permanent molar, ranking of cusp height is more variable. In small teeth, the metaconid is taller than the entoconid and the hypoconid is taller than the hypoconulid.

The contribution of cusp height to total crown height also differs markedly between the deciduous and permanent molar, reflecting the different partitioning of cusp formation relative to crown formation. In the fast growing second deciduous molar, biomineralization begins some 6 weeks after initiation of the tooth germ and horizontal and vertical distances between cusp tips are similar, reflecting the rapid growth in length and breadth as well as apically in this tooth while the cusps are forming. In the first permanent molars, biomineralization only begins some 4 months after the initiation of the tooth germ and continues for nearly three times as long as that of the deciduous molars. In the permanent molar height differences between cusp tips are small, relative to horizontal distances, indicating that the tooth germ is larger when biomineralization begins and that less growth takes place between the initiation of successive cusps. This may reflect the slower rate of growth of this tooth but is also expressed in a different spatial partitioning of cell proliferation within the tooth (Butler 1967, Kraus and Jordan 1965). In both large and small first molars, the hypoconid is finalized at an early stage of development, when it coalesces with the protoconid, while the size of the hypoconulid is dependent on that of the entoconid (Fig. 14.5).

Comparison of the DEJ and OES of deciduous and permanent molars demonstrates the contribution of the thickness of the enamel shell to tooth size and proportions. The volume as well as the shape of the tooth crown seen at the OES reflects the combined volume of the tooth defined at the DEJ that results from cell proliferation, plus the amount of enamel matrix laid down by a fixed number of differentiated cells-the ameloblasts. Obviously, the thicker the enamel, or the more it varies in different locations, the greater the extent of modification of the crown from the underlying template defined by the DEJ. Like the covering of snow that smoothes out jagged mountain peaks and fills in crevices, thick enamel obscures details present at the DEJ.

Examination of enamel thickness in different locations over the cusps shows that the thickness of the enamel shell varies over the surface of each cusp. Consequently cusps are more rounded and the topography seen at the OES is much smoother than that at the DEJ. At the same time, local variation in enamel thickness creates additional features such as marginal cuspules not represented at the DEJ. Enamel on the outer surface of all cusps is thicker than that on the inner slopes (Grine 2005; Keinan et al. 2006; Suwa and Kono 2005). It is thicker on the buccal cusps than on the lingual cusps and is usually thickest on the hypoconulid followed by the hypoconid, so that these cusps occupy a relatively larger area on the crown surface than at the DEJ. This applies even to the thin-enameled deciduous molars so that the rank order of cusp area at the OES may change from that measured at the DEJ (Table 14.2). However, the angle formed by the cusps to one another also affects intercusp distances. Where the cusps diverge from one another, distances at the OES are greater than those at the DEJ. Where there is little intercusp growth, cusps may be parallel or incline towards one another and distances between them at the OES are similar or smaller to those seen at the DEJ (Smith et al. 1997, 2000). This is demonstrated here by the differences shown between intercusp distances at the DEJ and OES of the dm2 and permanent molars in Table 14.1.

In the small M1a represented in Figs. 14.3 and 14.5 and Tables 14.1 and 14.2, the hypoconid and hypoconulid are very much smaller than other cusps at the DEJ, but the thicker enamel obscures this at the OES. In the permanent molars, the increase in hypoconid cusp area is more than double that of the metaconid. Moreover, when the hypoconulid is extremely reduced, the thicker enamel may obscure the boundary between it and the entoconid (Korenhof 1979; Kraus and Jordan 1965).

Table 14.1 Basal cusp area at DEJ and OES (mm^2)

Area at DEJ

Tooth	Prd	Mtd	Hyd	Etd	Hld
dm_2	3.6	5.43	3.27	5.55	2.62
M_1 a	4.89	6.52	2.7	6.02	1.72
M_1 b	5.85	8.72	4.12	9.72	3.15

Area at OES

Cusp	Prd	Mtd	Hyd	Etd	Hld
dm_2	13.59	13.7	12.42	12.06	14.73
M_1 a	21.63	21.23	15.01	18.9	11.77
M_1 b	20.66	22.04	17.06	20.96	17.33

Note area of buccal cusps changes more than area of lingual cusps because of thicker enamel. This changes cuspal proportions at the OES. Note that in both permanent molars the entoconid is larger than the hypoconid at the OES, while the deciduous tooth with a larger talonid maintains the conservative sequence with the hypoconid larger

Prd Protoconid, *Mtd* Metaconid, *Hyd* Hypoconid, *Etd* Entoconid, *Hld* Hypoconulid, M_1a is a small 1st permanent molar, M_1b is a large 1st permanent molar

Table 14.2 Distance between cusp tips (mm) at DEJ and OES, and maximum area (mm^2) at OES

Tooth	Area		Prd-Mtd	Prd-Hyd	Prd-Etd	Prd-Hld	Mtd-Hyd	Mtd-Etd	Mtd-Hld	Hyd-Etd	Hyd-Hld	Etd-Hld
dm_2	–	DEJ	3.51	3.58	6.46	6.41	5.00	4.71	6.80	4.64	2.92	3.94
dm_2	72.1	OES	3.65	3.93	7.12	7.18	5.65	5.20	7.67	5.60	3.45	4.75
M_1 a	–	DEJ	4.15	4.18	7.26	7.24	6.27	5.55	8.07	5.44	3.26	4.50
M_1 a	83.4	OES	4.85	3.64	7.45	6.89	6.35	5.67	8.42	5.58	3.29	4.97
M_1 b	–	DEJ	4.31	4.56	6.97	6.85	6.48	5.32	7.67	4.88	2.50	4.04
M_1 b	99.4	OES	5.28	4.78	7.45	7.16	7.40	5.99	8.68	5.02	2.51	4.31

For each tooth first row = DEJ, second row = OES. Tooth area in mm^2. Calculated from maximum convexity of crown

Note that the Prd-Hld intercusp distance is smaller than Mtd- Hld distance in all teeth, despite the variation in location of the Hld

Implications for Neandertal Tooth Formation

The results obtained using our three dimensional model show how differences in the partitioning of cell division and differentiation between cusp initiation and coalescence of individual cusps, are reflected in spatial geometry of the cusps at the DEJ and the extent to which the DEJ is modified on the crown surface by the enamel shell. They show differences in the partitioning of growth between deciduous and permanent molars as well as between large and small teeth of the same tooth class. They also show that the hypoconid is incorporated into the developmental sequence of the modified trigonid, while the early entoconid and hypoconulid coalescence means that these two cusps grow independently until the coalescence of the hypoconulid with the hypoconid. The location of the hypoconulid relative to the entoconid is therefore defined before its position relative to the hypoconid is finalized. Finally as shown here in Tables 14.1 and 14.2 and Fig. 14.5, differences in enamel thickness modify the crown surface defined at the DEJ, so that while the topography of the DEJ reflects epithelial-mesenchymal interactions, the epithelial derived ameloblasts "have the last word" in terms of fine-tuning enamel thickness and so crown form. These findings provide a developmental framework with which to interpret the significance of morphological features that distinguish Neandertal teeth from those of other fossil or contemporary hominins.

Neandertal deciduous and first permanent molars are characterized by thin enamel and taurodont roots (Macchiarelli et al. 2006; Molnar et al. 1993; Ramirez Rozzi 1996; Smith 1990; Zilberman and Smith 1992; Zilberman et al. 1992). The occlusal surface shows deep anterior fovea, a high frequency of mid-trigonid crests and a tendency for cusp tips to be more closely aligned on the occlusal surface than those of other hominins (Bailey 2002a, b; Bailey 2005). Radiographic studies of Neandertal deciduous teeth also indicate accelerated development and eruption relative to that of modern humans (Faerman et al. 1994; Legoux 1966; Skinner and Sperber 1982). However, the extent to which this accelerated growth pattern is maintained in the permanent dentition is unclear, since conflicting results have been published from both radiographic studies and those based on counts of perichymata or incremental lines (Dean et al. 1986; Thompson and Nelson 2000; Guatelli-Steinberg et al. 2005, 2007; Macchiarelli et al. 2006; Mann et al. 1990; Ramirez Rozzi and Bermudez De Castro 2004).

Enamel Thickness

The thin enamel that distinguishes Neandertal teeth means that the OES in permanent as well as deciduous Neandertal molars retains a more faithful resemblance to the underlying DEJ than the similar sized but thicker enameled molars of other hominins. This may account for the greater frequency of features such as midtrigonid crests and 6th and 7th cusps that are more commonly present at the DEJ of contemporary humans than at the OES where they tend to be obscured by the overlying enamel (Korenhof 1979; Kraus and Jordan 1965; Sasaki and Kanazawa 2000).

Cusp Pattern

Molar cusp pattern in *Homo sapiens sapiens* is directly linked to tooth size, with smaller teeth changing from the Dryopithecine 5Y pattern to 5+ and eventually 4+ as the hypoconulid rotates lingually and is eventually integrated with the entoconid. Bailey (2002b) reported that in anatomically modern *Homo* and Upper Paleolithic *Homo sapiens sapiens* cusp size rank was predominately protoconid>metaconid>hypoconid>entoconid>hypoconulid, but changed in contemporary teeth to protoconid>metaconid>entoconid>hypoconid>hypoconulid. These differences within fossil and modern *Homo sapiens sapiens* illustrate the changing cusp relations associated with reduction in tooth size. In the Neandertals, cusp pattern differs from that predicted from their large size, and clusters with that found for contemporary teeth, emphasizing yet again their unique pattern of development. Moreover these differences appear to be even more pronounced in the early developing deciduous teeth (Fig. 14.6).

The ratio of distances between cusp tips in deciduous teeth of Neandertals, anatomically modern *Homo* and a modern sample is shown in Table 14.3. In both anatomically modern *Homo* and contemporary deciduous teeth, the cusps are more evenly spaced than is the case for the Neandertal molars and the distance between the protoconid and hypoconulid is greater than that between the metaconid and hypoconulid. In the Neandertal molars, intercusp distances vary markedly and the shape of the tooth differs, with the hypoconulid internally located and much closer to the entoconid than in any of the other teeth. Placed within the developmental pattern shown by our model, this suggests a major shift in the partitioning of cell division and differentiation within the tooth germ, expressed in a striking reduction of the talonid.

The thin enamel of Neandertal molars also means that the contribution of enamel thickness to tooth volume is smaller than in similar sized molars of *Homo sapiens* or other hominins. This implies a larger dental papilla and so a larger surface area at the DEJ with more ameloblasts even though they produce less enamel. This makes the reduced functional activity of the ameloblasts shown by the thin enamel even more striking. It indicates a major difference in the expression of growth factors regulating cell division and differentiation of the inner enamel epithelium in Neandertals.

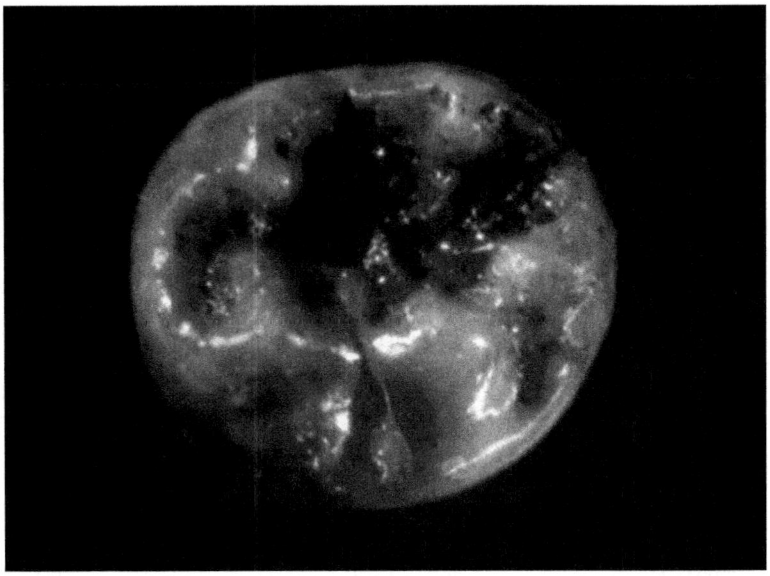

Fig. 14.6 Occlusal view of the second deciduous molar from Kebara 1. Note the internally located cusps and rounded outline of this tooth

Table 14.3 Intercusp distances measured at OES and standardized as % of Prd-Hld distance

	dm_2[a]	dm_2[b]	KEB I	KEB IV	SKHUL X
Prd-Mtd	51	58	63	56	59
Prd-Hyd	55	54	61	55	50
Prd-Etd	99	101	104	106	99
Prd-Hld	100	100	100	100	100
Mtd-Etd	72	70	64	68	58
Mtd-Hld	107	107	94	94	101
Hyd-Etd	78	81	73	79	73
Hyd-Hld	48	50	42	50	51
Etd-Hld	66	65	54	55	69

Neandertals: KEB I – Kebara I, KEB IV – Kebara IV. Anatomically modern human: SKHUL X. Modern specimens taken from archaeological collections where [a]dm_2 – measured from micro-CT reconstructions, [b]dm_2 mean value for 20 teeth. This and fossil specimens measured directly from occlusal surface using "Galai" image analyzer as described in Smith et al. 1995. Note: Similar ranking of intercusp distances measured directly from the occlusal surface of the dm2 and those computed using the 3D model. In the two Neandertal specimens, Prd-Hld, Prd-Etd ratios are reversed, with the Hld closer to the Etd and further from the Prd than in all other specimens

Root Formation

Root formation in Neandertals is also unique. In Neandertals the cervical region lacks the cervical constriction typical of *Homo sapiens sapiens* molars, while the common root stem is elongated and bifurcation of the roots is apically located (Kallay 1963). This is due to a delay in the change of direction of the cells of Hertwig's epithelial root sheath that define root form. While this condition does occur in modern humans as well as in other fossil hominids, it is not combined with thin enamel, but rather associated with short roots and so differs from the pattern seen in Neandertal molars that are larger with longer roots than most modern teeth (Bailey 2005).

Macchiarelli et al. (2006) reported that the Neandertal first permanent molar shows a slower initial phase of root development than modern teeth, but that later development is more rapid so that it completes development at the same time as modern teeth. Their findings, based on examination of incremental lines in the dentin of ground sections, provide independent confirmation of our analyses that suggest marked differences between Neandertal and *Homo sapiens sapiens* in the partitioning (timing and amount of growth) of successive developmental stages within the tooth germ.

Interpretation

We propose that the Neandertal molar complex comprising large teeth with a small occlusal area, a relatively small entoconid and hypoconulid, thin enamel and long taurodont roots demonstrates the presence of unique growth trajectories in Neandertal teeth. All of these dental features may be attributed to changes in signaling pathways that regulate the epithelial-mesenchymal interactions that form the teeth. These may specifically affect the epithelial component, since this gives rise to the enamel knots that define the timing and location of cusp initiation as well as the inner enamel epithelium that differentiates into enamel forming ameloblasts and Hertwig's epithelial root sheath. There is good evidence to show that many of the active signaling pathways affecting tooth formation are common to other organs dependent on epithelial-mesenchymal interactions, such as hair, sweat glands and lungs (Plikus et al. 2005). The changes seen in the Neandertal teeth may constitute one component of a broader distinctive pattern of development affecting the form and function of other organs in which epithelial derived tissues play a major role. Thus it may be time to evaluate the selective significance of the dental changes seen in the Neandertals, within a wider perspective than that of dental function.

Acknowledgements Research supported by Grant No. 032-5302 from the Israel Science Foundation.

References

Avishai, G., Muller, R., Gabet, Y., Bab, I., Zilberman, U., & Smith, P. (2004). New approach to quantifying developmental variation in the dentition using serial microtomographic imaging. *Microscopy Research and Technique, 65*(6), 263–269.

Bailey, S. E. (2002a). A closer look at Neanderthal postcanine dental morphology: The mandibular dentition. *The Anatomical Record, 269*(3), 148–156.

Bailey, S. E. (2002b). *Neandertal dental morphology: Implications for modern human origins*. Ph.D. thesis, Arizona State University, Tempe, Arizona.

Bailey, S. E. (2005). Diagnostic dental differences between Neandertals and Upper Paleolithic modern humans: Getting to the root of the matter. In E. Zadzinska (Ed.), *Current trends in dental morphology research* (pp. 201–210). Lodz: University of Lodz Press.

Bailey, S. E., & Hublin, J. J. (2006). Dental remains from the Grotte du Renne at Arcy-sur-Cure (Yonne). *Journal of Human Evolution, 50*(5), 485–508.

Butler, P. M. (1956). The ontogeny of molar patterns. *Biological Reviews, 31*, 30–70.

Butler, P. M. (1967). Comparison of the development of the second deciduous molar and first permanent molar in man. *Archives of Oral Biology, 12*(11), 1245–1260.

Butler, P. M. (2000). The relation of cusp development and calcification to growth In: J. Mayhall, & T. Heikkinen (Eds.), *Proceedings of the 11th International Symposium on Dental Morphology* (pp. 26–32). Oulu: Oulu University Press.

Dahlberg, A. (1961). Relationship of tooth size to cusp number and groove conformation of occlusal surface patterns of lower molar teeth. *Journal of Dental Research, 40*, 34–36.

Dahlberg, A. (1985). Ontogeny and dental genetics in forensic problems. *Forensic Science International, 30*, 163–176.

Dean, M. C., Stringer, C. B., & Bromage, T. G. (1986). Age at death of the Neanderthal child from Devil's Tower, Gibraltar and the implications for studies of general growth and development in Neanderthals. *American Journal of Physical Anthropology, 70*(3), 301–309.

Faerman, M., Kharitonov, V., Batsevich, B., Zilberman, U., & Smith, P. (1994). A Neanderthal infant from the Barakai Cave, Western Caucasus. *Journal of Human Evolution, 27*, 405–415.

Grine, F. (2005). Enamel thickness of deciduous and permanent molars in modern *Homo sapiens*. *American Journal of Physical Anthropology, 126*(1), 14–31.

Guatelli-Steinberg, D., Reid, D. J., Bishop, T. A., & Larsen, C. S. (2005). Anterior tooth growth periods in Neandertals were comparable to those of modern humans. *Proceedings of the National Academy of Sciences, 102*(40), 14197–14202.

Guatelli-Steinberg, D., Reid, D. J., & Bishop, T. A. (2007). Did the lateral enamel of Neandertal anterior teeth grow differently from that of modern humans? *Journal of Human Evolution, 52*(1), 72–84.

Gutiérrez, G., Sanchez, D., & Marin, A. (2002). A reanalysis of the ancient mitochondrial sequences recovered from Neandertal bones. *Molecular Biology and Evolution, 19*, 1359–1136.

Harvati, K. (2003). The Neanderthal taxonomic position: Models of intra- and inter-specific craniofacial variation. *Journal of Human Evolution, 44*(1), 107–132.

Hebsgaard, M. B., Wiuf, C., Gilbert, M. T., Glenner, H., & Willerslev, E. (2007). Evaluating Neanderthal genetics and phylogeny. *Journal of Molecular Evolution, 64*(1), 50–60.

Hrdlička, A. (1930). *The skeletal remains of early man* (Smithsonian Miscellaneous Collections, Vol. 83). Washington, DC: Smithsonian Institution.

Hublin, J. J. (1998). Climatic changes, paleogeography and the evolution of the Neandertals. In T. Akazawa, K. Aoki, & O. Bar-Yosef (Eds.), *Neandertals and modern humans in Western Asia* (pp. 295–310). New York: Plenum.

Hublin, J. J. (2011) Dental development and age at death of a Middle Paleolithic juvenile hominin from Obi-Rakhmat Grotto, Uzbekistan. In S. Condemi & G.-C. Weniger (Eds.), *Continuity and discontinuity in the peopling of Europe* (pp. 155–164). Dordrecht: Springer.

Jernvall, J., & Thesleff, I. (2000). Return of lost structure in the developmental control of tooth shape. In M. F. Teaford, M. M. Smith, & M. W. J. Ferguson (Eds.), *Development, function and evolution of teeth* (pp. 13–22). Cambridge: Cambridge University Press.

Jernvall, J., Keranen, S. V., & Thesleff, I. (2000). Evolutionary modification of development in mammalian teeth: Quantifying gene expression patterns and topography. *Proceedings of the National Academy of Sciences, 97*(26), 14444–14448.

Kallay, J. (1963). A radiographic study of the Neanderthal teeth from Krapina, Croatia. In D. R. Brothwell (Ed.), *Dental anthropology* (pp. 75–86). Oxford: Pergamon.

Keinan, D., Smith, P., & Zilberman, U. (2006). Microstructure and chemical composition of primary teeth in children with down syndrome and cerebral palsy. *Archives of Oral Biology, 51*(10), 836–843.

Korenhof, C. A. (1979). The evolution of the lower molar pattern and remnants of the trigonid crests in man. *Nederlands Tijdschrift voor Tandheelkunde, 86*(Suppl 17), 6–31.

Kraus, B. S. (1952). Morphological relationships between enamel and dentin surfaces of lower first molar tooth. *Journal of Dental Research, 31*, 248–256.

Kraus, B. S., & Jordan, R. E. (1965). *The human dentition before birth*. Philadelphia: Lea & Febiger.

Legoux, P. (1966). *Détermination de l'âge dentaire de fossiles de la lignée humaine*. Paris: Maloine.

Macchiarelli, R., Bondioli, L., Debenath, A., Mazurier, A., Tournepiche, J. F., Birch, W., & Dean, M. C. (2006). How Neanderthal molar teeth grew. *Nature, 444*(7120), 748–751.

Mann, A., Lampl, M., & Monge, J. (1990). Patterns of ontogeny in human evolution: Evidence from dental development. *Yearbook of Physical Anthropology, 33*, 11–150.

McCollum, M., & Sharpe, P. T. (2001). Evolution and development of teeth. *Journal of Anatomy, 199*(Pt 1–2), 153–159.

Molnar, S., Hildebolt, C., Molnar, I. M., Radovcic, J., & Gravier, M. (1993). Hominid enamel thickness: I. The Krapina Neandertals. *American Journal of Physical Anthropology, 92*(2), 131–138.

Pääbo, S., Poinar, H., Serre, D., Jaenicke-Despres, V., Hebler, J., Rohland, N., Kuch, M., Krause, J., Vigilant, L., & Hofreiter, M. (2004). Genetic analyses from ancient DNA. *Annual Review of Genetics, 38*, 645–679.

Plikus, M. V., Zeichner-David, M., Mayer, J. A., Reyna, J., Bringas, P., Thewissen, J. G., Snead, M. L., Chai, Y., & Chuong, C. M. (2005).

Morphoregulation of teeth: Modulating the number, size, shape and differentiation by tuning Bmp activity. *Evolution and Development, 7*(5), 440–457.

Ramirez Rozzi, F. (1996). Comment on the causes of thin enamel in Neandertals. *American Journal of Physical Anthropology, 99*(4), 625–626.

Ramirez Rozzi, F. V., & Bermudez De Castro, J. M. (2004). Surprisingly rapid growth in Neanderthals. *Nature, 428*(6986), 936–939.

Rosas, A., Martinez-Maza, C., Bastir, M., Garcia-Tabernero, A., Lalueza-Fox, C., Huguet, R., Ortiz, J. E., Julia, R., Soler, V., de Torres, T., Martinez, E., Canaveras, J. C., Sanchez-Moral, S., Cuezva, S., Lario, J., Santamaria, D., de la Rasilla, M., & Fortea, J. (2006). Paleobiology and comparative morphology of a late Neandertal sample from El Sidron, Asturias, Spain. *Proceedings of the National Academy of Sciences, 103*(51), 19266–19271.

Sasaki, K., & Kanazawa, E. (2000). Morphological traits on the dentino-enamel junction of lower deciduous molar series. In J. Mayhall, & T. Heikkinen (Eds.), *1998: Proceedings of the 11th International Symposium on Dental Morphology* (pp. 167–178). Oulu: Oulo University Press.

Skinner, M. F., & Sperber, G. H. (1982). *Atlas of radiographs of early man*. New York: Alan R. Liss.

Smith, P. (1990). Specialised features of Neanderthal teeth and the Latium Neanderthals. *Quaternaria Nova, 1*, 663–671.

Smith, F. H., Jankovic, I., & Karavanic, I. (2005). The assimilation model, modern human origins in Europe, and the extinction of Neandertals. *Quaternary International, 137*, 7–19.

Smith, P., Peretz, B., & Forte-Koren, R. (1995). The ontogeny of cusp morphology: The evidence from tooth germs. In R. Radlanski & H. Renz (Eds.), *Proceedings of the 10th international symposium on dental morphology* (pp. 49–53). Berlin: "M" Marketing Services.

Smith, P., Gomorri, J. M., Spitz, S., & Becker, J. (1997). Model for the examination of evolutionary trends in tooth development. *American Journal of Physical Anthropology, 102*(2), 283–294.

Smith, P., Gomori, J., Shaked, R., Haydenblit, R., & Joskowicz, L. (2000). A computerized approach to reconstruction of growth patterns in hominid molar teeth. In J. Mayhall & T. Heikkinen (Eds.), *Proceedings of the 11th international symposium on dental morphology* (pp. 388–397). Oulo: Oulo University Press.

Smith, P., Muller, R., Gabet, Y., & Avishai, G. (2007). A computerized model for reconstruction of dental ontogeny: A new tool for studying evolutionary trends in the dentition. In S. E. Bailey & J. J. Hublin (Eds.), *Dental perspectives on human evolution* (pp. 273–286). AA Dordrecht, The Netherlands: Springer.

Soficaru, A., Dobos, A., & Trinkaus, E. (2006). Early modern humans from the Pestera Muierii, Baia de Fier, Romania. *Proceedings of the National Academy of Sciences, 103*(46), 17196–17201.

Stringer, C. B. (2002). Modern human origins – progress and prospects. *Philosophical Transactions of the Royal Society, 357*, 563–579.

Stringer, C., Pälike, H., van Andel, T., Huntley, B., Valdes, P., & Allen, J. R. M. (2004). Climatic stress and the extinction of the Neanderthals. In T. H. van Andel, & W. Davies (Eds.), *Neanderthals and modern humans in the European landscape during the last glaciation* (pp. 233–240). Cambridge: McDonald Institute Monographs. chapter 13.

Suwa, G., & Kono, R. T. (2005). A micro-CT based study of linear enamel thickness in the mesial cusp section of human molars: Reevaluation of methodology and assessment of within-tooth, serial and individual variation. *Anthropological Science, 113*, 273–289.

Swindler, D. R., & McCoy, H. A. (1965). Primate odontogenesis. *Journal of Dental Research, 44*(SUPPL), 283–295.

Swindler, D. R., & Meekins, D. (1991). Dental development of the permanent mandibular teeth in the Baboon, *Papio cynocephalus*. *American Journal of Human Biology, 3*, 571–580.

Swindler, D. R., Orlosky, F. J., & Hendrickx, A. G. (1968). Calcification of the deciduous molars in baboons (*Papio anubis*) and other primates. *Journal of Dental Research, 47*(1), 167–170.

Thompson, J. L., & Nelson, A. J. (2000). The place of Neandertals in the evolution of hominid patterns of growth and development. *Journal of Human Evolution, 38*(4), 475–495.

Trinkaus, E., Moldovan, O., Milota, S., Bilgar, A., Sarcina, L., Athreya, S., Bailey, S. E., Rodrigo, R., Mircea, G., Higham, T., Ramsey, C. B., & van der Plicht, J. (2003). An early modern human from the Pestera cu Oase, Romania. *Proceedings of the National Academy of Sciences, 100*(20), 11231–11236.

Weiss, K. M., & Smith, F. H. (2007). Out of the veil of death rode the one million! Neandertals and their genes. *Bioessays, 29*(2), 105–110.

Winkler, L. A., Schwartz, J. H., & Swindler, D. R. (1996). Development of the orangutan permanent dentition: Assessing patterns and variation in tooth development. *American Journal of Physical Anthropology, 99*(1), 205–220.

Wolpoff, M. H., Hawks, J., Frayer, D. W., & Hunley, K. (2001). Modern human ancestry at the peripheries: A test of the replacement theory. *Science, 291*(5502), 293–297.

Zaho, Z., Weiss, K. M., & Stock, D. W. (2000). Development and evolution of dentition patterns and their genetic basis. In M. F. Teaford, M. M. Smith, & M. W. J. Ferguson (Eds.), *Development function and evolution of teeth* (pp. 152–172). Cambridge: Cambridge University Press.

Zilberman, U., & Smith, P. (1992). A comparison of tooth structure in Neanderthals and early *Homo sapiens sapiens*: A radiographic study. *Journal of Anatomy, 180*(Pt 3), 387–393.

Zilberman, U., Skinner, M., & Smith, P. (1992). Tooth components of mandibular deciduous molars of *Homo sapiens sapiens* and *Homo sapiens neanderthalensis*: A radiographic study. *American Journal of Physical Anthropology, 87*(3), 255–262.

Chapter 15
Endostructural Characterisation of the Regourdou 1 Neanderthal Proximal Arm: Bilateral Asymmetry and Handedness

Virginie Volpato, Christine Couture, Roberto Macchiarelli, and Bernard Vandermeersch

Abstract Cross-sectional properties of long bones are commonly used for reconstructing mechanical loading histories related to locomotion, subsistence strategies, manipulative behavior. In this respect, a significant degree of functional bilateral asymmetry of the proximal arm, likely related to unilateral activity levels, has been reported for Neanderthals.

Previous external analysis of the upper limb bones of the Regourdou 1 adult partial skeleton from Montignac-sur-Vézère, France, showed right side hypertrophy (Vandermeersch and Trinkaus 1995). In order to precise the amount of functional asymmetry and handedness characterizing this individual, we have investigated at 35%, 44%, 50%, and 65% of the shaft length the cross-sectional geometric properties of its humeri by means of synchrotron radiation microtomography (SR-μCT).

Present results support previous conclusion that Regourdou 1 was right-handed. Nonetheless, while a greater strength to compression, flexion, and torsion is shown by the right humerus, only a modest degree of right dominance characterizes this individual with respect to the available Neanderthal figures. Interestingly, the high-resolution three-dimensional-mapping of the humeral cortical bone volume documents a heterogeneous topographic pattern of structural asymmetry along the shaft.

V. Volpato (✉)
Abteilung Paläoanthropologie und Messelforschung, Forschungsinstitut und Naturmuseum, Senckenberganlage 25, 60325 Frankfurt am Main, Germany
e-mail: vvolpato@senckenberg.de

C. Couture and B. Vandermeersch
Lab. d'Anthropologie des Populations du Passé, UMR 5199 CNRS, Université Bordeaux 1, 33405 Talence Cedex, France
e-mail: c.couture@pacea.u-bordeaux1.fr; bvanderm@bio.ucm.es

R. Macchiarelli
Dép. de Préhistoire, UMR 5198, MNHN, 1, rue René Panhard, 75013 Paris, France
and
Dép. de Géosciences, Université de Poitiers, 40 av. du Recteur Pineau, 86022 Poitiers, France
e-mail: roberto.macchiarelli@univ-poitiers.fr

Keywords Humerus • Cross-sectional properties • Cortical thickness • Cortical volume • 3D topography

Introduction

Cross-sectional properties of long bone shaft are commonly used in (paleo)anthropology for reconstructing mechanical loading histories and patterns of activity related to locomotion, subsistence strategies, manipulative behaviors (Ruff et al. 1993, 1994; Trinkaus et al. 1994; Churchill and Schmitt 2002; Sládek et al. 2006; Stock 2006).

Available evidence from the fossil record shows that substantial functionally-related structural changes in diaphyseal morphology of the upper limb occurred through the Pleistocene (Ruff et al. 1993). Compared to the condition shown by early anatomically modern humans, the Neanderthal arm is characterized by a degree of robusticity suggesting higher levels of habitual manipulative loads (Trinkaus 1996; Churchill and Schmitt 2002). Likely related to unilateral activity levels and preferences, the humerus also evidences marked bilateral asymmetry, with usual right side hypertrophy (Trinkaus et al. 1994; Trinkaus 2000). In this respect, Neanderthals approximate the pattern of pronounced functional asymmetry currently shown by professional tennis-players (Trinkaus et al. 1994).

Regourdou 1 is a Neanderthal partial skeleton discovered in 1957 at Montignac-sur-Vézère, Dordogne, France (Bonifay and Vandermeersch 1962; Piveteau 1963–1965; Bonifay 1964). This young adult individual of undetermined sex preserves skeletal elements from both upper limbs suitable for accurate morpho-structural comparative analysis and functional interpretation (Vandermeersch and Trinkaus 1995). Nonetheless, while the right humerus is perfectly preserved, the left one is incomplete, lacking the head and consisting of two non-joining diaphyseal portions (the distal one not described by Vandermeersch and Trinkaus 1995).

Previous tomographic (Senut 1985) and radiographic investigation (Trinkaus et al. 1994) of the Regourdou 1 right humerus showed that its structural morphology fits the

Vertebrate Paleobiology and Paleoanthropology, DOI 10.1007/978-94-007-0492-3_15,

typical Neanderthal pattern, with marked reduction of the medullary area towards the distal third of the diaphysis. Additionally, the comparative analysis of the external morphology of the right and left upper limb bones revealed diaphyseal asymmetry, with right dominance for most dimensions (Vandermeersch and Trinkaus 1995).

In order to precise the polarity (right vs left side dominance) and the amount of functional asymmetry and handedness characterizing this individual, here we report the results of a high-resolution non-invasive analysis of the endostructural organisation and cross-sectional geometry of its humeri.

Methods

The bi- three-dimensional analyses of the Regourdou 1 humeri are based on a synchrotron radiation microtomographic record (SR-μCT) performed at the beamline ID 17 set at the European Synchrotron Radiation Facility, Grenoble, France (www.esrf.fr/UsersAndScience/Experiments/Imaging/ID17/; experiments SC-1587 and SC-1749).

The ESRF microtomographic system is characterized by continuous energy spectrum, high photon flux, intense monochromatic X-ray beam, nearly parallel projections, and small angular source size. Monoenergetic X-ray beams enable absolute linear attenuation coefficients to be measured and avoid the risk of beam hardening artefacts in the reconstruction of images from dense specimens such as mineralised fossils (Macchiarelli et al. 2006, 2007; Mazurier et al. 2006).

Scans of the two investigated specimens were performed at energy of 51 keV. Projections were taken each 0.35° (1,024/360°). Final sections have been reconstructed from sinograms and saved on a 32 bit floating point raw format at a voxel size of 350×350×350 μm/pixels. The final 8 bits volumes were elaborated by means of AMIRA v4.0 package (Mercury Computer Systems, Inc.).

On digital cross-sections taken at 35%, 44%, 50%, and 65% of the biomechanical humeral length (Fig. 15.1), the following parameters have been quantitatively measured/calculated: total area (TA, in mm^2); cortical area (CA, in mm^2); medullary area (MA, in mm^2); second moment of area about m-l axis (Ix, in mm^4); second moment of area about a–p axis (Iy, in mm^4); maximum second moment of area (Imax, in mm^4); minimum second moment of area (Imin, in mm^4); orientation of greatest bending rigidity (Theta, in degrees); polar second moment of area (J, in mm^4); bilateral asymmetry ([(max–min)/min] × 100) of CA, MA, J, Imax/Imin (shape asymmetry) (Trinkaus et al. 1994).

Cortical area measures the amount of compact bone in the diaphysis and relates to the resistance of the shaft to axial loadings. Second moments of area reflect the bending and the torsional strengths of the shaft under loading. The medullary cross-sectional area, which relates to CA, more directly reflects endosteal deposition/resorption patterns. As a whole, these variables measure the geometric contribution of cortical bone to diaphyseal strength (Trinkaus et al. 1994), while the cross-sectional shape of long bone shafts, expressed as ratio Imax/Imin, is used as barometers of activity patterns (Pearson 2001).

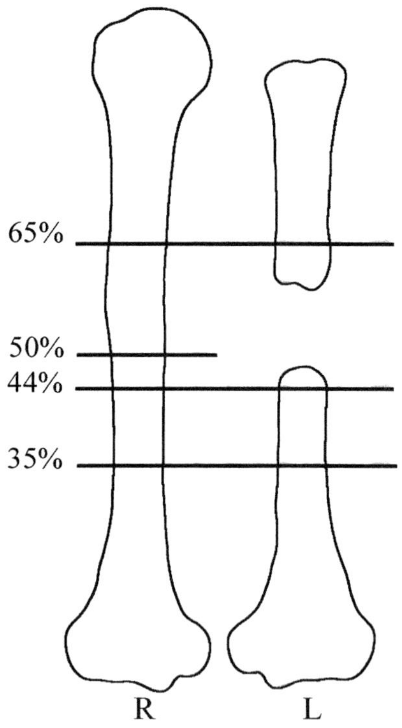

Fig. 15.1 Regourdou 1: sketch of the right (R) and left (L) humerus with indication of the cross-sections set at 35%, 44%, 50%, and 65% of the biomechanical length

Measurements of cross-sectional properties have been performed following Ruff (2002) running NIH Image by means of a Scion Image macro based on established geometric computations (www.hopkinsmedicine.org/FAE/mmacro.htm). Because of its incompleteness, Regourdou 1 left humerus has not been detailed at 50%.

The topographic variation of the cortical thickness between 20% and 80% of the diaphyseal length has been assessed following three-dimensional rendering and mapping. In the case of the right humerus (Fig. 15.2), the cortical volume (CV) has been quantified (in mm^3) for the distal and the proximal portions comprised between 35% and 50%, and 50–65%, respectively.

Results

Despite the incompleteness of the left humerus, bone endostructure is very well preserved in both specimens (Fig. 15.3), granting a high reliability of the digital measurements.

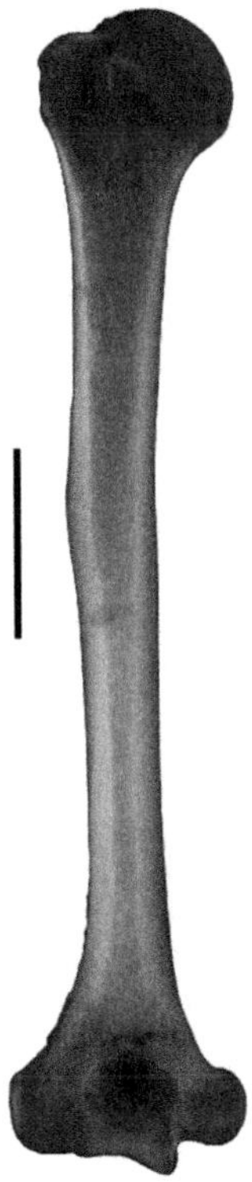

Fig. 15.2 SR-μCT-based 3D virtual modelling of the Regourdou 1 right humerus rendered in semi-transparency, showing thickening of the cortical bone towards the distal third of the diaphysis. Scale bar is 50 mm

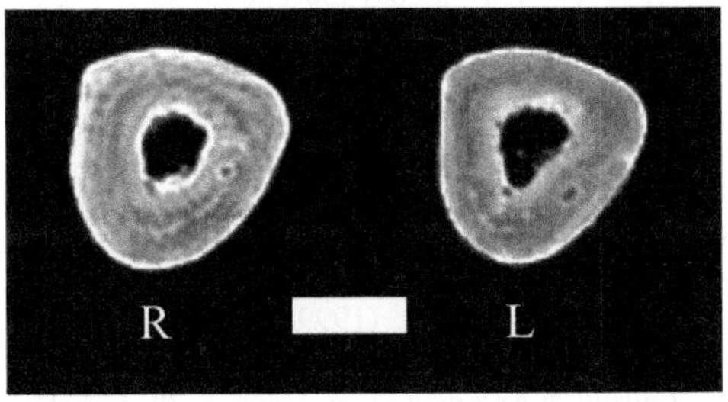

Fig. 15.3 Digital cross-sections of the right (R) and left (L) humerus at 35% of the biomechanical length. Scale bar is 10 mm

For all three comparable sections (35%, 44%, and 65%), the right humerus systematically shows the highest values for all investigated variables. The only exception is represented by the medullary area, which is larger on the left side at 35% (26.1 vs 21.3 mm^2; see Fig. 15.3) and at 44% (37 vs 30.3 mm^3).

In both humeri, the relatively thickest cortical bone is found distally (35–44%). At 35% of the biomechanical length, cortical bone occupies 90% and 93% of the total cross-sectional area on the left and right specimen, respectively. Similarly, the thinnest bone is found proximally (65%), the difference between the distal and the proximal portions reaching 11.6% on the left humerus.

As assessed on the shaft of a reference young adult humerus (nineteenth century, European origin) detailed by means of the same analytical tool (SR-μCT), the modern human condition is characterized by an absolutely and relatively thinner cortical bone distributed rather homogeneously through the shaft, while both Neanderthal and modern specimens share a topographic distribution pattern of volumetric thickening towards the distal end (Senut 1985; for additional comparative data on Middle and Upper Palaeolithic humans and living urban populations, see Trinkaus and Churchill 1999; Trinkaus 2006).

Table 15.1 Measures of bilateral asymmetry of the Regourdou 1 humerus assessed at 35%, 44%, and 65% of its diaphyseal biomechanical length

	CA	MA	J	Imax/Imin
35%	9.7	(–)18.3	14.6	1.9
44%	11.1	(–)18.2	16.1	4.0
65%	14.5	14.4	32.3	23.6

See "Methods" for the meaning of the variables. (–) indicating a left dominance

Percent cortical bone volume (%CV) measured on the Regourdou 1 right humerus for the portion 35–65% equals 83.15%, while a value of 50.45% is obtained for the modern reference specimen.

In Regourdou 1, the diaphyseal section is systematically more circular distally (Imax/Imin). As usually seen in modern humeri, the second moment of area (principal and anatomical axes) and the polar second moment of area (J) decrease distally on both Neanderthal specimens.

With the exception of the medullary area assessed at 35% and 44%, measures of bilateral asymmetry estimated for the variables CA, Imax/Imin, and J indicate right arm dominance in Regourdou 1, being the distal third of the humeral shaft (35%) the least asymmetrical portion (Table 15.1). The most asymmetrical value has been recorded for the polar second moment of area (32.3% at the level of proximal section). Notably, with respect to its counterside, the three-dimensional rendering of the cortical bone topography of the right humerus uniquely evidences a pattern of absolutely thicker bone located on the anterior and posterior aspects of its distal third, likely in relation to the insertion areas of the muscles *brachialis* and *triceps brachii*.

Discussion and Conclusions

On both humeri from the Regourdou 1 Neanderthal partial skeleton, the middle-upper shaft (65% to 44%) is characterized by a greater strength to bending in the mediolateral plane (even if the left side is incomplete), while the distal one (44% to 35%) shows a greater bending rigidity in the anteroposterior plane. As a whole, diaphyseal geometric properties of the shafts indicate a greater strength of the right proximal arm to the compressive axial loads, particularly on the distal part, and a greater rigidity to bending and torsion on the proximal portion. Accordingly, present evidence suggests a differential muscular development and mechanical loading pattern for this side, likely related to peculiar and/or more intensive manipulative activities (cf. Schmitt et al. 2003).

The bi-3D elaborations of the high-resolution microtomographic record of the two humeri support previous conclusion that Regourdou 1 was right-handed (Vandermeersch and Trinkaus 1995). Nonetheless, compared to the currently available Neanderthal figures (which are mostly radiographically-based), this individual displays only a modest degree of right dominance, close to the minimum values of the Neanderthal variation range reported so far (for the cortical area measured at 35%, Regourdou 1 shows 9.7% vs 8.4–45.9% for the Neanderthal whole sample; Trinkaus et al. 1994). As suggested by the ongoing high-resolution volumetric analysis of the endostructural morphology of the forearm bones from the same individual (for a preliminary assessment, see Volpato et al. 2005; Volpato 2007), this characteristics of Regourdou 1 may be related to sex variation in Neanderthals (Ben-Itzhak et al. 1988), even it should be noted that the least asymmetric among the Neanderthals is the male Kebara 2 (Trinkaus et al. 1994).

Acknowledgments Research developed within the EC TNT project (http://www.the-neanderthal-tools.org) in collaboration with the Musée d'Art et d'Archéologie du Périgord (Périgueux, France) and the European Synchrotron Radiation Facility (ESRF, Grenoble, France), supported by the French CNRS, the GDR 2152, the Région Poitou-Charentes. Special thanks to A. Bergeret, L. Bondioli, A. Bravin, F. Couturas, G. Marchesseau, V. Merlin-Anglade, A. Mazurier, C. Nemoz, C. Ruff, P. Tafforeau, G.-C. Weniger.

References

Ben-Itzhak, S., Smith, P., & Bloom, R. A. (1988). Radiographic study of the humerus in Neandertals and *Homo sapiens*. *American Journal of Physical Anthropology, 77*, 231–242.

Bonifay, E. (1964). La grotte du Régourdou (Montignac, Dordogne) Stratigraphie et industrie lithique moustérienne. *L'Anthropologie, 68*, 49–64.

Bonifay, E., & Vandermeersch, B. (1962). Dépôts rituels d'ossements d'ours dans le gisement moustérien de Régourdou (Montignac, Dordogne). *Comptes Rendus de l'Académie des Sciences, 255D*, 1635–1636.

Churchill, S. E., & Schmitt, D. (2002). Biomechanics in palaeoanthropology: Engineering and experimental approaches to the study of behavioural evolution in the genus. In C. S. Harcourt & B. R. Sherwood (Eds.), *New perspectives in primates evolution and behaviour* (pp. 71–102). Otley: Westbury Academic & Scientific Publishing.

Macchiarelli, R., Bondioli, L., Debénath, A., Mazurier, A., Tournepiche, J.-F., Birch, W., & Dean, C. (2006). How Neanderthal molar teeth grew. *Nature, 444*, 748–751.

Macchiarelli, R., Mazurier, A., & Volpato, V. (2007). L'apport des nouvelles technologies à l'étude des Néandertaliens. In B. Vandermeersch (Ed.), *Les Néandertaliens* (pp. 169–179). Paris: Comité des Travaux Historiques et Scientifiques.

Mazurier, A., Volpato, V., & Macchiarelli, R. (2006). Improved non-invasive microstructural analysis of fossil tissues by means of SR-microtomography. *Applied Physics A: Materials Science & Processing, 83*, 229–233.

Pearson, O. M. (2001). Postcranial remains and the origin of modern humans. *Evolutionary Anthropology, 9*, 229–247.

Piveteau, J. (1963–1965). La grotte du Régourdou (Dordogne), *paléontologie humaine. Annales de Paléontologie* (Vertébrés) *49*, 285–305; *50*, 155–194; *52*, 163–194.

Ruff, C. B. (2002). Long bone articular and diaphyseal structure in old world monkeys and apes I: Locomotor effects. *American Journal of Physical Anthropology, 119*, 305–342.

Ruff, C. B., Trinkaus, E., Walker, A., & Larsen, C. S. (1993). Postcranial robusticity in Homo I: Temporal trends and mechanical interpretation. *American Journal of Physical Anthropology, 91*, 21–53.

Ruff, C. B., Walker, A., & Trinkaus, E. (1994). Postcranial robusticity in Homo III: Ontogeny. *American Journal of Physical Anthropology, 93*, 35–54.

Schmitt, D., Churchill, S. E., & Hylander, W. L. (2003). Experimental evidence concerning spear use in Neandertals and early modern humans. *Journal of Archaeological Science, 30*, 103–114.

Senut, B. (1985). Computerized tomography of a Neanderthal humerus from Le Regourdou (Dordogne, France): Comparisons with modern man. *Journal of Human Evolution, 14*, 717–723.

Sládek, V., Berner, M., & Sailer, R. (2006). Mobility in central European Late Eneolithic and Early Bronze Age: Femoral cross-sectional geometry. *American Journal of Physical Anthropology, 130*, 320–332.

Stock, J. T. (2006). Hunter-gatherer postcranial robusticity relative to patterns of mobility, climatic adaptation, and selection for tissue economy. *American Journal of Physical Anthropology, 131*, 194–204.

Trinkaus, E. (1996). The humerus versus the femur: Changing patterns of diaphyseal robusticity across the late archaic to early modern human transition. *L'Anthropologie, 34*(3), 239–247.

Trinkaus, E. (2000). The "robusticity transition" revisisted. In C. Stringer, R. N. E. Barton, & C. Finlayson (Eds.), *Neanderthals on the Edge* (pp. 227–236). Oxford: Oxbow Books.

Trinkaus, E. (2006). The upper limb remains. In E. Trinkaus & J. A. Svoboda (Eds.), *Early modern human evolution in Central Europe* (pp. 327–372). Oxford: Oxford University press.

Trinkaus, E., & Churchill, S. E. (1999). Diaphyseal cross-sectional geometry of Near Eastern Middle Palaeolithic humans: The humerus. *Journal of Archaeological Science, 26*, 173–184.

Trinkaus, E., Churchill, S. E., & Ruff, C. B. (1994). Postcranial robusticity in Homo II: Humeral bilateral asymmetry and bone plasticity. *American Journal of Physical Anthropology, 93*, 1–34.

Vandermeersch, B., & Trinkaus, E. (1995). The postcranial remains of the Regourdou 1 Neandertal: The shoulder and arm remains. *Journal of Human Evolution, 28*, 439–476.

Volpato, V. (2007). Morphogenèse des propriétés texturales du tissu osseux et environnement biomécanique. *Caractérisation non invasive du réseau trabéculaire et de l'os cortical du squelette appendiculaire de Mammifères actuels et fossiles, Hominidés inclus*. Ph.D. dissertation, University of Poitiers, France.

Volpato, V., Couture, C., Vandermeersch, B., & Macchiarelli, R. (2005). Cortical structural geometry of the Regourdou 1 Neandertal upper limb: Bilateral asymmetry and handedness. *Bulletins et Mémoires de la Société d'Anthropologie de Paris, 17*, 288–289.

Chapter 16
A Three-Dimensional Look at the Neanderthal Mandible

Katerina Harvati, Nandini Singh, and Elisabeth Nicholson López

Abstract Mandibular traits that differentiate Neanderthals from modern humans include greater robusticity, a receding symphysis, a large retromolar space, a rounder gonial area, an asymmetric mandibular notch and a posteriorly positioned mental foramen in Neanderthals. These features have been shown to separate Neanderthals from modern humans in both non-metric and metric, including 3-D geometric morphometric, studies. However the degree to which these differences are related to size and function is still under discussion. The aim of this study is to further assess the effects of allometry and evaluate the influence of masticatory and paramasticatory activities on mandibular shape using a morphological integration approach. Data were collected in the form of three-dimensional coordinates of 27 landmarks, superimposed using generalized Procrustes analysis, and analyzed using canonical variates, regression and partial least squares analyses. Consistent with previous findings, Neanderthals are separated from modern human mandibles in the canonical variates analysis. However, partial least squares analysis indicates a similar pattern of integration for the two human groups, suggesting homology across modern humans and Neanderthals in the mandibular features examined. This finding does not support a paramasticatory hypothesis for Neanderthal mandibular shape, although it also does not refute this hypothesis. Finally, allometry was found to influence the expression of the retromolar gap.

Keywords Mandibular morphology • Variation • Integration • Function • Phylogeny • Neanderthals • Geometric morphometrics

K. Harvati (✉) and N. Singh
Department of Early Prehistory and Quaternary Ecology, Senckenberg Center for Human Evolution and Paleoecology, Eberhard Karls Universität Tübingen, Rümelinstrasse 23, 72070 Tübingen, Germany
e-mail: katerina.harvati@ifu.uni-tuebingen.de; singhnandini@gmail.com

E. Nicholson López
Northwestern University, Chicago, IL 60611-3008, USA
e-mail: lisaknlopez@gmail.com

Introduction

Differences between Neanderthal and modern human mandibular morphology have been extensively documented and include both archaic and presumed derived characters. Primitive features retained in Neanderthal mandibles include overall robusticity and a receding symphysis which results in the absence of a mental eminence or chin. Proposed derived traits include a posteriorly positioned mental foramen; a large retromolar space; a rounder, less angled gonial area; a mandibular notch that meets the condyle in a more medial position, resulting in a laterally expanded condyle; a shallow and asymmetric mandibular notch with a higher coronoid process than condyle; very deep submandibular and pterygoid fossae; and an oval-horizontal shape of the mandibular foramen. These features have been shown to successfully discriminate Neanderthals from modern human in both metric and non-metric analyses (Fig. 16.1; Boule 1911–1913; Coon 1962; Vandermeersch 1981; Stringer et al. 1984; Tillier et al. 1989; Condemi 1991; Rosas 1992, 2001; Franciscus and Trinkaus 1995; Creed-Miles et al. 1996; Arensburg and Belfer-Cohen 1998; Hublin 1998; Rak 1998; Jabbour et al. 2002; Rak et al. 2002; Trinkaus et al. 2003; Nicholson and Harvati 2006). However, the degree to which Neanderthal mandibular morphology is influenced by allometry and by masticatory adaptations is still a matter of discussion (see e.g. Humphrey et al. 1999; Nicholson and Harvati 2006).

Allometry

Recent studies have evaluated the role of size and allometry in producing Neanderthal-like mandibular morphology. Rosas and Bastir (2002) explored the allometric trends in the human craniofacial complex using 2D geometric morphometric techniques. They found that size-related variation in the mandible among modern humans occurs mainly in the vertical dimensions of the ramus. In a more recent study using the same methods, these authors (Rosas and Bastir 2004) evaluated the

S. Condemi and G.-C. Weniger (eds.), *Continuity and Discontinuity in the Peopling of Europe: One Hundred Fifty Years of Neanderthal Study*, Vertebrate Paleobiology and Paleoanthropology, DOI 10.1007/978-94-007-0492-3_16,

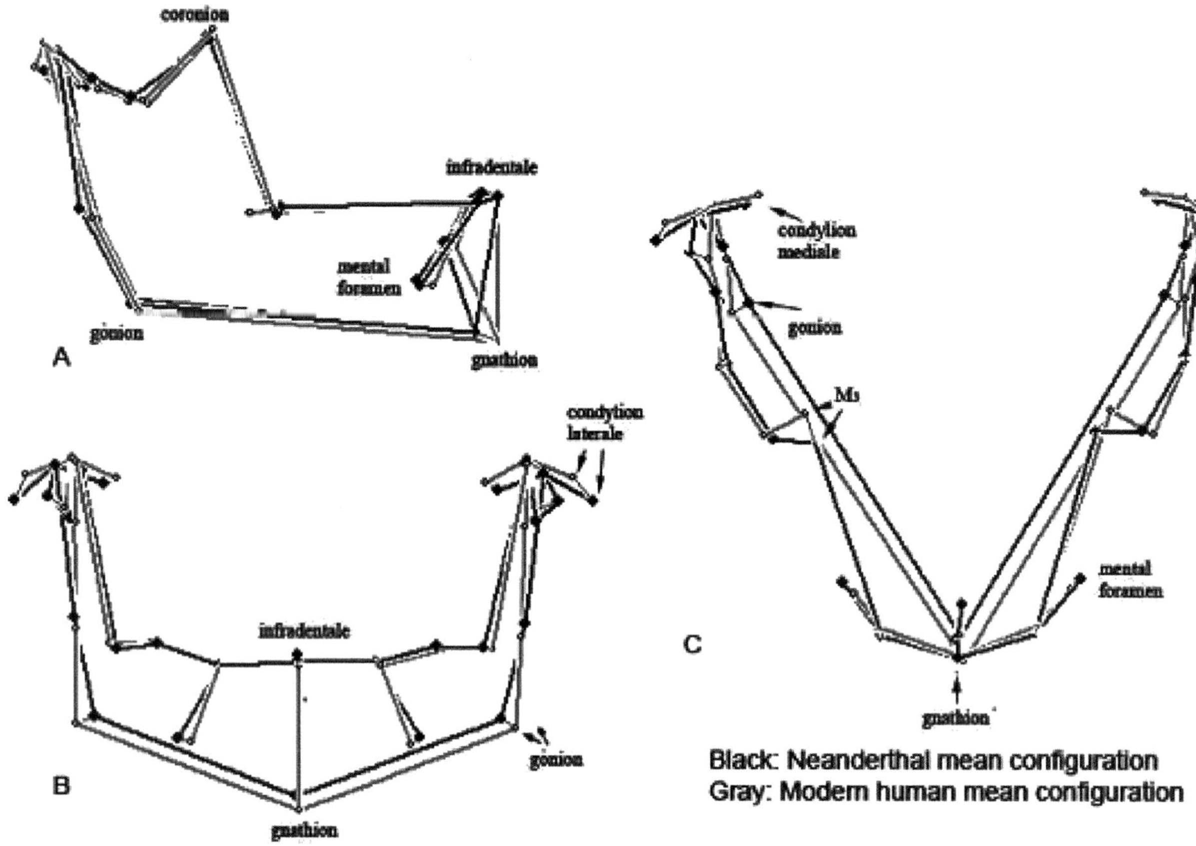

Fig. 16.1 Mean Neanderthal mandibular configuration superimposed with the mean modern human configuration (Adapted from Nicholson and Harvati 2006)

role of size in a large sample of hominoids, including modern and Pleistocene humans. They found a trend for greater retromolar spaces with increased mandibular size in Neanderthals, *H. heidelbergensis*, chimpanzees, gorillas, and modern humans, suggesting that this feature may not necessarily indicate Neanderthal affinities. This result agrees with those of Nicholson and Harvati (2006), who, using 3D geometric morphometric techniques, found that increased mandibular size in modern humans does not produce Neanderthal-like morphology with the exception of the retromolar space. All of these studies used regression of shape on centroid size to evaluate allometric shape differences. Recently, however, the logarithm of centroid size has been shown to be a more appropriate size variable for use in such analyses (see Mitteroecker et al. 2004), thus throwing these results into question.

Biomechanics

Several biomechanical hypotheses have been proposed to explain the combination of features exhibited by Neanderthal cranio-mandibular form (Linderholm and Wennstrom 1970; Smith 1983; Rak 1986; Demes 1987; Trinkaus 1987; Smith and Paquette 1989; Antón 1990, 1994a, 1996a; Spencer and Demes 1993; Lieberman et al. 2004; O'Connor et al. 2005). A number of these researchers have linked Neanderthal facial and mandibular morphology to paramasticatory adaptations.

Excessive attrition on the anterior dentition, differential occlusal wear patterns, along with the presence of degenerative disease in the temporo-mandibular joint of Neanderthals was attributed to heavy loading on the anterior dentition and commonly referred to as the "anterior dental loading hypothesis" (hereafter ADLH; Smith 1983; Rak 1986). Neanderthal facial morphology was therefore hypothesized to be an adaptive response to heavy paramasticatory activities, such as using "teeth-as-tools" (Brace et al. 1981; Smith 1983). An assumption of the ADLH is that those activities generated high-magnitude forces on the anterior dentition, leading to extensive occlusal loading and wear on the anterior teeth (Rak 1986; Demes 1987; Smith and Paquette 1989).

To test some of the assumptions made by the ADLH, Spencer and Demes (1993) conducted a quantitative study of the Neanderthal masticatory system. Their approach included pair-wise comparisons of Neanderthal facial morphology with that of Inuit and other Native American modern humans. Because the Inuit are known to produce intense incisal bite force, they served as a good comparative model for Neanderthals (Balicki 1970; Hylander 1977). Spencer and Demes' biomechanical analysis of bite-force production efficiency showed that Neanderthals share a number of morphological similarities with the Inuit that are possibly related to extensive usage of the anterior teeth. Moreover, pair-wise comparisons between Neanderthals and Middle Pleistocene hominins showed that Neanderthal anterior dental architecture was better adapted to repeated and high-level usage than other Pleistocene hominins'. Based on their findings, these authors concluded that the Neanderthal face was well-designed for and capable of high-force production. However, their study was based entirely on the face and did not include elements of the mandible.

Conversely, others have suggested that the facial morphology of Neanderthals lacked the ability to produce high anterior bite forces. According to Trinkaus (1987), large anterior dentition was not related to intensified usage of incisor teeth in Neanderthals. Their overall reduction in facial robusticity, posterior placement of the zygomatico-ramal region and anterior placement of the dentition further implied a lack of ability to produce high-level masticatory loads. According to Trinkaus (1987), this particular combination of posteriorly placed masticatory muscles and mid-facial prognathism reduced the potential of generating heavy bite forces in Neanderthals, thus refuting the assumptions made by the ADLH that Neanderthal cranio-mandibular morphology was mechanically advantageous for anterior tooth use.

This interpretation was supported by the work of Antón (1990, 1994a). She conducted a quantitative study on the amount and capability of bite force generated by Neanderthals to evaluate the extent to which these biomechanical factors contributed to the evolution of the Neanderthal face. Based on low estimates of bite force production and occlusal loading, Antón concluded that Neanderthals were less capable of and efficient at generating high-magnitude bite force than the modern humans in her comparative sample. In other words, she rejected the ADLH hypothesis as an explanation for the evolution of Neanderthal cranio-mandibular structure.

More recently, O'Connor et al. (2005) conducted a more comprehensive study on bite force capability and efficiency in Neanderthals and modern humans. Their overall results indicated that masticatory biomechanical explanations such as "bite force dissipation" do not adequately account for the underlying mechanisms driving Neanderthal facial and mandibular form, concurring with Trinkaus (1987) and Antón (1990, 1994a). However, their results differed from Antón's (1990, 1994a) in that they found less of a difference in force production efficiency between Neanderthals and modern humans than indicated by her study. Their findings further suggested that size did not affect force-production "efficiency" of the cranio-mandibular system, although force production "capability" was significantly correlated with overall size.

The majority of these studies have focused on the effects of biomechanical forces in the evolution of overall facial architecture in Neanderthals. And even though they reached diverging conclusions on the degree of impact of biomechanical factors on Neanderthal cranio-mandibular morphology, a point of agreement among the studies was the excessive and repetitive usage of the anterior dentition by Neanderthals.

This study does not propose to test biomechanical hypotheses contributing to Neanderthal mandibular features. Instead, it aims to examine whether increased dental attrition found in Neanderthals impacts the integration of the anterior and posterior regions of the mandible and, consequently, contributes to the shape variation seen between modern human and Neanderthal mandibles. Studies on mandibular patterns of integration are few and primarily on aspects of the mouse mandible (Cheverud et al. 1991; Leamy 1993; Klingenberg et al. 2003) with the exception of Bastir et al. (2005). Morphological integration is generally defined as the coordinated variation between units of a phenotype (Olson and Miller 1958; Klingenberg et al. 2001a). The pattern and degree of integration among anatomical units or subsets is correlated with the degree of developmental and functional relatedness among those subsets. For example, a subset of traits related due to masticatory activity is characterized as a functional unit.

So far no work has been done exclusively on morphological integration in the mandible of Pleistocene hominins and modern humans. Though exploring integration patterns, Bastir et al. (2005) included aspects of the cranium, with the primary objective of determining the degree of morphological integration between the cranium and mandible.

Extensive research on murine mandibles has contributed greatly to our general understanding of mandibular integration patterns and allows for further testing of similar functional hypotheses in a paleoanthropological context. Previous studies have shown that the alveolar (tooth bearing corpus) and ascending ramus are two key regions of variation in the mandible (Atchley and Hall 1991). Examining mandibular patterns of integration in modern humans and Neanderthals is a novel way of investigating the underlying processes that generate morphological variability in modern and fossil human mandibles.

Our Goals

The goals of the present study are: (1) To further explore the effect of size and allometry on modern human and Neanderthal mandibular shape using a size-shape analysis. Only the modern human sample was used in this analysis, as Neanderthals differ from modern humans both in shape and in their greater size. (2) To evaluate the influence of masticatory and paramasticatory behaviours on mandibular shape using a morphological integration approach to understanding the shape differences between Neanderthals and modern humans. By dividing the mandible into the anterior and posterior regions, we hypothesized that given the repetitive usage of the anterior dentition in Neanderthals, the pattern and degree of integration between the alveolar region and ascending ramus of Neanderthals will be different from other Pleistocene and modern humans.

Materials and Methods

Samples

Our modern human sample is a subset of that used by Nicholson and Harvati (2006) and comprises 141 mandibles (Table 16.1) from 10 relatively broad geographic regions These samples do not represent biological populations in the sense of demes. The fossil sample comprises two Middle Pleistocene European specimens, seven Neanderthals, seven Upper Paleolithic specimens from Europe and Asia, and two Late Pleistocene early anatomically moderns (Table 16.2). Where original specimens were unavailable for study, casts were measured from the collections of the Departments of Anthropology of the American Museum of Natural History and New York University, and of the Department of Human Evolution of the Max Plank Institute for Evolutionary Anthropology.

Table 16.1 Modern human samples from the American Museum of Natural History included in this study

Population	Specimens
Total	141
Oceania (Australia, New Guinea, and Tasmania)	18
Polynesia	18
Southeast Asia (Southeast Asia and China)	14
North Asia (Japan, Korea, Siberia, and Mongolia)	13
East Africa (Masai)	14
South Africa (Khoisan, Bantu)	8
Europe	26
South America	11
Central America (Central America and Mexico)	9
North America Arctic (Alaska, Greenland, and Northern Canada)	10

Specimens were labeled by species and population, with *H. neanderthalensis* and *H. heidelbergensis* assigned to two distinct species, each separate from *H. sapiens*. Upper Paleolithic humans were treated as a population of *H. sapiens*, as were the early anatomically modern humans. Only adult specimens, as determined by a fully erupted permanent dentition, and only mandibles preserving all 27 landmarks, were included in this study, limiting sample sizes for both recent and fossil groups. Sex was known only for few specimens in each sample, making an assessment of sexual dimorphism in mandibular shape impossible. Although the mandible is sexually dimorphic (see e.g. Morant et al. 1936; Martin 1936; Hrdlička 1940a, b; De Villiers 1968a, b; Hunter and Garn 1972; Humphrey et al. 1999), here we pooled sexes in our analyses due to the lack of secure sex assignments for either recent or fossil specimens.

Data

Twenty-seven landmarks, represented by 81 three-dimensional coordinates and selected to represent the overall shape of the mandible, were collected with a Microscribe 3DX digitizer

Table 16.2 Fossil sample used in this study[a]

Specimen	Location	Museum	Collected by	Species	Population
Mauer 1[b]	Germany	AMNH	EN	H	H
Arago 13[b]	France	NYU	KH	H	H
Montmaurin	France	MH	KH	N	N
La Ferrassie 1	France	MH	KH	N	N
Zafarraya[b]	Spain	MPI	KH	N	N
Krapina J[b]	Croatia	AMNH	EN	N	N
Amud 1[b]	Israel	AMNH	EN	N	N
Tabun 1	Israel	NHML	KH	N	N
Shanidar 1[b]	Iraq	AMNH	EN	N	N
Chancelade[b]	France	NYU	KH	S	UP
Isturitz 1950–4–1	France	IPH	KH	S	UP
Oberkassel 2[b]	Germany	AMNH	EN	S	UP
Grimaldi-Grotte-des-Enfants 6	Italy	AMNH	EN	S	UP
Skhul 5	Israel	PEA	KH	S	EAM
Qafzeh 9[b]	Israel	MPI	KH	S	EAM

[a] *AMNH* American Museum of Natural History, *IPH* Institut de Paléontologie Humaine, *MH* Musée de l'Homme, *MPI* Max Planck Institute, Leipzig, *NHML* Natural History Museum London, *NYU* New York University, *PEA* Peabody Museum, Harvard. *N* Neanderthal, H *H. heidelbergensis*, S *H. sapiens*. Fossil *H. sapiens* were divided into two samples: *UP* Upper Paleolithic human, *EAM* Early Anatomically Modern Human

[b] Indicate casts from the collections of the AMNH, MPI and NYU

Table 16.3 Definitions of landmarks used. The first 12 landmarks were collected from both right and left sides

Landmark	Definition
1. Gonion	The point along the rounded posterioinferior corner of the mandible where the line bisecting the angle between the body and the ramus would hit
2. Posterior ramus	The point at the posterior margin of the ramus at the level of the M_3
3. Condyle tip	The most superior point on the mandibular condyle
4. Condylion mediale	The most medial point on the mandibular condyle
5. Condylion laterale	The most lateral point on the mandibular condyle
6. Root of sigmoid process	The point where the mandibular notch intersects the condyle
7. Mandibular notch	The most inferior point on the mandibular notch
8. Coronion	The most superior point on the coronoid process
9. Anterior ramus	The point at the anterior margin of the ramus at the level of the M_3
10. M_3	The point on the alveolar bone just posterior to the midline of the third molar
11. Mental foramen	The point in the middle of the mental foramen
12. Canine	The point on the alveolar margin between the canine and the first premolar
13. Gnathion	The most inferior midline point on the symphysis
14. Infradentale	The midline point at the superior tip of the septum between the mandibular central incisors
15. Mandibular orale	The most superior midline point on the lingual side of the mandible between the two central incisors

(Table 16.3). Three of these were located on the midsagittal plane; the others consisted of 12 pairs of homologous points on the left and right sides. All recent human specimens were measured by EN, as were most of the casts of fossil specimens used. Some casts and all the original fossils were measured by KH (for inter- and intra-observer error assessment see Nicholson and Harvati 2006).

Because morphometric analyses do not accommodate missing data, and because many of the fossil specimens were incomplete, some data reconstruction was allowed. During data collection for specimens with minimal damage, landmarks were reconstructed by estimating the position of the structure of interest using the morphology of the preserved surrounding areas. Missing landmarks were further reconstructed by reflecting the right and left sides of the specimen. Incomplete specimens were least-squares superimposed with their reflected equivalents using Morpheus (Slice 1994–1999), and missing data were reconstructed from their homologous counterparts on the other side. Further reconstruction by substituting sample means was permitted for a few fossil specimens missing one or two landmarks on both sides.

Methods

The landmark coordinates were superimposed using generalized Procrustes analysis (GPA) in Morphologika (O'Higgins and Jones 2006). GPA superimposes the specimens landmark configurations by translating them to common origin, scaling them to unit centroid size (the square root of the sum of squared distances of all landmarks to the centroid of the object; the measure of size used here), and rotating them according to a best-fit criterion. This procedure removes 'size' as a factor (although size-related shape differences may remain). 'Shape' and 'size' can therefore be analyzed separately (Rohlf and Marcus 1993; Slice 1996; O'Higgins and Jones 1998, 2006).

Size-Shape Principal Components Analysis

Here we evaluate the effect of size on mandibular morphology by performing a principal components analysis on the Procrustes aligned coordinates and logarithm of centroid size, a more appropriate proxy variable for size than centroid size (see Mitteroecker et al. 2004), using the software Morphologika (O'Higgins and Jones 2006). Because the Pleistocene fossil humans are differentiated from modern humans by their larger size in addition to their shape differences, we conducted this analysis on the modern human sample alone. In this way we were able to evaluate whether increased size in the modern human mandible results in Neanderthal-like morphology.

Partial-Least Squares Analysis (PLS)

For the purpose of the analysis, the mandibular landmarks were divided into two subsets *a priori*, representing the alveolar (tooth-bearing) and ascending ramus (attached to the skull and muscles of mastication); this sub-division is based on previous work on the development and morphological integration of mouse mandibles (e.g. Atchley and Hall 1991; Klingenberg et al. 2003). Prior to conducting the PLS analysis, a GPA was performed on the subset of landmarks for the anterior and posterior regions separately. This was done in order to reduce possible effects of extra correlations from the original Procrustes fit.

A two-block PLS analysis of shape variables was conducted to examine co-variation patterns between the anterior and posterior parts of the mandible. This method finds pairs of axes, one axis per block of variables, which successively account for the maximum amount of covariance between the two sets of variables examined. Each PLS axis in one block is only correlated to the corresponding axis in the other block, but not to the other PLS axes in that set. Therefore, the patterns of covariance can be analyzed one pair of PLS axes at a time (Bookstein et al. 2003; Klingenberg et al. 2003).

In addition, permutation tests were conducted to assess the statistical significance of the observed correlations between blocks.

Results

Allometry

As expected, the first principal component was very strongly related to variation in size (see Fig. 16.2). Visualization along this axis (Fig. 16.3) allowed for an evaluation of the presence of Neanderthal-like morphology in large modern human mandibles. Larger modern human mandibles are characterized by a medio-laterally wider and antero-posteriorly shorter overall mandibular shape; a supero-inferiorly higher ramus and symphysis; a antero-posteriorly wider ramus; and a lightly more laterally projecting gonion. The features commonly described as "Neanderthal-like" among modern humans refer to a more anterior placement of the distal margin of the M_3, which results in a morphology? similar to the Neanderthal retromolar space, and secondly, a coronoid process that is higher than the condyle. The latter trait is exhibited by Neanderthal mandibles, which, however, display several additional differences in this area from modern humans (i.e. a shallow, asymmetric notch, a laterally expanded condyle, a more medial placement of the root of the sigmoid notch; see Rak 1998; Jabbour et al. 2002; Rak et al. 2002; Nicholson and Harvati 2006). Our results very closely match those found previously by Rosas and Bastir (2002, 2004) and Nicholson and Harvati (2006).

Co-variation Between the Anterior and Posterior Regions

For the pooled modern human and fossil PLS analysis, the first axis accounts for 64.91% of the total co-variance (Fig. 16.4), with a correlation of $r=0.62$. The P-value of the permutation test is not statistically significant ($P=0.75$),

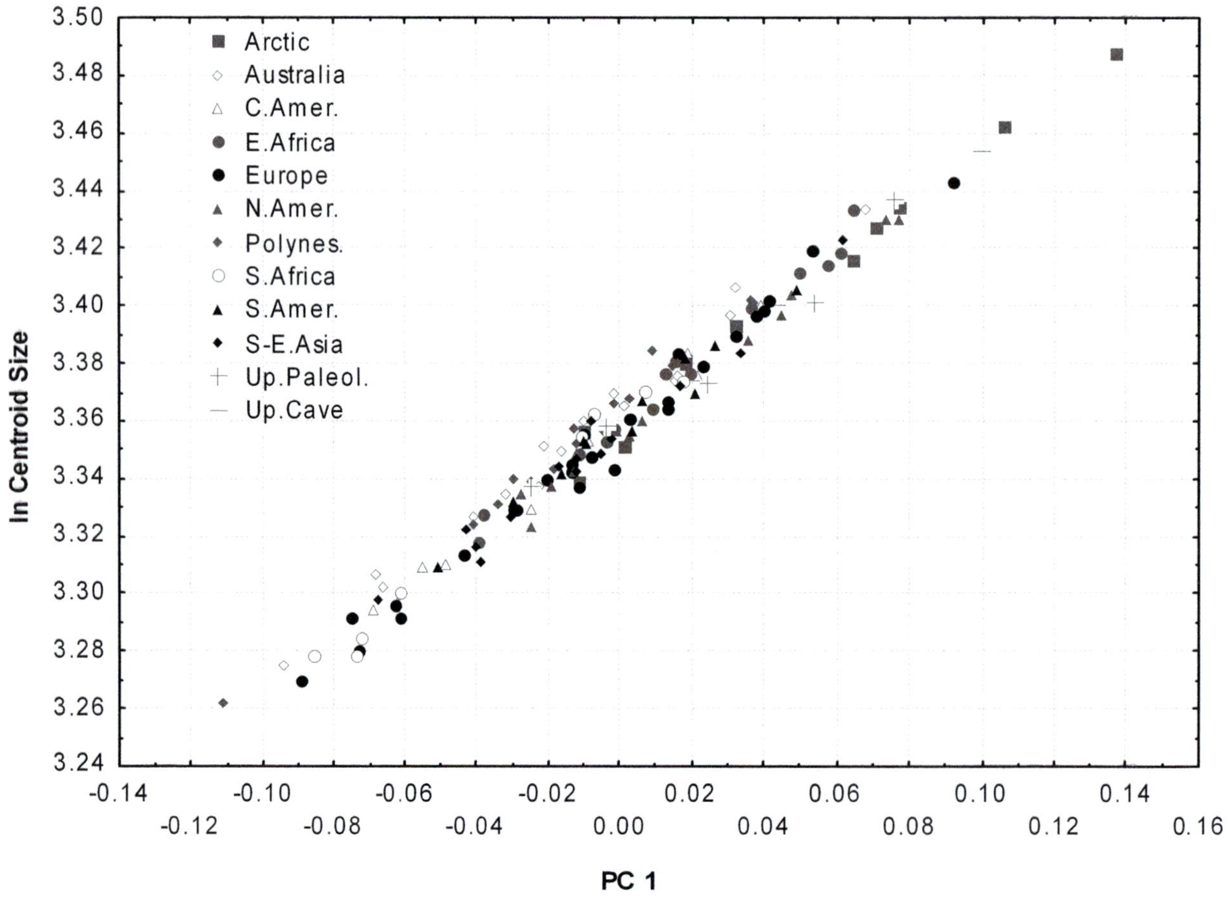

Fig. 16.2 PC 1 plotted against log centroid size in the size-shape analysis of the modern human sample

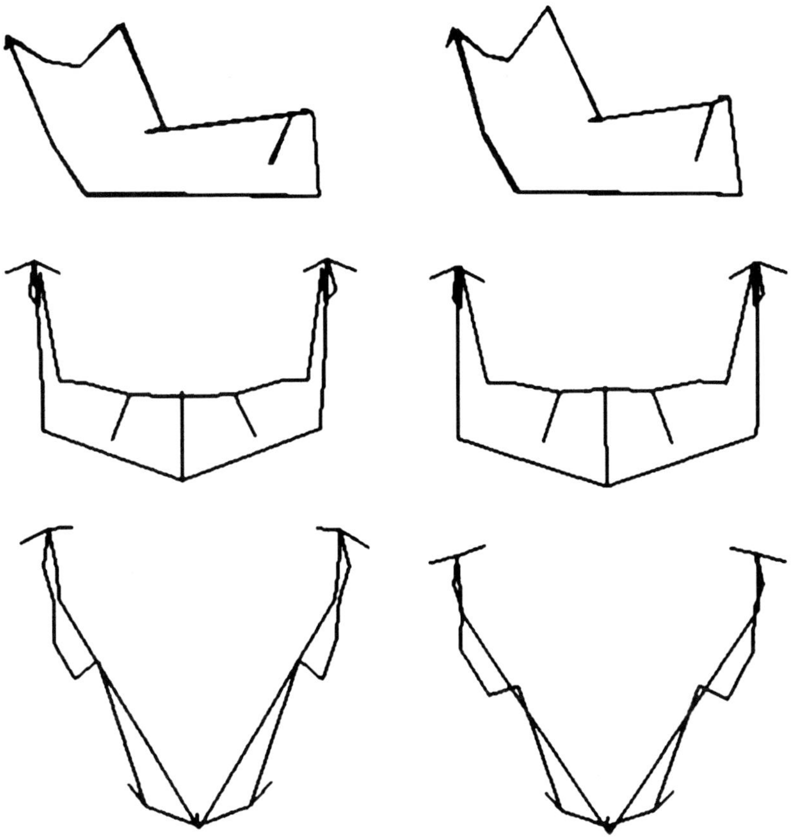

Fig. 16.3 Shape differences along PC 1, associated with small (*left*) and large (*right*) size among modern humans. Lateral (*top*), frontal (*middle*) and occlusal (*bottom*) views

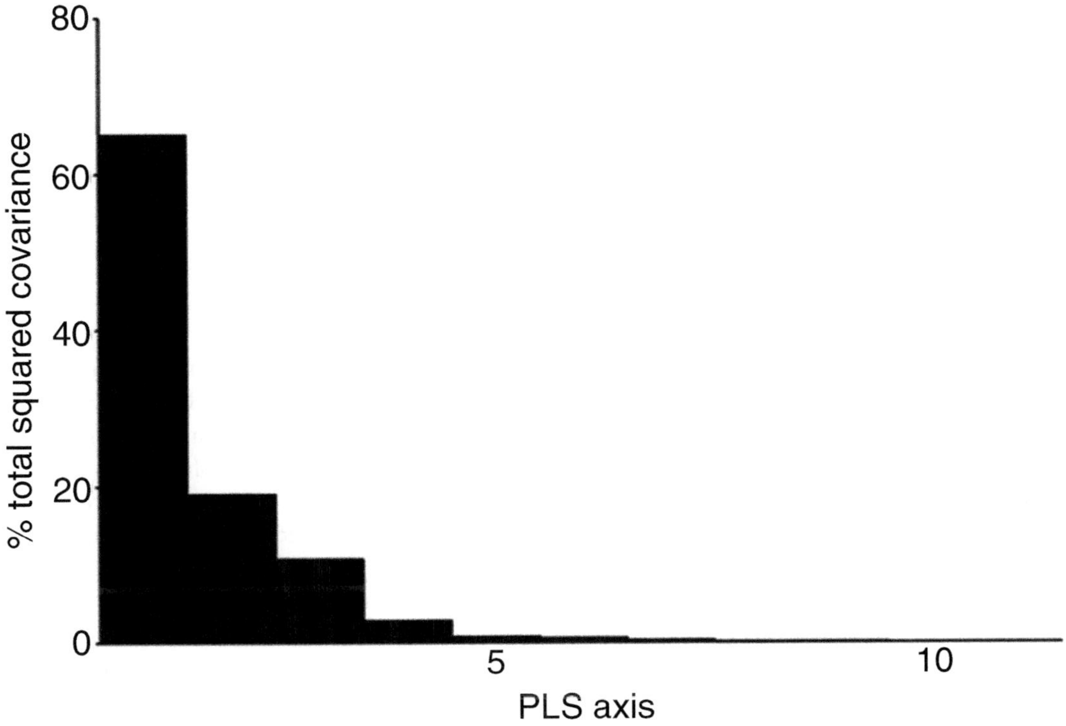

Fig. 16.4 Plot showing the distribution of total% co-variance on the respective PLS axes

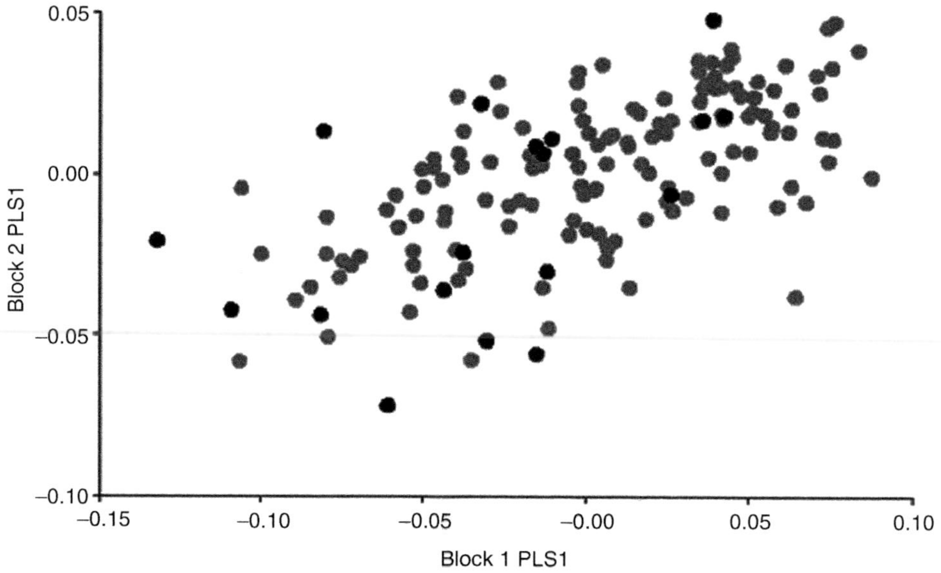

Fig. 16.5 Distribution of groups along PLS 1. Fossils are represented by the *black dots* and modern humans by *gray dots*

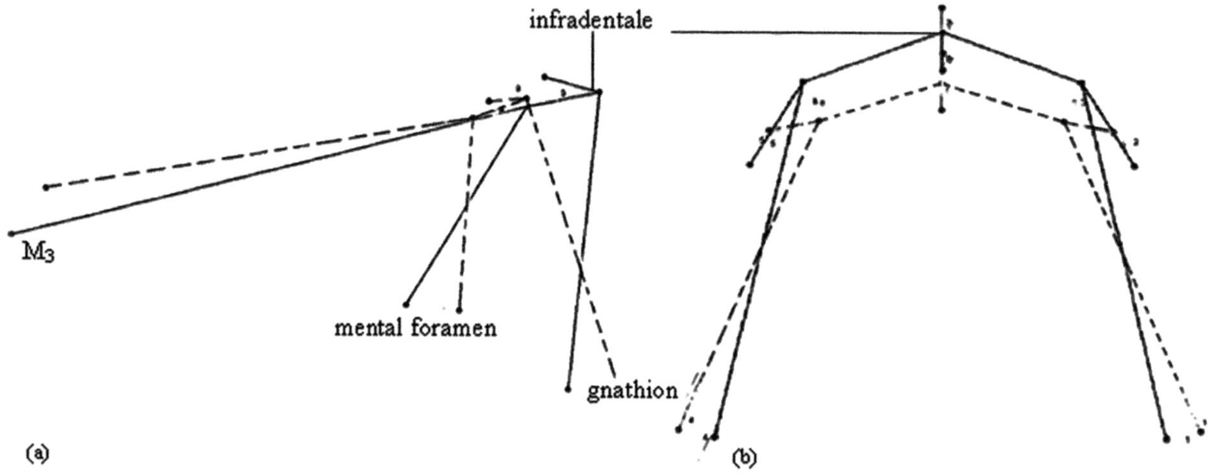

Fig. 16.6 Shape differences in block 1 are shown (**a**) in lateral view and (**b**) in superior view along PLS 1: high scores are represented by the *dashed line* and low scores by the *solid line*

suggesting a strong association between the two blocks. The trend in the distribution of groups along PLS I (Fig. 16.5) indicates a shared pattern of integration, with no clear separation between Neanderthals and other Pleistocene and modern humans. The distribution along this axis reflects modern human variation, with the fossils not only sharing the slope, but also falling within the range of this variation. Low scores on PLS I relate to shape changes associated with a more forwardly placed alveolar region, posteriorly retracted symphysis area, reduced distance between the M_3s and posteriorly placed mental foramen (Fig. 16.6). Corresponding changes in the posterior region relate to an antero-posteriorly wider ascending ramus, shallow mandibular notch and medially drawn in gonia (Fig. 16.7). High scores on this axis show a less anteriorly placed alveolar region, forwardly placed symphysis, wide posterior alveolar region (laterally expanded), and anteriorly placed mental foramen relative to the position of M_3 (Fig. 16.6). Related shape changes in the posterior region are associated with an antero-posteriorly narrow ascending ramus, deep mandibular notch and laterally expanded gonia (Fig. 16.7).

PLS II, which accounts for 19.8% of the total co-variance in the sample, with a moderate correlation coefficient of $r = 0.37$, shows a slight separation between Neanderthals and

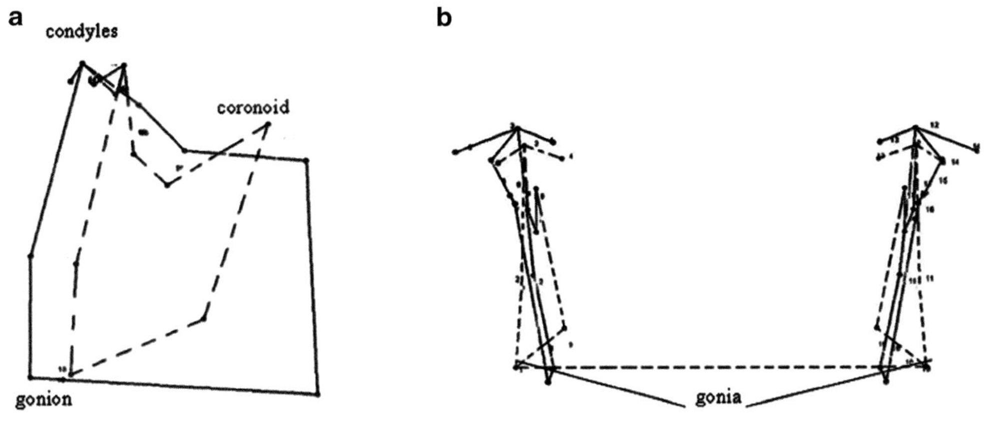

Fig. 16.7 Shape differences in block 2 are shown (**a**) in lateral view and (**b**) in posterior view along PLS 1; high scores are represented by the *dashed line* and low scores by the *solid line*

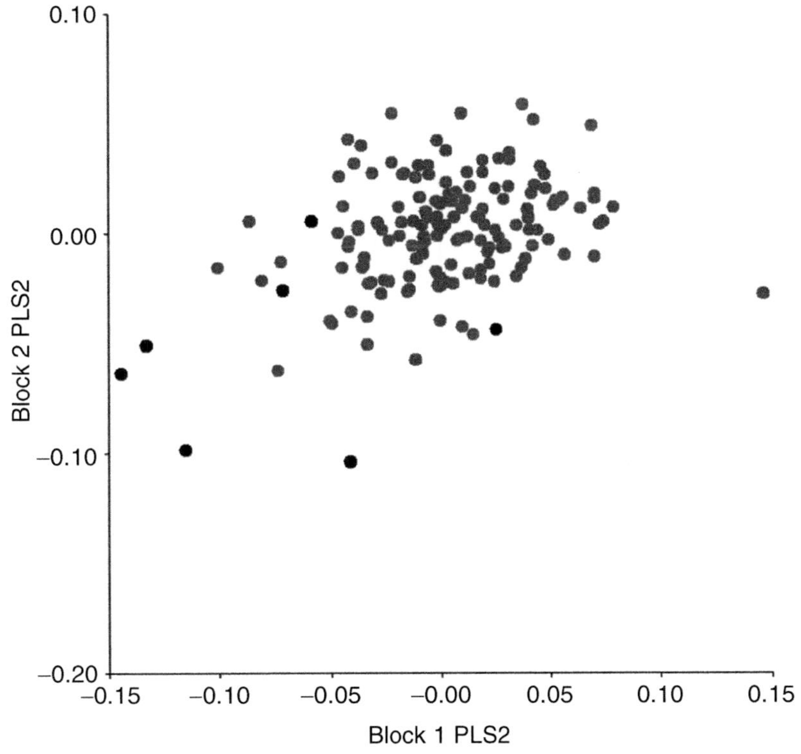

Fig. 16.8 Distribution of groups along PLS II. Neanderthals represented by *black dots* and other Pleistocene and modern humans by *gray dots*

other groups (Fig. 16.8). Neanderthals having lower scores on PLS II than the other taxa suggests a mean difference between the two groups. However, as mentioned earlier, the trend in the distribution of modern human and Neanderthal specimens along the regression line is similar, showing no distinct differences in the respective patterns of integration. Low scores on this axis relate to a posteriorly retracted lower symphysis and slightly laterally expanded alveolar region (indicated by position of distal M_3, suggesting a wider mandible (Fig. 16.9). Associated shape changes in the posterior part consist of a shallow and asymmetric mandibular notch with medially (inward) shifted gonia (Fig. 16.10) These shape differences correspond well with those described for Neanderthals and modern humans in the principal

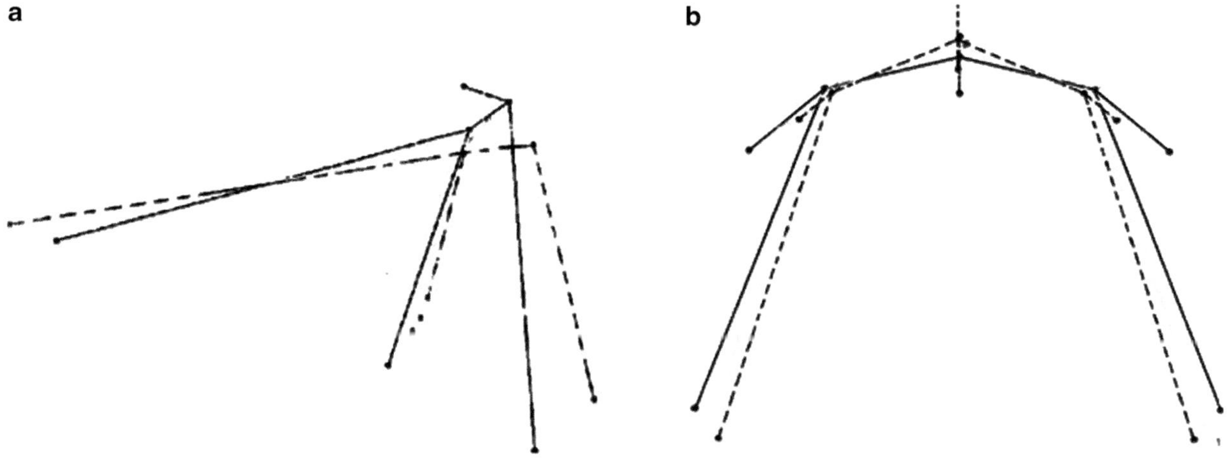

Fig. 16.9 Shape differences in block 1 are shown (**a**) in lateral view and (**b**) in superior view along PLS II: high scores are represented by the *dashed line* and low scores by the *solid line*

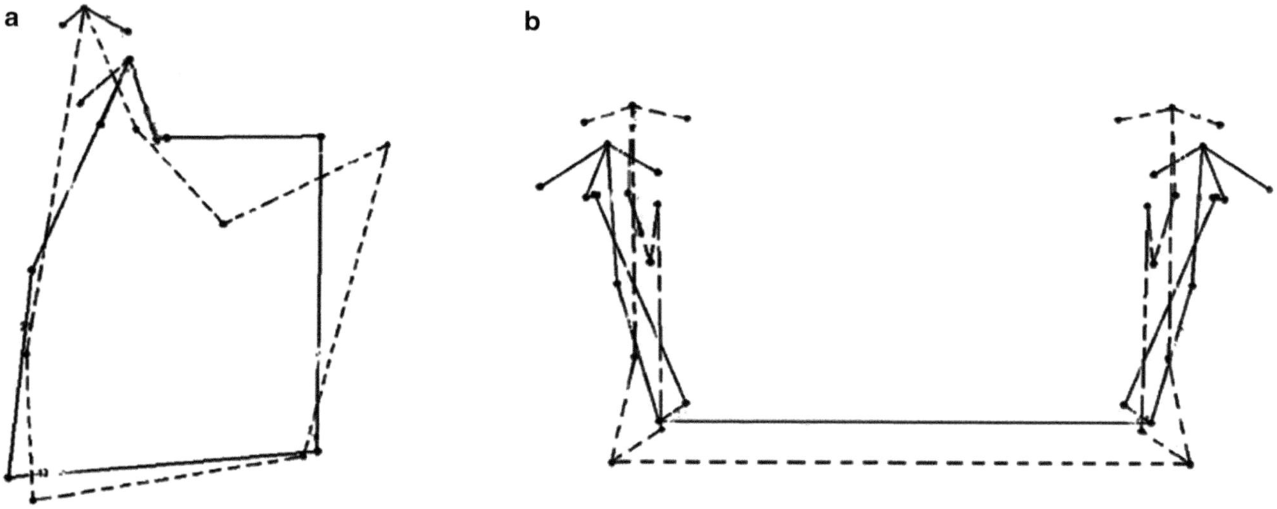

Fig. 16.10 Shape differences in block 2 are shown (**a**) in lateral view and (**b**) in posterior view along PLS II: high scores are represented by the *dashed line* and low scores by the *solid line*

components analysis of Nicholson and Harvati (2006). The modern humans and other Pleistocene fossils have higher values than the Neanderthals along PLS II. Majority of the shape changes in the anterior region consist of a forwardly placed lower symphysis, less projecting anterior alveolar region and anteriorly placed mental foramen (Fig. 16.9). Corresponding changes in the posterior region relate to a deep mandibular notch and postero-laterally expanded gonia (Fig. 16.10).

A second analysis was conducted on the fossil taxa alone. This was done in order to clarify subtle differences between the Pleistocene hominins that may have been obscured in the previous pooled-groups PLS analysis. The first PLS axis accounts for 44% of the total co-variation, with a correlation coefficient of $r = 0.58$ (Fig. 16.11). A low P-value ($P = 0.43$) of the anterior-posterior block permutation test indicates a high level of co-variation between the two parts. This analysis does not separate Neanderthals from other middle-late

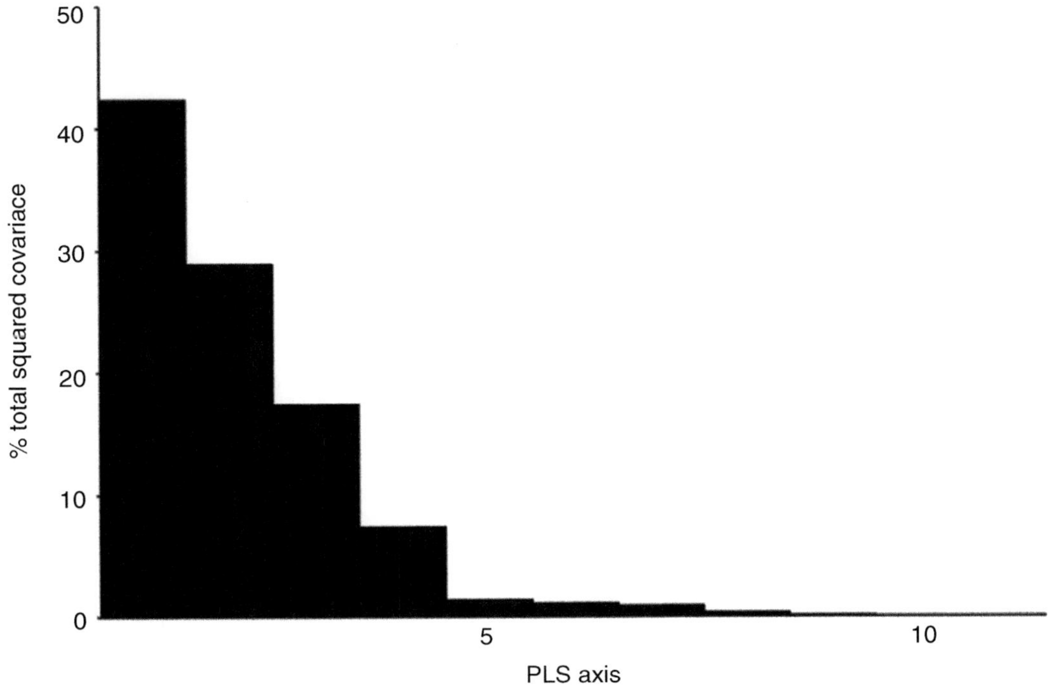

Fig. 16.11 Plot showing the distribution of total% co-variance on the respective PLS axes

Pleistocene hominins (Fig. 16.12). Given the small number of fossil specimens, these results do not reflect the variation within the group and little can be concluded about the pattern and degree of integration between the two mandibular parts. The higher PLS axes do not show any separation between the groups either and are, therefore, not discussed here.

Discussion and Conclusions

The results of the present study concur with previous findings (Rosas and Bastir 2004; Nicholson and Harvati 2006) suggesting that Neanderthal mandibular morphology is, for the most part, not the result of large size and therefore cannot be accounted for solely through allometric differences. The only Neanderthal-like features found to be related to increased centroid size in modern humans were the retromolar gap – also found in higher frequencies among the larger pre-Neanderthal specimens from Sima de los Huesos, Atapuerca (Rosas and Bastir 2004) – and the higher position of the coronoid process relative to the condyle. Other features commonly described as Neanderthal-like, including a less anteriorly projecting symphysis, an inwardly sloping ramus and a shallow mandibular notch, are in fact related to smaller centroid size among our sample of recent humans.

Given the extensive use of the anterior dentition in Neanderthals, we had hypothesized that the degree and patterns of integration in the alveolar region and ascending ramus of Neanderthals would be different from other Pleistocene and modern human groups. Our findings suggest that the activities involved in causing intense wear on the incisors and canines of Neanderthals did not influence the pattern of integration between the mandibular regions. Even though our objective was not to test biomechanical hypotheses, our findings concur with results from previous research (Antón 1990, 1994a; O'Connor et al. 2005) that show that evolution of Neanderthal cranio-mandibular morphology cannot be attributed to mechanical demands. Those findings further suggest that Neanderthals were similar to modern humans in their potential to produce high-magnitude bite forces.

Additional fossil specimens are needed to conduct a more thorough comparative analysis of mandibular integration patterns. Nonetheless, our results indicate that modern humans serve as a good model taxon for exploring mandibular integration in Pleistocene hominins. Such an approach could be expanded to include other anatomical subdivisions, which would further refine our interpretations.

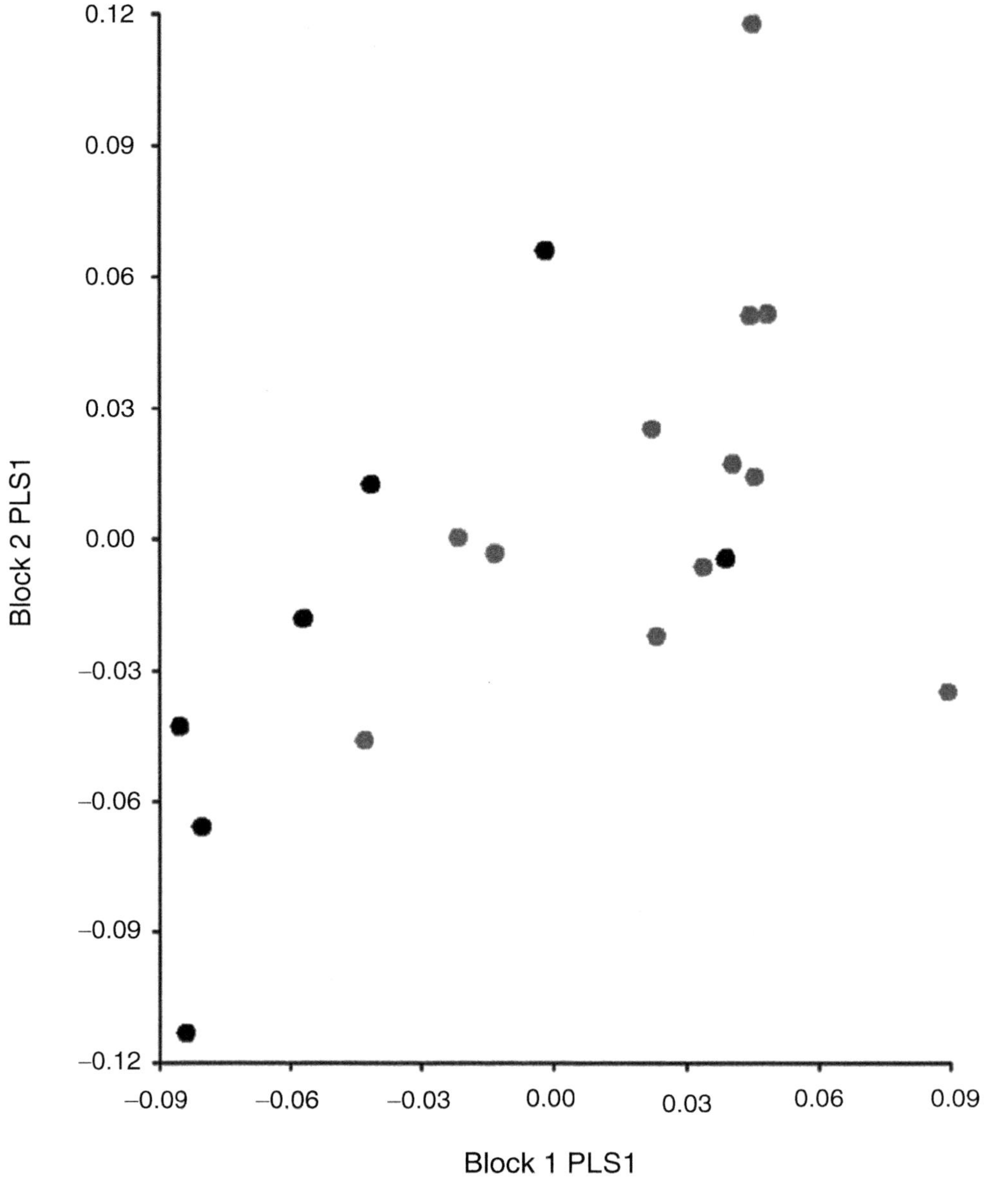

Fig. 16.12 Distribution of groups along PLS 1. Neanderthals are represented by the *black dots* and other fossils by *gray dots*

Acknowledgments We thank Chris Stringer, Rob Kruszynski, Henry and Marie-Antoinette de Lumley, Philippe Mennecier, Dominique Grimaud-Hervé, Ian Tattersall, Ken Mowbray and Jean-Jacques Hublin for allowing access to the specimens in their care and Chris Klingenberg and Philipp Gunz for assistance with the PLS analysis. We also thank the anonymous reviewers for their helpful comments and suggestions. This research was supported by the Centennial Scholars Program of Barnard College, New York University and the Max Planck Institute Society and the "EVAN" Marie-Curie training network MRTN-CT-019564. This is NYCEP morphometrics contribution number 33.

References

Antón, S. C. (1990). Neanderthals and the anterior dental loading hypothesis: A biomechanical evaluation of bite force production. *Kroeber Anthropology Socical Paper, 71–72*, 67–76.

Antón, S. C. (1994). Mechanical and other perspectives on Neanderthal craniofacial morphology. In R. S. Corruccini & R. L. Ciochon (Eds.), *Integrative paths to the past* (pp. 677–695). Englewood Cliffs: Prentice Hall.

Antón, S. C. (1996). Tendon associated bone features of the masticatory system in Neandertals. *Journal of Human Evolution, 31*, 391–408.

Arensburg, B., & Belfer-Cohen, A. (1998). *Sapiens* and Neandertals: Rethinking the Levantine Middle Paleolithic hominids. In T. Akazawa, K. Aoki, & O. Bar-Yosef (Eds.), *Neanderthals and modern humans in Western Asia* (pp. 311–322). New York: Plenum.

Atchley, W. R., & Hall, B. K. (1991). A model for development and evolution of complex morphological structures. *Biological Review, 66*, 101–157.

Balicki, A. (1970). *The Netsilik Eskimo*. Garden City, NJ: The Natural History Press.

Bastir, M., Rosas, A., & Sheets, H. D. (2005). The morphological integration of the hominoid skull: A partial-least squares and PC analysis with implications for European middle Pleistocene mandibular variation. In D. E. Slice (Ed.), *Modern morphometrics in physical anthropology* (pp. 265–282). New York: Kluwer/Plenum.

Bookstein, F. L., Gunz, P., Mitteroecker, P., Prossinger, H., Schaefer, K., & Seidler, H. (2003). Cranial integration in *Homo:* Singular warp analysis of the midsagittal plane in ontogeny and evolution. *Journal of Human Evolution, 44*, 167–187.

Boule, M. (1911–1913). L'homme fossile de la Chapelle-aux-Saints. *Ann Paléont, 6*, 11–172. 7, 21–56; 8, 1–70.

Brace, C. L., Ryan, A. S., & Smith, B. H. (1981). Comment on, "Tooth wear in La Ferrassie man". *Current Anthropology, 22*, 426–430.

Cheverud, J. M., Hartman, S. E., Richtsmeier, J. T., & Atchley, W. R. (1991). A quantitative genetic analysis of localized morphology in mandibles of inbred mice using finite element scaling. *Journal of Craniofacial Genetics and Developmental Biology, 11*, 122–137.

Condemi, S. (1991). Circeo I and variability among classic Neanderthals. In M. Piperno & G. Scichilone (Eds.), *The Circeo 1 Neanderthal skull: Studies and documentation*. Rome: Instituto Poligrafico e Zecca dello Stato.

Coon, C. S. (1962). *The origin of the races*. New York: Knopf.

Creed-Miles, M., Rosas, A., & Kruszynski, R. (1996). Issues in the identification of Neanderthal derivative traits at early post-natal stages. *Journal of Human Evolution, 30*, 147–153.

De Villiers, H. (1968a). *The skull of the South African Negro. A biometrical and morphometrical study*. Johannesburg: Witwatersrand University Press.

De Villiers, H. (1968b). Sexual dimorphism of the skull of the South African Bantu-speaking Negro. *South African Journal of Science, 64*, 118–124.

Demes, B. (1987). Another look at an old face: Biomechanics of the Neanderthal facial skeleton reconsidered. *Journal of Human Evolution, 16*, 297–303.

Franciscus, R. G., & Trinkaus, E. (1995). Determinants of retromolar space presence in Pleistocene *Homo* mandibles. *Journal of Human Evolution, 28*, 577–595.

Hrdlička, A. (1940a). Lower jaw. The gonial angle, I. The bigonial breadth. II. *American Journal of Physical Anthropology, 27*, 281–308.

Hrdlička, A. (1940b). Lower jaw further studies. *American Journal of Physical Anthropology, 27*, 383–467.

Hublin, J. J. (1998). Climatic changes, paleogeography and the evolution of Neanderthals. In T. Akazawa, K. Aoki, & O. Bar-Yosef (Eds.), *Neanderthals and modern humans in Western Asia* (pp. 295–310). New York: Plenum.

Humphrey, L. T., Dean, M. C., & Stringer, C. B. (1999). Morphological variation in great ape and modern human mandibles. *Journal of Anatomy, 195*, 491–513.

Hunter, W. S., & Garn, S. M. (1972). Disproportionate sexual dimorphism in the human face. *American Journal of Physical Anthropology, 36*, 133–138.

Hylander, W. L. (1977). The adaptive significance of Eskimo craniofacial morphology. In A. A. Dahlberg & T. M. Graber (Eds.), *Orofacial growth and development* (pp. 129–170). Paris: Mouton Publishers.

Jabbour, R. S., Richards, G. D., & Anderson, J. Y. (2002). Mandibular condyle traits in Neanderthals and other *Homo*: A comparative, correlative, and ontogenetic study. *American Journal of Physical Anthropology, 119*, 144–155.

Klingenberg, C. P., Badyaev, A. V., Sowry, S. M., & Beckwith, N. J. (2001). Inferring developmental modularity from morphological integration: Analysis of individual variation and asymmetry in bumblebee wings. *The American Naturalist, 157*, 11–23.

Klingenberg, C. P., Mebus, K., & Auffray, J.-C. (2003). Developmental integration in a complex morphological structure: How distinct are the modules in the mouse mandible. *Evolution and development, 5*(5), 522–531.

Leamy, L. (1993). Morphological integration of fluctuating asymmetry in the mouse mandible. *Genetica, 89*, 139–153.

Lieberman, D. E., Krovitz, G. E., Yates, F. W., Devlin, M., & St Claire, M. (2004). Effects of food processing on masticatory strain and craniofacial growth in a retrognathic face. *Journal of Human Evolution, 46*, 655–677.

Linderholm, H., & Wennstrom, A. (1970). Isometric bite force and its relation to general muscle force and body build. *Acta Odontologica Scandinavica, 28*, 679–689.

Martin, E. S. (1936). A study of an Egyptian series of mandibles with special reference to mathematical models of sexing. *Biometrika, 28*, 149–178.

Mitteroecker, P., Gunz, P., Bernhard, M., Schaefer, K., & Bookstein, F. L. (2004). Comparisons of cranial ontogenetic trajectories among great apes and humans. *Journal of Human Evolution, 46*, 679–698.

Morant, G. M., Collett, M., & Adyanthaya, N. K. (1936). A biometric study of the human mandible. *Biometrika, 28*, 84–122.

Nicholson, E., & Harvati, K. (2006). Quantitative analysis of human mandibular shape using three-dimensional geometric morphometrics. *American Journal of Physical Anthropology, 131*, 368–383.

O'Connor, C. F., Franciscus, R. G., & Holton, N. E. (2005). Bite force production capability and efficiency in Neanderthals and modern humans. *American Journal.of Physical Anthropology, 127*, 129–151.

O'Higgins, P., & Jones, N. (1998). Facial growth in *Cercocebus torquatus*: An application of three-dimensional geometric morphometric techniques to the study of morphological variation. *Journal of Anatomy, 193*, 251–272.

O'Higgins P., & Jones, N. (2006). *Tools for statistical shape analysis*. Heslington: Functional Morphology and Evolution Research Group, Hull York Medical School.

Olson, E. C., & Miller, R. L. (1958). *Morphological integration*. Chicago: University of Chicago Press.

Rak, Y. (1986). The Neanderthal: A new look at an old face. *Journal of Human Evolution, 15*, 151–164.

Rak, Y. (1998). Does any Mousterian cave present evidence of two hominid species? In T. Akazawa, K. Aoki, & O. Bar-Yosef (Eds.), *Neanderthals and modern humans in Western Asia* (pp. 353–366). New York: Plenum.

Rak, Y., Ginzberg, A., & Geffen, E. (2002). Does *Homo neanderthalensis* play a role in modern human ancestry? The mandibular evidence. *American Journal of Physical Anthropology, 119*, 199–204.

Rohlf, F. J., & Marcus, L. F. (1993). A revolution in morphometrics. *Trends in Ecology & Evolution, 8*, 129–132.

Rosas, A. (1992). *Ontogenia y filogenia de la mandíbula en la evolución de los homínidos. Aplicación de un modelo de morfogénesis a las mandíbulas fósiles de Atapuerca*. Ph.D. thesis, Complutense University, Madrid.

Rosas, A. (2001). Occurrence of Neanderthal features in mandibles from the Atapuerca-SH site. *American Journal of Physical Anthropology, 114*, 74–91.

Rosas, A., & Bastir, M. (2002). Thin-plate spline analysis of allometry and sexual dimorphism in the human craniofacial complex. *American Journal of Physical Anthropology, 117*, 236–245.

Rosas, A., & Bastir, M. (2004). Geometric morphometric analysis of allometric variation in the mandibular morphology of the hominids of Atapuerca, Sima de los Huesos site. *The Anatomical Record, 278A*, 551–560.

Slice, D. E. (1994–1999). Copyright. Morpheus et al., Software for morphometric research. Department of Ecology and Evolution, State University of New York, Stonybrook, New York.

Slice, D. E. (1996). Three-dimensional generalised resistant fitting and the comparison of least-squares and resistant fit residuals. In L. F. Marcus, M. Corti, A. Loy, G. J. P. Naylor, & D. Slice (Eds.), *Advances in morphometrics* (pp. 179–199). New York: Plenum.

Smith, F. H. (1983). Behavioral interpretations of changes in cranio-facial morphology across the archaic/modern *Homo sapiens* transition. *BAR International Series, 164*, 137–209.
Smith, F. H., & Paquette, S. P. (1989). The adaptive basis of Neanderthal facial form, with some thoughts on the natura of modern human origins. In E. Trinkaus (Ed.), *The emergence of modern humans: Biocultural adaptions in the later pleistocene* (pp. 181–210). Cambridge: Cambridge University Press.
Spencer, M. A., & Demes, B. (1993). Biomechanical analysis of masticatory system configuration in Neanderthals and Inuits. *American Journal of Physical Anthropology, 91*, 1–20.
Stringer, C. B., Hublin, J. J., & Vandermeersch, B. (1984). The origin of anatomically modern humans in Western Europe. In F. H. Smith & F. Spencer (Eds.), *The origins of modern humans: A world survey of the fossil evidence* (pp. 51–135). Liss: New York.
Tillier, A. M., Arensburg, B., & Duday, H. (1989). La mendibule et les dents. In O. Bar-Yosef & B. Vandermeersch (Eds.), *Le squelette Mousterien de Kebara 2* (pp. 97–111). Paris: CNRS Éditions.
Trinkaus, E. (1987). The Neanderthal face: Evolutionary and functional perpectives on a recent hominid face. *Journal of Human Evolution, 16*, 429–443.
Trinkaus, E., Moldovan, O., Milota, S., Bîlgăr, A., Sarcina, L., Athreya, S., Bailey, S., Rodrigo, R., Mircea, G., Migham, T., Ramsey, C. B., & van der Plicht, J. (2003). An early modern human from the Peştera cu Oase, Romania. *Proceedings of the National Academy of Sciences of the United States of America, 100*, 11231 11236.
Vandermeersch, B. (1981). *Les hommes fossiles de Qafzeh (Israel) Cahiers de paléontologie*. Paris: CNRS Éditions.

Chapter 17
Integration and Homology of "Chignon" and "Hemibun" Morphology

Philipp Gunz and Katerina Harvati

Abstract The presence of a weak occipital bun in some Upper Paleolithic European fossils is often cited as evidence for admixture between Neanderthals and anatomically modern humans, because the "chignon" morphology is considered by many to be a derived Neanderthal trait.

It is impossible, however, to split this morphology into "present" or "absent" character states (and thus "primitive" or "derived"); it rather varies in continuous degrees of expression. Furthermore the shape of the upper scale of the occipital bone is tightly integrated with the shape of the other bones forming the vault. To assess whether the "hemibun" of some Upper Paleolithic European crania should be considered evidence for possible hybridization, it is thus crucial to understand the integration of this morphology and whether this shape feature is homologous between modern humans and Neanderthals. Here we present a geometric morphometric analysis assessing the integration of the posterior midsagittal profile and the temporal bone quantitatively. We digitized 3-D coordinates of anatomical landmarks on the posterior vault and semilandmarks along a midsagittal curve from bregma to inion on 356 modern and archaic human crania. These points were converted into shape coordinates using Procrustes superimposition and then analyzed using the method of singular warps. The occurrence of an occipital bun is highly correlated with a flat parietal midline and an anteriorly positioned temporal bone. While Upper Paleolithic *Homo sapiens* cannot be distinguished from recent humans, archaic *Homo* fall outside the range of modern variation. The *pattern* of integration however, which accounts for ~30% of the total variation, is shared between modern humans and archaic *Homo*. Our results suggest that the occurrence of "hemibuns" in UPE should not be used as evidence for admixture between modern humans and Neanderthals, as this morphology is a predictable correlate of the relative position of the temporal bone and not an independent trait.

Keywords Procrustes analysis • Homo • Hybridization • Singular warps • Partial Least Squares • Homo neanderthalensis

Introduction

The "chignon" or "occipital bun" morphology, defined as a posterior projection or a great convexity of the upper scale of the occipital bone, is one of the features often cited as evidence for admixture between modern humans and Neanderthals. Because the "chignon" is considered by many to be a derived Neanderthal trait (Hublin 1988; Dean et al. 1998; see also Lieberman 1995) and because some Upper Palaeolithic crania from Europe exhibit a similar, yet less pronounced, morphology (sometimes called a "hemibun") some authors have argued that this indicates interbreeding between Neanderthals and early modern Europeans (Jelinek 1969; Genet-Varcin 1970; Vlcek 1970; Smith 1982, 1984; Bräuer 1989; Gambier 1997; Churchill and Smith 2000; Wolpoff et al. 2001; Smith et al. 2005). However, since the "chignon" is a morphological continuum, it is difficult to split into "present" or "absent" character states (Trinkaus and LeMay 1982), and therefore to be considered a trait in a cladistic sense. The putative evidence for interbreeding is thus not the presence of a derived Neanderthal discrete trait in Upper Paleolithic modern Europeans; instead, it is the simple assumption that a hybrid of two morphologically distinct populations, one with a distinct "chignon", the other one with a more globular cranium, would show an average morphology.

Using a geometric morphometric analysis of points measured along the midline of the posterior profile, we recently studied the chignon quantitatively and assessed its usefulness in separating Neanderthals from modern humans (Gunz and Harvati 2007; see also Harvati 2001; Harvati et al. 2002; Reddy et al. 2005). In that study we showed that it is not

P. Gunz (✉)
Department of Human Evolution, Max Planck Institute for Evolutionary Anthropology, Deutscher Platz 6, D-04103 Leipzig, Germany
e-mail: gunz@eva.mpg.de

K. Harvati
Department of Early Prehistory and Quaternary Ecology, Senckenberg Center for Human Evolution and Paleoecology, Eberhard Karls Universität Tübingen, Rümelinstrasse 23, 72070 Tübingen, Germany
e-mail: katerina.harvati@ifu.uni-tuebingen.de

S. Condemi and G.-C. Weniger (eds.), *Continuity and Discontinuity in the Peopling of Europe: One Hundred Fifty Years of Neanderthal Study*, Vertebrate Paleobiology and Paleoanthropology, DOI 10.1007/978-94-007-0492-3_17, © Springer Science+Business Media B.V. 2011

possible to metrically separate modern humans and archaic *Homo*, including Neanderthals, based on the midline *shape* of the upper scale of the occipital bone alone. It is rather the position and orientation of the occipital bone relative to the parietal and temporal bones, or to put it differently, the overall shape of the posterior half of the neurocranium, that constitutes the unique Neanderthal morphology. Our results concurred with previous work that this condition is the result of an integrated shape change and that differences in the shape of the posterior profile are probably related to differences in the rate of brain growth and timing of suture closure (Trinkaus and LeMay 1982; Lieberman 1995; Lieberman et al. 2000).

Our findings, however, differed from those of Lieberman et al. (2000), who observed a correlation between occipital bunning and narrow cranial bases in modern humans, and the inverse relationship in Neanderthals. Lieberman et al. (2000) tentatively concluded that bunning in Neanderthals and Upper Paleolithic Europeans is not homologous, and hence its presence in the latter does not indicate genetic continuity. Contrary to Lieberman et al. (2000), we (Gunz and Harvati 2007) failed to detect differences in integration between modern and archaic humans, and concluded that the "chignon" in Neanderthals and "hemibun" morphology in modern humans were homologous. In this chapter we revisit our previous analysis and elaborate on methodological details that received little attention in the original publication. Furthermore we discuss our method and results in light of the recent work on modularity and integration by Mitteroecker and Bookstein (2007).

Aim of This Study

Our goal was to reassess the hypothesis that Neanderthals and modern humans exhibit different patterns of cranial integration resulting in non-homologous posterior cranial projection. For this purpose we evaluated the influence of the shape of the posterior midsagittal profile and the cranial base on the expression of occipital bunning in our combined recent and fossil human sample. We quantified how tightly the bones forming the posterior vault are integrated and whether Neanderthals and modern humans follow different trajectories, or form distinct clusters along the same trajectory. We then repeated the analysis looking at the recent human and fossil samples separately.

In order to quantify the shape of the posterior vault we used geometric morphometrics based on coordinate data, measured with a 3-D digitizer on both temporal bones and the midsagittal profile. Using homologous points (semilandmarks) along the midsagittal curve enabled us to quantify the continuous variation in the expression of the midsagittal aspect of the "chignon" morphology. We explored the co-variation of the shape of the posterior vault and the cranial base (as represented by our parietal and temporal bone landmarks and semilandmarks) on the expression of occipital bunning using the method of singular warps (Bookstein et al. 2003). If modern humans and Neanderthals differed in their pattern of cranial integration, and hence did not share homologous occipital buns, as suggested by Lieberman et al. (2000), the two groups would not follow the same predictions of occipital shape based on the shape and relative positions of the parietal and temporal bones.

In this chapter we specifically address two questions that were not explored in full detail in our previous study of the same data-set: (1) whether the small sample of fossils biases the results, and (2) whether partitioning the midsagittal curve into two blocks exaggerates the correlation between these two blocks.

Because the number of extant modern humans is more than ten times the number of fossil specimens, one could easily imagine situations where the analysis of the pooled sample would impose the modern integrational pattern upon the fossils, even if the modern and fossil patterns were distinct. To check whether the excess of modern specimens would bias the results, we performed a separate analysis of all fossil specimens.

The second concern is slightly more technical: The midsagittal profile is a single curve, not two independent curves. Partioning it into two parts at an arbitrary point to look for covariation (as we did in Gunz and Harvati 2007), will naturally yield high correlations between these two blocks. Furthermore the semilandmarks along the posterior profile were allowed to slide along the curve so as to minimize the thin-plate spline bending energy *before* partitioning into blocks. Because these semilandmarks slide along the same curve, this sliding step potentially exaggerates correlations among them. In the present study we have tried to avoid those potential pitfalls by treating the midsagittal profile as one curve, thereby reducing the analysis from a three-block to a two-block design.

Materials and Methods

Samples

This study included a large comparative sample of recent humans ($n=326$) and Middle and Late Pleistocene fossil specimens (Table 17.1). Recent humans were represented by ten regional groups from a wide geographical range, including an Iberomaurusian series from Afalou and Taforalt (North Africa), dated to 14–8.5 ka (Chamla 1978). Only adult crania with fully erupted permanent dentition were measured. Most

Table 17.1 Sample of recent humans and fossil specimens used in this analysis

Fossil humans	($n=29$)
Middle-Late Pleistocene African and near Eastern fossil humans (MLA)	($n=7$)
Kabwe	
Ndutu	
Ngaloba	
Omo 2	
Qafzeh 9	
Skhul 5	
Singa	
Middle Pleistocene European Pre-Neanderthals (MPE)	($n=2$)
Reilingen	
Sima de los Huesos 5*	
'Classic' and Early Neanderthals (NE)	($n=10$)
Amud 1	
Guattari 1	
La Chapelle-aux-Saints	
La Ferrassie 1	
La Quina 5	
Saccopastore 1	
Shanidar 1*	
Spy 1 and Spy 10	
Tabun	
Eurasian Upper Paleolithic specimens (UP)	($n=10$)
Abri Pataud	
Chancelade*	
Cioclovina	
Cro Magnon 1,2	
Ein Gev	
Mladeč 1 and 5	
Předmostí 3* and 4*	
Recent humans	($n=326$)
1. African *(Mali, Kenya)*	($n=36$)
2. Andaman *(Andaman Islands)*	($n=30$)
3. Asian *(North China, Thailand)*	($n=38$)
4. Australian *(South Australia)*	($n=32$)
5. Inuit *(Alaska, Greenland)*	($n=42$)
6. European *(Austria, Greece, Italy, Germany, Yugoslavia)*	($n=45$)
7. Khoi-San *(South Africa)*	($n=28$)
8. Melanesian *(New Britain)*	($n=30$)
9. Middle Eastern *(Syria)*	($n=19$)
10. Iberomaurusian *(Afalou, Taforalt)*	($n=26$)

*Asterisks indicate specimens for which casts or stereolithographs were used

Table 17.2 Landmarks collected on the temporal, parietal and occipital bones

Temporal block (L+R)	Number of landmarks: 18
Porion	
Auriculare	
Mastoidiale	
Stylomastoid Foramen	
Most postero-lateral point of the jugular fossa	
Lateral origin of the petro-tympanic crest	
Most medial point of the petro-tympanic crest at the level of the carotid canal	
Most inferior point on the juxtamastoid crest (following Hublin 1978)	
Deepest point of the lateral margin of the articular eminence	
(root of the articular eminence)	
Midsagittal Block	Number of landmarks and semilandmarks: 37
Bregma	
20 semilandmarks between Bregma and Lambda	
Lambda	
14 semilandmarks between Lambda and Inion	
Inion	

See also Fig. 17.1

crania were only sexed based on anthropological criteria, so the sexes were pooled in the analysis. The fossil sample comprised ten Neanderthal specimens from Europe and the Near East; two European Middle Pleistocene specimens; seven Middle-Late Pleistocene fossils from Africa and the Near East; and ten Upper Paleolithic anatomically modern humans from Europe and the Near East (Table 17.2). In the few cases where the original specimens were not available, high quality casts or stereolithographs from the Anthropology Departments of the American Museum of Natural History and of New York University were measured.

Data

3-D coordinates of anatomical landmarks on the temporal, parietal and occipital bones as well as semilandmarks along the midsagittal from bregma to inion were collected by KH using a Microscribe 3DX digitizer. Each cranium was measured in a 'top' and 'bottom' orientation, which were later superimposed using four fiducial points, which were not used in the analysis. The midsagittal profile was digitized as closely spaced points along the curve. These points were then automatically resampled to yield equal point count on every specimen: we computed a cubic spline through the points measured

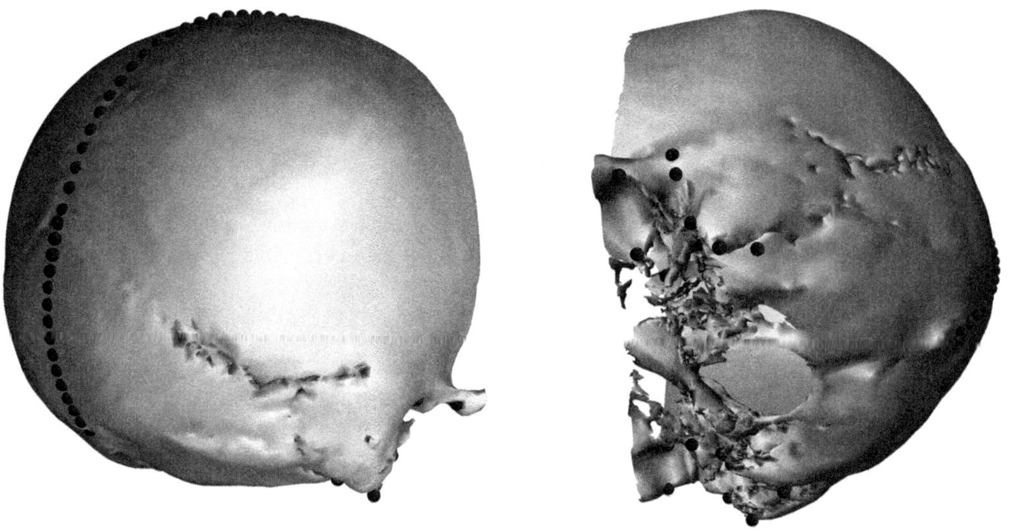

Fig. 17.1 Landmarks and semilandmarks used in this analysis. For the singular warps analysis the set of landmarks and semilandmarks was partitioned into two blocks: (1) the midsagittal profile from bregma to lambda to inion, (2) the landmarks on the temporal bones (both sides)

on the midsagittal curve and then placed 20 equidistantly spaced points between bregma and lambda, and 14 between lambda and inion. Since morphometric analyses do not accommodate missing data, some data reconstruction was necessary for the incomplete fossils. During data collection, and only for specimens with minimal damage, landmarks were reconstructed using anatomical information from the preserved surrounding areas. In cases where bilateral landmarks were missing on one side only, they were estimated by reflected relabeling (Mardia and Bookstein 2000; see Gunz 2005; Gunz et al. 2009 regarding application to missing data).

Incomplete specimens were least-squares superimposed with their reflected configurations in Procrustes space and missing data were reconstructed from their homologous counterparts on the other side. Landmarks on those specimens that had only one side of temporal bone preserved, or were one side was affected by distortion were reflected by mirroring across the empirical midplane (Reilingen, Předmostí 4, Kabwe, Omo2, LH18, Cro Magnon2, Spy10, Singa).

In limited instances (specimens Cro Magnon2, Ndutu, Omo2, Mladeč1, Saccopastore1), points missing on both sides of the cranium or on the midsagittal plane were estimated by minimizing the bending energy of the thin-plate spline between the incomplete specimen and the sample Procrustes average, following (Gunz 2005; Gunz et al. 2009).

The semilandmarks were iteratively allowed to slide along the midsagittal curve to minimize the bending energy of the thin-plate spline interpolation function computed between each specimen and the sample Procrustes average. We used the algorithm of Bookstein (1997; see also Gunz et al. 2005, 2009), which allows points to slide along tangents to the curve. These tangents were approximated for each semilandmark by converting the vector between the two neighboring points to unit length. Missing points were allowed to slide without constraining them to a curve ("full relaxation"). Spline-relaxation removes the effects of "digitizing error" in the tangent direction that results from the practical necessity of having to place the semilandmarks somewhere along the curve. After relaxation these semilandmarks can be treated in multivariate analyses as if they had been homologous points in the first place (Bookstein 1997; Bookstein et al. 1999; Gunz et al. 2005). Landmarks and slid semilandmarks were converted to shape coordinates by Procrustes superimposition (GPA: Gower 1975; Rohlf and Slice 1990). This procedure removes information about location and orientation from the raw coordinates and standardizes each specimen to unit centroid size, a size-measure computed as the square root of the summed squared Euclidean distances from each landmark to the specimen's centroid (Dryden and Mardia 1998).

Unlike in Gunz and Harvati (2007) the landmark and semilandmark data were partitioned into two blocks rather than three: temporal and midsagittal (Table 17.2, Fig. 17.1). The "midsagittal-block" included the landmarks and semilandmarks defining the posterior midline cranial profile from bregma to lambda to inion. The "temporal bone block" included landmarks on the basicranial portion of both right and left temporal bones (see Table 17.2). The temporal landmarks were measured on both sides in order to include information about the width of the cranial base.

Analyses

We studied the morphological integration of the cranial base and the posterior midline profile using the method of "singular warps" introduced by Bookstein et al. (2003) that quantifies and visualizes the covariation of anatomical regions. Blocks of landmarks were defined a priori; then the linear combinations of the original shape variables that provide the best mutual cross-prediction between these landmark-blocks were calculated. Usually a singular warps analysis with two blocks is calculated via a singular value decomposition of the cross-block covariance matrix (Bookstein et al. 2003). Here we used a more flexible algorithm suggested by Mitteroecker and Bookstein (2007), based on a "Wright-style" factor analysis (Wright 1932). The results are identical to the SVD approach, when the singular vectors are scaled appropriately. We will return to this numerical identity between these two approaches in the discussion section. Mitteroecker and Bookstein (2007) demonstrate that the singular vectors have to be scaled before joint visualisation. This scaling was performed in the analysis of Gunz and Harvati (2007) but not described in the paper. Higher singular warps were calculated using the same algorithm after projecting out the lower singular vectors for each block separately.

Singular warps are linear combinations of the original shape variables and can be visualized either as scores, or as deformations. We scaled the singular vectors following Mitteroecker and Bookstein (2007) and visualized the deformations of the temporal and midsagittal block together. To create the surface-morphs the landmarks and semilandmarks of one modern human specimen were used to warp the triangulated surface from the mean configuration in Procrustes space to this mean shape with different multiples of the singular vectors added. Note that the surface areas where there is no (semi)landmark information are just smoothly warped according to the thin-plate spline interpolation.

For data processing and analyses we used software routines written in Mathematica (© Wolfram Research) and "R" by PG together with Philipp Mitteroecker (University of Vienna).

Results

Singular Warps: Pooled Sample

The scores of the first singular warp (which explains ~30% of the total variation) are plotted in Fig. 17.2 (scores for midsagittal and temporal blocks together as well as separately for each group); the associated shape changes are visualized in Fig. 17.3. The integration of the midsagittal profile and the position and shape of the temporal bones is very tight (the correlation coefficient of the midsagittal and temporal scores $r > 0.89$); the points lie along a single trajectory. The fossil specimens do not diverge from this common pattern of integration. Looking at the scores for each block and "population" separately shows that the variation along these axes within each group is considerable. Neanderthals and the other fossil human groups have on average higher scores than the modern mean, but almost all modern groups have specimens that fall well within the Neanderthal range of variation.

Figure 17.3 visualizes the shape change associated with the first singular warp as a surface-morph. The singular vectors are scaled following Mitteroecker and Bookstein (2007) so that they can be visualized jointly. An anteriorly and superiorly placed cranial base predicts (and is predicted by) a flat parietal and a posterior projection of the occipital. Along the first singular warp, there are only subtle shape changes of the temporal bone; it is the relative position of the cranial base that changes. The higher singular warps were calculated after projecting out the lower singular vectors for every block separately. The plots are not shown here, as the associated shape changes do not involve occipital bunning.

Because the sample comprises so many more modern humans than fossil specimens, we checked carefully whether this shared trajectory could be an artifact due to the unequal sample sizes. There exists another potential pitfall with pooled samples: if populations had different means, then even if there were different patterns of integration among populations, looking at integration based on the uncentered pooled covariance matrix would create the impression of a common trajectory. To avoid this, we subtracted the respective group mean from each specimen before computing the covariance matrix for the singular warps.

Singular Warps: Fossil Sample

Figure 17.4 shows the scores of the first and second singular warp of the fossil sample. In essence the results stay the same as those for the pooled analysis; in the fossil sample the shape-changes described above for the pooled analysis are captured by the first two singular warps: the first singular warp is driven by the correlation of a flat parietal midline with a relatively wide cranial base ($r \sim 0.82$); the second singular warp entails posterior projection of the occipital midline, an anteriorly and superiorly positioned temporal bone and a relatively anterior placement of bregma (correlation $r > 0.62$). The scores of the first singular warp (upper left panel) show two clusters along the midsagittal scores. Only the second singular warp is associated with occipital bunning, and the corresponding scores (upper right panel) show no differences between anatomically modern humans and archaic *Homo* in these dimensions.

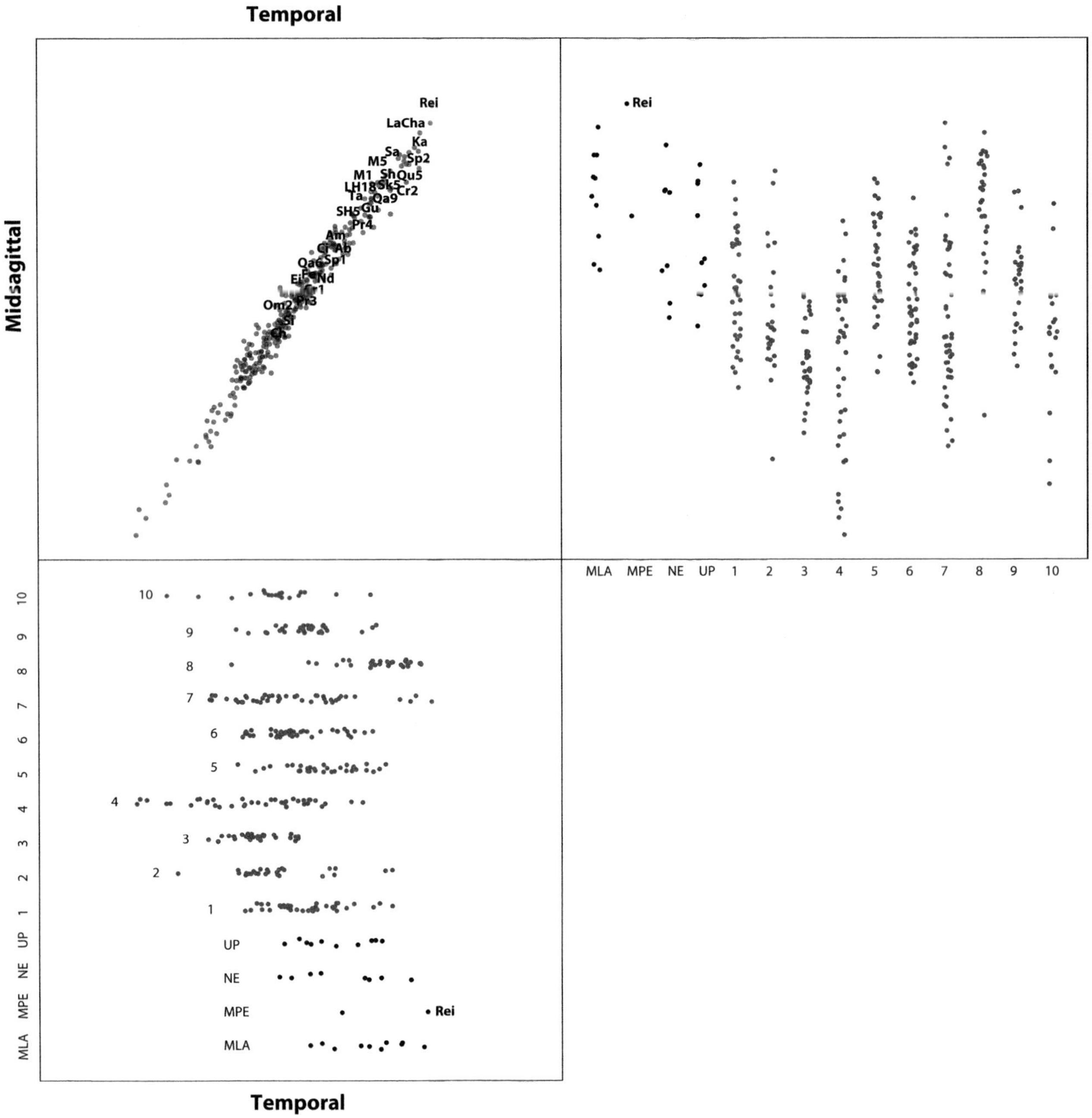

Fig. 17.2 Pooled sample; scores of the first singular warp of midsagittal and temporal block plotted against each other. These plots show how well one can predict the shape of one block from the other block. NE and MPE crania have higher scores than most modern humans for all blocks, but the integrational pattern is the same between groups. The scores are plotted for each group separately to demonstrate the considerable variability within geographic groups. The numbers correspond to the order of modern human populations in Table 17.1

Shape Regression

Because Lieberman et al. (2000) found occipital bunning in recent humans to be correlated with narrow cranial bases, we also looked at shape predictions for modern humans with *absolutely* narrower cranial bases (in contrast to the *relative* shape differences in Procrustes space discussed above). For all extant *Homo sapiens* crania we computed the absolute width of the cranial base as the length of the vector between left and right porion of the original coordinates before superimposition, then calculated the multivariate regression of all Procrustes shape coordinates on this width. When visualized by adding the regression slopes to a mean shape (Fig. 17.5), an absolute increase in cranial base width in recent humans predicts a projection of the occipital bone and a slightly flatter parietal. In the same way we predicted all shape variables

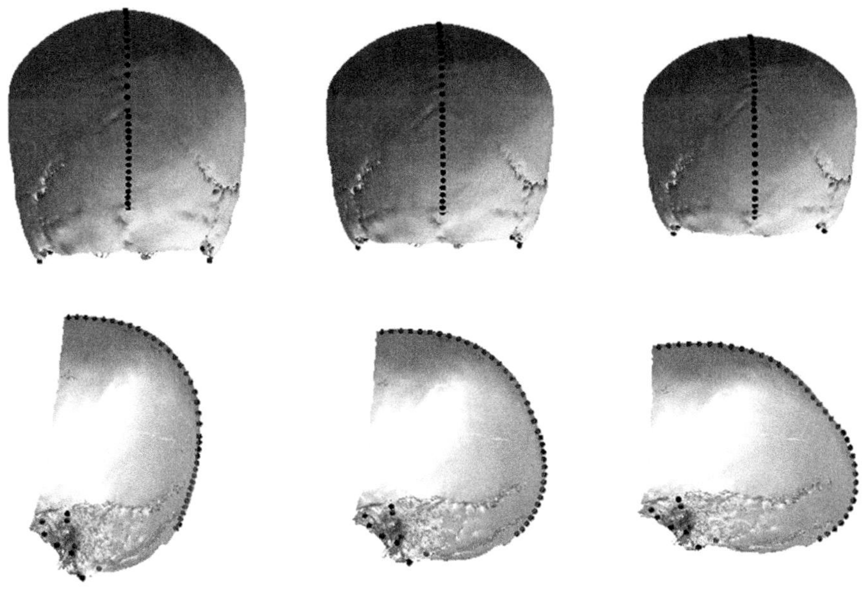

Fig. 17.3 Pooled sample; first singular warp. The shape dimensions associated with the scores shown in Fig. 17.2 can be visualized together. Here a modern human cranium's surface is deformed by subtracting (*left*) and adding (*right*) a multiple of the singular vectors to the mean shape (*middle*). The first singular warp contrasts a globular cranium with a more elliptical shape. Note that as occipital bunning increases, the cranial width does not change at all; the temporal bones get more superiorly and anteriorly placed and the mastoid decreases in size

by the logarithm of centroid size, as a crude proxy for endocranial volume. An increase in absolute size predicts a slightly more globular shape of the posterior vault and an occipital protuberance, but no occipital bun.

Discussion

The results of our singular warps analysis confirm our previous findings (Gunz and Harvati 2007), indicating that occipital bunning in modern humans and Neanderthals alike is associated with anteriorly and superiorly placed temporal bones/cranial bases (Figs. 17.3 and 17.4). The shape of the posterior midline profile and the relative position of the cranial base are tightly integrated; the scores of the two landmark blocks are highly correlated and fall along a common trajectory in those dimensions related to occipital bunning. These high correlations are unlikely to be caused by direct individual interactions of the bones; instead they are most likely spurious correlations because all bones are affected by a single factor, brain expansion, during ontogeny. Mitteroecker (2007) and Mitteroecker and Bookstein (2007) show that the results of a two-block partial least squares analysis and Wrigth (1932) factor analysis are identical (when scaled appropriately). Thus the results of the present analysis can be interpreted as loadings of the two landmark modules on a common factor.

There exists considerable shape variability within groups of modern humans, but the pattern of integration is shared among modern human populations, fossil modern human specimens and archaic *Homo*. While Neanderthals and archaic *Homo* consistently had higher scores than most modern humans, they followed the same linear trend when the two blocks were viewed together. Neanderthals thus have the amount of occipital bunning that one would predict for a human with such a supero-anteriorly positioned temporal bone and such a flat parietal. This is particularly apparent in the modern humans that have scores similar to archaic humans. The Upper Paleolithic specimens all plot within the modern human cloud. Note that – with the exception of the Andamenese (population #2) – every modern human group has outliers that fall within the Neanderthal variation along the midsagittal scores of the first singular warp (Fig. 17.2). This includes crania from Africa, Asia and the Middle East, hence no uniquely European pattern can be discerned.

Changing the singular warp analysis from the three-block design of Gunz and Harvati (2007) to a two-block design had no impact on the results. Analyzing the fossil specimens separately confirmed the conclusions of our original publication that the fossils and extant specimens follow the same shape prediction between the relative position of the temporal bones and the occipital profile. A posterior projection of the occipital bone is associated with a relatively anterior and superior position of the temporal bone. We would like to stress however, that this conclusion only pertains to the external aspects

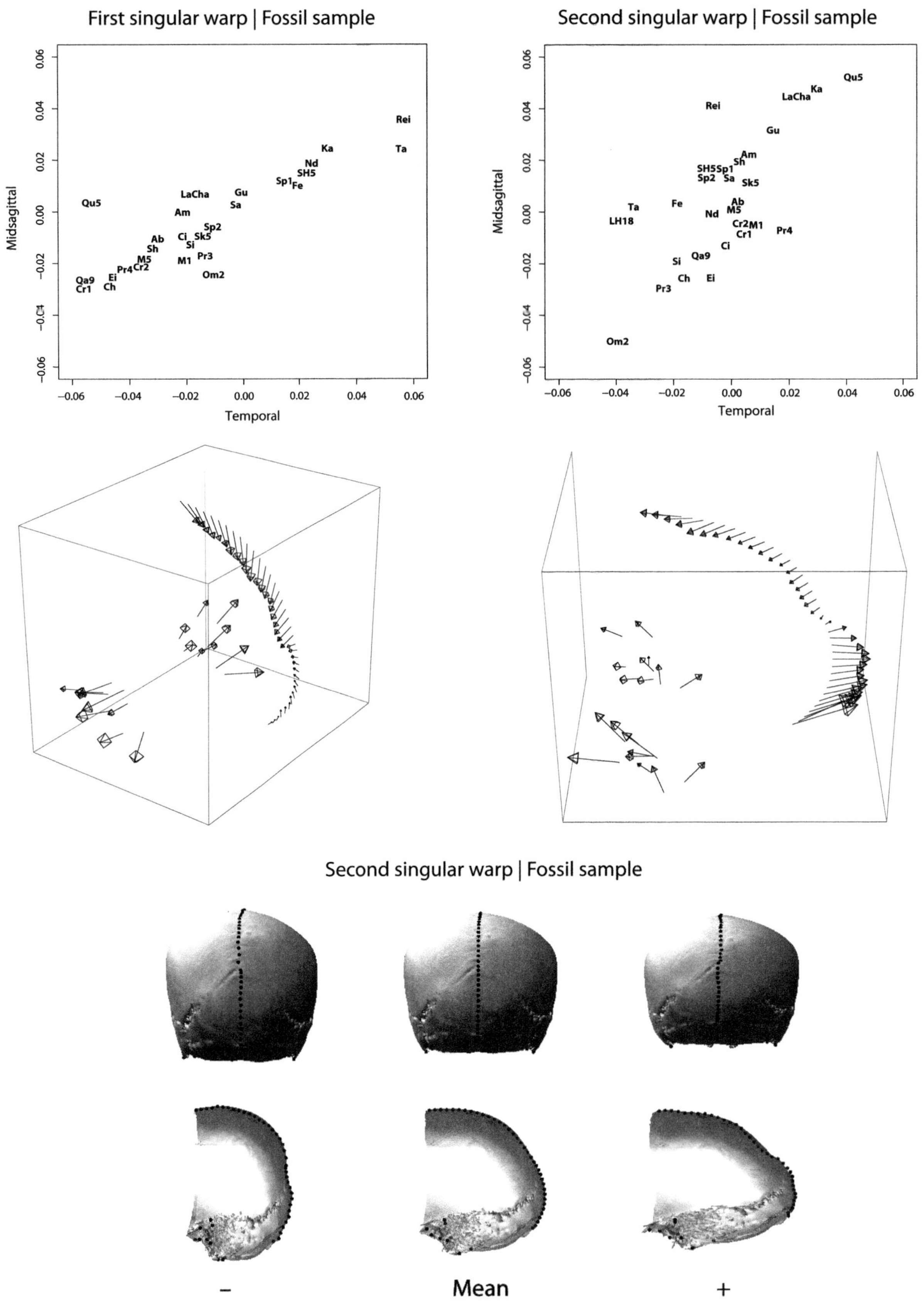

Fig. 17.4 Fossil sample; first and second singular warp. *Left:* Scores of the first singular warp computed from the fossil specimens only. There are two clusters along the midsagittal scores along a general trend. The bottom panel shows the associated shape change as vectors: increased width of the cranial base co-varies with a flat parietal profile. Note that there is almost no shape change in the occipital profile. *Right:* One common trajectory with complete overlap of modern and archaic humans. An projecting occipital bun is predicted by an anterior and superior placement of the cranial base. Figure 17.4b visualizes the same shape difference as a surface morph

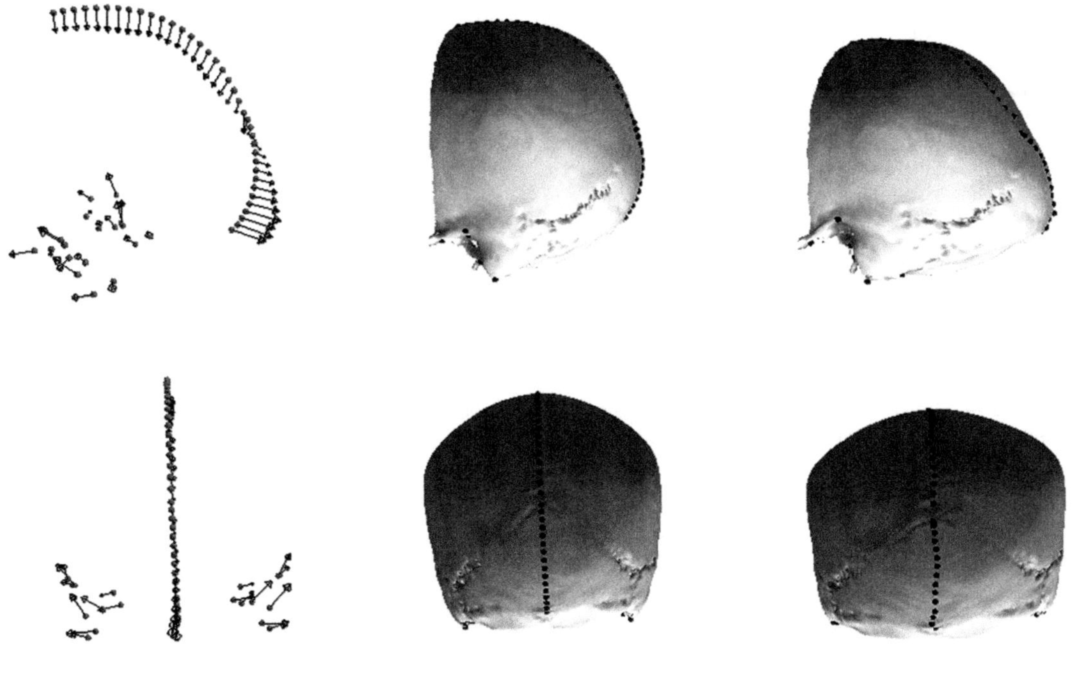

Fig. 17.5 Shape predictions for a wide cranial base in modern humans. Overall shape was regressed on the absolute length of the vector between left and right porion. An absolutely wider cranial base predicts a posterior projection of the midsagittal profile in extant modern humans. The regression slopes are visualized as vectors (*left*) from the mean shape towards an increase in cranial base width. Mean shape (*middle*) and positive extreme shape (*right*) are also shown as a surface morph

of the occipital bun in the midsagittal plane, as there were no endocranial landmarks used in this study.

Conclusions

We could not detect differences in integration between modern and fossil humans with respect to the external aspects of midsagittal posterior projection and relative width and position of the temporal bones. We thus consider the midsagittal aspect of the "chignon" morphology in Neanderthals to be homologous to bunning morphology in modern humans. The occurance of "hemibuns" in Upper Paleolithic modern Europeans should not be used as evidence for admixture between modern humans and Neanderthals, as this morphology is a predictable correlate of the relative position of the cranial base and not an independent trait.

Acknowledgments We thank all the curators and collections managers in several institutions across Europe, Africa and the USA, for kindly allowing access to both fossil and extant material used in this study. We thank Maximilian v. Harling for the CT scan used to create surface morphs, and Silvana Condemi for her efforts putting together this volume. We are also grateful to Jean-Jacques Hublin, Tim Weaver, Fred Bookstein, Philipp Mitteröcker, Markus Bastir, Susan Antón, Dan Lieberman and two anonymous reviewers for providing very helpful comments and suggestions. This research was funded in its various stages by grants to KH by the American Museum of Natural History; NYCEP; the Onassis and the CARE Foundations; and the U.S. National Science Foundation. Support was also provided by New York University, the Max Planck Institute for Evolutionary Anthropology and the "EVAN" Marie Curie Research Training Network MRTN-CT-019564. This is NYCEP morphometrics contribution number 34.

References

Bookstein, F. L. (1997). Landmark methods for forms without landmarks: Morphometrics of group differences in outline shape. *Medical Image Analysis, 1*(3), 225–243.

Bookstein, F. L., Schaefer, K., Prossinger, H., Seidler, H., Fiedler, M., Stringer, C. B., Weber, G. W., Arsuaga, J. L., Slice, D., Rohlf, F. J., et al. (1999). Comparing frontal cranial profiles in archaic and modern *Homo* by morphometric analysis. *Anatomical Record, 257*(6), 217–224.

Bookstein, F. L., Gunz, P., Mitteroecker, P., Prossinger, H., Schaefer, K., & Seidler, H. (2003). Cranial integration in *Homo*: Singular warps analysis of the midsagittal plane in ontogeny and evolution. *Journal of Human Evolution, 44*(2), 167–187.

Bräuer, G. (1989). The evolution of modern humans: A comparison between the African and non-African Evidence. In C. B. Stringer & P. Mellars (Eds.), *The human revolution* (pp. 123–154). Princeton: Princeton University Press.

Chamla, M. C. (1978). Le peuplement de l'Afrique duNord de l'Epipaleolithique a l' epoque actuelle. *L'Anthropologie, 82*, 385–430.

Churchill, S. E., & Smith, F. H. (2000). Makers of the Early Aurignacien of Europe. *American Journal of Physical Anthropology (Yrbk), 43*, 61–115.

Dean, D., Hublin, J. J., Holloway, R., & Ziegler, R. (1998). On the phylogenetic position of the pre-Neandertal specimen from Reilingen, Germany. *Journal of Human Evolution, 34*(5), 485–508.
Dryden, I., & Mardia, K. V. (1998). *Statistical shape analysis*. New York: Wiley.
Gambier, D. (1997). Modern humans at the beginning of the Upper Paleolithic in France; anthropological data and perspectives. In G. A. Clark & C. M. Willermet (Eds.), *Conceptual issues in modern human origins research* (pp. 117–131). New York: Aldine de Gruyter.
Genet-Varcin, E. (1970). Considérations morphologiques sur l'homme de Cro-Magnon. In G. Camps & G. Olivier (Eds.), *l'homme de Cro Magnon*. Paris: Arts et Métiers Graphiques.
Gower, J. C. (1975). Generalized procrustes analysis. *Psychometrika, 40*, 33–51.
Gunz, P. (2005). *Statistical and geometric reconstruction of hominid crania: Reconstructing australopithecine ontogeny*. Ph.D. dissertation, University of Vienna, Vienna.
Gunz, P., & Harvati, K. (2007). The Neanderthal "chignon": Variation, integration and homology. *Journal of Human Evolution, 52*, 262–274.
Gunz, P., Mitteroecker, P., & Bookstein, F. L. (2005). Semilandmarks in three dimensions. In D. E. Slice (Ed.), *Modern morphometrics in physical anthropology* (pp. 73–98). New York: Kluwer/Plenum.
Gunz, P., Mitteroecker, P., Neubauer, S. Weber, G. W., Bookstein, F. L. (2009). Principles for the virtual reconstruction of hominin crania. *Journal of Human Evolution, 57*(1), 48–62.
Harvati, K. (2001). *The Neanderthal problem: 3-D geometric morphometric models of cranial shape variation within and among species*. Ph.D. dissertation, City University of New York, New York.
Harvati, K., Reddy, D. P., & Marcus, L. F. (2002). Analysis of the posterior cranial profile morphology in Neanderthals and modern humans using geometric morphometrics. *American Journal of Physical Anthropology, S34*, 83.
Hublin, J. J. (1978). Apomorphic characters of Neanderthalian skull and their phylogenetic interpretation. *Comptes Rendus Hebdomadaires des Séances de L'Académie des Sciences SérieD, 287*(10), 923–926.
Hublin, J. J. (1988). Caractères dérivés de la région occipito-mastoïdienne chez les Néandertaliens. *L'Anatomie, 3*, 67–73.
Jelinek, J. (1969). Neanderthal man and *Homo sapiens* in Central and Eastern Europe. *Current Anthropology, 10*(5), 475.
Lieberman, D. E. (1995). Testing hypotheses about recent human-evolution from skulls – integrating morphology, function, development, and phylogeny. *Current Anthropology, 36*(2), 159–197.
Lieberman, D. E., Pearson, O. M., & Mowbray, K. M. (2000). Basicranial influence of overall cranial shape. *Journal of Human Evolution, 38*, 291–315.
Mardia, K. V., & Bookstein, F. L. (2000). Statistical assessment of bilateral symmetry of shapes. *Biometrika, 87*, 285–300.
Mitteroecker, P. (2007). *Modularity and Integration*. Ph.D. dissertation, University of Vienna, Vienna.
Mitteroecker, P., & Bookstein, F. L. (2007). The conceptual and statistical relationship between modularity and morphological integration. *Systematic Biology, 56*(5), 818–836.
Reddy, D. P., Harvati, K., & Kim, J. (2005). Alternative approaches to ridge-curve analysis using the example of the Neanderthal occipital bun. In D. Slice (Ed.), *Modern morphometrics in physical anthropology* (pp. 99–115). New York: Kluwer/Plenum.
Rohlf, F. J., & Slice, D. (1990). Extensions of the Procrustes method for the optimal superimposition of landmarks. *Systematic Zoology, 39*, 40–59.
Smith, F. H. (1982). Upper Pleistocene Hominid evolution in South–Central Europe: A review of the evidence and analysis of trends. *Current Anthropology, 23*, 667–703.
Smith, F. H. (1984). Fossil hominids from the Upper Pleistocene of Central Europe and the origin of modern Europeans. In F. H. Smith & F. Spencer (Eds.), *The origins of modern humans: A world survey of the fossil evidence* (pp. 211–250). New York: Liss.
Smith, F. H., Jankovič, I., & Karavanič, I. (2005). The assimilation model, modern human origins in Europe, and the extinction of Neandertals. *Quaternary International, 137*, 7–19.
Trinkaus, E., & LeMay, M. (1982). Occipital bunning among Later Pleistocene hominids. *American Journal of Physical Anthropology, 57*, 27–35.
Vlcek, E. (1970). Relations morphologiques des types humains fossiles de Brno et Cro-Magnon auPleistocene Supérieur d'Europe. In G. Camps & G. Olivier (Eds.), *L'Homme de Cro-Magnon* (pp. 59–72). Paris: Arts et Métiers Graphiques.
Wolpoff, M. H., Hawks, J., Frayer, D., & Hunley, K. (2001). Modern human ancestry at the peripheries: A test of the replacement theory. *Science, 291*, 293–297.
Wrigth, S. (1932). General, group and special size factors. *Genetics, 15*, 603–619.

Chapter 18
Virtual Synthesis of the Skull in Neanderthals by FESS

Ulrich Witzel

Abstract The measurement of strains in real skulls or the calculation of strain distribution in individual finite-element models is an inductive method that yields information about the stresses occurring in the a priori existing shape. In contrast, the approach taken here to determine the relationship between skull function and skull shape applies Wolff's law through a deductive technique of structure synthesis. This paper describes the application of this method in the virtual synthesis of a Neanderthal skull.

As a first step a non-specific homogeneous solid is constructed, giving the stresses ample volume to spread between points of force application and constraint. The FE software ANSYS 10 is used to form 10-noded tetrahedral finite elements with a maximum of 129,000 nodes. The initial conditions are the functional spaces for the brain, the eye openings, and the nose cavity. Further initial conditions are the muscle forces, and the placement of the dental arcade, including assumed bite and chewing forces and the spatial relationships of these with respect to each other. Enforcing equilibrium of forces, the primary 3D stress flows in each load case are summarized by a physiological superposition, which accumulates the highest value of compressive stress in each finite element. If the stress-free parts are eliminated and the summarized stress flows are maintained, a reduced model appears which is very similar to the real skull. This reduction of shape can be repeated iteratively and leads to a more exact form. The final FE-model is presented by using the CAD software CATIA V5 and the resultant cross-sections are compared with CT scans of a real Neanderthal skull.

Changes in the form of the dental arcade, its position relative to the braincase, the origins of muscles, or the volume of the brain lead to models that clearly resemble morphological differences between species or genera.

The deductive virtual synthesis of the typical skull of Neanderthals using the finite-element structure synthesis (FESS) technique demonstrates the direct correlation between functional loading and the biological structure and shape and can be used to test hypotheses regarding the relationship between structure and function during skull evolution.

Keywords Wolff's law • CT-scan • Skull shape • Functional loading • Bauraum

Introduction

In comparison to the skull of modern humans, Neanderthal skulls (Fig. 18.1) show some special characteristics:

- elongated braincase
- prominent maxilla
- no fossa canina
- wide nose opening rounded at the upper edge
- distinct brow ridges
- prominent occipital structure, torus occipitalis and so on.

How We Can Explain These Morphological Details with a Biomechanical Approach?

The anatomical structure of a Neanderthal skull is difficult to calculate in a simple classical manner in order to obtain information about form and function, therefore we make use of the finite-element method (FEM) as a modern numerical engineering tool for 3D calculations and powerful computer (Zienkiewicz 1971). Finite-element structure analysis (FESA) is used nowadays worldwide in paleontology (Rayfield 1998, 2004; Fastnacht et al. 2002) in a similar way to its application in engineering. This analysis enables engineers to control stability and tolerable stress distribution in designed buildings or machine elements and in a priori given FE-models of skulls paleontologists are able to explore what structures are optimised for as a result of evolution. This inductive approach subordinates the theory of morphogenesis

U. Witzel (✉)
Research Group of Biomechanics,
Ruhr-University Bochum, D 44780 Bochum, Germany
e-mail: Ulrich.Witzel@ruhr-uni-bochum.de

S. Condemi and G.-C. Weniger (eds.), *Continuity and Discontinuity in the Peopling of Europe: One Hundred Fifty Years of Neanderthal Study*, Vertebrate Paleobiology and Paleoanthropology, DOI 10.1007/978-94-007-0492-3_18,

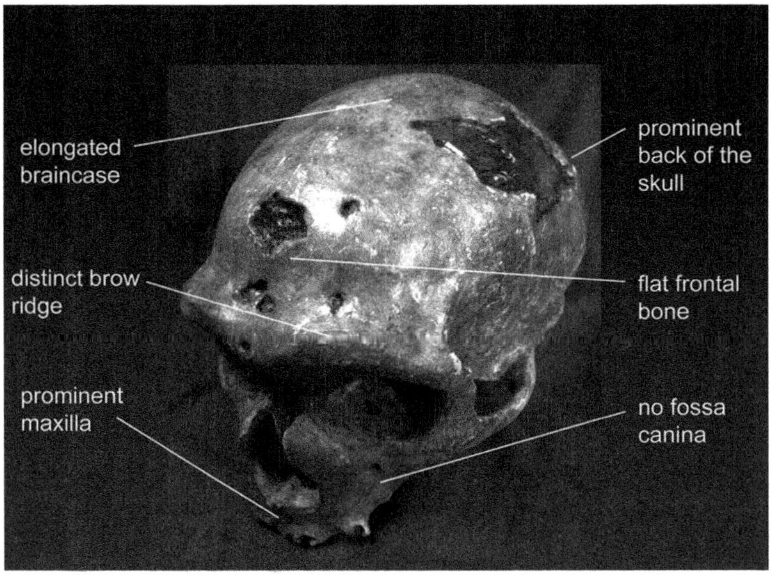

Fig. 18.1 Characterization of a Neanderthal skull. Cast of La Chapelle-aux-Saints

and does not provide precise explanations for the existence of bone in a specific position in the given model. This calculation technique does not facilitate the acquisition of new knowledge about bone characteristics which are currently not understood and does not allow the shape of the skull to be explained as an absolute necessity and an essential biological answer to mechanical loading (Wolff 1892); but we know that a mixture of social display and developmental history also determine the final form.

In this study we used the deductive technique of finite-element structure synthesis (FESS) in order to realize skull calculations. This method was established in 1985 by synthesizing a cross-section of the diaphysis of a human femur (Witzel 1985). A similar approach was later developed by Carter and Wong (1988). In general the method starts from a non-specific homogeneous body that offers the stresses ample volume for spreading between points of force application and constraint. External forces are applied to the FE model, low stress areas are iteratively removed, and the resulting shapes are compared with those observed in nature. In this way, the method is used to deduce skull form from a few initial and boundary conditions. Surely this makes FESS ideal for discovering which parts are explicable in terms of mechanics and which are not.

This method has been used to elucidate the mechanical reasons for the pneumatized spaces in the skulls of primates (Preuschoft et al. 2002) and for the prominent bony nose in humans and other primates (Witzel and Preuschoft 1999), as well as a virtual synthesis of the facial part of human and gorilla skulls (Witzel and Preuschoft 2002; Preuschoft and Witzel 2004).

More than 100,000 tetrahedral finite-elements are located in a closed package the so-called *Bauraum* (Witzel and Preuschoft 2005; Rossmann et al. 2005) which can be loaded by functional forces. For each finite-element deformation mechanical stress is calculated by FE software. Mechanical stresses in elements show a specific load relating force per area: Newton/mm^2 or MPa. These stresses are transformable in fields of different bone densities. The present article, based on an abstract of Witzel (2006), extends this method to the deductive synthesis of the skull in *Homo neanderthalensis.*

Materials and Methods

Finite-element structure synthesis with its deductive approach is normally based on a general theory but in this special case two theories offer further insights into skull biomechanics.

The first theory says that the structure of a skull is determined by powerful coupling structures with light-weight construction that connect functional spaces and loading regions such as the dental arcade, membrane fastenings and muscle insertions.

The relevant functional spaces contain the brain, the two eyes and the olfactory organ (Preuschoft et al. 2002; Witzel and Preuschoft 2002). To synthesize the skull of Homo *neanderthalensis,* a model is constructed that makes only six a priori assumptions: the brain space, the position of the two eye openings, the nasal cavity and the position of the dental arcade (Fig. 18.2b, c) with assumed bite and chewing forces, and the observed muscle insertions and membrane fastenings with the estimated activity of the muscles and membrane forces under consideration of the equilibrium of the whole system; surely another assumption are the spatial relationships of each of these with each other (Witzel and Preuschoft 2005). The stable three-point support, which is a basic necessity for FE calculations, is achieved by constraining the position of the both occipital condyles and the position of

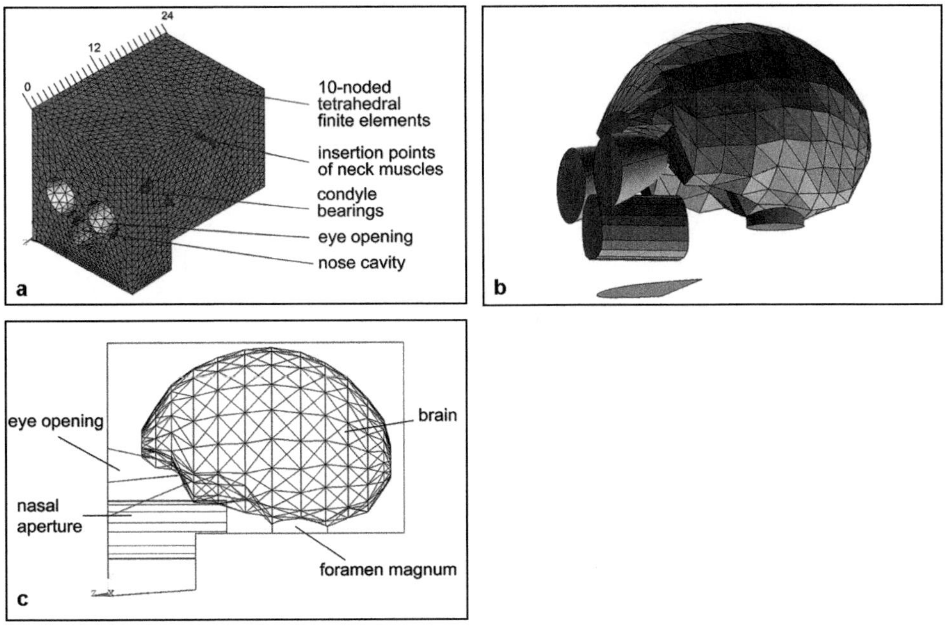

Fig. 18.2 *Bauraum* as an unspecific homogeneous solid with functional spaces as initial conditions. (**a**) *Bauraum*. (**b**) Functional spaces in a 3D presentation. (**c**) Lateral view

neck muscle insertions where reaction forces are applied: compression forces in the area of condyles and tensile forces in the attachment points of neck muscles.

The second theory focuses on the relationship between skull function and structure. With regard to the genetic facts bone is mechanically determined. That means the morphology of bone is known to be heavily influenced by mechanical loading history (Darwin 1859; Wolff 1892; Jansen 1920; Pauwels 1965; Frost 1988; Witzel 1993; Witzel and Hoffmann 1993; Witzel and Preuschoft 2002, 2005). Loading due to neuromuscular activity plays a key role in bone formation and subsequent development of form in ontogeny and phylogenetically bone form adapts according to loading (Wong and Carter 1990). Important in signalling loading for adaptation during life are hydraulic pressure and fluid flow which take place in Haversian and Volkmann canals in osteons (Imai et al. 1990; Qin L et al. 1999; Qin YX et al. 2003). To increase stability of osteons regarding compression forces they are twisted by collagenous fibres in several layers with different pitches (Ascenzi and Bonucci 1968; Pidaparti and Burr 1992). On account of this arrangements which arms, osteons with passive tension chords we found 24% higher compressive stiffness relative to theoretically assumed non-armed structures. With hydraulic filling of Harversian canals compressive stiffness increases up to 47% (DeMicheli and Witzel, not published).

These considerations indicate that bone represents an optimized compression structure. Accordingly it is reasonable to attempt virtual synthesis of bone form by modeling functional compression cases (Witzel 1985; Witzel and Preuschoft 1999, 2002, 2005; Preuschoft and Witzel 2004).

Table 18.1 Study data: Maximal bite forces of each tooth

	Load [N]
1st incisors	112
2nd incisors	115
Canine tooth	157
1st premolar	196
2nd premolar	212
1st molar	232
2nd molar	249
3rd molar	221

The modelling procedure begins with a 3D *Bauraum* (Fig. 18.2a), a cubical solid structure that envelops the total structure of the objects of interest. This *Bauraum* in natural scale is created by using ANSYS 10 and consists of 10-noded tetrahedral finite elements (FEs) with 129,000 nodes. Young's modulus is 17 GPa and the Poisson ratio is 0.3.

It is illogical to load the *Bauraum* simultaneously because this would be non-physiological and results in unrealistically high moments and stresses. So we have to simulate asynchronous physiological loading using a load case technique. Five load cases are dictated by bite forces:

1st: 4 incisors and 2 canine teeth,
2nd left and 3rd right: 2 premolars,
4th left and 5th right: 3 molars.

Using the strain gauge technique we measured the maximal bite forces of each tooth of a strong male person. The experiment was undertaken (Table 18.1) with the assumption that bite forces in Neanderthals are not very much

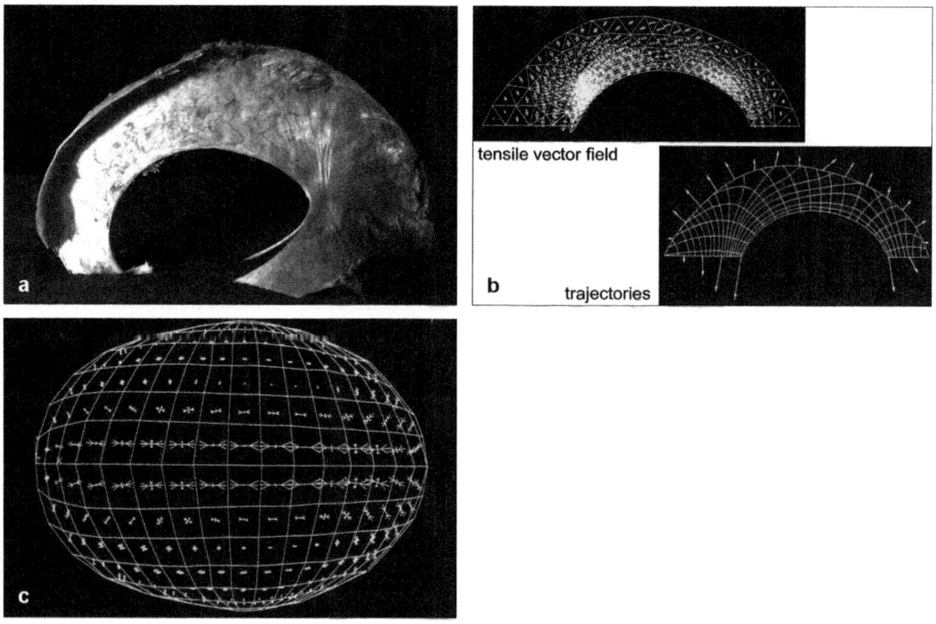

Fig. 18.3 Falx cerebri and calvaria. (**a**) Falx cerebri as part of dura mater; modern human preparation. (**b**) Biomechanics of falx cerebri. Tensile stresses: vector field and trajectories. (**c**) Compressive stress distribution in the calvaria as a function of lateral acceleration. Sagittal bridge formation (*high compressive stresses*) and parietal thinning (*low stress areas*)

Table 18.2 Study data: Tensile forces along the periphery of the falx cerebri

		Load [N]	
	Number	Fy	Fz
Frontal	1	4.98	0.44
	2	9.93	−1.22
	3	10.60	−6.62
	4	2.82	−5.30
	5	0.00	−6.00
	6	−2.54	−5.44
	7	−4.02	−5.73
	8	−5.66	−5.66
Top	9	−6.93	−4.00
	10	−7.52	−2.74
	11	−7.88	−1.39
	12	−10.47	0.73
	13	−10.04	3.07
	14	−9.09	5.25
	15	−9.66	8.70
Occipital	16	−9.32	11.11
	17	−7.49	14.70
	18	−1.26	14.45
	19	2.92	12.67
	20	6.72	8.04
	21	6.71	4.36
	22	7.31	3.25
	23	7.31	−3.25

higher than normal bite forces in living man (O'Connor et al. 2005).

The sixth load case models acceleration forces of the brain transmitted by the falx cerebri through its stable connection with calvaria (Fig. 18.3a) (Witzel and Hoffmann 1993; Witzel 1994). The lateral acceleration forces lead to tensile stresses in the falx membrane drawn as a vector field with directions indicated in Fig. 18.3b. Along the periphery of the falx membrane tensile forces (Table 18.2) result in compression within the calvarium, rather like the situation found in a bridge (Fig. 18.3c). Both lateral sides of the calvaria show low-stressed areas in the parietal bone which could explain parietal thinning in this region.

The primary 3D stress flows found for each load case are summarized by physiological superposition (not addition) because every finite element accumulates the highest value of compressive stress that occurs in one of the load cases. If during the synthesis of the skull the stress-free elements are eliminated, maintaining stress flows and reduced model appears that is similar to the real skull. This reduction of shape can be repeated iteratively and leads stepwise to more exact forms with each step. The final iteration is dictated by the reaching of a physiological stress in all synthesized structures. Changes in the form of the dental arcade, its position relative to the braincase or the origins of muscles, or the length or height of the skull, all lead to models that clearly resemble morphological differences between species and genera.

For the Neanderthal skull, the geometry of the *Bauraum* (Fig. 18.2a) was derived from the outer shape, the dental arcade, and the positions of joints taken from the cast of the skull of *Homo neanderthalensis* from La Chapelle-aux-Saints procured by H. Preuschoft, University Bochum. L. Bondioli, Roma, Museo Nazionale "L.Pigorini" made several CT-scans of *Homo neanderthalensis* (Guattari) available for us for comparison with our synthesis.

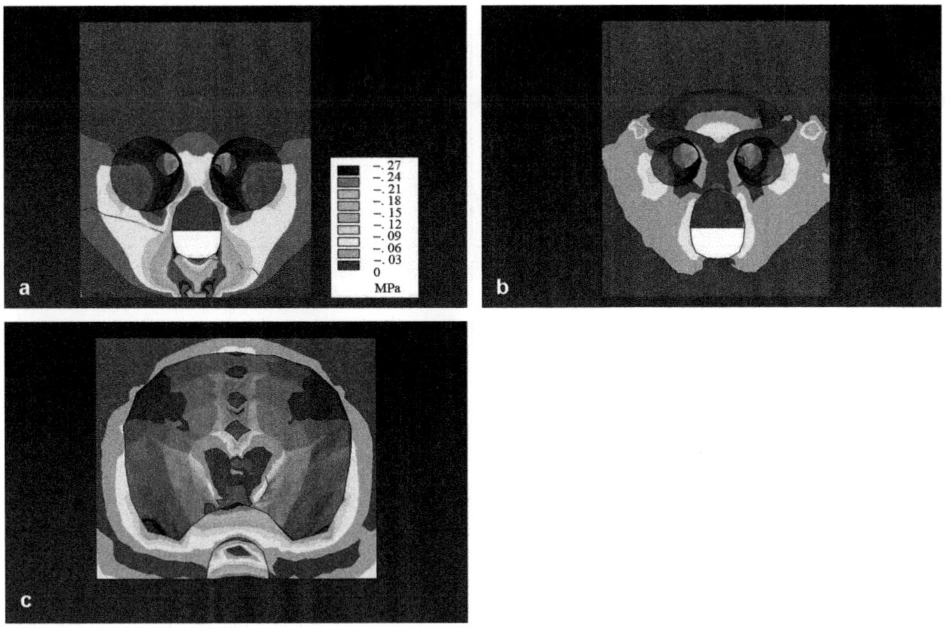

Fig. 18.4 Frontal view and two cross-sections with compressive stresses after first calculation. (**a**) Frontal view and stress legend, first load case. (**b**) 3rd cross-section, first calculation and first load case. (**c**) 7th cross-section, first calculation, 6th load case

The insertions of m. masseter, m. temporalis, the temporal fascia as the active tension chord of the zygomatic arch (Witzel et al. 2004) and the falx cerebri were used to locate the places of application of the muscle and membrane forces.

Before starting the FE calculation, bite and chewing forces, muscle forces and membrane forces were brought into equilibrium for each load case according to mechanical laws as follows:

$\Sigma F_{i\,x,y,z} = 0$, with F_{ix} = all Forces in the direction of x-coordinate, and so forth; and

$\Sigma M_{i\,x,y,z} = 0$, with M_{ix} = all Moments about x-coordinate, and so forth.

The models were constructed and calculated on a 2.4 GHz computer with 2 GB of RAM and a storage capacity of 240 GB. After the models were calculated, the postprocessor of the FE program was used to extract all principal stresses, including the von Mises stresses. Our explanation above and in particular Wolff´s law and Pauwel´s theory emphasize compressive stress. This was also the case in this study because tensile stresses are usually taken by collagenous fibres or neutralized by muscle forces, and in physiological cases the bending moments are zero (Sverdlova and Witzel 2010).

Results

Figure 18.4a shows the distribution of compressive stress on the frontal surface of the model after first calculation. Two incisors produce a high load in the maxillary region; the unloaded regions or those with little stress are the lower corners and the area above the eye openings. The stress distribution over the interior of the volume can be shown by cross-sections through the model. For example, cross-section 3 (see Fig. 18.2a for section number) gives information on stress distribution around the nose cavity and the eye openings in the first load case (Fig. 18.4b). The 7th cross-section through the skull (Fig. 18.4c) shows the influence of the loaded falx cerebri in the 6th load case. The calvaria, the temporal bone and the skull base are presented by compressive stresses.

The result of the first calculation is the physiological superposition of compressive stresses by load case technique and the realization of 23 cross-sections with their according stress distribution. If the non- or lightly stressed parts in all cross-sections using an arbitrarily selected threshold of −.06 N/mm² are eliminated, the equivalents of the remaining bony cross-sections are shown. Figure 18.5 presents cross-section 5 as an example (Fig. 18.5a) and cross-sections 5 and 6 with extrusion (Fig. 18.5b) to prepare a gradation-model shown in Fig. 18.5c.

To smooth this gradation-model we used CAD software CATIA V5. The merging together of all cross-sections leads to a 3D reconstruction of the reduced model which is presented in Fig. 18.6. The 3rd cross-section (Fig. 18.6a) and the 6th cross-section (Fig. 18.6b) show some remarkable details as incisura supraorbitalis, foramen infraorbitale, zygomatic arch and ethmoid bone but the first calculation did not resolve the frontal sinus, the inferior orbital rim and the maxillary sinus at this position. In Fig. 18.6c, d the reduced model possesses a maxillary sinus with hiatus and a sphenoidal sinus. Further back the auditory passage (porus acusticus externus

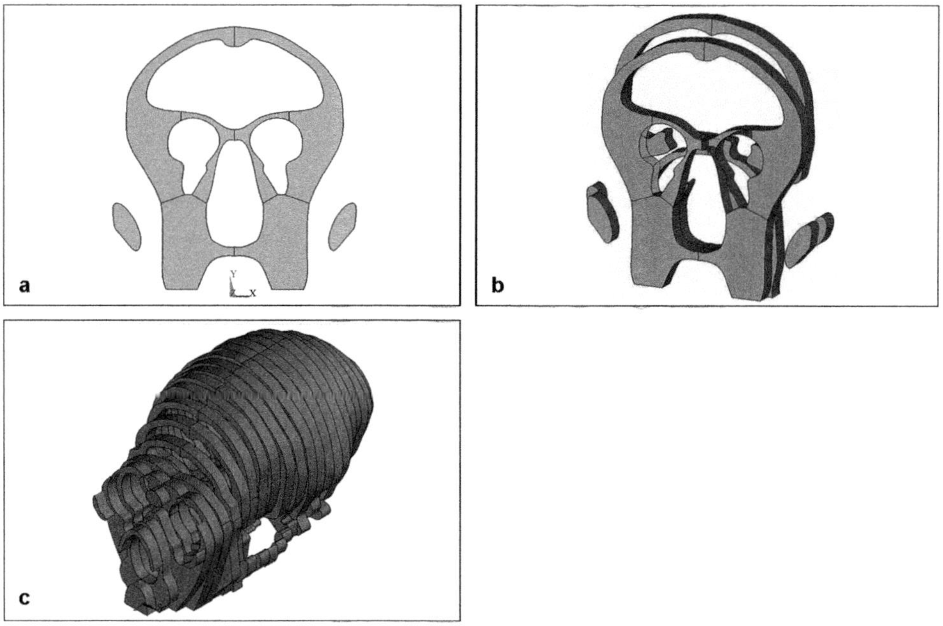

Fig. 18.5 Result of the first calculation including physiological summation of all load cases and elimination of non- or lightly stressed parts. Preparation of the gradation-model. (**a**) Cross-section 5. (**b**) Extrusion of cross-sections 5 and 6. (**c**) Gradation model. View from *top left*

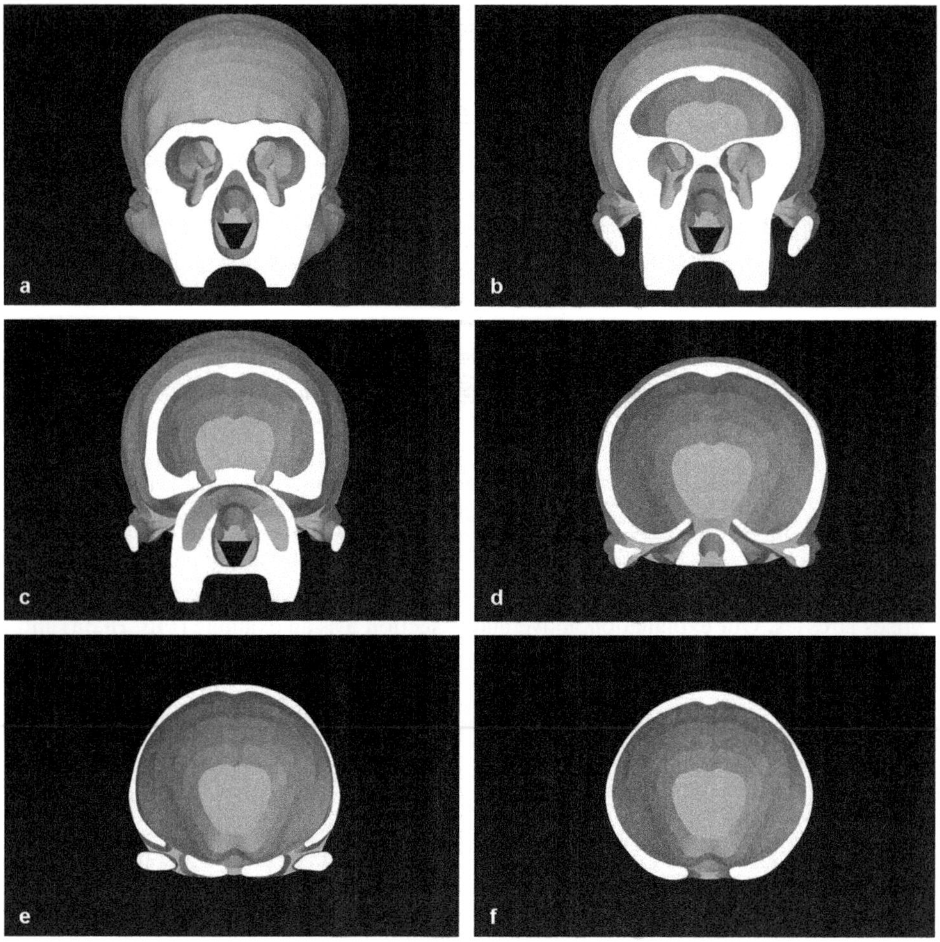

Fig. 18.6 Reduced model after first calculation using FESS. Presentation as a CAD model. (**a**) 3 rd cross-section and (**b**) 6th cross-section with incisura supraorbitalis, foramen infraorbitale, zygomatic arch and ethmoid bone. Restrictions: no frontal sinus, inferior orbital rim and maxillary sinus at this position. (**c**) 7th cross-section. Insight into maxillary sinus with hiatus. (**d**) 10th cross-section. Insight into sphenoidal sinus and on each side the foramen lacerum. (**e**) 14th cross-section. Auditory passage with porus acusticus externus and internus. Section through apertura externa canalis carotici. Restriction: no petrous bone. (**f**) 16th cross-section with foramen magnum

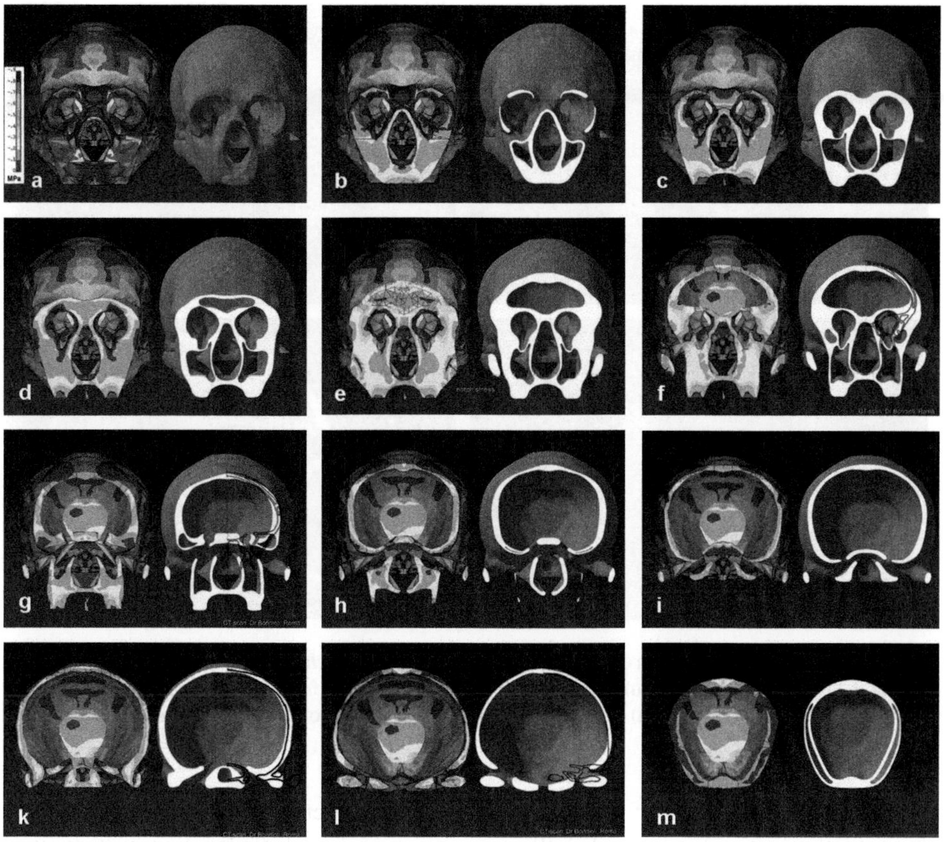

Fig. 18.7 Result of second calculation. Frontal view with compressive stress distribution (*left*) and as reduced CAD model (*right*). (**a**) pure frontal view. (**b**) 2nd cross-section with foramen supraorbitale, incisura frontalis, foramen infraorbitale, canalis infraorbitalis and maxillary sinus. Thinner bony roof of the nose and correction of the nose cavity. Restrictions: no inferior orbital rim, no spina nasalis anterior, no prominent glabella. (**c**) 3rd cross-section with foramen supraorbitale, foramen infraorbitale and maxillary sinus. Restriction: no inferior orbital rim. (**d**) 4th cross-section with frontal sinus. (**e**) 5th cross-section with the beginning of zygomatic arch. Restrictions: no tabula interna of the frontal sinus and no vertical segmentation. (**f**) 6th cross-section with brain cavity, parts of cancellous bone. Overlay of CT scan (Guattari). (**g**) 7th cross-section with canalis opticus, parts of cancellous bone, connection (hiatus) between nasal cavity and maxillary sinus. Overlay of CT scan (Guattari). (**h**) 8th cross-section with parietal thinning, fossa cranialis media, end of maxilla and lamina lateralis processus pterygoidei. (**i**) 9th cross-section with sella turcica, sulcus caroticus, foramen ovale. Restriction: no distal lamina of sinus sphenoidalis. (**k**) 11th cross-section with fossa hypophysialis and sinus sphenoidalis. Overlay of CT scan (Guattari). (**l**) 14th cross-section with porus acusticus externus et internus, temporomandibular joint, apertura externa canalis caroti, condylus occipitalis, and beginning of foramen magnum. Overlay of CT scan (Guattari). (**m**) 20th cross-section with crista occipitalis interna, protuberantia occipitalis interna and diploe (cancellous bone)

and internus) is synthesized (Fig. 18.6e) and the foramen magnum which was defined a priori is integrated into the skull base in Fig. 18.6f.

For the next step of reduction, this 3D model was meshed again with 10-noded tetrahedral finite-elements. The subsequent calculation with the same loading regime (load case technique) as in the first case yields the surprising results shown in Fig. 18.7. All cortical shells, the coupling structures for connecting functional spaces, change to a more concentrated walled construction. If the lightly stressed parts in all cross-sections with a threshold of −.2 N/mm² are eliminated, the synthesized structure very closely resembles the skull of *Homo neanderthalensis* in itself (Fig. 18.7a). In Fig. 18.7b–e and h the thinning effect of bony walls is demonstrated. Other thick structures, such as the skull base (Fig. 18.7g) and occipital skull wall (Fig. 18.7m) are reduced to a cortical-trabecular bone sandwich like a technical light-weight construction. Highly stressed areas are artefacts due to the notch (stress raiser) effect (Fig. 18.7e), or because there is a deficit of material.

Into Fig. 18.7f, g, k, and l CT scans of *Homo neanderthalensis* (Guattari) are projected.

This comparison with the result of FESS is very satisfactory and demonstrates the potential advantages of FESS.

The summary result of the second calculation using FESS is shown in Fig. 18.8a–d. Several 3D views of the virtual synthesized skull are presented using CAD techniques.

Discussion

The results of the described synthesis of the *Homo neanderthalensis* skull, as described above, demonstrate a direct correlation between reasonable functional loadings and form.

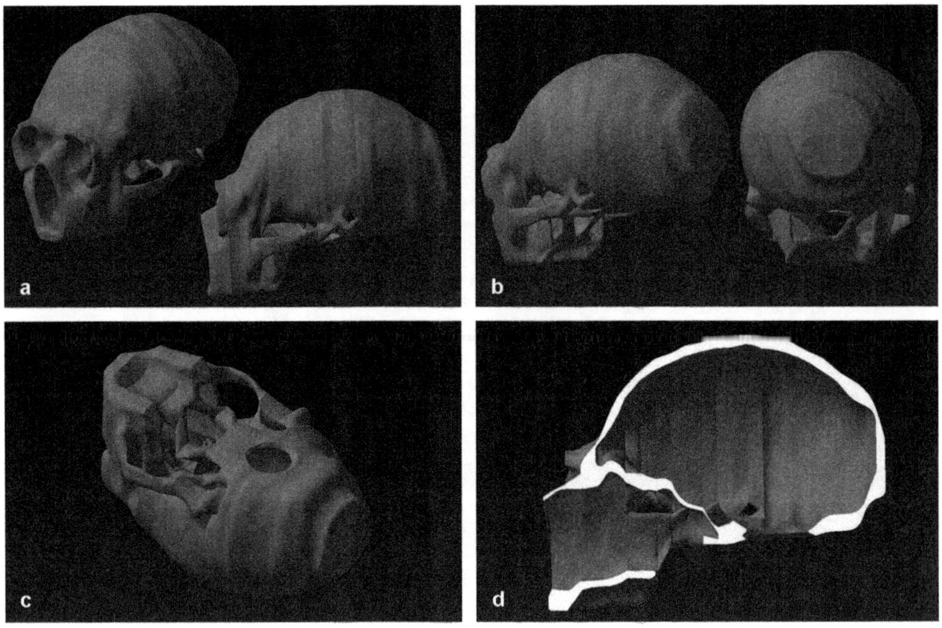

Fig. 18.8 Summarized result of the second calculation using FESS demonstrated in several views of the virtual synthesized skull of *Homo neanderthalensis* modelled by CAD software CATIA V5. (**a**) lateral views. (**b**) views from *behind*. (**c**) view from *below*. (**d**) sagittal section with brow ridges, sinus frontalis, palatinum and dental arcade. Restrictions: no tabula interna of frontal sinus and no distal lamina of sinus sphenoidalis

This synthesis suggests that the evolution of skull form in Neanderthals reflects natural selection for optimal skull construction, where optimality is defined as maximum strength with minimum material.

The synthesized skull represents thin-walled functional spaces with high stability and their coupling structures. Their function is to optimally bear loads from biting, chewing and head motion and to encapsulate the spaces for sense organs. The similarity between the stress flow and the shape of Neanderthal skull seems to indicate that the shape and structure of the skull is dependent on mechanical stresses. The skull shape is optimised to loading through mechanically influenced modelling and remodelling during ontogeny. The shape and structure of the Neanderthal skull, including several openings within it, are essentially predicted by biomechanical necessity.

In this study, the inferior orbital rim (Fig. 18.7c), the tabula interna of frontal sinus and the vertical segmentation in it (Fig. 18.7d, e), the upper closing of the maxillary sinus (Fig. 18.7g) and the distal lamina of sphenoidal sinus (Fig. 18.7i) are not synthesized. This is due to the fact that the FE software capacity in our university is limited (129,000 nodes instead of unlimited versions of other institutes or industry) and cannot model small structures such as very thin shells.

Furthermore, the brow ridges are not as distinct as in *Homo neanderthalensis* (La Chapelle-aux-Saints and Guattari) but almost identical to *Homo neanderthalensis* (La Ferrassie 1) and *Homo heidelbergensis* (Atapuerca 5) (Bookstein et al. 1999). If our synthesized model is not in full agreement with the shape in one fossil but in another, the most probable explanation is that our assumptions about loading do not fit perfectly. In future we should add mm. pterygoideus medialis and lateralis and pars profunda as the deep part of m. masseter.

This study again demonstrates the utility of finite-element structure synthesis (FESS) for the virtual synthesis of human skulls in order to test assumptions and hypotheses regarding the relationship between skull function and structure.

Acknowledgments The author is greatly obliged to Mrs J. Mannhardt (cand. ing.), Mr. V. Meimann (Dipl.-Ing.) and Mr. M. Neges (cand. ing.) who executed countless calculations and prints with incredible engagement. His special thanks go to Prof. H. Preuschoft and Prof. G.-C. Weniger for the procurement of skull castings and to Dr. L. Bondioli, Roma, Museo Nazionale "L. Pigorini" for several CT scans of the skull of *Homo neanderthalensis* (Guattari). The latter contact was given by Prof. G. W. Weber to whom he feels bound with great thanks. He would also like to thank an anonymous reviewer for his careful reading and constructive criticism of his article.

References

Ascenzi, A., & Bonucci, E. (1968). The compressive properties of single osteons. *The Anatomical Record, 161*, 377–392.

Bookstein, F., Schäfer, K., Prossinger, H., Seidler, H., Fieder, M., Stringer, C., Weber, G. W., Arsuaga, J. L., Slice, D. E., Rohlf, F. J., Recheis, W., Mariam, A. J., & Marcus, L. F. (1999). Comparing frontal cranial profiles in archaic and modern homo by morphometric analysis. *The Anatomical Record, 257*, 217–224.

Carter, D. R., & Wong, M. (1988). Mechanical stresses and endochondral ossification in the chondroepiphysis. *Journal of Orthopaedic Research, 6*, 148–154.

Darwin, C. (1859). The origin of the species. London. (1963). Die Entstehung der Arten. Philipp Reclam, Stuttgart.

Fastnacht, M., Hess, N., Frey, E., & Weiser, H. P. (2002). Finite-element analysis in vertebrate palaeontology. *Senckenbergiana Lethaea, 82*, 195–206.

Frost, H. M. (1988). Vital biomechanics: Proposed general concepts for skeletal adaptations to mechanical usage. *Calcified Tissue International, 42*, 145–156.

Imai, M., Shibata, T., Moriguchi, K., & Takada, Y. (1990). Fluid-path in the mandible and maxilla. *Okajimas Folia Anatomica Japonica, 67*(4), 243–247.

Jansen, M. (1920). *On Boneformation, its relation to tension and pressure*. London: Longmans, Green & Co.

O'Connor, C. F., Franciscus, R. G., & Holton, N. E. (2005). Bite force production capability and efficiency in Neanderthals and modern humans. *American Journal of Physical Anthropology, 127*, 121–151.

Pauwels, F. (1965). *Gesammelte Abhandlungen zur funktionellen Anatomie des Bewegungsapparates*. Berlin: Springer.

Pidaparti, R. M., & Burr, D. B. (1992). Collagen fibre orientation and geometry effects on the mechanical properties of secondary osteons. *Journal of Biomechanics, 25*(8), 869–880.

Preuschoft, H., & Witzel, U. (2004). Functional structure of the skull in hominoidea. *Folia Primatologica, 75*, 219–252.

Preuschoft, H., Witte, H., & Witzel, U. (2002). Pneumatized spaces, sinuses and spongy bones in the skulls of primates. *Anthropology Anzeiger, 60*, 67–79.

Qin, L., Mak, A. T., Cheng, C. W., Hung, L. K., & Chan, K. M. (1999). Histomorphological study on pattern of fluid movement in cortical bone in goats. *The Anatomical Record, 255*(4), 380–387.

Qin, X. Y., Kaplan, T., Saldanha, A., & Rubin, C. (2003). Fluid pressure gradients, arising from oscillations in intramedullary pressure, is correlated with the formation of bone and inhibition of intracortical porosity. *Journal of Biomechanics, 36*, 1427–1437.

Rayfield, E. J. (1998). Finite-element analysis of the snout of *Megalosaurus bucklandi. Journal of Vertebrate Paleontology, 18*, 71A.

Rayfield, E. J. (2004). Cranial mechanics and feeding in *Tyrannosaurus rex. Proceedings of the Royal Society of London. Series B: Biological Sciences, 271*, 1451–1459.

Rossmann, T., Witzel, U., & Preuschoft, H. (2005). Mechanical stress as the main factor in skull design of the fossile reptile *Proterosuchus* (Archosauria). In T. Rossmann & C. Tropea (Eds.), *Bionik, Aktuelle Forschungsergebnisse in Natur-Ingenieur- und Geisteswissenschaft* (pp. 517–528). Berlin: Springer.

Sverdlova, N., & Witzel, U. (2010). Principles of determination and verification of muscle forces in the human musculoskeletal system: Muscle forces to minimise bending stress, *Journal of Biomechanics, 43*, 387–396.

Witzel, U. (1985). Zur Biomechanik der ossaeren Schafteinbettung von Hueftendoprothesen, 5. Symposium ueber isoelstische RM-Hueftprothesen. Bettlach, Schweiz: Mathys.

Witzel, U. (1993). Biomechanische Untersuchungen am Verbundsystem des Binde- und Stützgewebes mit der Methode der finiten Elemente. In H. J. Pesch, H. Stoess, & B. Kummer (Eds.), *Osteologie aktuell* (Vol. VII, pp. 91–97). Berlin: Springer.

Witzel, U. (1994). Ueber die Spannungsverteilung in der Calvaria aufgrund physiologischer und traumatisierender Beschleunigungen des Schaedels. Osteologie, Band 3, Suppl 1, abstracts. Verlag Hans Huber, Bern, p. 50.

Witzel, U. (2006). Virtual synthesis of the skull in Neanderthals by FESS. In: v. Koenigswald, W., Litt, T. (Eds.), 150 years of Neanderthal discoveries. Abstract. Terra Nostra 2006/2, (p. 141) Berlin:Geo Union.

Witzel, U., & Hoffmann, P. (1993). Ueber den Einfluss der durch Beschleunigungskraefte belasteten Dura mater encephali auf die Spannungsverteilung in der Calvaria. Osteologie Band 2, Suppl 1, abstracts. (p. 37) Bernlin:Verlag Hans Huber

Witzel, U., & Preuschoft, H. (1999). The bony roof of the nose in humans and other primates. *Zoologischer Anzeiger, 238*, 103–115.

Witzel, U., & Preuschoft, H. (2002). Function-dependent shape characteristics of the human skull. *Anthropology Anzeiger, 60*, 113–135.

Witzel, U., & Preuschoft, H. (2005). Finite-element model construction for the virtual synthesis of the skulls in vertebrates: Case study of *Diplodocus. The Anatomical Record, 283A*, 391–401.

Witzel, U., Preuschoft, H., & Sick, H. (2004). The role of the zygomatic arch in the statics of the skull and its adaptive shape. *Folia Primatologica, 75*, 202–218.

Wolff, J. (1892). *Das Gesetz der transformation der Knochen*. Berlin: Hirschwald.

Wong, M., & Carter, D. R. (1990). A theoretical model of endochondral ossification and bone architectural construction in long bone ontogeny. *Anatomy and Embryology, 181*, 523–532.

Zienkiewicz, O. C. (1971). *The finite element method in engineering science*. London: McsGraw-Hill.

Chapter 19
Neandertal mtDNA from a Late Pleistocene Human Mandible from the Cova del Gegant (Spain)

Juan Luis Arsuaga, Rolf Quam, Joan Daura, Montserrat Sanz, Maria Eulàlia Subira, Love Dalén, and Anders Götherström

Abstract Over the last decade, D-loop fragments of mtDNA of varying lengths have been published from 12 Neandertal specimens. The fossils have been recovered from geographically diverse sites, ranging from the Caucasian mountains to the Iberian Peninsula. In the sequences from these fossils, some mutations, present even in the shortest fragments, characterize all published Neandertal sequences to date and are absent or very rare in *Homo sapiens*. Here we use some of these diagnostic genetic substitutions to support our recent morphological assessment of the Cova del Gegant mandible as representing a Neandertal. The short sequence obtained (52 bp) contains substitutions common to all Neandertals. To further verify the Neandertal nature of the Cova del Gegant sequence, it was compared with the 232 most similar *H. sapiens* sequences from the GenBank. NJ bootstrap values as well as Bayesian posterior probability for the sequence from Cova del Gegant clearly place it with other Neandertals. The variation within the fragment does not show any geographical structure, but there is one substitution (16243) that may covary with age to some degree. We conclude that the recovery of this short mtDNA fragment can be used as a diagnostic tool for taxonomic classification in European Upper Pleistocene fossil human specimens.

Keywords Morphology • Iberian Peninsula • NJ bootstrap • Taxonomic classification

J.L. Arsuaga
Departamento de Paleontología, Universidad Complutense de Madrid, Avenida Complutense s/n, 28040 Madrid, Spain
and
Centro UCM-ISCIII de Investigación sobre, Evolución y Comportamiento Humanos, c/Sinesio Delgado, 4, 28029 Madrid, Spain
e-mail: jlarsuaga@isciii.es

R. Quam
Centro UCM-ISCIII de Investigación sobre, Evolución y Comportamiento Humanos, c/Sinesio Delgado, 4, 28029 Madrid, Spain
and
Division of Anthropology, American Museum of Natural History, Central Park West at 79th St., New York, NY 10024-5192, USA
e-mail: rquam@amnh.org

J. Daura and M. Sanz
Grup de Recerca del Quaternari, SERP–Seminari d'Estudis i Recerques Prehistòriques, Dpto. de Prehistoria, Historia Antiga i Arqueologia, Universitat de Barcelona, c/Montalegre, 6, 08001 Barcelona, Spain
e-mail: grupquaternari@hotmail.com

M.E. Subira
Unitat d'Antropologia, Departament de Biologia Animal, de Biologia Vegetal i d'Ecologia, Edifici C, Universitat Autònoma de Barcelona, 08193, Bellaterra (Cerdanyola del Vallès) Barcelona, Spain
e-mail: Eulalia.Subira@uab.es

L. Dalén
Centro UCM-ISCIII de Investigación sobre, Evolución y Comportamiento Humanos, c/Sinesio Delgado, 4, 28029 Madrid, Spain
e-mail: ldalen@isciii.es

A. Götherström (✉)
Centro UCM-ISCIII de Investigación sobre, Evolución y Comportamiento Humanos, c/Sinesio Delgado, 4, 28029 Madrid, Spain
and
Department of Evolutionary Biology, Uppsala University, Norbyvägen 18D, SE-75236 Uppsala, Sweden
e-mail: Anders.Gotherstrom@ebc.uu.se

Introduction

The publication of the first Neandertal mtDNA sequence (Krings et al. 1997), from the type specimen, was considered a major breakthrough in the field of ancient DNA. However, it was not until 3 years later that a second mtDNA sequence appeared in the scientific press (Ovchinnikov et al. 2000). Knowledge of the Neandertal mtDNA sequence has made it easier to design Neandertal-specific experiments, as illustrated by the recent publication of three Neandertal sequences in the same year (Caramelli et al. 2006; Lalueza-Fox et al. 2006; Orlando et al. 2006). Currently, there are 12 known Neandertal mtDNA sequences (Krings et al. 1997; Ovchinnikov et al. 2000; Krings et al. 2000; Schmitz et al. 2002; Serre et al. 2004; Beauval et al. 2005; Caramelli et al. 2006; Lalueza-Fox et al. 2006; Orlando et al. 2006), and the increasing number of Neandertal sequences has led researchers to propose both phylogeographic and demographic hypotheses (Currat and Excoffier 2004; Lalueza-Fox et al. 2005;

S. Condemi and G.-C. Weniger (eds.), *Continuity and Discontinuity in the Peopling of Europe: One Hundred Fifty Years of Neanderthal Study*, Vertebrate Paleobiology and Paleoanthropology, DOI 10.1007/978-94-007-0492-3_19, © Springer Science+Business Media B.V. 2011

Excoffier 2006). Recently, the feasibility of recovering nuclear DNA and massive mtDNA from extremely well-preserved Neandertal specimens has been convincingly demonstrated (Green et al. 2006; Noonan et al. 2006; Green et al. 2008), as well as the possibility to type mitochondrial Single Nucleotide Polymorphisms from material excavated in earlier hominid contexts (Valdiosera et al. 2006), and these represent promising lines of future research.

Within the D-loop in the mitochondrial genome, there are substitutions and insertions/deletions (indels) that are present in Neandertals at a much higher frequency than in modern humans, some of which appear to be Neandertal-specific, being present in all known Neandertal sequences to date. Short mtDNA fragments, spanning some of these substitutions, have previously been used as a diagnostic tool to distinguish between *Homo neanderthalensis* and *H. sapiens* (Serre et al. 2004). A genetic test for Neandertal taxonomic status may also help to address the recent assertions of a Neandertal genetic contribution to the early *H. sapiens* inhabitants of Europe (Duarte et al. 1999; Smith et al. 2005; Trinkaus 2005; Soficaruet al. 2006). Here we analyze a short 52 bp D-loop mtDNA fragment (spanning positions 16210–16261) recovered from the Cova del Gegant human mandible to examine the genetic diversity within the known Neandertal sample and to complement the anthropological assessment of Neandertal affinities for the specimen (Daura et al. 2005).

Materials and Methods

The site of Cova del Gegant is located near the city of Sitges, some 40 km to the south of Barcelona along the Spanish Mediterranean coast. A human mandible was recovered by Santiago Casanova during excavations in 1952, but was not recognized until 2001 during a revision of the faunal collection. The material recovered from the same area of the site which yielded the human mandible consists of Upper Pleistocene faunal remains and Mousterian stone tools. The chronological interpretation of the site suggests a human occupation of the cave, and presumably the deposition of the human mandible, spanning the period from approximately 100-40 kya (Daura et al. 2005).

The mandible from Cova del Gegant has been described and compared with Pleistocene *Homo* specimens, and shows a suite of archaic anatomical features which clearly distinguish it from living humans (Daura et al. 2005). Some of these features represent the primitive condition for the genus *Homo*, such as the lack of a bony chin, and are seen even in the earliest members of this genus. However, several features associated with the derived midfacial prognathism in Neandertals are also present in Cova del Gegant, particularly the posterior placement of the anatomical structures of the lateral corpus. Further, the low position of the mental foramen, in the lower half of the mandibular corpus, also seems to be a feature which occurs in a higher frequency among Neandertals and their Middle Pleistocene precursors. Given that the cave deposits at the Cova del Gegant site from which the mandible derives contain exclusively Pleistocene faunal remains and Mousterian stone tools, the anatomy, chronology, and cultural association of the fossil specimen are consistent with its classification as a Neandertal (Daura et al. 2005).

DNA was extracted from the specimen using a silica spin column-based method (Malmström et al. 2005; Yang et al. 1998). However, two modifications were made: changing the SDS in the extraction buffer for 1 M urea, and letting the guanidium buffer with the DNA rest on the silica filter for an hour before spinning it through. The fragment was amplified with previously described primers (NL16209, NH16262, (Krings et al. 1997)). PCR (55 cycles) was carried out in 25 μl reactions containing 3U HotStarTaq (Qiagen), 200 μM of each dNTP, and 0.5 pMol of each primer. 5 μl of the extract was used as template and annealing temperature was 58°C. Sequencing was performed in both directions on a Megabace instrument according to the suppliers' recommendations. The amplification and sequencing was repeated once to confirm the sequence. The two negative extraction controls were processed parallel to the hominid sample, and the PCR reaction was controlled for specificity with ten modern human DNA extracts.

The 500 most similar *H. sapiens* sequences were downloaded from the GenBank (with BLAST). The reason behind this nonrandom selection was that if any H. sapiens sequences could be confused with Neandertal sequences, it would be the most similar ones. Those that lacked information (such as undetermined bases or missing data in the 5′ or 3′ end) were removed, leaving a total of 232 sequences. These, together with the seven published Neandertal sequences (Krings et al. 1997; Krings et al. 2000; Ovchinnikov et al. 2000; Schmitz et al. 2002; Caramelli et al. 2006; Lalueza-Fox et al. 2006; Orlando et al. 2006) containing the same information as we targeted in the specimen from Cova del Gegant, were selected for phylogenetic analysis. A single *Pan troglodytes* sequence was also used as an outgroup. A neighbor joining (NJ) phylogeny (Kimura-2,10,000 pseudoreplicas) as well as a maximum likelihood phylogeny (HKY, using a discrete gamma model with four categories, a gammashape of 1.108, a proportion of invariants of 0.240, and a transition/transversion ratio of 4:1000 pseudoreplicas) were constructed. As the fragment is short and contains little data, we constructed a second phylogeny using a likelihood method based on Bayesian statistics and using MrBayes (GTR nst = 6;10,000,000 generations with every 2000th generation sampled, burnin set for 1,000) to ensure a possible Neandertal clade supported by the NJ phylogeny was not method-dependent.

As phylogenetic reconstruction tests for evolutionary branching patterns, and generally demands a certain amount of data to provide fair support, we also performed a $\chi 2$ test with 500 randomly drawn *H. sapiens* D-loop sequences from the GenBank. The test was based on the C/A substitution in 16256, where Neandertal sequences display A, while all, or at least the vast majority of the *H. sapiens* sequences display C.

Results and Discussion

The fossil yielded a short mtDNA fragment (52 bp, spanning positions 16210–16261). The PCR products were sequenced in triplicate, forward as well as reverse. The quality of the sequence was good, and we could not detect any deviation from published Neandertal sequences (Fig. 19.1). None of the negative controls yielded PCR products, nor did the ten modern human samples yield any PCR products. Thus, we could not detect any signs of degraded bases or contamination, and the sequence appears to be of authentic ancient origin. The material is of considerable age, but it is still well within the temporal range of ancient DNA retrieval (Loreille et al. 2001; Willerslev et al. 2003; Valdiosera et al. 2006). In addition, the size of the targeted fragment appears to be of major importance (Noonan et al. 2006; Poinar et al. 2006; Valdiosera et al. 2006) and by targeting a short fragment, the possibility of successfully retrieving genetic material was maximized.

Several factors suggest that the mtDNA sequence recovered from the Cova del Gegant specimen is of Neandertal origin. The first one is the specificity of the reaction. The primers were designed to target Neandertal DNA (Krings et al. 1997). We further tested the specificity by trying to amplify modern human DNA of high quality. The only DNA extract amplifying with the PCR system was the one from the Cova del Gegant hominid specimen. Further, the sequence contains several substitutions (bp positions 16234, 16244, and 16256) which are ubiquitous in the published Neandertal sequences. The combination of these same three substitutions is not found in any of the *H. sapiens* individuals in our study, but they do each occur individually in very low frequencies among the modern humans in our sample. Had the Cova del Gegant sequence originated from an anatomically modern human, we would not expect to find all three of these substitutions.

The phylogenetic analysis indicates that the hominid has a genetic origin among Neandertals (Fig. 19.2). In comparison with the *H. sapiens* and the *Pan troglodytes* sequences (yielding a total of 15 polymorphic sites within the data set), the bootstrap value yielded in the NJ tree provided fair support (69%) for the Neandertal clade, including the hominid specimen from Cova del Gegant. The ML tree also supported the Neandertal clade (73%). When we constructed a Bayesian phylogeny, the posterior probability for the Neandertal clade, including Cova del Gegant, also provided strong support (100). Further, the $\chi2$ test indicated that the probability for the two sample groups belonging to the same population was very low ($p < 0.0001$). The *H. sapiens* clade was not as well resolved. The bootstrap support was lower for one complete human clade (NJ 59, ML 34), and we did not find the *H. sapiens* sequences in a distinct and well-supported sistergroup to the Neandertal clade. Judging from previous work (Krings et al. 1997; Krings et al. 1999; Ovchinnikov et al. 2000; Green et al. 2006; Noonan et al. 2006), this is probably due to the limited data included in the short length of the sequence fragment under study. Nevertheless, the Neandertal clade was supported, and the hominid specimen from the Cova del Gegant grouped within it.

The 52 bp fragment from the Cova del Gegant contained two sites which are polymorphic within Neandertals. One of

```
                16,210    16,220    16,230    16,240    16,250    16,260
CRS             ACAGCAATCAACCCTCAACTATCACACATCAACTGCAACTCCAAAGCCACCC
CDG             .............T......G...T.........A...........A.G...
Feldhofer 2     .............T......G...T.........A...........A.....
Mezmaiskaya     .............T......G...T.........A...........A.....
El Sidron       .............T......G...T.........A...........A.G...
Monti Lessini   .............T......G...T.........A...........A.G...
Vi75            .............T......G...T.........A...........A.G...
Feldhofer 1     .............T......G...T.........A...........A.G...
Scladina        .............T......G...T........CA...........A.....
Vindija 77      ~~~~~~~~~~~~~~~~~~~~~...T.........A...........A.G...
Vindija 80      ~~~~~~~~~~~~~~~~~~~~~...T.........A...........A.G...
Engis 2         ~~~~~~~~~~~~~~~~~~~~~...T.........A...........A.....
La Chapelle     ~~~~~~~~~~~~~~~~~~~~~...T.........A...........A.....
Les Rochers     ~~~~~~~~~~~~~~~~~~~~~...T.........A...........A.....
```

Fig. 19.1 Alignment of the 12 published Neandertal sequences (16210–16261 bp), the sequence from the Cova del Gegant specimen (CDG), and the modern human Cambridge Reference Sequence (CRS). The last five specimens in the alignment have so far only yielded a portion of the 52 bp spanned by the Cova del Gegant sequence. Two substitutions (16243 and 16258) are polymorphic within Neandertals, and one of them (16243) singles out the oldest specimen (Scladina)

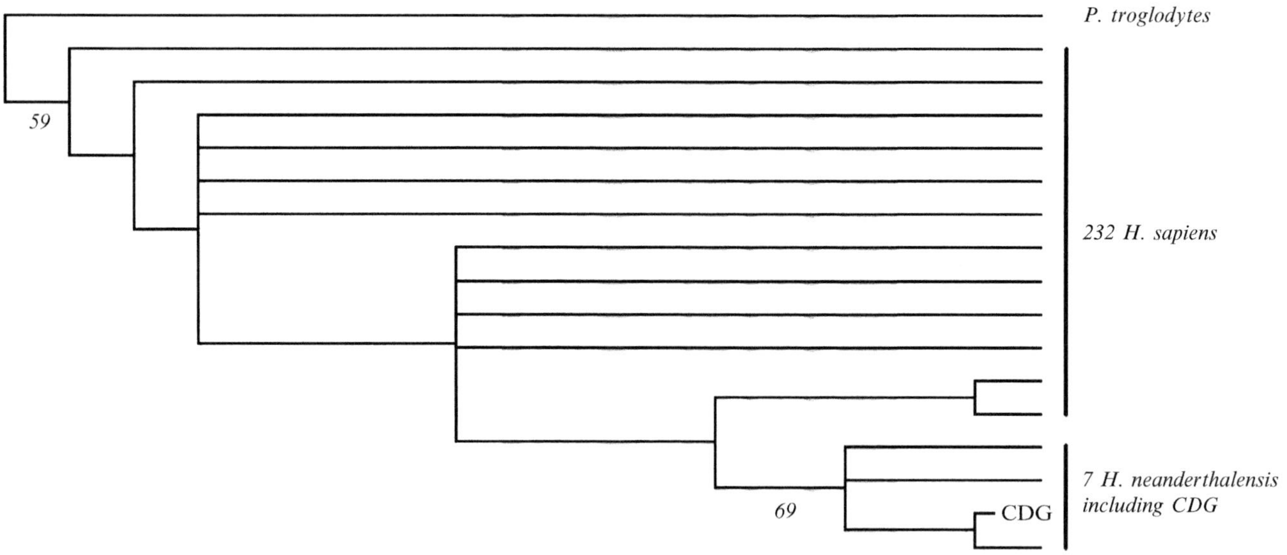

Fig. 19.2 NJ tree including 232 modern *H. sapiens* sequences, seven Neandertal sequences, the sequence from the hominid for Cova del Gegant (CDG), and a chimpanzee sequence as outgroup. The Neandertal sample includes only those specimens which preserved all 52 bp spanned by the Cova del Gegant sequence. The Neandertal clade, supported with a bootstrap of 69, was also supported with a posterior value (100) when a Bayesian tree was inferred from the same data as well as an ML value of 73 for the same data

them (16243) is only polymorphic in the oldest specimen published so far (Orlando et al. 2006), while the other (16258) is more variable. It has been suggested that this latter mutation indicates a deep split among Neandertals (Lalueza-Fox et al.2005). As such, it could potentially be used to infer a phylogeographic structure, spanning from east to west. However, more recently published sequences have shown that the substitution is recurrent (Caramelli et al. 2006), and it is therefore not possible to make such assertions based on it. Further, no phylogeographic structure of the kind is visible when more sequence data are analyzed (Excoffier 2006). To address this question more definitively, further Neandertal sequences from geographically diverse specimens are clearly necessary.

Substitution 16243 presents a more interesting feature. Thus far, only the sequence from the oldest specimen (Scladina) presents a deviating base in this position. Interestingly, the substitution seems to be derived in the Scladina specimen, since the younger Neandertal specimens, and Cova del Gegant, are similar to the majority of the *H. sapiens* and *P. troglodytes* sequences. The Scladina sequence also deviates from the other published Neandertal sequences in several other positions, and it has been suggested that this elevated variation may indicate population dynamics and replacement of Neandertal populations over time in the species' northern European distribution (Excoffier 2006). Nevertheless, the implications of the Scladina sequence can only be tested by obtaining further Neandertal mtDNA sequences from specimens spanning different time periods and geographical areas. The recent recovery of Middle Pleistocene mtDNA from the bear species *Ursus deningeri* at the site of the Sima de los Huesos in northern Spain (Valdiosera et al. 2006) suggests the feasibility of retrieving mtDNA from the *H. heidelbergensis* fossil human specimens at this same site as well. Successful mtDNA retrieval from *H. heidelbergensis*, considered to represent the ancestral population from which Neandertals later evolved (Arsuaga et al. 1993; Arsuaga et al. 1997), may provide new insights on the genetic origin and evolution of the Neandertal lineage.

In conclusion, the hominid specimen from Cova del Gegant is genetically a Neandertal, an assessment which complements the previous anthropological analysis (Daura et al. 2005). Amplifying and sequencing this short fragment of mtDNA may be a helpful tool for establishing the taxonomic affinities of more fragmentary fossil specimens where diagnostic anatomical regions are not preserved. The limited variation within this fragment may also be useful in further studies of Neandertal material and help to answer questions surrounding the phylogeography and evolutionary processes which shaped this Pleistocene population.

Acknowledgments We are especially grateful to X. Miret i Mestre for his work as director of the Arxiu Historic Municipal de Sitges and S. Casanova-Amunt for his work at the cave site. M.C. Ortega was responsible for the cleaning and restoring of the specimen. R. Quam was supported by a grant from the Fundación Duques de Soria/ Fundación Atapuerca and J. Daura by the Ministerio de Cultura, Educación y Deportes. This chapter is included in the research project Els Primers Pobladors de Garraf-Ordal, which is encompassed within the Projecte de Recerca el Plistocè Superior i l'Holocè a Catalunya, supported by the SGR2001-00007 Research Quality Group of the Generalitat de Catalunya, the 2006EXCAVA00012 project of Generalitat de Catalunya, and the HUM2004-600 project of the Ministerio de

Educación y Ciencia of Spain . This research was supported by the Ministerio de Ciencia y Tecnología of the Government of Spain, Project No. BOS-2003-08938-C03-01.

References

Arsuaga, J. L., Martinez, I., Gracia, A., Carretero, J. M., & Carbonell, E. (1993). Three new human skulls from the Sima de los Huesos Middle Pleistocene site in Sierra de Atapuerca, Spain. *Nature, 362*(6420), 534–537.

Arsuaga, J. L., Martinez, I., Gracia, A., & Lorenzo, C. (1997). The Sima de los Huesos crania (Sierra de Atapuerca, Spain). A comparative study. *Journal of Human Evolution, 33*(2–3), 219–281.

Beauval, C., Maureille, B., Lacrampe-Cuyaubere, F., Serre, D., Peressinotto, D., Bordes, J. G., Cochard, D., Couchoud, I., Dubrasquet, D., Laroulandie, V., Lenoble, A., Mallye, J. B., Pasty, S., Primault, J., Rohland, N., Pääbo, S., & Trinkaus, E. (2005). A late Neandertal femur from Les Rochers-de-Villeneuve, France. *Proceedings of the National Academy of Sciences USA, 102*(20), 7085–7090.

Caramelli, D., Lalueza-Fox, C., Condemi, S., Longo, L., Milani, L., Manfredini, A., de Saint, P. M., Adoni, F., Lari, M., Giunti, P., Ricci, S., Casoli, A., Calafell, F., Mallegni, F., Bertranpetit, J., Stanyon, R., Bertorelle, G., & Barbujani, G. (2006). A highly divergent mtDNA sequence in a Neandertal individual from Italy. *Current Biology, 16*(16), 630–632.

Currat, M., & Excoffier, L. (2004). Modern humans did not admix with Neanderthals during their range expansion into Europe. *Public Library of Sciences Biology, 2*(12), 421.

Daura, J., Sanz, M., Subira, M. E., Quam, R., Fullola, J. M., & Arsuaga, J. L. (2005). A Neandertal mandible from the Cova del Gegant (Sitges, Barcelona, Spain). *Journal of Human Evolution, 49*(1), 56–70.

Duarte, C., Mauricio, J., Pettitt, P. B., Souto, P., Trinkaus, E., van der Plicht, H., & Zilhao, J. (1999). The early Upper Paleolithic human skeleton from the Abrigo do Lagar Velho (Portugal) and modern human emergence in Iberia. *Proceedings of the National Academy of Sciences USA, 96*(13), 7604–7609.

Excoffier, L. (2006). Neandertal genetic diversity: A fresh look from old samples. *Current Biology, 16*(16), 650–652.

Green, R. E., Krause, J., Ptak, S. E., Briggs, A. W., Ronan, M. T., Simons, J. F., Du, L., Egholm, M., Rothberg, J. M., Paunovic, M., & Pääbo, S. (2006). Analysis of one million base pairs of Neanderthal DNA. *Nature, 444*(7117), 330–336.

Green, R. E., Malaspinas, A. S., Krause, J., Briggs, A. W., Johnson, P. L., Uhler, C., Meyer, M., Good, J. M., Maricic, T., Stenzel, U., Prüfer, K., Siebauer, M., Burbano, H. A., Ronan, M., Rothberg, J. M., Egholm, M., Rudan, P., Brajković, D., Kućan, Z., Gusić, I., Wikström, M., Laakkonen, L., Kelso, J., Slatkin, M., & Pääbo, S. (2008). A complete Neandertal mitochondrial genome sequence determined by high-throughput sequencing. *Cell, 134*, 416–426.

Krings, M., Stone, A., Schmitz, R. W., Krainitzki, H., Stoneking, M., & Pääbo, S. (1997). Neandertal DNA sequences and the origin of modern humans. *Cell, 90*(1), 19–30.

Krings, M., Geisert, H., Schmitz, R. W., Krainitzki, H., & Pääbo, S. (1999). DNA sequence of the mitochondrial hypervariable region II from the Neandertal type specimen. *Proceedings of the National Academy of Sciences USA, 96*(10), 5581–5585.

Krings, M., Capelli, C., Tschentscher, F., Geisert, H., Meyer, S., von Haeseler, A., Grossschmidt, K., Possnert, G., Paunovic, M., & Pääbo, S. (2000). A view of Neandertal genetic diversity. *Nature Genetics, 26*(2), 144–146.

Lalueza-Fox, C., Sampietro, M. L., Caramelli, D., Puder, Y., Lari, M., Calafell, F., Martinez-Maza, C., Bastir, M., Fortea, J., de la Rasilla, M., Bertranpetit, J., & Rosas, A. (2005). Neandertal evolutionary genetics: Mitochondrial DNA data from the iberian peninsula. *Molecular Biology and Evolution, 22*(4), 1077–1081.

Lalueza-Fox, C., Krause, J., Caramelli, D., Catalano, G., Milani, L., Sampietro, M. L., Calafell, F., Martinez-Maza, C., Bastir, M., Garcia-Tabernero, A., de la Rasilla, M., Fortea, J., Pääbo, S., Bertranpetit, J., & Rosas, A. (2006). Mitochondrial DNA of an Iberian Neandertal suggests a population affinity with other European Neandertals. *Current Biology, 16*(16), 629–630.

Loreille, O., Orlando, L., Patou-Mathis, M., Philippe, M., Taberlet, P., & Hänni, C. (2001). Ancient DNA analysis reveals divergence of the cave bear, *Ursus spelaeus*, and brown bear, *Ursus arctos*, lineages. *Current Biology, 11*(3), 200–203.

Malmström, H., Storå, J., Dalén, L., Holmlund, G., & Götherström, A. (2005). Extensive human DNA contamination in extracts from ancient dog bones and teeth. *Molecular Biology and Evolution, 22*(10), 2040–2047.

Noonan, J. P., Coop, G., Kudaravalli, S., Smith, D., Krause, J., Alessi, J., Chen, F., Platt, D., Pääbo, S., Pritchard, J. K., & Rubin, E. M. (2006). Sequencing and analysis of Neanderthal genomic DNA. *Science, 314*(5802), 1113–1118.

Orlando, L., Darlu, P., Toussaint, M., Bonjean, D., Otte, M., & Hänni, C. (2006). Revisiting Neandertal diversity with a 100,000 year old mtDNA sequence. *Current Biology, 16*(11), 400–402.

Ovchinnikov, I. V., Götherström, A., Romanova, G. P., Kharitonov, V. M., Liden, K., & Goodwin, W. (2000). Molecular analysis of Neanderthal DNA from the northern Caucasus. *Nature, 404*(6777), 490–493.

Poinar, H. N., Schwarz, C., Qi, J., Shapiro, B., Macphee, R. D., Buigues, B., Tikhonov, A., Huson, D. H., Tomsho, L. P., Auch, A., Rampp, M., Miller, W., & Schuster, S. C. (2006). Metagenomics to paleogenomics: Large-scale sequencing of mammoth DNA. *Science, 311*(5759), 392–394.

Schmitz, R. W., Serre, D., Bonani, G., Feine, S., Hillgruber, F., Krainitzki, H., Pääbo, S., & Smith, F. H. (2002). The Neandertal type site revisited: Interdisciplinary investigations of skeletal remains from the Neander Valley, Germany. *Proceedings of the National Academy of Sciences USA, 99*(20), 13342–13347.

Serre, D., Langaney, A., Chech, M., Teschler-Nicola, M., Paunovic, M., Mennecier, P., Hofreiter, M., Possnert, G., & Pääbo, S. (2004). No evidence of Neandertal mtDNA contribution to early modern humans. *Public Library of Sciences Biology, 2*(3), E57.

Smith, F., Janković, I., & Karavanić, I. (2005). The assimilation model, modern human origins in Europe and the extinction of Neandertals. *Quaternary International, 137*, 7–19.

Soficaru, A., Dobos, A., & Trinkaus, E. (2006). Early modern humans from the Pestera Muierii, Baia de Fier, Romania. *Proceedings of the National Academy of Sciences USA, 103*(46), 17196–17201.

Trinkaus, E. (2005). Early modern humans. *Annual Review of Anthropology, 34*, 207–230.

Valdiosera, C., Garcia, N., Dalen, L., Smith, C., Kahlke, R. D., Liden, K., Angerbjorn, A., Arsuaga, J. L., & Götherström, A. (2006). Typing single polymorphic nucleotides in mitochondrial DNA as a way to access Middle Pleistocene DNA. *Biology Letters, 2*(4), 601–603.

Willerslev, E., Hansen, A. J., Binladen, J., Brand, T. B., Gilbert, M. T., Shapiro, B., Bunce, M., Wiuf, C., Gilichinsky, D. A., & Cooper, A. (2003). Diverse plant and animal genetic records from Holocene and Pleistocene sediments. *Science, 300*(5620), 791–795.

Yang, D. Y., Eng, B., Waye, J. S., Dudar, J. C., & Saunders, S. R. (1998). Technical note: Improved DNA extraction from ancient bones using silica-based spin columns. *American Journal of Physical Anthropology, 105*(4), 539–543.

Chapter 20
Towards Neanderthal Paleogenomics

David Caramelli, Lucio Milani, Roscoe Stanyon, and Carles Lalueza Fox

Abstract Recent advances in ancient DNA technology have made it possible to recover DNA from paleontological remains allowing the scientist to go back in time studying the genetic relationships between Humans and Neandertals. However, the field is fraught with technical pitfalls and needs stringent criteria to ensure the reliability of results, particularly when human and Neanderthal remains are studied.

Keywords Paleogenetics • Ancient DNA • Evolution

Thirteen years ago, in 1997, the dream of recovering DNA from Neanderthal remains was finally achieved by the publication of the first mitochondrial hypervariable region I sequence, from the famous Neanderthal specimen: Feldhofer 1, from the Neander Valley, in Germany (Krings et al. 1997). Analysis of this sequence suggested that there was no archaic Neanderthal mtDNA in the current European mtDNA pool and that the mtDNA last common ancestor of modern humans and Neanderthals dated to between 300,000 and 700,000 years ago.

Since this date, more Neanderthal samples have been sequenced and we now have the hypervariable region I sequences (or its fragments) from 13 individuals from all over Europe: two from Spain (El Sidron Cave), two from France (Rochers de Veilleneuve and La Chappelle-aux Saint), two from Belgium (Scladina and Engis 2), two from Germany (Feldhofer 1 and 2), one from Italy (Monte Lessini), three from Croatia (Vindija 75, 77, and 80), and one from the Caucasus (Mezmaiskaya Cave, southern Russia) (Beauval et al. 2005; Caramelli et al. 2006; Krings et al. 1997, 2000; Lalueza-Fox et al. 2005, 2006; Orlando et al. 2006; Ovchinnikov et al. 2000; Schmitz et al. 2002; Serre et al. 2004). These additional data confirmed the original conclusions from Feldhofer 1 that Neanderthals contributed little or no mtDNA to the gene pool of modern humans. If there was a small contribution, it probably was erased by genetic drift or by the continuous influx of modern human DNA into the Neanderthal gene pool.

These data now make it possible to generate some preliminary hypotheses about Neanderthal population dynamics, even if only relatively recent fossils have been examined (only Scladina dates to about 100,000 years ago, being the rest 50,000–60,000 years more recent). Regrettably, we have only seven sequences extensive enough (more than 300 base pairs) to allow robust phylogenetic tests. Right now, there is no genetic information on West Asian Neanderthals or from samples older than 100,000 years (Excoffier 2006).

The sequences from El Sidrón 1252, those found in Croatia (Vindija 75 and 80) and Feldhofer 1 and 2 from Germany, are very similar one each other. All share the haplotype with the nucleotide positions 16078 G and 16154 C. The cluster Feldhofer 1, Vindija 75–80, and el Sidrón 1252 also share the nucleotide position 16258 G (few other substitutions are exclusive of one on another of these sequences). The most recent common ancestor of these closely related sequences was estimated by coalescent methods to around 130,000 years ago, close to the end of the isotopic stage 6 glacial maximum (Lalueza-Fox et al. 2006). Therefore, it seems to be a relatively recent genetic group of sequences that most likely reflect a population expansion that took place late in Neanderthal's history. Since the genetic diversity can predate population expansions, it is not sure that these sequences show an expansion that took place immediately after the glacial maximum; it could be, in fact, much younger. The fact that the oldest sequence (around 100,000 years), that from Scladina, is quite divergent and clusters at the root of the Neandertal mtDNA genealogy seems to support this view.

In any case, the basal position of Monti Lessini, Mezmaiskaya, and Scladina in the mtDNA genealogy suggests that before that glacial maximum Neanderthals were

D. Caramelli (✉), L. Milani, and R. Stanyon
Dipartimento di Biologia Evoluzionistica, Laboratori di Antropologia, via del Proconsolo 12, 50122 Firenze, Italy
e-mail: david.caramelli@unifi.it; lucio.milani@unifi.it; roscoe.stanyon@unifi.it

C. Lalueza Fox
Institute of Evolutionary Biology (CSIC-UPF), Barcelona (Spain), Barcelona, Spain
e-mail: carles.lalueza@upf.edu

S. Condemi and G.-C. Weniger (eds.), *Continuity and Discontinuity in the Peopling of Europe: One Hundred Fifty Years of Neanderthal Study*, Vertebrate Paleobiology and Paleoanthropology, DOI 10.1007/978-94-007-0492-3_20, © Springer Science+Business Media B.V. 2011

genetically – and probably, morphologically also (see Rosas et al. 2006) – more variable. Climatic changes brought on by the glaciation may have constrained Neanderthals to Southern European refugia (Italy, Iberia, and the Balkans), and reduced their genetic variability. After the glacial maximum, the Central and Northern European regions may have been recolonizated by populations that originated from some Southern refugia, and that represent only a subsample of the original mitochondrial diversity. It seems, therefore, that there was some level of phylogeographic structure within Neandertals. The last Neanderthal populations could be divided into three lineages: East European with the Mezmaiskaya sample, Central and Western European Neanderthals, and the Italian Neanderthals, that is perhaps a form isolated by the Alps since before 130,000 years ago. More samples will be needed to confirm this scenario and perhaps discover new lineages by better exploring the Southern and Eastern range of the Neanderthals (Excoffier 2006).

Other hypotheses about the Neanderthal genetic contribution to early modern humans are possible comparing the mitochondrial DNA because of its particular characteristics of transmission and mutation; in particular the hypervariable region I sequences of this genome extract from ancient *Homo sapiens*. An example is the Cro Magnon sample dating to 24,700 years ago from Paglicci, Italy (Caramelli et al. 2003). Sequences from this specimen were indistinguishable from modern Europeans, and therefore, could not be distinguished from putative modern human contaminants even if exhaustive protocols of authentication (Cooper and Poinar 2001; Hofreiter et al. 2001) were followed. Serre et al. (2004) assumed that it is impossible to authenticate any anatomically modern human (AMH) sequence obtained from archaeological specimens. He did however confirm the absence of Neanderthal-specific mtDNA sequences from five European early modern human specimens. Since coalescence theory indicates that the (inferred) modern human mtDNA sequences of the five AMH specimens are unlikely to exactly match the five to seven ancestral lineages of modern populations, this effectively doubles the number of modern human mtDNA lineages known to exist in the Late Pleistocene. This value was used with population genetic models to calculate that the maximum Neanderthal genetic contribution to early modern humans is likely to have been between 25% and 0.1% (Serre et al. 2004; Currat and Excoffier 2004).

These conclusions are based only on mtDNA, a single, maternally inherited locus, and are therefore limited. Recently, the recovery of nuclear DNA from Neanderthal remains (Green et al. 2006) has opened new horizons. The publication of about a million base pair sequences that were extracted from a 38,000-year-old Neanderthal from Croatia (Vindija 80) represent a major success in human ancient DNA research (Green et al. 2006; Noonan et al. 2006) although there have been revisions of results lowering the initial enthusiasm (Wall and Kim 2007). These studies used two distinct, advanced technical processes and were carried out in two different laboratories. In both cases, they were able to extract and amplify the broken and tiny DNA fragments present in the remains. The Vindija 80 sample was selected among others because this extract was free of contaminating modern human DNA. Sequences from autosomes and both sex chromosomes showed that the bone specimen is from a male individual, this providing the first Neanderthal genetic sexing. According to these nuclear sequence data, the time of most recent common ancestor of Neanderthals and modern humans was about 700,000 years ago for Noonan et al. (2006) and about 500,000 for Green et al. (2006) (the latter using a larger fraction of the genome). These dates are similar to the mtDNA estimates, both in these samples (461–825 kyr) (Krings et al. 2000) and in those derived from earlier work (317–741 kyr) (Krings et al. 1997). The best estimate for the demographic split (i.e. the point beyond which no gene flow would have occurred) between the ancestors of modern humans and those of Neanderthals is 370,000 years ago. The two publications are contradictory on the possibility of admixture between Neanderthals and modern humans. Noonan et al. (2006) indicate that there is no evidence for admixture, or it is extremely improbable, while Green et al. (2006) based on high frequency of derived human SNPs in the Neanderthal genome suggest that this observation is compatible with a limited level of gene flow between modern humans and Neanderthals. Since the X chromosome shows a higher Neanderthal–modern human divergence than the autosomes, this gene flow could have been of modern human males and Neanderthal females. But this inference strongly depends on the Neanderthal population size estimates and on the divergence dates. The shallow genomic divergence dates obtained not only can be compatible with some degree of modern humans inbreeding, but also with the possibility of long-term contacts between Neanderthals and other archaic human species (for instance, during the arrival of Mode 3 industry to the European continent, 300,000 years ago). It could well be that the speciation events that lead to the Neanderthal lineage lasted several hundreds of thousands of years and that this long process shaped the Neanderthal genome.

The main criticisms of this work are directed to the results from Green et al. In particular, the reanalyses of these data showed that they are not consistent with each other and point to serious problems with the data quality, possibly due to modern human contaminants and high rate of sequencing errors (Wall and Kim 2007). The problems of contamination are the main limitations for all studies on ancient DNA.

More precise evidences on the inbreeding debate will be possible when more genomic data could be obtained, not only from Vindija but also from other Neanderthal fossils such as those from El Sidrón, whose date (43,000 years ago, Rosas et al. 2006) clearly predates the arrival of modern humans to Western Europe. In addition, the genomic results should provide clues about the peculiar morphologic characteristics and other evolutionary traits of the Neanderthals.

References

Beauval, C., Maureille, B., Lacrampe-Cuyaubere, F., Serre, D., Peressinotto, D., Bordes, J. G., Cochard, D., Couchoud, I., Dubrasquet, D., Laroulandie, V., et al. (2005). A late Neandertal femur from Les Rochers-de-Villeneuve, France. *Proceedings of the National Academy of Sciences of the United States of America, 102*, 7085–7090.

Caramelli, D., Lalueza-Fox, C., Vernesi, C., Lari, M., Casoli, A., Mallegni, F., Chiarelli, B., Dupanloup, I., Bertranpetit, J., Barbujani, G., et al. (2003). Evidence for a genetic discontinuity between Neandertals and 24,000-year-old anatomically modern Europeans. *Proceedings of the National Academy of Sciences of the United States of America, 100*, 6593–6597.

Caramelli, D., Lalueza-Fox, C., Condemi, S., Longo, L., Milani, L., Manfredini, A., de Saint Pierre, M., Adoni, F., Lari, M., Giunti, P., et al. (2006). A highly divergent mtDNA sequence in a Neandertal individual from Italy. *Current Biology, 16*, R630–R632.

Cooper, A., & Poinar, H. N. (2001). Ancient DNA do it right or not at all. *Science, 18*(2001), 1139.

Currat, M., & Excoffier, L. (2004). Modern humans did not admix with Neanderthals during their range expansion into Europe. *PLoS Biology, 2*, 2264–2274.

Excoffier, L. (2006). Neandertal genetic diversity: A fresh look from old samples. *Current Biology, 16*(16), R650–R652.

Green, R. E., Krause, J., Ptak, S. E., Briggs, A. W., Ronan, M. T., Simons, J. F., et al. (2006). Analysis of one million base pairs of Neanderthal DNA. *Nature, 444*, 330–336.

Hofreiter, M., Serre, D., Poinar, H. N., Kuch, M., & Pääbo, S. (2001). Ancient DNA. *Nature Reviews Genetics, 2*, 353.

Krings, M., Stone, A., Schmitz, R. W., Krainitzki, H., Stoneking, M., & Pääbo, S. (1997). Neandertal DNA sequences and the origin of modern humans. *Cell, 90*, 19–30.

Krings, M., Capelli, C., Tschentscher, F., Geisert, H., Meyer, S., von Haeseler, A., Grossschmidt, K., Possnert, G., Paunovic, M., & Pääbo, S. (2000). A view of Neandertal genetic diversity. *Nature Genetics, 26*, 144–146.

Lalueza-Fox, C., Sampietro, M. L., Caramelli, D., Puder, Y., Lari, M., Calafell, F., Martinez-Maza, C., Bastir, M., Fortea, J., de la Rasilla, M., et al. (2005). Neandertal evolutionary genetics: Mitochondrial DNA data from the Iberian Peninsula. *Molecular Biology and Evolution, 22*, 1077–1081.

Lalueza-Fox, C., Krause, J., Caramelli, D., Catalano, G., Milani, L., Sampietro, L., Calafell, F., Martínez-Maza, C., Bastir, M., García-Tabernero, A., et al. (2006). Mitochondrial DNA of an Iberian Neandertal suggests a population affinity with other European Neandertals. *Current Biology, 16*, R629–R630.

Noonan, J., Coop, G., Kudaravalli, S., Smith, D., Krause, J., Alessi, J., et al. (2006). Sequencing and analysis of Neanderthal genomic DNA. *Science, 314*, 1113–1118.

Orlando, L., Darlu, P., Toussaint, M., Bonjean, D., Otte, M., & Hänni, C. (2006). Revisiting Neandertal diversity with a 100,000 year old mtDNA sequence. *Current Biology, 16*, R400.

Ovchinnikov, I. V., Götherström, A., Romanova, G. P., Kharitonov, V. M., Lidén, K., & Goodwin, W. (2000). Molecular analysis of Neandertal DNA from the northern Caucasus. *Nature, 404*, 490–493.

Rosas, A., Martinez Maza, C., Bastir, M., Garcia-Tabernero, A., Lalueza Fox, C., Huguet, R., Ortiz, J. E., Julia, R., Soler, V., De Torres, T., Martinez, E., Canaveras, J. C., Sanchez Moral, S., Cuezva, S., Lario, J., Santamaria, D., De La Rasilla, M., & Fortea, J. (2006). Paleobiology and comparative morphology of a late Neandertal sample from El Sidron, Asturias, Spain. *Proceedings of the National Academy of Sciences of the United States of America, 103*, 19266–19271.

Schmitz, R. W., Serre, D., Bonani, G., Feine, S., Hillgruber, F., Krainitzki, H., Pääbo, S., & Smith, F. H. (2002). The Neandertal type site revisited: Interdisciplinary investigations of skeletal remains from the Neander Valley, Germany. *Proceedings of the National Academy of Sciences of the United States of America, 99*, 13342–13347.

Serre, D., Langaney, A., Chech, M., Teschler-Nicola, M., Paunovic, M., Mennecier, P., Hofreiter, M., Possnert, G., & Pääbo, S. (2004). No evidence of Neandertal mtDNA contribution to early modern humans. *PLoS Biology, 2*, 313–317.

Wall, J. D., & Kim, S. K. (2007). Inconsistencies in Neanderthal genomic DNA sequences. *PLoS Genetics, 3*, 1862–1866.

Chapter 21
Twelve Years of Neandertal Genetic Discoveries: State-of-the-Art and Future Challenges

Ludovic Orlando and Catherine Hänni

Abstract The first mitochondrial DNA sequence from a Neandertal specimen was recovered in 1997. Now the completion of the whole Neandertal genome has been announced to be completed in the forthcoming months. About one million nucleotides of nuclear DNA have already been sequenced and so far 15 Neandertal specimens have delivered authentic mitochondrial sequences. This information has helped us to better understand the evolution of the Neandertal gene pool over space and time and to address the long-standing question of possible admixture with their modern human relatives. This chapter reviews current knowledge on NeandertalDNA sequences and presents future challenges related to Neandertal genomics.

Keywords Ancient DNA • mtDNA • Comparative genomics • Pyrosequencing • 2-round Multiplex • SNP

Introduction

The year 1997 was not like any other for ancient DNA (aDNA) researchers. First, in April, the analysis of a large repertoire of insects trapped in Oligocene Dominican amber (and Quaternary African copal resin) failed to recover any trace of authentic DNA (Austin et al. 1997; Sykes 1997). This study came a couple of years after the so-called dinosaur DNA molecules were recognized as modern human DNA contaminants (Woodward et al. 1994; Zischler et al. 1995), and the bell had definitely tolled for the fantasy of DNA preservation over millions of years. Nevertheless, this year gave rise to one of the most important advances of the field as well: the very first DNA fragments gathered from a Paleolithic human being, in this case, a Neandertal (Krings et al. 1997). Technically, the challenge was difficult to take up since the sample (from Feldhofer) was part of the first Neandertal specimen ever discovered and as such was most certainly contaminated by the DNA of generations of anthropologists. Even Ötzi, the 5,000-year-old Iceman discovered in the Austro-Italian Alps in September 1991, showed significant levels of contamination and multiple human genetic signatures (Handt et al. 1996). However, for the Feldhofer specimen, rigorous controls supported that the eight overlapping fragments recovered in the mitochondrial hypervariable region 1(HVR-I) were all endogenous to the Neandertal specimen (Krings et al. 1997).

Comparison with modern humans and chimpanzees revealed that the Neandertal haplotype was more similar to humans than chimpanzees but fall outside the range of modern human genetic diversity. In a phylogenetic tree, Neandertal and modern humans appeared as two separate lineages that diverged about half-a-million years ago (Krings et al. 1997). This first sensational result was later confirmed through the analysis of a supplemental 340-bp in the mitochondrial hypervariable region II (HVR-II, Krings et al. 1999). Yet, as long as the result relied on the analysis of only one bone, skeptics could still favor the alternative view that the Feldhofer bone was heavily contaminated by a very unusual modern HVR-I variant in the last 140 years and therefore could give no relevant indication relative to the real Neandertal sequences (Lindahl 1997). However, 12 years later, we now have 14 supplemental Neandertal HVR-I sequences, covering the last 70,000 years of their history and large parts of their geographic range (from Iberia to Central Asia – and surprisingly Siberia see below), and all point in the same direction: the divergence between Neandertals and modern humans well predated the origin of the current mitochondrial diversity of modern humans (Ovchinnikov et al. 2000; Krings et al. 2000; Schmitz et al. 2002; Serre et al. 2004; Lalueza-Fox et al. 2005; Orlando et al. 2006; Caramelli et al. 2006; Lalueza-Fox et al. 2006; Krause et al. 2007a, b).

L. Orlando (✉)
Centre for GeoGenetics, Natural History Museum of Denmark, University of Copenhagen, Oster Voldgade 5-7, Kobenhavn K 1350, Denmark
e-mail: Lorlando@snm.ku.dk

C. Hänni
Institut de Génomique Fonctionnelle de Lyon, IFR 128 Biosciences Lyon-Gerland, CNRS UMR 5242, INRA, UCB Lyon1, Ecole Normale Supérieure de Lyon, Université de Lyon, 46 Allée d'Italie, 69364 Lyon Cedex 07, France
e-mail: catherine.hanni@ens-lyon.fr

S. Condemi and G.-C. Weniger (eds.), *Continuity and Discontinuity in the Peopling of Europe: One Hundred Fifty Years of Neanderthal Study*, Vertebrate Paleobiology and Paleoanthropology, DOI 10.1007/978-94-007-0492-3_21,

These sequences now allow us to study Neandertals for themselves (and not solely for their relationships with modern humans) and to start to deliver precious information on their demographic stability over time, the dynamics of their gene pool with regards to global climatic changes or their possible migration patterns over their geographic range (Excoffier 2006). Concomitantly, new high-throughput sequencing technologies have made it possible to get access to the complete sequence of the Neandertal genome, and several Neandertal genome projects are now ongoing (Green et al. 2006; Noonan et al. 2006). These projects offer a unique opportunity to get large amounts of sequence information of nuclear DNA, which is responsible for the phenotypic specificities of Neandertals. Comparison with the complete genomes of chimpanzees and modern humans promises to shed light on the specific genetic modifications that make us all human. Therefore, a century and a half after their discovery, Neandertals continue to play a key role in understanding the very origin of our humanity. Here we will first review what is currently known about the levels of putative admixture between Neandertals and modern humans and about the evolution of genetic diversity of Neandertal populations over space and time. Then we will discuss the future challenges related to Neandertal paleogenomics.

Part One – Neandertal and Modern Humans: Levels of Admixture

The Onset: The Feldhofer HVR-I Haplotype

When the first piece of DNA was recovered from the Neandertal-type specimen found in 1856 in Feldhofer (western Germany) (Krings et al. 1997), the paleoanthropologist Dan Lieberman (Rutgers University, New Brunswick, New Jersey) declared to *Science* that "[it was the] *proof that there is a God who likes paleoanthropology*" (interviewed by Kahn and Gibbons 1997: 177). First of all, because of the technical challenge: very short overlapping fragments were amplified to cover 379-bp in the mitochondrial HVR-I, and both DNA–damage induced mutations and endogenous sequences from contemporary contaminant sequences from modern humans were filtered through a cloning and sequencing procedure (Krings et al. 1997). Amino-acid racemization analysis, multiple independent extractions and duplication of the results by a second laboratory warranted that the putative Neandertal sequence could be considered as really endogenous and authentic.

What was most important, however, was that apparently the Neandertal sequence fell outside the statistical range of modern human variation: whereas all possible pairs of 994 modern human haplotypes (i.e., different sequences found among 478 Africans, 510 Europeans, 494 Asians, 167 Native Americans, and 20 Australasians) showed only 8.0±3.1 substitutions, the 994 pairs consisting of the Neandertal haplotype and one modern human haplotype exhibited 3.4 times more differences (27.2±2.2 substitutions). Together, these two distributions showed virtually no overlap (only 0.002% of modern human pairs were more distant than the minimal difference between Neandertal and modern humans). Furthermore, though Neandertals and modern humans cohabitated in Europe between 28 and 42 KY BP (Finlayson et al. 2006; Mellars 2006), the Feldhofer haplotype exhibited no specific affinity with modern Europeans. Lastly, the Neandertal fell outside the group of modern humans in phylogenetic analyses and the age of the most recent common ancestor (MRCA) to both lineages could be estimated at 550,000–690,000 years ago (based on molecular clock calibration of four to five millions year for the chimp–*Homo* divergence), which is long before the first fossil evidence of modern humans (McDougall et al. 2005). These findings were later corroborated by the DNA sequence of the mitochondrial HVR-II of the same specimen (Krings et al. 1999). As consistent with the "Rapid replacement" model, these observations stimulated considerable debate (and numerous critics, as discussed below) regarding the origins of modern humans.

A Point-by-Point Answer to the Critics

Critic 1: A Possible Numt?

Admittedly, the Feldhofer sequence was endogenous, but as regards as its extreme divergence to contemporary modern human haplotypes, it could well not come from a mitochondrion origin. Insertions of mitochondrial DNA in the nuclear genome occur in the course of evolution. After duplication, the inserted nuclear mitochondrial DNA (numt) evolves independently from the genuine mitochondrial copy and may accumulate extensive mutations. This phenomenon is well known in humans and Primates (Zischler et al. 1995; Van der Kuyl et al. 1995) and is particularly prone to hamper the reconstructions of the evolutionary relationships among individuals and populations (Thalmann et al. 2004). Therefore, if the Feldhofer sequence was a numt, such extensive differences with contemporary mitochondrial sequences would be expected. However, using PCR primers specific to the putative Neandertal haplotype under conditions allowing less than one copy to be detected, Krings et al. (1997) demonstrated that the Neandertal-like sequence was completely absent from the genomes of contemporary modern human populations. This unambiguously indicated that the Neandertal haplotype derived from an individual whose lineage is not represented in the current modern human mtDNA gene pool, rather than from a numt.

Critic 2: An Unusual and Degraded Modern Human Haplotype?

The Feldhofer specimen was discovered in 1856 and should have experienced a high level of DNA contamination from modern human DNA. As a matter of fact, numerous modern human sequences were found among the clones sequenced by Krings et al. (1997). In these circumstances, the putative Neandertal sequence could be seen as a modern human haplotype either divergent (and unknown) (Hawks and Wolpoff 2001) or that experienced large degradation during the taphonomic process (Gutierrez et al. 2002). Alternatively, if the Feldhofer sequence was truly of Neandertal origin, a sequence of a second Neandertal specimen should exhibit a similar mutational pattern with regards to modern human haplotypes. However, adding this new piece to the puzzle would not be easy because preliminary analyses of amino-acid racemization ratio and amino-acid, nitrogen, and DNA contents of faunal and human bones from three key Neandertal sites (Zafarraya, Krapina, and La Chaise) suggested extensive diagenetic degradation and poor (if any) DNA preservation (Cooper et al. 1997). The so-expected second Neandertal sequence finally came from a 29,000-year-old infant specimen from Mezsmaïskaya (Caucasus) in March 2000 (Ovchinnikov et al. 2000). Along 345 base pairs sequenced, the HVR-I from Mezsmaïskaya was found closely related, but not identical, to that of the Feldhofer specimen.

This definitely destroyed the possibility that either sequence was an artifact or the product of contamination, but surprisingly not the hypothesis that modern humans and Neandertal intermixed. For Hawks and Wolpoff (2001), the Neandertal affiliation of the Mezsmaïskaya infant is doubtful for anatomical, cultural, and chronological reasons. If the buried infant were actually intrusive into an older Mousterian layer as suggested by these authors, it should be considered as a post-Neandertal modern human. Its HVR-I haplotype would then give clear evidence of persistence of Neandertal-like mtDNA sequences in later populations, rather than confirming a proposed isolation of Neandertal and modern human populations (Hawks and Wolpoff 2001). However, there is now compelling evidence that Neandertals persisted in isolated refuges as late as 28,000 years BP (Finlayson et al. 2006). In addition, the first radiocarbon date at 29,000 years BP for the buried Mezsmaïskaya infant is now dismissed and considered to be due to contamination by modern carbon (Skinner et al. 2005). Standard electron spin resonance datations indicate that would be at least 36,000 old (Skinner et al. 2005), suggesting the specimen cannot be viewed as a post-Neandertal modern human.

Critic 3: Unknown Neandertal Sequences Might be More Modern-Like

It is clear that Neandertals were a diverse group of hominoids, possibly structured both at the geographic and temporal scales. The particular Neandertal sequences from Feldhofer and Mezsmaïskaya specimens might be at one extreme of a diverse spectrum in Neandertals that could include other, more modern-like sequences. Hence that other Neandertals should be examined was a prerequisite to further test genetic interbreeding with modern humans. Today, partial HVR-I sequences are available for 13 supplemental Neandertal specimens (Table 21.1). Though some indicate significant

Table 21.1 List of the available Neandertal mtDNA HVR-I sequences

Specimen	Location	Age (KYBP)	Length (bp)	Reference
Feldhofer 1	Neander valley, Germany	40	379	Krings et al. 1997
Mezmaïskaya	Sukhoy Kurdzhips river, Northern Caucasus	29–36	345	Ovchinnikov et al. 2000
Feldhofer 2	Neander valley, Germany	40	357	Schmitz et al. 2002
Vindija 75	Vindija, North of Croatia	42	357	Krings et al. 2000
Vindija 77	Vindija, North of Croatia	40	31	Serre et al. 2004
Vindija 80	Vindija, North of Croatia	38	31	Serre et al. 2004
La Chapelle aux Saints	Dordogne, France	40–50	31	Serre et al. 2004
Rochers de Villeneuve	Vienne, France	41	31	Beauval et al. 2005
Engis 2	Meuse basin, Belgium	40–50	31	Serre et al. 2004
El Sidron	Asturias, North of Spain	43	303	Lalueza-Fox et al. 2005
El Sidron	Asturias, North of Spain	43	48	Lalueza-Fox et al. 2006
Scladina	Meuse, Belgium	100	123	Orlando et al. 2006
Monte Lessini	Riparo Mezzena, Northern Italy	50	378	Caramelli et al. 2006
Teshik Tash	Uzbekistan, Central Asia	75	190	Krause et al. 2007a
Okladnikov	Altaï mountains, Siberia	30–38	345	Krause et al. 2007a

BP and bp stand for Before Present and base pair, respectively diversity among Neandertal populations (Orlando et al. 2006; Caramelli et al. 2006; Krause et al. 2007a), all of them cluster in a monophyletic group separate from contemporary modern humans. Therefore, over the last 70,000 years of their evolutionary history, European Neandertal populations remained more closely related to one another than to any modern human populations, suggesting low (if any) admixture.

Critic 4: Absence of Regional Affinity Does Not Preclude Multiregionalism

The fact that Neandertal sequences show no preferential regional affinity with Europeans has been generally taken as an argument in favor of the rapid replacement model for the origin of modern humans. This interpretation can be refuted though (Relethford 2001a, b). It is true that the rapid replacement model predicts no regional affinity between Neandertals and modern humans from Europe, as in this scenario, the former have been replaced without intermixing by the latter. But multiregional evolution is *not* independent regional evolution as archaic human populations across the Old World (including the European Neandertals) remained interconnected by gene flow (Relethford 1999). Yet migration matrix theory shows that accumulated ancestry of a given population is a function of both the pattern of gene flow and the number of generations. Providing there are low levels of continued gene flow between different archaic human populations and sufficient time for convergence to equilibrium, then the accumulated Neandertal ancestry for any kind of modern human population is expected to be the same (Relethford 2001a, b). Therefore, the observed lack of regional affinity is not exclusive for multiregionalism as a model of modern human origins.

Critic 5: Admixture Level Is Underestimated as Modern-Like Sequences Are Discarded as Contaminants

PCR experiments have clearly shown human mtDNA sequences can be virtually retrieved from almost every ancient animal specimen (Hofreiter et al. 2001b), often at levels exceeding the amount of authentic endogenous DNA (Malmström et al. 2005). Therefore, any sequence taken from a Neandertal fossil that looks like modern humans is considered as contamination. In contrast, any sequence similar to the Feldhofer haplotype is taken as genuine (Serre et al. 2004). Consequently, checking for modern-like sequences in Neandertal remains offers poor (if any) chance of detecting possible admixture events. However, HVR-I from all Neandertals sequenced so far exhibit striking differences with the sequences of extant modern human worldwide. Finding Neandertal-like mtDNA sequences in early human remains could not possibly be interpreted as contaminant by-products but would unambiguously be the hallmark of interbreeding. Forty specimens of early modern human fossils (Abri Pataud, Cro-Magnon, and La Madeleine, France; Mladec, Czech Republic; Sandalja and Verternica, Croatia) were analyzed for amino-acid contents, and five were suggestive of endogenous DNA preservation (Serre et al. 2004). They consisted both in typical modern humans (La Madeleine, Cro-Magnon, and Abri-Pataud) and more controversial specimens, sometimes regarded as transitional between modern humans and Neandertals (e.g., Mladec). No Neandertal-like but only modern-like human sequences could be detected, supporting no (or low) genetic contribution of Neandertals to the modern human gene pool (Serre et al. 2004).

Critic 6: In-Silico Analyses Might Have Serious Flaws

Along the two mitochondrial hypervariable regions (HVR-I and HVR-II), many nucleotide positions are mutational hotspots while others are rather stable over time (Excoffier and Yang 1999; Tamura 2000). Yet mutational hotspots give rise to excessive homoplasic changes that may mislead phylogenetic reconstructions (Innan and Nordborg 2002). Accurate estimation of the nucleotide substitution model (e.g., considering both a proportion of invariable sites and correcting for heterogeneity in substitution rates) is thus a prerequisite for phylogenetic reconstructions relying on HVR sequence information. However, such appropriate models have not necessarily been used to infer the phylogenetic relationships between Neandertals (Feldhofer, Mezsmaïskaya and Vindija-75) and modern humans (Krings et al. 1997, 1999; Ovchinnikov et al. 2000; Krings et al. 2000). Using maximum-likelihood principles for model selection and parameter estimation, Gutierrez et al. (2002) have re-evaluated the phylogenetic support for Neandertal branching off the modern human cluster. First, Neandertals nest inside modern humans as long as the whole HVR-I and HVR-II sequence information is considered, and according to taxon and sequence sampling, all alternative topologies receive similar likelihood support (Gutierrez et al. 2002). Second, if Neandertal and modern humans could be supported as different species, then the most frequent quartet between two modern humans, one Neandertal and one chimp would be the one clustering the two modern humans altogether, as Krings et al. (1997) emphasized. However, this methodology is very prone to long-branch attraction (a process clustering long branches together regardless of their true phylogenetic relationships). Given the relatively longer branches of chimps and Neandertals in quarter puzzling analyses, Gutierrez et al. (2002) have shown that the support for a difference between modern humans and Neandertals has been overstated. The same was true for pair-wise distance distributions because a lot of the human–human pairs consisted of individuals that actually belonged to the same population. Correcting for over-representations and thus providing an unbiased estimation of human HVR-I diversity, it appeared that 1.6% of human–human pairs exhibited larger distances than the closest human–Neandertal pair (contra 0.002% in the analysis of Krings et al. 1997). Strikingly, 27% of human–Neandertal pairs were closer than the largest distance between modern humans, suggesting considerable overlap between both distributions (Gutierrez et al. 2002).

However, scrupulous analyses of parsimony informative sites along HVR-I and HVR-II have revealed eight stable synapomorphies that set apart Neandertals from modern humans (Knight 2003). Interestingly, none of them have been found to be susceptible to *postmortem* damage (Gilbert et al. 2003), therefore providing consistent support for taxonomic differences between Neandertals and modern humans. Moreover, network reconstructions and accurate phylogenetic reconstructing based either on partial HVR-I sequence information (Orlando et al. 2006; Hebsgaard et al. 2007) or genomic data (see below; Green et al. 2006; Noonan et al. 2006) have now confirmed Neandertals and modern humans as two distinct phylogenetic lineages (Fig. 21.1).

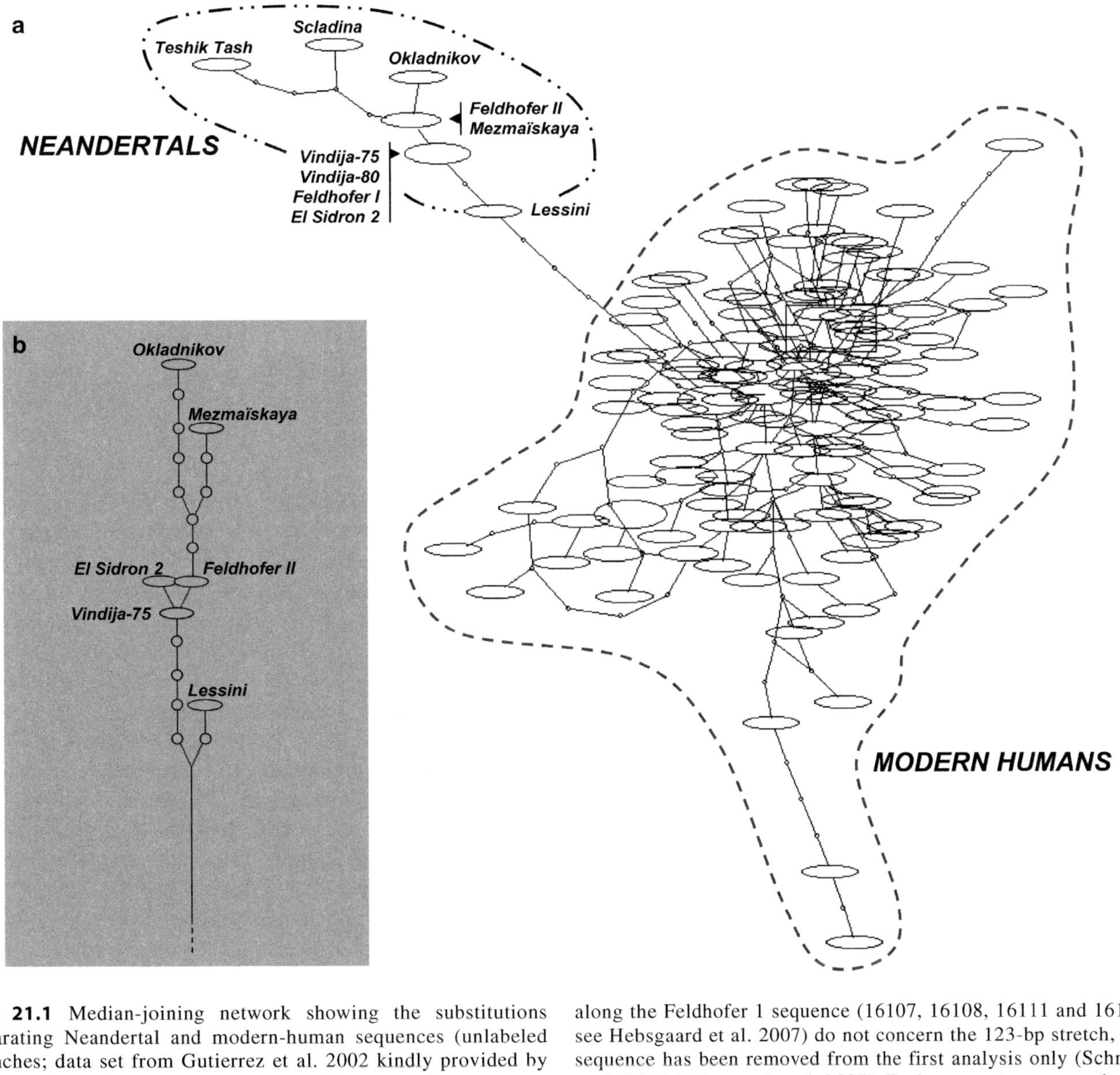

Fig. 21.1 Median-joining network showing the substitutions separating Neandertal and modern-human sequences (unlabeled branches; data set from Gutierrez et al. 2002 kindly provided by G. Gutierrez). (**a**) 111-bp. (**b**) 303-bp. The networks were constructed with TCS (Clement et al. 2000) considering gaps as a 5th state. As the positions that show likely *postmortem* substitutions along the Feldhofer 1 sequence (16107, 16108, 16111 and 16112, see Hebsgaard et al. 2007) do not concern the 123-bp stretch, this sequence has been removed from the first analysis only (Schmitz et al. 2002; Hebsgaard et al. 2007). Each segment corresponds to a single mutational event and each oval represents one specific haplotype

Critic 7: Current Modern Humans Are Not the Contemporaries of Neandertals

Neandertals have not been compared with contemporary but extant modern humans. Yet using coalescent theory and assuming different demographic scenarios (e.g., constant population size, population expansion), the mtDNA of all modern humans can be traced back to a very low number of ancestral lineages around 20,000–30,000 years ago (Nordborg 1998; Cooper et al. 2004). Therefore, the major part of the genetic diversity present in current modern human populations has been shaped after the cohabitation with Neandertals. Indeed, though early Neolithic farmers have certainly contributed little to the current European gene pool (Haak et al. 2005), population contractions in glacial refugia followed by expansion episodes have most probably erased a substantial part of the human diversity that prevailed at the time of Neandertal and modern human cohabitation as shown by mtDNA (Richards et al. 2000; Torroni et al. 2001), Y-chromosome (Semino et al. 2000), or SNPs diversity (Currat and Excoffier 2005). Recent surveys of the regional distribution of a genome-wide set of SNPs (Marth et al. 2003, 2004) and of a *NF1*-region subset of SNPs (Schmegner et al. 2005) are consistent with a demographic collapse in the demography of European and Asian populations around 40 KYA (size recovery over the last 10,000 generations would have created about up to one third of the current diversity) (Marth et al. 2004). Furthermore, given that some mtDNA lineages are associated with propensity for bioenergetic disorders, the regional variation in mtDNA sequences of extant modern humans may have been shaped by natural selection, some variants being preferentially selected as adaptative to northern climates (Mishmar et al. 2003; Ruiz-Pesini et al. 2004). In this context, it is still possible that Neandertals have contributed to the gene pool of modern humans at the time of their cohabitation but that their sequences were later eliminated through demographic events (Nordborg 1998) or selective sweeps (Hawks and Wolpoff 2001). Coalescence simulations even indicate that the available HVR-I data set cannot statistically reject interbreeding rates as high as 25% (Cooper et al. 2004).

Some efforts have been made at circumventing the drift/selection problem by sequencing the HVR-I of modern human fossils contemporary with Neandertals (Caramelli et al. 2003). The sequences of the two specimens (coming from the Paglicci cave, Italy, and dated around 23–25 KYBP) were found very similar to those of current Europeans but drastically divergent with regards to those of Neandertals (Caramelli et al. 2003). Nevertheless, even though the study fulfilled the most stringent standards for authenticating ancient DNA sequences, doubts have later been raised as for the validity of these sequences which appeared like mosaic recombinations of human mtDNA from different origins (Bandelt 2005).

Rather than gathering DNA sequences from early modern humans, Currat and Excoffier (2004) have decided to use realistic models to provide accurate estimates of the interbreeding rate between Neandertals and modern humans. Previous models assumed an instantaneous mixing of Neandertal and modern human populations (Nordborg 1998; Serre et al. 2004). In the new sophisticated model, Europe is subdivided into small territories harboring one Neandertal and one modern human subpopulation and is settled progressively from the Near East by modern humans. Concomitant with their geographic expansion, modern humans undergo a demographic growth, resulting in the progressive replacement of Neandertals. Archeological and paleodemographic information was used to calibrate different parameters of the model, such as the estimated duration of the replacement process or the speed of the geographic expansion. In striking contrast with previous reports, the maximum initial input of Neanderthal genes into the Paleolithic European population, given it is no longer present in current populations today, is estimated to 0.09% at best (Currat and Excoffier 2004). Thus, though models cannot be taken for reality, most convincing simulations have so far left little place for interbreeding, given the absence of Neandertal-like mtDNA haplotype in current modern human populations.

Neandertal Genomics

Levels of Admixture

Taken overall, the Neandertal HVR-I sequences do suggest an early divergence of *Homo neanderthalensis* from *Homo sapiens* lineages and suggest a minimal contribution of interbreeding. Of course, because mtDNA comes only from the mother, the mitochondrial evidence leaves the possibility open that Neandertal fathers contributed to the modern nuclear gene pool. Hopefully, new technological advances (multiplex-PCR and next-generation sequencing technologies, such as Roche: 454 Life Sciences and Illumina: Solexa) now furnish the opportunity to gather genomic data and therefore to take a look at both parental contributions for the first time (Green et al. 2006; Noonan et al. 2006; Krause et al. 2006, 2007b; Römpler et al. 2006; see Part 3 for in-depth developments).

If Neandertals and modern humans admixed, a substantial part of the Neandertal nuclear alleles should match European alleles (SNPs). Consequently, checking if those sites that specifically derived in European populations are present at the derived state in the Neandertal genome as well offers a direct way to gauge admixture levels (note that here, derived means different from the chimpanzee reference, therefore considered as ancestral). Interestingly, 35 human SNPs are present among the 65,520 nucleotide information reported in Noonan et al. (2006), but only 3 of them are present at the derived state in the Neandertal genome. While two of them are derived among Yorubas (and consequently are not related to the cohabitation process), the third one is a C→T mutation and could correspond a DNA damage by-product. Ancient DNA molecules indeed consist of short fragments (50–150 nucleotides long) that include subtle changes in sequence information; the most prevalent of these DNA lesions is Cytosine deamination in Uracil and generates GC→AT artifactual mutations (Hofreiter et al. 2001a; Stiller et al. 2006; Briggs et al. 2007; Brotherton et al. 2007). Therefore, in this data set, the survey of genomic human SNPs shows no evidence of admixture (Noonan et al. 2006). In addition, simulations were computed to further test the likelihood of a single admixture event 40,000 years ago. Though the absence of contribution receives maximal probabilistic support, interbreeding rates as high as 20% cannot be excluded at the 5% level due to low sequence coverage (Noonan et al. 2006).

In striking contrast with this observation is the one-million nucleotide (1-Mb) data set showing as much as 33% of 786 human HapMap SNPs at the derived state in the Vindija-80 individual (Green et al. 2006; note that (1) this value is about 30% if 318 Perlegen SNPs are added and drops to 21.8% when only the short fragments (<50 bp) of this data set are considered, see below: "Part 3. Contamination concerns" and Wall and Kim (2007) for further discussion). This may suggest substantial gene flow between modern humans and Neandertals, corroborating what the analysis of the pattern of linkage disequilibrium (LD) in contemporary human DNA sequences (Plagnol and Wall 2006), or what the age of haplogroup D of the gene microcephalin (MCPH1; Evans et al. 2006; reviewed in Hawks et al. 2007) have recently suggested. It is noteworthy, however, that the alleles are not found at the derived state solely among HapMap Europeans (e.g., HapMap alleles were genotyped among 210 individuals coming from 4 different populations: Yoruba, Japan, Northern Europeans, and Chinese). Furthermore, even though not present in the Vindija-80 individual, one should not forget that other Neandertals could still have carried the ancestral alleles. That Neandertals and modern humans share derived alleles would then reflect either convergence (homoplasy) or polymorphism inheritance (plesiomorphy) from an ancestral gene pool but not interbreeding. Clearly, more extensive sequencing of the Neandertal genome from Vindija-80 and other individuals is still necessary to further address the long-standing admixture hypothesis. In particular, contrasting genomic data from Neandertals that lived before or at the time the cohabitation period with modern humans would help to identify those SNPs that derived at the time of the cohabitation, i.e., to identify putative genetic markers of admixture.

Timing the MRCA and the Divergence of Neandertal and Modern Human Lineages

So far, the 1-Mb genomic data has revealed a total of 41 sequences dispersed over the mtDNA genome, exhibiting 39 Neandertal specific substitutions (Green et al. 2006). A majority of them (25) have been double-checked by two-round multiplex PCR (both rounds work as classical PCRs except that in the first one, multiple pairs of primers are included in order to restore DNA fragments of the different targets to amplifiable levels; the second step consists in amplifying each target separately in a specific reaction, allowing the recovery of sequence information from many loci from a rather limited amount of DNA extract; see Römpler et al. 2006) and have been used for timing the divergence between Neandertal and modern humans at 461–825 KYA (given 4.7–8.4 million years for the human–chimpanzee split) (Green et al. 2006).

Neandertal pyrosequencing reads cover also about one-million nucleotides of the nuclear genome (on average 3.61 bases per 10,000 bases for autosomes; 2.18 and 1.62 bases per 10,000 bases for X and Y chromosomes, respectively). Reads of at least 30 nucleotides were unambiguously aligned with the human and the chimp genomes. Potential inaccurate chimp positions were filtered from the alignment (the chimp genome project is still ongoing and some regions exhibit poor sequence coverage), leaving a total of 750,989 positions. More than 98% of them are conserved over the three hominids, but 434 nucleotides are human-specific and therefore postdate the MRCA of the human and Neanderthal sequences (Green et al. 2006). Additionally, 10,167 positions have mutated either along the chimp lineage or along the human lineage after the chimp–human divergence but before the Neandertal–modern human split (Green et al. 2006). Assuming constant rates of DNA sequence change, an average human–chimpanzee divergence time at 6.5 million years, and correcting for potential biases in divergence estimates, this implies a human–Neandertal DNA sequence divergence time of 465–569 KY (note that this estimate shifts to around 700 KY as long as only the short fragments (<50 bp) are analyzed; this would provide a better estimate, in agreement to what Noonan et al. (2006) reported; see below: "Part 3. Contamination concerns" and Wall and Kim (2007) for further discussion).

It is worth mentioning that such a method does not provide a direct estimate of the split but rather the average time at which the Neanderthal and human reference sequences began to diverge in the common ancestral population. As such, it gives an upper limit for the mean time of the split as population partitioning postdated the origin of the polymorphisms. Using their 65,250 bp data set of Neandertal sequences and the polymorphism data of the HapMap project), Noonan et al. (2006) have been able to provide a preliminary estimate for the human–Neanderthal population time split. The method relies upon explicit population model simulations (10,000 individuals in the ancestral population, no gene flow after instantaneous split, substitution rates, etc.) and likelihood estimations. Briefly, the likelihood of the data, i.e., the probability of getting the sequence alignment given the population model is computed for different split times and the maximum of likelihood is used to identify the most probable time for the split. Interestingly, the most-likely estimate (at around 370 KYA) largely predates the emergence of anatomically modern humans, first documented in the fossil records at 195 KYA (McDougall et al. 2005). The method still provides large confidence intervals though (170–670 KYA), but ongoing Neandertal genome sequencing will certainly provide more precise estimates in the future. For the time being, it is noteworthy that a congruent split time of 400 KYA might be inferred from the fossil record (J.J. Hublin, cited in Green et al. 2006). Furthermore, using the one-million nucleotide data set and 370 KYA as a split time, computer simulations under a likelihood-framework estimate a maximum of 3,000 individuals (rather than supposed 10,000 in Noonan et al. (2006) simulations) for the ancestral population of both modern humans and Neandertals (Green et al. 2006).

Part Two – Neandertal Populations: Evolution of Genetic Diversity Over Space and Time

Preliminary Cautions

So far, only a small number of Neandertal specimens have been analyzed at the genetic level. HVR-I sequences of 15 specimens have been published. Though closely related to each other, but clearly different from their modern and early human counterparts, a few of them have started to reveal substantial levels of genetic diversity among Neandertals. The geographic distribution of Neandertal haplotypes now allows us to draw a first picture of population relationships and subdivisions (Caramelli et al. 2006). Moreover, the genetic diversity of Neandertals can now be followed at the mtDNA level over their last 70,000 years of existence and suggests possible changes over time (Orlando et al. 2006).

This part is devoted to present these new advances. Though interesting, they should be taken as preliminary for the following reasons. Firstly, the available genetic information is reduced in a few individuals and while some sequences are large (>300 bp; Feldhofer 1&2, Mezsmaïskaya, Vindija-75 and 80, El Sidron 2, Monte Lessini and Okladnikov), others are restricted to short (about 120 nucleotides, Scladina) or extremely short HVR-I fragments (about 30 nucleotides, Vindija-77, La-Chapelle-aux-Saints, Rochers-de-Villeneuve and Engis 2). Secondly, so far no genetic information is available from Middle Eastern Neandertals. The specimens from Amud (Israel, Serre et al. 2004) and Dederyieh (Syria, Serre et al. 2004) have been analyzed though but exhibited poor DNA contents according to amino-acid racemization ratios. Furthermore, the mtDNA sequence from the Okladnikov specimen has clearly demonstrated that Neandertals inhabitated the Altaï region, suggesting a much greater geographic distribution than previously acknowledged (Krause et al. 2007a). Finally, only two specimens older than 70 KYA have delivered some genetic information so far (Scladina and Teshik Tash), which makes any inference of the long-term changes in Neandertal genetic diversity very speculative. It is therefore highly recommended to wait for the analysis of additional individuals to draw more robust conclusions. But undoubtedly, as noted by Laurent Excoffier (Excoffier 2006:R652), "*the focus now switch from the mere relationship between Neandertals and modern humans to more Neandertal specific questions, such as the dynamics of their extension and retreat in space and time.*"

The Geographic Genetic Diversity of Neandertals

The sequence of the Scladina specimen (100 KYA) was the first to document a greater genetic diversity among European Neandertals by showing 3 novel substitutions with other previously reported Neandertal sequences over 123 nucleotides (Orlando et al. 2006). Soon after, the characterization of the whole HVR-I from the Monte Lessini specimen (Italy, 50 KYA) revealed another very divergent mtDNA haplotype (Caramelli et al. 2006). The Teshik Tash and Okladnikov haplotypes have revealed additional genetic diversity among Neandertals (Krause et al. 2007a). Median-spanning network and phylogenetic analyses revealed that, while clearly belonging to the Neandertal group, the Monte Lessini haplotype branched near the root of the Neandertal genealogy (Caramelli et al. 2006; see Fig. 21.1a and b). By contrast, the sequences from El Sidron (Northern Spain), Vindija (Croatia), and Feldhofer (Germany) were found very similar and clustered altogether (cluster I) (Lalueza-Fox et al. 2006). Intermediate between this cluster and Monte Lessini was the Mezmaïskaya (Caucasus) haplotype (note that the Scladina haplotype cannot be attributed to either of these groups since the 123-nucleotides sequenced so far do not include the positions that structure the genealogy; Fig. 21.1a and b).

This pattern has been described as compatible with a geographic subdivision of Neandertal populations into three clades at 40–50 KYA: one spanning from Spain to Germany and Croatia (cluster I), and two others respectively centred in Italy (Monte Lessini) and the Caucasus (Mezsmaïskaya). According to coalescence simulations, cluster I originated around 130±30 KYA. This could suggest that the El Sidrón, Feldhofer, and Vindija Neandertals belong to a group of Neandertals that expanded from a hypothetical southern glacial refugium after a demographic collapse associated with the Riss-glacial maximum (Lalueza-Fox et al. 2006). In turn, the Monte Lessini and Mezsmaïskaya haplotypes, which are both clearly distinct from the cluster I, would be the representatives of earlier Neandertal lineages (Caramelli et al. 2006; Excoffier 2006).

We should note however that such a model was drawn before Teshik Tash and Okladnikov haplotypes were characterized and we can now add further details/revisions. First, though the haplotype from Monte Lessini still appears at an ancestral position in the network, the haplotypes from Mezsmaïskaya, Okladnikov, and Teshik Tash (and Scladina) are certainly the most derived (Fig. 21.1a and b). They would most probably be the representatives of more recent lineages that dispersed over large geographic areas. Second, the haplotypes from Scladina and Teshik Tash exhibit striking similarities, which possibly suggests large geographic dispersion from a common ancestral population. But as most of the information relies on short sequences, it is basically impossible to infer at the present time any definitive population connection; homoplasic mutations may indeed cause a chance similarity. However, the presence of Neandertal haplotypes in the Altaï region (Siberia) raises the interesting possibility that they dispersed even farther to the east, in Mongolia or China.

Neandertal Genetic Diversity Through Time

The Scladina sequence represents the most ancient Neandertal sample analyzed at the DNA sequence level (100 KYA) (Orlando et al. 2006). Though limited to 123-nucleotides, it offers a unique opportunity to document either drastic changes or long-time stability of the Neandertal mtDNA-pool by comparison with other Neandertal sequences (29–50 KYBP). The Scladina sequence has nonetheless revealed an intriguing feature: younger Neandertal sequences appear more similar to modern humans than older Neandertal sequences (Orlando et al. 2006; Fig. 21.1a). Interestingly, such a pattern is still observed if the Scladina and the Teshik Tash sequences (>70 KYA) are compared to younger Neandertal sequences (<50 KYA).

Selection or genetic drift (random sorting of Neandertal lineages) could be responsible for this pattern. If both Neandertals and modern humans carried similar adaptative mutations, selection acting under similar environmental constraints between 70 and 50 KYA might have driven convergent evolution of their mtDNA (Excoffier 2006). This selective sweep would have made recent Neandertals (<50 KYA) more human-like than ancient (>70 KYA). Alternatively, demographic bottlenecks could have neutrally eliminated the less expanded haplotypes (i.e., the most derived) from the Neandertal mtDNA gene pool between 70 and 50 KYA (Orlando et al. 2006; Fig. 21.1a). Whether the cold isotope stage 4 (74–60 KYA; Ambrose 1998) could have promoted such a demographic decline will require further sequence data, in particular from specimens that lived between 50 and 100 KYA.

Finally, there is no doubt that the four Neandertal specimens that have been characterized at the HVR-I level since June 2006 have helped to better characterize the real extent of the Neandertal genetic diversity (see Fig. 21.1a: these four haplotypes are the most diverse – the less central haplotypes). Overall, the mean genetic pairwise distance among Neandertals appears within the range of the one seen among present day Europeans or Asians (Krause et al. 2007a). Interestingly, it remains significantly lower than the diversity present in current African populations.

Part Three – *Homo Sapiens* in Light of the Complete Genome of *Homo Neanderthalensis*

Rationale

The Neandertal genome sequence offers a real opportunity to give insights into the very genetic changes that make us human. The first complete draft of the human and chimpanzee genomes were published respectively in 2001 (The International Human Genome Sequencing Consortium 2001) and 2005 (The Chimpanzee Sequencing and Analysis Consortium 2004). About 35 million nucleotide differences have been identified so far by comparative genomics, but only part of these differences have occurred along the human lineage (the other part is chimp-specific).

One way to pinpoint the repertoire of nucleotidic changes that promoted the emergence of modern humans is to check for nucleotides conserved between chimps and gorillas (and possibly other great apes) but divergent in humans. For instance, recent genomic screens of chimp and modern human genomes have recently identified regions more repeated in humans than in chimps (and other mammals) (Cheng et al. 2005). Interestingly, some of these regions correspond to proteic domains that are expressed in the neocortex and might be involved in differences in the cognitive capabilities between the two species (Popesco et al. 2006). The determination of the number of repetitions found in the

Neandertal genome could provide the exact timing of the gene expansion along the *Homo* lineage. The same holds true for 992 recently identified noncoding sequences located preferentially in the vicinity of genes involved in neuronal cell adhesion, which remained conserved in the course of Primate evolution but that accumulated substantial nucleotide change along the human lineage (Prabhakar et al. 2006).

Likewise, the Neandertal genome will offer the unique opportunity to check which part of the coding-loci exhibiting excess in nonsynonymous mutations along the human lineage (Hellman et al. 2003; reviewed in Enard and Pääbo 2004; Harris and Meyer 2006) has been positively selected since the Neandertal–modern human split. Among them, the FOXP2 gene (a forkhead transcription factor involved in larynx development and language ability; see below) might reveal if the expansion of modern humans was driven by the appearance of a proficient spoken language (Enard et al. 2002). All in all, comparative genomics of chimp, Neandertal and modern humans promises to provide the complete list of the functionally important genetic changes that gave rise to our species. However, this achievement is by no mean straightforward, as the processing of ancient DNA requires numerous controls.

Getting the Neandertal Genome

An Unexpected Dream

Recovering even short pieces of ancient sequence information is most often the result of very extensive work. This is principally due to the nature of ancient DNA molecules that are highly fragmented and chemically modified (Höss et al. 1996). The classical strategy consists in amplifying a short DNA fragment by PCR and in subcloning the amplification by-products before sequencing a great number of clones. As each cell houses thousands of mitochondria but only one nucleus and as very few templates are preserved during the taphonomic process, paleogeneticists have generally focused on mitochondrial genes. Thus, nuclear genes received so far only little attention (Greenwood et al. 1999; Jaenicke-Desprès et al. 2003). Furthermore, considering the level of fragmentation of aDNA molecules (a given fragment is around 50–150 nucleotides long) and the total length of the human genome (about three billions nucleotides), collecting large amounts of nuclear information would require hundreds of millions PCRs. Consequently, that the complete genomic sequence of an individual from the past could be determined was once beyond paleogeneticists's dreams.

Technological Breakthrough

Recent high-throughput technological breakthroughs have made this dream a realistic objective though. In brief, the methodology relies on a three-step procedure: first, aDNA fragments are used as templates for constructing DNA libraries. Then, the clones (or the single stranded fragments according to the methodology selected) undergo massive parallel sequencing, and finally, the sequences are processed *in silico*. Sequences of bacterial, microbial, and environmental origin are filtered through comparison with sequences available in worldwide databanks, and only primate-like sequences are conserved for further analyses. Overlapping regions are assembled into larger DNA scaffolds that one can align onto the human or the chimp genome to check for human-specific (or Neandertal) nucleotidic changes. In 2000, a much more straightforward strategy (relying on Southern-blot hybridization experiments) was used to directly get information on the structural differences of Neandertals and modern humans genomes (Scholz et al. 2000). However, further investigations showed the inadequacy of such a method (Geigl 2001; Bachmann 2001). Therefore, it is now clear that precious fossils such as early hominoids should be destroyed to be analyzed by high-throughput genomic methods rather than unreliable procedures such as southern hybridization.

Contamination Concerns

The three-step strategy was first used for gathering genomic data from skeletal remains of a 40,000 year old cave bear (Noonan et al. 2005) and a 27,000 year old mammoth (Poinar et al. 2006). Only a small fraction (1–6%) of cave bear genomic sequences could be attributed to cave-bear (Noonan et al. 2005), and at the same time, 30–100 times fewer sequences consisted in DNA of human origin. Though it was better preserved by the optimal environmental conditions of the northern region of the Artic circle (about 45% of the sequences belonged to the mammoth), the mammoth specimen also still exhibited a significant proportion of human DNA contaminants (about a 30th of the mammoth sequences; Poinar et al. 2006). Though not critical as long as animal DNA is studied, this problem becomes a major concern for Neandertal genomic survey since one expects only a few (if any) differences between Neandertal and modern human DNA fragments. Therefore, the only way to warrant that modern human contaminants are not be taken for Neandertal genomic fragments is to check for Neandertal samples with no (or low) contamination levels. Since the mitochondrial HVR-I from Neandertals exhibit specific mutations with regards to modern human sequences, the most straightforward way to assess for human contamination is to use primers that coamplify Neandertal and modern human HVR-I fragments and to determine the proportion of Neandertal and modern human sequences after amplification. One bone (named Vindija-80), found in 1980 in the Vindija cave (Croatia) and dated at 38,310 ± 2,130 BP, exhibited 94–99% of Neandertal sequences in the assay, suggesting very low levels of modern human contamination.

Filtering for Quality

Sequencing Errors

As no sequencing method is error-free, the quality of the Neandertal sequences generated through pyrosequencing needed to be assessed first, even more so for the 454 technology which provides hundreds of thousands of reads per run (even very low error-rates would insert significant uncertainty in the Neandertal genome). A straightforward strategy to circumvent sequencing errors is to take advantage of reads that overlap the same region in order to filter for poor-matching positions (low-quality scores). Then, if pyrosequencing errors occur at no significant level, these filtered regions should exhibit identical sequences if sequenced with the traditional Sanger method. Noonan et al. (2006) systematically compared pyrosequences and Sanger electrophoregrams along 6,200 nucleotides and found very little discordance (about 1 nucleotide every 1,000) between both sequence technologies, suggesting pyrosequencing does not introduce significantly more errors than the Sanger method.

Artifactual Mutations

Pyrosequencing of ancient DNA fragments generates another systematic type of error though. As pyrosequence reactions stem from just one original single-stranded template molecule, every damaged base might be misread, leading to misincorporations of substantial artifactual mutations (DNA-damage induced errors). Large genome coverage (>3X for every haploid stock, i.e., 6X for our diploid genome) is promised to solve this problem by providing multiple reads of the same genomic region (different templates are not likelihood to carry the same kind of damages). However, such a strategy will be very costly. The specific amplification of Neandertal loci showing sequence divergence with humans (or chimps) through multiplex PCRs would offer a more simple way to correct for artifactual substitutions. This strategy has been successfully used to correct the mtDNA reads of the one-million nucleotide survey of the Neandertal genome: 14 of 34 mutated sites were actually artifactual and explained the 2.5-fold excess in Neandertal branch length (Green et al. 2006).

Surprisingly, pyrosequencing also offers the unique opportunity to distinguish every different type of nucleotide misincorporation for the following reason (Stiller et al. 2006; Gilbert et al. 2007; Briggs et al. 2007). Let us focus on a Neandertal specific site where, for instance, Neandertals exhibit a T where human and chimp genomes carry a C. The pyrosequencing reaction might have either sequenced this strand (T) or the complementary (A). Therefore, the number of sites where Neandertals carry a T and modern humans a C (C → T) should be equal to the number of sites where Neandertals carry a A and modern humans a G (G → A). The only reason for any discrepancy between pairs of substitutions (C → T *versus* G → A) is that the substitution was carried by only one strand and therefore was a *post-mortem* degradation by-product. As expected for extant organisms, such pairs of substitutions for chimp-specific sites (or human specific sites) were similar, suggesting these differences were not degradation derivatives. As far as the Neandertal specific sites were concerned, excessive C → T changes were found (relative to G → A), suggesting that 12% of Neandertal specific sites were actually artifactual (Green et al. 2006). This overall excess of mutations due to artifactual changes can be statistically removed from global estimates of divergence time or ancestral population size. The correction is however less straightforward for describing the precise repertoire of genetic changes that occurred specifically along the Neandertal genome. For instance, multicoverage after specific fishing of these regions with complementary probes (as suggested in Noonan et al. (2006)) or through a single primer extension (SPEX)-based approach (Brotherton et al. 2007) could provide direct experimental strategies to evaluate their accuracy. Fortunately, Briggs et al. (2007) have quantitatively shown that *postmortem* degradations most probably affects the Cytosines that are located at the 5′-extremities of the aDNA strands, leaving open the possibility that the reliability of a neandertal specific change could be estimated according to its position in sequencing reads. Whatever the strategy, such corrections appear as a prerequisite.

Contamination Concerns

Wall and Kim (2007) emphasized several discrepancies between the two currently available Neandertal genomic data sets, though both are derived from the same specimen (Vindija-80). In particular, according to the 454-data set (1-Mb), the level of divergence between Neandertal and modern humans is about the same as between two modern human ethnic groups (while it is greater according to the Noonan et al. (2006) data set). Moreover, Neandertal specific mutations exhibit higher levels of DNA damage induced mutations (C→T and G→A; as expected for ancient molecules) in the former data set than in the latter (Wall and Kim 2007). This pattern most probably reflects large contamination of the 454-data set (possibly up to 78%) with long fragments of modern human DNA since these discrepancies vanish as long as only small fragments (<50 bp) are analyzed. Therefore, in addition to PCR-assays developed to identify the Neandertal extracts free of contamination, further experimental procedures are crucial to minimize contamination postextraction along the different steps of the 454 process (Briggs et al. 2007). First, ssDNA libraries must be performed in a laboratory specifically equipped and exclusively dedicated to work on aDNA, and second, the DNA libraries corresponding to different extracts should be generated with different primers that could act as barcodes to monitor possible cross-contamination *a posteriori*.

Focusing on Candidate Genes

So far, shotgun genomic sequencing has still delivered no information relative to possible phenotypic features of Neandertals. Two-round multiplex has however been used to gather information about their possible language capabilities and skin color by specific amplification of two candidate genes (FOX-P2 and MC1R, respectively; Krause et al. 2007b; Lalueza-Fox et al. 2007).

In modern humans, FOX-P2 deficiencies lead to a limitation in orofacial movements (associated with reduced word-repetition abilities) and Broca's aphasia (Vargha-Khadem et al. 2005). Moreover, two nonsynonymous substitutions have been fixed in the modern human lineage, possibly in relation to a selective advantage as suggested by the extended haplotype homozygosity at the FOX-P2 locus present in modern human populations (Enard et al. 2002). Coalescent simulations based on these polymorphisms have estimated 200 KY as the possible time for the onset of the positive selective sweep, suggesting that these cognitive abilities were confined solely to our species. The genotyping of Neandertal individuals for these two positions offers the opportunity to test if the mutations (and possibly the cognitive abilities they have driven) were already present in the last common ancestor to modern humans and Neandertals. Obviously, contamination with modern DNA would inevitably lead to the identification of derived mutations in the Neandertal FOX-P2 gene – but interestingly for all other loci as well – whereas those loci that derived early in the course of modern human evolution should be genotyped as ancestral in absence of contamination. Therefore, determining the ancestral versus derived status of such loci gives a way to monitor contamination at the nuclear level. Although a large fraction of the amplification attempts gave no result, eight such loci (located in the X and Y, and autosomal chromosomes 1 and 8) revealed reproducible ancestral states for two El Sidron Neandertal specimens (El Sidron 1253 and 1351c), excluding contamination of the Neandertal extracts at the nuclear level. Surprisingly, both Neandertals exhibited the same mutations as in the modern human FOX-P2 gene (Krause et al. 2007b; note that the same is found for polymorphic sites located in the upstream intrinsic region of the gene, revealing that the identity with modern human is true at the whole haplotypic level). Consequently, it leaves no ambiguity that Neandertals carried the same FOX-P2 mutations as modern humans. Of course, language emergence could not rely exclusively on two isolated mutations only but whatever their exact function has played in this complex and integrative process, it concerned both modern humans and Neandertals. Ongoing *in vivo* experiments should help to determine their precise role in Neandertal cognitive abilities as well as further genotyping of other loci possibly involved in language acquisition or processing (e.g., ASPM and Microcephalin; Dediu and Ladd 2007).

The second nuclear Neandertal gene that has been genotyped is MC1R (Lalueza-Fox et al. 2007). This gene encodes for a cell transmembrane receptor that is involved in the balance between eumelanin and pheomelanin synthesis. Allelic variants that lead to red hair and pale skin have been described among modern populations. However, the Arg307Gly mutation variant, which has been observed in the Monte Lessini and El Sidron 1252 specimens, has never been described in more than 3,700 modern humans, suggesting that there is almost no chance it could be a contaminant by-product. However, after transfection in human fibroblasts (COS-7 cells), expression and correct localization at the plasmidic membrane, this MC1-R variant exhibited significantly reduced levels of activation after α-MSH activation As a result, and assuming they were homozygous for this allelic variant, the two neandertal individuals would most probably have been red haired and pale skinned (Lalueza-Fox et al. 2007).

These two last studies demonstrate how aDNA could come as a complement to physical anthropology approaches by providing direct access to characters that do not fossilize and thereby completing our knowledge and representation modes of ancient humans. There is no doubt that other phenotypically interesting candidate genes will be genotyped soon.

Beyond the Neandertal Genome

The reason for differences between species does not simply rely on genomic differences. Differences at the transcriptional level/timing/pattern of specific genes may result in large phenotypic differences. For instance, a recent survey of micro-RNA tissue contents (small RNA molecules interfering with gene expression) revealed large differences among the human and chimpanzee brains (Berezikov et al. 2006). Microarray experiments revealing gene expression patterns at the transcriptional level have clearly identified the brain and testis as key organs involved in chimp–human differences (Khaitovitch et al. 2005). Besides, epigenetic modifications of nucleotidic bases (e.g., cytosine methylation) or histones (e.g., acetylation/deacetylation) play a critical role in this phenomenon without sequence divergence (Eckhardt et al. 2006). The knowledge of the Neandertal genome will not provide such kinds of information.

Conclusion

The recovery of the Neandertal genome has now started and is announced to be completed within the forthcoming months. Given the current capacity of the 454 sequencing platform, a first complete coverage of the genome (1X) from a single individual will take 6,000 supplemental runs and about 20-g of fossil material (Green et al. 2006). A large genome coverage (>6X) is required yet to correct for possible sequence mistakes. Improvements in the technique are therefore essential before the whole project can possibly be achieved. Interestingly, the latest generations of high-throughput sequencing platforms already provide larger sequence information per run (250 Mb for the Roche GS-FLX and 1 Gb for the Ilumina Genetic Analyzer).

Even though, the Neandertal genome project clearly shows the way forward. By defining the standard procedure, it opens the hunt for large-scale DNA retrieval of other ancient hominids from the Upper Pleistocene. Obviously, Neandertal individuals coming from interspersed geographic regions would be interesting to look at since they would provide a clearer picture on the origins and diversity of Neandertals. Above all, they would reveal what is the exact part of their genomic specificities that is attributable to real interspecific differences and not to individual polymorphism. For certain, a few other ancient hominids will be serious candidates for genomic studies so long as contamination and DNA degradation reactions have occurred at extremely low levels: the Neandertal-like specimens from China or the Pestera-Muierii fossils for instance (showing a mosaic of modern human and archaic Neandertal features; Soficaruet al. 2006) but also the southern-African early *sapiens* (Grun et al. 1990) or even the late hobbits from Flores (Brown et al. 2004; Morwood et al. 2005) or pre-Neandertals (Valdiosera et al. 2006). By doing so, ancient DNA will place the "Rapid Replacement" versus "Multiregionalism" debate in a broader context than the single Neandertal-*sapiens* admixture possibility and promises to give invaluable insights on our phylogenetic tree.

The last decade has known the emergence of Neandertal paleogenetics. The next will be devoted to Stone Age genomics!

Acknowledgments We are indebted to Dr. Marina Faerman, Dr. Colin Smith, and one supplemental anonymous reviewer for critical reading of the manuscript and helpful comments.

References

Ambrose, S. H. (1998). Late Pleiste human population bottlenecks, volcanic winter, and differentiation of modern humans. *Journal of Human Evolution, 34*, 623–651.

Austin, J. J., Ross, A. J., Smith, A. B., Fortey, R. A., & Thomas, R. H. (1997). Problems of reproducibility – does geologically ancient DNA survive in amber-preserved insects? *Proceedings of the Royal Society of London, Biological Sciences, 264*, 467–474.

Bachmann, L. (2001). Reply to Geigl. *American Journal of Human Genetics, 68*, 290–291.

Bandelt, H. J. (2005). Mosaics of ancient mitochondrial DNA: Positive indicators of nonauthenticity. *European Journal of Human Genetics, 13*, 1106–1112.

Beauval, C., Maureille, B., Lacrampe-Cuyaubere, F., Serre, D., Peressinotto, D., Bordes, J. G., Cochard, D., Couchoud, I., Dubrasquet, D., Laroulandie, V., Lenoble, A., Mallye, J. B., Pasty, S., Primault, J., Rohland, N., Pääbo, S., & Trinkaus, E. (2005). A late Neandertal femur from Les Rochers-de-Villeneuve, France. *Proceedings of the National Academy of Sciences of the United States of America, 102*, 7085–7090.

Berezikov, E., Thuemmler, F., van Laake, L. W., Kondova, I., Bontrop, R., Cuppen, E., & Plasterk, R. H. (2006). Diversity of micro RNAs in human and chimpanzee brain. *Nature Genetics, 38*, 1375–1377.

Briggs, A. W., Stenzel, U., Johnson, P. L., Green, R. E., Kelso, J., Prüfer, K., Meyer, M., Krause, J., Ronan, M. T., Lachmann, M., & Pääbo, S. (2007). Patterns of damage in genomic DNA sequences from a Neandertal. *Proceedings of the National Academy of Sciences of the United States of America, 104*, 14616–14621.

Brotherton, P., Endicott, P., Sanchez, J. J., Beaumont, M., Barnett, R., Austin, J., & Cooper, A. (2007). Novel high-resolution characterization of ancient DNA reveals C>U-type base modification events as the sole cause of post mortem miscoding lesions. *Nucleic Acids Research, 35*, 5717–5728.

Brown, P., Sutikna, T., Morwood, M. J., Soejono, R. P., Jatmiko, Saptomo, E. W., & Due, R. A. (2004). A new small-bodied hominin from the Late Pleistocene of Flores, Indonesia. *Nature, 431*, 1055–1061.

Caramelli, D., Lalueza-Fox, C., Vernesi, C., Lari, M., Casoli, A., Mallegni, F., Chiarelli, B., Dupanloup, I., Bertranpetit, J., Barbujani, G., & Bertorelle, G. (2003). Evidence for a genetic discontinuity between Neandertals and 24,000-year-old anatomically modern Europeans. *Proceedings of the National Academy of Sciences of the United States of America, 100*, 6593–6597.

Caramelli, D., Lalueza-Fox, C., Condemi, S., Longo, L., Milani, L., Manfredini, A., de Saint Pierre, M., Adoni, F., Lari, M., Giunti, P., Ricci, S., Casoli, A., Calafell, F., Mallegni, F., Bertranpetit, J., Stanyon, R., Bertorelle, G., & Barbujani, G. (2006). A highly divergent mtDNA sequence in a Neandertal individual from Italy. *Current Biology, 2006*(16), R630–R632.

Cheng, Z., Ventura, M., She, X., Khaitovich, P., Graves, T., Osoegawa, K., Church, D., DeJong, P., Wilson, R. K., Pääbo, S., Rocchi, M., & Eichler, E. E. (2005). A genome-wide comparison of recent chimpanzee and human segmental duplications. *Nature, 437*, 88–93.

Clement, M., Posada, D., & Crandall, K. (2000). TCS: A computer program to estimate gene genealogies. *Molecular Ecology, 9*, 1657–1660.

Cooper, A., Poinar, H. N., Pääbo, S., Radovcic, J., Debenath, A., Caparros, M., Barroso-Ruiz, C., Bertranpetit, J., Nielsen-Marsh, C., Hedges, R. E., & Sykes, B. (1997). Neandertal genetics. *Science, 277*, 1021–1024.

Cooper, A., Drummond, A. J., & Willerslev, E. (2004). Ancient DNA: Would the real Neandertal please stand up? *Current Biology, 14*, R431–R433.

Currat, M., & Excoffier, L. (2004). Modern humans did not admix with Neanderthals during their range expansion into Europe. *PLoS Biology, 2*, e421.

Currat, M., & Excoffier, L. (2005). The effect of the Neolithic expansion on European molecular diversity. *Proceedings of the Royal Society of London, Biological Sciences, 272*, 679–688.

Dediu, D., & Ladd, R. (2007). Linguistic tone is related to the population frequency of the adaptative haplogroups of two brain size genes, ASPM and Microcephalin. *Proceedings of the National Academy of Sciences of the United States of America, 104*, 10944–10949.

Eckhardt, F., Lewin, J., Cortese, R., Rakayan, V. K., Attwood, J., Burger, M., Burton, J., Cox, T. V., Davies, R., Down, T. A., Haefliger, C., Horton, R., Howe, K., Jackson, D. K., Kunde, J., Koenig, C., Liddle, J., Niblett, D., Otto, T., Pettett, R., Seemann, S., Thompson, C., West, T., Rogers, J., Olek, A., Berlin, K., & Beck, S. (2006). DNA methylation profiling of human chromosomes 6, 20 and 22. *Nature Genetics, 38*, 1378–1385.

Enard, W., & Pääbo, S. (2004). Comparative primate genomics. *Annual Review of Genomics and Human Genetics, 5*, 351–378.

Enard, W., Przeworski, M., Fisher, S. E., Lai, C. S., Wiebe, V., Kitano, T., Monaco, A. P., & Pääbo, S. (2002). Molecular evolution of FOXP2, a gene involved in speech and language. *Nature, 418*, 869–872.

Evans, P. D., Vallender, E. J., & Lahn, B. T. (2006). Molecular evolution of the brain size regulator genes CDK5RAP2 and CENPJ. *Gene, 375*, 75–79.

Excoffier, L. (2006). Neandertal genetic diversity: A fresh look from old samples. *Current Biology, 16*, R650–R652.

Excoffier, L., & Yang, Z. (1999). Substitution rate variation among sites in mitochondrial hypervariable region I of humans and chimpanzees. *Molecular Biology and Evolution, 16*, 1357–1368.

Finlayson, C., Pacheco, F. G., Rodriguez-Vidal, J., Fa, D. A., Gutierrez Lopez, J. M., Santiago Perez, A., Finlayson, G., Allue, E., Baena Preysler, J., Caceres, I., Carrion, J. S., Fernandez Jalvo, Y., Gleed-Owen, C. P., Jimenez Espejo, F. J., Lopez, P., Lopez Saez, J. A., Riquelme Cantal, J. A., Sanchez Marco, A., Guzman, F. G., Brown, K., Fuentes, N., Valarino, C. A., Villalpando, A., Stringer, C. B., Martinez Ruiz, F., & Sakamoto, T. (2006). Late survival of Neanderthals at the southernmost extreme of Europe. *Nature, 443*, 850–853.

Geigl, E. M. (2001). Inadequate use of molecular hybridization to analyze DNA in Neanderthal fossils. *American Journal of Human Genetics, 68*, 287–291.

Gilbert, M. T., Willerslev, E., Hansen, A. J., Barnes, I., Rudbeck, L., Lynnerup, N., & Cooper, A. (2003). Distribution patterns of postmortem damage in human mitochondrial DNA. *American Journal of Human Genetics, 72*, 32–47. *Erratum in*: 72, 779.

Gilbert, M. T., Binladen, J., Miller, W., Wiuf, C., Willerslev, E., Poinar, H., Carlson, J. E., Leebens-Mack, J. H., & Schuster, S. C. (2007). Recharacterization of ancient DNA miscoding lesions: Insights in the era of sequencing-by-synthesis. *Nucleic Acids Research, 35*, 1–10.

Green, R. E., Krause, J., Ptak, S. E., Briggs, A. W., Ronan, M. T., Simons, J. F., Du, L., Egholm, M., Rothberg, J. M., Paunovic, M., & Pääbo, S. (2006). Analysis of one million base pairs of Neanderthal DNA. *Nature, 444*, 330–336.

Greenwood, A. D., Capelli, C., Possnert, G., & Pääbo, S. (1999). Nuclear DNA sequences from late Pleistocene megafauna. *Molecular Biology and Evolution, 16*, 1466–1473.

Grun, R., Beaumont, P. B., & Stringer, C. B. (1990). ESR dating evidence for early modern humans at border cave in South Africa. *Nature, 344*, 537–539.

Gutierrez, G., Sanchez, D., & Marin, A. (2002). A reanalysis of the ancient mitochondrial DNA sequences recovered from Neandertal bones. *Molecular Biology and Evolution, 19*, 1359–1366.

Haak, W., Forster, P., Bramanti, B., Matsumura, S., Brandt, G., Tanzer, M., Villems, R., Renfrew, C., Gronenborn, D., Alt, K. W., & Burger, J. (2005). Ancient DNA from the first European farmers in 7500-year-old Neolithic sites. *Science, 310*, 1016–1018.

Handt, O., Krings, M., Ward, R. H., & Pääbo, S. (1996). The retrieval of ancient human DNA sequences. *American Journal of Human Genetics, 59*, 368–376.

Harris, E. E., & Meyer, D. (2006). The molecular signature of selection underlying human adaptations. *American Journal of Physical Anthropology, 43*, 89–130.

Hawks, J., & Wolpoff, M. H. (2001). Paleoanthropology and the population genetics of ancient genes. *American Journal of Physical Anthropology, 114*, 269–272.

Hawks, J., Cochran, G., Harpending, H. C., & Lahn, B. T. (2007). A genetic legacy from archaic Homo. *Trends in Genetics, 24*, 19–23.

Hebsgaard, M. B., Wiuf, C., Gilbert, M. T., Glenner, H., & Willerslev, E. (2007). Evaluating Neanderthal genetics and phylogeny. *Journal of Molecular Evolution, 64*, 50–60.

Hellman, I., Zöllner, S., Enard, W., Ebersberger, I., Nickel, B., & Pääbo, S. (2003). Selection on human genes as revealed by comparisons to chimpanzee cDNA. *Genome Research, 13*, 831–837.

Hofreiter, M., Jaenicke, V., Serre, D., Haeseler, A. V. A., & Pääbo, S. (2001a). DNA sequences from multiple amplifications reveal artifacts induced by cytosine deamination in ancient DNA. *Nucleic Acids Research, 29*, 4793–4799.

Hofreiter, M., Serre, D., Poinar, H. N., Kuch, M., & Pääbo, S. (2001b). Ancient DNA. *Nature Reviews Genetics, 2*, 353–359.

Höss, M., Jaruga, P., Zastawny, T. H., Dizdaroglu, M., & Pääbo, S. (1996). DNA damage and DNA sequence retrieval from ancient tissues. *Nucleic Acids Research, 24*, 1304–1307.

Innan, H., & Nordborg, M. (2002). Recombination or mutational hot spots in human mtDNA? *Molecular Biology and Evolution, 19*, 1122–1127.

Jaenicke-Despres, V., Buckler, E. S., Smith, B. D., Gilbert, M. T., Cooper, A., Doebley, J., & Pääbo, S. (2003). Early allelic selection in maize as revealed by ancient DNA. *Science, 302*, 1206–1208.

Kahn, P., & Gibbons, A. (1997). DNA from an extinct human. *Science, 277*, 176–178.

Khaitovitch, P., Hellmann, I., Enard, W., Nowick, K., Leinweber, M., Franz, H., Weiss, G., Lachmann, M., & Pääbo, S. (2005). Parallel patterns of evolution in the genomes and transcriptomes of humans and chimpanzees. *Science, 309*, 1850–1854.

Knight, A. (2003). The phylogenetic relationship of Neandertal and modern human mitochondrial DNAs based on informative nucleotide sites. *Journal of Human Evolution, 44*, 627–632.

Krause, J., Dear, P. H., Pollack, J. L., Slatkin, M., Spriggs, H., Barnes, I., Lister, A. M., Ebersberger, I., Pääbo, S., & Hofreiter, M. (2006). Multiplex amplification of the mammoth mitochondrial genome and the evolution of elephantidae. *Nature, 439*, 724–727.

Krause, J., Orlando, L., Serre, D., Viola, B., Prüfer, K., Richards, M. P., Hublin, J. J., Hänni, C., Derevianko, A. P., & Pääbo, S. (2007a). Neanderthals in central Asia and Siberia. *Nature, 449*, 902–904.

Krause, J., Lalueza-Fox, C., Orlando, L., Enard, W., Green, R. E., Burbano, H. A., Hublin, J. J., Hänni, C., Fortea, J., de la Rasilla, M., Bertranpetit, J., Rosas, A., & Pääbo, S. (2007b). The derived FOXP2 variant of modern humans was shared with Neandertals. *Current Biology, 17*, 1908–1912.

Krings, M., Stone, A., Schmitz, R. W., Krainitzki, H., Stoneking, M., & Pääbo, S. (1997). Neandertal DNA sequences and the origin of modern humans. *Cell, 90*, 19–30.

Krings, M., Geisert, H., Schmitz, R. W., Krainitzki, H., & Pääbo, S. (1999). DNA sequence of the mitochondrial hypervariable region II from the Neandertal type specimen. *Proceedings of the National Academy of Sciences of the United States of America, 96*, 5581–5585.

Krings, M., Capelli, C., Tschentscher, F., Geisert, H., Meyer, S., von Haeseler, A., Grossschmidt, K., Possnert, G., Paunovic, M., & Pääbo, S. (2000). A view of Neandertal genetic diversity. *Nature Genetics, 26*, 144–146.

Lalueza-Fox, C., Sampietro, M. L., Caramelli, D., Puder, Y., Lari, M., Calafell, F., Martinez-Maza, C., Bastir, M., Fortea, J., de la Rasilla, M., Bertranpetit, J., & Rosas, A. (2005). Neandertal evolutionary genetics: Mitochondrial DNA data from the Iberian peninsula. *Molecular Biology and Evolution, 22*, 1077–1081.

Lalueza-Fox, C., Krause, J., Caramelli, D., Catalano, G., Milani, L., Sampietro, M. L., Calafell, F., Martinez-Maza, C., Bastir, M., Garcia-Tabernero, A., de la Rasilla, M., Fortea, J., Pääbo, S., Bertranpetit, J., & Rosas, A. (2006). Mitochondrial DNA of an Iberian Neandertal suggests a population affinity with other European Neandertals. *Current Biology, 16*, R629–R630.

Lalueza-Fox, C., Römpler, H., Caramelli, D., Stäubert, C., Catalano, G., Hughes, D., Rohland, N., Pilli, E., Longo, L., Condemi, S., de la Rasilla, M., Fortea, J., Rosas, A., Stoneking, M., Schöneberg, T., Bertranpetit, J., & Hofreiter, M. (2007). A melanocortin 1 receptor allele suggests varying pigmentation among Neanderthals. *Science, 318*, 1453–1455.

Lindahl, T. (1997). Facts and artifacts of ancient DNA. *Cell, 90*, 1–3.

Malmstrom, H., Stora, J., Dalen, L., Holmlund, G., & Gotherstrom, A. (2005). Extensive human DNA contamination in extracts from ancient dog bones and teeth. *Molecular Biology and Evolution, 22*, 2040–2047.

Marth, G., Schuler, G., Yeh, R., Davenport, R., Agarwala, R., Church, D., Wheelan, S., Baker, J., Ward, M., Kholodov, M., Phan, L., Czabarka, E., Murvai, J., Cutler, D., Wooding, S., Rogers, A., Chakravarti, A., Harpending, H. C., Kwok, P. Y., & Sherry, S. T. (2003). Sequence variations in the public human genome data reflect a bottlenecked population history. *Proceedings of the National Academy of Sciences of the United States of America, 100*, 376–381.

Marth, G. T., Czabarka, E., Murvai, J., & Sherry, S. T. (2004). The allele frequency spectrum in genome-wide human variation data reveals signals of differential demographic history in three large world populations. *Genetics, 166*, 351–372.

McDougall, I., Brown, F. H., & Fleagle, J. G. (2005). Stratigraphic placement and age of modern humans from Kibish, Ethiopia. *Nature, 433*, 733–736.

Mellars, P. (2006). A new radiocarbon revolution and the dispersal of modern humans in Eurasia. *Nature, 439*, 931–935.

Mishmar, D., Ruiz-Pesini, E., Golik, P., Macaulay, V., Clark, A. G., Hosseini, S., Brandon, M., Easley, K., Chen, E., Brown, M. D., Sukernik, R. I., Olckers, A., & Wallace, D. C. (2003). Natural selection shaped regional mtDNA variation in humans. *Proceedings of the National Academy of Sciences of the United States of America, 100*, 171–176.

Morwood, M. J., Brown, P., Jatmiko, Sutikna, T., Saptomo, E. W., Westaway, K. E., Due, R. A., Roberts, R. G., Maeda, T., Wasisto, S., & Djubiantono, T. (2005). Further evidence for small-bodied hominins from the Late Pleistocene of Flores, Indonesia. *Nature, 437*, 1012–1017.

Noonan, J. P., Hofreiter, M., Smith, D., Priest, J. R., Rohland, N., Rabeder, G., Krause, J., Detter, J. C., Pääbo, S., & Rubin, E. M. (2005). Genomic sequencing of Pleistocene cave bears. *Science, 309*, 597–599.

Noonan, J. P., Coop, G., Kudaravalli, S., Smith, D., Krause, J., Alessi, J., Chen, F., Platt, D., Pääbo, S., Pritchard, J. K., & Rubin, E. M. (2006). Sequencing and analysis of Neanderthal genomic DNA. *Science, 314*, 1113–1118.

Nordborg, M. (1998). On the probability of Neanderthal ancestry. *American Journal of Human Genetics, 63*, 1237–1240.

Orlando, L., Darlu, P., Toussaint, M., Bonjean, D., Otte, M., & Hänni, C. (2006). Revisiting Neandertal diversity with a 100,000 year old mtDNA sequence. *Current Biology, 16*, R400–R402.

Ovchinnikov, I. V., Gotherstrom, A., Romanova, G. P., Kharitonov, V. M., Liden, K., & Goodwin, W. (2000). Molecular analysis of Neanderthal DNA from the northern Caucasus. *Nature, 404*, 490–493.

Plagnol, V., & Wall, J. D. (2006). Possible ancestral structure in human populations. *PLoS Genetics, 2*, e105.

Poinar, H. N., Schwarz, C., Qi, J., Shapiro, B., Macphee, R. D., Buigues, B., Tikhonov, A., Huson, D. H., Tomsho, L. P., Auch, A., Rampp, M., Miller, W., & Schuster, S. C. (2006). Metagenomics to paleogenomics: Large-scale sequencing of mammoth DNA. *Science, 311*, 392–394.

Popesco, M. C., Maclaren, E. J., Hopkins, J., Dumas, L., Cox, M., Meltesen, L., McGavran, L., Wyckoff, G. J., & Sikela, J. M. (2006). Human lineage-specific amplification, selection, and neuronal expression of DUF 1220 domains. *Science, 313*, 1304–1307.

Prabhakar, S., Noonan, J. P., Pääbo, S., & Rubin, E. M. (2006). Accelerated evolution of conserved noncoding sequences in humans. *Science, 314*, 786.

Relethford, J. H. (1999). Models, predictions, and the fossil record of modern human origins. *Evolutionary Anthropology, 8*, 7–10.

Relethford, J. H. (2001a). Absence of regional affinities of Neandertal DNA with living humans does not reject multiregional evolution. *American Journal of Physical Anthropology, 115*, 95–98.

Relethford, J. H. (2001b). Ancient DNA and the origin of modern humans. *Proceedings of the National Academy of Sciences of the United States of America, 98*, 390–391.

Richards, M., Macaulay, V., Hickey, E., Vega, E., Sykes, B., Guida, V., Rengo, C., Sellitto, D., Cruciani, F., Kivisild, T., Villems, R., Thomas, M., Rychkov, S., Rychkov, O., Rychkov, Y., Golge, M., Dimitrov, D., Hill, E., Bradley, D., Romano, V., Cali, F., Vona, G., Demaine, A., Papiha, S., Triantaphyllidis, C., Stefanescu, G., Hatina, J., Belledi, M., Di Rienzo, A., Novelletto, A., Oppenheim, A., Norby, S., Al-Zaheri, N., Santachiara-Benerecetti, S., Scozari, R., Torroni, A., & Bandelt, H. J. (2000). Tracing European founder lineages in the near eastern mtDNA pool. *American Journal of Human Genetics, 67*, 1251–1276.

Römpler, H., Rohland, N., Lalueza-Fox, C., Willerslev, E., Kuznetsova, T., Rabeder, G., Bertranpetit, J., Schoneberg, T., & Hofreiter, M. (2006). Nuclear gene indicates coat-color polymorphism in mammoths. *Science, 313*, 62.

Ruiz-Pesini, E., Mishmar, D., Brandon, M., Procaccio, V., & Wallace, D. C. (2004). Effects of purifying and adaptive selection on regional variation in human mtDNA. *Science, 303*, 223–226.

Schmegner, C., Hoegel, J., Vogel, W., & Assum, G. (2005). Genetic variability in a genomic region with long-range linkage disequilibrium reveals traces of a bottleneck in the history of the European population. *Human Genetics, 118*, 276–286.

Schmitz, R. W., Serre, D., Bonani, G., Feine, S., Hillgruber, F., Krainitzki, H., Pääbo, S., & Smith, F. H. (2002). The Neandertal type site revisited: Interdisciplinary investigations of skeletal remains from the Neander Valley, Germany. *Proceedings of the National Academy of Sciences of the United States of America, 99*, 13342–13347.

Scholz, M., Bachmann, L., Nicholson, G. J., Bachmann, J., Giddings, I., Ruschoff-Thale, B., Czarnetzki, A., & Pusch, C. M. (2000). Genomic differentiation of Neanderthals and anatomically modern man allows a fossil-DNA-based classification of morphologically indistinguishable hominid bones. *American Journal of Human Genetics, 66*, 1927–1932.

Semino, O., Passarino, G., Oefner, P. J., Lin, A. A., Arbuzova, S., Beckman, L. E., De Benedictis, G., Francalacci, P., Kouvatsi, A., Limborska, S., Marcikiae, M., Mika, A., Mika, B., Primorac, D., Santachiara-Benerecetti, A. S., Cavalli-Sforza, L. L., & Underhill, P. A. (2000). The genetic legacy of paleolithic *Homo sapiens sapiens* in extant Europeans: A Y chromosome perspective. *Science, 290*, 1155–1159.

Serre, D., Langaney, A., Chech, M., Teschler-Nicola, M., Paunovic, M., Mennecier, P., Hofreiter, M., Possnert, G., & Pääbo, S. (2004). No evidence of Neandertal mtDNA contribution to early modern humans. *PLoS Biology, 2*(3), E57.

Skinner, A. R., Blackwell, B. A., Martin, S., Ortega, A., Blickstein, J. I., Golovanova, L. V., & Doronichev, V. B. (2005). ESR dating at Mezmaiskaya Cave, Russia. *Applied Radiation and Isotopes, 62*, 219–224.

Soficaru, A., Dobos, A., & Trinkaus, E. (2006). Early modern humans from the Pestera Muierii, Baia de Fier, Romania. *Proceedings of the National Academy of Sciences of the United States of America, 103*, 17196–17201.

Stiller, M., Green, R. E., Ronan, M., Simons, J. F., Du, L., He, W., Egholm, M., Rothberg, J. M., Keates, S. G., Ovodov, N. D., Antipina, E. E., Baryshnikov, G. F., Kuzmin, Y. V., Vasilevski, A. A., Wuenschell, G. E., Termini, J., Hofreiter, M., Jaenicke-Despres, V., & Pääbo, S. (2006). Patterns of nucleotide misincorporations during enzymatic amplification and direct large-scale sequencing of ancient DNA. *Proceedings of the National Academy of Sciences of the United States of America, 103*, 13578–13584. *Erratum in*: 103,14977.

Sykes, B. (1997). Really ancient DNA. Lights turning red on amber. *Nature, 386*, 764–765.

Tamura, K. (2000). On the estimation of the rate of nucleotide substitution for the control region of human mitochondrial DNA. *Gene, 259*, 189–197.

Thalmann, O., Hebler, J., Poinar, H. N., Pääbo, S., & Vigilant, L. (2004). Unreliable mtDNA data due to nuclear insertions: A cautionary tale from analysis of humans and other great apes. *Molecular Ecology, 13*, 321–335.

The Chimpanzee Sequencing and Analysis Consortium. (2004). Initial sequence of the chimpanzee genome and comparison with the human genome. *Nature, 437*, 69–87.

The International Human Genome Sequencing Consortium. (2001). Initial sequencing and analysis of the human genome. *Nature, 409*, 860–921.

Torroni, A., Bandelt, H. J., Macaulay, V., Richards, M., Cruciani, F., Rengo, C., Martinez-Cabrera, V., Villems, R., Kivisild, T., Motopalu, E., Parik, J., Tolk, H. V., Tambets, K., Forster, P., Karger, B., Francalacci, P., Rudan, P., Janicijevic, B., Rickards, O., Savontaus, M. L., Huoponen, K., Laitinen, V., Koivumaki, S., Sykes, B., Hickey, E., Novelletto, A., Moral, P., Sellitto, D., Coppa, A., Al-Zaheri, N., Santachiara-Benerecetti, A. S., Semino, O., & Scozzari, R. (2001). A signal, from human mtDNA, of postglacial recolonization in Europe. *American Journal of Human Genetics, 69*, 844–852.

Valdiosera, C., Garcia, N., Dalen, L., Smith, C., Kahlke, R. D., Liden, K., Angerbjorn, A., Arsuaga, J. L., & Gotherstrom, A. (2006). Typing single polymorphic nucleotides in mitochondrial DNA as a way to access Middle Pleistocene DNA. *Biology Letters, 2*, 601–603.

van der Kuyl, A. C., Kuiken, C. L., Dekker, J. T., & Goudsmit, J. (1995). Phylogeny of African monkeys based upon mitochondrial 12S rRNA sequences. *Journal of Molecular Evolution, 40*, 173–180.

Vargha-Khadem, F., Gadian, D. G., Copp, A., & Mishkin, M. (2005). FOXP2 and the neuroanatomy of speech and language. *Nature Reviews Neurosciences, 6*, 131–138.

Wall, J. D., & Kim, S. K. (2007). Inconsistencies in Neanderthal genomic DNA sequences. *PLoS Genetics, 3*, 1862–1866.

Woodward, S. R., Weyand, N. J., & Bunnell, M. (1994). DNA sequence from Cretaceous period bone fragments. *Science, 266*, 1229–1232.

Zischler, H., Höss, M., Handt, O., von Haeseler, A., van der Kuyl, A. C., & Goudsmit, J. (1995). Detecting dinosaur DNA. *Science, 268*, 1192–1193.

Chapter 22
Radiocarbon Dating the Middle to Upper Palaeolithic Transition: The Demise of the Last Neanderthals and the First Appearance of Anatomically Modern Humans in Europe

Olaf Jöris, Martin Street, Thomas Terberger, and Bernhard Weninger

Abstract Only a precise chronological/stratigraphical framework can enable an understanding of the dynamics of change underlying the replacement of Neanderthals by Anatomically Modern Humans and the emergence of what are recognized as Upper Palaeolithic technologies and behaviour. This paper therefore examines the European radiocarbon-based chronometric record for the period between ca. 40.0 and 30.0 ka ^{14}C BP with reference to the stratigraphic evidence. The following testable hypotheses are proposed:

^{14}C ages for remains of the latest Neanderthals will regularly date to older than 38.0 ka ^{14}C BP. While at present the oldest direct dates for remains of Anatomically Modern Humans are <ca. 35.0 ka ^{14}C BP, AMH possibly appear in Europe as early as ca. 38.0 ka ^{14}C BP.

^{14}C will date Final Middle Palaeolithic "transitional" industries (leaf-point industries, Chatelperronian, Uluzzian) to between ca. 41.0 and 38.0 ka ^{14}C BP, and possibly as young as 35.0/34.0 ka ^{14}C BP.

Initial and Early Upper Palaeolithic "transitional" industries (Bachokirian, Bohunician, Protoaurignacian, Kostenki 14, level IVb) will date to between ca. 39.0 and 35.0 ka ^{14}C BP.

The earliest Aurignacian (I) will not significantly pre-date ca. 35.0 ka ^{14}C BP, whereas the earliest appearance of Aurignacian figurative art will not date earlier than 32.5 ka ^{14}C BP.

O. Jöris (✉) and M. Street
Forschungsbereich Altsteinzeit des, Römisch-Germanischen Zentralmuseums, Schloß Monrepos, D-56567 Neuwied, Germany
e-mail: joeris@rgzm.de; street@rgzm.de

T. Terberger
Lehrstuhl für Ur- und Frühgeschichte, Historisches Institut, Ernst Moritz Arndt Universität Greifswald, Hans-Fallada-Strasse 1, D-17489 Greifswald, Germany
e-mail: terberge@uni-greifswald.de

B. Weninger
Institut für Ur- und Frühgeschichte, Universität zu Köln, ^{14}C-Labor, Weyertal 125, D-50923 Köln, Germany
e-mail: b.weninger@uni-koeln.de

Keywords Radiometric data • Stratigraphic context • Chronological framework • Transitional industries • Aurignacien • Replacement

Introduction

The replacement of western Eurasian Neanderthals (Fig. 22.1) by Anatomically Modern Humans (AMH) ultimately descended from African hominins (Fig. 22.2) and the change from Middle Palaeolithic (MP) to Upper Palaeolithic (UP) technologies in the same region both mark important developments in human history and cultural evolution. They are among the most debated issues in palaeoanthropology and Palaeolithic archaeology and the frequently heated discussion by specialists has often been dominated by the assumption that they are synchronous and causally linked.

Schools of thought in the debate range from the hypothesis that behaviour is clearly species-related (e.g. Lahr and Foley 1998, 168; cf Foley and Lahr 1997), with the implication that replacement of MP technologies by UP ones was a "revolutionary" process that started in a certain core area and is directly equated with the replacement of Neanderthals by AMH (e.g. Bar-Yosef 1998; cf Bar-Yosef 2001), to the proposal that observed cultural changes are trans-specific, with aspects of UP technology being invented independently by Neanderthals and AMH (d'Errico 2003; cf Zilhão and d'Errico 1999; Zilhão 2006a) or adopted by the former following contact with the latter (e.g. Hublin et al. 1996; Mellars 2000).

In both the anthropological and the archaeological discussion it is clear that meaningful models can only be created within a reliable chronological framework defined by highly accurate stratigraphic records and chronometric age determinations. However, it is increasingly obvious that the basis for constructing a chronology for the period under consideration (ca. 40.0–30.0 ka ^{14}C BP) is flawed and it appears that major contextual and methodological problems have been underestimated. This paper will discuss the evaluation and re-evaluation of critical chronometric

Vertebrate Paleobiology and Paleoanthropology, DOI 10.1007/978-94-007-0492-3_22,

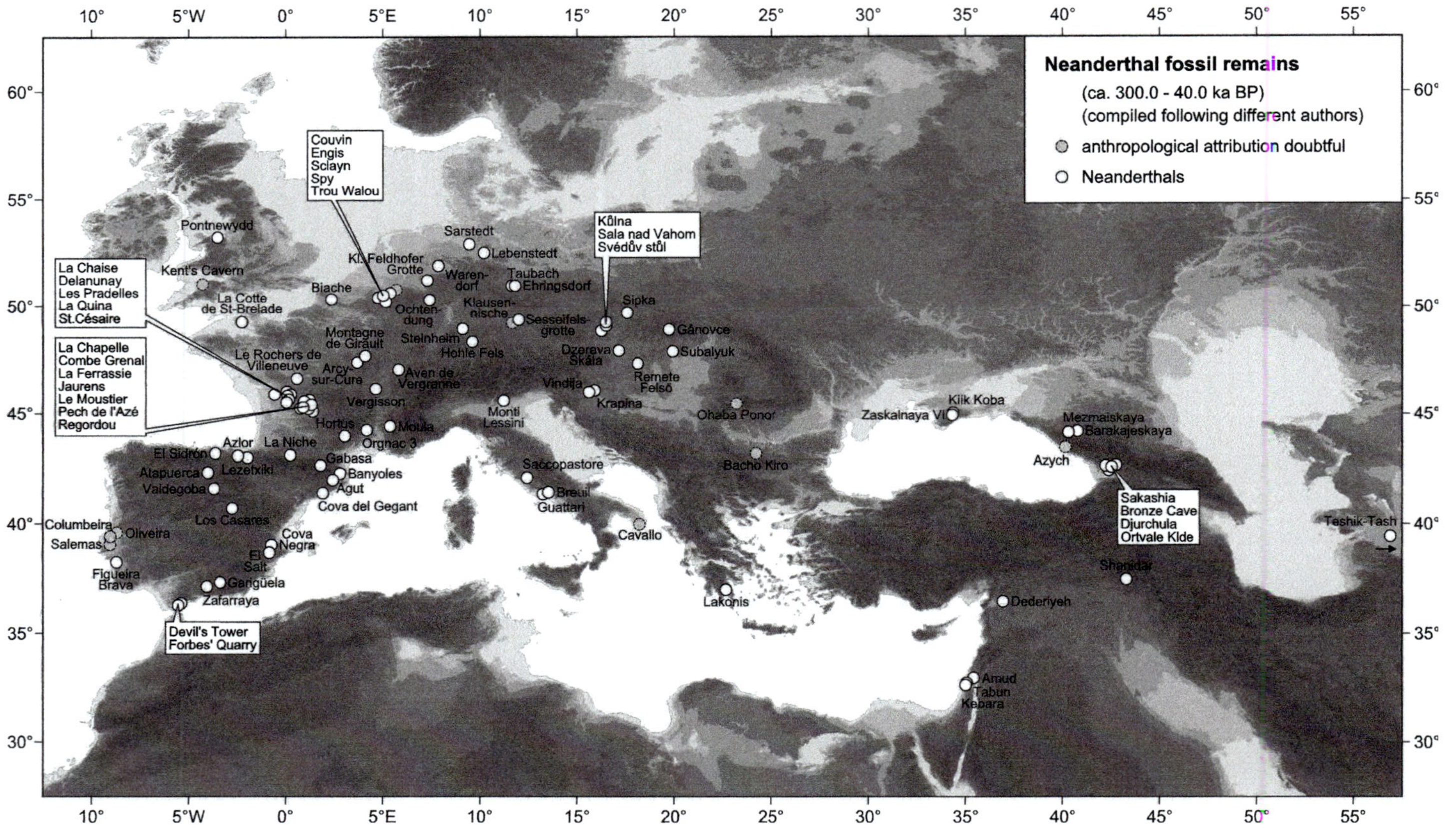

Fig. 22.1 Neanderthal remains in Europe (Modified from Jöris, 2005). Map based on SRTM data; sea level lowered by 75 m

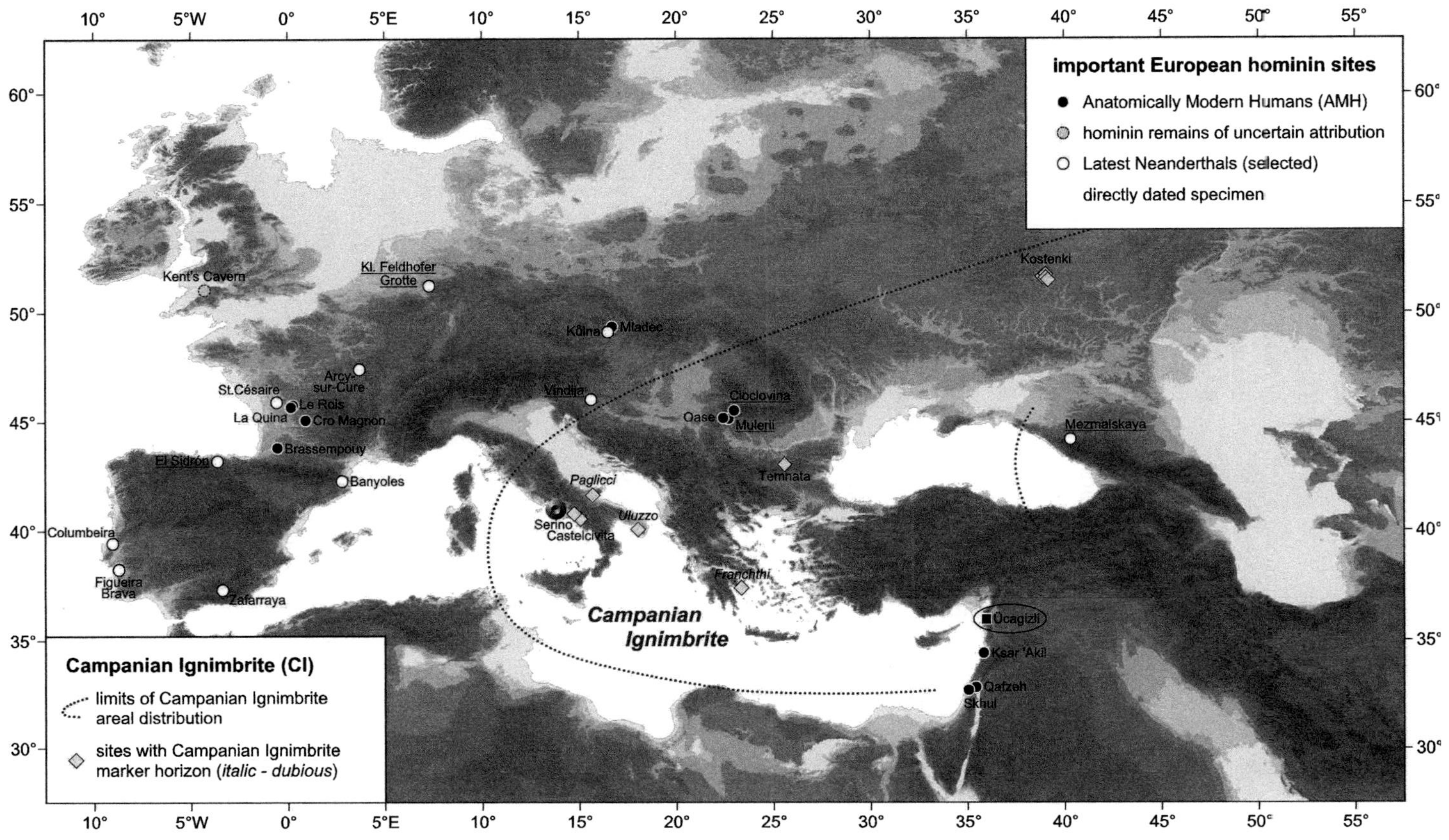

Fig. 22.2 Key sites with hominin remains around the Middle to Upper Palaeolithic transition (after different authors) and sites associated with the Campanian Ignimbrite (CI) marker horizon (modified from Giaccio et al. 2006; Pyle et al. 2006). Map based on SRTM data; sea level lowered by 75 m

records[1], taking into account recent advances in radiocarbon dating and interpretation, and address their relevance for specific problems in both the physical anthropological and cultural records.

Dating the Transition – Old Problems, Recent Trends

In this discussion radiocarbon dating provides the backbone for current models regarding the absolute chronometric scale of the two developments, due to both its wide applicability and precision. Other radiometric evidence is still applied quite unsystematically and many methods have only limited suitability or offer only relatively poor precision due to large standard deviations. Additionally, it has long been recognized that the radiocarbon time scale requires absolute age calibration before meaningful comparisons with ages scaled to calendar years can be made. This is most clearly the case for age estimates derived from other radiometric dating methods (e.g. U/Th-series, TL), which directly supply absolute ages. Unfortunately, the systematic application of radiocarbon age calibration for the pre-Holocene period is still not generally accepted.

Sample Context

Regardless of the dating technique applied, sample provenance and choice are among the most important parameters for assessing the quality of dates. The relevance of a sample is too often merely assumed, for example by the spatial proximity of the dated object to the supposedly associated archaeological/anthropological context, but without objective examination of the taphonomic environment. The relevance of a date is a function of both the secure stratigraphic context in which a sample was recovered and the association of the sample with human presence or activity within this context. To allow a meaningful interpretation of a dated sample its context should be guaranteed by rigorous recovery and documentation and its archaeological/palaeoanthropological relevance clearly demonstrated by unambiguous alteration due to human action. Against this background, dates on samples from older excavations in particular need to be treated with caution especially if they are based on large bulk samples.

[1]This paper identifies dates obtained by the radiocarbon method with the format "^{14}C BP" for actual radiocarbon results and "ka ^{14}C BP" for summary or abbreviated radiocarbon results. Similarly, specific calibrated radiocarbon dates are given as e.g. "41,000 cal BP" and abbreviated calibrated radiocarbon ages as "41.0 ka cal BP".

The introduction of the AMS ^{14}C method made possible the measurement of much smaller amounts (<50 mg) of material (Hedges and van Klinken 1992), allowing direct dating of single objects with potentially better taphonomic and stratigraphic control. However, problems arise even with AMS dates if they are not interpreted rigorously:

1. In many cases the relevance of direct dates on hominid fossils for archaeological questions is uncertain.
2. The relevance of direct dates on "significant" MP or UP organic artefacts for hominin evolution or replacement at the transition is uncertain due to their unknown association with a specific hominin type.
3. Possible stratigraphic displacement of the now often very small sampled objects can mean that direct dates on "associated" material (e.g. cut-marked bone, charcoal) are misleading for the interpretation of both the archaeological and anthropological records.

Sample Contamination and Pre-treatment

The MP–UP transition and the time of the last Neanderthals/first AMH are close to the limit of radiocarbon dating and we are therefore dealing with an intensification of methodological problems. Generally speaking, these are due to difficulties in extracting and purifying the often extremely small amounts of indigenous organic carbon from bone samples and from difficulties to chemically address suitable organic components (Hedges and van Klinken 1992). However, continual improvements in ^{14}C measurement technology and sample pre-treatment have enabled laboratories to produce increasingly older – probably more accurate – ages for samples close to the technical limits of radiocarbon dating. Despite these unquestionable improvements we should be aware that we are only now beginning to appreciate these problems. Further improvements and corrections are to be expected.

Due to the small amount of measurable residual ^{14}C, dates beyond an age of ca. 30.0 ka ^{14}C BP have increasingly high standard deviations and often cannot provide sufficiently fine temporal resolution to answer specific questions. Due to the magnified effects of even minute amounts of intrusive carbon, we are confronted with an increase in unreliable measurements. Sample chemistry becomes even more important with sample age and the sensitivity of different materials to contamination is amplified disproportionately in the period beyond 30.0 ka ^{14}C BP. All in all, effects which may cause divergent measurements are still difficult to interpret.

These problems are well-known to scientists working with radiocarbon data in this age range. We judge that the problem is nevertheless widely underestimated. It is a major methodological step between critically rejecting the one or the other outlying radiocarbon date and accepting that the majority of measurements in this age-range may be erroneous. This problem is further complicated since divergent data from the same stratigraphical context may, on the one hand, call into question the integrity of the layer/context but may on the other hand result from reported extreme oscillations in past atmospheric radiocarbon levels (e.g. Beck et al. 2001). According to some researchers the synchronization of the different time scales currently in use remains problematic and a standard ^{14}C calibration curve beyond 26.0 kacal BP cannot yet be accepted (van der Plicht et al. 2004; cf Reimer et al. 2004; Bronk Ramsey et al. 2006). However, it is in fact quite possible to scale ^{14}C measurements from a variety of climate archives (Shackleton et al. 2004; cf Shackleton 2005; Fairbanks et al. 2005; Hughen et al. 2006; Chiu et al. 2007) via marine synchronisms against a synthetical U/Th-based calendric Greenland ice core age model, allowing the construction of an extended Glacial ^{14}C calibration curve (Fig. 22.3; CalPal-2007$_{Hulu}$; Weninger and Jöris 2008; http://www.calpal.de).

Another possible reason for contamination may be the specific location of a sample within a site. For example, bones from MP horizons inside the Sesselfelsgrotte rock shelter produced radiocarbon ages between ca. 48.0 and 40.0 ka ^{14}C BP in consistent stratigraphical order. Samples from equivalent strata outside the rock shelter drip line gave significantly younger ages, all lying around 37.0–34.0 ka ^{14}C BP (Richter 2004; cf Mellars 2006a), probably due to contamination with younger carbon (cf Mellars 2006a).

The problem of sample contamination is highlighted by progress in methods of sample pre-treatment and has consequences for inter-laboratory comparability. Although comparisons are routinely carried out (e.g. Scott 2003), it is rare for the same archaeological/anthropological specimen to be cross-dated by different laboratories. In the case of the UP burials of Sun'gir dates produced by Oxford and Arizona disagree by several thousand radiocarbon years (cf Pettitt and Bader 2000; Kuzmin et al. 2004), possibly due to differences in sample pre-treatment (Kuzmin et al. 2004).

A significant factor in this context is the choice of sample material. Because of the high yield of carbon (ca. >50 %) charcoal was originally the preferred material for dating the MP and early UP. Nevertheless, even though bone provides much less datable carbon, it has become increasingly important since the advent of AMS ^{14}C dating due to its normally greater availability at sites and frequently more direct contextual relevance, leading to an increased emphasis on dating bone. However, systematic comparison of charcoal and bone dates for the European late MP and early UP documents significant differences between the two series (cf Jöris et al. 2003, 2006). In the case of charcoal, dates for the MP cluster consistently before 38.0 ka ^{14}C BP with only few younger results, while dates for the early UP are mainly younger than 38.0 ka ^{14}C BP. This is in contrast with the much larger series of available bone dates, which suggests extremely wide age ranges for both the MP and UP, with age overlaps between them far beyond any statistical dating errors. In our view, charcoal dates represent the more reliable evidence and do not support an extended (>10 ka ^{14}C BP) coexistence of the MP and the UP, but instead indicate a temporal succession between the two periods.

In qualification of the last statement, recent developments in the purification of bone collagen by advanced methods of "ultrafiltration" pre-treatment (Bronk Ramsey et al. 2004) have made it possible to remove far more younger contaminant carbon from bone samples than previously, which has already had important chronological implications for a number of English sites, producing higher ages on bone specimens previously dated without this pre-treatment (Higham et al. 2006b; Jacobi et al. 2006). The new results are in accordance with the observation, that measurements on bone samples (in the past) have a strong tendency to turn out younger than charcoal dates. This new pre-treatment technique will in future ensure higher reliability of measurements on bone samples at the limit of ^{14}C dating.

In the absence of an agreed standard pre-treatment procedure for bone samples and due to incomplete protocols between different laboratories for sample carbon content and carbon–nitrogen ratio, it is often difficult to evaluate the reliability of dating results. For now we must accept that an unknown proportion of radiocarbon results measured in the past may be unreliable and in the case of many bone samples (much) too young.

The Hominin Record

The oldest fossil AMH are known from Africa at the Ethiopian sites of Kibish (Omo I & II) and Herto, dated to ca. 195.0 ka (McDougall et al. 2005; Trinkaus 2005) and ca 160.0 ka (White et al. 2003) respectively. The younger AMH remains from burials at the sites of Skhul and Qafzeh in Israel are estimated at between 135.0 and 100.0 ka (Grün et al. 2005), suggesting that African populations had expanded northwards by this time. This is consistent with genetic analysis of both mitochondrial and nuclear recent human DNA, which suggests that all modern *Homo sapiens* descend from a small founder population located at ca. 130,000 ka in southern or eastern Africa (Forster 2004). This initial AMH incursion into the Levant appears to have been temporary since younger hominins in the region (Amud, Kebara, Dederiyeh; e.g. Akazawa et al. 1998; Hovers 2006) are morphologically clearly identified as Neanderthals of western Eurasian origin.

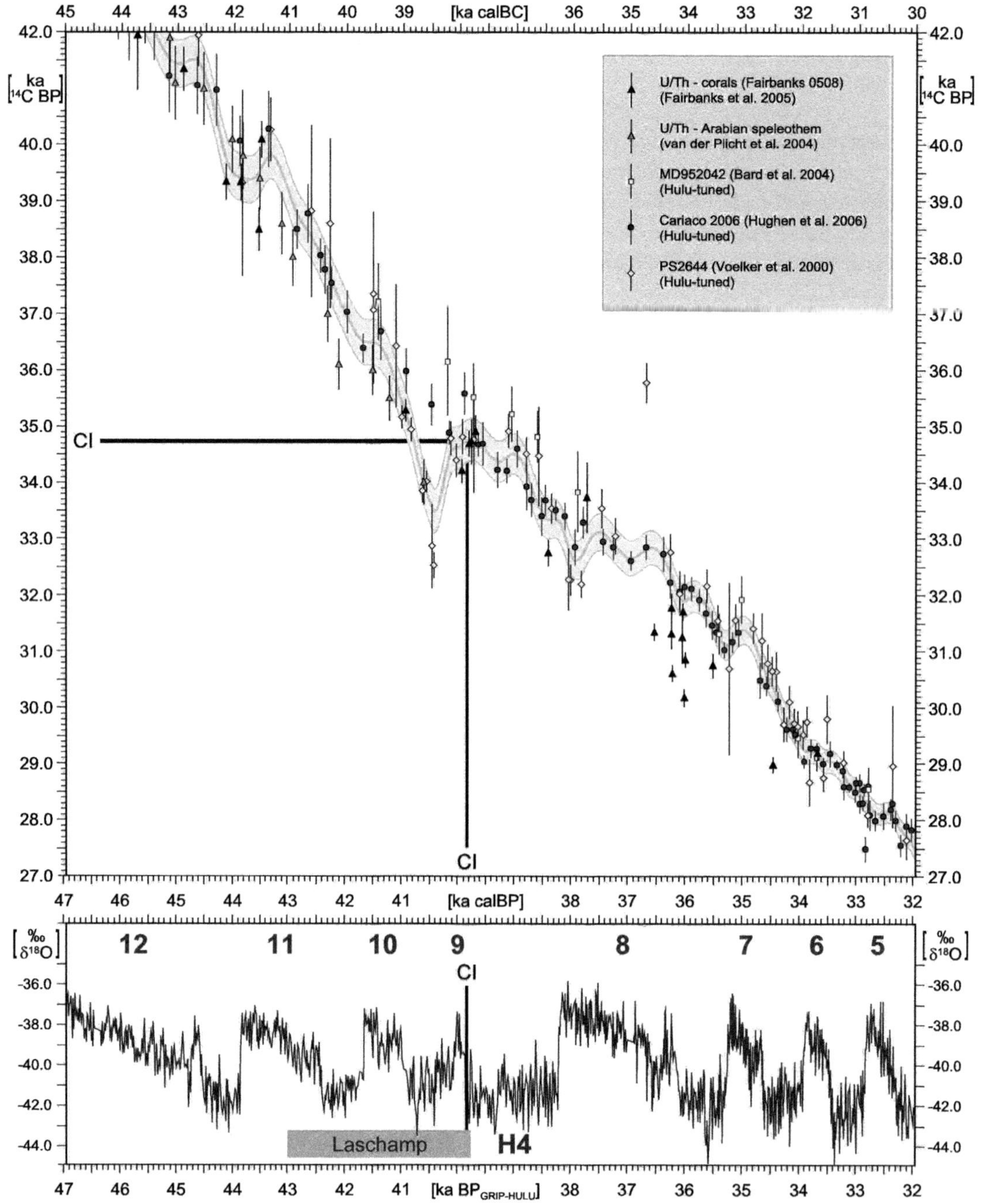

Fig. 22.3 Radiocarbon calibration records around the Middle to Upper Palaeolithic transition between 42.0 and 27.0 ka ^{14}C BP and 47.0–32.0 ka calBP (45.0–30.0 ka calBC) respectively, shown against the background of climatic change recorded in the Greenland GRIP ice core (data from Johnsen et al. 2001), scaled against the U/Th-dated Hulu-chronology (Wang et al. 2001) via synchronization of $\delta^{18}O$ signatures (Weninger and Jöris 2008), with interstadial oscillations labelled in grey (cf Johnsen et al. 1992). Over the entire period plotted, the presented datasets show overall agreement with the available paired ^{14}C vs. U/Th data on corals (Chiu et al. 2007; Fairbanks et al. 2005) when scaled to "Hulu" as the common age-model, allowing for construction of a synthetic curve for Glacial radiocarbon calibration: CalPal-2007$_{Hulu}$ (cf http://www.calpal.de). The curve indeed contains long plateaux and inversions within which massive age distortions will be produced, while other parts of the curve, including the section before 35.0 ka ^{14}C BP, are steep and allow precise radiocarbon age calibration. Note that the calibration spline is running through dates from Fairbanks et al. (2005), Bard et al. (2004, Hulu-tuned), Hughen et al. (2006, Hulu-tuned) and Voelker et al. (2000, Hulu-tuned), excluding apparent outliers (Weninger and Jöris 2008). Graph produced with CalPal (www.calpal.de), spline stiffness: 15. H4 – Heinrich event 4. CI – Campanian Ignimbrite marker horizon (after Fedele et al. 2002, 2003; Giaccio et al. 2006). Laschamp – Laschamp geomagnetic excursion (after Giaccio et al. 2006; cf Voelker et al. 2000; Southon 2004)

Genetic and Palaeoanthropological Evidence

A chronologically distant separation of the Neanderthal and AMH lineages is shown by genetic studies of Neanderthal remains revealing fundamental differences between both their mitochondrial and nuclear aDNA and that of recent humans. It also appears probable that Neanderthal mtDNA made no major contribution to the gene pool of modern *Homo sapiens*, suggestive of a rapid replacement scenario (Excoffier 2006; Serre et al. 2004; Forster and Matsumura 2005; Macaulay et al. 2005). The hypothesis of replacement of Neanderthals by AMH is supported by recent studies of Neanderthal nuclear aDNA (Noonan et al. 2006; Lambert and Millar 2006). However, there has also been criticism of an exclusively African origin of humans (Eswaran et al. 2005) based on anthropological and aDNA data from the Mezmaiskaya infant burial (Hawks and Wolpoff 2001). Independently, limited genetic transfer between the two lineages has been admitted as possible (Green et al. 2006).

It has also been suggested that morphological analyses of some fossil hominins provide evidence for intermixture of Neanderthals and AMH. The well preserved Peştera cu Oase 2 AMH cranium from Romania, dated to ca. 35.0 ka ^{14}C BP, has been described as presenting "an unusual mosaic of features", including some found "principally among the Neandertals" (Rougier et al. 2007, 1,169) which could, on one interpretation, be viewed as due to "admixture with Neandertal populations as ... modern humans spread through western Eurasia" (Rougier et al. 2007, 1,169). This scenario might imply that the absence of genetic evidence for admixture is due purely to subsequent loss of genetic diversity in the resulting population (vid. Zilhão 2006a, 8). This viewpoint has been established against the background of the much younger Mid-Upper Palaeolithic (MUP) AMH child burial from Lagar Velho in Portugal. Here, apparently Neanderthal skeletal traits have been claimed as evidence for very late Neanderthal survival at the western edge of Europe (Zilhão and Trinkaus 2002; cf Finlayson et al. 2006; Zilhão and Pettitt 2006). We view this sceptically given the fact that the skeleton is that of an immature individual.

Directly Dated Hominin Fossils

The duration of any contact between Neanderthals and AMH has been variously suggested to have been extremely short (Currat and Excoffier 2004; but see Zilhão 2006a, 4–6), or on the order of several thousands of years, allowing for the gradual diffusion of AMH into Eurasia and extended coexistence with Neanderthals (Zilhão 2006b). Of particular importance in this context are the few directly dated specimens of Neanderthals and AMH (Table 22.1; Fig. 22.4). Below we discuss those results known at the time of the 2006 Neanderthal Anniversary Meeting in Bonn; for Neanderthals these comprise two uranium-series measurements for Vindija and 17 direct radiocarbon dates from five different sites, for pre-MUP AMH 11 direct radiocarbon measurements were available.

Neanderthals

The oldest direct measurement of a Neanderthal is on a charred bone from the MP layer 7a at the Kůlna cave in the Czech Republic (Mook 1988) with an age of 45,660 + 2,850/– 2,200 ^{14}C BP (GrN-6060; Table 22.1; Fig. 22.4). Direct dated Neanderthal remains from the type locality in the Düssel valley, Germany (Schmitz et al. 2002), and from the cave of El Sidrón in Cantabrian Spain (Lalueza-Fox et al. 2005) produced radiocarbon ages ranging from 40.8 to 37.3 ka ^{14}C BP which cluster tightly around 39.8 ka ^{14}C BP and 38.4 ka ^{14}C BP, respectively (Table 22.1). A similar radiocarbon age of 38,310 ± 2,130 ^{14}C BP was obtained for the Vindija Vi-80 fossil from layer G3 (Serre et al. 2004), while dating a different specimen from the same layer gave an infinite age estimate of >42,000 ^{14}C BP (Ua-13873; Krings et al. 2000; Table 22.1; Fig. 22.4). These four sites together demonstrate late Neanderthals in different parts of Europe shortly before 38.0 ka ^{14}C BP.

Appreciably younger are seven direct radiocarbon dates from layer G1 at Vindija and a single date for Mezmaiskaya (Russia) which fall between ca. 32.4 and 28.0 ka ^{14}C BP (Table 22.1; Fig. 22.4). Initial results (<30.0 ka ^{14}C BP) from these sites were interpreted as evidence for prolonged survival of Neanderthals in South-eastern and Eastern Europe, implying co-existence with contemporary early AMH in immediately adjacent regions (e.g. Smith et al. 1999).

The remarkably young Vindija dates OxA-8295 (28,020 ± 360 ^{14}C BP) and OxA-8296 (29,080 ± 400 ^{14}C BP) for a Neanderthal parietal (Vi-208) and mandible (Vi-207) from layer G1 were in apparent stratigraphical association with a mixture of MP and early UP technological elements (but see discussion in Ahern et al. 2004; discussion in Karavanić 2000). However, new results for both these specimens re-dated at the same laboratory following ultrafiltration pre-treatment are appreciably older (Table 22.1; Fig. 22.4). It is concluded that the "true" ages of the specimens "should be in the vicinity of ca. 32,000 ^{14}C BP or slightly older" (Higham et al. 2006a, 555). Independent U-series measurements on Vi-207 produced dates of 46.7 ± 7.0 ka BP (U/Pa) and 51.0 ± 8.8 ka BP (U/Th; Table 22.1; Karavanić and Smith 1998),

Table 22.1 Results from direct radiometric dating of European hominin remains

Layer	Method	No.	Lab.-No.	^{14}C age	STD+	STD–	$\delta^{13}C$	Material	Species	Skeletal element	Assoc. ind.	Source
Kůlna, CZ												
7a	^{14}C		GrN-6060	45,660	2850	2200		Bone, charred	Neanderthal		Micoquian	Mook (1988)
Kleine Feldhofer Grotte (Neandertal), D												
	^{14}C	1	ETH-19660	39,240	670	670	−20.0	Bone	Neanderthal: "NN1" (burial ?)	Right humeral shaft	Micoquian?	Schmitz et al. (2002)
	^{14}C	2	ETH-20981	39,900	620	620	−19.6	Tooth	Neanderthal: "Nean 1"	Milk tooth	Micoquian?	Schmitz et al. (2002)
	^{14}C	3	ETH-19661	40,360	760	760	−18.8	Bone	Neanderthal: "NN4" (burial ?)	Right tibia shaft	Micoquian?	Schmitz et al. (2002)
	^{14}C	*(1–3)*	*Weighted mean*	*39,797*	*390*	*390*	*0.72*	*t-Value*				
El Sidrón, E												
	^{14}C	1	Beta-192066	37,300	830	830		Bone	Neanderthal: "sample 2"		Mousterian	Lalueza-Fox et al. (2005)
	^{14}C	2	Beta-192067	38,240	890	890		Tooth	Neanderthal: "sample 3"		Mousterian	Lalueza-Fox et al. (2005)
	^{14}C	3	Beta-192065	40,840	1200	1200		Tooth	Neanderthal: "sample 1"		Mousterian	Lalueza-Fox et al. (2005)
	^{14}C	*(1–3)*	*Weighted mean*	*38,369*	*542*	*542*	*1.88*	*t-Value*				
Vindija, CRO												
G1	^{14}C AG		OxA-8296	29,080	400	400	−20.5	Bone	Neanderthal: "Vi-207"	Mandible	MP/UP	Smith et al. (1999)
G1	^{14}C AG		OxA-X-2082–10	29,100	360	360	−22.8	Bone	Neanderthal: "Vi-207"	Mandible	MP/UP	Higham et al. (2006)
G1	^{14}C AF > 30 kDa		OxA-X-2089–07	32,400	1800	1800	−24.6	Bone	Neanderthal: "Vi-207"		MP/UP	Higham et al. (2006)
G1	U-Pa			46,000	7000	7000		Bone	Neanderthal: "Vi-207"	Mandible	MP/UP	Karavanić et al. (1998)
G1	U-Th			51,000	8800	8800		Bone	Neanderthal: "Vi-207"	Mandible	MP/UP	Karavanić et al. (1998)
G1	^{14}C AG		OxA-8295	28,020	360	360	−19.5	Bone	Neanderthal: "Vi-208"	Parietal	MP/UP	Smith et al. (1999)
G1	^{14}C AG		OxA-X-2082–09	29,200	360	360	−19.8	Bone	Neanderthal: "Vi-208"	Parietal	MP/UP	Higham et al. (2006)
G1	^{14}C AF < 30 kDa		OxA-X-2094–10	31,390	220	220	−19.5	Bone	Neanderthal: "Vi-208"		MP/UP	Higham et al. (2006)
G1	^{14}C AF > 30 kDa		OxA-X-2089–06	32,400	800	800	−20.2	Bone	Neanderthal: "Vi-208"		MP/UP	Higham et al. (2006)
G3	^{14}C		Ua-?	38,310	2130	2130		Bone	Neanderthal: "Vi-80"		Mousterian	Serre et al. (2004)
G3/h-203	^{14}C		Ua-13873	>42,000				Bone	Neanderthal: "Vi-75"		Mousterian	Krings et al. (2000)
Mezmaiskaya Cave, RU												
3D	^{14}C		Ua-14512	29,195	965	965		Bone	Neanderthal infant (burial)	Rib fragment	E-Europ. Micoquian	Ovchinnikov et al. (2000)
Kent's Cavern, GB												
	^{14}C		OxA-1621	30,900	900	900		Bone	Hominin: "Kent's Cavern 4"	Maxilla	?	Jacobi (1999)
Pestera cu Oase, RO												
	^{14}C	1	GrA-22810	34,290	970	870	−19.0	Bone	AMH: "Oase 1"	Mandible	No assoc. ind.	Trinkaus et al. (2003)
	^{14}C	2	OxA-11711	>35,200			−18.7	Bone	AMH: "Oase 1"	Mandible	No assoc. ind.	Trinkaus et al. (2003)
	^{14}C	*(1–2)*	*Weighted mean*	*34,950*	*990*	*890*						Trinkaus et al. (2003)

Mladeč, CZ												
2	^{14}C	1	VERA-2736	26,330	170	170	–24.6	Bone	AMH: "Mladeč 25c"	Ulna	Aurignacian	Wild et al. (2005)
2	^{14}C	2	VERA-3075	30,680	380	360	–20.1	Tooth collagen	AMH: "Mladeč 8"	Left M2, mesial-buccal root	Aurignacian	Wild et al. (2005)
2	^{14}C	3	VERA-3073	31,190	400	390	–19.1	Tooth	AMH female: "Mladeč 1"	Rigth M2, distal half of the crown	Aurignacian	Wild et al. (2005)
2	^{14}C	4	VERA-3074	31,320	410	390	–20.6	Tooth	AMH female: "Mladeč 2"	Left M3, distal half of the crown	Aurignacian	Wild et al. (2005)
2	^{14}C	5	VERA-3076A	31,500	420	400	–19.7	Tooth collagen (white)	AMH: "Mladeč 9a"	Right maxillary canine	Aurignacian	Wild et al. (2005)
2	^{14}C	6	VERA-3076B	27,370	230	230	–19.7	Tooth collagen (brown)	AMH: "Mladeč 9a"	Right maxillary canine	Aurignacian	Wild et al. (2005)
	^{14}C	*(2–5)*	*Weighted mean*	*31,150*	*210*	*210*	*1.09*	*t-Value*				
Pestera Muierilor de la Baia de Fier, RO												
	^{14}C		LuA-5228	30,150	800	800		Bone collagen	AMH	Mandible (with material from cranium, scapula, tibia diaphysis)	?	Paunescu (2001)
Pestera Mare-Cioclovina, RO												
	^{14}C		LuA-5229	29,000	700	700		Bone collagen	AMH	Cranium	Aurignacian	Paunescu (2001)
Cro Magnon, F												
				Not directly dated					AMH (burial)			
	^{14}C		Beta-157439	27,680	270	270		Shell	*Littorina* sp.	Perforated shell (grave good)	Aurignacian	Henry-Gambier (2002)

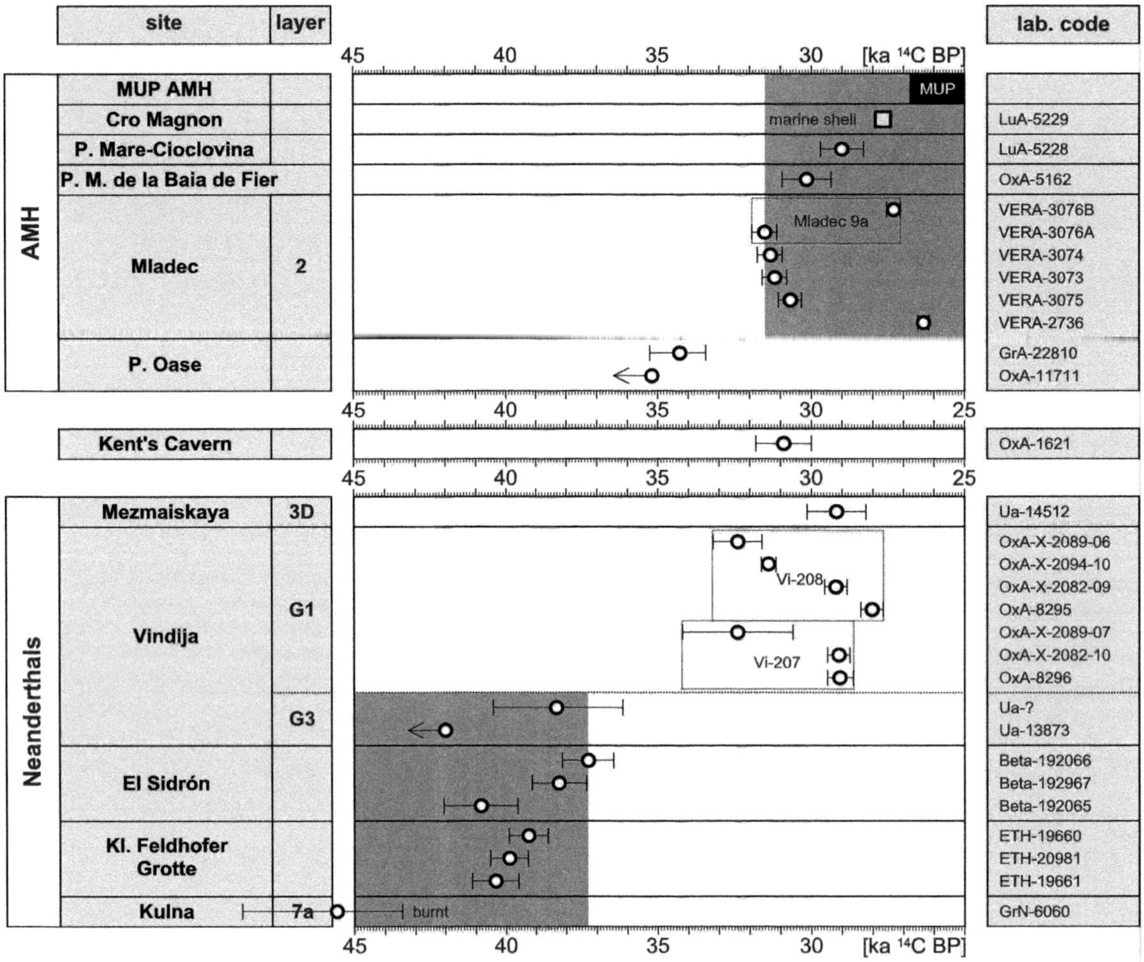

Fig. 22.4 Direct radiocarbon age determinations for hominin remains from around the Middle to Upper Palaeolithic transition (Compiled after different authors: see text), given with 1σ-standard deviation. Dates on bone (*circles*); dates on shell (*squares*); *AMH* Anatomically Modern Humans, *MUP* Mid-Upper Palaeolithic; *Dark grey* shading: most likely dating range

indicating a much higher age for this fossil. Furthermore, bones of *Ursus spelaeus* from the same level date to between ca. 46.8 and 18.3 ka ^{14}C BP (Higham et al. 2006a), suggesting that the layer probably contains a highly heterogeneous secondary mix of different material (cf Zilhão and d'Errico 1999).

At Mezmaiskaya cave, direct radiocarbon measurement of the Neanderthal infant burial in layer 3D gave a result of 29,195 ± 965 ^{14}C BP (Ua-14512; Ovchinnikov et al. 2000; Table 22.1; Fig. 22.4), although another radiocarbon measurement from the same layer gave an infinite age of >45,000 ^{14}C BP (LE-3841; Golovanova 1998). The younger age is also contradicted by a number of radiocarbon measurements from the overlying levels 2B – 1C, ranging from ca. 40.7 to 30.0 ka ^{14}C BP (Golovanova 1998). These data suggest contamination of sample Ua-14512 and on the evidence of ESR dating a true age of ca. 40.0 ka BP has been proposed (Skinner et al. 2005). Further to the disputed age of the specimen, there has also been criticism of its anthropological attribution (Hawks and Wolpoff 2001).

Anatomically Modern Humans

In recent years the record of European fossil AMH has undergone major revision and supposedly key specimens, including those from Vogelherd and several other hominin remains have lost their relevance (Terberger et al. 2001; Terberger and Street 2003; Conard et al. 2004; Trinkaus 2005; Street et al. 2006). Only a few AMH specimens directly dated to before the MUP are of significance for the present question (Table 22.1; Churchill et al. 2000), among them remains from Romania which shed important light on the timing of the arrival of AMH in Europe (Table 22.1; Fig. 22.4).

Of the latter, the Peştera cu Oase 1 mandible produced radiocarbon ages of 34,290 + 970/– 870 ^{14}C BP (GrA-22810) and >35.2 ka ^{14}C BP (OxA-11711), with a mean age estimate of 34,950+990/− 890 ^{14}C BP providing the earliest reliable dating evidence for AMH in Europe (Trinkaus et al. 2003). However, the finds have no archaeological context and provide no information on their cultural association. Direct dates from two other Romanian sites, Peştera Muierilor de la Baia de Fier and Peştera Mare-Cioclovina, produced radiocarbon ages of 30,150±800 ^{14}C BP (LuA-5228) and 29,000±700 ^{14}C BP (LuA-5227) respectively (Paunescu 2001). Both fossils may be associated with an Aurignacian lithic industry (Paunescu 2001).

Further to the North-west, AMH specimens from the Mladeč cave in the Czech Republic associated with an Aurignacian facies characterized by bone points of the eponymous Mladeč type (Hahn 1988a; Svoboda 2000, 2001) are directly dated by six radiocarbon measurements (Wild et al. 2005; Table 22.1; Fig. 22.4). One was taken on bone (Mladeč 25c ulna) with an extremely low $\delta^{13}C$ content and resulted in 26,330±170 ^{14}C BP (VERA-2736). All other dates derive from teeth. Although one age determination on apparently contaminant brown coloured collagen produced a result of 27,370±230 ^{14}C BP (VERA-3076B) the remaining four measurements lie close together (Wild et al. 2005) with a weighted mean of 31,150 ± 210 ^{14}C BP (Table 22.1). A calcite layer may have originally sealed the fossil bearing horizon had already been radiocarbon dated to ca. 34.5 ka ^{14}C BP, confirming the high age of the Mladeč fossils (Svoboda et al. 2002).

Another direct dated hominin in Aurignacian context is a maxilla fragment from Kent's Cavern (horizon 4), Great Britain, with an age of 30,900±900 ^{14}C BP (OxA-1621; Hedges et al. 1989; Table 22.1; Fig. 22.4). Although an attempt to re-date the maxilla fragment after ultrafiltration failed, new dates on other stratified material bracketing the find horizon of the maxilla suggests that the OxA-1621 result can be regarded as a minimum age estimate only. The "true" radiocarbon age of the hominin is expected to fall between ca. 37.0 and 35.0 ka ^{14}C BP (Higham et al. 2006b). Although recent publications tend to classify the specimen as AMH, its fragmentary state means that there is some doubt about its identification (Jacobi et al. 2006).

Although not directly dated by radiocarbon, an AMH skeleton from the Cro-Magnon site in France is mentioned here since this find defines the eponymous early modern human type believed to be associated with the Aurignacian. The context of the burial was radiocarbon dated to 27,680±270 ^{14}C BP on a perforated *Littorina* sp. shell from the grave fill (Beta-157439; Table 22.1; Fig. 22.4), which would attribute it to the early MUP (Henry-Gambier 2002). This interpretation is complicated by the fact that the Aurignacian forms the top of the sequence at the rock shelter where MUP layers are completely absent (Gambier 1989). Furthermore, the *Littorina* date from Cro-Magnon can give only an approximate age for the burial since marine-derived samples require correction for reservoir effects which are difficult to evaluate (cf Mellars 2004).

Summary

At present the "transitional" European fossil hominin record can be summarised as follows:

- Neanderthal fossils are reliably dated only to before ca. 38.0 ka ^{14}C BP.
- AMH fossils in Europe are not reliably dated to before ca. 35.0 ka ^{14}C BP but the association of AMH with the Aurignacian can probably be accepted (e.g. Churchill and Smith 2000; Paunescu 2001; Trinkaus et al. 2003; Bailey and Hublin 2005; Trinkaus 2005; Wild et al. 2005; cf Henri-Gambier et al. 2004).
- With the exception of an isolated tooth from Kostenki 14, level IVb, discussed as AMH (Sinitsyn 2003:91), the interval 38.0–35.0 ka ^{14}C BP has so far not provided unambiguous anthropological evidence in Europe for either Neanderthals or early AMH, leaving room for discussion of interactions such as cultural and/or genetic transfer. In the Near East the suggested date for a burial of an AMH, "Egbert" (Bergman and Stringer 1989) from an Early Ahmarian context at Ksar'Akil would fall into this gap (cf Mellars 2004; Zilhão 2007).

The Middle to Upper Palaeolithic Transition

The course of research in Western Europe led to the distinction of the MP and UP as two monolithic blocks, believed to be essentially different and implicitly associated with Neanderthals and AMH respectively.

It is today recognized that the later MP is heterogeneous and comprises diverse, regionally different assemblages, probably indicative of distinct spatio-chronological entities such as the western European Moustérien de Tradition Acheuléen (MTA; cf Soressi 2002), the Central European Keilmessergruppen (Bosinski 1967; Jöris 2004) or the Eastern European Micoquian and various further industries on the Crimean peninsula, Russian Plain and Caucasus (e.g. Chabai 2003; Monigal 2006; Usik et al. 2006; Sinitsyn 2003). From all these cultural contexts only Neanderthal skeletal remains are known.

The first European technocomplex generally accepted as being fully UP in character is the Aurignacian (Figs. 22.5, 22.13, and 22.19; cf Bar-Yosef and Zilhao 2006). By the early Aurignacian I phase we find all of the traits regarded as

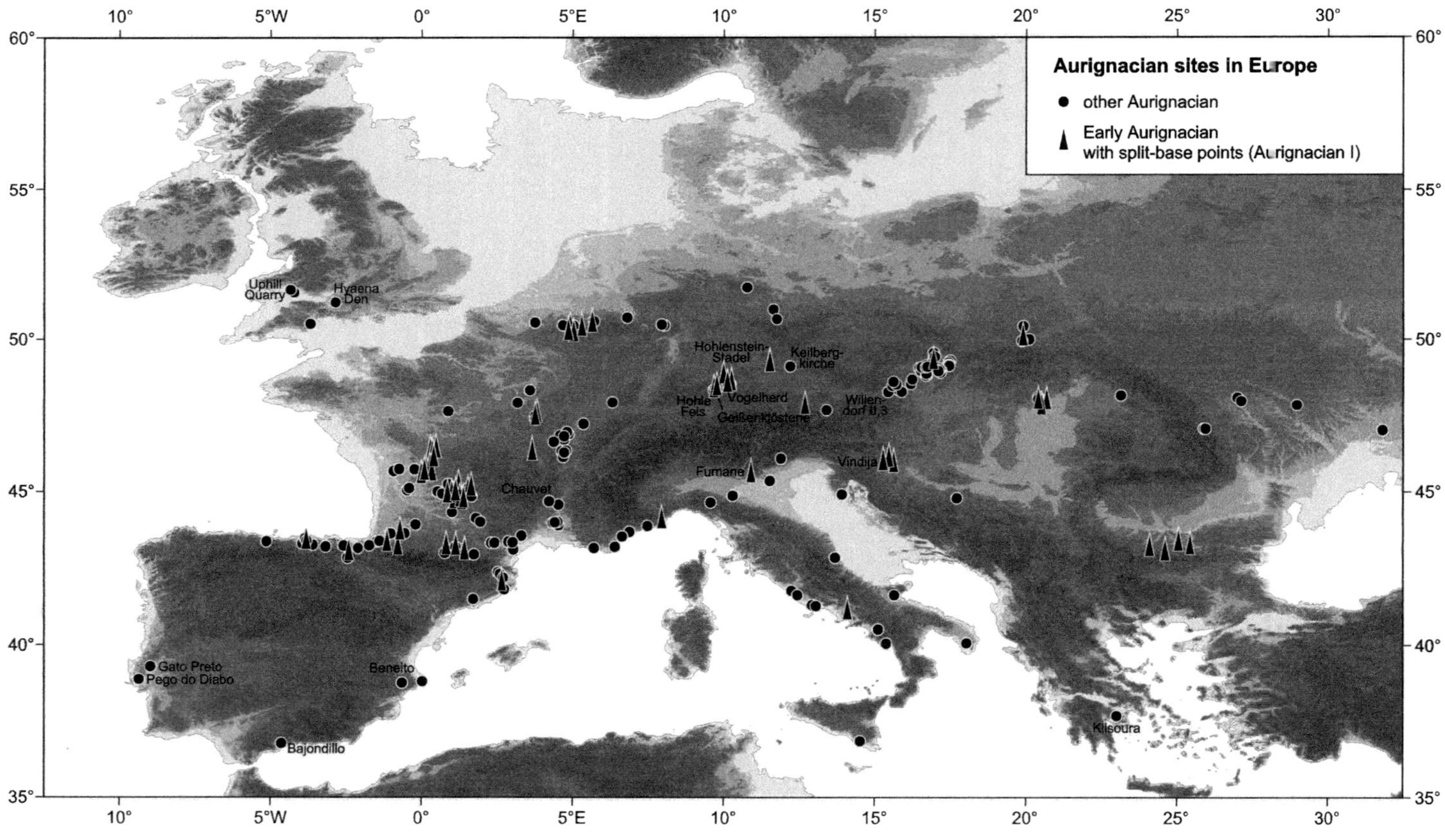

Fig. 22.5 European Aurignacian industries (Compiled after different authors). Map based on SRTM data; sea level lowered by 75 m

reflecting "behavioural modernity", i.e. highly specific lithic tool-kits, systematic production of (backed) bladelets, elaborated bone and ivory technology, personal ornaments, figurative art, music and – probably – ritual/religious beliefs (cf Dunbar 2004). Furthermore, the Aurignacian is sporadically associated with AMH, leading many researchers to believe that the spatio-temporal patterns of Aurignacian appearance map the spread of AMH into Europe (Bolus and Conard 2001; Conard and Bolus 2003; Mellars 2004, 2006a, b; Bar-Yosef 2001).

"Transitional" Industries

Detailed stratigraphic observation and techno-/typological studies, in combination with improvements in absolute dating, show that change from the MP to the UP in western Eurasia, broadly placed within the period 40.0–30.0 ka ^{14}C BP, is complex in both space and time (Fig. 22.6). A wide variety of assemblages between the Aurignacian and the later MP are characterized by elements regarded by some researchers as reflecting "behavioural modernity". Lithic assemblages regularly reveal the systematic production of blades and the presence of higher frequencies of specific tools such as end-scrapers, burins or backed pieces and the presence of these "progressive" items in such assemblages has led to their being grouped together under the term of "transitional" industries, even though there is no consensus on the definition of this term (cf Bar-Yosef 2006a).

On one interpretation of the term these industries would be regarded as "transitional" only because they bridge the period between the preceding (clearly) MP and the subsequent UP assemblages. Another interpretation of the term is steered by an evolutionist concept of cultural change, with the implication that these assemblages represent some form of developmental stage between the MP and the UP.

For example, the south-western European Chatelperronian, characterized *inter alia* by curve-backed pieces, was initially considered to be fully UP (e.g. de Sonneville-Bordes 1960) but later claimed to display MP features (Leroi-Gourhan 1968; Guilbaud 1993; Zilhão and d'Errico 1999; d'Errico et al. 1998). In the latter case it is argued that the Chatelperronian arises regionally from the latest MTA, with both groups having a similar geographical distribution (cf Bosinski 1989, 1990).

Similarly, it is believed that other "transitional" industries with curve-backed pieces – among them the Uluzzian of the Italian Peninsula (Palma di Cesnola 1982, 129) and the Layer V industry from Klisoura Cave 1 in Greece (Koumouzelis et al. 2001a, b) – may also be the result of filiation from a regional late MP, though others see a clear break between the Uluzzian and the preceding Italian MP (e.g. Gioia 1990).

To the North of the high alpine mountain chains, Northwestern and Central European leaf-point industries or Blattspitzengruppen (cf Freund 1952) are characterized by bifacially worked, foliate tools (leaf points and blade points: e.g. Jacobi 1990) and variously described as Lincombian (Campbell 1977; but see objection in Jacobi 1990), Jerzmanovician, Altmühlgruppe/"Ranisian", Szeletian etc. (e.g. Allsworth-Jones 1986; Bolus 2004; Jacobi et al. 2006, 567), would emerge from the regional late MP Keilmessergruppen (Bosinski 1967; Jöris 2004; cf Richter 1997; Uthmeier 1998, 2004). The assemblage of layer F (Zone 4) of the Weinberghöhlen close to Mauern in southern Germany (von Koenigswald et al. 1974) probably best demonstrates this filiation.

On this interpretation, those "transitional" industries which are most probably derived from a regional MP substrate should parsimoniously be regarded only as a diverse, more elaborate and differentiated "Final Middle Palaeolithic" (FMP) and we see no need to interpret them as "transitional" in the sense of foreshadowing the "behavioural modernity" of the UP. With the exception of Neanderthal remains from two French Chatelperronian contexts at St. Césaire (Lévêque and Vandermeersch 1980; but see also discussion in Bar-Yosef 2006b; Morin et al. 2005) and the Grotte du Renne at Arcy-sur-Cure (Leroi-Gourhan 1958; Bailey and Hublin 2006) none of these industries are associated with significant hominin remains.

Close in time to the described foliate and backed point industries occur South-eastern European assemblages characterized by a significant blade component and "UP" tool types. Such "Bachokirian" assemblages at the Bulgarian sites Bacho Kiro cave, level 11, and Temnata TD-I and TD-V, layer 4, appear to represent a break with previous regional traditions and have often been perceived as the earliest evidence for the "behavioural modernity" of the UP in the region (e.g. Kozłowski 1982).

Originally assigned to the early UP, they were seen as close to but distinct from the Aurignacian (e.g. Kozłowski 1992, 2006). Recent studies of the lithic technology of these laminar assemblages show that they differ in important details from the Aurignacian and are close to an evolved Levallois Mousterian (Tsanova and Bordes 2003), with particular parallels to a group of Near Eastern assemblages (e.g. Teyssandier 2005, 2006) generally classed together as "Initial Upper Palaeolithic" (IUP; e.g. Bar-Yosef 2003; Kuhn 2003, 2004). If one postulates an origin of the Bachokirian within these industries it might be logical to designate it too as IUP. On this interpretation, the Bachokirian might now be classified as a "transitional" industry, which, although clearly rooted in the (not necessarily regional) MP, already foreshadows some UP technological traditions. The Bohunician

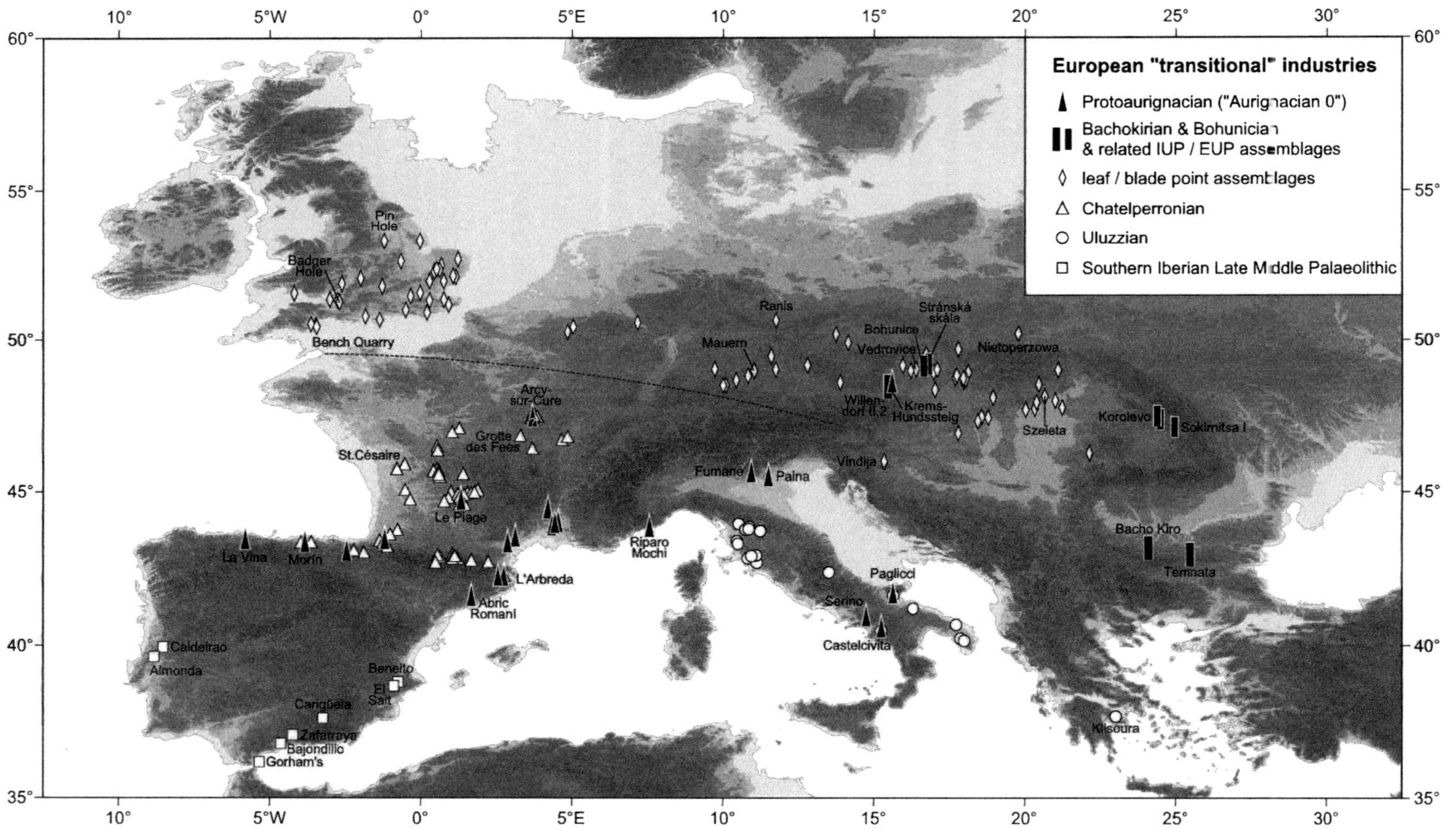

Fig. 22.6 European "transitional" industries (Modified from Jöris et al. 2003; 2006; compiled after different authors: see text). Map based on SRTM data; sea level lowered by 75 m. *IUP* Initial Upper Palaeolithic, *EUP* Early Upper Palaeolithic. The *broken line* divides the FMP "transitional" industries of northen Europe (leaf-point assemblages) from those of southern Europe (Chatelperronian, Uluzzian, Klisoura); cf Figs. 22.8 and 22.9

of eastern Central Europe is interpreted in a similar way (Škrdla 1996, 2003; Svoboda 2003; Svoboda and Bar-Yosef 2003).

Around the northern Mediterranean a very early appearance of the UP has been suggested on the basis of dates for assemblages designated "Aurignacian 0", "Protoaurignacian" (Broglio and Laplace 1966; Laplace 1966) or "Fumanian" (Mellars 2006b, 175) at sites such as L'Arbreda (Bischoff et al. 1989; Canal i Roquet and Carbonell i Roura 1989; Ortega Cobos et al. 2005) and Abric Romaní (Bischoff et al. 1994; Canal i Roquet and Carbonell i Roura 1989) in Catalonian Spain or Grotta di Fumane and Riparo Mochi in northern Italy (Broglio 1996, 2000, 2001; Laplace 1977). However, here too, detailed analysis suggests close technological similarities with assemblages of Levallois Mousterian type and a close relationship between these Mediterranean industries and broadly contemporaneous Early Upper Palaeolithic (EUP) industries in the Near East such as the Ahmarian (e.g. Ohnuma 1988; Boëda and Muhesen 1993; Marks 1993; Bourguignon 1996; Kuhn et al. 1999; Bar-Yosef 2000, 2003).

Although many authors propose that pre-Aurignacian European IUP and EUP blade assemblages may represent the input of Near Eastern technologies and ideas by AMH (Bar-Yosef 2000; Tostevin 2000; Škrdla 2003; Svoboda 2003; Svoboda and Bar-Yosef 2003), Nicolas Teyssandier (2006, 14) cautions that "The question remains totally open whether or not these typo-technological similarities represent technical convergences, diffusion of ideas or human migrations".

The Campanian Ignimbrite Marker Horizon

Radiocarbon data close to the MP–UP boundary are particularly sensitive with regard to sample provenance and association and their integrity must, whenever possible, be referenced against the stratigraphic sequence. In effect, continuous stratigraphic sequences are one of the most important tools for the critical evaluation of the chronological depth of cultural change involving the "transitional" industries.

The volcanic deposits of the Campanian Ignimbrite (CI) mega-eruption that took place in the south Italian Phlegrean Fields some 39,300 years ago (de Vivo et al. 2001; cf Giaccio et al. 2006; Southon 2004) provide the most valuable marker horizon for the evaluation of the chronostratigraphy of the MP–UP transition in southern and eastern Europe (Fig. 22.7). Beyond the southern Italian occurrences, the distal Y5 ash layer is widespread in the eastern

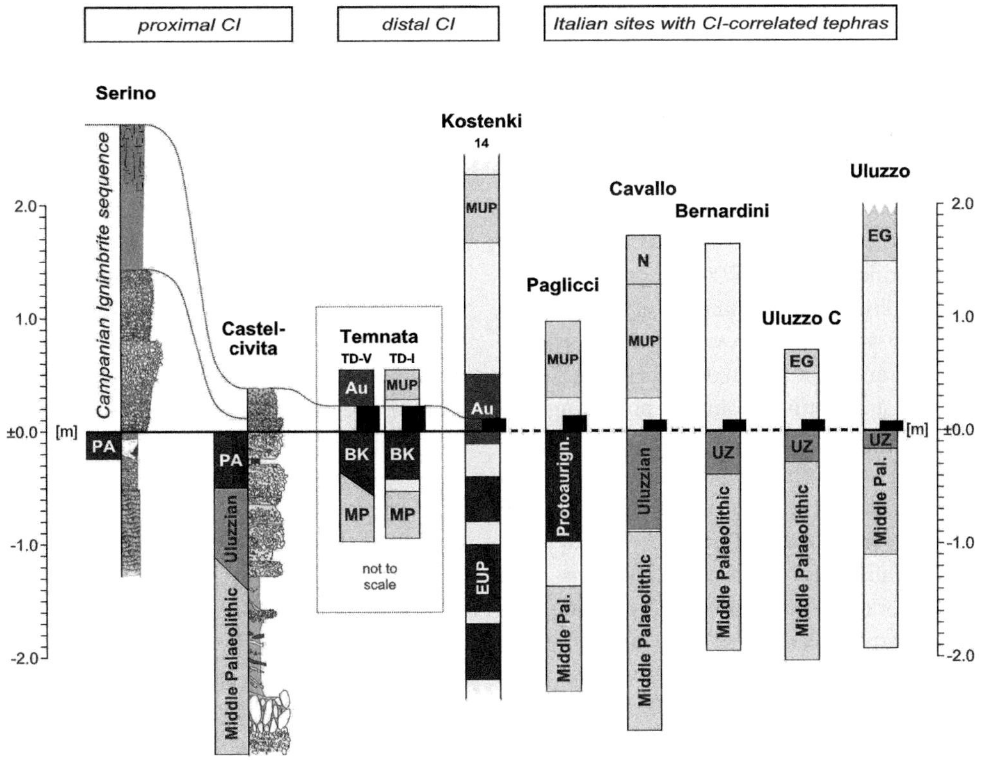

Fig. 22.7 Key-stratigraphies providing the Campanian Ignimbrite (Y5 tephra) marker horizon and correlated equivalents (*black*) in relation to the archaeological stratigraphical record (*MP* Middle Palaeolithic, *UZ* Uluzzian, *BK* Bachokirian, *PA* Protoaurignacian, *Au* Aurignacian, *MUP* Mid-Upper Palaeolithic, *EG* Epigravettian, *N* Neolithic). Modified after Fedele et al. (2002) and Giaccio et al. (2006). Note: For the attribution of Temnata see Teyssandier (2006)

Mediterranean, in south-eastern Europe at the site of Temnata in Bulgaria (Fedele et al. 2002, 2003; Giaccio et al. 2006) and as far East as the Kostenki-Borschevo region along the Don River in Russia, where several occurrences of the tephra are known (Anikovich et al. 2007; Pyle et al. 2006).

Mediterranean $\delta^{18}O$-records fix the ash layer stratigraphically immediately before the Heinrich 4 event (Ton-That et al. 2001; cf Fig. 22.3). Synchronisms of these records with Greenland ice core chronologies further allow the CI event to be linked to an extreme peak in volcanic sulphur recorded around 40.0 ka $_{GISP2}$ in the GISP2 ice core at the very end of Greenland Interstadial (GI) 9 (Giaccio et al. 2006; Zielinski et al. 1996, 1997). In Mediterranean records the Y5 ash post-dates the Laschamp geomagnetic excursion, which is identified by rock magnetic parameters and ^{10}Be flux, allowing for further tightening of both the CI-eruption and the Laschamp excursion with Greenland ice core chronologies (Giaccio et al. 2006; roughly dating between ca. 43.0 and 40.0 ka $_{GRIP-HULU}$: cf Voelker et al. 2000; Southon 2004).

That the geomagnetic event must have resulted in enhanced levels of atmospheric ^{14}C is probably indicated by radiocarbon measurements obtained at high stratigraphical resolution from the Mediterranean CT85-5 core (Giaccio et al. 2006) and by radiocarbon age distortions recorded earlier in the North Atlantic PS2644 core (Voelker et al. 2000). The latter documents offsets of up to 8,500 years between calendar ages and uncalibrated radiocarbon dates (Fig. 22.3; cf Weninger and Jöris 2004, 2008; cf www.calpal.de). Nevertheless, new data (Hughen et al. 2006; cf Fairbanks et al. 2005; Chiu et al. 2007) do not show such age distortions in the radiocarbon time scale. Archaeological sites from the MP-UP "transition" stratified below CI-deposits broadly date within the age-range of 39.1–30.1 ka ^{14}C BP (Giaccio et al. 2006; Anikovich et al. 2007).

In the long stratigraphy of the Russian site of Kostenki 14 (Markina Gora) the Y5 volcanic ash overlies two horizons, "Cultural Layer IVb" and the "Horizon of Hearths", the assemblages of which are distinct from the preceding MP and earlier than Aurignacian level first documented in stratigraphical positions within the ash horizon at the site. The pre-Aurignacian layers are assigned to an EUP (Anikovich et al. 2007) and differ from the MP in their lithic technology (production of blades and bladelets from prismatic cores), elaborated bone and ivory working, the presence of personal ornaments in the form of perforated marine shells (*Columbella* sp.) imported over distances of at least 500 km, and – possibly – figurative art (Sinitsyn 2003; Anikovich et al. 2007). The EUP at this site is dated to ca. 36,167 ± 176 ^{14}C BP (weighted mean of three dates from Layer IVb), with an oldest measurement of 36,540 ± 270 ^{14}C BP (GrA-15961), and 35,964 ± 121 ^{14}C BP (weighted mean of five dates from the "Horizon of Hearths"), with an oldest measurement of 37,240 ± 430 ^{14}C BP (GrA-10948).

At Temnata in Bulgaria the Y5 tephra covers Bachokirian laminar assemblages. A series of radiocarbon measurements, mostly of charcoal, provided dates bracketed between 39,100 ± 1,800 ^{14}C BP (OxA-5169; TD-I: 4B) and 36,900 ± 1,300 ^{14}C BP (OxA-5173; TD-V: 4B) (Ginter et al. 1996). Two TL measurements on burnt flint also date these levels (Table 22.4).

At the Italian cave of Castelcivita, CI tephra covers a long stratigraphy extending from the MP, followed by Uluzzian levels which are themselves covered by Protoaurignacian horizons. Moreover, at the open-air site of Serino some 50 km east of the CI eruption centre, thick CI pumice layers, ashes and pyroclastic flow deposits sealed an archaeological layer attributed to the Protoaurignacian (Fedele et al. 2002, 2003; Giaccio et al. 2006). At the Grotta Paglicci on the Gargano peninsula a Protoaurignacian layer is sealed by tephra which has been equated with CI deposits, although doubts have recently been expressed about this correlation (Giaccio et al. 2006, 366).

At the southern Italian cave sites of Cavallo, Uluzzo, Uluzzo C and Bernardini tephra equated with the Y5 ash directly overlie Uluzzian layers, which themselves overlay MP horizons (Giaccio et al. 2006). Although Uluzzian assemblages are only poorly dated radiometrically, their superstratification in many cases by CI tephra provides an excellent *terminus ante quem* for the age of the Uluzzian in the southern half of the Italian Peninsula.

According to the southern Italian record, it appears certain that the MP–UP "transition" includes an Uluzzian phase followed by industries ascribed to the Protoaurignacian, which on stratigraphical grounds must be fixed shortly before 40,000 calendar years ago. The Uluzzian assemblages should therefore be slightly older than the CI eruption which in radiocarbon terms is ca. 34.8–34.7 ka ^{14}C BP (cf Fig. 22.3).

In summary, MP, Uluzzian and Protoaurignacian assemblages have never been demonstrated to overlie CI-deposits. If the former two groups are suggested to have been made by Neanderthals, the survival of this hominin on the central and eastern Mediterranean peninsulas (Kuhn and Bietti 2000) later than 39.6 ka $_{HULU}$ appears increasingly unlikely.

Curve-Backed Piece Industries

Unlike at some Italian sites, where Uluzzian industries are associated with the CI marker horizon, the Klisoura Cave 1 (Greece) has no such independent stratigraphic control. Here the Uluzzian-like inventory of level V stratigraphically pre-dates a long Aurignacian sequence. The Protoaurignacian is not present at the site. The Aurignacian of the next overlying levels IV and IIIg/IIIe' is dated to 32,400 ± 600 ^{14}C BP (Gd-10562) and 34,700 ± 1,600 ^{14}C BP (Gd-7892) respectively, whereas the level V assemblage gave infinite dates of >30,800 ^{14}C BP (Gd-10715) and >31,100 ^{14}C BP (Gd-10714)

Table 22.2 Selection of results from radiometric dating of Chatelperronian and Uluzzian assemblages

Layer	Method	No.	Lab.-No.	^{14}C age	STD+	STD–	^{13}C	Material	Species	Skeletal element	Assoc. ind.	Source
Valiña, E												
IV	^{14}C		GrN-20833	31,730	2800	2110		Bone			Chatelperronian	Llana Rodríguez and Soto Barreiro (1991); Fortea Pérez (1996)
IV	^{14}C		GrA-3014	32,600	250	250		Bone			Chatelperronian	Llana Rodríguez and Soto Barreiro (1991); Fortea Pérez (1996)
IV	^{14}C		GrN-17729	34,800	1900	1500		Bone			Chatelperronian	Llana Rodríguez and Soto Barreiro (1991); Fortea Pérez (1996)
Labeko Koba, E												
IX (middle)	^{14}C		Ua-3325	29,750	740	740					Chatelperronian	Arrizabalaga (1995); Barandiarán Maetzu (1996)
IX (base)	^{14}C		Ua-3034	26,575	505	505					Chatelperronian	Arrizabalaga (1995); Barandiarán Maetzu (1996)
IX (base)	^{14}C		Ua-3324	34,215	1265	1265					Chatelperronian	Arrizabalaga (1995); Barandiarán Maetzu (1996)
Ekain, E												
Xb (I)	^{14}C		I-11056	>30,600				Bone			Chatelperronian	Davies (2000)
Brassempouy, Grande Galerie 2, F												
2g sup.	^{14}C		Gif-8172	31,690	810	810		Bone collagen			Chatelperronian	Bon et al. (1998); Mellars (2000)
Abri Dubalen (Brassempouy), F												
EBC2	^{14}C		GifA-101045	36,130	690	690		Bone			Chatelperronian	Zilhão (2007)
Caune de Belvis, F												
7	^{14}C		AA-7390	35,425	1140	1140		Bone			Chatelperronian	Bazile (1984)
Le Flageolet, Grotte XVI, F												
B	^{14}C	1	GifA-95581	35,000	1200	1200		Bone			Chatelperronian	Zilhão and d'Errico (1999)
B	^{14}C	2	AA-2997	38,100	1670	1670		Bone			Chatelperronian	Zilhão and d'Errico (1999)
B	^{14}C	3	AA-2674	>39,000				Bone			Chatelperronian	Zilhão and d'Errico (1999)
B	*^{14}C*	*(1–2)*	*Weighted mean*	*36,056*	*975*	*975*	*1.06*	*t-Value*			*Chatelperronian*	
Roc de Combe, F												
8	^{14}C	1	GifA-101264	39,540	970	970		Bone	K9		Chatelperronian	Zilhão (2007)
8	^{14}C	2	GifA-101266	40,000	1300	1300		Bone	K9		Chatelperronian	Zilhão (2007)
8	^{14}C	3	GifA-101265	45,100	2100	2100		Bone	K9		Chatelperronian	Zilhão (2007)
8	*^{14}C*	*(1–2)*	*Weighted mean*	*39,705*	*777*	*777*	*0.20*	*t-value*			*Chatelperronian*	
10	^{14}C		OxA-1264	31,000	750	750		Bone			Chatelperronian	Hedges et al. (1990)
10	^{14}C		OxA-1443	38,000	2000	2000		Bone			Chatelperronian	Hedges et al. (1990)

(continued)

Table 22.2 (continued)

Layer	Method	No.	Lab.-No.	^{14}C age	STD+	STD–	^{13}C	Material	Species	Skeletal element	Assoc. ind.	Source
Châtelperron, F												
B5	^{14}C	1	OxA-13622	39,150	600	600		Bone			Chatelperronian	Gravina et al. (2005)
B5	^{14}C	2	OxA-14320	39,240	380	380		Bone			Chatelperronian	Gravina et al. (2005)
B5	^{14}C	3	OxA-13621	40,650	600	600		Bone			Chatelperronian	Gravina et al. (2005)
B5	^{14}C	*(1–3)*	*Weighted mean*	*39,534*	*283*	*283*	*1.68*	*t-Value*			*Chatelperronian*	
Roche-au-Loup, F												
5 (b)	^{14}C		Gif-2414	>40,000				Charcoal			Chatelperronian	Delibrias et al. (1976)
La Quina, aval, F												
4	^{14}C		OxA-10261/ Ly-1367	35,950	450	450		Bone			Chatelperronian	Zilhão (2007)
Les Cottés (St. Pierre de Maille), F												
G1	^{14}C		GrN-4333	33,300	500	500		Teeth			Chatelperronian	Vogel and Waterbolk (1967)
G2	^{14}C		GrN-4510	31,900	430	430		Bone			Chatelperronian	Vogel and Waterbolk (1967)
Combe Saunière, F												
X	^{14}C		OxA-6504	33,000	900	900		Bone			Chatelperronian	Mellars (1999)
X	^{14}C		OxA-6503	35,900	1100	1100		Bone			Chatelperronian	Mellars (1999)
X	^{14}C		OxA-6503 (tripeptide)	38,100	1000	1000		Bone			Chatelperronian	Mellars (1999)
Grotte du Renne, Arcy-sur-Cure, F												
VIII	^{14}C		Ly-5062	32,000	1200	1200					Chatelperronian	David et al. (2001)
VIII	^{14}C		Ly-2163	33,000	1400	1400		Bone (collagen)			Chatelperronian	David et al. (2001)
VIII	^{14}C		GrN-1736	33,500	400	400		Ash/burnt bones			Chatelperronian	David et al. (2001)
VIII	^{14}C		GrN-1742	33,860	250	250		Bone, residue (charred)			Chatelperronian	David et al. (2001)
IX	^{14}C		L-340-C	15,700	400	400		Bone (charred)			Chatelperronian	David et al. (2001)
IX	^{14}C		Ly-5063	31,500	1200	1200					Chatelperronian	David et al. (2001)
IX	^{14}C		OxA-3465	45,100	2800	2800		Bone	A15 (86)19'		Chatelperronian	David et al. (2001)
X	^{14}C		L-340-D	15,350	400	400		Bone (charred)			Chatelperronian	David et al. (2001)
Xa	^{14}C		OxA-8450/ Ly-893	25,820	280	280		Bone	Y11	Mammoth	Chatelperronian	David et al. (2001)
Xb	^{14}C		GrN-4216	24,500	360	360		Bone, burnt ("extra")			Chatelperronian	David et al. (2001)
Xb	^{14}C		GrN-4251	25,500	800	800		Bone, residue (burnt)			Chatelperronian	David et al. (2001)
Xb	^{14}C		Ly-5064	25,550	380	380					Chatelperronian	David et al. (2001)
Xb	^{14}C		OxA-3464	33,820	720	720		Bone splinter			Chatelperronian	David et al. (2001)
Xb1	^{14}C		OxA-9122/ Ly-1055	33,400	600	600		Bone	Y10	Horse or reindeer	Chatelperronian	David et al. (2001); Stevens and Hedges (2004)
Xb1	^{14}C		OxA-8451/ Ly-894	38,300	1300	1300		Bone	Y10	Horse	Chatelperronian	David et al. (2001); Stevens and Hedges (2004)

Xb2	^{14}C	OxA-8452/ Ly-895	34,450	750	750	Bone	Y11	Mammoth or horse	Chatelperronian	David et al. (2001); Stevens and Hedges (2004)
Xc	^{14}C	OxA-8533/ Ly-896	31,300	600	600	Bone	Y11	Mammoth	Chatelperronian	David et al. (2001)
Castelcivita, I										
rsa"	^{14}C	F-71	32,470	650	650	Bones (charred)			Uluzzian	Gambassini (1997)
rpi	^{14}C	GrN-13985	33,330	430	430	Bone			Uluzzian	Gambassini (1997)
rpi	^{14}C	F-106	>34,000			Bones (charred)			Uluzzian	Azzi and Gulisano (1979)
pie	^{14}C	F-107	33,220	780	780	Bones (charred)			Uluzzian	Azzi and Gulisano (1979); Gambassini (1997)
Grotta La Cala, I										
tg.14	^{14}C	OxA-6265	29,120	300	300	Charcoal			Uluzzian	Hedges et al. (1998)
Klisoura, GR										
V	^{14}C	Gd-10715	>30,800			Bone	Hearth 53		"Uluzzian-like"	Koumouzelis et al. (2001a, b)
V	^{14}C	Gd-10714	>31,100			Bone	Hearth 42		"Uluzzian-like"	Koumouzelis et al. (2001a, b)
V	^{14}C	GifA-99168	40,010	740	740	Bone (burnt)	Hearth 42		"Uluzzian-like"	Koumouzelis et al. (2001b)

obtained from the two hearths no. 53 and 42 respectively (Koumouzelis et al. 2001a). More recently, a measurement on a (burnt) bone from hearth 42 in level V produced an age of 40,010 ± 740 ^{14}C BP (Koumouzelis et al. 2001b).

In western Europe, the Chatelperronian is dated by a fairly large number of radiocarbon results, the majority of them from French sites. Where the dated material is known, all but one of these are on bone. In general, the Chatelperronian radiocarbon dataset is very heterogeneous, with regard to both the laboratories involved and the date of sample submission, and contains measurements ranging from ca. 45.0 to 25.0 ka ^{14}C BP (Table 22.2; Fig. 22.8; cf compilation in Zilhão 2007), many of them with very large standard deviations.

The old and young extremes of this range are clearly erroneous but the validity and significance of dates falling between ca. 41.0 and 31.0 ka ^{14}C BP must be discussed. A first group of older radiocarbon ages lies in the range of ca. 40.7–38.0 ka ^{14}C BP. Most of these results have been obtained in the past several years, but the significance of a number of these dates is lessened by their large standard deviations. The weighted mean of two radiocarbon measurements for layer B at Le Flageolet, Grotte XVI, is 36,056 ± 956 ^{14}C BP, with a further infinite measurement of >39,800 ^{14}C BP (AA-2674) (Rigaud 2001). Layer X at Combe Saunière has a date of 38,100 ± 1,000 ^{14}C BP (OxA-6503 tripeptide), although two other dates (OxA-6503, OxA-6504) are much younger (Mellars 1999). Three radiocarbon measurements for level B5 of the Grotte des Fées (Châtelperron type site) have a weighted mean of 39,534 ± 283 ^{14}C BP (Gravina et al. 2005) but the contextual association of the radiocarbon samples from layers B4-B1 at the Grotte des Fées remains unclear (Zilhão et al. 2006). Older still are dates for level 8 at Roc-de-Combe (weighted mean of three radiocarbon measurements: 39,705 ± 777 ^{14}C BP; cf Zilhão 2007) and a single date on charcoal from layer 5(b) at Roche-au-Loup (Gif-2414: >40,000 ^{14}C BP; cf Delibrias et al. 1976).

The most comprehensive series of dates for the Chatelperronian comes from the site of Grotte du Renne (Arcy-sur-Cure: Table 22.2; Fig. 22.8). Here, the stratigraphic record covers a continuous sequence from the Late MP, through the Chatelperronian (levels X–VIII) to the Aurignacian with numerous radiocarbon dates from locations throughout the sequence (David et al. 2001). The oldest Chatelperronian at Grotte du Renne (level X) provided a number of dates, many of which are impossibly young (L-340-D; GrN-4216; GrN-4251; Ly-5064; OxA-8450/Ly-893). Other dates from this level (OxA-8533/Ly-896; OxA-9122/Ly-1055; OxA-3464; OxA-8452/Ly-895) form a more plausible cluster around 34.0 to 33.0 ka ^{14}C BP, however, it might be questioned whether a much older age for the Chatelperronian at Grotte du Renne is implied by the date of 38,300 ± 1,300 ^{14}C BP on a horse bone from layer Xb1 (OxA-8451/Ly-894), which resembles results from Grotte des Fées (level B5) and Roc-de-Combe (level 8). The middle Chatelperronian level IX at Grotte du Renne has two widely discrepant dates (OxA-3465 and L-340 C) with a third result (Ly-5063) also seemingly too young for the context. Dates for the youngest level VIII form a close group between 33,860 ± 250 ^{14}C BP (GrN-1742) and 32,000 ± 1,200 ^{14}C BP (Ly-5062).

The Grotte du Renne is important for the question of late Neanderthal survival (from level Xb; Hublin et al. 1996; cf Bailey and Hublin 2006) and due to the highest frequency of personal ornaments attributed to the Chatelperronian (cf discussion in Zilhão 2007). However, the radiocarbon dataset at this site is far from optimal and cannot resolve the debate on the age of the Chatelperronian. Most results can only be taken as minimum estimates for the age of specific layers.

The only Chatelperronian inventories represented in Cantabrian Spain are from level 10 at Cueva Morín, the small assemblages from Labeko Koba IX and Ekain X, and those from El Pendo and A Valiña, where there are stratigraphical problems (Maíllo Fernández 2003; Montes Barquín and Sanguino González 2001). At Cueva Morín and Labeko Koba the Chatelperronian assemblages are overlain by Protoaurignacian levels. Protoaurignacian level 8 at Cueva Morín is dated to 36,590 ± 1,100 ^{14}C BP (GifA-96263; Maíllo Fernández et al. 2001), providing a minimum age for the underlying Chatelperronian level 10 at the site. From Catalonia only a few Chatelperronian points are known from L'Arbreda, Cova Pau and Reclau Viver (cf Canal i Roquet and Carbonell i Roura 1989: 337; Zilhão 2006b). At L'Arbreda these are stratigraphically associated with the Protoaurignacian, implying that they may be as old as ca. 38.0 ka ^{14}C BP.

Leaf Point Industries

Stratigraphic dating evidence for the leaf point industries is generally poor. A limited amount of information is provided by certain cave sites such as the Szeleta cave in Hungary (Simán 1995), or at the Weinberghöhlen at Mauern (von Königswald et al. 1974) and the Ilsenhöhle in Ranis (Hülle 1977), Germany, but it must be concluded that no well developed and accurately dated stratigraphies are available (Bolus 2004). Nonetheless, what restricted information does exist leaves no doubt that leaf point industries were always recovered from below any UP layers present.

The number of radiocarbon dates relevant for the leaf point industries is restricted, but they nevertheless form a quite consistent group between ca. 40.0 and 37.5 ka ^{14}C BP (Table 22.3; Fig. 22.9). Five charcoal dates measured for the Szeletian site of Vedrovice V in the Czech Republic span the period between 39,500 ± 1,100 ^{14}C BP (GrN-12375) and 35,150 ± 650 ^{14}C BP (GrN-15513; Valoch 1996; Svoboda et al. 1996; Table 22.3; Fig. 22.9) with a weighted mean of

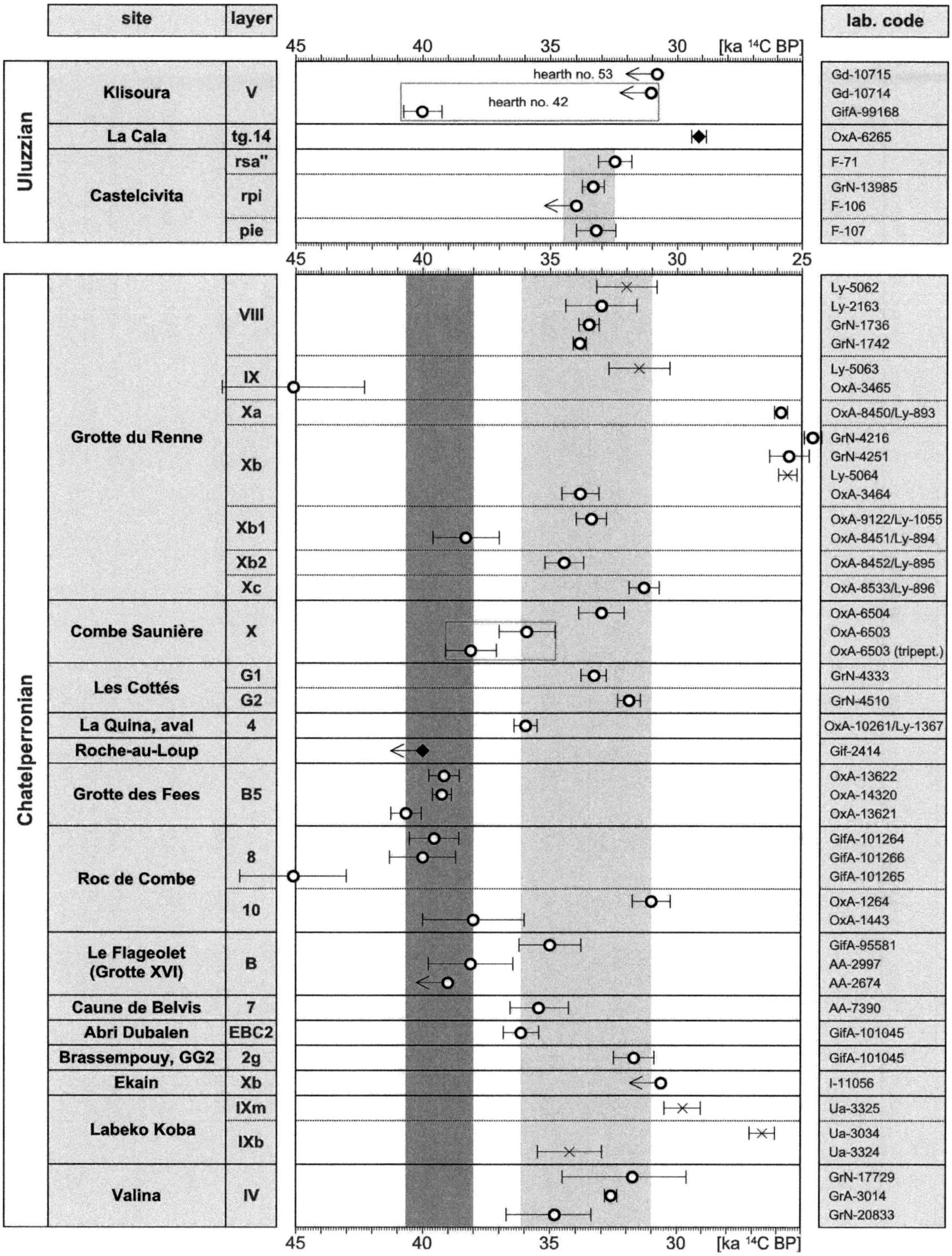

Fig. 22.8 Selection of results from radiometric dating of Chatelperronian and Uluzzian assemblages (see: Table 22.2), given with 1σ-standard deviation. Dates on bone (*circles*: open – without traces of hominin activity); dates on charcoal (*diamonds*); unknown material (*asterixes*); *Dark grey* shading: most likely dating range; *Light grey* shading: Divergent dates

ca. 37,100 ± 350 ¹⁴C BP. Radiocarbon dating of charcoal from the leaf point sites of Nietoperzowa (level 6), in southern Poland, and Certova Pec in Slovakia (Vogel and Waterbolk 1964) produced dates of ca. 38.5 ka ¹⁴C BP (GrN-2181) and 38.4 ka ¹⁴C BP (pooled mean of two measurements) respectively (Table 22.3; Fig. 22.9), but the precision of these pioneering dates is lessened by their high standard deviations. Radiocarbon measurements on bone from the Szeletian

Table 22.3 Selection of results from radiometric dating of leaf/blade point assemblages (only pre-Aurignacian contexts)

Layer	Method	No.	Lab.-No.	^{14}C age	STD+	STD–	^{13}C	Material	Species	Skeletal element	Assoc. ind.	Source
Nietoperzowa, PL												
6	^{14}C		GrN-2181	38,500	1240	1240		Charcoal			Szeletian	Vogel and Waterbolk (1964)
Vedrovice V, CZ												
4	^{14}C	1	GrN-17261	30,170	300	300		Charcoal	Hearth ca 15 cm above cultural layer		Szeletian	Valoch (1996)
4	^{14}C	2	GrN-15513	35,150	650	650		Charcoal	10/C		Szeletian	Valoch (1996)
4	^{14}C	3	GrN-15514	37,600	800	800		Charcoal	17/I		Szeletian	Valoch (1996)
4	^{14}C	4	GrN-12374	37,650	550	550		Charcoal	17/P		Szeletian	Valoch (1996)
4	^{14}C	5	GrN-12375	39,500	1100	1100		Charcoal	11/O		Szeletian	Valoch (1996)
4	^{14}C	6	GrN-19105	>39,500				Charcoal	13/S-T		Szeletian	Valoch (1996)
4	^{14}C	7	GrN-19106	47,250	3700	2500		Charcoal	Underlying cultural layer (14/Z)		Szeletian	Valoch (1996)
4	*^{14}C*	*(3–6)*	*Weighted mean*	*37,096*	*352*	*352*	*2.63*	*t-Value*			*Szeletian*	
Čertova Pec (Radosina), SLO												
	^{14}C	2	?	38,320	2480	2480		Charcoal?			Szeletian	Svoboda et al. (1996)
−1.9–2.0m	^{14}C	1	GrN-2438	38,400	2800	2100		Charcoal			Szeletian	Vogel and Waterbolk (1964)
	^{14}C	*(1–2)*	*Weighted mean*	*38,355*	*1856*	*1856*	*0.01*	*t-Value*			*Szeletian*	
Szeleta Cave, HU												
7	^{14}C		GrN-5130	32,620	400	400		Bone	Light grey layer (–3m)		Szeletian	Allsworth-Jones (1978)
4	^{14}C		?	32,500	420	420		Bone?			Szeletian	Allsworth-Jones (1978)
4	^{14}C		GXO-197	>41,700				Bone?			Szeletian	Allsworth-Jones (1978)
3	^{14}C		GrN-6058	43,000	1100	1100		Bone	Dark brown layer (–6m)		Szeletian	Allsworth-Jones (1978)
Badger Hole, Wookey Hole, GB												
	^{14}C - AF*		OxA-11963	36,000	450	450	−20.1	Bone			Blade point	Jacobi et al. (2006)
Bench Quarry, Brixham, GB												
	^{14}C - AC		OxA-1620	34,500	1400	1400	−21.0	Bone			Leaf point	Jacobi et al. (2006)
	^{14}C - AI		OxA-5961	32,500	1200	1200	−18.5	Bone			Leaf point	Jacobi et al. (2006)
	^{14}C - AF*		OxA-13512	36,800	450	450	−18.4	Bone			Leaf point	Jacobi et al. (2006)
	^{14}C - AG*		OxA-13324	37,500	900	900	−18.5	Bone			Leaf point	Jacobi et al. (2006)
Pin Hole, Creswell Crags, GB												
	^{14}C - AI		OxA-4754	37,800	1600	1600	−17.3	Bone			Leaf point	Jacobi et al. (2006)

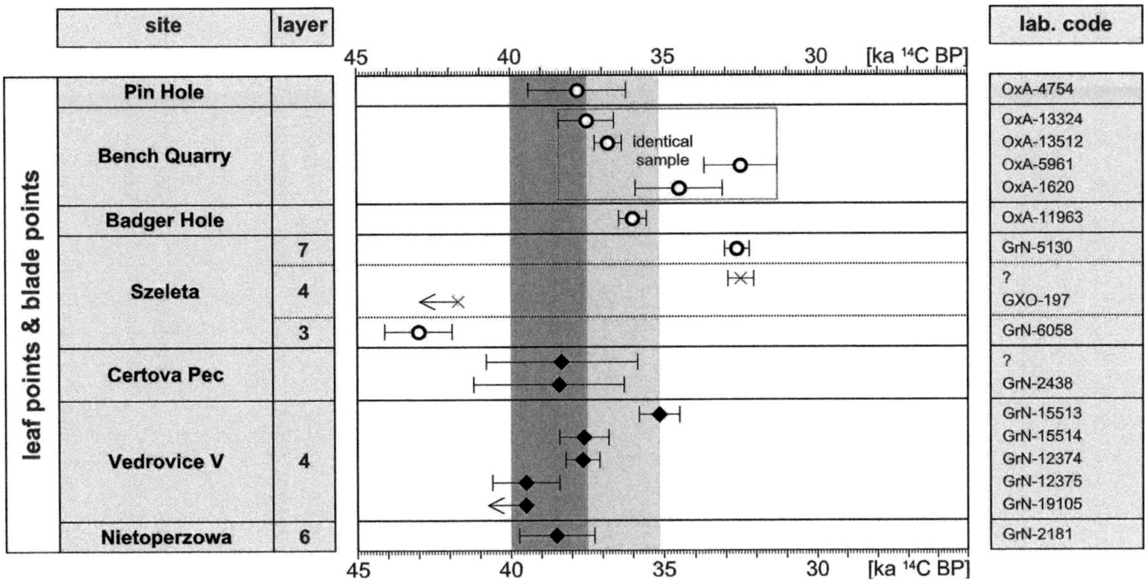

Fig. 22.9 Selection of results from radiometric dating of leaf/blade point assemblages (only pre-Aurignacian contexts; see: Table 22.3), given with 1σ-standard deviation. Dates on bone (*circles*: open – without traces of hominin activity); dates on charcoal (*diamonds*); unknown material (*asterisk*); *Dark grey* shading: most likely dating range; *Light grey* shading: less likely dating range

horizons 3 to 7 of Szeleta cave in Hungary produced highly divergent dates, which only assign the Szeletian to the broad age range of the MP–UP transition (Table 22.3; Fig. 22.9).

To date, no reliable absolute dates are available for southern and eastern German leaf point sites. Recently published ^{14}C AMS dates for the Ilsenhöhle at Ranis (Grünberg 2006) are unable to clarify the absolute age of the foliate point assemblage at this site, since results obtained for the supposedly distinct Szeletian, Aurignacian and Gravettian contexts 2, 3 & 4 place all three entities indiscriminately within the period 33.5–27.0 ka ^{14}C BP. The early date of excavations (Hülle 1977) and the probably disturbed stratigraphy might be reasons why the sampled bones do not reliably date specific cultural units.

Good evidence has been obtained from British sites with leaf- and blade-point industries (Campbell 1977; Jacobi 1990, 1999), where the dating of closely associated material has changed ideas dramatically during the last few years. Whereas British blade point sites were initially assigned to ca. 29.0–27.0 ka ^{14}C BP. (Aldhouse-Green and Pettitt 1998), recently measured more precise AMS dates provide evidence for a much older age (Jacobi et al. 2006).

Series of AMS dates were obtained from the sites of Pin Hole, Badger Hole and Bench Quarry following newly established ultrafiltration sample pre-treatment techniques (Bronk Ramsey et al. 2004) (Table 22.3; Fig. 22.9). At Pin Hole a large series of radiocarbon dates assigns the MP occupation to between ca. 58.8 and 40.7 ka ^{14}C BP, which is in accordance with an AMS date ca. 37,760 ± 340 ^{14}C BP (OxA-11980) for a reindeer antler found "seemingly above the distribution of the MP artefacts" (Jacobi et al. 2006, 563). A comparable result of 37,800 ± 1,600 ^{14}C BP (OxA-4754) was measured for a bone found in close association with a leaf point in brecciated sediment overlying the MP levels. At Bench Quarry two AMS dates of 34,500 ± 1,400 ^{14}C BP (OxA-1620) and 32,500 ± 1,200 ^{14}C BP (OxA-5961; Table 22.3; Fig. 22.9) had been previously obtained for a hyena dentary found in close association with a leaf point. However, two new AMS results on the same bone suggest an appreciably earlier time range of 37.5–36.8 ka ^{14}C BP (Jacobi et al. 2006, 568). A recently obtained AMS-date for Badger Hole (OxA-11963) is somewhat younger and may date the blade points here to 36,000 ± 450 ^{14}C BP (Table 22.3; Fig. 22.9).

In summary, the British blade point assemblages are now dated to between 37.8 and >36.0 ka ^{14}C BP. This range is in overall agreement with the (generally slightly older) dates for the continental leaf point assemblages and compares well with the most robust radiocarbon age determinations from Vedrovice V, Certova Pec and Nietoperzowa.

Early Laminar Industries

Although at a few key-sites early laminar lithic assemblages (e.g. Bachokirian, Protoaurignacian) can be stratigraphically fixed below CI-tephra deposits (see above), other important Italian, French and northern Spanish sites with horizons containing laminar lithic industries lie outside the distal dispersal area of this independent marker horizon (Pyle et al. 2006). It is nevertheless possible to examine the radiometric age of some of the crucial Central European and northern Mediterranean assemblages in the context of their stratigraphies.

Table 22.4 Selection of results from radiometric dating of the Initial Upper Palaeolithic (IUP) of Üçağızlı, Bachokirian and Bohunician assemblages, and from the Aurignacian of Keilberg-Kirche and Willendorf II (KS 3)

Layer	Method	No.	Lab.-No.	^{14}C age	STD+	STD−	^{13}C	Material	Location		Assoc. ind.	Source
Keilberg-Kirche, D												
	^{14}C	1	KN-4691	37,500	1250	1250		Charcoal			Aurignacian	Uthmeier (1996)
	^{14}C	2	KN-4690	37,500	1450	1450		Charcoal			Aurignacian	Uthmeier (1996)
	^{14}C	3	KN-4692	38,600	1200	1200		Charcoal			Aurignacian	Uthmeier (1996)
	^{14}C	*(1–3)*	*Weighted mean*	*37,922*	*743*	*743*	*0.48*	*t-Value*			*Aurignacian*	
Willendorf II, AU												
KS 3	^{14}C		GrN-11192	34,100	1200	1000		Charcoal			Aurignacian	Haesaerts (1990)
KS 3	^{14}C		GrA-896	37,930	750	750		Charcoal			Aurignacian	Damblon et al. (1996)
KS 3	^{14}C		GrN-17805	38,880	1530	1280		Charcoal			Aurignacian	Damblon et al. (1996)
Stránská-skála, CZ												
5	^{14}C	1	GrN-12297	38,200	1100	1100		Charcoal	III-1	Upper soil	Bohunician	Svoboda (2003)
5	^{14}C	2	GrN-12298	38,500	1400	1200		Charcoal	III-2	Upper soil	Bohunician	Svoboda (2003)
5	^{14}C	3	AA-32059	37,900	1100	1100		Charcoal	IIId	Upper soil	Bohunician	Svoboda (2003)
5	^{14}C	4	GrN-11504	34,530	830	740		Charcoal	IIId	Upper soil	Bohunician	Svoboda (2003)
5	^{14}C	5	AA-32061	35,080	830	830		Charcoal	IIId	Upper soil	Bohunician	Svoboda (2003)
5	^{14}C	6	GrN-11808	35,320	320	300		Charcoal	IIId	Upper soil	Bohunician	Svoboda (2003)
5	^{14}C	7	AA-32060	37,270	990	990		Charcoal	IIId	Upper soil	Bohunician	Svoboda (2003)
5	^{14}C	8	AA-41477	34,530	770	770		Charcoal	IIIc	Upper soil	Bohunician	Svoboda (2003)
5	^{14}C	9	AA-41480	34,680	820	820		Charcoal	IIIc	Upper soil	Bohunician	Svoboda (2003)
5	^{14}C	10	AA-41478	36,350	990	990		Charcoal	IIIc	Upper soil	Bohunician	Svoboda (2003)
5	^{14}C	11	AA-41476	36,570	940	940		Charcoal	IIIc	Upper soil	Bohunician	Svoboda (2003)
5	^{14}C	12	AA-32059	37,900	1100	1100		Charcoal	IIIc	Upper soil	Bohunician	Svoboda (2003)
5	^{14}C	13	AA-32058	38,300	1100	1100		Charcoal	IIIc	Redeposited lower soil		Svoboda (2003)
4	^{14}C	14	GrN-12298	41,300	3100	2200		Charcoal	IIIa	Redeposited lower soil		Svoboda (2003)
		(1–12)	*Weighted mean*	*35,726*	*213*	*213*	*2.21*	*t-Value*			*Bohunician*	
Bohunice-Kejbyly, CZ												
4a	^{14}C		Q-1044	40,173	1200	1200		Charcoal	I	Lower soil	Bohunician	Valoch (1976)
4a	^{14}C		GrN-6802	41,400	1400	1200		Charcoal	II	Lower soil	Bohunician	Valoch (1976)
Bohunice-cihelna, CZ												
4a	^{14}C		GrN-6165	42,900	1700	1400		Charcoal		Below soil of Bohunice culture	No assoc. ind.	Valoch (1996)
4a	^{14}C		GrN-16920	36,000	1100	1100		Charcoal		Below soil of Bohunice culture	No assoc. ind.	Damblon et al. (1996)
Korolevo II, RU												
II	^{14}C		GIN-2774	38,500	1000	1000		Charcoal			Bohunician-type	Usik et al. (2006)

Bacho Kiro, BG											
11	^{14}C	1	OxA-3212	34,800	1150	1150		Tooth		Bachokirian	Hedges et al. (1994)
11	^{14}C	2	OxA-3183	37,650	1450	1450		Charcoal		Bachokirian	Hedges et al. (1994)
11	^{14}C	3	OxA-3213	38,500	1700	1700		Bone		Bachokirian	Hedges et al. (1994)
11	^{14}C	4	GrN-7545	>43,000				Charcoal		Bachokirian	Mook (1982)
11	^{14}C	*(1–3)*	*Weighted mean*	*36,471*	*796*	*796*	*1.19*	*t-Value*		*Bachokirian*	
Temnata, BG											
4B	^{14}C	1	OxA-5171	38,200	1500	1500		Charred (mostly charcoal)	TD-I (interior)	Bachokirian	Ginter et al. (1996)
4B	^{14}C	2	OxA-5170	38,800	1700	1700		Charred (mostly charcoal)	TD-I (interior)	Bachokirian	Ginter et al. (1996)
4B	^{14}C	3	OxA-5169	39,100	1800	1800		Charred (mostly charcoal)	TD-I (interior)	Bachokirian	Ginter et al. (1996)
4B	^{14}C	*(1–3)*	*Weighted mean*	*38,642*	*954*	*954*	*0.25*	*t-Value*		*Bachokirian*	
4B	TL		GdTL-255	46,000	8000	8000		Burnt flint	TD-I (interior)	Bachokirian	Ginter et al. (1996)
4C	TL		GdTL-256	45,000	7000	7000		Burnt flint	TD-I (interior)	Bachokirian	Ginter et al. (1996)
4B	^{14}C		OxA-5173	36,900	1300	1300		Charred (mostly charcoal)	TD-V (interior)	Bachokirian	Ginter et al. (1996)
4B (or 4C?)	^{14}C		OxA-5172	38,300	1800	1800		Charred (mostly charcoal)	TD-V (interior)	Bachokirian	Ginter et al. (1996)
Üçağizli, T											
G	^{14}C		AA-37626	39,100	1500	1500		Charcoal		IUP	Kuhn et al. (2001)
H	^{14}C	1	AA-37623	33,040	1400	1400		Charcoal		IUP	Kuhn et al. (2001)
H	^{14}C	2	AA-35261	35,670	730	730		Charcoal		IUP	Kuhn et al. (2001)
H	^{14}C	3	AA-27995	38,900	1100	1100		Charcoal		IUP	Kuhn et al. (2001)
H	^{14}C	4	AA-27994	39,400	1200	1200		Charcoal		IUP	Kuhn et al. (2001)
H	^{14}C	5	AA-37625	41,400	1100	1100		Charcoal		IUP	Kuhn et al. (2001)
H	^{14}C	*(2–5)*	*Weighted mean*	*38,036*	*487*	*487*	*2.80*	*t-Value*		*IUP*	

At the Bulgarian site of Bacho-Kiro three radiocarbon dates for layer 11, one each on charcoal, tooth and bone, fall between 38,500±1,700 ^{14}C BP (OxA-3213) and 34,800 ± 1,150 ^{14}C BP (OxA-3212) with a weighted mean of 36,471±796 ^{14}C BP (Table 22.4; Fig. 22.10). A further charcoal measurement provided an infinite result of >43,000 (GrN-7545). At Temnata in Bulgaria the Bachokirian laminar assemblages are dated by a series of radiocarbon measurements, mostly on charcoal, and by two TL measurements (Table 22.4; Fig. 22.10). TD-V (interior) dates to 38,300± 1,800 ^{14}C BP (OxA-5172) and 36,900 ± 1,300 ^{14}C BP (OxA-5173) (Ginter et al. 1996) and TD-I (interior) to 38,642±954 ^{14}C BP by a weighted mean of three results (Table 22.4; Fig. 22.10). TL-measurements on burnt flint place the Bachokirian levels at around 46.0±8.0 ka BP (GdTL-255) and 45.0±7.0 ka BP (GdTL-256; Ginter et al. 1996).

At the Czech site of Stránská skála Bohunician levels were dated by 14 radiocarbon measurements on charcoal (Damblon et al. 1996; Valoch 1996; Svoboda 2003). Seven results from the "upper paleosol" locations III-1, III-2, IIId range from 38,500+1,400/– 1,200 ^{14}C BP (GrN-12298) to 34,530+830/– 740 ^{14}C BP (GrN-11504), while five measurements from the "upper paleosol" at Stránská skála IIIc fall between 37,900±1,100 ^{14}C BP (AA-32059) and 34,530±770 ^{14}C BP (AA-41477) (Table 22.4; Fig. 22.10). The 12 results give a weighted mean of 35,726±213 ^{14}C BP.

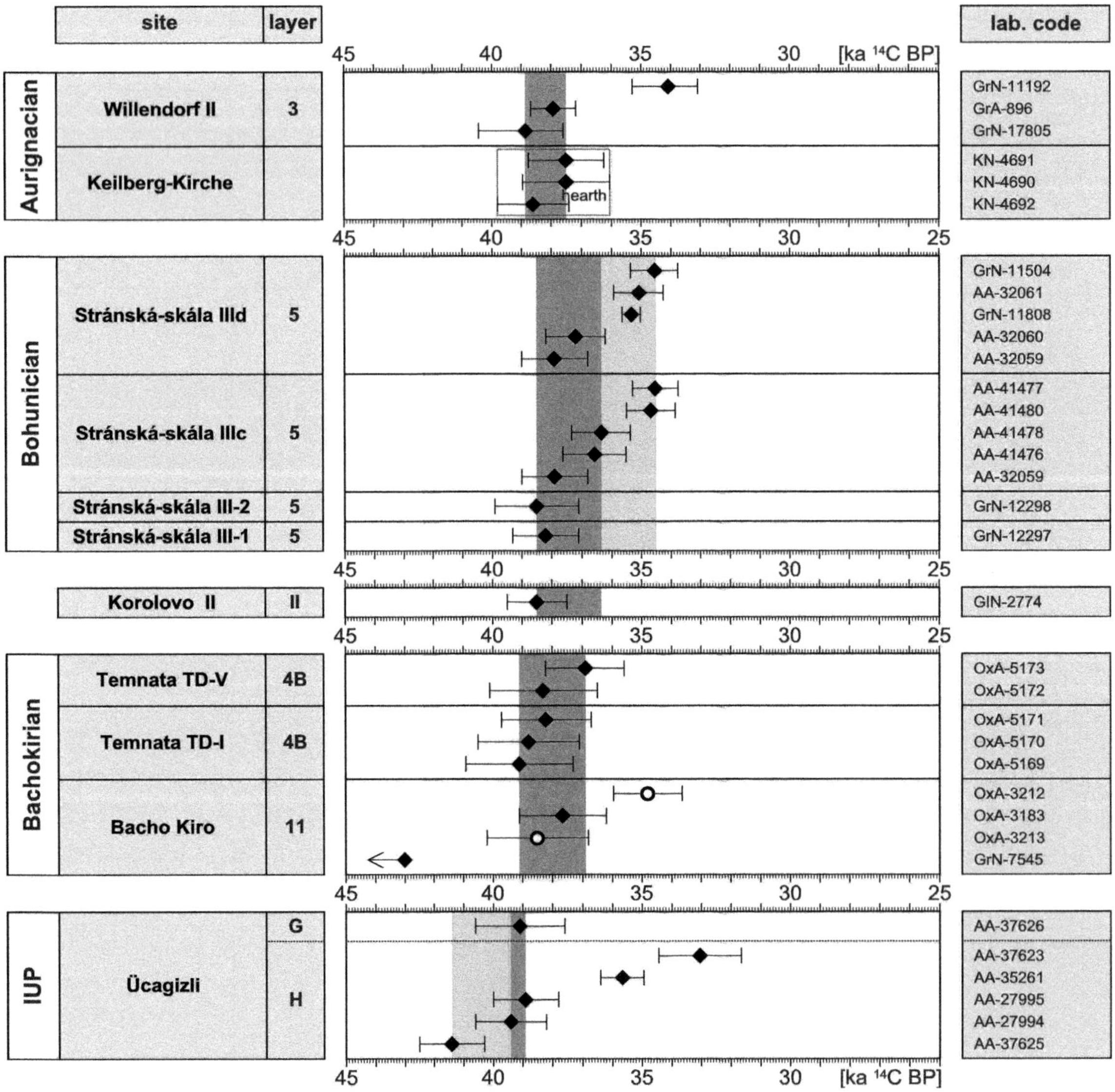

Fig. 22.10 Selection of results from radiometric dating of the Initial Upper Palaeolithic (IUP) of Üçağizli, Bachokirian and Bohunician assemblages, and from the early Aurignacian of Keilberg-Kirche and Willendorf II (KS 3) (see: Table 22.4), given with 1σ-standard deviation. Dates on bone (*circles*: open – without traces of hominin activity); dates on charcoal (*diamonds*); *Dark grey* shading: most likely dating range; *Light grey* shading: less likely dating range

For the Czech site of Bohunice-Kejbyly I a single radiocarbon date on charcoal from the "lower soil" horizon may date the Bohunician assemblage to 40,173 ± 1,200 ^{14}C BP (Q-1044; Valoch 1976). Again, at Bohunice-Kejbyly II a charcoal specimen from the "lower soil" horizon gave a result of 41,400 + 1,400/– 1,200 ^{14}C BP (GrN-6802; Valoch 1976; Table 22.4; Fig. 22.10). In both cases it must be considered whether the dated charcoal might represent somewhat older material reworked into the soil.

In contrast to earlier interpretations, the assemblage of Korolevo II, layer II, in Transcarpathian Ukraine, dated to 38,500 ± 1,000 ^{14}C BP (GIN-2774; Usik et al. 2006; Table 22.4; Fig. 22.10) is now also identified as a "Bohunician-type" industry. Assumed to be of similar age is the site of Korolevo I, I-a, which is attributed to a non-Aurignacian EUP (Usik et al. 2006). The blade-dominated industry from area A, level 3, at Sokirnitsa I, which is dated by a highly consistent series of charcoal radiocarbon measurements to around 38,880 ± 110 ^{14}C BP (KI-10837; Usik et al. 2006), also falls within this time range.

Protoaurignacian – "Aurignacien 0" – "Fumanian"

It had long been recognized that lithic material from the Austrian site Krems-Hundssteig (Strobl 1901; Strobl and Obermaier 1909) shows a number of specific typological similarities with Mediterranean "Aurignacian" sites now recognized as forming a separate Protoaurignacian group (Laplace 1970; Hahn 1977; Broglio 2000; Demidenko 2002). A recent study of the Hundssteig material confirms that at least some elements of the large, probably mixed assemblage can indeed be assigned by technological and typological criteria to the Protoaurignacian of the northern Mediterranean (Teyssandier 2003, 2006).

The largest series of dates for the Protoaurignacian comes from layer A2 at Grotta di Fumane in Northern Italy (Fig. 22.12). Here, the Protoaurignacian assemblage overlies a long MP sequence with dates on charcoal samples between ca. 42.0 ka ^{14}C BP (layer A11) and slightly older than 33.0 ka ^{14}C BP (layer A4). Radiocarbon measurements for horizon A2 itself scatter between 36.8 and 31.3 ka ^{14}C BP with few apparent outliers (e.g. UtC-1774: 40,000 + 4,000/–3,000 ^{14}C BP) and probably allow the distinction of a series older than ca. 35.4–33.6 ka ^{14}C BP from a younger one. The best contextual information for the Protoaurignacian is provided by Hearth S14 from the interior of the cave. Of seven charcoal samples taken here, five measurements give a pooled mean of 34,164 ± 281 ^{14}C BP with the oldest measurement from this feature dating to 36,800 + 1,200/–1,400 ^{14}C BP (UtC-2688). The fact that some of the Protoaurignacian radiocarbon measurements from the Grotta di Fumane produced dates significantly older than those of the underlying MP horizons has been interpreted as resulting from extreme variations of past atmospheric ^{14}C levels at the time of the Laschamp geomagnetic excursion (Giaccio et al. 2006; see above).

At the north-western Italian site of Riparo Mochi and at the Grotta Paglicci (Mussi et al. 2006; see above) radiocarbon dates place the Protoaurignacian close to 35.0 ka ^{14}C BP (Table 22.5; Fig. 22.11), with the three oldest Mochi dates forming a tight cluster between 35.7 and 34.7 ka ^{14}C BP and layer 24 at Paglicci dated to 34,300 + 900/–800 ^{14}C BP (UtC-?).

In south-western France, two radiocarbon measurements on charcoal obtained for the Protoaurignacian assemblage of layer 4d at Isturitz (Zilhão and d'Errico 1999) have a weighted mean of 35,490 ± 413 ^{14}C BP; (Table 22.5; Fig. 22.11). The oldest date (Ly-1898: >35,800 ^{14}C BP) for Aurignacian level G at Tournal provides a *terminus ante quem* for the Protoaurignacian assemblage in the underlying levels F and C (Tavoso 1976).

In the case of Abric Romaní in Catalonia G. Laplace and N. Soler suggested that the small assemblage from layer 2, might be a palimpsest of different occupations, one Aurignacian and another during the late Gravettian, because of similarities between six backed tools and Gravettian points (cited in Canal i Roquet and Carbonell i Roura 1989; cf Carbonell et al. 1994). This "Aurignacian" layer from the initial excavations of A. Romaní between 1909 and 1929 is today recognized as Protoaurignacian. Layer 2 (now labelled "level A") is still partly preserved; it produced "abundant faunal remains, dispersed charcoal and artifacts" (Bischoff et al. 1994: 544) and most likely represents a living floor sealed by overlying travertine. Five of the seven radiocarbon measurements on charcoal from remnant level A in three different locations (Table 22.5; Fig. 22.11) give a weighted mean of 36,644 ± 373 ^{14}C BP, with individual dates ranging from 37,290 ± 990 ^{14}C BP (AA-7395) to 36,390 ± 629 ^{14}C BP (pooled from AA-8037A and AA-8037B). These dates contrast greatly with two significantly younger measurements from the New Zealand Waikato radiocarbon laboratory (NZA-1817; NZA-1818), possibly due to contamination with more recent carbon. The laboratory also produced a date of 36,590±640 ^{14}C BP (NZA-2311) which is contained in the weighted mean given above. The Abric Romaní dates are stratigraphically consistent with the radiocarbon age of a charcoal sample (USGS-2839: 36,600 ± 1,300 ^{14}C BP) embedded in travertine dated to between 42.9 and 39.1 ka (corrected) U/Th BP (Bischoff et al. 1994; Carbonell et al. 1994).

The Abric Romaní Protoaurignacian assemblage closely resembles that of L'Arbreda, especially in the presence of Dufour bladelets (Carbonell et al. 2000, 18). The Protoaurignacian horizon at of L'Arbreda produced a consistent group of radiocarbon age determinations significantly older than 36.5 ka ^{14}C BP (Bischoff et al. 1989; Canal i Roquet and Carbonell i Roura 1989; Table 22.5; Fig. 22.11). Four charcoal samples taken from

Table 22.5 Selection of results from radiometric dating of Protoaurignacian assemblages and the Early Upper Palaeolithic (EUP) of Kostenki 14.

Layer	Method	No.	Lab.-No.	^{14}C age	STD+	STD–	^{13}C	Material	Location	Assoc. ind.	Source
La Viña, E											
XIII	^{14}C		Ly-6390	36,300	750	750		Charcoal		Protoaurignacien	Fortea Pérez (1999)
Cueva Morín, E											
8	^{14}C		GifA-96263	36,590	1100	1100		Charcoal		Protoaurignacien	Maillo et al. (2001)
Isturitz, F											
4d	^{14}C	1	GifA-98233	34,630	560	560		Bone, burnt	V1	Protoaurignacien	Zilhão and d'Errico (1999)
4d	^{14}C	2	GifA-98232	36,510	610	610		Bone, burnt	U27	Protoaurignacien	Zilhão and d'Errico (1999)
4d	*^{14}C*	*(1–2)*	*Weighted mean*	*35,490*	*413*	*413*	*1.39*	*t-Value*		*Protoaurignacien*	
Abric Romaní, E											
2 (A)	^{14}C	1	AA-8037A	35,400	810	810		Charcoal		Protoaurignacien	Bischoff et al. (1994)
2 (A)	^{14}C	2	AA-8037B	37,900	1000	1000		Charcoal		Protoaurignacien	Bischoff et al. (1994)
2 (A)	^{14}C	3	NZA-2311	36,590	640	640		Charcoal		Protoaurignacien	Bischoff et al. (1994)
2 (A)	^{14}C	4	AA-6608	36,740	920	920		Charcoal		Protoaurignacien	Bischoff et al. (1994)
2 (A)	^{14}C	5	AA-7395	37,290	990	990		Charcoal		Protoaurignacien	Bischoff et al. (1994)
2 (A)	*^{14}C*	*(1–5)*	*Weighted mean*	*36,644*	*374*	*374*	*1.39*	*t-Value*		*Protoaurignacien*	
2 (A)	U-series			43,000	1000	1000		Travertine		Protoaurignacien	Bischoff et al. (1994)
L'Arbreda, E											
H (B1)	^{14}C	1	OxA-3730	35,480	820	820		Bone		Protoaurignacien	Bischoff et al. (1989)
H (B1)	^{14}C	2	OxA-3729	37,340	1000	1000		Bone		Protoaurignacien	Bischoff et al. (1989)
H (B1)	^{14}C	3	AA-3779	37,700	1000	1000		Charcoal		Protoaurignacien	Bischoff et al. (1989)
H (B1)	^{14}C	4	AA-3780	37,700	1000	1000		Charcoal		Protoaurignacien	Bischoff et al. (1989)
H (B1)	^{14}C	5	AA-3782	38,700	1200	1200		Charcoal		Protoaurignacien	Bischoff et al. (1989)
H (B1)	^{14}C	6	AA-3781	39,900	1300	1300		Charcoal		Protoaurignacien	Bischoff et al. (1989)
H (B1)	*^{14}C*	*(3–6)*	*Weighted mean*	*38,307*	*552*	*552*	*1.13*	*t-Value*		*Protoaurignacien*	
Grotte Tournal, F											
G	^{14}C		Ly-1898	>35,800				Charcoal		Protoaurignacien	Evin et al. (1983)
Esquicho-Grapaou, F											
SLC 1b	^{14}C		MC-2161	34,540	2000	2000		Charcoal		Protoaurignacien	Delibrias and Evin (1980)
Riparo Mochi, I											
G	^{14}C	1	OxA-3588	32,280	580	580		Charcoal		Protoaurignacien	Mussi et al. (2006)
G	^{14}C	2	OxA-3589	33,400	750	750		Charcoal		Protoaurignacien	Mussi et al. (2006)
G	^{14}C	3	OxA-3590	34,680	760	760		Charcoal		Protoaurignacien	Mussi et al. (2006)
G	^{14}C	4	OxA-3592	34,870	800	800		Charcoal		Protoaurignacien	Mussi et al. (2006)
G	^{14}C	5	OxA-3591	35,700	850	850		Charcoal		Protoaurignacien	Mussi et al. (2006)
G	*^{14}C*	*(3–5)*	*Weighted mean*	*35,045*	*462*	*462*	*0.68*	*t-Value*		*Protoaurignacien*	

Grotta di Fumane, I											
D6	^{14}C		OS-5872	37,100	240	240		Marine shell (personal ornament)		Aurignacian	Giaccio et al. (2006)
A2	^{14}C	1	OS-5999	32,000	90	90		Marine shell (personal ornament)		Protoaurignacian	Giaccio et al. (2006)
A2	^{14}C	2	OS-5871	32,700	140	140		Marine shell (personal ornament)		Protoaurignacian	Giaccio et al. (2006)
A2	*^{14}C*	*(1–2)*	*Weighted mean*	*32,205*	*76*	*76*	*3.11*	*t-Value*		*Protoaurignacien*	
A2	^{14}C	1	UtC-2048	36,500	600	600		Charcoal	Near hearth S14	Protoaurignacian	Giaccio et al. (2006)
A2	^{14}C	2	OxA-6566	31,900	1100	1100		Charcoal	Hearth S14, top	Protoaurignacian	Giaccio et al. (2006)
A2	^{14}C	3	OxA-8052	34,120	460	460		Charcoal	Hearth S14, top	Protoaurignacian	Giaccio et al. (2006)
A2	^{14}C	4	UtC-2688	36,800	1200	1400		Charcoal	Hearth S14, level A	Protoaurignacian	Giaccio et al. (2006)
A2	^{14}C	5	UtC-2689	35,400	1100	1300		Charcoal	Hearth S14, level B1	Protoaurignacian	Giaccio et al. (2006)
A2	^{14}C	6	UtC-2690	34,200	900	900		Charcoal	Hearth S14, level B2	Protoaurignacian	Giaccio et al. (2006)
A2	^{14}C	7	OxA-6465	31,620	500	500		Charcoal	Hearth S14, base	Protoaurignacian	Giaccio et al. (2006)
A2	^{14}C	8	OxA-8053	33,640	440	440		Charcoal	Hearth S14, base	Protoaurignacian	Giaccio et al. (2006)
A2	*^{14}C*	*(3–6, 8)*	*Weighted mean*	*34,164*	*281*	*281*	*2.14*	*t-Value*	*Hearth S14*	Protoaurignacian	
Grotta di Paina, I											
9	^{14}C	2	UtC-2042	37,900	800	800		Bone		Protoaurignacien	Mussi et al. (2006)
9	^{14}C	1	UtC-2695	38,600	1400	1800		Bone		Protoaurignacien	Mussi et al. (2006)
9	*^{14}C*	*(1–2)*	*Weighted mean*	*38,015*	*731*	*731*	*0.31*	*t-Value*		Protoaurignacien	
Paglicci, I											
24	^{14}C	1	UtC-?	34,300	900	800		Charcoal	B2-1	Protoaurignacien	Mussi et al. (2006)
Krems-Hundssteig, AU											
Brown layer	^{14}C		KN-654	35,500	2000	2000		Charcoal		Protoaurignacien	Hahn (1977)
Kostenki 14, RU											
IVb	^{14}C	1	OxA-9568	32,600	280	280		Charcoal		EUP	Sinitsyn (2003)
IVb	^{14}C	2	GrA-13302	34,940	630	630		Charcoal		EUP	Sinitsyn (2003)
IVb	^{14}C	3	OxA-9569	35,280	330	330		Charcoal		EUP	Sinitsyn (2003)
IVb	^{14}C	4	GrA-15957	36,040	250	250		Charcoal		EUP	Sinitsyn (2003)
IVb	^{14}C	5	GrA-15961	36,540	270	270		Charcoal		EUP	Sinitsyn (2003)
IVb	*^{14}C*	*(2–5)*	*Weighted mean*	*35,970*	*155*	*155*	*1.89*	*t-Value*		EUP	
IVb	IRSL		UIC-1128	47,730	3480	3480				EUP	Anikovich et al. (2007)
Hor. hearths	^{14}C	1	GrA-15958	35,330	240	240		Charcoal		EUP	Sinitsyn (2003)
Hor. hearths	^{14}C	2	GrA-15962	35,870	250	250		Charcoal		EUP	Sinitsyn (2003)
Hor. hearths	^{14}C	3	GrA-15965	36,010	250	250		Charcoal		EUP	Sinitsyn (2003)
Hor. hearths	^{14}C	4	GrA-15956	36,320	270	270		Charcoal		EUP	Sinitsyn (2003)
Hor. hearths	^{14}C	5	GrA-10948	37,240	430	430		Charcoal		EUP	Sinitsyn (2003)
Hor. hearths	*^{14}C*	*(1–5)*	*Weighted mean*	*35,964*	*121*	*121*	*2.86*	*t-Value*		*EUP*	
Hor. hearths	IRSL		UIC-749	34,170	2700	2700				EUP	Sinitsyn (2003)
Hor. hearths	IRSL		UIC-748	44,880	3580	3580				EUP	Sinitsyn (2003)

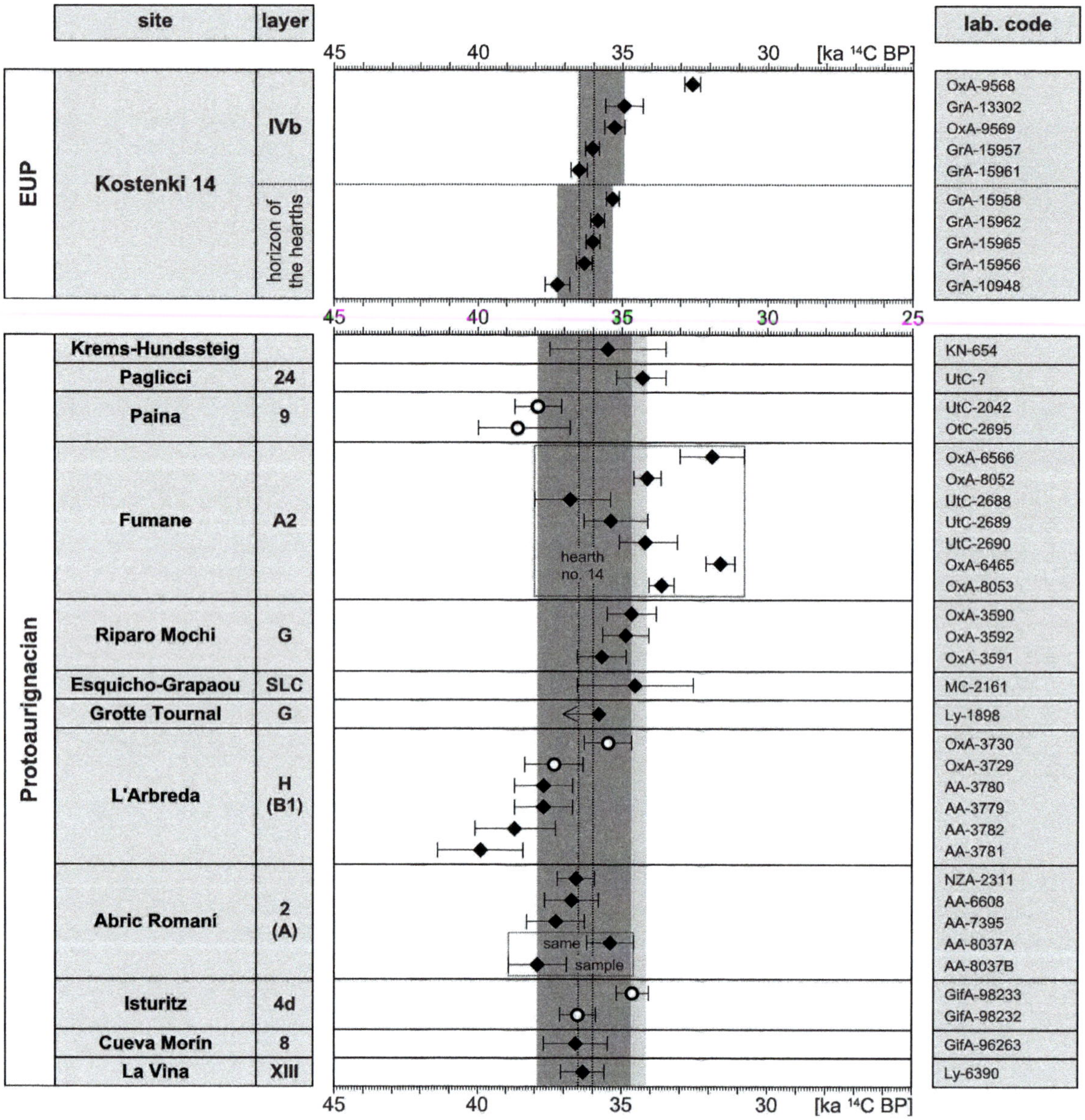

Fig. 22.11 Selection of results from radiometric dating of Protoaurignacian assemblages and the Early Upper Palaeolithic (EUP) of Kostenki 14 (see: Table 22.5), given with 1σ-standard deviation. Dates on bone (*circles*: open – without traces of hominin activity); dates on charcoal (*diamonds*); unknown material (*asterisks*); The *dotted lines* divide more secure evidence of personal ornaments <36.0 ka ^{14}C BP from less secure evidence >36.5 ka ^{14}C BP. *Dark grey* shading: most likely dating range; *Light grey* shading: less likely dating range

a 5 cm spit (level B1[H], 5.50–5.55 m below surface) in square E2 (BE 111) immediately sheltered by the travertine wall (Bischoff et al. 1989) produced a weighted mean of 38,307±552 ^{14}C BP. The samples came from the very base of the Protoaurignacian deposits, immediately next to the main profile published by Canal i Roquet and Carbonell i Roura (1989). A radiocarbon measurement on bone assigned to the same cultural unit produced a somewhat younger age (OxA-3730: 35,480±820 ^{14}C BP), but another measurement on bone from a few metres away (CE 103) dates level (H) to 37,340±1,000 ^{14}C BP (OxA-3729). A measurement (Gif-6422) obtained earlier is considered to be too young.

It has been discussed whether the L'Arbreda series of dates gives the age of the Protoaurignacian occupation itself or simply provides a *terminus post quem* (Zilhão and d'Errico 1999; Zilhão 2006b). Nevertheless, the consistent radiometric dating and stratigraphic evidence from both L'Arbreda and Abric Romaní strongly indicates that the earliest Protoaurignacian in the North of the Iberian Peninsula dates between 38.0 and 37.0 ka^{14}C BP. An equally early appearance of the Protoaurignacian may be indicated in northern Italy at the Grotta di Paina (Veneto), where two dates on bone from the base of archaeological level 9 provide a *terminus post quem* of 38,600+1,400/−1,800 ^{14}C BP

(UtC-2695) and 37,900±800 ^{14}C BP (UtC-2042) (Table 22.5; Fig. 22.11; Mussi et al. 2006).

The Emergence of the Aurignacian

In view of the interpretation of several early laminar industries (Bachokirian, Bohunician and Protoaurignacian/"Aurignacian 0") as entities technologically distinct from the early Aurignacian (Aurignacien I) (e.g. Teyssandier 2003, 2005, 2006), assemblages of the former type are excluded from the following discussion.

Many authors argue that the appearance of "innovative" Aurignacian technology in western Eurasia can be equated with the expansion of AMH into the region from the Near East. This dispersal has been seen variously as occurring across a broad front (Bocquet-Appel and Demars 2000) or by penetration along specific major axes such as the Danube (Conard and Bolus 2003) and the Don (Anikovich et al. 2007, 225) river systems.

In the "Danube Corridor" hypothesis (Conard and Bolus 2003) radiocarbon dates significantly older than 35.0 ka ^{14}C BP for the Austrian site of Willendorf II, cultural layer (*Kulturschicht*: KS) 3, and the Aurignacian assemblages from the southern German sites Keilberg-Kirche and Geißenklösterle (layers IIIb–IIa) have been interpreted as reflecting the precocious movement of AMH into the region.

Willendorf II in Lower Austria has provided one of the most important archaeological sequences for the second half of OIS 3 in Central Europe and chronostratigraphy at the site is well established (Damblon et al. 1996; Fig. 22.14), with the basal archaeological horizons (KS 1–KS 4) falling at the MP–UP transition. The youngest of these, KS 4, is assigned unquestionably to the Aurignacian with radiocarbon dates between ca. 32.0 and 30.0 ka ^{14}C BP. Basal KS 1 yielded only three undiagnostic artefacts, while the very small laminar assemblage from KS 2 (32 pieces) lacks "typical Aurignacian or transitional forms" (Haesarts and Teyssandier 2003, 144; Teyssandier et al. 2006, 247).

Of crucial interest to the present question is the interpretation of the finds from KS 3, which is clearly stratified below KS 4 and has traditionally been attributed to the (early) Aurignacian (e.g. Felgenhauer 1959; Broglio and Laplace 1966; Hahn 1977; Teyssandier 2003). This view was challenged by Zilhão and d'Errico (1999) who are sceptical of the radiometric age estimate for the layer and also question the archaeological attribution of the small inventory of only 43 artefacts (Felgenhauer 1959). However, detailed study of the KS 3 assemblage supports the view that the artefacts "present some technical particularities that are very close to what is seen in the Aurignacian technocomplex" and concludes that "some specimens are very similar to certain pieces from layer 4" (Haesaerts and Teyssandier 2003, 146). The discovery of hitherto unpublished artefacts from the Szombathy, Bayer and Obermaier excavations has enlarged the KS 3 assemblage to a total of some 500 lithic pieces and their preliminary technological and typological evaluation supports their interpretation as fully Aurignacian (Nigst 2004, 2006).

Criticism has been made of the radiometric age of KS 3 (Zilhão and d'Errico 1999), for which three radiocarbon measurements on charcoal are available (Damblon et al. 1996). Two dates of 38,880+1,530/−1,280 ^{14}C BP (GrN-17805) and 37,930±750 ^{14}C BP (GrA-896) provide a weighted mean slightly older than 38.1 ka ^{14}C BP, while a third result is significantly younger (GrN-11192: 34,100+1,200/−1,000 ^{14}C BP). Whereas most recent studies (Haesaerts and Teyssandier 2003; Nigst 2006) tend to accept the age of around 38.0 ka ^{14}C BP for Willendorf II/KS 3, Zilhão and d'Errico (1999) suggested the possibility that the older samples may be reworked and that GrN-11192 represents the "true" age of the assemblage. New AMS-dates on charcoal samples taken during recent fieldwork (www.willendorf-project.org) may further clarify the age of Willendorf II/KS 3.

Further to the West, at Keilberg-Kirche, close to Regensburg in Bavaria, lithic material was recognized and collected over several decades from an extended surface find scatter under ploughing. A small (4 m²) rescue excavation in 1987 recovered burnt bone and charcoal from what was probably an *in situ* hearth, together with 215 artefacts, among them typically Aurignacian forms and burnt pieces (Uthmeier 1996, 1998, 2004). The site was documented in greater detail by a subsequent (1991) sondage program over an area of some 25 m by 10 m which recovered a further 721 artefacts (Uthmeier 1996). Charcoal samples from the 1987 excavation provided three statistically identical radiocarbon measurements (KN-4692: 38,600±1,200 ^{14}C BP; KN-4690: 37,500 ± 1,450 ^{14}C BP; KN-4691: 37,500 ± 1,250 ^{14}C BP) with a weighted mean of ca. 37,922±743 ^{14}C BP (Table 22.4), i.e. approximately the same age as the older age estimate for Willendorf II, KS 3. The presence of a few, not very diagnostic but probably MP artefacts in some test pits (Uthmeier 1996) led Zilhão and d'Errico (1999) to suggest that the hearth might in fact be linked to an older (pre-Aurignacian) occupation phase which left almost no other traces (Zilhão and d'Errico 1999). Weakening this argument is the fact that during the 1991 field work typical Aurignacian artefacts were documented in spatial association with patches of burnt loess immediately around the hearth uncovered earlier. The most parsimonious interpretation remains that the age of the Aurignacian assemblage is given by the dates for the hearth (Uthmeier 1996).

An important stratigraphic sequence for the Central European Aurignacian is that at the Geißenklösterle cave in

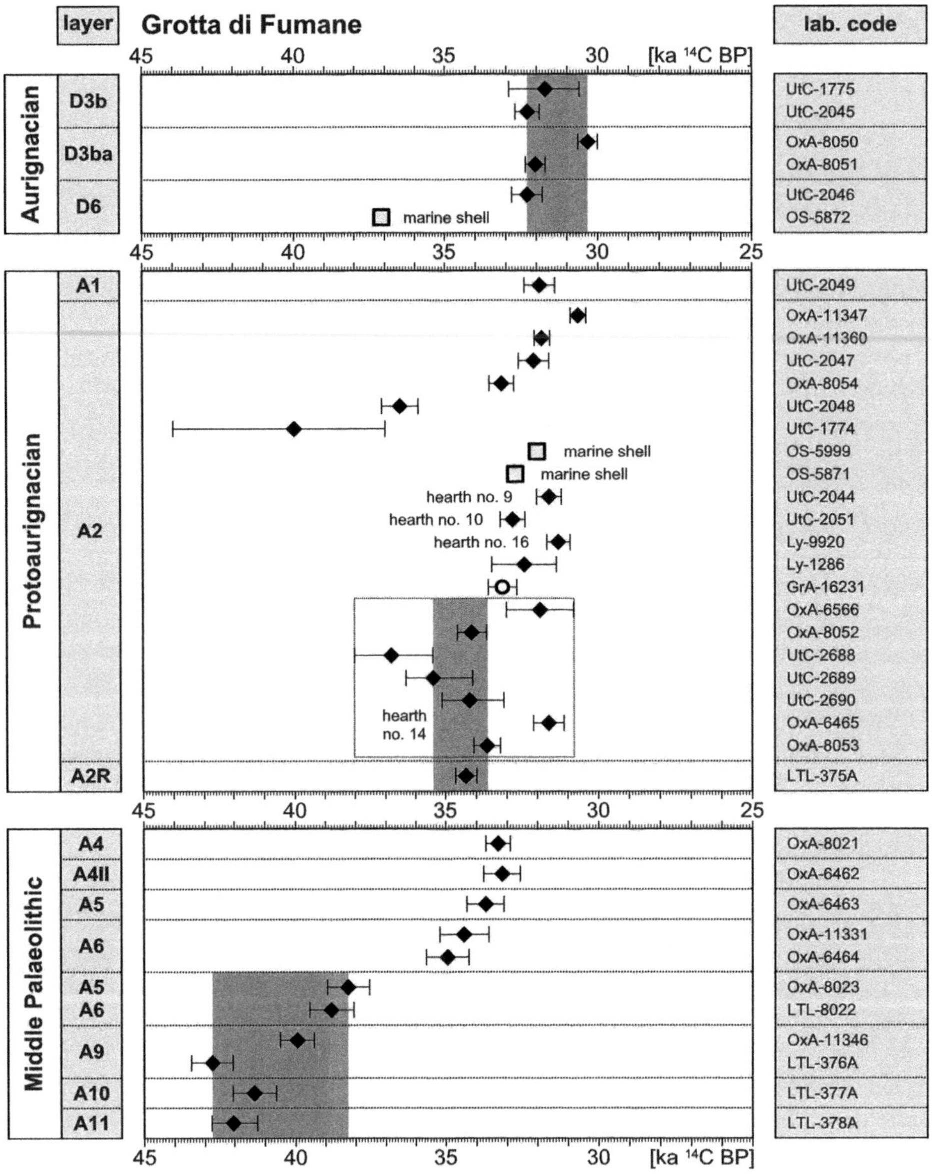

Fig. 22.12 Radiocarbon age determinations for the Middle to Upper Palaeolithic transition at Grotta di Fumane, Italy, given with 1σ-standard deviation. Dates on bone (*circles*: open – without traces of hominin activity); dates on charcoal (*diamonds*); dates on shell (*squares*); *Dark grey* shading: most likely dating range

south-western Germany (Hahn 1988b; Conard et al. 2003a). In his site monograph Joachim Hahn (1988b) carefully reconstructed archaeological horizons (AH) based on refits of the lithic material between different geological horizons (GH). Despite doubts as to the validity of Hahn's assemblages (Zilhão and d'Errico 1999), recent excavations with even more detailed recording of finds and further comprehensive refitting work have confirmed the overall reliability of the Geißenklösterle stratigraphical sequence and the integrity of the specific archaeological inventories (Conard et al. 2003a, 2006).

Hahn distinguished the Aurignacian AH II, which provided all the figurative art, from the stratigraphically earlier AH III. He suggested that horizon AH II can be compared with the French early Aurignacien I (particularly due to similarities of the organic technology) but that the interpretation of AH III is more difficult. Hahn originally saw no clear parallels with the Mediterranean Protoaurignacian and instead considered the possibility of a progressive development within "the Aurignacian". AH III would then represent a phase of the Aurignacian preceding the "classical Aurignacien I", being without Dufour bladelets and possessing only few bone tools and different types of pendants (Hahn 1988b, 246). Probably not for technological/typological reasons and perhaps influenced by very old radiocarbon dates for AH III obtained in the early 1990s, Hahn subsequently

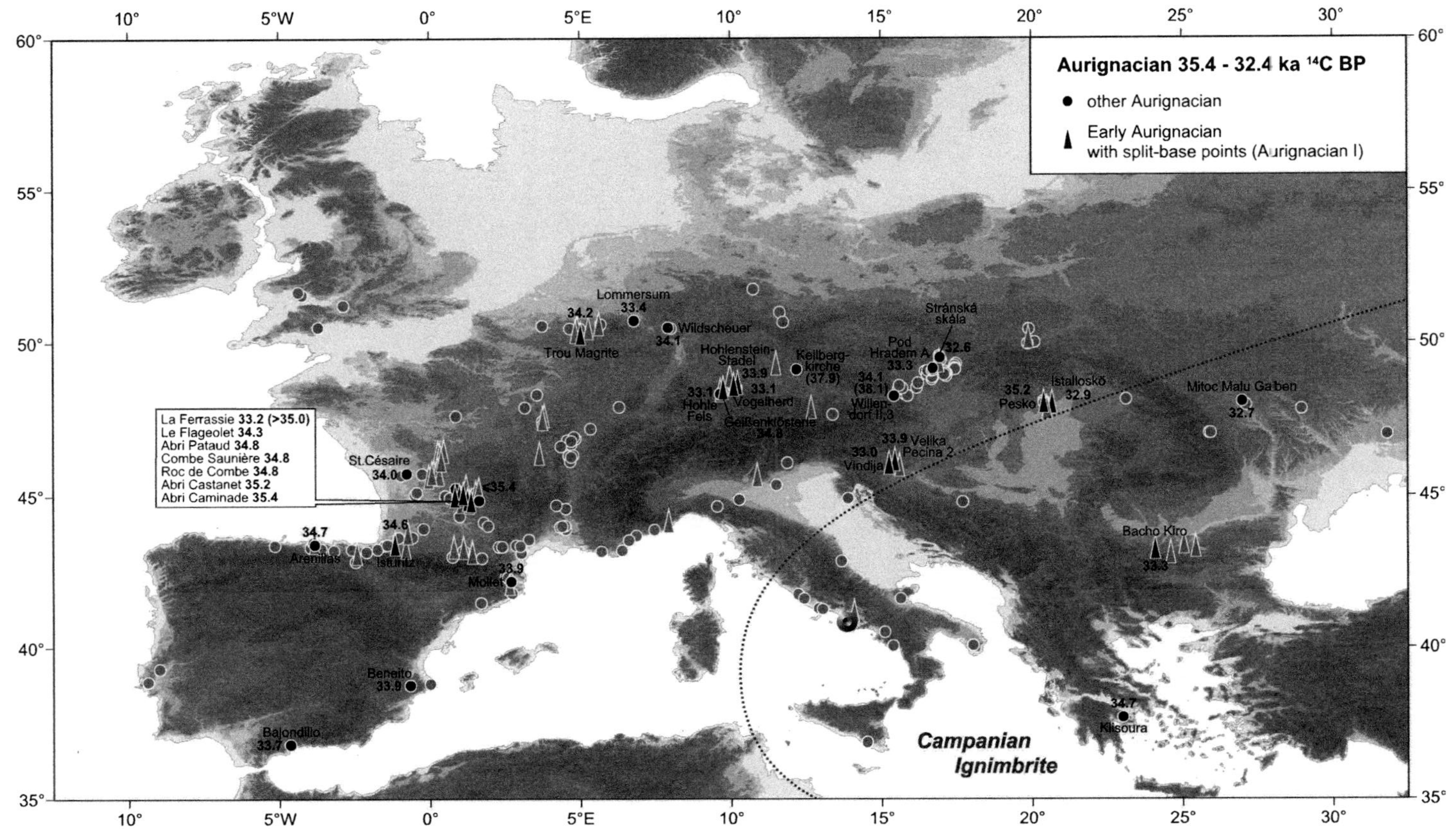

Fig. 22.13 Earliest radiocarbon dated evidence for European Aurignacian industries (Compiled after different authors). Map based on SRTM data; sea level lowered by 75 m

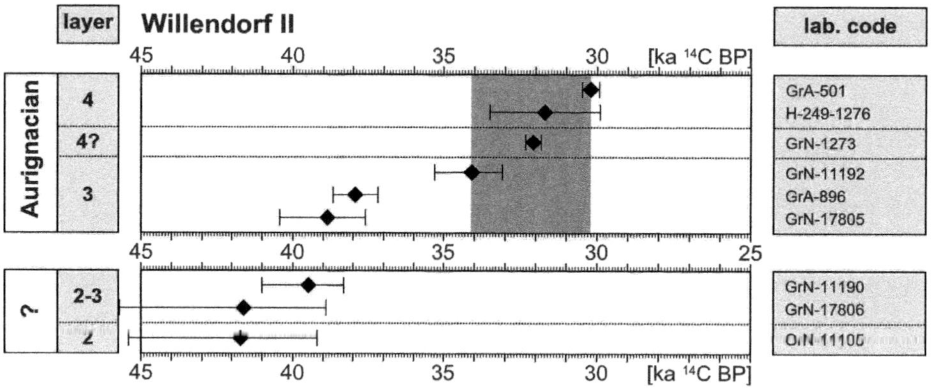

Fig. 22.14 Radiocarbon age determinations for the Middle to Upper Palaeolithic transition at Willendorf II, Austria, given with 1σ-standard deviation. Dates on charcoal (*diamonds*); *Dark grey* shading: most likely dating range

did use the term Protoaurignacian (Hahn 1995). This alteration of the initial designation of AH III almost certainly lies behind later doubts about the fully Aurignacian character of this horizon and attempts to assign it to other complexes e.g. a "Pre-Aurignacian" or "Protoaurignacian" (Kozłowski and Otte 2000; Zilhão and d'Errico 1999, 2003a, b). More recent analyses of technological aspects of the lithic and organic material indeed confirm the fully Aurignacian character of the entire Geißenklösterle sequence from horizons AH IIIb–IIa (e.g. Bolus 2003; Teyssandier 2005, 2006). Teyssandier and Liolios (2003) conclude that in AH III "The technical processes involved in bone working … are unsophisticated" and that in the case of ivory "implemented processes are not as complex as in reconstructed level II". Nonetheless, antler working in AH III "documents the use of a technique specific to the Aurignacian" and there are good arguments for the "existence of Aurignacian levels predating the split-based point facies" (Teyssandier and Liolios 2003).

The age of the Geißenklösterle Aurignacian layers remains a matter of discussion since the large series of 34 AMS and conventional radiocarbon measurements (on bone and charred bone) from this sequence spans the entire period between 40,200±1,600 ^{14}C BP (OxA-4595) and 28,640+380/−360 ^{14}C BP (KIA-8962) (Fig. 22.15). They are seen as providing the earliest evidence for the Aurignacian in Central Europe (Hahn 1995; Bolus and Conard 2001; Conard and Bolus 2003; Conard et al. 2003a), an interpretation supported by TL results of ca. 40.0 ka BP on burnt artefacts from AH III (Richter et al. 2000).

Nonetheless, in their pan-European evaluation of the age of the Aurignacian, Zilhão and d'Errico (2003b), conclude that "the Aurignacian of the Swabian Jura in fact is no earlier than ca. 36.5 ka" (Zilhão and d'Errico 2003b, 328), while for Verpoorte even this "more moderate age of 36.5 ka, the critical boundary for the Early Aurignacian, … is problematic" (Verpoorte 2005, 271). While confirming the fully Aurignacian character of the AH III assemblage, Teyssandier (2005) also rejects the supposed early dating and suggests a younger age of ca. 35.5–33.0 ka ^{14}C BP (with AH II lying between 33.0 and 31.5 ka ^{14}C BP).

It is important to note that the Geißenklösterle radiocarbon results are quite often inconsistent with their stratigraphic provenance (Conard et al. 2003a, 173). Conard and colleagues explain these age-distortions as being due to a "Middle to Upper Palaeolithic dating anomaly" (cf Conard and Bolus 2003) indications for which have been claimed in a speleothem record from the Bahamas (Beck et al. 2001) or in the "young" radiocarbon measurements from layers below the Campanian Ignimbrite (see above, Fig. 22.3).

In fact, the majority of radiocarbon dates for AH IIIb–AH IIa falls between ca. 35.0 and 29.5 ka ^{14}C BP and is bracketed between results of 34,800+290/−280 ^{14}C BP (KIA-13074) from AH IIIa and 29,800±240 ^{14}C BP (KIA-8960) from AH IIb. The only exceptions are an apparent outlier in AH IIIb (KIA-8962), which falls into the age range of the MUP sequence above the Aurignacian, and seven dates significantly older than 35.0 ka ^{14}C BP scattered throughout the sub-horizons of AH III and AH II (one each from AH IIIb, IIIa, IIb and IIa, and three from AH III). Six of the seven dates obtained in the early 1990s (Hahn 1995) have large standard deviations (at least ±1,000 ^{14}C BP and as much as ±3,560 ^{14}C BP) and are therefore far less significant for any age estimate of the Geißenklösterle Aurignacian.

Only one of the seven measurements before 35.0 ka ^{14}C BP (KIA-16032: 36,560+410/−390 ^{14}C BP) was obtained in recent years. This is on a roe deer metacarpal with an impact mark and is the oldest acceptable date from the Geißenklösterle sequence with direct relevance for past hominin actions. However, the species determination of the sample gives grounds for scepticism regarding its relevance for the Aurignacian, since the habitat demands of *Capreolus* are totally different from those of the typical glacial fauna comprising most of the archaeozoological assemblage of AH III & II (Niven 2003; cf Conard et al. 2006). Possibly

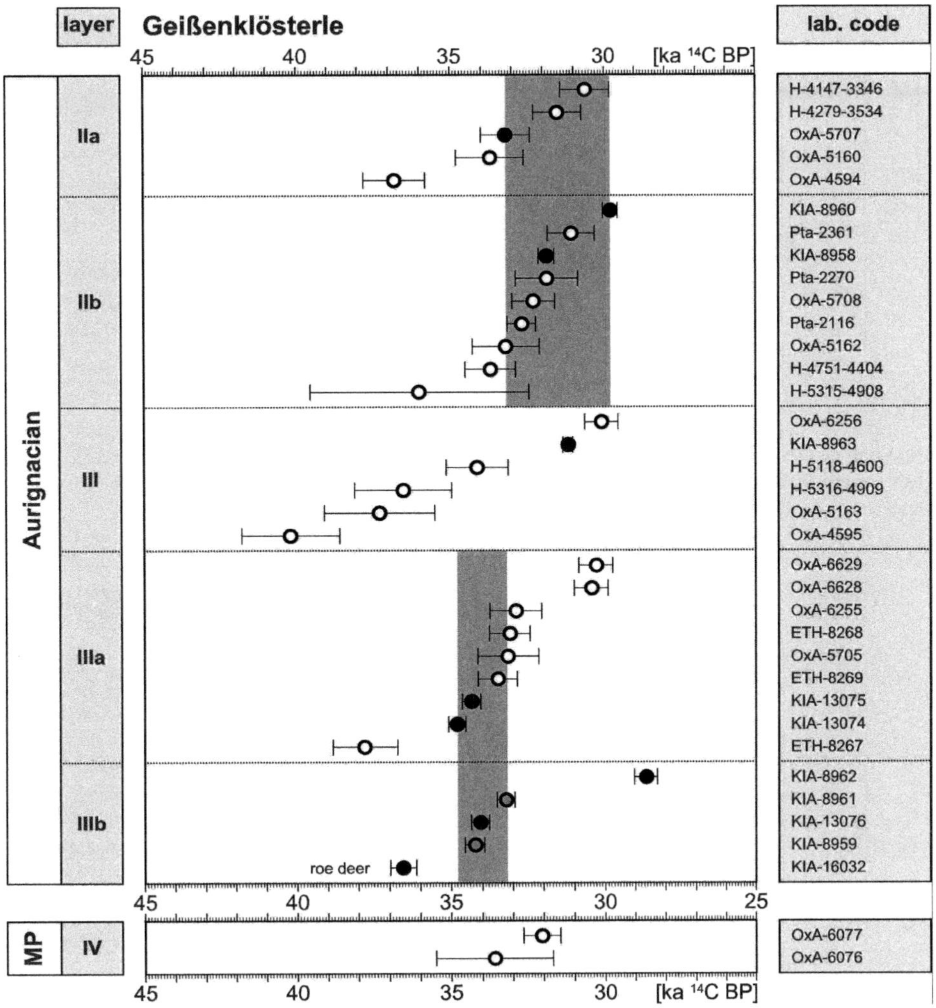

Fig. 22.15 Radiocarbon age determinations for the Middle Palaeolithic (MP) and Aurignacian layers of Geißenklösterle cave, Germany, given with 1σ-standard deviation. Dates on bone (*circles*: open – without traces of hominin activity; *grey fill* – "fresh break"; *black fill* – hominin modification [e.g. cut-marked and/or impact scars]); *Dark grey* shading: most likely dating range

the specimen represents material from an older (interstadial) event at the site (cf Conard et al. 2006; Table 1) reworked into AH IIIb and therefore of questionable cultural relevance (cf Zilhão and d'Errico 2003b, 335).

To summarize, the best estimate for the age of the Geißenklösterle Aurignacian is between ca. 35.0 and 29.5 ka ^{14}C BP. If the dates are selected for their relevance for hominin activity and correlated with stratigraphy they might be distinguished into an older cluster between ca. 34.8 and 33.2 ka ^{14}C BP (for AH III) and a younger one around 33.2–29.8 ka ^{14}C BP (for AH II).

The Aurignacian assemblages from Hohlenstein-Stadel (Fig. 22.16), Vogelherd (Fig. 22.17) and Hohle Fels at Schelklingen (Fig. 22.18) consistently produced radiocarbon dates >29.5 ka ^{14}C BP and <34.0 ka ^{14}C BP. Only two samples, one each from Vogelherd and Hohlenstein-Stadel, produced dates >35.0 ka ^{14}C BP. Sample KIA-8950 (36,910+490/–460 ^{14}C BP) from Hohlenstein-Stadel dates an unmodified metatarsal of elk (*Alces alces*) unearthed 19 m inside the cave in "spit 7" during Wetzel's initial excavations at the site. In the absence of any fine stratigraphic recording the relevance of this date for the age of the Aurignacian must be viewed sceptically, particularly since a date for a reindeer bone with a "fresh break" from the same spit (KIA-8949: 33,920+310/– 300 ^{14}C BP) is in full accordance with other dates available for the Lone valley Aurignacian (cf Niven 2006; cf Conard and Bolus 2006). At Vogelherd, a single sample on a cut-marked bovid or equid long bone fragment from level V gave a result of 35,810±710 ^{14}C BP (PL-0001337A), whereas all other measurements have turned out significantly younger than 34.0 ka ^{14}C BP (Conard and Bolus 2003; Conard et al. 2003b).

Seen critically, the radiocarbon record for the Swabian Aurignacian (cf Verpoorte 2005) agrees well with the dating of other German Aurignacian sites such as Lommersum (Hahn 1989) and Wildscheuer (Pettitt et al. 1998).

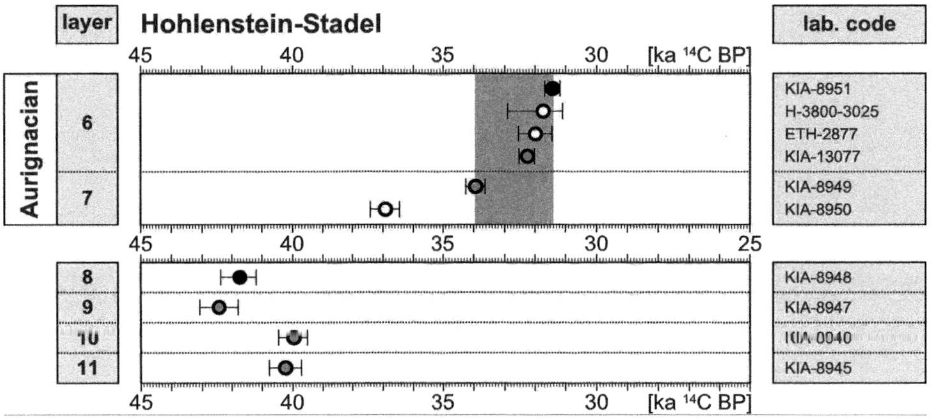

Fig. 22.16 Radiocarbon age determinations for the Aurignacian and underlying layers of Hohlenstein-Stadel cave, Germany, given with 1σ-standard deviation. Dates on bone (*circles*: open – without traces of hominin activity, *grey fill* – "fresh break"; *black fill* – hominin modification [e.g. cut-marked and/or impact scars]); *Dark grey* shading: most likely dating range

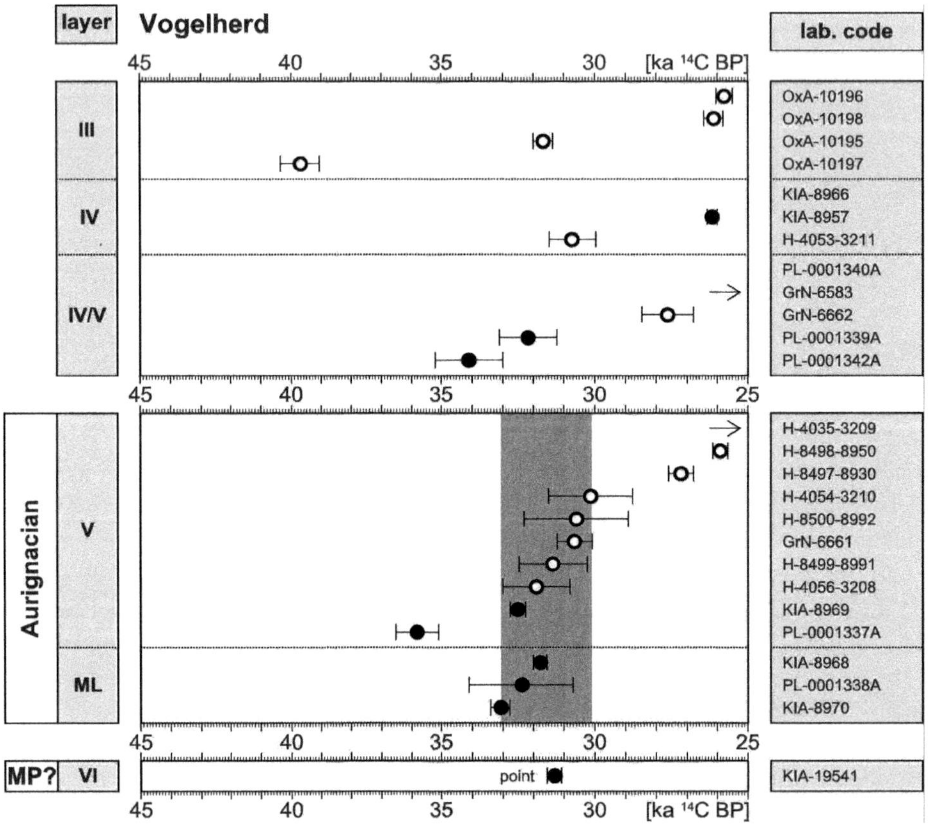

Fig. 22.17 Radiocarbon age determinations for the Middle Palaeolithic (MP), Aurignacian and overlying layers of Vogelherd cave, Germany, given with 1σ-standard deviation. Dates on bone (*circles*: open – without traces of hominin activity, *black fill* – hominin modification [e.g. cut-marked and/or impact scars]); *Dark grey* shading: most likely dating range

While the Central European dating evidence has been discussed here in detail, the evidence for an Aurignacian presence prior to 36.5 ka ^{14}C BP is everywhere rare and disputed (see discussion in Zilhão and d'Errico 1999; cf Djindjian et al. 2003). Although we find no good arguments against the validity of the radiocarbon dates of samples from Willendorf II, KS 3, and Keilberg-Kirche, their relevance for the age of the Central European Aurignacian can be regarded with scepticism in view of the overall European evidence dating the Aurignacian to a period younger than ca. 35.0 ka ^{14}C BP.

High ages for south-western European Aurignacian assemblages can be met with similar scepticism: At Caminade-Est, bone from the Aurignacian I assemblage with split-based points (layer F) is radiocarbon dated to 35,400 ± 1,100 ^{14}C BP

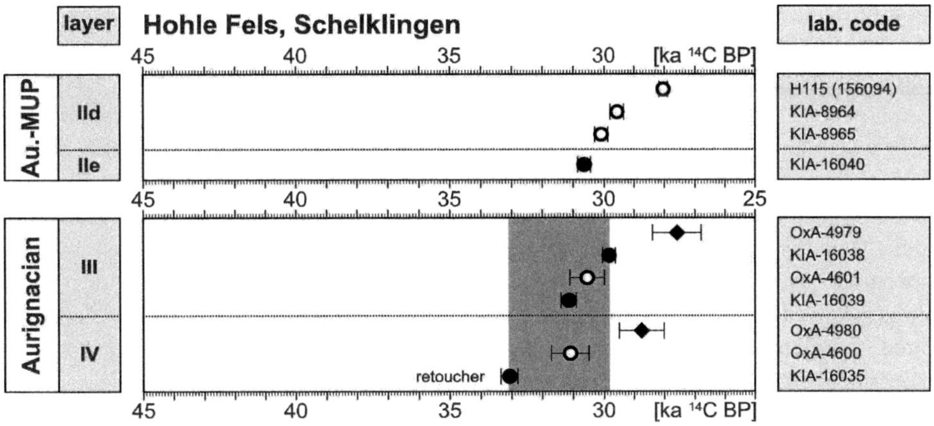

Fig. 22.18 Radiocarbon age determinations for the Aurignacian and Aurignacian to Mid-Upper Palaeolithic (MUP) layers of Hohle Fels cave, Germany, given with 1σ-standard deviation. Dates on bone (*circles*: open – without traces of hominin activity; *black fill* – hominin modification [e.g. cut-marked and/or impact scars]); *Dark grey* shading: most likely dating range

(GifA-97186; Rigaud 2001). A date of 37,200 ± 150 ^{14}C BP (GifA-97185) was obtained on bone from layer G, which underlies the Aurignacian I horizon. GifA-97185 cannot be interpreted as dating the Aurignacian since "systematic refitting work carried out by J.-G. Bordes (1998, 2000) demonstrated that around 30% of the archaeological material included in this layer comes from the underlying Mousterian deposits, to which the dated sample could conceivably be related" (Zilhão and d'Errico 1999, 17). Similarly, while the upper part of the basal layer (TIII) at Reclau Viver contains an Aurignacian I assemblage with split-based points (Canal i Roquet and Carbonell i Roura 1989), the single radiocarbon measurement of 40,000 ± 1,400 ^{14}C BP (OxA-3727) is from the base of this layer and may also date reworked bone irrelevant to the Aurignacian context (Zilhão and d'Errico 1999).

In summary, a critical evaluation of the radiocarbon database for the Aurignacian shows its quasi simultaneous appearance across the whole of Europe at ca. 35.0 ka ^{14}C BP (Fig. 22.13). This corresponds to the radiocarbon age of the CI eruption and it is tempting to speculate on a causal association between the two phenomena.

The Emergence of "Behavioural Modernity"

A number of apparently innovative features appears in the European Palaeolithic record at the time under investigation, some of which have been invested with particular significance in the discussion of the emergence of the Aurignacian, "behavioural modernity" and the question of Neanderthal – AMH abilities and interactions. As shown by a number of recent detailed studies, the Aurignacian is not monolithic but heterogeneous and use of this term may artificially unite what are diverse chronological, technological, social and possibly ethnic units (cf Vanhaeren and d'Errico 2006). Many of the elements which supposedly characterize the Aurignacian are regionally exclusive and absent at many sites identified with this entity. Additionally, some phenomena regarded as typical for the Aurignacian are also found associated with non-Aurignacian, probably earlier, contexts.

Innovative features at this time with potentially fundamental significance for the question of "behavioural" (or even cognitive) modernity (cf d'Errico 2003) include the first appearance of personal ornaments, of figurative art and of standardized bone, antler and ivory weapon technologies.

Personal Ornaments

The occurrence of personal ornaments is generally seen as a milestone in the emergence of "behavioural modernity" and recent reviews of their first appearance in the archaeological record (Alvárez Fernández 2006; Zilhão 2007) should be seen against the background of controversial discussions of their presence in pre-Aurignacian contexts. The use of personal ornament is regarded by some researchers as the prerogative of AMH; on this view its apparent presence in "transitional" contexts is interpreted as due to contamination (e.g. White 2001). An opposing position sees the presence of ornaments in "transitional" contexts as evidence for their independent invention by indigenous Neanderthals (d'Errico et al. 1998; d'Errico 2003). A third hypothesis would accept that while Neanderthals possessed the cognitive skills implicit in the use of personal ornaments, they only adopted these as a result of acculturation following contact with AMH (e.g. Hublin et al. 1996; Mellars 2000).

Evidence for personal ornaments in "transitional" contexts is generally limited, with the majority of claimed associations from the Chatelperronian and related complexes with curve-backed pieces, from the pre-Aurignacian EUP of Eastern Europe and from the Protoaurignacian of the northern Mediterranean.

A large proportion of the personal ornament attributed to the Chatelperronian is made up of grooved or perforated teeth of a wide range of mammal species (after Alvárez Fernández 2006; Zilhão 2007). Other categories of personal ornament include grooved and carved ivory (Grotte du Renne) and modified shells.

Zilhão attributes 33 items of personal ornament from the Grotte du Renne to the Chatelperronian levels (Zilhão 2007: Table 3). The uppermost level VIII contains four teeth (one perforated, three grooved) and level IX two grooved teeth, one perforated tooth and one perforated reindeer phalange. The 25 specimens from the basal Chatelperronian level X comprise four fossil shells (three perforated, one grooved), two fragments of ivory rings, two grooved and one perforated reindeer bone and 16 teeth (three perforated, 13 grooved). White (2001) is very sceptical of the attribution of the Grotte du Renne ornaments to the Chatelperronian, criticising the integrity of site stratigraphy and pointing out that the supposedly Chatelperronian ornaments are manufactured using identical techniques to those found in the overlying Aurignacian layers. The personal ornaments found in the Chatelperronian layers of the Grotte du Renne would thus best be explained as intrusive from the overlying Aurignacian of level VII (White 2001), an interpretation followed by Alvárez Fernández (2006) but contested by Zilhão (2007).

Perforated and grooved teeth are also reported from Chatelperronian levels at Roche au Loup (two perforated bovine incisors and one grooved reindeer incisor), Roc de Combe level 8 (a perforated lynx canine may actually derive from a mixed context; Rigaud 2001), Grotte des Fées (Châtelperron) level B4 (two perforated teeth of fox and a feline) and Quincay (three perforated canines of fox, one of wolf and two of red deer). At the latter site contamination from a younger context is excluded by Zilhão due to the absence of overlying layers. He also notes the use of a technique of manufacture (abrasion and piercing) identical to that employed at the Grotte du Renne (Zilhão 2007).

Finds of shell ornament in Chatelperronian context are rare. Apart from the four specimens described from the Grotte du Renne, two beads of fossil *Turritella temprina* from Caune de Belvis (Taborin 1993) have a poor contextual association (Zilhão 2007), while several "*Dentalium* beads" found within a Chatelperronian layer at St. Césaire and apparently associated with a Neanderthal burial still remain unpublished (Zilhão and d'Errico 1999, 47).

In Italy shell ornament and unmodified shells were recovered from Uluzzian levels at Grotta del Cavallo. The lowermost archaic Uluzzian level E-III contained only fragments of unmodified scaphopods (*Dentalium entalis*) whereas the uppermost Uluzzian levels E-I and D contained perforated specimens of shells of *Cyclonassa neritea* and *Columbella rustica*. It is considered highly probable that the perforated shells represent contamination from overlying Aurignacian or Romanellian layers (Gioia 1990; Zilhão 2007), in which case *Dentalium* is possibly the only species associated with the Uluzzian. Furthermore, Alvárez Fernández (2006) suggests that the unmodified *Dentalium* shells from Cavallo may not represent true ornaments but were simply collected as curiosities.

At Klisoura Cave 1 in Greece, the layer V assemblage with curve-backed pieces was associated with *Dentalium* shells of two species and some fragments of *H. figulina* shells (Koumouzelis et al. 2001a, 533), although it is not made clear whether any of these specimens are artificially modified.

From Central Europe personal ornaments are reported in the form of a spindle-shaped bone pendant (Kozłowski 1982; Figure 6, 2) and two pierced teeth of bear and fox at Bacho Kiro (Bulgaria) in association with the laminar industry of level 11 (Kozłowski 1982). Further west at Willendorf II (Austria) a perforated shell of a fossil gastropod is reported from KS 2 (Felgenhauer 1959). The association of an ivory disc with artificial central hole and a needle like bone point from the Ilsenhöhle at Ranis (Germany) with the leaf-point assemblage at the site (Hülle 1977) was already contested by Hahn (1977, 103) who believed they may derive from contamination from the Aurignacian level.

Contrasted with the weak evidence for personal adornment from the "transitional" industries described above, ornaments in the form of (marine) shell beads, often in large numbers, and, sporadically, animal teeth are regularly described for the Protoaurignacian in far more plausible contexts (cf Alvárez Fernández 2006; Zilhão 2007).

The northern Italian "Protoaurignacian" site of Grotta di Fumane (Broglio and Gurioli 2004; Broglio and Dalmeri 2005) provided the largest assemblage of 650 marine shells, half of them modified as beads, while 240 marine shell beads were recovered at Riparo Mochi (Kuhn and Stiner 1998; Zilhão 2007). At both sites a large range of taxa was identified (Vanhaeren and d'Errico 2006), although at Fumane "the large majority of the shells are *Homalopoma sanguineum*" (Zilhão 2007). The dominant taxon at Riparo Mochi is *Cyclope neritea* (29%) followed by *H. sanguineum* (16%), while at Castelcivita beads of *Homalopoma* were the only ornaments present (Vanhaeren and d'Errico 2006). In Mediterranean France the Protoaurignacian at the Abri Rothschild also produced a large number of beads (almost 400), the overwhelming majority made of marine shells of a range of taxa (Zilhão 2007).

In Protoaurignacian contexts bead ornaments other than marine shell are insignificant. From Fumane the only other possible ornaments are three red deer incisors with roots grooved for suspension, while at Riparo Mochi only one pierced tooth (incisor of a small carnivore) is present. Four beads of soft stone or bone found at the latter site are carved to resemble red deer maxillary canine teeth (Stiner 1999).

The stratigraphic attribution of the Fumane personal ornaments to the Protoaurignacian has been questioned by Zilhão, who suggests that "direct radiocarbon dates on three such shell ornaments from different species prove contemporaneity with the Aurignacian occupation" (Zilhão 2007). The situation seems somewhat more complicated, since these results in fact document an inversion of the radiometric age of the specimens relative to the stratigraphy. Two direct dates on shell ornaments assigned to Protoaurignacian level A2 (OS-5999: 32,700 ± 140 ^{14}C BP; OS-5999: 32,000 ± 90 ^{14}C BP) fall within the chronometric range of charcoal dates from both Aurignacian level D and the majority of measurements from Protoaurignacian levels A1 and A2, whereas the third measurement on marine shell from the basal Aurignacian horizon D6 produced an older result of 37,100 ± 240 ^{14}C BP (OS-5872) more appropriate for the Protoaurignacian (Fig. 22.12). This inconsistency and potential reservoir effects on marine shell render the interpretation of radiometric dates on these finds relative to their archaeological context particularly difficult.

Unlike at sites where the intrusion of material from overlying Aurignacian layers into Protoaurignacian levels cannot entirely be excluded, 17 perforated beads made from marine shells of five taxa and "a few" *Dentalium* at La Laouza in Mediterranean France must be associated with the single Protoaurignacian horizon present (Taborin 1993; Zilhão 2007). Other French Protoaurignacian sites have provided smaller quantities of personal ornaments, sometimes from insecure contexts (cf Zilhão 2007).

Protoaurignacian ornaments are rare in Spain (Alvárez Fernández 2006; Zilhão 2007). In northern Catalonia the Protoaurignacian level H of L'Arbreda (Maroto 1994) produced only three *Dentalium* fragments, one fragment of *H. sanguineum*, three other shell fragments of marine species and a single pierced specimen of *Trivia pulex*, while Abric Romaní Protoaurignacian level 2(A) produced personal ornaments of scaphopod shells and an atrophied red deer canine tooth (Bischoff et al. 1994; Alvárez Fernández 2006).

That personal ornaments of marine shells do not simply reflect the proximity of a Protoaurignacian site to the sea is demonstrated by their considerable presence at Grotta di Fumane, at that time ca. 200 km from the Ligurian coast. Even more convincing is their presence in central and eastern European EUP contexts.

Personal ornaments from the site of Krems-Hundssteig in Austria, more than 600 km from the nearest contemporary coastline, include 128 perforated mollusc shells, among them *Clanculus*, *Columbella*, *Cyclope*, *Dentalium*, *Melanopsis*, and a pendant of nephrite, all of which are possibly associated with a Protoaurignacian assemblage (Hahn 1972, 1977; Vanhaeren and d'Errico 2006).

On the Don River in southern Russia the oldest occupation horizon at Kostenki 14 (level IVb) produced a double-perforated ornament of a mollusc shell identified as Columbellidae alongside worked ivory, bone and antler and associated with a tooth attributed to an AMH (Sinitsyn 2003, 91). The shell is apparently "derived from a source no closer than the Black Sea (i.e. transported >500 km)" (Anikovich et al. 2007, 225).

With the exception of the small amount of evidence from appreciably older contexts (cf Vanhaeren et al. 2006) not directly relevant to the period under consideration here, personal ornaments from layer H at Üçağizli in south-eastern Turkey (e.g. Kuhn et al. 2001; Stiner 2003; Stiner et al. 2003) are the earliest securely dated specimens outside Africa. Layer H is dated to between 41.4 and 35.7 ka ^{14}C BP with a weighted mean of four dates of 38,036 ± 487 ^{14}C BP, an age supported by a radiocarbon measurement (AA-37626: 39,100 ± 1,500 ^{14}C BP) from the next higher level of the cave (Kuhn et al. 2001; Table 22.4). Less well dated shell beads from IUP levels at the Lebanese site of Ksar'Akil may be even older (Kuhn 2004; Kuhn et al. 2001).

Despite the insecure or undemonstrated association of marine shell ornaments at a few Protoaurignacian sites, the evidence for their presence in the time range ca. 36.5/36.0 to 35.0 ka ^{14}C BP is, in contrast to the "transitional" industries, remarkably robust (Table 22.5, Fig. 22.11). If dates as old as 38.0 ka ^{14}C BP for the Protoaurignacian at L'Arbreda and Abric Romaní (Table 22.5) should be confirmed, the appearance of personal adornments at the western end of the Mediterranean would fall very close to their presence at Üçağizli (e.g. Stiner 2003; Stiner et al. 2003). It might be significant that the earliest and most convincing European evidence for personal ornaments is linked with probably intrusive assemblages of ultimately Near Eastern IUP origin.

In this case, the practically simultaneous appearance of personal ornaments manufactured on marine shells in association with laminar lithic assemblages would imply a rapid diffusion of this concept around the northern Mediterranean, into the Danube Basin and areas adjacent to the Black Sea and, in the case of inland localities, directly demonstrates the transfer of specimens themselves. This probably reflects the emergence of novel large scale social networks at this time, most plausibly created by newly arrived human groups, the identity of which might be indicated by the apparently AMH tooth from Kostenki 14, level IVb, and the AMH burial from Ksar'Akil ("Egbert": Bergman and Stringer 1989) referred to above.

Figurative Art

Among the most convincing evidence for fully developed modern human behaviour in the western Eurasian UP is the appearance of figurative art. In the discussion of its emergence certain regions of Western and Central Europe are of key importance (Fig. 22.19). These have provided examples of carved figurines (southern Central Europe), painted images (several sites in southern France, the southern German Geißenklösterle and the northern Italian Grotta di Fumane) and picked engravings (southern France: Style 1 after Leroi-Gourhan 1971). We do not accept the suggestion of a pre-Aurignacian age for the Russian site of Sungir, with its burials, elaborate associated personal ornaments and figurative art (Bosinski 1989, 1990) since direct radiocarbon dating of the AMH skeletons by two laboratories clearly demonstrates their MUP age (Pettitt and Bader 2000; Kuzmin et al. 2004).

In south-western Germany Aurignacian levels at Geißenklösterle, Vogelherd, Hohle Fels and Hohlenstein-Stadel have produced the oldest known complex portable figurative artworks. At Geißenklösterle the younger Aurignacian horizon AH II dates to between 33.2 and 29.8 ka ^{14}C BP, with similar radiocarbon ages for other regional Aurignacian horizons with figurative portable art (Figs. 22.15–22.18). A small stone figure, supposedly female, from Stratzing (Krems-Rehberg, Austria) also dates to the Aurignacian (Neugebauer-Maresch 1989).

Portable figurative art is unknown in Aurignacian contexts outside southern Central Europe, but it is now generally accepted (*pace* Züchner 1996, 2003; Pettitt and Bahn 2003) that parietal figurative art appears at approximately the same time, with Grotte Chauvet providing the best known and most convincing evidence (Valladas et al. 2001, 2005; Table 22.6). Radiocarbon direct dating of charcoal from paintings in the "Salle Hillaire" produced results between 32,410 ± 720 ^{14}C BP (GifA-95132) and 29,670 ± 950 ^{14}C BP (GifA-98160) which are in complete agreement with other evidence for pre-MUP hominin presence within the cave (cf Table 22.6).

The radiocarbon dating of Grotte Chauvet revolutionized ideas on the age and context of the appearance of parietal art but intentionally coloured fragments of limestone were already known from Aurignacian horizons at several caves and rock shelters, among these Abri Blanchard (Breuil 1952; Delluc and Delluc 1978), Abri Pataud (Movius 1975) and the base of horizon AH II (AH IIb) at Geißenklösterle (Hahn 1988b). To the south of the Alps, painted limestone blocks showing one (fragment II) or possibly two (fragment IV) anthropomorphic figures (Broglio et al. 2005, 46–47) were recovered at the Grotta di Fumane in north-eastern Italy from an Aurignacian horizon dated to between 32.3 and 30.3 ka ^{14}C BP (Giaccio et al. 2006; Fig. 22.12).

Limestone blocks recovered from Aurignacian layers in south-western France reveal figurative depictions deeply picked into their surfaces which were classed together by André Leroi-Gourhan (1971) under his "Style 1". When stratigraphic control is adequate, they appear to be associated with the later Aurignacian and date younger than ca. 32.0 ka ^{14}C BP (Zilhão 2007).

Aurignacian Bone, Antler and Ivory Weapon Technology

Compared with organic artefacts from Châtelperronian or EUP contexts (e.g. Grotte du Renne: d'Errico et al. 2003; Kostenki 14, level IVb: Sinitsyn 2003), Aurignacian organic technology is more elaborated and aimed at the production of quite standardized tool-types (cf Albrecht et al. 1972; Liolios 2006).

During recent years series of radiocarbon measurements have been directly obtained on bone, antler and ivory projectile points such as the split-based points characteristic of Aurignacian I or points with massive bases of so called "Mladeč" type regarded as typical for the later Aurignacian (e.g. Hahn 1988a, c; Hofreiter and Pacher 2004). Dates fall into a fairly narrow range between 32,470 + 270/– 260 ^{14}C BP (KIA-19551; Bolus and Conard 2006) for a massive-base bone point from layer XIV at the Brillenhöhle in the Swabian Jura and 29,210 ± 210 ^{14}C BP (OxA-13048; Grünberg 2006) on a Mladeč point from the Hermannshöhle in Saxony-Anhalt, Germany (Table 22.7). Only one result on a distal fragment of a bone point from Aurignacian level V at Sirgenstein in Baden-Württemberg is younger (KIA-13082: 26,730 + 170/–160 ^{14}C BP; Bolus and Conard 2006), but the specimen does not permit a more precise typological attribution. A point from the supposedly MP level VI of Vogelherd (Riek 1934) was dated to 31,310 + 240/–230 ^{14}C BP (KIA-19541) and falls within the range of the other directly dated Aurignacian-type specimens (Bolus and Conard 2006). Due to the limited spatial distribution and stratigraphical integrity of level VI and the small amount of recovered archaeological material a cultural attribution to the MP cannot be verified. A context radiocarbon measurement (GifA-101459: 32,650 ± 540 ^{14}C BP) suggests that an ivory point with massive base from the "Galerie des Mégacéros" at Grotte Chauvet (Valladas et al. 2005) is marginally older than the directly dated Central European specimens (Table 22.7). It thus appears that the emergence of an elaborated use of bone, antler and ivory for the production of standardized projectiles of Mladeč type falls at around 32.5 ka ^{14}C BP in both Central and south-western Europe.

Although Bolus and Conard (2006) point out that the stratigraphical distinction between older (Aurignacien I) split-based and younger Aurignacian massive-based points recognizable in south-western Europe (cf Djindjian et al. 2003) is not reflected by the direct dates for Central European

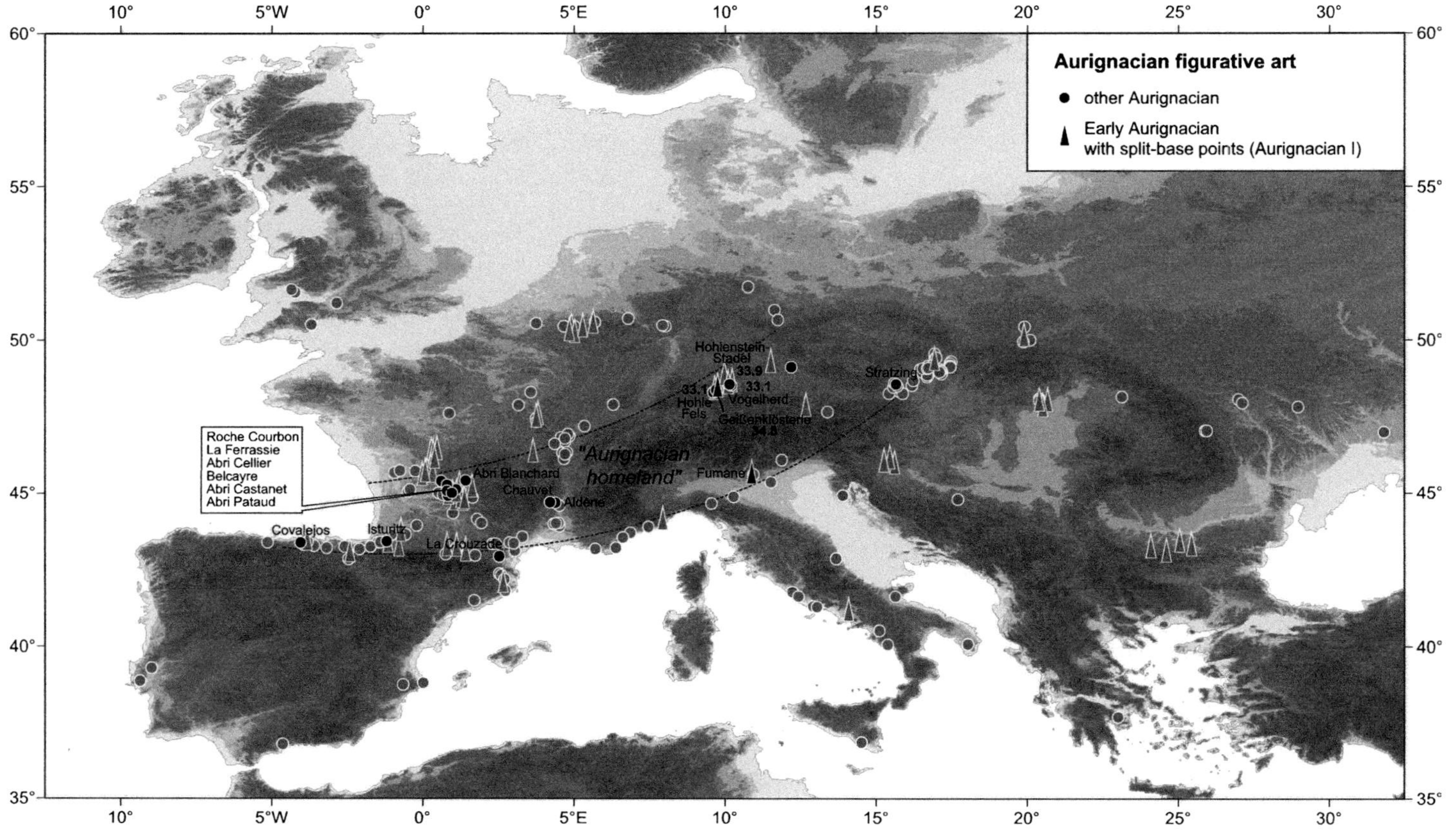

Fig. 22.19 Evidence of Aurignacian figurative art compiled after different authors ("Aurignacian homeland" following Bar-Yosef, 2006b). Map based on SRTM data; sea level lowered by 75 m

Table 22.6 Results from radiocarbon dating of Grotte Chauvet (only pre-Mid-Upper Palaeolithic data >> 27,500^{14}C BP)

Layer	Method	No.	Lab.-No.	^{14}C age	STD+	STD–	^{13}C	Material	Location	Assoc. ind.	Source
Chauvet, F											
Paintings	^{14}C	1	GifA-95126	30,940	610	610			Painting of rhinoceros, left ("Salle Hillaire: Panneau des Chevaux")		Valladas et al. (2005)
	^{14}C	2	GifA-95133	30,790	600	600			Painting of rhinoceros, right ("Salle Hillaire: Panneau des Chevaux")		Valladas et al. (2005)
	^{14}C	3	GifA-95132	32,410	720	720			Painting of rhinoceros, right ("Salle Hillaire: Panneau des Chevaux")		Valladas et al. (2005)
	^{14}C	4	GifA-96065	30,230	530	530			Painting of aurochs ("Salle Hillaire: Panneau des Chevaux")		Valladas et al. (2005)
	^{14}C	5	GifA-98157	20,790	340	340		Carbon fraction humic fraction	Painting of horse ("Salle Hillaire")		Valladas et al. (2001)
	^{14}C	6	GifA-98160	29,670	950	950			Painting of horse ("Salle Hillaire")		Valladas et al. (2001)
	^{14}C	7	GifA-96063	31,350	620	620			Painting of Megaloceros ("Galerie des Mégacéros")		Valladas et al. (2005)
	^{14}C	8	GifA-95128	30,340	570	570			Painting of large bison ("Salle du Fond")		Valladas et al. (2005)
	^{14}C	*(1–4,6–8)*	*Weighted mean*	*30,811*	*237*	*237*	*2.11*	*t-Value*			
"Artefacts"	^{14}C	1	GifA-101459	32,650	540	540		Carbonaceous earth	Carbonaceous clay with ivory point ("Galerie des Mégacéros")	Aurignacian	Valladas et al. (2005)
	^{14}C	2	GifA-99809	32,360	490	490			On the bloc with cave bear cranium ("Salle du Crâne")		Valladas et al. (2005)
	^{14}C	3	GifA-99810	31,390	420	420			On the bloc with cave bear cranium ("Salle du Crâne")		Valladas et al. (2005)
	^{14}C	4	GifA-99811	32,600	490	490			On the bloc with cave bear cranium ("Salle du Crâne")		Valladas et al. (2005)
	^{14}C	*(1–4)*	*Weighted mean*	*32,159*	*240*	*240*	*1.59*	*t-Value*			
Traces of fire	^{14}C	1	GifA-102567	30,980	410	410		Carbonaceous earth	("Galerie du Cierge")		Valladas et al. (2005)
	^{14}C	2	GifA-101462	31,180	400	400		Carbonaceous earth	("Salle Hillaire")		Valladas et al. (2005)
	^{14}C	3	GifA-101461	32,130	460	460		Carbonaceous earth	("Salle du Crâne")		Valladas et al. (2005)

	^{14}C	4	GifA-102565	31,060	400	400	Carbonaceous earth	("Salle du Croisillons")	Valladas et al. (2005)
	^{14}C	5	GifA-101456	31,560	490	490	Carbonaceous earth	("Salle du Croisillons")	Valladas et al. (2005)
	^{14}C	6	GifA-99238	31,430	420	420	Carbonaceous earth	("Salle des Mégacéros")	Valladas et al. (2005)
	^{14}C	7	GifA-99239	29,740	390	390	Carbonaceous earth	("Salle des Mégacéros")	Valladas et al. (2005)
	^{14}C	8	GifA-99768	31,910	390	390	Carbonaceous earth	("Salle des Mégacéros": carbonaceous zone 1)	Valladas et al. (2005)
	^{14}C	9	GifA-99769	31,520	360	360	Carbonaceous earth	("Salle des Mégacéros": carbonaceous zone 2)	Valladas et al. (2005)
	^{14}C	10	GifA-99770	31,860	380	380	Carbonaceous earth	("Salle des Mégacéros": carbonaceous zone 3)	Valladas et al. (2005)
	^{14}C	11	GifA-99771	32,220	400	400	Carbonaceous earth	("Salle des Mégacéros": carbonaceous zone 4)	Valladas et al. (2005)
	^{14}C	12	GifA-99775	32,080	430	430	Carbonaceous earth	("Salle des Mégacéros": carbonaceous zone 6)	Valladas et al. (2005)
	^{14}C	13	GifA-99773	31,020	350	350	Carbonaceous earth	("Salle des Mégacéros": carbonaceous zone 7)	Valladas et al. (2005)
	^{14}C	14	GifA-99774	32,500	400	400	Carbonaceous earth	("Salle des Mégacéros": carbonaceous zone 8)	Valladas et al. (2005)
	^{14}C	15	GifA-99778	32,900	490	490	Carbonaceous earth	("Salle des Mégacéros": carbonaceous zone 10)	Valladas et al. (2005)
	^{14}C	16	Ly-6878	29,000	400	400	Carbonaceous earth	("Salle des Mégacéros": below painting of horse)	Valladas et al. (2005)
Below calcitic concretions or collapsed blocks	^{14}C	1	GifA-101458	33,100	590	590		("Salle Hillaire")	Valladas et al. (2005)
	^{14}C	2	GifA-102573	30,020	350	350		("Salle Hillaire")	Valladas et al. (2005)
	^{14}C	3	GifA-102574	31,350	440	440		("Salle Hillaire")	Valladas et al. (2005)
	^{14}C	4	GifA-102566	30,560	370	370		("Salle du Croisillons")	Valladas et al. (2005)
	^{14}C	5	GifA-102568	29,180	340	340		("Salle du Croisillons")	Valladas et al. (2005)
	^{14}C	6	GifA-99778	31,020	370	370		("Salle des Mégacéros")	Valladas et al. (2005)

Table 22.7 Results from direct radiometric dating of organic projectiles

Layer	Method	No.	Lab.-No.	^{14}C age	STD+	STD–	^{13}C	Material	Species	Skeletal element	Assoc. ind.	Source
Brillenhöhle, D												
XIV	^{14}C		KIA-19550	30,400	240	230		Bone collagen	Bone point, split-based (?)		Aurignacian	Bolus and Conard (2006)
XIV	^{14}C		KIA-19550	32,110	480	450		Residual fraction	Bone point, split-based (?)		Aurignacian	Bolus and Conard (2006)
XIV	^{14}C		KIA-19551	32,470	270	260		Bone collagen	Bone point, massive base		Aurignacian	Bolus and Conard (2006)
Chauvet, F												
						Not directly dated						
	^{14}C		GifA-101459	32,650	540	540		Ivory carbonaceous earth	Point, massive base carbonaceous clay with ivory point ("Galerie des Mégacéros")		Aurignacian	Valladas et al. (2005)
											Aurignacian	Valladas et al. (2005)
Hermannshöhle (Rübeland), D												
	^{14}C		OxA-13048	29,210	210	210	−17.5	Antler/bone?	Point, massive base (Mladec Type)		MP & UP	Grünberg (2006)
Hyaena Den, GB												
	^{14}C AI		OxA-3451	24,600	300	300	−20.1	Ion-exchanged gelatin	Bone or antler point		Aurignacian	Jacobi et al. (2006)
	^{14}C AF*		OxA-13803	31,550	340	340	−19.2	Ion-exchanged gelatin	Bone or antler point		Aurignacian	Jacobi et al. (2006)
Potočka Zijalka, SLO												
	^{14}C		VERA-2526	29,560	270	270		Bone	Bone point, massive base: "PZ-128"	Proximal fragment	Aurignacian	Hofreiter and Pacher (2004)
	^{14}C		VERA-2525	29,740	330	310		Bone	Bone point, massive base: "PZ-126"	Proximal fragment	Aurignacian	Hofreiter and Pacher (2004)
	^{14}C		VERA-2524	29,760	330	310		Bone	Bone point, massive base: "PZ-121"	Proximal fragment	Aurignacian	Hofreiter and Pacher (2004)
	^{14}C		VERA-2522	30,140	340	330		Bone	Bone point, massive base: "PZ-59"	Almost complete distal fragment	Aurignacian	Hofreiter and Pacher (2004)
	^{14}C		VERA-2521	31,080	370	360		Bone	Bone point: "PZ-54"		Aurignacian	Hofreiter and Pacher (2004)
	^{14}C		VERA-2523	31,490	350	340		Bone	Bone point, massive base: "PZ-112"	Proximal fragment	Aurignacian	Hofreiter and Pacher (2004)

Sirgenstein, D											
V	^{14}C	KIA-13082	26,730	170	160		Bone	Bone point	Distal fragment	Aurignacian	Conard and Bolus (2003)
Tischoferhöhle, AU											
	^{14}C	KIA-19543	32,010	510	480		Bone collagen	Bone point: "T-139"	Distal fragment	Aurignacien	Bolus and Conard (2006)
	^{14}C	KIA-19543	31,220	400	380		Residual fraction	Bone point: "T-139"	Distal fragment	Aurignacien	Bolus and Conard (2006)
	^{14}C	KIA-19544	31,530	210	200		Bone collagen	Bone point, split-based: "T-143"		Aurignacien	Bolus and Conard (2006)
	^{14}C	KIA-19544	30,250	360	340		Residual fraction	Bone point, split-based: "T-143"		Aurignacien	Bolus and Conard (2006)
	^{14}C	KIA-19545	29,500	200	200		Antler collagen	Antler point, massive base (?): "T-137"		Aurignacien	Bolus and Conard (2006)
Trou de la Mère, F											
	^{14}C	Beta-150312	29,490	190	190		Antler (Rangifer)	Antler point, split-based (?)		Aurignacien	Brou (1997, 2001)
Uphill Quarry, GB											
	^{14}C AG	OxA-8408	28,080	360	360	−17.3	Antler	Antler point: "BRSMG Ce 16476"	Distal fragment	Aurignacian	Jacobi and Pettitt (2000)
	^{14}C AF	OxA-13716	31,730	250	250	−17.5	Antler	Antler point: "BRSMG Ce 16476"	Distal fragment	Aurignacian	Jacobi et al. (2006)
Vindija, CRO											
G1	U-Pa		30,000	5,000	5,000		Bone	Bone point, split-based		MP/UP	Karavanic et al. (1998)
G1	U-Th		45,000	6,000	6,000		Bone	Bone point, split-based		MP/UP	Karavanic et al. (1998)
Vogelherd Cave, D											
VI	^{14}C	KIA-19541	31,310	240	230		Bone collagen	Bone point, massive base		Middle Palaeolithic	Bolus and Conard (2006)

Aurignacian projectiles, unambiguous direct dating evidence for split-based points is restricted to the single specimen T-143 from Tischoferhöhle (Bolus and Conard 2006) securely attributed to this type (cf Table 22.7).While the younger Geißenklösterle horizon AH II contains a larger number of diverse projectile forms of both ivory and reindeer antler, including 11 split-based antler points, the only specimens from the underlying AH III assemblage are three ivory points with plain bases (Hahn 1988b, Plate 35).

The distal fragment of an antler point from Uphill Quarry and a rod-like point from Hyaena Den (both Somerset, England) were re-dated following ultrafiltration pre-treatment (Jacobi et al. 2006). The new results of 31,730 ± 250 ^{14}C BP (OxA-13716) and 31,550 ± 340 ^{14}C BP (OxA-13803) respectively are appreciably older than previous measurements (OxA-8408: 28,080 ± 360 ^{14}C BP, Jacobi and Pettitt 2000; OxA-3451: 24,500 ± 300 ^{14}C BP, Hedges et al. 1996) and firmly place the English projectile points within the time range of Aurignacian points in continental Europe.

The Relationship of the Aurignacian and the "Transitional" Industries

The possibility that the "makers of the early Aurignacian of Europe", identified as AMH (Churchill and Smith 2000), would have met the last Neanderthals has led to heated discussions in the past. Any such debate must examine critically the radiometric and stratigraphic dating evidence for potential contemporaneity.

Interstratification?

In south-eastern and Central Europe, all stratified sequences containing MP or "transitional" and Aurignacian or other UP assemblages show an unambiguous succession with the former underlying the latter and no suggestion of any interstratification. Indeed, sites in the Swabian Jura often show a clear hiatus between MP and UP horizons (Conard et al. 2006) and it seems clear that central European Aurignacian assemblages postdate the MP and/or "transitional" (e.g. leaf point) industries by an appreciable margin. Further to the East, radiocarbon dates between ca. 38.9 and 37.5 ka ^{14}C BP from Willendorf II, KS 3 and Keilberg-Kirche may provide the earliest evidence for the Aurignacian in Central Europe. The Aurignacian lithic industries of this region might derive from earlier laminar traditions further to the East such as the Bachokirian at Temnata (Kozłowski 2006, 33).

In Western Europe stratigraphic evidence generally shows that the Chatelperronian lies between Late MP and Protoaurignacian levels around the Pyrenees and in Mediterranean France (Djindjian et al. 2003) and between Late MP and Aurignacian I levels in the rest of France (Djindjian et al. 2003; cf Bosinski 1989, 1990; Demars 1996).

Interstratification of the Chatelperronian and Aurignacian has been suggested at El Pendo in northern Spain and at the French sites of Roc de Combe and Le Piage, although these arguments have been convincingly refuted (Zilhão and d'Errico 1999, 2003b; Rigaud 2001; Djindjian et al. 2003). Although interstratification of Aurignacian and Chatelperronian in the northern part of Le Piage was rejected on grounds of vertical mixing (e.g. d'Errico et al. 1998), new analyses have demonstrated a valid stratigraphic sequence in the southern part of the site (Bordes 2006). Here, remnants of Mousterian and Chatelperronian layers are overlain by three horizons originally attributed to the Aurignacian. Still more recent claims for interstratification within the sequence of the Grotte des Fées (Châtelperron) (Gravina et al. 2005) have been challenged and it is claimed that stratigraphical problems were underestimated and typological arguments are questionable (Zilhão and Pettitt 2006, 7; Zilhão et al. 2006). In a reply to Zilhão et al. (2006), Mellars et al. (2007) insist that an ephemeral Aurignacian presence is documented stratigraphically between two Chatelperronian layers at the type site. They interpret two consistent series of radiocarbon results on bone as dating the older and the younger Chatelperronian occupations respectively and conclude that the Aurignacian presence at the site "could easily date from as late as, say, 36,000–37,000 BP" (Mellars et al. 2007 3662).

We see no convincing evidence for interstratification of Chatelperronian and Aurignacian assemblages in Western Europe and suggest that, as in Central Europe, any possible interaction between indigenous Neanderthals and intrusive AMH would considerably pre-date the appearance of Aurignacien I assemblages. In this context it is interesting that at both Le Piage (Bordes 2003, 2006) and, possibly, at Arcy-sur-Cure (Bon 2003 2006) details of the lithic technology may indicate *in situ* evolution of the Aurignacian from the preceding Protoaurignacian. Interestingly, it also has been suggested that the Le Piage K assemblage shows affinities with the underlying Chatelperronian (Bordes 2006, 165), thereby opening up the possibility that there is at least some degree of continuity between the Chatelperronian and the Aurignacian.

Southern European Refugia?

Against the background of the hypothesis of AMH westward migration through the Danube valley or along the northern Mediterranean coastline there has been much discussion of the possibility of Neanderthal survival on southern European peninsulas appreciably later than to the north. This has led to the concept of refugial situations on the Iberian, Italian and Crimean peninsulas and in the Caucasus region.

Since the Iberian Peninsula lies at the western extremity of the supposed trajectory of AMH expansion it has been suggested that the Aurignacian would logically arrive here later than in regions further to the East. In fact, as shown above, radiocarbon dates for both the Aurignacian and the Protoaurignacian in the northern part of the peninsula are no younger than elsewhere (Figs. 22.6 and 22.13). In the southern part of the Iberian Peninsula the situation is different: Neither Protoaurignacian nor Aurignacian I sites with characteristic split-based points are known south of the Ebro River, suggesting that only later Aurignacian industries are present (Vega Toscaño 1990; Zilhão 1993, 2006b).

Although the radiometric evidence for this period and region is particularly poor, radiocarbon dates from Cova Beneito (AA-1388: 33,900 ± 1,100 ^{14}C BP) in the region of Valencia and Bajondillo (Ua-17150: 33,690 ± 1,195 ^{14}C BP) in Andalucia (Zilhão 2006b) might suggest an Aurignacian presence only marginally younger than to the North, in which case the absence of split-based points in southern Iberia could simply show the presence of a particular Aurignacian facies rather than a chronological stage within the Aurignacian succession. The absence of split-based points from the southern Iberian Aurignacian has been explained by some authors as a response to the absence of reindeer in the region and consequent limited availability of antler, the main material used for their manufacture (Bernaldo de Quirós et al. 2001, quoted in Olszewski and Dibble 2006, 364–366).

The timing of the demise of the last Neanderthals on the Iberian Peninsula is hotly disputed. While some authors argue that the severe cold of Heinrich event 4 (Greenland Stadial 9; Fig. 22.3) led to the final extinction of Neanderthals (Giaccio et al. 2006; cf d'Errico and Sánchez Goñi 2003, 2004; contra Finlayson et al. 2004), the existence of several young dates for Late MP assemblages and a small number of Neanderthal remains from the southern half of the Iberian Peninsula has led others to propose later survival, with Neanderthals finally going extinct as recently as ca. 35,000 calendar years ago, more or less equivalent to GI 7 (Zilhão 2006b).

Radiocarbon dates from the interior of Gorham's Cave, Gibraltar, have led Finlayson et al. (2006) to propose an even later survival of the MP and the last Neanderthals at the southern tip of the Iberian Peninsula until as recently as 28.0 and probably even 24.0 ka ^{14}C BP. This would correspond to Greenland Stadial (GS) 5 or younger if compared with calibration data sets available today (cf Fig. 22.3), however the context and relevance of the dated samples for the age of the MP levels at Gorham's Cave were immediately questioned (Zilhão and Pettitt 2006).

Systematic evaluation of the available radiocarbon data shows that the Late Middle Palaeolithic (LMP) of the southern Iberian Peninsula is in fact poorly dated (Jöris et al. 2003; cf Vaquero et al. 2006), with a chronology often based on measurements obtained several decades ago. They often display high standard deviations, are frequently contradicted by results obtained by other dating methods (e.g. Zafarraya: Hublin et al. 1995) and may derive from poorly recorded contexts (Jöris et al. 2003; cf Vaquero et al. 2006). The most recent compilation of radiometric age determinations for the MP–UP transition on the Iberian Peninsular cannot resolve the observed problem (Zilhão 2006b). Potential dating evidence for a LMP presence in southern Iberia comes from the open-air site of El Salt in Andalucia. Here, a stratigraphically coherent series of TL dates places the older levels XII to IX between 59.1 ± 8.9 and 43.2 ± 3.3 ka TL BP., with younger level VIII beginning "a partir del 40.000 B.P." (Galván Santos et al. 2006, 132). Still younger LMP levels VI and V produced five teeth assigned to Neanderthals.

During the period from ca. 38.0 to 34.0 ka ^{14}C BP we can neither confirm late Neanderthal survival on the southern Iberian Peninsula nor do we recognize a significant delay in the first occurrence of the Aurignacian. As pointed out above, the LMP of the region is still poorly dated, while the earliest radiometric evidence for the Aurignacian can be found as early as 33.7 ka ^{14}C BP along the Spanish Mediterranean coastline as far south as Andalucia, making its appearance here more or less contemporaneous with that in the northern part of Iberia. These observations have implications for the "Ebro frontier hypothesis" (Zilhão 1993, 1997, 1998, 2000, 2006a, b; d'Errico et al. 1998; Zilhão and Trinkaus 2002; Vega Toscaño 1993) with claims for the co-existence of Neanderthals and AMH over as many as 7,000 calendar years (Zilhão 2006b).

The scarcity of well dated archaeological evidence in southern Iberia during the crucial period from ca. 38.0 to 34.0 ka ^{14}C BP might suggest plausible alternative scenarios. For example, regional population shifts due to climatic and environmental change at a millennial to centennial scale may have partially or totally emptied the region of hominins between the Late MP and the arrival of the Aurignacian (Jöris et al. 2003).

Such a scenario could possibly also explain the absence of the Chatelperronian and Protoaurignacian to the South of the Ebro. The dense presence of both in south-western France and their extension onto the northern Iberian Peninsula suggests that these were probably the regions and contexts in which contact between Neanderthals and AMH would have taken place (see above). While the nature and consequences of such contact remain obscure, a scenario of brief (in terms of hominin generations still not inconsiderable) interaction and influence between adaptive human groups appears more plausible than that of a long independent survival of an innovative "French/Cantabrian" (e.g. d'Errico et al. 2003) and an extremely conservative "Iberian" (e.g. Zilhão 2006b) Neanderthal population geographically separated from each other by intrusive AMH avoiding contact with both.

During recent years, claims for the late survival of Neanderthals and the MP have also been made for the Crimean Peninsula and the Caucasus region. The hypothesis is difficult to demonstrate in Eastern Europe since "there are very few sites with a clear stratigraphic sequence which includes a Micoquian, a transitional industry, and an acceptable Upper Paleolithic" (Marks and Monigal 2000, 213).

Important for this discussion is the stratigraphy at the collapsed rock-shelter of Buran-Kaya III on the Crimean Peninsula, reported to document a sequence in which an "Eastern Szeletian" EUP "transitional" industry (level C) is overlain by a Late MP "Kiik-Koba" Micoquian (level B/B1), which is itself covered by UP layers (Marks 1998; Marks and Monigal 2000, 2004; Monigal 2001, 2004, 2006).

Although the total excavated area covered less than 30 m² (Marks 1998) it is difficult to correlate later descriptions of the site stratigraphy (Marks 1998; Marks and Monigal 2000) with that given by the discoverer and first excavator of the site (Yanevich et al. 1996). A fixed point appears to be the equation of cultural layer 7.2 "undoubtedly analogous to the so-called Kiik-Kobian" (Yanevich et al. 1996, 318) with "Kiik-Koba-type Micoquian" level B/B1 (Monigal 2006; Yanevich et al. 1997, 85). Confusing is the fact that this layer was initially described stratified below a layer in which "were found some bifacially worked points similar to those of the Streletskaya culture" (Yanevich et al. 1996, 317), whereas in later publications the "Kiik-Koba-type Micoquian" is reported to lie above "the Streletskaya-like occupation" (Monigal 2006, 202) of level C, now attributed to the "early Upper Palaeolithic" (Monigal 2006, 205) but originally classified as "Eastern Szeletian" (Marks and Monigal 2000). In view of the complete absence of "truncated tools, burins, or perforators" and the lack of "evidence for the use of any purposeful blade technology" (Marks and Monigal 2000, 218) we cannot agree that the assemblage is "generically UP" (Monigal 2006, 205) and see no reason to doubt the original interpretation of level C as a "transitional" assemblage. Parallels for the "bifacial trapezoidal microliths" of level C might be found in the disc-shaped, often bifacially retouched microlithic "groszaki" of Central Europe Micoquian assemblages (cf Hillgruber 2006) which, although morphologically distinct, are produced by a similar chaine opératoire.

Radiocarbon dates on bone from Buran-Kaya III (Yanevich et al. 1996; Pettitt 1998; Marks and Monigal 2000) have been argued to conform to the stratigraphic superposition of a MP assemblage of level B/B1 above the Streletskaya-like industry of level C. In fact, taken together, the dates are in a number of cases inverted relative to stratigraphy. For example, a date of 32,710±940 ^{14}C BP (OxA-4130) for the Kiik-Koba Micoquian of level 7.2 (Yanevich et al. 1996) contrasts with two measurements of 28,840±460 ^{14}C BP (OxA-6673) and 28,520 ± 460 ^{14}C BP (OxA-6674) for the Kiik-Koba Micoquian of level B/B1 (Pettitt 1998) preferred by Marks and Monigal (2000). The younger dates are in fact indistinguishable from OxA-4128 (28,700±650 ^{14}C BP) obtained for level 6.10 which overlies the Kiik-Koban assemblage and contains "points similar to those of the Streleskaya culture" (Yanevich et al. 1996, 317). We suggest that doubts concerning "the unknowable association between each date and its archaeological correlate" (Marks and Monigal 2000, 221) may apply generally to the Buran-Kaya III samples and propose that the two results for the overlying Gravettian (OxA-6882: 30,740±460 ^{14}C BP) and Aurignacian (OxA-6990: 34,400±1,200 ^{14}C BP) and that for the "Eastern Szeletian" of level C (OxA-6868: 36,700 ± 1,500 ^{14}C BP) (Marks and Monigal 2000, 221) at present provide the only convincing radiometric evidence for the chronology of the site.

The Mezmaiskaya Neanderthal burial from the northern Caucasus has been discussed above and its supposedly young direct date questioned. Further south, in the Georgian Republic the cave of Ortvale Klde provides the best dated sequence of the MP–UP transition within the wider region (e.g. Adler et al. 2006). Although the lowermost UP (non-Aurignacian) layer 4d has not yet provided radiocarbon dates, the basal part of the immediately overlying layer 4c produced a series of charcoal dates ranging between 34,600±600 ^{14}C BP (RTT-4213) and 33,700 ± 620 ^{14}C BP (AA-45864; Bar-Yosef et al. 2006) with a weighted mean of 34,188±328 ^{14}C BP (AA-45864, RRR-4212, RTT-4213, RTT-4214). The uppermost part of the youngest MP horizon at Ortvale Klde, layer 5, produced dates of 39,275±1,200 ^{14}C BP (RTT-3826a) and 37,770±1,000 ^{14}C BP (RTT-3826b) on the same sample of bone (Adler 2002) with a weighted mean of 38,387±768 ^{14}C BP. The age of the earliest UP at Ortvale Klde (layer 4d) can thus be broadly estimated to lie between ca. 38.0 and 35.0 ka ^{14}C BP. These results speak in favour of a clear chronological separation of distinct MP and UP assemblages and provide no evidence for continuity or "transition" between the two.

As in the Caucasus, a clear chronological separation of the MP and UP can be demonstrated on the Italian Peninsula, with the difference that here they are separated by the "transitional" Uluzzian. Advantageous in southern Italy is the possibility of an independent control of the radiocarbon evidence due to the presence of the CI stratigraphic marker (ca. 40.0 ka BP_{GISP2}; Giaccio et al. 2006) discussed above. Since neither MP nor Uluzzian assemblages have ever been shown to overlay these deposits, a final Neanderthal refugium on the Italian peninsula (cf Kuhn and Bietti 2000) dating to significantly after ca. 35.0 ka ^{14}C BP (40.0 kacal BP; Fig. 22.3) is to be rejected.

Conclusions

Examination of the available radiometric data against their stratigraphic context and the consideration of taphonomic criteria show that many ^{14}C ages higher than ca. 30.0 ka ^{14}C BP are problematic. The authors propose that ongoing re-evaluation of the database and refinement of radiocarbon dating procedures will lead to major revisions of many presently widespread assumptions concerning the course of the MP–UP "transition" in western Eurasia.

Proposed evidence for the survival of Neanderthals significantly later than ca. 38.0 ka ^{14}C BP on the southern European peninsulas (Zilhão 2006b) or in other supposed geographical refugia is here considered improbable. On current evidence it can be concluded that Neanderthal disappearance was, on the contrary, a more rapid process which probably started around 39.0–37.0 ka ^{14}C BP. The first appearance of AMH in Europe falls at around this time and, at latest by ca. 35.0 ka ^{14}C BP, these newcomers were the only European hominins. This scenario contradicts models for both a long coexistence of Neanderthals and AMH and the persistence of long-term geographical borders during the process of AMH expansion (e.g. Vega Toscaño 1993; Zilhão 1993). The chain of events within this period might be summarized as follows:

Industries such as the Chatelperronian and the Uluzzian in the southern part of Europe and the leaf or blade point industries to the North are interpreted in this paper as a "Final Middle Palaeolithic" (FMP), ultimately derived from regional late MP substrates. The only hominin evidence from these contexts is provided by Neanderthal remains from Arcy-sur-Cure and Saint Césaire, both associated with the Chatelperronian.

Available data place Chatelperronian assemblages broadly between ca. 41.0 and 38.0 ka ^{14}C BP, although a large number of appreciably younger measurements falls between ca. 36.0 and 31.0 ka ^{14}C BP, with a group of results clustered around 34.0–33.0 ka ^{14}C BP (Fig. 22.20). Many of the younger dates can certainly be explained by effects of sample contamination or poor contextual integrity and younger outliers within otherwise homogeneous series of dates should not be quoted in support of late Neanderthal survival.

Similarly, the limited radiocarbon evidence for the Uluzzian might suggest ages as young as 32.0–31.0 ka ^{14}C BP. However, at Castelcivita the Uluzzian is stratigraphically overlain by the Campanian Ignimbrite (and at other sites by tephras correlated with this) and, at least in the former case, must pre-date ca. 40.0 ka BP_{HULU} (~ 34.8/34.7 ka ^{14}C BP; Fig. 22.3). Given the similarity of the Klisoura V assemblage in Greece to the Uluzzian, the single finite date for the former site (GifA-99168: 40,010 ± 740 ^{14}C BP) would also support a higher age for these eastern Mediterranean industries with curve-backed pieces.

When stratigraphical controls are available it can frequently be demonstrated (cf Giaccio et al. 2006) that the true age of some radiocarbon dates as young as 32.5 ka ^{14}C BP is appreciably older (Fig. 22.3). Nevertheless, these results may represent technically accurate measurements, the discrepancy being due to an age inversion for the critical period around 40.5 ka BP_{HULU} towards the end of the Laschamps geomagnetic excursion. In the absence of stratigraphic or other controls such dates remain impossible to evaluate, creating an unknown error factor.

In the case of the blade and leaf point assemblages of northern Europe the bulk of the radiocarbon dating evidence places them between 40.0 and 37.5 ka ^{14}C BP. A few results marginally older than 35.0 ka ^{14}C BP might imply a younger survival of these industries.

In contrast to the described "transitional" industries interpreted as a FMP phenomenon, a second group with specific laminar lithic production strategies displays technological similarities with the Near Eastern Initial Upper Palaeolithic and is best interpreted as of exogenous origin. Radiocarbon dates for the Bachokirian of south-eastern Europe at the type locality and at Temnata cluster tightly between ca. 39.1 and 36.9 ka ^{14}C BP. The Bachokirian overlaps in time with the Bohunician to the North, which covers the time span ca. 38.5–36.4 ka ^{14}C BP, although some dates are as young as ca. 34.5 ka ^{14}C BP. The Bohunician resembles the Bachokirian but contains foliate points with affinities to FMP leaf-point industries.

It is unclear whether the appearance of the Bachokirian and the Bohunician represents the spread of AMH populations into Europe or reflects the adoption of their technology by pre-existing Neanderthal populations (cf Teyssandier 2006, 14). In the first scenario, the presence of leaf points in the Bohunician would imply adoption of Neanderthal technology by AMH at the northern periphery of their expansion. In the second scenario, the Bachokirian and the Bohunician show the adoption of AMH technology by indigenous hominins, but leaf-points survived as an element of the lithic assemblage in the northern region in which they had developed. In a more differentiated model the Bachokirian might have been made by AMH, whereas the Bohunician could be the result of transferral of AMH technology to Neanderthals further to the North.

The Protoaurignacian ("Aurignacian 0") found along the northern Mediterranean is characterized by a lithic technology for the production of both blades and bladelets within the same chaine opératoire. The Protoaurignacian is technologically and typologically distinct from earlier industries in the region in which it occurs. The first evidence of marine shell ornaments outside Africa and the Near East is regularly found

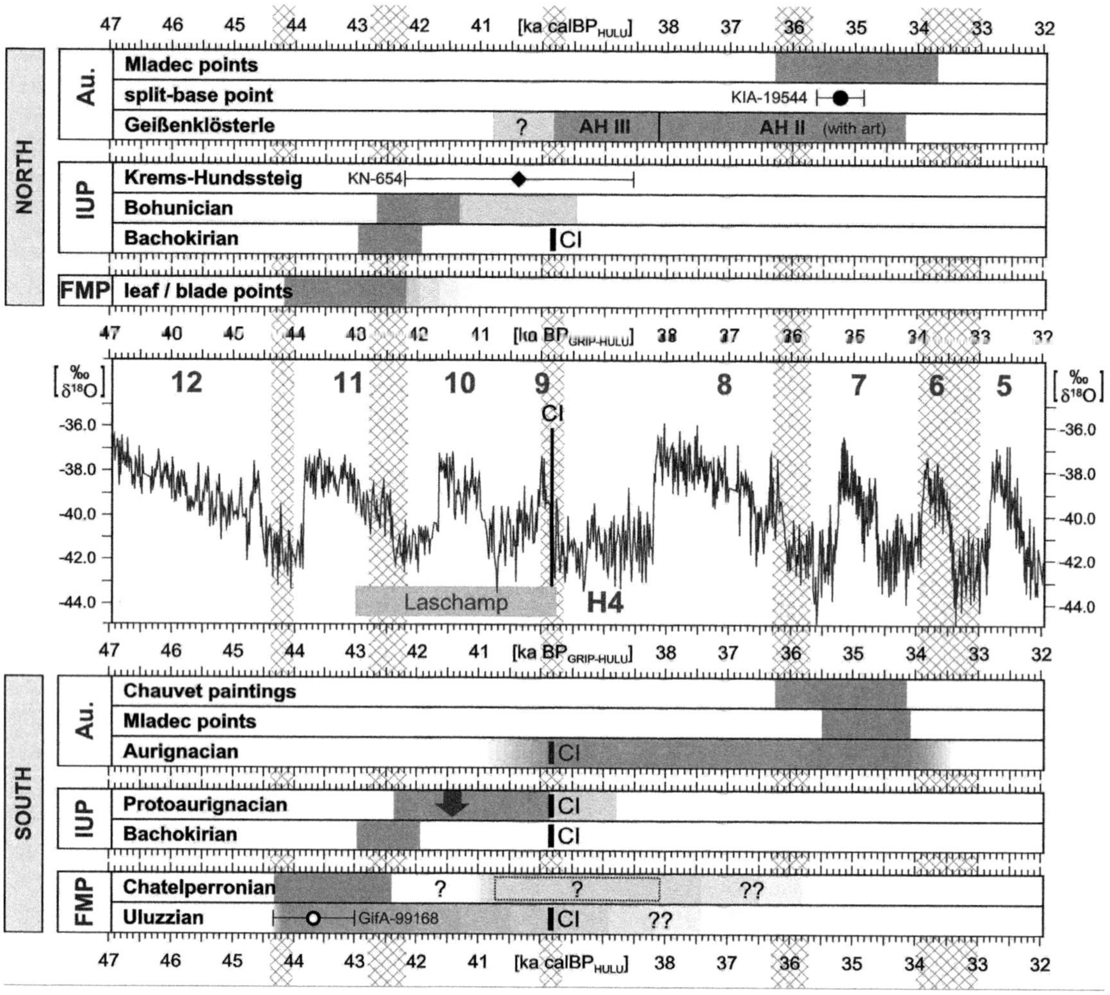

Fig. 22.20 Compilation of calibrated radiocarbon dating ranges for "transitional" and Aurignacian industries in the northern and southern regions of Europe. Hatched bars separate the most probable dating ranges of FMP "transitional" industries (centred around GI 11), IUP "transitional" industries (42.5 ka cal BP_{HULU} – CI event), the early Aurignacian (CI event - 36.0 ka cal BP_{HULU}) and the younger Aurignacian (ca. 36.0–33.5 ka cal BP_{HULU}). For further explanation see text and captions for Figs. 22.3 and 22.15

in Protoaurignacian contexts. In combination, this suggests that the Protoaurignacian may have been made by AMH spreading rapidly westwards (from the Near East?) into Europe while maintaining long distance social networks.

The Protoaurignacian dates to ca. 37.9–34.7 ka ^{14}C BP or only slightly younger. In calendar years it first appears within GS 11 around 42.0 ka_{HULU} and persists until the CI event (Fig. 22.20). The Early Upper Palaeolithic at Kostenki 14, level IVb has provided similar radiocarbon dates to the Protoaurignacian, falling mainly between ca. 36.5 (or possibly 37.2) and 35.0 ka ^{14}C BP. By analogy with the Protoaurignacian, the sparse evidence for marine shell ornament at Kostenki may suggest that the EUP here is due to a simultaneous movement of AMH onto the Russian Plain.

The next well defined stage in the proposed sequence of cultural transition is fully UP in nature – the Aurignacian (here excluding assemblages attributed to the Protoaurignacian/"Aurignacian 0"). Wherever secure stratigraphic control is present the Aurignacian consistently overlies both FMP and IUP/EUP "transitional" assemblages. In Eastern Europe the Aurignacian lies above tephra of the Campanian Ignimbrite eruption (Temnata) and is found at Kostenki mixed with Y5 CI tephra in a cryoturbated layer formed during GS 9 (~ Heinrich 4 cold interval) dated to ca. 40.0 ka BP_{HULU} (Fig. 22.3). On the combined evidence the earliest Aurignacian assemblages across Europe do not significantly pre-date 35.0 ka ^{14}C BP and are more or less contemporary with the CI eruption and the onset of GS 9 (~34.8/34.7 ka ^{14}C BP; Fig. 22.13).

Traditionally, the appearance of the Aurignacian in Europe has often been equated with the arrival of AMH in the region. As discussed above, the first arrival of AMH here probably pre-dates the Aurignacian by as much as 3000–4000 radiocarbon years and any contact between them and indigenous Neanderthals would probably have involved the makers of the Protoaurignacian and/or Bachokirian rather than the Aurignacian (pace Floss 2003).

A plausible alternative explanation for the apparently sudden appearance of the Aurignacian might be offered by the practically simultaneous impacts of the CI mega-eruption and the GS 9 climatic decline on the ecosystem (cf Fedele et al. 2002), which may have acted as catalysts for population shifts leading to social/technological innovation within the pre-existing IUP/EUP.

While it can thus be argued that Aurignacian laminar lithic technology and shell bead personal ornaments (as a proxy for social networks) are "inherited" from earlier assemblages such as the Protoaurignacian or Bachokirian (and ultimately the eastern Mediterranean IUP), other features seem to be innovative and intrinsic only to specific chronological and geographical facies of the Aurignacian. The apparent pan-European lithic technological unity of the Aurignacian would therefore be complemented by diverse other cultural phenomena reflecting diachronic and regional social and ethnic differences. Features characteristic of the second half of the Aurignacian are, for example, Mladeč projectile points, which are directly dated consistently to 32.5–29.2 ka ^{14}C BP, or figurative art, dated at Grotte Chauvet to ca. 32.4–29.7 ka ^{14}C BP and somewhat before this in the case of the earliest portable art in south-western Germany. This is in agreement with contextual dates for Style 1 art in southwestern France (cf Zilhão 2007). These dates equate to period between the end of GI 8 and the onset of GI 6 (Fig. 22.20).

Despite the different medium involved (portable ivory figurines in the southern German Aurignacian/parietal art at Grotte Chauvet), parallels between the motifs involved were recognized soon after the discovery and radiometric dating of the Chauvet paintings. This is particularly true of the animal species depicted, most impressively the large carnivore species lion and bear. It might, however, also be considered whether anthropomorphic depictions in frontal view also belong to this iconic repertoire. There are at least similarities between the pose of the carved figure on a small ivory plaque from Geißenklösterle IIb (Hahn 1988b, Figure 89) and the red painted anthropomorph(s) from Grotta di Fumane (Broglio et al. 2005). Possibly other figures shown standing erect, such as the carving of a bear from Geißenklösterle IIa (Hahn 1988b, Figure 91) or the Stratzing figurine (Neugebauer-Maresch 1989) project a similar message. Several further anthropomorphic motifs, ranging from southwestern German carvings (Vogelherd IV anthropomorph: Riek 1934, Pl. IIa; Hohlenstein-Stadel "Löwenmensch": Hahn 1970; Hohle Fels IV "Löwenmensch": Conard 2003) to the Grotte Chauvet "bison man" therianthrope (Clottes 2001) might all point to the existence of a shared ideological content in the Aurignacian at this period.

We suggest that this particular Aurignacian facies with standardized mammoth ivory and reindeer antler weapon technology and figurative art featuring anthropomorphs/therianthropes and a particular suite of animal depictions, reflects the emergence of hitherto unseen "memetic complexity" within a specific Western/Central European core region which can possibly be equated with the "Aurignacian Homeland" (Bar-Yosef 2006b).

References

Adler, D. S. (2002). *Late Middle Palaeolithic patterns of lithic reduction, mobility, and land use in the Southern Caucasus*. Unpubl. Ph.D dissertation, Harvard University.

Adler, D. S., Belfer-Cohen, A., & Bar-Yosef, O. (2006). Between a rock and a hard place: Neanderthal-modern human interactions in the southern Caucasus. In N. J. Conard (Ed.), *When Neanderthals and modern humans met* (pp. 165–187). Tübingen: Kerns Verlag.

Ahern, J. C. M., Karavanic, I., Paunovic, M., Jankovic, I., & Smith, F. H. (2004). New discoveries and interpretations of hominid fossils and artifacts from Vindija Cave, Croatia. *Journal of Human Evolution, 46*, 27–67.

Akazawa, T., Aoki, K., & Bar-Yosef, O. (Eds.). (1998). Neandertals and modern humans in western Asia. Colloquium Tokio 1995. New York and London.

Albrecht, G., Hahn, J., & Torke, W. G. (1972). *Merkmalanalyse von Geschoßspitzen des mitteren Jungpleistozäns in Mittel- und Osteuropa* (Archaeologica Venatoria, Vol. 2). Tübingen: Verlag Kohlhammer.

Aldhouse-Green, S., & Pettitt, P. B. (1998). Paviland Cave: Contextualizing the "Red Lady". *Antiquity, 72*, 756–772.

Allsworth-Jones, P. (1986). *The Szeletian and the transition from Middle to Upper Palaeolithic in Central Europe*. Oxford: Clarendon.

Alvárez Fernández, E. (2006). *Los objetos de adorno-colgantes del Paleolítico superior del Mesolítico en la Cornisa Cantábrica y en el Valle de del Ebro*. Ed. Universidad de Salamanca. Colección Vítor 195. Salamanca.

Anikovich, M. V., Sinitsyn, A. A., Hoffecker, J. F., Holliday, V. T., Popov, V. V., Lisitsyn, S. N., Forman, S. L., Levkovskaya, G. M., Pospelova, G. A., Kuz'mina, I. E., Burova, N. D., Goldberg, P., Macphail, R. I., Giaccio, B., & Praslov, N. D. (2007). Early Upper Paleolithic in Eastern Europe and implications for the dispersal of modern humans. *Science, 315*, 223–315.

Arrizabalaga, A. (1995). *La industria lítica del Paleolítico Superior en el Oriente cantábrico*. Tesis doctoral, Universidad dl País Vasco.

Azzi, C. M., & Gulisano, F. (1979). Florence Radiocarbon dates IV. *Radiocarbon, 21*, 353–357.

Bailey, S., & Hublin, J. J. (2005). Who made the early Aurignacian? Reconsideration of the Brassempouy dental remains. *Bulletins et Mémoires de la Société d'Anthropologie de Paris, 17*, 115–121.

Bailey, S., & Hublin, J. J. (2006). Dental remains from the Grotte du Renne at Arcy-sur-Cure (Yonne). *Journal of Human Evolution, 50*, 485–508.

Barandiarán Maetzu, I. (1996). Le Paléolithique supérieur au Pays Basque et dans le Bassin de L'Ebre (1990–1995). In M. Otte (Ed.), *Le Paléolithique supérieur européen*. Bilan Quinquennal 1991–1996 (pp. 319–322). UISPP-Colloquium Forlí 1996. E.R.A.U.L. 76. Liège.

Bard, E., Rostek, F., & Ménot-Combes, G. (2004). Radiocarbon calibration beyond 20, 000 ^{14}C yr B.P. by means of planktonic foraminifera of the Iberian Margin. *Quaternary Research, 61*, 204–214.

Bar-Yosef, O. (1998). On the nature of transitions: The Middle to Upper Palaeolithic and the Neolithic revolution. *Cambridge Archaeological Journal, 8*, 141–163.

Bar-Yosef, O. (2000). The Middle and Upper Paleolithic in Southwest Asia and neighbouring regions. In O. Bar-Yosef & D. Pilbeam (Eds.), *The geography of Neandertals and modern humans in*

Europe and the Greater Mediterranean (pp. 107–156). Colloquium Cambridge (Mass.) 1997. Peabody Museum Bulletin 8. Cambridge (Mass.).

Bar-Yosef, O. (2001). Dating the transition from the Middle to Upper Paleolithic. In J.-N. Barrandon, P. Guibert, & V. Michel (Eds.), *Datation. XXIe rencontres internationals d'archeologie et d'histoire d'Antibes* (pp. 279–293). Antibes: Editions APDCA.

Bar-Yosef, O. (2003). Away from home: Prehistoric colonizations, exchanges and diffusions in the Mediterranean basin. In B. Vandermeersch (Ed.), *Echanges et diffusion dans la préhistoire méditerranéenne* (pp. 71–81). Paris: CTHS.

Bar-Yosef, O. (2006a). Defining the Aurignacian. In O. Bar-Yosef & J. Zilhão (Eds.), *Towards a definition of the Aurignacian*. Proceedings of the Symposium held in Lisbon, Portugal, June 25–30, 2002. Instituto Português de Arqueologia.Trabalhos de Arqueologia 45. Artes Graficas, pp. 11–18.

Bar-Yosef, O. (2006b). Neanderthals and modern humans: A different interpretation. In N. J. Conard (Ed.), *When Neanderthals and modern humans met. Tübingen publications in prehistory* (pp. 467–482). Tübingen: Kerns Verlag.

Bar-Yosef, O., & Zilhao, J. (Eds.). (2006). *Towards a definition of the Aurignacian*. Proceedings of the Symposium held in Lisbon, Portugal, June 25–30, 2002, Trabalhos de Arqueologia 45. Istituto Português de Arqueologia, Lisbon.

Bar-Yosef, O., Belfer-Cohen, A., & Adler, D. S. (2006). The implications of the Middle-Upper Paleolithic chronological boundary in the Caucasus to Eurasian Prehistory. *L'Anthropologie, XLIV*, 49–60.

Bazile, F. (1984). Les industries du Paléolithique supérieur en Languedoc oriental. *L'Anthropologie, 88*, 77–88.

Beck, J. W., Richards, D. A., Edwards, R. L., Silverman, B. W., Smart, P. L., Donahue, D. J., Hererra-Osterheld, S., Burr, G. S., Calsoyas, L., Jull, A. J. T., & Biddulph, D. (2001). Extremely large variations of atmospheric ^{14}C concentration during the Last Glacial period. *Science, 292*, 2453–2458.

Bergman, C. A., & Stringer, C. B. (1989). Fifty years after: Egbert, an early Upper Palaeolithic juvenile from Ksar Akil, Lebanon. *Paleorient, 15*, 99–111.

Bernaldo de Quirós, F., Cabrera, V., Lloret, M., & Pike-Tay, A. (2001). New kids on the block? Some comments on the Middle-Upper Paleolithic transition in Cantabrian Spain. In M. A. Hays & P. T. Thacker (Eds.), *Questioning the answers: Re-solving fundamental problems of the Early Upper Paleolithic*. Papers from a Symposium held at the Society for American Archaeology Meeting in Chicago. British Archaeological Reports (International Series) *1005*, 27–38.

Bischoff, J. L., Soler, N., Maroto, J., & Julià, R. (1989). Abrupt Mousterian/Aurignacian boundary at c. 40 ka bp: Accelerator C^{14} dates from L'Arbreda Cave (Catalunya, Spain). *Journal of Archaeological Science, 16*, 563–576.

Bischoff, J. L., Ludwig, K., Garcia, J. F., Carbonell, E., Vaquero, M., Stafford, T. W., & Jull, A. J. T. (1994). Dating of the Basal Aurignacian Sandwich at Abric Romani (Catalunya, Spain) by radiocarbon and uranium-series. *Journal of Archaeological Science, 21*, 541–551.

Bocquet-Appel, J.-P., & Demars, P.-Y. (2000). Neanderthal contraction and modern human colonization of Europe. *Antiquity, 74*, 544–552.

Boëda, É., & Muhesen, S. (1993). Umm el Tlel (El Kowm, Syrie): Étude préliminaire des industries lithiques du Paléolithique moyen et supérieur 1991–1992. *Cahiers de l'Euphrate, 7*, 47–91.

Bolus, M. (2003). The cultural context of the Aurignacian of the Swabian Jura. In J. Zilhão & F. d'Errico (Eds.), *The Chronology of the Aurignacian and of the Transitional Technocomplexes. Dating, Stratigraphies, Cultural Implications* (Trabalhos de Arqueologia, Vol. 33, pp. 153–164). Lisboa: Istituto Português de Arqueologia.

Bolus, M. (2004). Der Übergang vom Mittel-zum Jungpaläolithikum in Europa.Eine Bestandaufnahme unter besonderer Berücksichtigung Mitteleuropas. *Germania, 82*, 1–54.

Bolus, M., & Conard, N. (2001). The late Middle Palaeolithic and earliest Upper Palaeolithic in Central Europe and their relevance for the out of Africa hypothesis. *Quaternary International, 75*, 29–40.

Bolus, M., & Conard, N. (2006). Zur Zeitstellung von Geschoßspitzen aus organischen Materialien im späten Mittelpaläolithikum und Aurignacien. *Archäologisches Korrespondenzblatt, 26*, 1–15.

Bon, F. (2006). A brief overview of Aurignacian cultures in the context of the industries of the transition from the Middle to the Upper Paleolithic. In O. Bar-Yosef & J. Zilhão (Eds.), *Towards a definition of the Aurignacian*. Proceedings of the Symposium held in Lisbon, Portugal, June 25–30, 2002. Instituto Português de Arqueologia. Trabalhos de Arqueologia 45. Istituto Português de Arqueologia, Artes Graficas, pp. 133–144.

Bon, F., Ferrier, C., Gambier, D., & Gardère, P. (1998). Gisement de Brassempouy (Landes): Les recherches de 1995 à 1997, bilan et perspectives. *Bulletin de la Société Borda*, 203–222.

Bordes, J. G. (1998). *L'Aurignacien en Périgord – analyse des données. Un example d'application d'une méthode de quantification des remontages d'intérêt stratigraphique, DEA*. Dissertation, Universität Bordeaux 1.

Bordes, J.-G. (2000). La séquence aurignacienne de Caminade revisitée: l´apport des raccords d´intérêt stratigraphique. *Paléo, 12*, 387–407.

Bordes, J.-G. (2003). Lithic taphonomy of the Châtelperronian/ Aurignacian interstratifications in Roc de Combe and Le Piage (Lot, France). In J. Zilhão & F. d'Errico (Eds.), *The Chronology of the Aurignacian and of the Transitional Technocomplexes. Dating, Stratigraphies, Cultural Implications* (Trabalhos de Arqueologia, Vol. 33, pp. 223–244). Lisboa: Istituto Português de Arqueologia.

Bordes, J.-G. (2006). News from the West: A reevaluation of the classical Aurignacian sequence of the Périgord. In O. Bar-Yosef & J. Zilhão (Eds.), *Towards a definition of the Aurignacian*. Proceedings of the Symposium held in Lisbon, Portugal, June 25–30, 2002. Instituto Português de Arqueologia.Trabalhos de Arqueologia 45. Istituto Português de Arqueologia, Artes Graficas, pp. 147–171.

Bosinski, G. (1967). *Die mittelpaläolithischen Funde im westlichen Mitteleuropa. Fundamenta A/4*. Köln - Graz: Böhlau Verlag.

Bosinski, G. (1989). Die große Zeit der Eiszeitjäger. Europa zwischen 40 000 und 10 000 v. Chr. *Jb Römisch Germanischen Zentralmuseums Mainz, 34*, 3–139.

Bosinski, G. (1990). *Homo sapiens. L'histoire des chasseurs du Paléolithique supérieur en Europe (40.000–10.000 av. J.-C.)*. Paris: Éditions Errance.

Bourguignon, L. (1996). Un Moustérien tardif sur le site d'Umm el Tlel (Bassin d'el Khowm, Syrie)? Exemples des niveaux II base et III2A. In E. Carbonell & M. Vaquero (Eds.), *The Last Neandertals, the first anatomically modern humans. A tale about the Human diversity culture change and human evolution. The crises at 40 KA BP* (pp. 317–336). Colloquium Capellades 1995. Barcelona.

Breuil, H. (1952). *Quatre cents siècles d'art pariétal. Les cavernes ornées de l'âge du renne*. Montignac: Centre d'études et de documentation préhistoriques.

Broglio, A. (1996). Le punte a dorso del Protoaurignaziano mediterraneo: i reperti della Grotta di Fumane (Prealpi venete). In A. Montet-White, A. Palma di Cesnola & K. Valoch (Eds.), *Colloquium 12: The Origin of the Gravettian. U.I.S.P.P., Section 6: The Upper Palaeolithic*. 13th international congress of prehistoric and protohistoric sciences, Forlí 1996. ABACO, Forlí, pp. 237–248.

Broglio, A. (2000). Considerations on the Aurignacian industries from Krems-Hundssteig in the Middle Danube Basin and Fumane in the Adige Basin. In Z. Mester & A. Ringer (Eds.), *À la recherche de l'homme préhistorique*. Volume commémoratif de Miklos Gábori et de Veronika Gábori-Csánk. E.R.A.U.L. 95. Liège, pp. 311–319.

Broglio, A. (2001). Discontinuity between the Mousterian and the Aurignacian: The archaeological sequence from Grotta di Fumane in the Veneto Prealp. In B. Ginter, B. Drobniewicz, B. Kazior, M. Nowak & M. Poltowicz (Eds.), *Problems of the Stone Age in the old world*. Jubilee book dedicated to Professor Janusz K. Kozłowski on his 40th scientific work in Jagiellonian University. Kraków, pp. 119–129.

Broglio, A., & Dalmeri, G. (2005). Pitture Paleolitiche nelle prealpi venete. Grotta di Fumane e Riparo Dalmeri. Memorie del Museo Civico di Storia Naturale di Verona, 2. Serie. Sezione Scienze dell'Uomo 9. Preistoria Alpina, Nr. Speciale. Cierre, Verona.

Broglio, A., & Gurioli, F. (2004). The symbolic behaviour of the first modern humans: The Fumane cave evidence (Venetian pre-Alps). In M. Otte (Ed.), *La Spiritualité*. Actes du colloque de la commission 8 de l'UISPP (Paléeolithique supéerieur), Liège, 10–12 décembre 2003, Études et Recherches Archéologiques de l'Université de Liège 106, Liège, pp. 97–102.

Broglio, A., & Laplace, G. (1966). Etudes de typologie analytique des complexes leptolithiques de l'Europe centrale.I. Les complexes aurignacoïdes de la Basse Autriche. *Rivista di Scienze Preistoriche, 21*, 60–121.

Broglio, A., Cremaschi, M., Peresani, M., de Stefani, M., Bertola, S., Gurioli, S., Marini, D., & Anastasio, G. (2005). Le Pietre dipinte dell'aurignaziano. In A. Broglio & G. Dalmeri (Eds.), *Pitture Paleolitiche nelle prealpi venete. Grotta di Fumane e Riparo Dalmeri* (pp. 38–63). Verona: Museo Civico di storia naturale di Verona.

Bronk Ramsey, C., Higham, T., Bowles, A., & Hedges, R. (2004). Improvements to the pretreatment of bone at Oxford. *Radiocarbon, 46*, 155–163.

Bronk Ramsey, C., Buck, C. E., Manning, S. W., Reimer, P., & van der Plicht, H. (2006). Developments in radiocarbon calibration for archaeology. *Antiquity, 80*, 783–798.

Campbell, J. B. (1977). *The Upper Palaeolithic of Britain. A Study of Man and Nature in the Late Ice Age* (2nd ed.). Oxford: Clarendon.

Canal i Roquet, J., & Carbonell i Roura, E. (1989). *Catalunya paleolitica*. Girona: Patronat Francesc Eiximenis.

Carbonell, E., Giralt, S., & Vaquero, M. (1994). Abric Romani Capellades, Barcelone, Espagne: Une Importante Séquence anthropisée du Pléistocène supérieur. *Bulletin de la Société Préhistorique Française, 91*, 47–55.

Carbonell, E., Vaquero, M., Maroto, J., Rando, J. M., & Mallol, C. (2000). A geographic perspective on the Middle to Upper Paleolithic transition in the Iberian Peninsula. In O. Bar-Yosef & D. Pilbeam (Eds.), *The geography of Neandertals and modern humans in Europe and the Greater Mediterranean* (pp. 5–34). Cambridge: Harvard University.

Chabai, V. (2003). The chronological and industrial variability of the Middle to Upper Palaeolithic transition in eastern Europe. In J. Zilhão & F. d'Errico (Eds.), *The chronology of the Aurignacian and of the transitional technocomplexes. Dating, stratigraphies, cultural implications*. Proceedings of Symposium 6.1 of the 14. Congress of the U.I.S.P.P., Liège 2001. Trabalhos de Arquelogia 33. Istituto Português de Arqueologia, Lissabon, pp. 71–86.

Chiu, T-Ch, Fairbanks, R. G., Cao, L., & Mortloc. (2007). Analysis of the atmospheric ^{14}C record spanning the past 50,000 years derived from high-precision $^{230}Th/^{234}U/^{238}U$, $^{231}Pa/^{235}U$ and ^{14}C dates on fossil corals. *Quaternary Science Reviews, 26*, 18–36.

Churchill, S. E., & Smith, F. H. (2000). Makers of the early Aurignacian of Europe. *Yearbook of Physical Anthropology, 43*, 61–115.

Churchill, S., Formicola, V., Holliday, T., Holt, B., & Schumann, B. (2000). The Upper Palaeolithic population of Europe in an evolutionary perspective. In W. Roebroeks, M. Mussi, J. Svoboda & K. Fennema (Eds.), *Hunters of the Golden Age. The mid Upper Palaeolithic of Eurasia 30,000–20,000 BP* (pp. 31–57). Leiden: University Leiden.

Clottes, J. (Ed.). (2001). *La Grotte Chauvet: L'Art des Origines*. Paris: Seuil.

Conard, N. J. (2003). Palaeolithic ivory sculptures from southwestern Germany and the origins of figurative art. *Nature, 426*, 830–832.

Conard, N. J., & Bolus, M. (2003). Radiocarbon dating the appearance of modern humans and timing of cultural innovations in Europe: New results and new challenges. *Journal of Human Evolution, 44*, 331–371.

Conard, N. J., & Bolus, M. (2006). The Swabian Aurignacian and its place in European Prehistory. In: O. Bar-Yosef & J. Zilhao (Eds.), *Towards a definition of the Aurignacian*. Proceedings of the Symposium held in Lisbon, Portugal, June 25–30, 2002, Trabalhos de Arqueologia 45. Istituto Português de Arqueologia, Lisbon, pp. 211–239.

Conard, N., Dippon, G., & Goldberg, P. (2003a). Chronostratigraphy and archaeological context of the Aurignacian deposits at Geißenklösterle Cave. In J. Zilhão & F. d'Errico (Eds.), *The chronology of the Aurignacian and of the transitional technocomplexes: dating, stratigraphies, cultural implications* (Trabalhos de Arqueologia, Vol. 33, pp. 165–178). Lisboa: Istituto Português de Arqueologia.

Conard, N. J., Niven, L. B., Mueller, K., & Sturart, A. J. (2003b). The Chronostratgraphy of the Upper Palaeolithic Deposits at Vogelherd. *Mitteilungen der Gesellschaft für Urgeschichte, 12*, 73–86.

Conard, N. J., Grootes, P. M., & Smith, F. H. (2004). Unexpectedly recent dates for human remains from Vogelherd. *Nature, 430*, 198–201.

Conard, N. J., Bolus, M., Goldberg, P., & Münzel, S. (2006). The Last Neanderthals and first modern humans in the Swabian Jura. In N. J. Conard (Ed.), *When Neanderthals and modern humans met* (pp. 305–341). Tübingen: Kerns Verlag.

Currat, M., & Excoffier, L. (2004). Modern Humans did not admix with Neanderthals during their range expansion into Europe. *PLoS Biology, 2*, 2264–2274.

de Sonneville-Bordes, D. (1960). *Le paléolithique supérieur en Périgord*. Bordeaux: Delmas.

De Vivo, B., Rolandi, G., Gans, P. B., Calvert, A., Bohrson, W. A., Spera, F. J., & Belkin, H. E. (2001). New constraints on the pyroclastic eruptive history of the Campanian volcanic Plain (Italy). *Mineralogy and Petrology, 73*, 47–65.

d'Errico, F. (2003). The invisible frontier. A multiple species model for the origin of behavioral modernity. *Evolutionary Anthropology, 12*, 188–202.

d'Errico, F., & Sánchez Goñi, M. F. (2003). Neanderthal extinction and the millennial scale climatic variability of OIS 3. *Quaternary Science Reviews, 22*, 769–788.

d'Errico, F., & Sánchez Goñi, M. F. (2004). A Garden of Eden for the Gibraltar Neandertals? A reply to Finlayson et al. *Quaternary Science Reviews, 23*, 1210–1216.

d'Errico, F., Zilhão, J., Baffier, D., Julien, M., & Pelegrin, J. (1998). Neanderthal acculturation in Western Europe? A critical review of the evidence and its interpretation. *Current Anthropology, 39* (Supp.), 1–44.

d'Errico, F., Julien, M., Liolios, D., Vanhaeren, M., & Baffier, D. (2003). Many awls in our argument. Bone tool manufacture and use in the Chatelperronian and Aurignacian levels of the Grotte du Renne at Arcy-sur-Cure. In J. Zilhão & F. d'Errico (Eds.), *The chronology of the Aurignacian and of the transitional technocomplexes: Dating, stratigraphies, cultural implications* (Trabalhos de Arqueologia, Vol. 33, pp. 247–270). Lisboa: Istituto Português de Arqueologia.

Damblon, F., Haesarts, P., & van der Plicht, J. (1996). New datings and considerations on the Chronology of Upper Palaeolithic sites in the Great Eurasiatic Plain. *Préhistoire Européenne, 9*, 177–231.

David, F., Connet, N., Girard, M., Lhomme, V., Miskovsky, J.-C., & Roblin-Jouve, A. (2001). Le Châtelperronien de la grotte du Renne à Arcy-sur-Cure (Yonne).Données sédimentologiques et chronologiques. *Bulletin de la Société Préhistorique Française, 98*, 207–230.

Davies, W. (2000). Archaeology Database. http://www.esc.cam.ac.uk/oistage3.

Delibrias, G., & Evin, J. (1980). Sommaire des datations ^{14}C concernant la préhistoire en France. II. Dates parues de 1974 à 1978. *Bulletin de la Société Préhistorique Française, 77*, 215–224.

Delibrias, G., Guillier, M.-Th., Evin, J., Thommeret, J., & Thommeret, Y. (1976). Datations absolues des dépôts quaternaires et des sites préhistoriques par la méthode du Carbone 14. In H. de Lumley, (Ed.), *La préhistoire française I* (pp. 1499–1514). Paris.

Delluc, B., & Delluc, G. (1978). Les manifestations graphiques aurignaciennes sur support rocheux des environs des Eyzies (Dordogne). *Gallia Préhistoire, 21*, 213–332.

Demars, P.-Y. (1996). La place du Piage et de Roc de Combe (Lot) dans la transition du Paléolithique moyen au Paléolithique supérieur.Bulletin Préhistorique du Sud-Ouest. *Nouvelles Etudes, 3*, 11–35.

Demidenko, Y. E. (2002). The European Early Aurignacian of Krems-Dufour type industries: A view from eastern Europe. *Préhistoire Européenne, 16–17*, 147–162.

Djindjian, F., Kozlowski, J., & Bazile, F. (2003). Europe during the early Upper Palaeolithic (40000–30000 BP): A synthesis. In J. Zilhão & F. d'Errico (Eds.), *The chronology of the Aurignacian and of the transitional technocomplexes dating, stratigraphies, cultural implications* (Trabalhos de Arqueologia, Vol. 33, pp. 29–48). Lisboa: Istituto Português de Arqueologia.

Dunbar, R. (2004). *The human story. A new history of mankind's evolution*. London: Faber & Faber.

Eswaran, V., Harpending, H., & Rogers, A. R. (2005). Genomics refutes an exclusively African origin of humans. *Journal of Human Evolution, 49*, 1–18.

Evin, J., Marechal, J., & Marien, G. (1983). Lyon natural radiocarbon measurements IX. *Radiocarbon, 25*, 59–128.

Excoffier, L. (2006). Neandertal genetic diversity: A fresh look from old samples. *Current Biology, 16*, 650–652.

Fairbanks, R. G., Mortlock, R. A., Chiu, T.-C., Cao, L., Kaplan, A., Guilderson, T. P., Fairbanks, T. W., Bloom, A. L., Grootes, P. M., & Nadeau, M.-J. (2005). Radiocarbon calibration curve Spanning 0 to 50, 000 Years B.P. based on paired ^{230}Th/^{234}U/^{238}U and ^{14}C dates on Pristine corals. *Quaternary Science Reviews, 24*, 1781–1796.

Fedele, F. G., Giaccio, B., Isaia, R., & Orsi, G. (2002). Ecosystem impact of the Campanian Ignimbrite Eruption in Late Pleistocene Europe. *Quaternary Research, 57*, 420–424.

Fedele, F. G., Giaccio, B., Isaia, R., & Orsi, G. (2003). The Campanian Ignimbrite eruption, Heinrich Event 4, and the Palaeolithic change in Europe: A high-resolution investigation. In A. Robock & C. Oppenheimer (Eds.), *Volcanism and the Earth's atmosphere. Geophys. Mon. 139* (pp. 301–325). Washington (DC): American Geophysical Union.

Felgenhauer, F. (1959). Willendorf in der Wachau. Monographie der Paläolith-Fundstellen I–VII. Mitteilungen der Prähistorischen Kommission der Österreichischen Akademie der Wissenschaften 8 & 9, 1956–1959. Wien: Rohrer.

Finlayson, C., Fa, D. A., Finlayson, G., Giles Pacheco, F., & Rodríguez Vidal, J. (2004). Did the Moderns kill off the Neanderthals? A reply to d'Errico & Sánchez Goñi. *Quaternary Science Reviews, 23*, 1205–1209.

Finlayson, C., Giles Pacheco, F., Rodríguez-Vidal, J., Fa, D. A., Guiterrez López, J. M., Santiago Pérez, A., Finlayson, G., Allue, E., Baena Preysler, J., Cáceres, I., Carrión, J. S., Fernández- Jalvo, Y., Gleed-Owen, C. P., Jimenez Espejo, F. J., López, P., López Sáez, J. A., Riquelme Cantal, J. A., Sánchez Marco, A., Giles Guzman, F., Brown, K., Fuentez, N., Valarino, C. A., Villalpando, A., Stringer, C. B., Martinez Ruiz, F., & Sakamoto, T. (2006). Late survival of Neanderthals at the southernmost extreme of Europe. *Nature* (advance online publication, 13 September, doi:10.1038/nature05195).

Floss, H. (2003). Did they meet or not? Observations on Châtelperronian and Aurignacian settlement patterns in eastern France. In J. Zilhão & F. d'Errico (Eds.), *The chronology of the Aurignacian and of the transitional technocomplexes dating, stratigraphies, cultural implications* (Trabalhos de Arqueologia, Vol. 33, pp. 273–287). Lisboa: Istituto Português de Arqueologia.

Foley, R. A., & Lahr, M. M. (1997). Mode 3 technologies and the evolution of modern humans. *Cambridge Archaeological Journal, 7*, 3–36.

Forster, P. (2004). Ice Ages and the mitochondrial DNA chronology of human dispersals: A review. *Philosophical Transactions of the Royal Society of London, Series B, 359*, 255–264.

Forster, P., & Matsumura, S. (2005). Did early humans go north or south? *Science, 308*, 965–966.

Fortea Pérez, J. (1996). Le Paléolithique supérieur en Espagne: Galice et Asturies (1991–1995). In M. Otte (Ed.), *Le Paléolithique supérieur européen. Bilan quinquennal 1991–1996* (pp. 329–344). E.R.A.U.L. 76. Liège.

Fortea Pérez, J. (1999). Abrigo de La Viña. Informe y primera valoración de las campañas de 1995 a 1998. *Excavationes Arqueológicas en Asturias 1995–1998* (pp. 31–41). Oviedo.

Freund, G. (1952). *Die Blattspitzen des Paläolithikums in Europa*. Bonn: Quartär-Bibliothek1.

Galván Santos, B., Hernández Gómez, C. M., Francisco Ortega, M. I., & Rodríguez Rodríguez, A. (2006). Datos para la caracterización del final del Musteriense en los valles de Alcoi. In: V. Cabrera Valdés, F. Bernaldo de Quirós Guidotti, J. M. Maíllo Fernández (Eds.), *En el centario de la Cueva de el Castillo: El ocaso de los Neandertales* (pp.127–141). Cantabria: UNED.

Gambassini, P. (Ed.). (1997). *Il Paleolitico di Castelcivita, culture e ambiente. Materiae no.5*. Naples: Electa.

Gambier, D. (1989). Fossil hominids from the Early Upper Palaeolithic (Aurignacian) of France. In P. Mellars & Ch. C. Stringer (Eds.), *The human revolution* (pp. 194–211). Edinburgh: Edinburgh University Press.

Giaccio, B., Hajdas, I., Peresani, M., Fedele, F. G., & Isaia, R. (2006). The Campanian Ignimbrite tephra and its relevance for the timing of the Middle to Upper Palaeolithic shift. In N. J. Conard (Ed.), *When Neanderthals and Modern Humans met* (pp. 343–375). Tübingen: Kerns Verlag.

Ginter, G., Kozlowski, J. K., Laville, H., Sirakov, N., & Hedges, R.E.M. (1996). Transition in the Balkans: News from the Temnata Cave, Bulgaria. In E. Carbonell & M. Vaquero (Eds.), *The last Neandertals, the first anatomically modern humans* (pp. 169–200). Tarragona: Universitat Rovira & Virgili.

Gioia, P. (1990). An aspect of the transition between Middle and Upper Paleolithic in Italy: The Uluzzian. In C. Farizy (Ed.), *Paléolithique moyen récent et Paléolithique supérieur ancien en Europe* (Mémoires du Musée de Préhistoire de l'Ile de France, Vol. 3, pp. 241–250). Nemours: Édition A.P.R.A.I.F.

Golovanova, L.V. (1998). Anthropological finds from the Palaeolithic sites of the northern Caucasus. In: C. Peretto & C. Giunchi (Eds.), *Atti del 13. congresso*, Vol. 2. U.I.S.P.P. Colloquium Forlì 1996 (pp. 397–400). ABACO, Forlì.

Gravina, B., Mellars, P., & Bronk Ramsey, C. (2005). Radiocarbon dating of interstratified Neanderthal and early modern human occupations at the Chatelperronian type site. *Nature, 438*, 51–56.

Green, R. E., Krause, J., Ptak, S. E., Briggs, A. W., Ronan, M. T., Simons, J. F., Du, L., Egholm, M., Rothberg, J. M., Paunovic, M., & Svante Pääbo, S. (2006). Analysis of one million base pairs of Neanderthal DNA. *Nature, 444*, 330–336.

Grün, R., Stringer, C., McDermott, F., Nathan, R., Porat, N., Robertson, S., Taylor, L., Mortimer, G., Eggins, S., & McCulloch,

M. (2005). U-series and ESR analyses of bones and teeth relating to the human burials from Skhul. *Journal of Human Evolution, 49*(3), 316–334.

Grünberg, J. M. (2006). New AMS dates for palaeolithic and Mesolithic camp sites and single finds in Saxony-Anhalt and Thuringia (Germany). *Proceedings of the Prehistoric Society, 72*, 95–112.

Guilbaud, M. (1993). Debitage from the Upper Castelperronian level at Saint-Césaire: Methodological approach and implications for the transition from Middle to Upper Paleolithic. In F. Léveque, A. Backer, & M. Guilbaud (Eds.), *Context of a Late Neandertal* (pp. 37–58). Madison: Prehistory Press.

Haesaerts, P, & Teyssandier, N. (2003). The early Upper Paleolithic occupations of Willendorf (Lower Austria): A contribution to the chronostratigraphic and cultural context of the beginning of the Upper Paleolithic in central Europe. In: J. Zilhão & F. d'Errico (Eds.), *The chronology of the Aurignacian and of the transitional technocomplexes: dating, stratigraphies, cultural implications*. Proceedings of Symposium 6.1 of the XIVth Congress of the U.I.S.P.P.. Trabalhos de Arqueologia; 33. Istituto Português de Arqueologia, Lisboa, pp. 133–151.

Hahn, J. (1970). Die Stellung der männlichen Statuette aus dem Hohlenstein-Stadel in der jungpaläolithischen Kunst. *Germania, 48*, 1–12.

Hahn, J. (1972). Aurignacian signs, pendants and art objects in central and eastern Europe. *World Archaeology, 3*, 252–266.

Hahn, J. (1977). *Aurignacien. Das ältere Jungpaläolithikum in Mittel- und Osteuropa. Fundamenta Reihe A 9*. Köln - Wien: Böhlau Verlag.

Hahn, J. (1988a). 1. Fiche sagaie à base simple de tradition aurignacienne. In: H. Delporte, J. Hahn, L. Mons, G. Pinçon & D. de Sonneville-Bordes (Eds.), *Fiches typologiques de l'industrie osseuse préhistorique. Paléolithique supérieur*. Cahier I: Sagaies, 1–17. Aix-en-Provence.

Hahn, J. (1988b). *Die Geißenklösterle-Höhle im Achtal bei Blaubeuren I. Fundhorizontbildung und Besiedlung im Mittelpaläolithikum und im Aurignacien. Forschungen und Berichte zu Vor- und Frühgeschichte in Baden-Württemberg 26*. Stuttgart: Konrad Theiss.

Hahn, J. (1988c). 2. Fiche sagaie à base fendue. In H. Delporte, J. Hahn, L. Mons, G. Pinçon & D. de Sonneville-Bordes (Eds.), *Fiches typologiques de l'industrie osseuse préhistorique. Paléolithique supérieur*. Cahier I: Sagaies, 1–21. Aix-en-Provence.

Hahn, J. (1989). *Genese und Funktion der jungpaläolithischen Freilandstation Lommersum im Rheinland*. Rheinische Ausgrabungen 29. Cologne.

Hahn, J. (1995). Neue Beschleuniger-^{14}C-Daten zum Jungpaläolithikum in Südwestdeutschland. *Eiszeitalter und Gegenwart, 45*, 86–92.

Hawks, J., & Wolpoff, M. H. (2001). Brief communication: Paleoanthropology and the population genetics of ancient genes. *American Journal of Physical Anthropology, 114*, 269–272.

Hedges, R.E.M, & van Klinken, G.J. (1992). A review of current approaches in the pretreatment of bone for radiocarbon dating by AMS. In A. Long, & R. S. Kra (Eds.), Proceedings of the 14th International ^{14}C conference. *Radiocarbon, 34*, 279–291.

Hedges, R. E. M., Housley, R. A., Law, I. A., & Bronk Ramsey, C. (1989). Radiocarbon dates from the Oxford AMS system. Archaeometry Datelist 9. *Archaeometry, 31*, 207–234.

Hedges, R. E. M., Housley, R. A., Law, I. A., & Bronk Ramsey, C. (1990). Radiocarbon dates from the Oxford AMS system. Archaeometry Datelist 10. *Archaeometry, 32*, 101–108.

Hedges, R. E. M., Housley, R. A., Bronk Ramsey, C., & van Klinken, G. J. (1994). Radiocarbon dates from the Oxford AMS system: Archaeometry Datelist 18. *Archaeometry, 36*, 337–374.

Hedges, R. E. M., Pettitt, P. B., Bronk Ramsey, C., & van Klinken, G. J. (1996). Radiocarbon dates from the Oxford AMS system: Archaeometry datelist 22. *Archaeometry, 38*, 391–415.

Hedges, R. E. M., Pettitt, P. B., Bronk Ramsey, C., & van Klinken, G. J. (1998). Radiocarbon dates from the Oxford AMS system: Archaeometry datelist 26. *Archaeometry, 40*, 437–455.

Henry-Gambier, D. (2002). Les fossiles de Cro-Magnon (Les Eyzies-de-Tayac, Dordogne): Nouvelles donées sur leur position chronologique et leur attribution culturelle. *Paléo, 14*, 201–204.

Henry-Gambier, H., Maureille, B., & White, R. (2004). Vestiges humains des niveaux de l'Aurignacien ancien du site de Brassempouy (Landes). *Bulletins et Mémoires de la Société d'Anthropologie de Paris, n.s., 16*, 49–87.

Higham, T., Bronk Ramsey, C., Karavanic, I., Smith, F. H., & Trinkaus, E. (2006a). Revised direct radiocarbon dating of the Vindija G1 Upper Palaeolithic Neandertals. *Proceeding of the National Academy of Sciences, 103*, 553–557.

Higham, T. F. G., Jacobi, R. M., & Bronk Ramsey, C. (2006b). AMS radiocarbon dating of ancient bone using ultrafiltration. *Radiocarbon, 48*, 179–195.

Hillgruber, F. (2006). The Middle Palaeolithic stone artefacts from the site in the Neander Valley. In R. W. Schmitz (Ed.), *Neanderthal 1856–2006* (Rheinische Ausgrabungen, Vol. 58, pp. 111–144). Mainz: Philipp von Zabern.

Hofreiter, M., & Pacher, M. (2004). Using ancient DNA to elucidate Raw material Origin of Bone points from Potocka zijalka (Slovenia): preliminary Results. In M. Pacher, V. Pohar & G. Rabeder (Eds.), *Potocka zijalka – palaeontological and archaeological results of the excavation campaigns 1997–2000* (pp. 201–210). Mitt. Komm. Quartärforsch. Österr. Akad. Wiss. 13. Wien.

Hovers, E. (2006). Neanderthals and modern humans in the Middle Paleolithic of the Levant: What kind of interaction? In N. J. Conard (Ed.), *When Neanderthals and Modern Humans met* (pp. 65–85). Tübingen: Kerns Verlag.

Hublin, J.-J., Barroso Ruiz, C., Medina Lara, P., Fontugne, M., & Reyss, J.-L. (1995). The Mousterian site of Zafarraya (Andalucia, Spain): Dating and implications on the palaeolithic peopling processes of Western Europe. *Comptes Rendus de l'Academie de Sciences de Paris série IIa, 321*, 931–937.

Hublin, J.-J., Spoor, F., Braun, M., Zonneveld, F. W., & Condemi, S. (1996). A late Neanderthal associated with Upper Palaeolithic artefacts. *Nature, 381*, 224–226.

Hughen, K., Southon, J., Lehman, S., Bertrand, C., & Turnbull, J. (2006). Marine-derived ^{14}C calibration and activity record for the past 50,000 years updated from the Cariaco Basin. *Quaternary Science Reviews, 25*, 3216–3227.

Hülle, W. M. (1977). *Die Ilsenhöhle unter Burg Ranis/Thüringen*. Stuttgart/New York: Eine paläolithische Jägerstation #Verlag.

Jacobi, R. M. (1990). Leaf-points and the British early Upper Palaeolithic. In J. K. Kozlowski (Ed.), *Feuilles de Pierre. Les industries à pointes foliacées du paléolithique supérieur européen* (pp. 271–289). Actes du Colloque de Cracovie 1989. E.R.A.U.L. 42. Liége.

Jacobi, R. M. (1999). Some observations on the British Earlier Upper Palaeolithic. In W. Davies & R. Charles (Eds.), *Dorothy Garrod and the progress of the Palaeolithic. Studies in the prehistoric archaeology of the Near East and Europe* (pp. 35–40). Oxford: Oxbow Books.

Jacobi, R. M., & Pettitt, P. B. (2000). An Aurignacian point from Uphill Quarry (Somerset) and the earliest settlement of Britain by Homo sapiens. *Antiquity, 74*, 513–518.

Jacobi, R. M., Higham, T. F. G., & Bronk Ramsey, C. (2006). AMS radiocarbon dating of Middle and Upper Palaeolithic bone in the British Isles: Improved reliability using ultrafiltration. *Journal of Quaternary Science, 21*, 557–573.

Johnsen, S. J., Clausen, H. B., Dansgaard, W., Fuhrer, K., Gundestrup, N., Hammer, C. U., Iversen, P., Jouzel, J., Stauffer, B., & Steffensen, J. P. (1992). Irregular glacial interstadials recorded in a new Greenland ice core. *Nature, 359*, 311–313.

Johnsen, S. J., Dahl-Jensen, D., Gundestrup, N., Steffensen, J. P., Clausen, H. B., Miller, H., Masson-Dulmeotte, V., Sveinbjørndottir,

A. E., & White, J. (2001). Oxygen isotope and palaeotemperature records from six Greenland ice-core stations: Camp Century, Dye-3, GRIP GISP2, Renland and NorthGRIP. *Journal of Quaternary Science, 16/4*, 299–307.

Jöris, O. (2004). Zur chronostratigraphischen Stellung der spätmittelpaläolithischen Keilmessergruppen.Der Versuch einer kulturgeographischen Abgrenzung einer mittelpaläolithischen Formengruppe in ihrem europäischen Kontext. *Berichte der Römisch-Germanischen Kommission, 84*, 49–153.

Jöris, O. (2005). Aus einer anderen Welt – Europa zur Zeit des Neandertalers. In N. J. Conard, St Kölbl, & W. Schürle (Eds.), *Vom Neandertaler zum Modernen Menschen. Alb und Donau – Kunst und Kultur 46* (pp. 47–70). Ostfildern: Thorbecke Verlag.

Jöris, O., Álvarez Fernández, E., & Weninger, B. (2003). Radiocarbon evidence of the Middle to Upper Palaeolithic transition in southwestern Europe. *Trabajos de Prehistoria, 60*, 15–38.

Jöris, O., Street, M., Terberger, Th, & Weninger, B. (2006). Dating the transition. In W. von Koenigswald & Th Litt (Eds.), *150 Years of Neanderthal Discoveries* (2006/2nd ed., pp. 68–73). Terra Nostra: Early Europeans – Continuity & Discontinuity.

Karavanić, I. (2000). Olschewian and appearance of bone technology in Croatia and Slovenia. In J. Orschiedt & G.-C. Weniger (Eds.), *Neanderthals and modern humans - Discussing the transition: Central and Eastern Europe from 50,000–30,000 B.P* (Wissenschaftliche Schriften des Neanderthal Museums, Vol. 2, pp. 159–168). Mettmann: Neanderthal Museum.

Karavanić, I., & Smith, F. H. (1998). The Middle/Upper Paleolithic interface and the relationship of Neanderthals and early modern humans in the Hrvatsko Zagorje, Croatia. *Journal of Human Evolution, 34*, 223–248.

Koumouzelis, M., Ginter, B., Kozlowski, J. K., Pawlikiwski, M., Bar-Yosef, O., Albert, R. M., Litynska-Zajac, M., Stworzewicz, E., Wojtal, P., Lipecki, G., Tomek, T., Bochenski, Z. M., & Pazdur, A. (2001a). The Early Upper Palaeolithic in Greece: The excavations in Klisoura Cave. *Journal of Archaeological Science, 28*, 515–539.

Koumouzelis, M., Kozlowski, J., Escutenaire, C., Sitlivy, V., Sobczyk, K., Valladas, H., Tisnerat-Laborde, N., Wojtal, P., & Ginter, B. (2001b). La fin du Paléolithique moyen et le début du Paléolithique supérieur en Grèce: la séquence de la grotte 1 de Klissoura. *L´Anthropologie, 105*, 469–504.

Kozłowski, J. K. (Ed.). (1982). *Excavation in the Bacho Kiro Cave (Bulgaria), final Report*. Warsaw: Polish Scientific Publishers.

Kozłowski, J. K. (1992). The Balkans in the Middle and Upper Palaeolithic: The gate to Europe or a Cul-de-sac. *Proceedings of the Prehistoric Society, 58*, 1–20.

Kozłowski, J.K. (2006). A dynamic view of Aurignacian technology. In O. Bar-Yosef & J. J. Zilhao (Eds.), *Towards a definition of the Aurignacian*. Proceedings of the Symposium held in Lisbon, Portugal, June 25–30, 2002, Trabalhos de Arqueologia 45. Istituto Português de Arqueologia, Lisbon, pp. 21–34.

Kozłowski, J. K., & Otte, M. (2000). The formation of the Aurignacian in Europe. *Journal of Anthropological Research, 56*, 513–534.

Krings, M., Capelli, C., Tschentscher, F., Geisert, H., Meyer, S., von Haeseler, A., Grossschmidt, K., Possnert, G., Paunovic, M., & Pääbo, S. (2000). A view of Neandertal genetic diversity. *Nature Genetics, 26*, 144–146.

Kuhn, S. L. (2003). In what sense is the Levantine Initial Upper Palaeolithic a "transitional" industry? In J. Zilhão & F. d'Errico, (Eds.), *The chronology of the Aurignacian and of the transitional technocomplexes: dating, stratigraphies, cultural implications*. Proceedings of Symposium 6.1 of the XIVth Congress of the U.I.S.P.P. Trabalhos de Arqueologia 33. Istituto Português de Arqueologia, Lisbon, pp. 61–69.

Kuhn, S. L. (2004). From Initial Upper Paleolithic to Ahmarian at Üçagizli Cave, Turkey. *L'Anthropologie, 42*(3), 249–262.

Kuhn, S. L., & Bietti, A. (2000). The Late Middle and Early Upper Paleolithic in Italy. In O. Bar-Yosef & D. Pilbeam (Eds.), *The geography of Neandertals and modern humans in Europe and the Greater Mediterranean* (pp. 49–76). Cambridge Mass: Harvard University Press.

Kuhn, S. L., & Stiner, M. (1998). The earliest Aurignacian of Riparo Mochi (Liguria, Italy). *Current Anthropology, 39*, 175–189.

Kuhn, S. L., Stiner, M. C., & Güleç, E. (1999). Initial Upper Palaeolithic in south-central Turkey and its regional context: A preliminary report. *Antiquity, 73*, 505–517.

Kuhn, S. L., Stiner, M. C., Reese, D. S., & Güleç, E. (2001). Ornaments of the earliest Upper Paleolithic: New insights from the Levant. *Proceedings of the National Academy of Sciences of the United States of America, 98*, 7641–7646.

Kuzmin, Y.V., Burr, G.S., Jull, A.J.T., & Sulerzhitsky, L.D. (2004) AMS ^{14}C age of the Upper Palaeolithic skeletons from Sungir site, Central Russian Plain. *Nuclear Instruments and Methods in Physics Research B*, 223–224, 731–734.

Lahr, M. M., & Foley, R. A. (1998). Towards a theory of modern human origins: Geography, demography, and diversity in recent human evolution. *Yearbook of Physical Anthropology, 41*, 137–176.

Lalueza-Fox, C., Lourdes Sampietro, M., Caramelli, D., Puder, Y., Lari, M., Calafell, F., Martinez-Maza, C., Bastir, M., Fortea, J., de la Rasilla, M., Bertranpetit, J., & Rosas, A. (2005). Neandertal evolutionary genetics: Mitochondrial DNA data from the Iberian Peninsula. *Molecular Biology and Evolution, 22*, 1077–1081.

Lambert, D. M., & Millar, C. D. (2006). Ancient genomics is born. *Nature, 444*, 275–276.

Laplace, G. (1966). Recherches sur l'origine et l'évolution des complexes leptolithiques. *École Française de Rome, Mélanges d'Archéologie et d'Histoire* (4th ed.). Paris: Éditions de Boccard.

Laplace, G. (1970). L'industrie de Krems Hundssteig et le problème de l'origine des complexes aurignaciens. In K. Gripp (Ed.), *Archäologische Beiträge. Frühe Menschheit und Umwelt I. Fundamenta A/2* (pp. 242–297). Köln Wien: Böhlau Verlag.

Laplace, G. (1977). Il Riparo Mochi ai Balzi Rossi di Grimaldi (Fouilles 1938–1949): Les industries leptolithiques. *Rivista di Scienze Preistoriche, 32*, 3–131.

Leroi-Gourhan, A. (1958). Étude des restes humains, fossiles provenant des grottes d'Arcy-sur-Cure. *Annales de Paleontologie, 44*, 87–148.

Leroi-Gourhan, A. (1968). Le petit racloir châtelperronien. *La préhistoire, problèmes et tendances* (pp. 275–282). Paris: CNRS.

Leroi-Gourhan, A. (1971). *Préhistoire de l'art occidental*. Paris: Éditions d'art Lucien Mazenod.

Lévêque, F., & Vandermeersch, B. (1980). Paléontologie humaine: Découverte de restes humaines dans un niveau Castelperronien à Saint-Césaire (Charente Maritime). *Comptes Rendues de l'Académie Nationale des Sciences, D 291*, 187–189.

Liolios, D. (2006). Reflections on the role of bone tools in the definition of the Early Aurignacian. In O. Bar-Yosef & J. Zilhao (Eds.), *Towards a definition of the Aurignacian*. Proceedings of the Symposium held in Lisbon, Portugal, June 25–30, 2002. Trabalhos de Arqueologia 45. Lisbon, pp. 37–51.

Llana Rodríguez, C., & Soto Barreiro, M.J. (Eds.). (1991). Cova da Valiña (Castroverde, Lugo). Un xacemento do Paleolítico Superior inicial en Galicia (Campañas de 1987 e 1988). Arqueoloxia/ Investigacíon 5. Xunta de Galicia.

Macaulay, V., Hill, C., Achilli, A., Rengo, C., Clarke, D., Meehan, W., Blackburn, J., Semino, O., Scozzari, R., Cruciani, F., Taha, A., Shaari, N. K., Raja, J. M., Ismail, P., Zainuddin, Z., Goodwin, W., Bulbeck, D., Bandelt, H. J., Oppenheimer, S., Torroni, A., & Richards, M. (2005). Single, rapid coastal settlement of Asia revealed by analysis of complete mitochondrial genomes. *Science, 308*, 1034–1036.

Maíllo Fernández, J.M. (2003). *La transición Paleolítico Medio-Superior en Cantabria: análisis tecnológico de la industria lítica de Cueva Morín*. Doctoral thesis, UNED Madrid.

Maíllo Fernández, J. F., Valladas, H., Cabrera Valdés, V., & Bernaldo de Quiros, F. (2001). Nuevos dataciones para el Paleolítico superior de

Cueva Morin (Villanueva de Villaescusa, Cantabria). Espacio, Tiempo y Forma Ser. 1. *Prehistoria y Arqueología, 14*, 145–150.

Marks, A. E. (1993). The Early Upper Paleolithic: The view from the Levant. In H. Knecht, A. Pike-Tay, & R. White (Eds.), *Before Lascaux. The complex record of the Early Upper Paleolithic* (pp. 5–21). *London & Tokio: Boca Raton, Ann Arbor*.

Marks, A. E. (1998). A new Middle to Upper Paleolithic "Transitional" assemblage from Buran-Kaya III, level C: A preliminary report. In M. Otte (Dir.), *Préhistoire d'Anatolie* (Vol. I, pp. 353–366). Colloquium Liège 1997. E.R.A.U.L. 85. Liège.

Marks, A., & Monigal, K. (2000). The Middle to Upper Palaeolithic interface at Buran-Kaya III, Eastern Crimea. In J. Orschiedt & G.-C. Weniger (Eds.), *Neanderthals and modern humans - Discussing the transition: Central and Eastern Europe from 50,000–30,000 B.P* (Wissenschaftliche Schriften des Neanderthal Museums, Vol. 2, pp. 212–226). Mettmann: Neanderthal Museum.

Marks, A. E., & Monigal, K. (2004). Origins of the European Upper Palaeolithic, seen from Crimea: Simple myth or complex reality? In P. J. Brantingham, S. L. Kuhn, & K. W. Kerry (Eds.), *The Early Upper Palaeolithic beyond Western Europe* (pp. 64–79). Berkeley and Los Angeles: University of California Press.

Maroto, J. (1994). *El pas del paleolític mitjá al paleolític superior a Catalunya i la seva interpretació dins del context geogràfic franco-ibric*. Ph.D. dissertation, University of Girona, Girona.

McDougall, I., Brown, F. H., & Fleagle, J. G. (2005). Stratigraphic placement and age of modern humans from Kibish, Ethiopia. *Nature, 433*, 733–736.

Mellars, P. A. (1999). The Neanderthal problem continued. *Current Anthropology, 40*, 341–350.

Mellars, P.A. (2000). Châtelperronian chronology and the case for Neanderthal/Modern Human "Acculturation" in Western Europe. In C. B Stringer, R. N. E. Barton, & J. C. Finlayson (Eds.), *Neanderthals on the edge*. Papers from a conference marking the 150th anniversary of the Forbes´ Quarry discovery, Gibraltar (pp. 33–39). Oxford: Oxbow Books.

Mellars, P. A. (2004). Neanderthals and the modern human colonization of Europe. *Nature, 432*, 461–465.

Mellars, P. A. (2006a). A new radiocarbon revolution and the dispersal of modern humans in Eurasia. *Nature, 439*, 931–935.

Mellars, P. A. (2006b). Archeology and the Dispersal of Modern Humans in Europe: Deconstructing the "Aurignacian". *Evolutionary Anthropology, 15*, 167–182.

Mellars, P., Gravina, B., & Bronk Ramsey, C. (2007). Confirmation of Neanderthal/modern human interstratification at the Chatelperronian type-site. *Proceedings of the National Academy of Sciences of the United States of America, 104*, 3657–3662.

Monigal, K. (2001). The Eastern Szeletian at Buran Kaya III (Crimea, Ukraine) and its Place in the Middle to Upper Palaeolithic Transition. In M. A. Hays & P. T. Thacker (Eds.), *Questioning the answers: Re-solving fundamental problems of the Early Upper Palaeolithic*. British Archaeological Reports (International Series) 1005 (pp. 51–64). Oxford.

Monigal, K. (2004). The Lithic Assemblage from Buran-Kaya III Level C. In V. P. Chabai, K. Monigal, & A. E. Marks (Eds.), *The Middle Palaeolithic and Early Upper Palaeolithic of Eastern Crimea* (pp. 57–78). E.R.A.U.L. 104. Liège.

Monigal, K. (2006). Transit lounge of Eastern Europe: Multicultural crimea during the late Middle Paleolithic and early Upper Paleolithic. In N. J. Conard (Ed.), *When Neanderthals and modern humans met* (pp. 189–211). Tübingen: Kerns Verlag.

Montes Barquín, R., & Sanguino González, J. (Eds.). (2001). *La cueva de El Pendo: actuaciones arqueológicas 1994–2000*. Santander: Monografías Arqueológicas de Cantabria.

Mook, W. G. (1982). Radiocarbon dating. In J. K. Kozlowski (Ed.), *Excavation in the Bacho Kiro Cave (Bulgaria): Final Report, 168* (p. 168). Warsaw: Polish Scientific Publishers.

Mook, W. G. (1988). Radiocarbon-Daten aus der Kůlna-Höhle. In K. Valoch,(Ed.), *Die Erforschung der Kůlna-Höhle 1961–1976*. Anthropos N.S. 16. Brno, pp. 285–286.

Morin, E., Tsanova, T., Sirakov, N., Rendu, W., Mallye, J.-B., & Lévêque, F. (2005). Bone refits in stratified deposits: Testing the chronological grain at Saint-Césaire. *Journal of Archaeological Science, 32*, 1083–1098.

Movius, H. (Ed.) (1975). *Excavation of the Abri Pataud, Les Eyzies (Dordogne)*. American School of Prehistoric Research Bulletin 30. Cambridge: Harvard University.

Mussi, M., Gioia, P., & Negrino, F. (2006). Ten small sites: The diversity of the Italian Aurignacien. In O. Bar-Yosef & J. Zilhao (Eds.), *Towards a definition of the Aurignacian. Proceedings of the Symposium held in Lisbon, Portugal, June 25–30, 2002. Trabalhos de Arqueologia 45* (pp. 189–209). Lisbon: Istituto Português de Arqueologia.

Neugebauer-Maresch, C. (1989). Zum Neufund einer weiblichen Statuette bei den Rettungsgrabungen an der Aurignacien-Station Stratzing/ Krems-Rehberg, Niederösterreich. *Germania, 67*, 551–559.

Nigst, P. R. (2004). Neue alte Funde aus Willendorf II, Schicht 3. *Archäologie Österreichs, 15/1*, 23–25.

Nigst, P. R. (2006). The first modern humans in the Middle Danube area? New evidence from Willendorf II (Eastern Austria). In N. J. Conard (Ed.), *When Neanderthals and modern humans met* (pp. 269–304). Tübingen: Kerns Verlag.

Niven, L. (2003). Patterns of subsistence and settlement during the Aurignacian of the Swabian Jura. In J. Zilhão & F. d´Errico (Eds.), *The chronology of the Aurignacian and of the transitional technocomplexes. Dating, stratigraphies, cultural implications*. Proceedings of Symposium 6.1 of the 14. Congress of the UISPP, Liège 2001. Trabalhos de Arquelogia 33. Lissabon, pp. 199–211.

Niven, L. (2006). *The Palaeolithic occupation of the Vogelherd Cave. Implications for the subsistance bahavior of late Neanderthals and early modern humans*. Tübingen: Tübingen Publications in Prehistory, Kerns Verlag.

Noonan, J. P., Coop, G., Kudaravalli, S., Smith, D., Krause, J., Alessi, J., Chen, F., Platt, D., Pääbo, S., Jonathan, K., Pritchard, J. K., & Rubin, E. M. (2006). Sequencing and analysis of Neanderthal genomic DNA. *Science, 314*, 1113–1118.

Ohnuma, K. (1988). Ksar Aqil, Lebanon: A technological study of the Earlier Upper Paleolithic Levels of Ksar Aqil, Levels XXV–XIV. British Archaeological Reports (International Series) 426. Oxford.

Olszewski, D. I., & Dibble, H. L. (2006). To be or not to be Aurignacian: the Zagros Upper Paleolithic. In O. Bar-Yosef & J. Zilhão (Eds.), *Towards a definition of the Aurignacian*. Proceedings of the Symposium held in Lisbon, Portugal, June 25–30, 2002. Instituto Português de Arqueologia.Trabalhos de Arqueologia 45, 355–373, Artes Graficas.

Ortega Cobos, D., Soler Masferrer, N., & Maroto Genover, J. (2005). La production des lamelles pendant l'Aurignacien archaïque dans la Grotte de l'Arbreda: Organisation de la production, variabilité des méthodes et des objectifs. In F. Le Brun-Ricalens, J.-G. Bordes, & F. Bon (Eds.), *Productions lamellaires attribuées à l´Aurignacien. Chaînes opératoires et perspectives technoculturelles*. Actes du XIVe congrès de l´UISPP September 2001 Session 6, Upper Palaeolithic, Symposium C6.7. ArchéoLogiques 1. Faber Mersch, Grand Duché du Luxemburg, pp. 359–373.

Ovchinnikov, I. V., Götherström, A., Romanova, G. P., Kharitonov, V. I., Lidén, K., & Goodwin, W. (2000). Molecular analysis of Neanderthal DNA from the northern Caucasus. *Nature, 404*, 490–493.

Palma di Cesnola, A. (1982). Le leptolithique archaique en Italie. *Aurignacien et Gravettien en Europe*. E.R.A.U.L. 13 (1) (pp. 119–143). ABACO, Forlí.

Paunescu, A. (2001). *Paleoliticul si mezoliticul din spatiul transilvan*. Bukarest: Academia Romana Institutul de Arheologie "Vasile Parvan".

Pettitt, P. B. (1998), Middle Palaeolithic and Early Upper Palaeolithic Crimea: The radiocarbon chronology. In: M. Otte (Ed.), *Préhistoire d'Anatolie* (Vol. I, pp. 329–338). Colloquium Liège 1997. E.R.A.U.L. 85. Liège.

Pettitt, P. B., & Bader, N. O. (2000). Direct AMS radiocarbon dates for the Sungir mid Upper Palaeolithic burials. *Antiquity, 74*, 269–270.

Pettitt, P., & Bahn, P. (2003). Current problems in dating Palaeolithic cave art: Candamo and Chauvet. *Antiquity, 77*, 134–141.

Pettitt, P. B., Street, M., & Terberger, T. (1998). Comments on the dating of Wildscheuer Cave. In: R. E. M. Hedges, P. B. Pettitt, C. Bronk Ramsey, & G. J. van Klinken (Eds.), Radiocarbon dates from the Oxford AMS system. Archaeometry Datelist 26. *Archaeometry, 40*, 441–443.

Pyle, D. M., Ricketts, G. D., Margari, V., van Andel, T. H., Sinitsyn, A. A., Praslov, N. D., & Lisitsyn, S. (2006). Wide dispersal and deposition of distal tephra during the Pleistocene 'Campanian Ignimbrite/Y5' eruption, Italy. *Quaternary Science Reviews, 25*, 2713–2728.

Reimer, P. J., Baillie, M. G. L., Bard, E., Bayliss, A., Warren Beck, J., Bertrand, C. J. H., Blackwell, P. G., Buck, C. E., Burr, G. S., Cutler, K. B., Damon, P. E., Edwards, R. L., Fairbanks, R. G., Friedrich, M., Guilderson, T. P., Hogg, A. G., Hughen, K. A., Kromer, B., McCormac, G., Manning, S., Bronk Ramsey, C., Reimer, R. W., Remmele, S., Southon, J. R., Stuiver, M., Talamo, S., Taylor, F. W., van der Plicht, J., & Weyhenmeyer, C. E. (2004). INTCAL04 terrestrial radiocarbon age calibration, 0–26 cal kyr BP. *Radiocarbon, 46*, 1029–1058.

Richter, J. (1997). Sesselfelsgrotte III. Der G-Schichten-Komplex der Sesselfelsgrotte. *Zum Verständnis des Micoquien*. Quartär-Bibliothek 7. Saarbrücken.

Richter, J. (2004). Die ^{14}C-Daten aus der Sesselsfelsgrotte und die Zeitstellung des Micoquien/M.M.O. *Germania, 80*, 1–22.

Richter, D.-J., Waiblinger, J., Rink, W. J., & Wagner, G. A. (2000). Thermoluminescence, electron spin resonance and ^{14}C-dating of the Late Middle and early Upper Palaeolithic site of Geissenklösterle cave in southern Germany. *Journal of Archaeological Science, 27*, 71–89.

Riek, G. (1934). *Die Eiszeitjägerstation am Vogelherd im Lonetal 1: Die Kulturen*. Tübingen: Heine Verlag.

Rigaud, J.-Ph. (2001). A propos de la contemporanéité du Castelperronien et de l´Aurignacien ancien dans le nord-est de l´Aquitaine: une révision des données et ses implications. In J. Zilhao, Th. Aubry, & A. F. Carvalho (Eds.), Les premiers hommes modernes de la Péninsula Ibérique. Actes de Colloque de la Commission VIII de l´UISPP. Colloquium Vila Nova de Foz Côa 1998. Trabalhos de Arqueologia 17. Lissabon, pp. 61–68.

Rougier, H., Milota, S., Rodrigo, R., Gherase, M., Sarcina, L., Moldovan, O., Zilhão, J., Constantin, S., Franciscus, R. G., Zollikofer, C. P. E., Ponce de León, M., & Trinkaus, E. (2007). Peştera cu Oase 2 and the cranial morphology of early modern Europeans. *PNAS, 104*(4), 1165–1170.

Schmitz, R. W., Serre, D., Bonani, G., Feine, S., Hillgruber, F., Krainitzki, H., Pääbo, S., & Smith, F. H. (2002). The Neandertal type site revisited: Interdisciplinary investigations of skeletal remains from the Neander Valley, Germany. *Proceedings of the National Academy of Sciences of the United States of America, 99*, 13342–13347.

Scott, E. M. (2003). The third international radiocarbon intercomparison (TIRI) and the fourth international radiocarbon intercomparison (FIRI), 1990–2002. Results, Analyses, and Conclusions. *Radiocarbon, 45*(2), 135–408.

Serre, D., Langaney, A., Chech, M., Teschler-Nicola, M., Paunovic, M., Mennecier, Ph. Hofreiter, M., Possnert, G., & Pääbo, S. (2004). No evidence of Neandertal mtDNA contribution to early modern humans. *PLoS Biology, 2*, 313–317.

Shackleton, N. (2005). The Absolute Chronology of the Last Glacial. Geophysical Research Abstracts 7, 05213 (SRef-ID: 1607-7962/EGU05-A-05213).

Shackleton, N. J., Fairbanks, R. G., Chiu, T.-C., & Parrenin, F. (2004). Absolute calibration of the Greenland time scale: Implications for Antarctic time scales and for Δ^{14}C. *Quaternary Science Reviews, 23*, 1513–1522.

Simán, K. (1995). La grotte Szeleta et le Szélétien. Les industries à pointes foliacées d'Europe centrale. Actes du Colloque commémoratif international, Miskolc 1991. Paléo-Supplement 1. Les-Eyzies-de-Tayac, pp. 37–43.

Sinitsyn, A. (2003). The most ancient sites of Kostenki in the context of the Initial Upper Palaeolithic of northern Eurasia. In J. Zilhão & F. d'Errico (Eds.), *The chronology of the Aurignacian and of the transitional technocomplexes: dating, stratigraphies, cultural implications* (Trabalhos de Arqueologia, Vol. 33, pp. 89–107) Lisboa: Istituto Português de Arqueologia. doi:Trabalhos de Arqueologia.

Skinner, A. R., Blackwell, B. A. B., Martin, S., Ortega, A., Blickstein, J. L. B., Golovanova, L. V., & Doronichev, V. B. (2005). ESR dating at Mezmaiskaya Cave, Russia. *Applied Radiation and Isotopes, 62*, 219–224.

Škrdla, P. (1996). The Bohunician reduction strategy. *Quaternaria Nova, 6*, 93–107.

Škrdla, P. (2003). Comparison of Boker Tachtit and Stránská skála MP/UP transitional industries. *Journal of the Israel Prehistoric Society, 33*, 37–73.

Smith, F. H., Trinkaus, E., Pettitt, P. B., Karavanic, I., & Paunovic, M. (1999). Direct radiocarbon dates for Vindija G1 and Velika Pecina Late Pleistocene hominid remains. *Proceedings of the National Academy of Sciences of the United States of America, 96*, 12281–12286.

Soressi, M. (2002). Le Moustérien de tradition acheuléenne du sud-ouest de la France. Discussion sur la signification du faciès à partir de l´étude comparée de quatre sites: Pech-de-l´Azé I, Le Moustier, La Rochette et la Grotte XVI. Dissertation, University of Bordeaux.

Southon, J. (2004). A radiocarbon perspective on Greenland ice-core chronologies: Can we use ice-cores for ^{14}C calibration? *Radiocarbon, 46*, 1239–1259. In: P. Reimer (Ed.), *IntCal04: Calibration Issue*.

Stiner, M. C. (1999). Palaeolithic mollusc exploitation at Riparo Mochi (Balzi Rossi, Italy): Food and ornaments from the Aurignacian through Epigravettian. *Antiquity, 73*, 735–754.

Stiner, M. C. (2003). "Standardization" in Upper Paleolithic ornaments at the coastal sites of Riparo Mochi and Üçagizli Cave. In J. Zilhão & F. d'Errico (Eds.), *The chronology of the Aurignacian and of the transitional technocomplexes: Dating, stratigraphies, cultural implications* (Trabalhos de Arqueologia, Vol. 33, pp. 49–59). Lisbon: Istituto Português de Arqueologia.

Stiner, M. C., Pehlevan, C., Sagir, M., & Özer, I. (2003). Zooarchaeological studies at Üçagızlı Cave: Preliminary results on Paleolithic subsistence and shell ornaments. *Araşterma Sonuçları Toplantısı, 17*, 29–36.

Street, M., & Terberger, T. (2003). New evidence for the chronology of the Aurignacian and the question of Pleniglacial settlement in western central Europe. In J. Zilhão & F. d'Errico (Eds.), *The chronology of the Aurignacian and of the transitional technocomplexes. Dating, stratigraphies, cultural implications* (Trabalhos de Arqueologia, Vol. 33, pp. 213–221). Lisboa: Istituto Português de Arqueologia.

Street, M., Terberger, T., & Orschiedt, J. (2006). A critical review of the German Paleolithic hominin record. *Journal of Human Evolution, 51*, 551–579.

Strobl, J. (1901). *Von der diluvialen Fundstelle auf dem Hundssteig in Krems*. Mitteilungen der Anthropologischen Gesellschaft Wien 31. Wien.

Strobl, J., & Obermaier, H. (1909). Die Aurignacienstation von Krems (Niederösterreich). *Jahrbuch für Altertumskunde, 3*, 129–148.

Svoboda, J. (2000). The depositional context of the Early Upper Palaeolithic human fossils from the Koneprusy (Zlatky kun) and Mladeč Caves, Czech Republic. *Journal of Human Evolution, 38*, 523–536.

Svoboda, J. (2001). Mladeč and other caves in the Middle Danube region: early modern humans, late Neandertals, and projectiles. In J. Zilhao, Th. Aubry, & A. F. Carvalho, (Eds.), *Les premiers hommes modernes de la Péninsula Ibérique*. Actes de Colloque de la Commission VIII de l'U.I.S.P.P.. Kolloquium Vila Nova de Foz Côa 1998. Trabalhos de Arqueologia 17 (pp. 45–60). Lissabon: Istituto Português de Arqueologia.

Svoboda, J. (2003). The Bohunician and the Aurignacian. In J. Zilhão & F. d'Errico (Eds.), *The chronology of the Aurignacian and of the transitional technocomplexes. Dating, stratigraphies, cultural Implications* (Trabalhos de arqueologia, Vol. 33, pp. 123–131). Lisbon: Instituto português de arqueologia.

Svoboda, J., & Bar-Yosef, O. (Eds.). (2003). *Stránská skála. Origins of the Upper Paleolithic in the Brno Basin, Moravia, Czech Republic. American School of Prehistoric Research Bulletin 47*. Cambridge: Harvard University Press.

Svoboda, J., Lozek, V., & Vlcek, E. (1996). *Hunters between East and West. The Palaeolithic of Moravia*. New York: Plenum.

Svoboda, J., van der Plicht, J., & Kuzelka, V. (2002). Upper Palaeolithic and Mesolithic human fossils from Moravia and Bohemia (Czech Republic): Some new ^{14}C dates. *Antiquity, 76*, 957–962.

Taborin, Y. (1993). Shells of the French Aurignacian and Périgordian. In H. Knecht, A. Pike-Tay, & R. White (Eds.), *Before Lascaux. The complex record of the Early Upper Paleolithic* (pp. 211–227). Boca Raton, Ann Arbor, London & Tokio: CRC Press.

Tavoso, A. (1976). La grotte Tournal ou grande grotte de Bize (Bize, Aude). Provence et Languedoc méditerranéen sites paléolithiques et néolithiques. Livret-Guide de l'excursion C 2 du IX Congrès U.I.S.P.P., Nice, pp. 232–239.

Terberger, Th., & Street, M. (2003). Jungpaläolithische Menschenreste im westlichen Mitteleuropa und ihr Kontext. In J. M Burdukiewicz, L. Fiedler, W.-D. Heinrich, A. Justus & E. Brühl (Eds.), Erkenntnisjäger. Kultur und Umwelt des frühen Menschen. Festschrift für Dietrich Mania. Veröffentlichungen des Landesamtes für Archäologie Sachsen-Anhalt – Landesmuseum für Vorgeschichte 57. Halle (Saale), pp. 579–592.

Terberger, Th., Street, M., & Bräuer, G. (2001). Der menschliche Schädelrest aus dem Elbmündungsgebiet bei Hahnöfersand und seine Bedeutung für die Steinzeit Norddeutschlands. *Archäologisches Korrespondenzblatt, 31*, 521–526.

Teyssandier, N. (2003). Les débuts de l'Aurignacien en Europe. Discussion à partir des sites de Geissenklösterle, Willendorf II, Krems-Hundssteig et Bacho Kiro. Ph.D. thesis, Université de Paris X-Nanterre.

Teyssandier, N. (2005). Neue Perspektiven zu den Anfängen des Aurignacien. *Mitteilungen der Gesellschaft für Urgeschichte, 14*, 11–24.

Teyssandier, N. (2006). Questioning the first Aurignacian: Mono- or multicultural phenomenon during the formation of the Upper Paleolithic in Central Europe and the Balkans? *L'Anthropologie, 44*, 9–29.

Teyssandier, N., & Liolios, D. (2003). Defining the earliest Aurignacian in the Swabian Alp: The relevance of the technological study of the Geissenklösterle (Baden-Württemberg, Germany) lithic and organic productions. In J. Zilhão & F. d'Errico (Eds.), *The chronology of the Aurignacian and of the transitional technocomplexes. Dating, stratigraphies, cultural implications* (Trabalhos de Arqueologia, Vol. 33, pp. 179–196). Lisboa: Instituto português de arqueologia.

Teyssandier, N., Bolus, M., & Conard, N.J. (2006). The Early Aurignacian in central Europe and its place in a European perspective. In O. Bar-Yosef, & J. Zilhão (Eds.), *Towards a definition of the Aurignacian*. Proceedings of the Symposium held in Lisbon, Portugal, June 25–30, 2002. Trabalhos de Arqueologia 45 (pp. 241–256). Artes Graficas: *Instituto Português de Arqueologia*.

Ton-That, T., Singer, B., & Paterne, M. (2001). 40Ar/39Ar dating of latest Pleistocene (41ka) marine tephra in the Mediterranean Sea: Implications for global climate records. *Earth and Planetary Science Letters, 184*, 645–658.

Tostevin, G. B. (2000). The Middle to Upper Paleolithic transition from the Levant to Central Europe: In situ development or diffusion? In J. Orschiedt & G.-C. Weniger (Eds.), *Neanderthals and modern humans - Discussing the transition: Central and Eastern Europe from 50,000–30,000 B.P* (Wissenschaftliche Schriften des Neanderthal Museums, Vol. 2, pp. 92–111). Mettmann: Neanderthal Museums.

Trinkaus, E. (2005). Early modern humans. *Annual Review of Anthropology, 34*, 207–230.

Trinkaus, E., Moldovan, O., Milota, S., Bîlgăr, A., Sarcina, L., Athreya, S., Bailey, S. E., Rodrigo, R., Mircea, G., Higham, T., Bronk Ramsey, C., & van der Plicht, J. (2003). An early modern human from the Pestera cu Oase, Romania. *Proceedings of the National Academy of Sciences of the United States of America, 100*, 11231–11236.

Tsanova, T., & Bordes, J.-G. (2003). Contribution au débat sur l'origine de l'Aurignacien: principaux résultats d'une étude technologique de l'industrie lithique de la couche 11 de Bacho Kiro. In T. Tsonev & E. Montagnari Kokelj (Eds.), *The humanized mineral world: Towards social and symbolic evaluation of prehistoric technologies in south eastern Europe*. E.R.A.U.L. 103 (pp. 41–50). Liège.

Usik, V. I., Monigal, K., & Kulakovskaya, L. (2006). New perspectives on the Transcarpathian Middle to Upper Paleolithic boundary. In N. J. Conard (Ed.), *When Neanderthals and Modern Humans met* (pp. 213–232). Tübingen: Kerns Verlag.

Uthmeier, T. (1996). Ein bemerkenswertes frühes Inventar des Aurignacien von der Freilandfundstelle "Keilberg-Kirche"bei Regensburg. *Archäologisches Korrespondenzblatt, 26*, 233–248.

Uthmeier, T. (1998). *Micoquien, Aurignacien und Gravettien in Bayern – Eine regionale Studie zum Übergang vom Mittel- zum Jungpaläolithikum*. Dissertation, University of Cologne.

Uthmeier, T. (2004): Micoquien, Aurignacien und Gravettien in Bayern. Eine regionale Studie zum Übergang vom Mittel- zum Jungpaläolithikum. Archäologische Berichte 18. Bonn.

Valladas, H., Tisnérat, N., Arnold, M., Évin, J., & Oberlin, Ch. (2001). Les dates des fréquentations. In J. Clottes (Ed.), *La Grotte Chauvet.L'art des origines* (pp. 32–33). Paris: Publisher Seuil.

Valladas, H., Tisnérat-Laborde, N., Cachier, H., Kaltnecker, É., Arnold, M., Oberlin, C., & Évin, J. (2005). Bilan des datations carbone 14 effectuées sur des charbons de bois de la Grotte Chauvet. *Bulletin de la Société Préhistorique Française, 102*, 109–113.

Valoch, K. (1976). Die altsteinzeitliche Fundstelle in Brno-Bohunice. Studie Archeologického ústvu CSAV v Brně 4, 1. Prag: Akademia.

Valoch, K. (1996). Das Mittelwürm in den Lössen Südmährens und seine paläolithischen Kulturen. *Eiszeitalter und Gegenwart, 46*, 54–64.

van der Plicht, J., Beck, J. W., Bard, E., Baillie, M. G. L., Blackwell, P. G., Buck, C. E., Friedrich, M., Guilderson, T. P., Hughen, K. A., Kromer, B., McCormac, F. G., Bronk Ramsey, C., Reimer, P. J., Reimer, R. W., Remmele, S., Richards, D. A., Southon, J. R., Stuiver, M., & Weyhenmeyer, C. E. (2004). NOTCAL04 – Comparison/Calibration ^{14}C records 26–50ka cal BP. *Radiocarbon, 46*, 1225–1238.

Vanhaeren, M., & d'Errico, F. (2006). Aurignacian ethno-linguistic geography of Europe revealed by personal ornaments. *Journal of Archaeological Science, 33*, 1105–1128.

Vanhaeren, M., d'Errico, F., Stringer, C., James, S. L., Todd, J. A., & Mienis, H. K. (2006). Middle Paleolithic shell beads in Israel and Algeria. *Science, 312*, 1785–1788.

Vaquero, M., Maroto, J., Arrizabalga, A., Baena, J., Baquedano, E., Carrión, E., Jordà, J. F., Martinón, M., Menéndez, M., Montes, R., & Rosell, J. (2006). The Neandertal–modern human meeting in Iberia: A critical view of the cultural, geographical and chronological data. In N. J. Conard (Ed.), *When Neanderthals and modern humans met* (pp. 419–439). Tübingen: Kerns Verlag.

Vega Toscaño, L.G. (1990). La fin du Paléolithique moyen au sud de l'Espagne: ses implications dans le contexte de la péninsule ibérique. In C. Farizy (Ed.), *Paléolithique moyen récent at Paléolithique supérieur ancien en Europe. Ruptures et transitions: examen critique des documents archéologiques*. Colloquium Nemours 1988. Mém. du Musée de Préhist. d'Ile de France 3 (pp. 169–176). Nemours: A.P.R.A.I.F.

Vega Toscaño, L. G. (1993). El tránsito del Paleolítico Medio al Paleolítico Superior en el sur de la Península Ibérica. In V. Cabrera Valdés (Ed.), *El origen del Hombre Moderno en el suroeste de Europa* (pp. 147–170). Madrid: UNED.

Verpoorte, A. (2005). The first modern humans in Europe? A closer look at the dating evidence from the Swabian Jura (Germany). *Antiquity, 79*, 269–279.

Voelker, A. H. L., Grootes, P. M., Nadeau, M.-J., & Sarnthein, M. (2000). ^{14}C levels in the Iceland Sea from 25–53 kyr and their link to the earth's magnetic field intensity. *Radiocarbon, 42*, 437–452.

Vogel, J. C., & Waterbolk, H. T. (1964). Groningen Radiocarbon dates V. *Radiocarbon, 6*, 349–369.

Vogel, J. C., & Waterbolk, H. T. (1967). Groningen radiocarbon dates VII. *Radiocarbon, 9*, 107–155.

von Koenigswald, W., Müller-Beck, H., & Pressmar, E. (1974). Die Archäologie und Paläontologie in den Weinberghöhlen bei Mauern (Bayern) – Grabungen 1937–1967. *Archaeologia Venatoria* (Vol. 3). Tübingen: Archaeologia Venatoria.

Wang, Y. J., Cheng, H., Edwards, R. L., An, Z. S., Wu, J. Y., Shen, C.-C., & Doralee, J. A. (2001). High-resolution absolute-dated late Pleistocene monsoon record from Hulu Cave, China. *Science, 294*, 2345–2348.

Weninger, B., & Jöris, O. (2004). Glacial Radiocarbon Age Calibration: The CALPAL Program. In Th. Higham,, Ch. Bronk Ramsey,, & Cl. Owen (Eds.), *Radiocarbon and archaeology*. Proceedings of the 4th Symposium, Oxford 2002. Oxford University School of Archaeology Monograph 62 (pp. 9–15). Oxford: Oxbow Books.

Weninger, B., & Jöris, O. (2008). A ^{14}C calibration curve fort he last 60ka: The Greeenland-Hulu U/Th timescale and its impact on understanding the Middle to Upper Paleolithic transition in Western Eurasia. In D. S. Adler & O. Jöris (Eds.), Setting the record straight: Toward a systematic chronological understanding of the Middle to Upper Paleolithic Boundary in Eurasia. *Journal of Human Evolution, 55*(5), 772–781

White, R. (2001). Personal ornaments from the Grotte du Renne at Arcy-sur-Cure. *Athena Review, 2*, 41–46.

White, T. D., Asfaw, B., DeGusta, D., Gilbert, H., Richards, G. D., Suwa, G., & Howell, F. C. (2003). Pleistocene Homo sapiens from Middle Awash, Ethiopia. *Nature, 423*, 742–747.

Wild, E. M., Teschler-Nicola, M., Kutschera, W., Steier, P., Trinkaus, E., & Wanek, W. (2005). Direct dating of Early Upper Palaeolithic human remains from Mladeč. *Nature, 435*, 332–335.

Yanevich, A. A., Stepanchuk, V. N., & Cohen, V. (1996). Buran-Kaya III and Skalistiy Rockshelter: Two new dated Late Pleistocene sites in the Crimea. *Préhistoire Européenne, 9*, 315–324.

Yanevich, A. A., Marks A. E., & Uerpmann H.-P. (1997). A bone handle from Buran-Kaya III: The earliest known in the Crimea. *Préhistoire Européenne, 11*, 85–87. Liège.

Zielinski, G. A., Mayewski, P. A., Meeker, L. D., Whitlow, S., Twickler, M. S., & Taylor, K. (1996). An 100, 000-year record of explosive volcanism from the GISP2 (Greenland) ice core. *Quaternary Research, 45*, 109–118.

Zielinski, G. A., Mayewski, P. A., Meeker, L. D., Grönvold, K., Germani, M. S., Whitlow, S. I., Twickler, M. S., & Taylor, K. C. (1997). Volcanic aerosol records and tephrochronology of the Summit, Greenland, ice cores. *Journal of Geophysical Research, 102*(C12), 26. 625–26,640.

Zilhão, J. (1993). Le passage du Paléolithique moyen au Paléolithique supérieur dans le Portugal. In V. Cabrera Valdés (Ed.), *El origen del hombre moderno en el suroeste de Europa. Colloquium Madrid 1991* (pp. 127–145). Madrid: Universidad Nacional de Educación a Distancia.

Zilhão, J. (1997). *O Paleolítico Superior da Estremadura portuguesa*. Lisbon: Colibri.

Zilhão, J. (1998). The extinction of Iberian Neandertals and its implications for the origins of modern humans in Europe. In F. Facchini, A. Palma di Cesnola, M. Piperno, & C. Peretto, (Eds.), XIII international congress of prehistoric and protohistoric sciences. Proceedings, Vol. 2, Abaco, Forli, pp. 299–312. Uploaded from "Mediterranean Prehistory Online" December 22, 1998. http://www.med.abaco-mac.it/articles/doc/006.htm

Zilhão, J. (2000). The Ebro frontier: A model for the late extinction of Iberian Neanderthals. In C. B. Stringer, R. N. E. Barton, & C. Finlayson (Eds.), *Neanderthals on the Edge* (pp. 111–121). Oxford: Oxbow Books.

Zilhão, J. (2006a). Genes, fossils, and culture. An overview of the evidence for Neandertal-modern human interaction and admixture. *Proceedings of the Prehistoric Society, 72*, 1–20.

Zilhão, J. (2006b). Chronostratigraphy of the Middle-to-Upper Paleolithic transition in the Iberian Peninsula. *Pyrenae, 37*, 7–84.

Zilhão, J. (2007). The emergence of ornaments and art: An archaeological perspective on the origins of "behavioral modernity". *Journal of Archaeological Research* Original Paper, online, Springer

Zilhão, J., & d'Errico, F. (1999). The chronology and taphonomy of the earliest Aurignacian and its implications for the understanding of Neandertal extinction. *Journal of World Prehistory, 13*, 1–68.

Zilhão, J., & d'Errico, F. (2003a). An Aurignacian Garden of Eden in southern Germany? An alternative interpretation of the Geissenklösterle and a critique of the Kulturpumpe model. *Paléo, 15*, 69–86.

Zilhão, J., & d'Errico, F. (2003b). The chronology of the Aurignacian and transitional technocomplexes. Where do we stand? In J. Zilhão, & F. d'Errico (Eds.), The chronology of the Ausrignacian and of the transitional technocomplexes: dating, stratigraphies, cultural implications. Proceedings of Symposium 6.1 of the XIVth Congress of the U.I.S.P.P. Trabalhos de Arqueologia 33. Instituto Português de Arqueologia, Lisboa, pp. 313–349.

Zilhão, J., & Pettitt, P. (2006). On the new dates for Gorham's Cave and the late survival of Iberian Neanderthals. *Before Farming, 2006/3*, 11–19.

Zilhão, J., & Trinkaus, E. (Eds.). (2002). *Portrait of the artist as a child. The Gravettian human skeleton from the Abrigo do Lagar Velho and its archaeological context* (Trabalhos de Arqueologia, Vol. 22). Lisboa: Instituto Português de Arqueologia, Lisbon.

Zilhão, J., d'Errico, F., Bordes, J.-G., Lenoble, A., Texier, J.-P., & Rigaud, J.-P. (2006). Analysis of Aurignacian interstratification at the Châtelperronian-type site and implications for the behavioral modernity of Neandertals. *Proceedings of the National Academy of Sciences of the United States of America, 103*(33), 12643–12648.

Züchner, Ch. (1996). The Chauvet Cave: Radiocarbon versus archaeology. *International Newsletter on Rock Art, 13*, 25–27.

Züchner, Ch. (2003). La cueva Chauvet y el problema del arte auriñaciense y gravetiense. In R. de Balbín Behrmann & P. Bueno Ramírez (Eds.), El arte prehistórico desde los inicios del siglo XXI. Primer symposium internacional de arte prehistórico de Ribadesella. Kolloquium Ribadesella 2002. Ribadesella, pp. 41–51.

Chapter 23
Archaeological, Paleontological and Genomic Perspectives on Late European Neandertals at Vindija Cave, Croatia

Ivor Janković, Ivor Karavanić, James C.M. Ahern, Dejana Brajković, Jadranka Mauch Lenardić, and Fred H. Smith

Abstract The Late Pleistocene sediments of Vindija Cave, NW Croatia, yield paleontological and archaeological finds that have an important role in understanding the patterns of late Neandertal/early modern human interaction and succession. The youngest securely dated Neandertal remains in Europe, found in association with an archaeological assemblage exhibiting a mixture of Middle and Upper Paleolithic elements, come from the G complex of this site. Assessments of both old and new data, as well as the results of newer analyses on the finds from Vindija are discussed in the light of the new revision of chronostratigraphic sequences of several important European sites. Further, the new genomic data obtained from the Vindija sample is included in a discussion on the problems, results and patterns of Late Pleistocene hominin evolution *and possible population interaction* patterns between Neandertals and early modern humans in Europe.

Keywords Modern human origins • Neandertals • Human evolution • Upper Paleolithic

I. Janković (✉)
Institute for Anthropological Research, Amruševa 8, 10000 Zagreb, Croatia
e-mail: ivor@inantro.hr

I. Karavanić
Department of Archaeology, University of Zagreb, I. Lučića 3, 10000 Zagreb, Croatia
e-mail: ikaravan@ffzg.hr

J.C.M. Ahern
Department of Anthropology (3431), University of Wyoming, 1000 E. University Ave, Laramie, WY 82071, USA
e-mail: jahern@uwyo.edu

D. Brajković and J.M. Lenardić
Institute for Quaternary Paleontology and Geology, Croatian Academy of Sciences and Arts, A. Kovačića 5, 10000 Zagreb, Croatia
e-mail: dejbraj@hazu.hr; jml@hazu.hr

F.H. Smith
Department of Sociology and Anthropology, Illinois State University, IL, USA
e-mail: fsmith@ilstu.edu

Site Background

Vindija has yielded a large sample of human skeletal remains and extensive evidence of their cultural behavior. Particularly significant is the evidence that relates to the latest phase of Neandertal occupation of Europe and specifically to the several millennia of possible overlap between late Neandertals and early modern Europeans. There is ample discussion of this critical period, and the importance of Vindija for this discussion has been repeatedly emphasized, most recently by Karavanić and Smith (1998), Smith et al. (1999), Wolpoff (1999), Churchill and Smith (2000), Ahern et al. (2004) and Janković et al. (2006). These works have noted the evidence for both biological and cultural interaction between late Neandertals and early modern Europeans and have suggested that the former evidence indicates genetic exchange between these populations. Even though some recent assessments of "mixing" during this time frame (Zilhão 2006) fail to give Vindija its due importance in this regard, the evidence marshaled here shows that Vindija continues to be one of the very best documentations of contact and interaction between these two late Pleistocene human populations.

Much of the reason Vindija has provided so much valuable information has to do with the nature of the site itself. Vindija is a large cave located in the rugged, semi-mountainous Hrvatsko Zagorje region of Croatia (Malez 1979). Its large size (~50 m in length, 28 m in width, and ~20 m in height) resulted in extensive accumulation of Quaternary sediments (>12 m) and created conditions favorable to the preservation of bone, including organic materials within bone. The latter has determined Vindija's critical role in the recovery of Neandertal genomic data (Krings et al. 2000; Serre et al. 2004; Green et al. 2006; Noonan et al. 2006). As exciting as the genetic data are, no less significant are the large samples of pertinent hominin fossil remains and the distinctive assemblage of Paleolithic stone and bone artifacts from Vindija that fall in the time span of ~43–31 ka, a timeframe critical to understanding the emergence of modern people in Europe.

S. Condemi and G.-C. Weniger (eds.), *Continuity and Discontinuity in the Peopling of Europe: One Hundred Fifty Years of Neanderthal Study*,
Vertebrate Paleobiology and Paleoanthropology, DOI 10.1007/978-94-007-0492-3_23,

The history of discovery and excavations at Vindija, as well as the discussion of its stratigraphy has been published in numerous papers (see Malez 1979, 1983; Malez and Rukavina 1979; Malez et al. 1980; Wolpoff et al. 1981; Paunović et al. 2001; Ahern et al. 2004; Janković et al. 2006 and references therein). Likewise, the archaeological, faunal, and hominin samples have been the subject of several detailed analyses (see Malez et al. 1980; Wolpoff et al. 1981; Smith et al. 1985; Karavanić 1993, 1995; Blaser et al. 2002; Ahern et al. 2004; Brajković 2005; Janković et al. 2006 and references cited there). Since analysis of the initial Vindija remains, mainly during the 1980s, several additional hominin fossils have been identified from the faunal collections (Smith and Ahern 1994; Ahern et al. 2004), and the archaeological and faunal assemblage has been the subject of several new detailed analyses (Miracle 1991; Karavanić 1993, 1995; Karavanić and Smith 1998; Blaser et al. 2002; Ahern et al. 2004; Brajković 2005). Although, modern human skeletal and cultural remains have been identified in the upper stratigraphic levels, it is the material from the G complex that provides critical information pertinent to understanding the fate of the Neandertals.

All of the Neandertal skeletal remains come from the complex G, with a possible exception of the Vi - 11.52 mandibular ramus that could be from the older layer I (Ahern et al. 2004). The archaeological sample from the complex G is quite interesting. Finds from the older part of the sequence (G_3) represent Mousterian with some Upper Paleolithic elements present, while the assemblage from the younger part (G_1) provides an even more intriguing picture (Karavanić and Smith 1998; Ahern et al. 2004; Janković et al. 2006). In this layer, three massive bone points (so-called Mladeč points), as well as other bone artifacts have been found. The most distinctive find from the G_1 is an Aurignacian-like split base bone point (Vi - 3437) (Fig. 23.1) found in association with a Neandertal mandible (Vi - 207). These bone tools represent distinctly Upper Paleolithic tool types, although bone points are also found in other late Mousterian contexts in Central Europe. Likewise, the stone tools from the G_1 layer exhibit a mixture of Mousterian and Upper Paleolithic types (Karavanić and Smith 1998) (Fig. 23.2), and one bifacial stone point made from non-local stone shows similarities to the Szeletian industry of nearby Hungary (Figs. 23.2; (4)). Therefore it remains unclear whether the archaeological

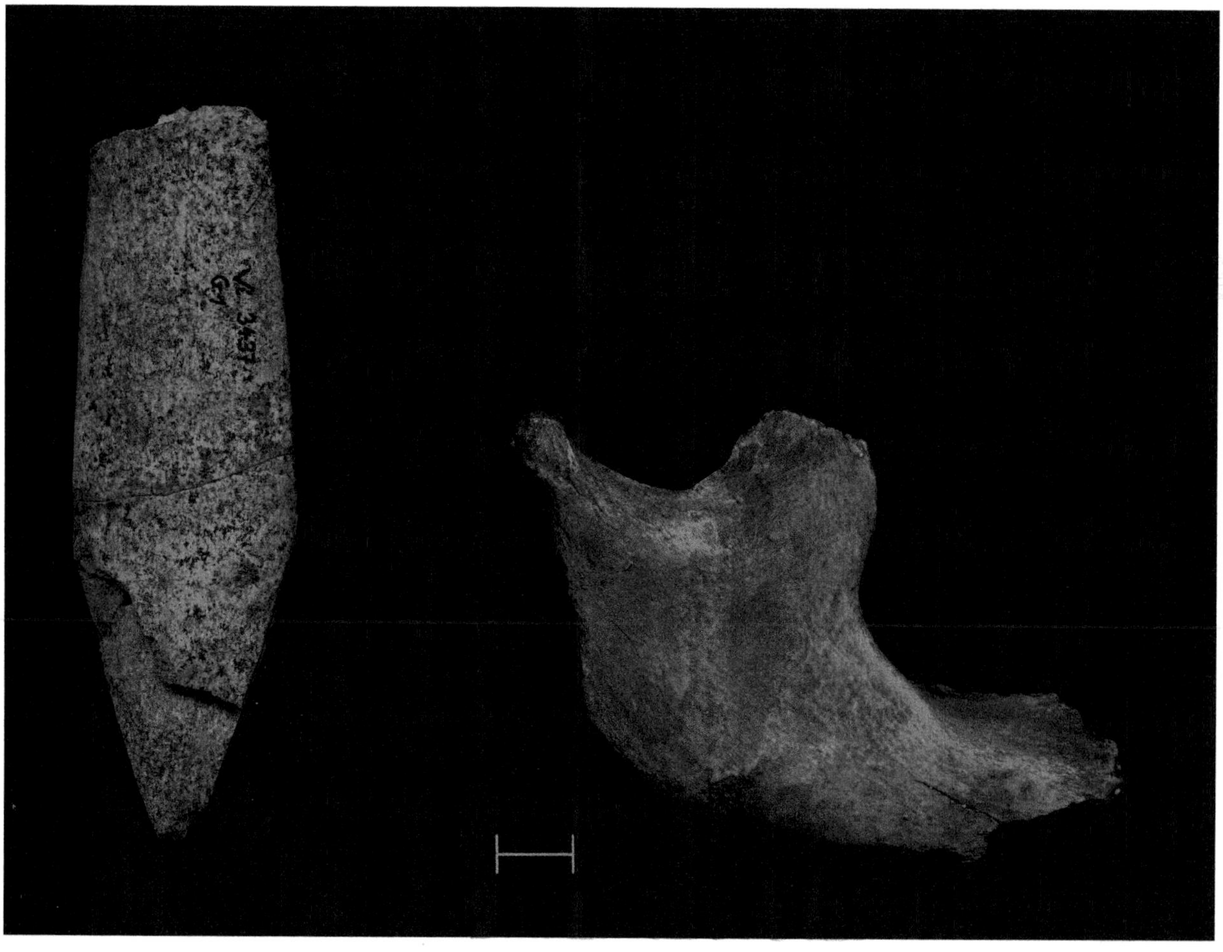

Fig. 23.1 Split-base bone point Vi 3437 and hominin mandible Vi 207

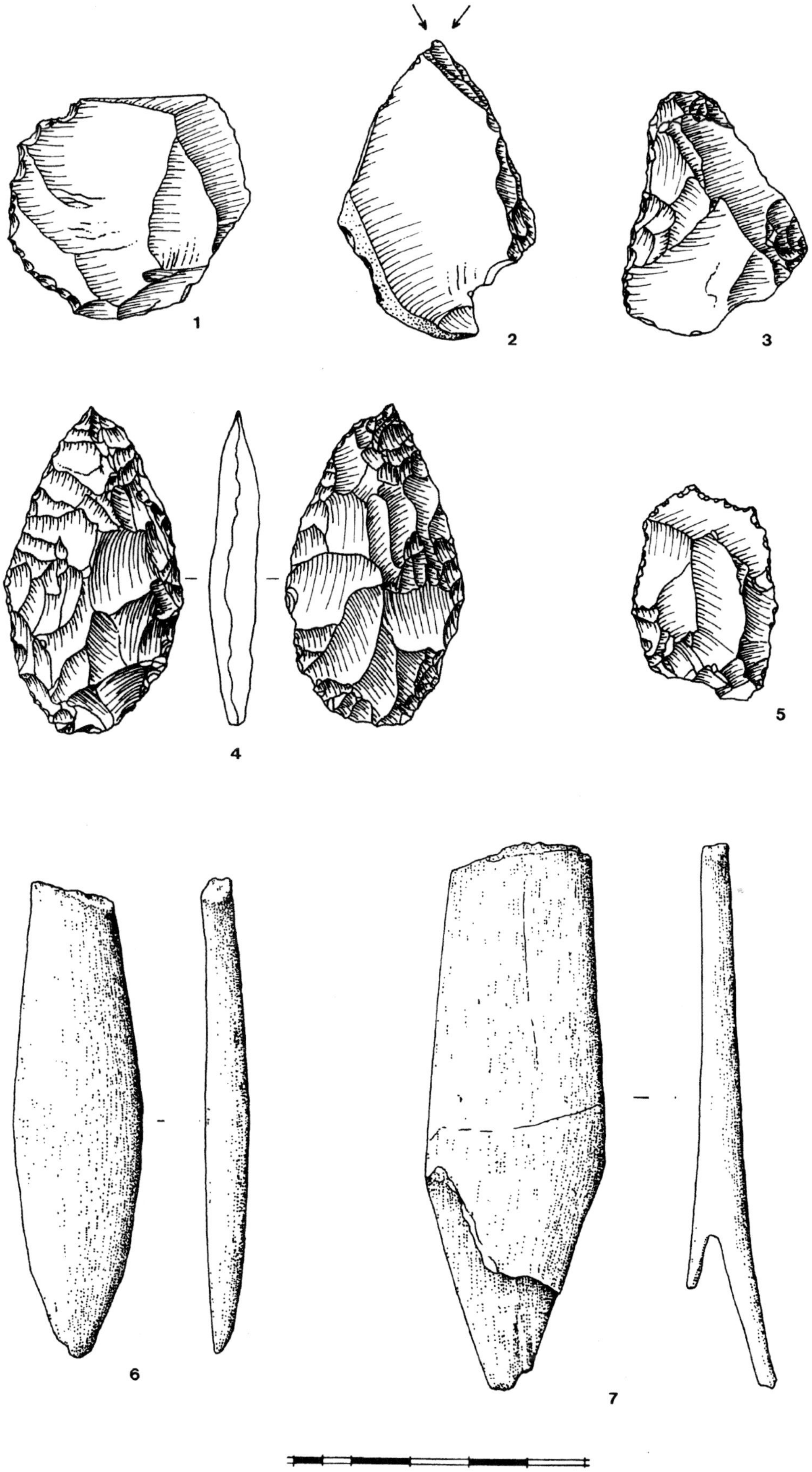

Fig. 23.2 Selected artefacts from Vindija level G1: *1*. probable a pseudo-tool (previously published as denticulated piece), *2*. burin. *3*. sidescraper. *4*. leaf-shaped bifacial piece, *5*. flake with marginal retouch on distal end (previously published as an endscraper on flake), *6*. massive base bone point, split base bone points (Modified after Karavanić 1995: Fig. 3; drawing by Marta Perkić)

material from this layer represents the late Mousterian with some Upper Paleolithic elements, the Aurignacian, or some other initial Upper Paleolithic ("transitional") industry (Miracle 1998; Karavanić and Smith 1998; Churchill and Smith 2000; Svoboda 2001; Janković et al. 2006). Another possibility is that the material from level G_1 was artificially mixed (D'Errico et al. 1998; Zilhão and D'Errico 1999a). This possibility cannot be excluded totally, but our reasons for considering it highly unlikely will be discussed below.

Vindija Faunal Sample

During the late Pleistocene, Vindija cave was situated on the southern edge of the Alpine ice sheet, which at the times of the glacial maximum covered the Alps. However, Vindija also lies near the edge of the Pannonian Plain, which explains the steppe elements in the classical forest faunal community during the OIS 2 and 3. As the majority of the Vindija finds are faunal, the zooarchaeological sample from this site has been studied at numerous times (Malez and Rukavina 1979; Malez and Ullrich 1982; Malez 1988; Paunović et al. 2001; Paunović and Smith 2002). With a better understanding of the taphonomy of the site and more detailed studies of specific taxa, new patterns emerged (Janković et al. 2006: pp 459).

After recent revision some ungulate taxa have been removed from the faunal list (Janković et al. 2006). The following species are present in layer G_3: red deer, elk, giant deer and aurochs, while in layer G_1 representatives are chamois, roe deer and Merck's rhinoceros. These taxa can tolerate a wide range of conditions, although none of them are considered to be particularly indicative of cold conditions. In contrast to Malez's original interpretation of the Vindija faunal assemblages, there are no indications of major climatic fluctuations within complex G. In the lower portion of complex F (layers F/d + F/d/d) ibex and bison are present and the roe deer is missing. In the remainder of complex F and E layer the presence of reindeer is confirmed. These faunal changes suggest somewhat cooler and perhaps more arid conditions in layers E and complex F relative to the G complex, although the continued presence of Merck's rhinoceros in complex E + F suggests either a relatively more temperate phase during the accumulation of these complexes, or a mosaic of biotopes in the vicinity of the cave that could support taxa with contrasting ecological tolerances.

New data (Janković et al. 2006) show that strong climatic fluctuations are not reflected in the composition of the larger mammal assemblages. Neither layers G3 nor G1 stand out in terms of faunal composition when compared to the rest of the site. The Neandertal remains in layers G3 and G1 appear to have been deposited during relatively temperate conditions – probably a more temperate interstadial/oscillation within OIS 3. The ungulate remains from the G complex were either deposited only during more temperate conditions within OIS 3, perhaps reflecting the fact that Neandertals visited Vindija only during more temperate periods, or that climates did not fluctuate dramatically in northern Croatia during OIS 3. There are indications from layer E and the F complex of somewhat cooler conditions that correspond with the onset of stadial conditions in OIS 2. Nevertheless, even at the time of the LGM during the deposition of layer E, the region around Vindija could still support steppe/woodland rhinoceros, aurochs, elk, and giant deer.

Chronometric Dating and the Earliest Modern Humans in Europe

Given the crucial period covered by particularly by the G complex at Vindija, the chronology of this complex must be as precise as possible. This is important in part because several key fossils at other localities in Europe have been redated by more precise methods, which has led to exclusion of several specimens previously held to be among the earliest modern humans in Europe from the debate (e.g., Vogelherd and Velika Pećina, now dated to the Neolithic, see Smith et al. 1999; Conard et al. 2004). However, probably more significant is the fact that we are dealing with a relatively short time period for potential Neandertal-early modern interactions to occur at Vindija. Dating of various stratigraphic layers of this site has been attempted several times, but not without problems (see Wild et al. 2001). Despite these problems, Neandertal remains from later G_1 were directly dated by AMS and yielded a date of 28–29 ka, thus making them the youngest Neandertals in Europe (Smith et al. 1999). The new technique of ultrafiltration of collagen samples in AMS dating offers more precise dating, and this technique has been applied at Vindija. This AMS method results in dates of 32,000–33,000 ka or the same G_1 fossils (Higham et al. 2006a, b).

It is now critical that the same method be applied to other crucial specimens of approximately the same time period, both late Neandertals and early anatomically modern humans in Europe. Otherwise the critical chronological details of the overlap between these two populations in Europe will remain unresolved. The main problem with radiocarbon dating is a high error margin for material older than about 30 kyr. Newer techniques, such as AMS ultrafiltration, add to the accuracy of dating and make these methods less destructive (see Mellars 2006). However when the time of overlap is expected to be several thousand years at best, the error margin can be still unacceptably high. Another problem is the large fluctuation in atmospheric radiocarbon over time (Conard and Bolus 2003; Hugen et al. 2004). Further, many specimens from this crucial time period (e.g. Mladeč, Kostenki, etc.) are likely to be older than reported (Kozłowski 1996; Trinkaus 2005). Therefore, the redating of the Vindija specimens does not necessarily shorten the temporal overlap between indigenous European Neandertals and anatomically modern newcomers.

At present, radiocarbon dating establishes remains from Kent's Cavern, England, Brassempouy and La Quina in France, Kostenki in Russia, Oase, Cioclovina and Baia de Fier in Romania, and Mladeč in the Czech Republic as the oldest modern human fossil specimens in Europe (Trinkaus 2005). However, there are problems with all of these sites. For example, the Kent's Cavern 4 maxillary fragment (Keith 1927) was found below the layer containing what was described as "Aurignacoid" industry (Garrod 1926; Oakley et al. 1971; Hedges et al. 1989; Churchill and Smith 2000), making the association of archeological industry and human fossil questionable. The fossil was directly dated to around 31 ka (Hedges et al. 1989; Stringer 1990), but it may be as old as 35–37 ka (Jacobi et al. 2006). Although this specimen was described as modern in morphology, its fragmentary state makes this assesment uncertain (Trinkaus 2005) and new analyses are still in progress. Similar problems exist with the other specimens listed and are detailed elsewhere (Churchill and Smith 2000; Janković et al. 2006).

Recently, human fossils from several Romanian sites have been directly dated (Olariu et al. 2004). A cranial vault, mandible, tibia and scapula were found in 1952 in Peştera Muierii (Nicolăescu-Popşlor 1968), of which the postcranial remains have been dated to 30,150 ± 700 ^{14}C years ka in 2001 (Păunescu 2001). In 2005, Higham and colleagues (Higham et al. 2006) obtained the dates of 29,930 ± 170 and 29,110 ± 190 ka for the Muierii 1 cranium and Muierii 2 temporal bone, which is in general agreement with the date prevoiusly published for the Muierii postcranial remains. Thus, Soficaru and colleagues published a more detailed description of the human remains from the site (Soficaru et al. 2006). Four bones (cranium, mandible, scapula and tibia) come from the Galeria Musteriană and most likely represent a single individual, collectively refered to as Muierii 1, while for two additional bones (temporal and fibula) the exact context is uncertain. The overal morphological pattern of the Muierii 1 remains is modern. However, like other early modern humans in Europe (Oase, Cioclovina, Mladeč, etc.) they exhibit a number of archaic features that are common in Neandertals. These include large interorbital breadth, relatively flat frontal arc, certain mandibular traits, bunning on the occipital bone, and narrow scapular glenoid fossa (Soficaru et al. 2006). The archaeological finds from the site have been described as Mousterian, while the upper layers contain some type of Upper Paleolithic industry. Unfortunately, the association of archaeological industries with the hominin skeletal remains are unclear.

The skull from Cioclovina cave now dated to around 28,510 ± 170 ka (Soficaru et al. 2006) has been described by Rainer and Simionescu (1942) as "*Homo sapiens fosillis*... with Neanderthalian characters", and although it is morphologically modern in overall gestalt, its supraorbital region is very robust and there is remnant bunning on the occipital bone (Smith 1984; Churchill and Smith 2000; Trinkaus 2005). Whether this pattern results from a Neandertal contribution is still debated. Harvati and colleagues, for example, have recently suggested exclusively modern affinities for this specimen (Harvati et al. 2007).

The most recently-recovered finds come from Peştera cu Oase in Romania are dated to around 35 ka, but unfortunately have no archaeological associations (Trinkaus et al. 2003a, b; Trinkaus 2005; Rougier et al. 2007). Trinkaus and colleagues (2003a, b) and Rougier and colleagues (2007) note the presence of several archaic features on these otherwise anatomically modern specimens (e.g. relatively large juxtamastoid eminence and robust and laterally oriented zygomatic bones and large molars in Oase 2). At least one feature (the horizontal-oval mandibular foramen form on the left ramus of Oase 1 mandible) is unknown in modern humans predating the Oase remains but is common in Neandertals and some of the later modern humans in Europe (Smith 1978; Trinkaus et al. 2003a, b).

Wild and colleagues (2005) have recently directly dated the early modern skeletal remains from Mladeč in the Czech Republic to ~31 ka. Although an association with the Aurignacian was previously suggested, the exact nature of the deposition at the site is uncertain (see Churchill and Smith 2000) and while Mladeč type bone points were found, the lithic material is scarce (and relatively non-diagnostic), and the split base bone points that are common in other Aurignacian-like industries of the earliest Central/Eastern European Upper Paleolithic are absent (Churchill and Smith 2000). Therefore, the question of whether these tools represent an early Aurignacian-like (transitional) industry, or Aurignacian *sensu stricto*, remains open. In addition, as in the Oase sample, several archaic features are seen in some of the fundamentally modern Mladeč specimens. These include occipital bunning in Mladeč 3, 5 and 6, and robust supraorbital regions in Mladeč 5 and 6, as well as large palatal and dental dimensions and some other anatomical details in the sample (see Smith 1982, 1984; Smith et al. 1989, 2005; Frayer 1986, 1992, 1997; Kidder et al. 1992; Frayer et al. 1993; Wolpoff 1999; Churchill and Smith 2000; Wild et al. 2005), all features that are common in earlier Neandertal populations.

Early Upper Paleolithic of Europe

Traditional equation of Neandertals to the Middle Paleolithic (Mousterian and its variants) and modern humans to the Upper Paleolithic no longer can be accepted uncritically for the European archaeological record. Between circa 50 and 30 ka in Western, Central and Eastern Europe there is a mosaic of industries that follow the Middle Paleolithic period, many of which are limited in space and time (Churchill and Smith 2000; Usik et al. 2006). These are sometimes referred to as "transitional industries" or the Initial Upper Paleolithic and include the Châtelperronian of France and northern Spain, Szeletian and Jankovichian of central and parts of eastern Europe, Uluzzian of Italy (Tuscany, Calabria, southern Adriatic part, Uluzzo Bay, etc.), Streletskian of eastern Europe, Jerzmanowician of eastern Germany and Poland, Altmühlian of southern Germany, Bohunician of Czech Republic, Brynzeny and Kostenki Szeletian of Russia, Bachokirian of Bulgaria, and several other unnamed or site-specific assemblages from Poland, Slovakia, Czech Republic and Romania, in which various elements of Mousterian appear alongside the Upper Paleolithic types or types produced using technology commonly associated with the Upper Paleolithic. As reviewed by us elsewhere (Janković et al. 2006) these industries seem to have their origin in local Mousterian variants and no abrupt change can be seen. Except for documented associations of Neandertal remains and Châtelperronian artifacts from La Roche à Pierrot at St. Cesaire and Grotte du Renne at Arcy-sur-Cure (Leroi-Gourhan 1958; Lévêque and Vandermeersch 1980; Hedges et al. 1994; Hublin et al. 1996) there are no diagnostic hominin fossils associated with any of these earliest Upper Paleolithic finds (Gambier 1989, 1997; Churchill and Smith 2000). Thus, even if we accept the earliest Aurignacian as an industrial complex that has its origins outside this area (Kozłowski and Kozłowski 1975; Kozłowski and Otte 2000) (which is far from proven) and attribute it to anatomically modern newcomers (for which there are no known hominin/industrial associations) we are left with the problem of who is responsible for these Initial Upper Paleolithic assemblages.

In our opinion, typological thinking has played a significant role in the acceptance of the Aurignacian as a single widespread complex that is unequivocally associated with the spread of morphologically modern humans into Europe (Miracle 1998; Karavanić and Smith 1998; Churchill and Smith 2000). We believe that, in light of the currently available evidence (or the lack thereof), this view should be carefully reexamined. The rationale for our view can be summarized as follows.

Numerous studies (Pradel 1966; Allsworth-Jones 1986, 1990; Clark and Lindly 1989; Gioia 1988; Harrold 1989; Rigaud 1989; Kozlowski 1990, 1996; Anikovich 1992; Cabrera Valdés et al. 1997; Straus 1997; Golovanova and Doronichev 2003) show that the earliest Upper Paleolithic ("transitional") industries in Europe develop within the local framework from (and including various elements of) the Mousterian complex. Thus the earliest distinctly Upper Paleolithic industry associated with anatomically modern humans should be the Aurignacian, purportedly brought into Europe as moderns move into the continent (Mellars 2006). Authors vary in opinion on whether and how much influence these modern newcomers and their culture had on the technology and behavior of late Neandertals. But in this model, the Aurignacian is regarded throughout Europe as a single imported complex that can be recognized in the archaeological record by the appearance of certain tool types and automatically assigned to anatomically modern populations.

While this sounds simple enough, it is not. First, detailed archaeological studies show that several tool types (especially bone tools) used as indicative of Aurignacian are in fact commonly found in various aforementioned "transitional" industries (Tode et al. 1953; Valoch 1972; Allsworth-Jones 1986; Svoboda 1993; Miracle 1998). Further, the Early Aurignacian differs from the Late Aurignacian (Miracle 1998), indicating that significant change occurs within the Aurignacian. In fact, these differences are similar in degree to those differentiating the Aurignacian and the Initial Upper Paleolithic. Finally, there are great differences between assemblages of typical Aurignacian from Western Europe, and those of Central/Eastern Europe (Oliva 1993; Karavanić and Smith 1998; Miracle 1998; Svoboda 2004).

All this makes it clear that there may be a different pattern of behavioral, and most likely populational change in Western vs. Central/Eastern Europe, and probably in differing regions within these broad regions. While this latter transition (whatever the mode of it) was more abrupt in Western Europe, evidence suggests a more gradual pattern for Central and Eastern areas of this region (or at least parts of it). Therefore, following from this archaeological pattern, we believe it is quite likely that some Neandertal populations had a significant role in the formation of early modern European gene pool (via assimilation into anatomically more modern populations), while other Neandertal groups had none. As Voisin (2006) and many others before him argued, Neandertal features are not uniformly spread across the Neandertal range. Models borrowed from contemporary studies of living animal, especially bird, distribution, such as "ring species model," may actually provide testable models for Neandertal and modern human interaction patterns in different geographical areas.

Finally, the origin of the Aurignacian is far less certain than is widely suggested (e.g. Mellars 2006). Although its origins were commonly seen in the Middle Eastern assemblages, some authors trace its initial rise in several independent centers in Europe, particularly Central Europe (see for example Oliva 1989; Svoboda 2004; Straus 2005). This explanation makes more sense if the Aurignacian is not a single widespread complex but actually represents different Early Upper Paleolithic assemblages that share several tool types (previously considered to be indicative of a single industrial complex). In this light there is no need to see these industries as a product of a single population. This also raises possibilities of different explanations for shared similarities (trade, influence, population mixing, etc.). However, we should bear in mind that population contacts differ in their pattern. Interbreeding and peaceful coexistence, trade, and sharing of knowledge might dominate some of these interactions; while in other circumstances, different patterns of contact, including more hostile interactions, might occur. Therefore, the same model of interaction is not likely to apply to all of Europe.

The G Layer of Vindija Cave and the Modern Human Peopling of Europe

As noted above, Vindija has an important place in the understanding of the initial anatomically modern human peopling of Europe. The significance of the association of Neandertal remains with an Upper Paleolithic industry has been a subject of a considerable debate (Karavanić and Smith 1998, 2000; D'Errico et al. 1998; Straus 1999; Zilhão and D'Errico 1999a, b), as it has been argued that the association is in fact artificial and the result of the cryoturbation that has been noted in some parts of the cave or bioturbation caused by denning activity of cave bear. Such mixing is always a possibility in Pleistocene cave sites, and the situation is exacerbated by the fact that the period in question here is characterized by substantial erosion of deposits from caves. Despite this, we contend that the arguments presented in favor of artificial mixing for much of the material at Vindija are weakened by careful consideration of data.

The excavations at Vindija have been criticized because current standards of excavation and documentation techniques were not followed. However, these problems do not invalidate the majority of the information available for Vindija. While important data were certainly lost (such as debitage or smaller non-diagnostic fragmentary bones etc.), the majority of recognizable tools, bones, and larger pieces of debitage were collected and recorded according to stratigraphic unit. Thus, while certainly far from perfect, contextual data on the Vindija fossil and archaeological finds are far more reliable and informative than is often suggested.

Cryoturbation, while present at the site (Malez and Rukavina 1975), has not been noted for the part of the cave where the associated Neandertal mandible (Vi 207) and split-based bone point (Vi 3437) (Fig. 23.2) were found (Wolpoff et al. 1981; Karavanić and Smith 1998, 2000; Ahern et al. 2004). Further, G_1 consists of characteristic reddish clay, easily recognizable and distinct from sediments from the older G_{2-5} and younger F complexes. This reddish clay was embedded in both the Vi-207 mandible and the Vi-3437 bone point and can still be observed on another massive-based bone point (Vi-3439) from this stratigraphic layer. In a recent paper, Ahern and colleagues (2004) reported additional Neandertal remains, one of which (a proximal radial shaft: Vi 13.8) has embedded reddish clay sediment that is characteristic of layer G_1. Neandertal attribution of this specimen (Ahern et al. 2004) is suggested by the strong curvature of the shaft and the medial orientation of radial tuberosity (Trinkaus and Churchill 1988; Churchill 1994). The presence of additional Neandertal specimens from layer G_1 further undercuts the claim for artificial mixing of layers and further supports the Neandertal association with the G_1 Upper Paleolithic industry.

Technologically related differences are also evident among the various Vindija layers. In older layers (unit K) typical Mousterian tools predominate and there is a clear evidence of the use of Levallois technology that is common in most European Mousterian assemblages. The most abundant raw material in unit K is local quartz (Kurtanjek and Marci 1990; Blaser et al. 2002), and flake technology predominates in tool production. Level G_3 presents a mixture of typical Mousterian tools, such as sidescrapers, but there are also Upper Paleolithic types of stone tools (such as endscrapers), and alongside flake technology, bifacial and blade technology was used in production of tools from this layer. It is important to note that no evidence of Levallois technology is seen in Vindija layer G_3 (Karavanić and Smith 1998). There is also evidence of more selective use of raw material, as there are more tools made from chert in this layer (Kurtanjek and Marci 1990; Blaser et al. 2002; Ahern et al. 2004).

The level G_1 assemblage shows an even more pronounced shift towards the use of higher quality raw material (i.e., chert) compared to the older layers of the site, and there are no tools made on quartz (Kurtanjek and Marci 1990; Blaser et al. 2002; Ahern et al. 2004). Upper Paleolithic elements among the stone tools are more abundant than in layer G_3, and bone points from G_1 layer represent a new distinctly Upper Paleolithic element that is not seen in any of the older layers (Karavanić 1993, 1995; Karavanić and Smith 1998).

All hominins from the Vindija G complex can be recognized as a belonging to Neandertal populations on the basis of their overall morphological gestalt. However, most of the commonly noted "Neandertal features" do not represent autapomorphies, but are instead either plesiomorphic characters inherited from preceding archaic hominins or shared with contemporary and/or post-Neandertal populations (Frayer 1992; Trinkaus 2005). Still, it is clear that there are many temporal and geographic differences. Several studies have shown that later Neandertals differ in morphological details from earlier "classic" members of this population, for instance in the reduction of facial dimensions and projection (Wolpoff et al. 1981; Smith and Ranyard 1980; Smith 1982, 1984; Smith et al. 1985; Ahern 1998; Ahern et al. 2004) as well as in other details of their anatomy. Generally, these changes are in the direction of early modern Europeans and

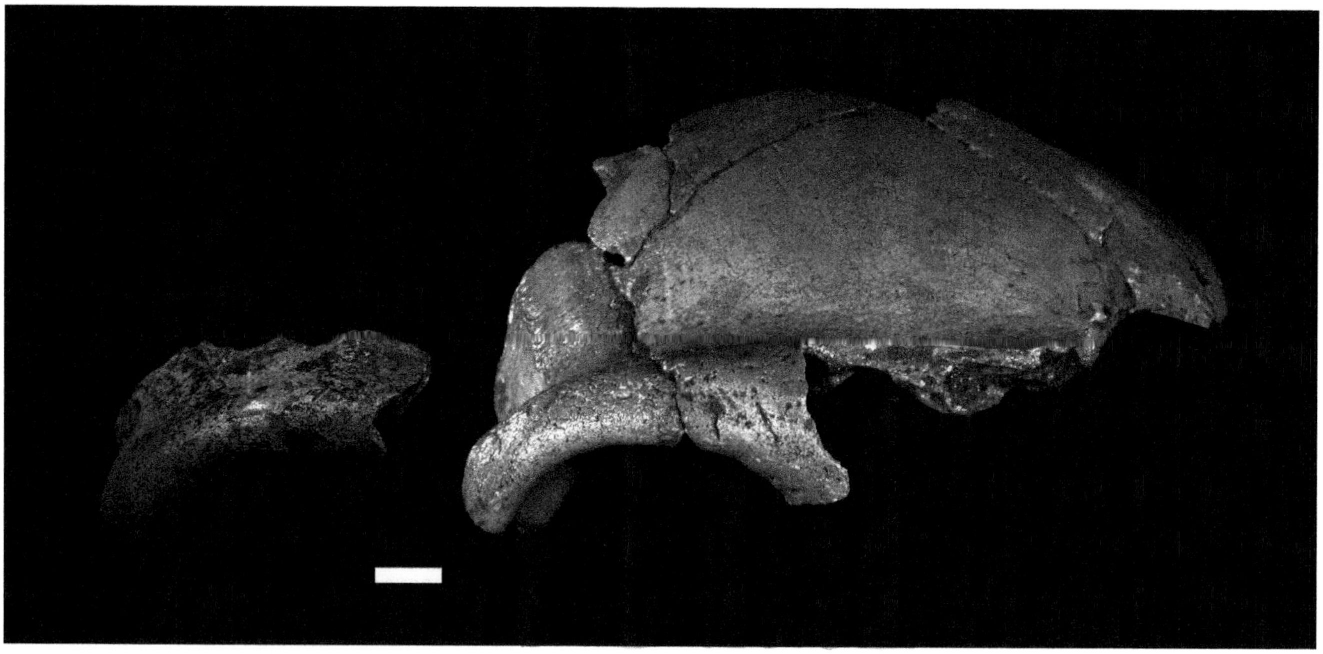

Fig. 23.3 Comparison of Vindija 202 (*left*) and Krapina 4 (*right*) frontal bones (Photo: J.C.M. Ahern)

have been interpreted as evidence of Neandertal change in the morphological direction of early modern humans (Wolpoff et al. 1981; Smith 1994; Ahern et al. 2004). One possible explanation for this is assimilation of early modern biological influences into late Neandertal populations. Recently, Hublin and Bailey (2006) argued that the difference between "classic" Neandertals and later, more gracile ones is not necessarily a result of interbreeding with anatomically modern populations, but can be explained by variation within the Neandertal sample, or, alternatively, by behavioral changes. While something like this is not impossible, it is very hard to see why the more gracile Neandertal morph is represented exclusively in the timeframe of potential overlap with anatomically modern humans. We regard it as more than just shear coincidence that this pattern of change in both morphology and culture does not occur except toward the end of the Neandertal timespan.

This more gracile morphological pattern is seen in the Vindija G Neandertals, as shown by several studies, especially on the supraorbital, facial and mandibular material (Smith and Ranyard 1980; Smith 1982, 1992; Ahern 1998). Analyses reveal the intermediate position of the Vindija supraorbitals, both in projection and shape compared to the older Krapina sample (Fig. 23.3). The Vindija supraorbital tori have relatively greater degrees of pinching above the orbits compared to the earlier Neandertals (Smith and Ranyard 1980; Smith 1984, 1994). Recent study of a newly reconstructed partial cranial vault from the G_3 level, comprised of supraorbital and frontal fragments (Vi 284, Vi 230, Vi 255, Vi 256), again suggests anatomical change in the direction of anatomically more modern morphology (Ahern et al. 2004) similar to that previously suggested for the other relatively complete frontal bone, Vi 261-277-278 (Malez et al. 1980). Change in the direction toward a more modern human pattern is also seen in the Vindija mandibular and maxillary sample, suggesting facial reduction. The Vindija maxillae have significantly narrower noses and shorter anterior alveolar processes that other Neandertals (Smith 1992); and the Vindija mandibles have more vertical symphyses than earlier Neandertals and exhibit incipient anterior basal projections, though not a true modern human chin (Wolpoff et al. 1981; Smith 1982, 1994; Ahern and Smith 1993; Kesterke and Ahern 2007). This observed gracility and change in shape is demonstrably not due to body size (Trinkaus and Smith 1995) or age and/or sex bias in the sample (Ahern 1998; Ahern et al. 2002; Ahern and Smith 2004; Kesterke and Ahern, 2007). We believe these changes, including those in level $G_{3,}$ are due to genetic interaction with early modern humans. As level G_3 dates to ~38–43 ka this would have to result from gene flow in this region prior to our ability to clearly document it. However, we suspect that modern humans did indeed make small-scale incursions into Europe prior to their ability to successfully adapt to and colonize the continent.

Genetics and Modern Human Origins Debate

Although genetic was applied to the study of human evolution earlier (e.g. Howells 1976), the initial analyses of recent human mitochondrial DNA suggested an exclusively African origin for all modern humans (Cann et al. 1987). More recently several authors have emphasized that such genetic results could be explained in different ways (Relethford 1999, 2001a, b; Harpending and Rogers 2000; Harpending and Eswaran 2005 and references therein). Moreover, mtDNA results do not seem to be in total agreement with results obtained from other parts of the genome, as several studies have shown a considerable depth for some non-African genetic polymorphisms (Harding 1997; Harding et al. 1997, 2000; Yu et al. 2001; Templeton 2002, 2005; Eswaran et al. 2005). Most recently, a published study on the microcephalin (MHPC1) haplogroup D shows that it most likely originated in some archaic human group and introgressed into the modern human gene pool much later, approximately 37 ka (Evans et al. 2006). Thus these analyses suggest some archaic non-African contributions to the modern human gene pool.

Newer analyses of mtDNA isolated directly from Neandertal specimens have added another dimension to the debate (Krings et al. 1997, 1999, 2000; Ovchinnikov et al. 2000; Schmitz et al. 2002; Serre et al. 2004; Lalueza-Fox et al. 2005, 2006; Orlando et al. 2006; Caramelli et al. 2006). Although these sequences are different from those of living humans, various processes (e.g. population expansions, migrations, bottlenecks etc.) could cloud our insight into how past events affect modern human gene pools, and many haplotypes of mtDNA could have been lost over time (Adcock et al. 2001; see also Relethford 2001a, b). Among Neandertal samples that yielded mtDNA, several Vindija specimens were included (Krings et al. 2000; Serre et al. 2004) and were reported to fall outside both contemporary modern human, as well as Upper Paleolithic hominin ranges. However, ancient DNA was extracted from Vindija fossils that are morphologically undiagnostic. Specifically these include the undiagnostic specimen Vi 77, the Vi 203, a proximal 5th metatarsal, designated Vi 75 in the genetic study by Serre et al. (2004) and Vi 33.16, a short segment of probable tibia shaft, designated Vi 80 in the Serre et al.'s genetic analysis in 2004. Vi 203 and Vi 33.16 specimens were dated by AMS radiocarbon, providing date estimates of >42 ka and 38.3±2.1 ka respectively. These dates would tend to support attribution of both specimens to the earlier G_3 level at Vindija, but the standard error on the Vi 33.16 fragment is so high that it could derive from as late as just over 34 ka.

A more meaningful insight into the question of whether or not Neandertals and anatomically modern humans interbred could be provided by extraction of DNA from the earliest modern humans in Europe. This has been accomplished for a total of seven early modern European specimens, all of which yield haplotypes that fall within the recent human range and outside that of Neandertals (Caramelli et al. 2003; Serre et al. 2004). While this seems compelling, assessment of probabilities demonstrates that a Neandertal contribution of <25% cannot be excluded on the basis of currently available data (Serre et al. 2004). Thus, despite claims to the contrary, none of the currently available genetic data excludes the possibility of some archaic Eurasian, including possibly Neandertal, contribution to early modern human gene pools outside of Africa. If such contributions occurred, we would not expect to find them in all pertinent samples. Rather we would expect such contributions to be relatively rare and to be found only in some samples.

The next step in genetic research of archaic humans has been aimed at a more complete understanding of the Neandertal genome as a whole. The studies of genomic DNA (Green et al. 2006; Noonan et al. 2006) of a presumably Neandertal individual are of particular interest, as they are based on the Vi 33.16 sample from Vindija. The reported sequences from this specimen are, according to these authors, exceptionally free of contamination. However, there are several problems relating to this specimen. It is a small piece (70×24 mm) of mammalian tubular bone that cannot be identified as Neandertal on the basis of its morphology. As noted above, the specimen's mtDNA sequence was previously reported as Neandertal-like (Serre et al. 2004). However a Neandertal-like sequence alone is not a conclusive proof of its Neandertal status as some degree of overlap between Neandertals and recent humans is seen in the pairwise comparisons used to contrast Neandertal and modern sequences (Krings et al. 1997). The AMS date of 38,310±2,130 ka (Serre et al. 2004) for this specimen is in general agreement with its G3 layer provenience but considering the relatively high error estimate of the date, the 95% confidence interval would include the earliest millennium that early anatomically modern human fossils are known from Europe (Trinkaus et al. 2003a, b; Trinkaus 2005; Soficaru et al. 2006; Zilhão 2006). Thus, while the non-diagnostic Vi-33.16 may be from a Neandertal, it is not certain that it is. Furthermore, Green and colleagues' data (Green et al. 2006) provide additional interesting results. If Vi-33.16 is indeed from a Neandertal, it shows that about 30% of SNP (single nucleotide polymorphism) derived alleles are shared with modern humans. This is best explained by gene flow between some Neandertal and some early modern populations. Thus it may well be that the Vi 33.16 specimen represents an admixed individual rather than a Neandertal. A series of genomic samples from other Neandertals is needed to address this issue, but these need to come from specimens with diagnostic Neandertal anatomical features and ideally from a time period before the possibility of the presence of early modern people in Europe.

Thus, distinction of Neandertals at the species level (if a strict biological species concept emphasizing reproductive isolation is used) is not conclusively demonstrated by current evidence (Nordborg 1998; Relethford 2001a, b; Relethford and Jorde 1999, Krings et al. 2000; Eshwaran 2002; Gutierrez et al. 2002; Serre et al. 2004; Eswaran et al. 2005; Smith et al. 2005). Any molecular analysis dealing with the question of Neandertal and anatomically modern human interaction must take into account these demographic changes and aforementioned factors. Even then, known problems such as small sample size and difficulties with extraction and contamination of DNA must continue to be recognized and acknowledged. The recent publication of a reconstructed Neandertal genome by Green and colleagues (2010) changes much of the generally accepted interpretations and views on Neandertals. According to this research, based largely on three fragments of long bone from Vindija, Neandertals contributed 1-4% of the genetic makeup of modern Eurasians. This may have actually been a larger contribution in, at least some, early modern populations. Thus Neandertals are our ancestors, at least in part.

Conclusions

Vindija cave has provided data critical to the assessment of the interaction between late Neandertals and early modern Europeans that derives from archaeology, human paleontology and ancient DNA. No other site has provided this breadth of crucial data. Vindija contributes the youngest securely dated Neandertal skeletal remains in Europe and a large sample of human fossil remains exhibiting a morphology potentially representing assimilation of early modern influences into a late Neandertal population. In addition, these remains have been found in association with an archaeological assemblage exhibiting Upper Paleolithic, combined with Mousterian, elements. Given all this and its geographic location, the Vindija remains are particularly crucial for the understanding of the initial modern human peopling of Europe and the nature of the Neandertal demise. We argue that the association of an early Upper Paleolithic industry with late Neandertals at Vindija is not likely to be a result of artificial mixing of specimens from different strata, but rather that these artifacts are reasonably considered to be products of the Vindija Neandertals. Although relatively similar archaeological samples in Europe have traditionally been regarded as Aurignacian and automatically assigned to anatomically modern humans, we believe that many of earliest Upper Paleolithic assemblages are in fact derived from the local Mousterian, and the question of which population is responsible for the production of these assemblages remains open. As Zilhão (2006) recently stated, there is no "modern human behavior" vs. "Neandertal behavior". Both Neandertals, and anatomicaly modern humans display different behavioral patterns at various sites throughout their geographic and temporal distribution. Site by site analysis is the only way through which we can hope to reach a better understanding of the complex and variable pattern of behaviour and adaptations of any prehistoric people.

The so-called transitional or Initial Upper Paleolithic industries such as the Uluzzian of Italy and Szeletian of Hungary and adjacent areas were quite likely a product of local Neandertal groups, as they have their origin in preceeding local Mousterian. In Europe at least, only Neandertals have been associated with Mousterian assemblages. Likewise, the only clear association of hominin remains with the Initial Upper Paleolithic thus far has been Neandertals with the Châtelperronian (at Arcy-sur-Cure and St. Cesaire, see Lévêque and Vandermeersch 1980; Hublin et al. 1996). Although it can be argued that the anatomically modern newcomers are the likely producers of the earlier distinctly Upper Paleolithic industry of Europe (later Aurignacian, or Aurignacian *sensu stricto*), this still remains to be proven. However if, as we argue, Aurignacian should no longer be considered a single Pan-European industrial complex, but rather represents a number of local early Upper Paleolithic assemblages, the association of Neandertals and early Upper Paleolithic would not be so surprising.

We hold that the "initial" Upper Paleolithic industry at Vindija is not Aurignacian *sensu stricto*, but one of the many "transitional" industry assemblages documented for Europe. This suggestion is supported by the presence of significant Mousterian types, one bifacial stone point typical of Szeletian, as well by significant differences in the assemblage compared to Western European sites (Montet-White 1996; Karavanić and Smith 1998; Miracle 1998; Karavanić 2000; Svoboda 2004). While we cannot equate industry with specific biological populations, the simplest explanation would be that late Neandertals developed at least some of these "transitional" industries. Further, we sugest, in accord with Zilhão (2006), that the Aurignacian sequence at various key sites, especially in Central and Eastern Europe, must be re-examined. Specificially it must be determined whether these are in fact Aurignacian *sensu stricto*, or if they also exhibit localized variations that are more accurately considiered to represent another "transitional" industry. If the later proves to be the case, the association of the split-base bone point (and therefore the Upper Paleolithic sequence) and late Neandertals at Vindija should not come as a surprise at all.

The first modern people to come to Europe were almost certainly small groups, and it is unclear how much they contributed to the later modern human groups (e.g. Gravettians, Magdalenians, and later groups). Therefore we must bear in mind that it is not only the issue of Neandertal genetic contribution to the initial anatomically modern newcomers, but also the relation of these first groups to the later modern humans that needs to be taken into account. At present, most genetic data is in agreement with paleoanthropological, archaeological, and paleontological studies pointing to a complex pattern of population dynamics in Europe (and elsewhere) at the end of the Pleistocene. Unfortunately the relatively short time frame of the populational overlap between late Neandertals and early moderns, possible differential site use, and numerous other factors (including sedimentation rates, preservation of the sediment which is erroding more quickly than forming, relatively short time frame of the populational overlap, etc.) will result in rare preservation of such evidence. Therefore, the Vindija G complex, especially the G_1 layer, is a rare and important find; and despite the problems with the site, the evidence it yields for "mixing" of early modern and late Neandertal populations must be considered crucial for our understanding of late Pleistocene human evolution of Europe.

Studies on the Vindija cave anthropological, archaeological and paleontological material are by no means over. New dating, DNA and various other skeletal analyses, as well as the recently recognized new hominins allow for a better insight into the human evolutionary past. There are many questions still to be answered and still more to be created by these answers. No doubt the material from Vindija will have a crucial part in answering some of them.

Acknowledgements The authors would like to thank several institutions and individuals for their support of various types of analyses on the Vindija material over the years. To the Ministry of Science, Education and Sports of the Republic of Croatia (project no. 196-1962766-2740 and 130-0000000-0871), the Fulbright Foundation, the Alexander von Humboldt Foundation, and the University of Wyoming we extend our gratitude for their financial support. We are also grateful to Dr. Silvana Condemi for the invitation to contribute to this volume. We are also greatly indebted to our many colleagues who have discussed the Vindija material and its implications with us on numerous occasions. Of course, any errors in facts or interpretation remain our responsibility.

References

Adcock, G., Dennis, E., Easteal, S., Huttley, G., Jermin, L., Peacock, W., & Thorne, A. (2001). Mitochondrial DNA sequences in ancient Australians: Implications for modern human origins. *Proceedings of the National Academy of Sciences of the United States of America, 98*, 537–542.

Ahern, J. C. M. (1998). *Late Pleistocene frontals of the Hrvatsko Zagorje: An analysis of intrapopulational variation among South Central European Neanderthals*. Ph.D dissertation, University of Michigan.

Ahern, J. C., & Smith, F. H. (1993). The transitional nature of the late Neanderthal mandibles from Vindija cave, Croatia. *American Journal of Physical Anthropology, 16*, 47.

Ahern, J., & Smith, F. H. (2004). Adolescent archaics or adult moderns? Le Moustier 1 as a model for estimating the age at death of fragmentary supraorbital fossils in the modern human origins debate. *Homo, 55*, 1–19.

Ahern, J. C. M., Hawks, J. D., & Lee, S. H. (2002). The late Neandertal supraorbital fossils from Vindija cave, Croatia: A biased sample? *Journal of Human Evolution, 43*, 419–432.

Ahern, J. C. M., Karavanić, I., Paunović, M., Janković, I., & Smith, F. H. (2004). New discoveries and interpretations of hominid fossils and artifacts from Vindija cave, Croatia. *Journal of Human Evolution, 46*, 27–67.

Allsworth-Jones, P. (1986). The Szeletian: Main trends, recent results, and problems for resolution. In M. Day, R. Foley, & Wu Rukang (Eds.), *The Pleistocene perspective* (pp. 1–25). Southampton: Papers of the World Archaeological Congress.

Allsworth-Jones, P. (1990). The Szeletian and stratigraphic succession in Central Europe and adjacent areas: Main trends, recent results and problems for resolution. In P. Mellars (Ed.), *The emergence of modern humans: An archaeological perspective* (pp. 160–242). Ithica, NY: Cornell University Press.

Anikovich, M. (1992). Early Upper Paleolithic industries of Eastern Europe. *Journal of World Prehistory, 6*, 205–245.

Blaser, F., Kurtanjek, D., & Paunovic, M. (2002). L'industrie du site neandertalien de la grotte de Vindija (Croatie): Une revision des matieres premieres lithiques: The industry from neandertalian site of the cave Vindija (Croatia): Revision of the lithic raw materials. *L'Anthropologie, 106*, 387.

Brajković, D. (2005). *Korelacija tafodema skupine ungulata iz gornjopleistocenskih sedimenata špilja: Vindija, Velika pećina i Veternica u sjeveriozapadnoj Hrvatskoj*. Ph.D dissertation, University of Zagreb, Zagreb.

Cabrera Valdés, V., Hoyos Gómez, M., & Dequirós, F. B. (1997). The transition from the middle to the Upper Paleolithic in the cave of El Castilo (Cantabria, Spain). In G. A. Clark & C. M. Willermet (Eds.), *Conceptual issues in modern human origins research* (pp. 177–188). New York: Aldine De Gruyter.

Cann, R. L., Stoneking, M., & Wilson, A. C. (1987). Mitochondrial DNA and human evolution. *Nature, 325*, 31–36.

Caramelli, D., Lalueza-Fox, C., Vernesi, C., Lari, M., Casoli, A., Mallegni, F., Chiarelli, B., Dupanloup, I., Bertranpetit, J., & Barbujani, G. (2003). Evidence for a genetic discontinuity between Neadertals and 24,000-year-old anatomically modern Europeans. *Proceedings of the National Academy of Sciences of the United States of America, 100*, 6593–6597.

Caramelli, D., Lalueza-Fox, C., Condemi, S., Longo, L., Milani, L., Manfredini, A., de Saint Pierre, M., Adoni, F., Lari, M., Giunti, P., Ricci, S., Casoli, A., Calafell, F., Mallegni, F., Bertranpetit, J., Stanyon, R., Bartorelle, G., & Barbijani, G. (2006). A highly divergent mtDNA sequence in a Neandertal individual from Italy. *Current Biology, 16*, R630–R632.

Churchill, S. E. (1994). *Human upper body evolution in the Eurasian later Pleistocene*. Ph.D dissertation. University of New Mexico, Albuquerque.

Churchill, S. E., & Smith, F. H. (2000). Makers of the early Aurignacian of Europe. *Yearbook of Physical Anthropology, 43*, 61–115.

Clark, G. A., & Lindly, J. M. (1989). The case of continuity: Observations on the biocultural transition in Europe and Western Asia. In P. Mellars & C. B. Stringer (Eds.), *The human revolution: Behavioural and biological perspectives on the origin of modern humans* (pp. 626–676). Princeton, NJ: Princeton University Press.

Conard, N. J., & Bolus, M. (2003). Radiocarbon dating the appearance of modern humans and timing of cultural innovations in Europe: New results and new challenges. *Journal of Human Evolution, 44*, 331–371.

Conard, N. J., Grootes, P. M., & Smith, F. H. (2004). Unexpectedly recent dates for human remains from Volgelherd. *Nature, 430*, 198–201.

D'Errico, F., Zilhão, J., Julien, M., Baffier, D., & Pelegrin, J. (1998). Neanderthal acculturation in Western Europe? A critical review of the evidence and its interpretation. *Current Anthropology, 39*, 1–44.

Eshwaran, V. (2002). A diffusion wave out of Africa. *Current Anthropology, 43*, 749–774.

Eswaran, V., Harpending, H., & Rogers, A. R. (2005). Genomics refutes an exclusively African origin of humans. *Journal of Human Evolution, 49*, 1–18.

Evans, P. D., Mekel-Bobrov, N., Vallender, E. J., Hudson, R. R., & Lahjn, B. T. (2006). Evidence that the adaptive allele of brain size gene *microcephalin* introgressed into *Homo sapiens* from an archaic *Homo* lineage. *Proceedings of the National Academy of Sciences (USA), 103*, 18178–18183.

Frayer, D. W. (1986). Cranial variation at Mladec and the relationship between Mousterian and Upper Paleolithic hominids. In V. V. Novotny & A. Mizerova (Eds.), *Fossil man: New facts, new ideas. Papers in honor of Jan Jelinek's life anniversary* (pp. 243–256). Brno: Anthropos.

Frayer, D. W. (1992). The persistence of Neandertal features in post-Neandertal Europeans. In G. Bräuer & F. H. Smith (Eds.), *Continuity or replacement. Controversies in Homo sapiens evolution* (pp. 179–188). Rotterdam: A. A. Balkema.

Frayer, D. W. (1997). Perspectives on Neanderthals as ancestors. In G. A. Clark & C. M. Willermet (Eds.), *Conceptual issues in modern human origins research* (pp. 220–234). New York: Aldine De Gruyter.

Frayer, D. W., Wolpoff, M. H., Thorne, A. G., Smith, F. H., & Pope, G. G. (1993). Theories of modern human origins: The paleontological test. *American Anthropologist, 95*, 14–50.

Gambier, D. (1989). Fossil hominids from the early Upper Paleolithic (Aurignacian) of France. In P. Mellars & C. B. Stringer (Eds.), *The human revolution* (pp. 194–211). Edinburgh: Edinburgh University Press.

Gambier, D. (1997). Modern humans at the beginning of the Upper Paleolithic in France. In G. A. Clark & C. M. Willermet (Eds.), *Conceptual issues in modern human origins research* (pp. 117–131). NewYork: Aldine De Gruyter.

Garrod, D. A. E. (1926). *The Upper Palaeolithic in Britain*. Oxford: Oxford University Press.

Gioia, P. (1988). Problems related to the origins of Italian Upper Paleolithic: Uluzzian and Aurignacian. In J. K. Kozlowski (Ed.), *La mutation* (pp. 71–101). Liege: Etudes et Recherches Archeologiques de l'Universite de Liege.

Golovanova, L. V., & Doronichev, V. B. (2003). The Middle Paleolithic of the Caucasus. *Journal of World Prehistory, 17*, 71–140.

Green, E. R., Krause, J., Ptak, S. E., Briggs, A. W., Ronan, M. T., Simons, J. F., Du, L., Egholm, M., Rothberg, J. M., Paunović, M., & Pääbo, S. (2006). Analysis of one million base pairs of Neanderthal DNA. *Nature, 444*, 330–336.

Gutierrez, G., Sanchez, D., & Marin, A. (2002). A reanalysis of the ancient mitochondrial DNA sequences recovered from Neandertal bones. *Molecular Biology and Evolution, 19*, 1359–1366.

Green, R. E., Krause, J., Briggs, A. W., Maričić, T., Stenzel, U., Kircher, M., Patterson, N., Li, H., Zhai, W., Fritz, M. H., Hansen, N. F., Durand, E. Y., Malaspinas, A. S., Jensen, J. D., Marques-Bonet, T., Alkan, C., Prufer, K., Meyer, M., Burbano, H. A., Good, J. M., Schultz, R., Aximu-Petri, A., Butthof, A., Hober, B., Hoffner, B., Siegemund, M., Weihmann, A., Nusbaum, C., Lander, E. S., Russ, C., Novod, N., Affourtit, J., Egholm, M., Verna, C., Rudan, P., Brajković, D., Kučan, Z., Gušić, I., Doronichev, V. B., Golovanova, L. V., Lalueza-Fox, C., de la Rasilla, M., Fortea, J., Rosas, A., Schmitz, R. W., Johnson, P. L., Eichler, E. E., Falush, D., Birney, E., Mullikin, J. C., Slatkin, M., Nielsen, R., Kelso, J., Lachmann, M., Reich, D., & Paabo. S. (2010). A draft sequence of the Neandertal genome. *Science, 328*, 710–722.

Harding, R. (1997). Lines of descent from mitochondrial Eve: An evolutionary look at coalescence. In P. Donnely & S. Tavaré (Eds.), *Progress in population genetics and human evolution* (pp. 15–31). New York: Springer.

Harding, R., Fullerton, S., Griffiths, R., Bond, J., Cox, M., Schneider, J., Moulin, D., & Clegg, J. (1997). Archaic African and Asian lineages in the genetic ancestry of modern humans. *American Journal of Human Genetics, 60*, 777–789.

Harding, R., Eugene, H., Ray, A., Ellis, N., Flanagan, N., Todd, C., Dixon, C., Sajantila, A., Jackson, I., Birch-Machin, M., & Rees, J. (2000). Evidence for variable selective pressures at MC1R. *American Journal of Human Genetics, 66*, 1351–1361.

Harpending, H., & Eswaran, V. (2005). Tracing modern human origins. *Science, 309*, 1995–1997.

Harpending, H., & Rogers, A. (2000). Genetic perspectives on human origins and differentiation. *Annual Review of Genomics and Human Genetics, 1*, 361–385.

Harrold, F. B. (1989). Mousterian, Châtelperronian and early Aurignacian in Western Europe: Continuity or discontinuity? In P. Mellars & C. Stringer (Eds.), *The human revolution: Behavioural and biological perspectives on the origin of modern humans* (pp. 677–713). Princeton, NJ: Princeton University Press.

Harvati, K., Gunz, P., & Grigorescu, D. (2007). Cioclovina (Romania): Affinities of an early modern European. *Journal of Human Evolution, 53*, 732–746.

Hedges, R. E. M., Housley, R. A., Law, I. A., & Bronk, C. R. (1989). Radiocarbon dates from the Oxford AMS system: Archaeometry datelist 9. *Archaeometry, 31*, 207–234.

Hedges, R. E. M., Housley, R. A., Bronk-Ramsey, C., & Van Klinken, G. J. (1994). Radiocarbon dates from the Oxford AMS system. *Archaeometry, 36*, 337–374.

Higham, T. F. G., Jacobi, R. M., & Bronk Ramsey, C. (2006a). AMS radiocarbon dating of ancient bone using ultrafiltration. *Radiocarbon, 48*, 179–195.

Higham, T., Ramsey, C. B., Karavanić, I., Smith, F. H., & Trinkaus, E. (2006b). Revised direct radiocarbon dating of the Vindija Upper Paleolithic Neandertals. *Proceedings of the National Academy of Sciences of the United States of America, 103*, 553–557.

Howells, W. (1976). Explaining the modern man: Evolutionists versus migrationists. *Journal of Human Evolution, 5*, 577–596.

Hublin, J.-J., & Bailey, S. E. (2006). Revisiting the last Neandertals. In N. J. Conrad (Ed.), *When Neanderthals and modern humans met* (pp. 105–128). Tübingen: Kerns Verlag. Tübingen Publications in Prehistory.

Hublin, J. J., Spoor, F., Braun, M., Zonnenveld, F., & Condemi, S. (1996). A late Neanderthal associated with Upper Paleolithic artefacts. *Nature, 381*, 224–226.

Hugen, K., Lehman, S., Southan, J., Overpeck, J., Marchal, O., Herring, C., & Turnbull, J. (2004). ^{14}C activity and global carbon cycle changes over the past 50,000 years. *Science, 303*, 202–207.

Jacobi, R. M., Higham, T. F. G., & Bronk Ramsey, C. (2006). AMS radiocarbon dating of the Middle and Upper Palaeolithic bone in the British Isles. *Journal of Quaternary Science, 21*, 557–573.

Janković, I., Karavanić, I., Ahern, J. C. M., Brajković, D., Mauch Lenardić, J., & Smith, F. H. (2006). Vindija cave and the modern human peopling of Europe. *Collegium Antropologicum, 30*, 315–319.

Karavanić, I. (1993). Gornjopaleolitičke kamene i koštane rukotvorine iz špilje Vindije. *Opuscula Archaeologica, 17*, 53–163.

Karavanić, I. (1995). Upper Paleolithic occupation levels and late-occuring Neandertal at Vindija cave (Croatia) in the context of Central Europe and the Balkans. *Journal of Anthropological Research, 51*, 9–35.

Karavanić, I. (2000). Olshewian and appearance of bone technology in Croatia and Slovenia. In J. Ortschiedt & G.-C. Weniger (Eds.), *Neanderthals and modern humans – Discussing the transition. Central and Eastern Europe from 50.000–30.000 B.P* (pp. 159–168). Mettmann: Neanderthal Museum.

Karavanić, I., & Smith, F. H. (1998). The Middle/Upper Paleolithic interface and the relationship of Neanderthals and early modern humans in the Hrvatsko Zagorje, Croatia. *Journal of Human Evolution, 34*, 223–248.

Karavanić, I., & Smith, F. H. (2000). More on the Neanderthal problem: The Vindija case. *Current Anthropology, 41*, 838–840.

Keith, A. (1927). Report on a fragment of a human jaw found at depth on (10 $^{1}/_{2}$ ft) 3.2 m. in the cave earth of the vestibule of Kent's

Cavern. *Transactions and Proceedings of the Torquay Natural History Society, 5*, 21–22.
Kesterke, M. J., & Ahern, J. C. M. (2007). Is the late Neadertal mandibular sample from Vindija cave (Croatia) biased? *Collegium Antropologicum, 31*, 365–373.
Kidder, J. H., Smith, F. H., & Jantz, R. L. (1992). Defining modern humans: A multivariate approach. In G. Bräuer & F. H. Smith (Eds.), *Continuity or replacement: Controversies in Homo sapiens evolution* (pp. 157–177). Rotterdam: A. A. Balkema.
Kozlowski, J. K. (1990). A multiaspectual approach to the origins of the Upper Palaeolithic in Europe. In P. Mellars (Ed.), *The emergence of modern humans: An archaeological perspective* (pp. 419–437). Ithaca, NY: Cornell University Press.
Kozłowski, J. K. (1996). Cultural context of the last Neandertals and early modern humans in the Central-Eastern Europe. In O. Bar-Yosef, L. L. Cavalli-Sforza, R. J. March, & M. Piperno (Eds.), *The lower and middle Paleolithic* (pp. 205–218). Forli: International Union of Prehistoric and Protohistoric Science.
Kozłowski, J. K., & Kozłowski, S. K. (1975). *Praziejie Europy od XL do IV tysiaclecia p.n.e.* Warsaw: Panstwowe wydawniciwo naukowe.
Kozłowski, J. K., & Otte, M. (2000). The formation of Aurignacian in Europe. *Journal of Anthropological Research, 56*, 513–534.
Krings, M., Stone, A., Schmitz, R. W., Krainitzki, H., Stoneking, M., & Pääbo, S. (1997). Neandertal DNA sequences and the origin of modern humans. *Cell, 90*, 19–30.
Krings, M., Geisert, H., Schmitz, R. W., Krainitzki, H., & Pääbo, S. (1999). DNA sequence of the mitochondrial hypervariable region II from the Neandertal type specimen. *Proceedings of the National Academy of Sciences of the United States of America, 96*, 5581–5585.
Krings, M., Capelli, C., Tschentscher, F., Geisert, H., Meyer, S., Von Haeseler, A., Grossschmidt, K., Possnert, G., Paunović, M., & Pääbo, S. (2000). A view of Neandertal genetic diversity. *Nature Genetics, 26*, 144–146.
Kurtanjek, D., & Marci, V. (1990). Petrografska istraživanja paleolitskih artefakata spilje Vindije. *Rad Jugoslavenske akademije znanosti i umjetnosti, 449*, 227–238.
Lalueza-Fox, C., Lourdes Sampietro, M., Caramelli, D., Puder, Y., Lari, M., Calafell, F., Martinez-Maza, C., Bastir, M., Fortea, J., de la Rasilla, M., Bertranpetit, J., & Rosas, A. (2005). Neandertal evolutionary genetics: Mitochondrial DNA data from the Iberian peninsula. *Molecular Biology and Evolution, 22*, 1077–1081.
Lalueza-Fox, C., Krause, J., Caramelli, D., Catalano, G., Milani, L., Lourdes Sampietro, M., Calafell, F., Martinez-Maza, C., Bastir, M., Garcia-Tabernero, A., de la Rasilla, M., Fortea, J., Pääbo, S., Bertranpetit, J., & Rosas, A. (2006). Mitochondrial DNA of an Iberian Neandertal suggests a population affinity with other European Neandertals. *Current Biology, 16*, R629–R630.
Leroi-Gourhan, A. (1958). Etude sur des restes humains provenant des grottes d'Arcy-sur-Cure. *Annales de Paleontologie, 44*, 87–147.
Lévêque, F., & Vandermeersch, B. (1980). Découverte de restes humains dans un niveau castelperronien à Saint-Cesaire (Charente Maritime). *Comptes Rendus de l'Académiedes Sciences, D291*, 187–189.
Malez, M. (1979). Nalazišta paleolithskog i mezolitskog doba u Hrvatskoj. In. A. Benac (Ed.), Praistorija jugoslavenskih zemalja, volume 1 (pp. 227–276). Sarajevo: Svjetlost.
Malez, M. (1983). *Razvoj kvartara, fosilnog čovjeka i njegovih materijalnih kultura na tlu Sjeverne Hrvatske.* Varaždin: Jugoslavenska akademija znanosti i umjetnosti. Posebni otisak iz knjige "Varaždinski Zbornik".
Malez, V. (1988). Pleistocenska orintofauna iz spilje Vindije u sjeverozapadnoj Hrvatskoj. *Radovi Zavoda za znanstveni rad Jugoslavenske akademije znanosti i umjetnosti Varaždin, 2*, 31–203.
Malez, M. & Rukavina, D. (1975). Krioturbacijske pojave u gornjopleistocenskim naslagama peiine Vindije kod Donje Voie u sjeverozapadnoj Hrvatskoj. *Rad JAZU, 371*, 245–265.
Malez, M., & Rukavina, D. (1979). Položaj naslaga spilje Vindije u sustavu članjenja Kvartara šireg podurčja Alpa. *Rad Jugoslavenske akademije znanosti i umjetnosti, 383*, 187–218.
Malez, M., & Ullrich, H. (1982). Neuere palaanthropologische Untersuchungen am Matrial aus her Hohle Vindija (Kroatien, Jugoslawien). *Palaeontologia Jugoslavica, 29*, 1–44.
Malez, M., Smith, F. H., Radovčić, J., & Rukavina, D. (1980). Upper Pleistocene hominids from Vindija, Croatia, Yugoslavia. *Current Anthropology, 21*, 365–367.
Mellars, P. (2006). A new radiocarbon revolution and the dispersal of modern humans in Eurasia. *Nature, 439*, 931–935.
Miracle, P. (1991). Carnivore dens or carnivore hunts? A review of Upper Pleistocene mammalian assemblages in Croatia and Slovenia. *Rad Jugoslavenske akademije znanosti i umjetnosti, 458*, 196–219.
Miracle, P. T. (1998). The spread of modernity in Paleolithic Europe. In K. Omoto & P. V. Tobias (Eds.), *Origins and past of modern humans: Towards reconciliation* (pp. 171–187). Singapore: World Scientific.
Montet-White, A. (1996). *Le Paléolithique en ancienne Yougoslavie.* Grenoble: Jérome Millon.
Nicolăescu-Popşlor. (1968). Les homes fossils en Roumanie. *Proceedings of the VIIIth International Congress of Anthropological and Ethnological Sciences*, Moscow 3 (1968) 381–386.
Noonan, J. P., Coop, G., Kudaravalli, S., Smith, D., Krause, J., Alessi, J., Chen, F., Platt, D., Pääbo, S., Pritchard, J. K., & Rubin, E. M. (2006). Sequencing and analysis of Neanderthal genomic DNA. *Science, 314*, 1113–1118.
Nordborg, M. (1998). On the probability of Neanderthal ancestry. *American Journal of Human Genetics, 63*, 1237–1240.
Oakley, K. P., Campbell, B. G., & Molleson, T. I. (1971). *Catalogue of fossil hominids. Part II: Europe.* London: British Museum.
Olariu, A., Skog, G., Hellborg, R., Stenström, K., Faarinen, M., & Persson, P. (2004). *Dating of two Paleolithic human fossil bones from Romania by accelerator mass spectrometry.* Report Wp1 IDRANAP.
Oliva, M. (1989). The early Aurignacian industry from Vedrovice II (Southern Moravia) and the question of the Aurignacian origins. *L'Anthropologie, 27*, 251–264.
Oliva, M. (1993). The Aurignacian in Moravia. In H. Knecht, A. Pike-Tay, & R. White (Eds.), *Before Lascaux. The complex record of the early Upper Paleolithic* (pp. 37–56). Boca Raton: CRC Press.
Orlando, L., Darlu, P., Toussaint, M., Bonjean, D., Otte, M., & Hänni, C. (2006). Revisiting Neandertal diversity with a 100,000 year old mtDNA sequence. *Current Biology, 16*, R400–R402.
Ovchinnikov, I. V., Götherström, A., Romanova, P. G., Kharitonov, V. M., Lidén, K., & i Goodwin, W. (2000). Molecular analysis of Neandertal DNA from the Northern Caucasus. *Nature, 404*, 490–493.
Păunescu, A. (2001). *A Paleoliticul şi Mezoliticul din Spaţiul Transilvan.* Bucuresti: Editura AGIR.
Paunović, M., & Smith, F. H. (2002). Taphonomy of lower vertebrates from Vindija cave (Croatia): Delicacy on Neandertal table of animal prey? *Journal of Human Evolution, 42*, A27.
Paunović, M., Jambrešić, G., Brajković, D., Malez, V., & Mauch Lenardić, J. (2001). Last glacial settlement of Croatia. *Acta Geologica, 26*, 27–70.
Pradel, L. (1966). Transition from Mousterian to Perigordian: Skeletal and industrial. *Current Anthropology, 7*, 33–50.
Rainer, F., & Simionescu, I. (1942). Sur le premier crâne d'homme paléolithique trouvé en Roumanie. *Analele Academiei Romane, 17*, 489–507.
Relethford, J. (2001a). *Genetics and the search for modern human origins.* New York: Wiley.
Relethford, J. H. (2001b). Absence of regional affinities of Neandertal DNA with living humans does not reject multiregional evolution. *American Journal of Physical Anthropology, 115*, 95–98.

Relethford, J. H., & Jorde, L. B. (1999). Genetic evidence for larger African population size during recent human evolution. *American Journal of Physical Anthropology, 108*, 251–260.

Rigaud, J.-P. (1989). From the middle to the Upper Paleolithic: Transition or convergence? In E. Trinkaus (Ed.), *The emergence of modern humans* (pp. 142–153). Cambridge: Cambridge University Press.

Rougier, H., Milota, S., Rodrigo, R., Gherase, M., Sarcină, L., Moldovan, O., Zilhão, J., Constantin, S., Franciscus, R. G., Zollikofer, C. P. E., Ponce de León, M., & Trinkaus, E. (2007). Peştera cu Oase 2 and the cranial morphology of early moderrn Europeans. *Proceedings of the National Academy of Sciences of the United States of America, 104*, 1165–1170.

Schmitz, R., Serre, D., Bonani, G., Feine, S., Hillgruber, F., Kraintzki, H., Pääbo, S., & Smith, F. (2002). The Neadertal type site revisited: Interdisciplinary investigations of skeletal remains from the Neander Valley, Germany. *Proceedings of the National Academy of Sciences of the United States of America, 99*, 13342–13347.

Serre, D., Langaney, A., Chech, M., Teschler-Nicola, M., Paunovic', M., Hofreiter, M., Possnert, G., Paabo, S., & Mennecier, P. (2004). No evidence of Neandertal mtDNA contribution to early modern humans. *PLoS Biology, 2*, 313–317.

Smith, F. H. (1978). Evolutionary significance of the mandibular foramen area in Neandertals. *American Journal of Physical Anthropology, 48*, 523–531.

Smith, F. H. (1982). Upper Pleistocene hominid evolution in South-Central Europe: A review of the evidence and analysis of trends. *Current Anthropology, 23*, 667–703.

Smith, F. H. (1984). Fossil hominids from the Upper Pleistocene of Central Europe and the origin of modern Europeans. In F. H. Smith & F. Spencer (Eds.), *The origin of modern humans: A world survey of the fossil evidence* (pp. 137–209). New York: Alan R. Liss.

Smith, F. H. (1992). The role of continuity in modern human origins. In G. Braüer & F. H. Smith (Eds.), *Continuity or replacement? Controversies in Homo sapiens evolution* (pp. 145–156). Rotterdam: Balkema.

Smith, F. H. (1994). Samples, species, and speculations in the study of modern human origins. In M. H. Nitecki & D. V. Nitecki (Eds.), *Origins of anatomically modern humans* (pp. 227–249). New York: Plenum Press.

Smith, F. H., & Ahern, J. C. (1994). Additional cranial remains from Vindija cave, Croatia, Yugoslavia. *American Journal of Physical Anthropology, 93*, 275–280.

Smith, F. H., & Ranyard, G. C. (1980). Evolution of the supraorbital region in Upper Pleistocene fossil hominids from South-Central Europe. *American Journal of Physical Anthropology, 53*, 589–610.

Smith, F. H., Boyd, D. C., & Malez, M. (1985). Additional Upper Pleistocene human remains from Vindija cave, Croatia, Yugoslavia. *American Journal of Physical Anthropology, 68*, 375–383.

Smith, F. H., Falsetti, A. B., & Donnelly, S. M. (1989). Modern human origins. *Yearbook of Physical Anthropology, 32*, 35–68.

Smith, F. H., Trinkaus, E., Pettitt, P. B., Karavanić, I., & Paunović, M. (1999). Direct radiocarbon dates for Vindija G1 and Velika Pećina Late Pleistocene hominid remains. *Proceedings of the National Academy of Sciences of the United States of America, 96*, 12281–12286.

Smith, F. H., Janković, I., & Karavanić, I. (2005). The assimilation model, modern human origins in Europe, and the extinction of Neandertals. *Quaternary International, 137*, 7–19.

Soficaru, A., Doboş, A., & Trinkaus, E. (2006). Early modern humans from the Peştera Muierii, Baia de Fier, Romania. *Proceedings of the National Academy of Sciences of the United States of America, 103*, 17196–17201.

Straus, L. G. (1997). The Iberian situation between 40,000 and 30,000 years B.P. in light of European models of migration and convergence. In G. A. Clark & C. M. Willermet (Eds.), *Conceptual issues in modern human origins research* (pp. 235–252). NewYork: Aldine De Gruyter.

Straus, L. G. (1999). The Neanderthal problem continued. *Current Anthropology, 40*, 352–355.

Straus, L. G. (Ed.). (2005). *Armageddon or entete? The demise of the European Neadertals in isotope stage 3 (special issue of quaternary international 137 (1))*. Oxford: Elsevier.

Stringer, C. B. (1990). British Isles. In R. Orban (Ed.), *Hominid remains: An update. British Isles and Eastern Germany* (pp. 1–40). Brussels: Université Libre de Bruxelles.

Svoboda, J. (1993). The complex origin of the Upper Paleolithic in the Czech and Slovak Republics. In H. Knecht, A. Pike-Tay, & R. White (Eds.), *Before Lascaux. The complex record of the early Upper Paleolithic* (pp. 23–36). Boca Raton: CRC Press.

Svoboda, J. (2001). Mladeč and other caves in the Middle Danube region: Early modern humans, late Neandertals, and projectiles. In J. Zilhão, T. Aubry, & A. F. Carvalho (Eds.), *Les premiers hommes modernes de la Péninsule Ibérique* (pp. 45–60). Lisboa: Actes du colloque de la Commission VIII de l'UISPP.

Svoboda, J. A. (2004). Continuities, discontinuities, and interactions in early Upper Paleolithic technologies. A view from the Middle Danube. In P. J. Brantingham, S. L. Kuhn, & K. W. Kerry (Eds.), *The early Upper Paleolithic beyond Western Europe* (pp. 30–49). Berkeley: University of California Press.

Templeton, A. (2002). Out of Africa again and again. *Nature, 416*, 45–51.

Templeton, A. (2005). Haplotype trees and modern human origins. *Yearbook of Physical Anthropology, 48*, 33–59.

Tode, A., Preul, F., Richter, K., Selle, W., Pfaffenberg, K., Kleinschmidt, A., & Guenther, E. (1953). Die Untersuchung der paläeolitischen Freilandstation von Salzgitter-Lebenstedt. *Eiszentaler und Gegenwart, 3*, 144–220.

Trinkaus, E. (2005). Early modern humans. *Annual Review of Anthropology, 34*, 207–230.

Trinkaus, E., & Smith, F. H. (1995). Body size of the Vindija Neandertals. *Journal of Human Evolution, 28*, 201–208.

Trinkaus, E., & Churchill, S. E. (1988). Neandertal radial tuberosity orientation. *American Journal of Physical Anthropology, 75*, 15–21.

Trinkaus, E., Milota, S., Rodrigo, R., Mircea, G., & Moldovan, O. (2003a). Early modern human cranial remains from the Pestera cu Oase, Romania. *Journal of Human Evolution, 45*, 245–253.

Trinkaus, E., Moldovan, O., Milota, S., Bilgar, A., Sarcina, L., Athreya, S., Bailey, S. E., Rodrigo, R., Mircea, G., Higham, T., Ramsey, C. B., & Plicht, J. C. D. (2003b). An early modern human from the Pestera cu Oase, Romania. *Proceedings of the National Academy of Sciences of the United States of America, 100*, 11231–11236.

Trinkaus, E., Zilhão, J., Rougier, H., Rodrigo, R., Milota, Ş., Gherase, M., Sarcina, L., Moldovan, O., Băltean, I. C., Codrea, V., Bailey, S. A., Franciscus, R. G., Ponce de Leon, M., & Zollikofer, C. P. E. (2006). The Peştera cu Oase and early modern humans in Southeastern Europe. In N. J. Conrad (Ed.), *When Neanderthals and modern humans met* (pp. 145–164). Tübingen: Kerns Verlag. Tübingen Publications in Prehistory.

Usik, V. I., Monigal, K., & Kulakovskaya, L. (2006). New perspectives on the Transcarpathian Middle to Upper Paleolithic boundary. In N. J. Conrad (Ed.), *When Neanderthals and modern humans met* (pp. 213–232). Tübingen: Kerns Verlag. Tübingen Publications in Prehistory.

Valoch, K. (1972). Rapports entre Le Paléolithique Moyen et le Paléolithique Supérieur en Europe Centrale. In F. Bordes (Ed.), *The Origins of Homo sapiens* (pp. 161–171). Paris: UNESCO.

Voisin, J.-L. (2006). Speciation by distance and temporal overlap: A new approach to understanding Neanderthal evolution. In K. Harvati & T. Harrison (Eds.), *Neanderthals revisited: New approaches and perspectives* (pp. 307–322). New York: Springer.

Wild, E. M., Paunović, M., Rabeder, G., Steffan, G., & Steier, P. (2001). Age determination of fossil bones from the Vindija Neanderthal site in Croatia. *Radiocarbon, 43*, 1021–1028.

Wild, E. M., Terschler-Nicola, M., Kutschera, W., Steier, P., Trinkaus, E., & Wanek, W. (2005). Direct dating of early Upper Paleolithic human remains from Mladeč. *Nature, 435*, 332–335.

Wolpoff, M. H. (1999). *Paleoanthropology* (2nd ed.). New York: McGraw-Hill.

Wolpoff, M. H., Smith, F. H., Malez, M., Radovčić, J., & Rukavina, D. (1981). Upper Pleistocene human remains from Vindija cave, Croatia, Yugoslavia. *American Journal of Physical Anthropology, 54*, 499–545.

Yu, N., Zhao, Z., Fu, Y., Sambuughin, N., Ramsay, M., Jenkins, T., Leskien, E., Patthy, L., Jorde, L., Kuromori, T., & Li, W. (2001). Global patterns of human DNA sequence variation in a 10-kb region on chromosome 1. *Molecular Biology and Evolution, 18*, 214–222.

Zilhão, J. (2006). Neandertals and moderns mixed, and it matters. *Evolutionary Anthropology, 15*, 183–195.

Zilhão, J., & D'Errico, F. (1999a). Reply in the Neanderthal problem continued. *Current Anthropology, 40*, 355–364.

Zilhão, J., & D'Errico, F. (1999b). The chronology and taphonomy of the earliest Aurignacian and its implications for the understanding of Neanderthal extinction. *Journal of World Prehistory, 13*, 1–68.

Chapter 24
Late Neandertals and Early Modern Humans in Europe, Population Dynamics and Paleobiology

Erik Trinkaus

Abstract On the sesquicentennial of the discovery at the Kleine Feldhofer Grotte, it is appropriate to reassess the nature of the biological transition between late Neandertals and the earliest modern humans in Europe. An assessment of the latter sample shows a predominantly modern human morphological pattern, but the persistent and varied presence of distinctive Neandertal features and/or archaic traits lost or rare in the ancestral Mid Upper Paleolithic modern humans. These traits includee variably present marked frontal flattening, large occipital buns, large juxtamastoid eminences, suprainiac fossae, wide mandibular rami, asymmetrical mandibular notch, medial notch position, mandibular foramen bridging, molar megadontia and incisor shoveling. Moreover, the later European Mid Upper Paleolithic (Gravettian) sample exhibits persistence of some of these and other Neandertal/archaic features. These combine to indicate a non-trivial Neandertal contribution to the gene pools of the early modern humans in Europe. At the same time, paleobiological assessment of late Neandertals and early modern humans in Europe indicates a mosaic of functional anatomical changes, involving the faces, humeri and femora of late Neandertals and the dentition, scapula, femora and talus of early modern humans. These data combine to indicate that the period between the Middle Paleolithic Neandertals and the Mid Upper Paleolithic modern humans was a complex mosaic in terms of population dynamics and behavioral patterns. Simple models of an abrupt behavioral and phylogenetic transition for this period in Europe should be abandoned.

Keywords Late Pleistocene • Human paleontology • Cranium • Dentition • Mandible • Femur

E. Trinkaus (✉)
Department of Anthropology, Washington University, Campus Box 1114, St. Louis, MO 63130, USA
e-mail: trinkaus@wustl.edu

Introduction

It is appropriate in 2006 to talk about Neandertals and modern humans, since this is the sesquicentennial of the discovery at the Kleine Feldhofer Grotte (Schmitz 2006), the approximate centennial of the classic papers by Schwalbe (1901), Klaatsch (1901) and Cunningham (1908) on the Feldhofer remains, the centennial of the Krapina monograph (Gorjanović-Kramberger 1906), the approximate centennial of the Le Moustier and Boufia Bonneval (La Chapelle-aux-Saints) discoveries (Boule 1908, 1911–1913; Klaatsch and Hauser 1909), the golden anniversary of a more upright position for the Neandertals (Arambourg 1955; Straus and Cave 1957; Toerien 1957), and the silver anniversary of the first proposal for an Out-of-Africa model of modern human emergence (Trinkaus 1981; Bräuer 1982). It is therefore appropriate that we reassess the nature of the human biological and behavioral transition which took place in the northwestern Old World between ≈ 45,000 and ≈ 35,000 years ago.

Not only have the past decades provided us with a greater appreciation of the biology and behavior of the late archaic humans that we call the Neandertals, but they have greatly increased our knowledge of early modern humans, the modern humans who were contemporaneous with, and immediately subsequent to, those Neandertals. Since the issues concern both the Neandertals and their issue, it is appropriate to review what we know about both of these groups, late Neandertals and early modern humans.[1] To do otherwise would be an attempt to solve one side of an equation while ignoring the other side.

[1] The term "modern" is used in a strictly cladistic sense, indicating an abundance of the uniquely derived features (and overall morphological pattern) of extant (i.e., modern) humanity (see Trinkaus 2006a for a list). The term "archaic" implies only the presence of the overall morphological pattern of, and an abundance of ancestral features of, earlier Pleistocene *Homo*. They are used without implications for the behavioral patterns or capabilities of the groups in question.

S. Condemi and G.-C. Weniger (eds.), *Continuity and Discontinuity in the Peopling of Europe: One Hundred Fifty Years of Neanderthal Study*, 315
Vertebrate Paleobiology and Paleoanthropology, DOI 10.1007/978-94-007-0492-3_24, © Springer Science+Business Media B.V. 2011

The Human Evolutionary Background

Of concern here is the western Old World ancestral situation of 50,000–100,000 years ago, since this is the one from which both late Neandertals and subsequent European early modern humans must have emerged. It consisted of regional groups of humans evolving out of the Middle Pleistocene human regional variation.

The earliest modern humans, given current fossil evidence, were an eastern and northeastern African phenomenon, represented principally by the Herto and Omo-Kibish remains >100,000 BP (Day and Stringer 1982; White et al. 2003; McDougall et al. 2005). They temporarily spread into extreme southwestern Asia and left their dead at Qafzeh and Skhul around 80,000–100,000 BP, but otherwise they appear to have remained geographically east and northeast African until ≈ 50,000 BP. This sample is joined by the similarly aged or younger fragmentary elements from Aduma, Bouri and Haua Fteah (Tobias 1967; Haile-Selassie et al. 2004). The Early Upper Paleolithic, ≈42,000 BP Nazlet Khater 2 skeleton (Thoma 1984; Crevecoeur and Trinkaus 2004; Crevecoeur 2008) may be of relevance, even though it exhibits archaic features previously lost in the earlier Middle Paleolithic modern humans and is approximately contemporaneous with the earliest modern humans in Europe, those from the Peştera cu Oase (Trinkaus et al. 2003; Rougier et al. 2007).[2]

Other purported "modern" humans from Africa, including the Middle Stone Age (MSA) remains from southern Africa and the Middle Paleolithic (including Aterian) remains from northwest Africa, are of questionable relevance. The southern African MSA remains, from sites such as Klasies River Mouth, Blombos Cave, Die Kelders, and Pinnacle Point, present a mosaic of distinctly archaic and possibly "modern" features, with the archaic features dominating. By any appropriate criteria, they are *not* modern humans, although the possibly "modern" features may represent minimal southward gene flow from eastern African early modern humans (Smith 1993; Lam et al. 1996; Trinkaus 2005 and references therein). The northwest African sample represents regional late archaic humans of uncertain age, geographically isolated in northwestern Africa and little different from regional Middle Pleistocene remains (Vallois and Roche 1958; Ferembach 1976; Trinkaus 2005); claims to the contrary confuse the absence of Neandertal features in the sample with purported modern human affinities. In addition, the ≈ 36,000 BP modern human Hofmeyr cranium (Grine et al. 2007) postdates the spread of modern humans across most of Europe, and it is therefore not directly relevant to the late archaic to early modern human transition in Europe.

The earliest securely dated modern humans from eastern Eurasia, the Niah Cave and Tianyuandong partial skeletons (Brothwell 1960; Barker et al. 2007; Shang and Trinkaus 2010) are both ≈ 40,000 years old, contemporaneous with the earliest modern Europeans [include the incomplete 110,000 BP Zhirendong fossils (Liu et al. 2010) older eastern Eurasian modern human specimens and Zhiren 3 remains present]. Even though the Tianyuan 1 archaic features and the Niah Cave cranium appears distinctly Australomelanesian (as opposed to African), their geographical position vis-à-vis the earliest modern Europeans minimizes their relevance to the transition in Europe.

The earliest modern humans, referred to here as the "Middle Paleolithic modern humans" (MPMH), are best represented paleontologically by the Qafzeh-Skhul sample, since it preserves all portions of the hard tissue for multiple individuals, all ages and both sexes. However, its full morphological range is supplemented by the east and northeast African members of this group. Relative to Early and Middle Pleistocene archaic *Homo*, the ancestral African modern humans present a large suite of uniquely derived features (Trinkaus 2006a), most of which have long been used to delimit modern humans from archaic humans. The details of the list may be altered, but the plethora of derived traits for the sample cannot be denied. Moreover, given the abundance of work to document that they are not Neandertals (Vandermeersch 1981; Tillier 1999; White et al. 2003; Haile-Selassie et al. 2004), this sample is almost devoid of traits that could be considered distinctive, derived, Neandertal features.

The Neandertals were a western Eurasian phenomenon, variably extending into extreme southwestern Asia. They emerged gradually within that geographical range during the later Middle Pleistocene and probably graded clinally with late archaic human populations further east in Eurasia. Their late Middle and Late Pleistocene lineage is well documented across this range and is the best known example of the gradual accumulation of regional features in human paleontology. The Neandertals are well known from early last glacial western Eurasian remains and multiple burials. They also have a set of uniquely derived features (Trinkaus 2006a), although a more modest number than the early modern humans when then underlying biology, and not multivariate confusions of unique combinations of aspects, is taken into account.

It is these two groups (pre-50,000 BP ancestral African modern humans and pre-45,000 BP Neandertals) which re-present the potential ancestral lineages for late Neandertals and western Eurasian early modern humans. It is therefore with respect to the biology of these two samples that the biology of those later early modern humans must be compared.

[2] All ages are given in calendar years, with radiocarbon ages "calibrated"using CalPal v.1.4 (www.calpal.de). Any uncertainties in the calibration of radiocarbon ages >30,000 ^{14}CBP are mitigated by the approximate nature of the dates used here.

Human Population Dynamics at the European Transition

The biological transition in Europe involved several features. There was a late survival of Neandertals, especially in Iberia south of the Pyrenees but also probably in more central regions of Europe and possibly northwestern Europe. There was a relatively late spread of modern humans, from east to west across Europe. The earliest modern humans are ≈ 40,000 years old, in southeastern Europe (Trinkaus et al. 2003, 2006b; Rougier et al. 2007). They are not documented further west until ≈ 37,000 BP (Dujardin 2003; Henry-Gambier et al. 2004; Wild et al. 2005), and they probably did not reach the Atlantic peripheries until ≈ 35,000 BP (Walker et al. 2008). The morphological contrasts between the late Neandertals and early modern humans in Europe are readily apparent and well documented. The issue remains as to the nature of the population dynamics during this transition – was it simply replacement, variable admixture or what?

So why does it matter what the population dynamics were? Some seem concerned with the whether there are Neandertals genes still among us? This nineteenth century question is neither appropriate nor relevant in the twenty-first century. Some want to know the nature of archaic-modern human reproduction, or what was the pattern of sex in the Pleistocene. I expect that they had it. Mate choice was probably limited given probable low populations densities, and people are known to be liberal in their sexual tastes. Perhaps of more relevance are the behavioral contrasts and similarities implied by one reproductive pattern or another. How "human" were the Neandertals, and what do inferred reproductive patterns tell us about how the two populations saw each other? Did they perceive each other as appropriate mates, in the context of widely dispersed small foraging populations? And finally, how special are we really? How closely linked are we to "nature" (or to less "human" groups such as the Neandertals)?

The Earliest European Modern Humans: The Sample and the Issue

The earliest modern Europeans are represented by a series of fossils, between ≈ 40,000 and ≈ 34,000 years ago. The majority of the information comes from Cioclovina, Mladeč, Muierii and Oase, with additional data from Brassempouy, La Quina Aval and Les Rois. Fortunately, many of these sites now have good radiocarbon dates, with four of them (Cioclovina, Mladeč, Muierii and Oase [Trinkaus et al. 2003; Wild et al. 2005; Soficaru et al. 2007; Rougier et al. 2007; Doboş et al. 2010]) providing direct dates on the human remains, and Brassempouy having a series of dates clearly associated with the human remains (Henry-Gambier et al. 2004). The Fontana Nuova and Fossellone remains are not included, given their derivations from disturbed deposits and hence uncertain ages (Mallegni and Segre-Naldini 1992; Chilardi et al. 1996).[3]

Given this sample, what unusual features do we find among them, assuming that they are primarily descendant from the highly derived MPMH. In this, the potential ancestral groups are the MPMH and the Neandertals. Any uniquely derived features should come from the group possessing them. Archaic features lost in one of these groups should come from the other group if it still possesses them, even if they are not uniquely derived (i.e., are ancestral) in that group. The alternative is a reversal, a possibility for one or two traits but not for a plethora of them.

In this, it needs to be emphasized (since this is frequently misunderstood) that the appearance of a purported "Neandertal" feature in recent (Holocene) human samples, or even in Late Pleistocene modern humans that long post-date the transition in Europe, has *no* bearing on this issue. The earlier Upper Paleolithic modern humans are not descended from more recent humans, and they therefore could not have inherited these traits from those recent human populations.

It also needs to be emphasized that a trait that is used to distinguish Neandertals from Late Pleistocene samples of modern humans is still the same trait if it appears in an otherwise modern human fossil specimen. For example, if arctic body proportions are used to distinguish the Neandertals in a Late Pleistocene context (e.g., Hublin 1983,1998, following Trinkaus 1981; Holliday 1997), such body proportions cannot then be denied as having phylogenetic valence when inconvenient (e.g., Hublin 2000, commenting on Duarte et al. 1999). The same applies to a series of craniomandibular traits, including suprainiac fossae, lingular bridging of the mandibular foramen, shovel-shaped incisors, and others. To do otherwise indicates a morphological double standard for phylogenetic convenience.

It is also apparent that the European earliest modern humans are indeed "modern" in their overall morphological pattern. From among the various derived modern human traits previously listed (Trinkaus 2006a), this sample (given preservation of features) shares an abundance of these traits with the MPMH and/or more recent humans. These features include: relative parietal expansion, supraorbital torus absence, infrequent platycephaly, prominent parietal bosses, laterally bulbous mastoid processes, labyrinthine morphology, reduced facial length, distinct canine fossae, narrow nasal aperture, projecting tuber symphyseos, femoral pilaster,

[3] Unless otherwise noted, morphological assessments are based on personal study of the original remains; references that provide further documentation, when available, are included.

reduced medial femoral buttress, and angular tibial and fibular diaphyses. There are other derived modern humans traits present in the sample, but they are variably present. Consequently, the overwhelming majority of the ancestry of these European earliest modern humans is "modern" and probably from the MPMH. Any Neandertal contribution to their ancestry is likely to minor but, from the considerations above, not necessarily trivial to our understanding of Late Pleistocene human evolution.

Therefore, simply documenting that the overall morphological affinities (especially morphometrically) of these fossils is with modern humans, as opposed to with the Neandertals, simply misses the point. No one is arguing that these fossils are other than basically modern in their morphology. No one is arguing that they are F1 hybrids. No one is arguing that they are some kind of Cuisinart-style genetic blend of equal parts of Neandertal and modern human morphology; such an argument would have been silly in the early twentieth century, and it is absurd in the twenty-first century. The question is whether, within the context of an overall modern human morphological gestalt, there is evidence of some degree, however modest, of Neandertal ancestry.

The concern is therefore not whether they are modern humans or even whether they are some average of Neandertal and modern human morphology, something intermediate between the two. The issue is whether, in the context of a basically modern human morphological pattern, there are epigenetic (i.e., near genetic, *sensu* Berry and Berry (1967)) Neandertal features and/or archaic traits lost in the MPMH.

The Earliest European Modern Humans: The Morphology

Even though their overall morphological pattern is clearly the derived pattern of modern humans, these earliest European modern humans exhibit a suite of morphological features that are generally unusual relative to the ancestral African modern humans. The traits involve a series of cranial, mandibular and dental features. Since many, if not most, of these characteristics are likely to be the result of complex multiple-locus genetic variation, they are treated in much the same way that haplotypes are considered in similarly complex molecular genetic systems.

Oase 2 has an exceptionally long and flat frontal sagittal arc, combined with a strongly curved parietal arc (Rougier et al. 2007; Fig. 24.1). The Cioclovina 1 and Muierii 1 crania exhibit the same tendency, although less pronounced than Oase 2. Nazlet Khater 2 has a relatively flat frontal arc, but it lacks the combination with a highly curved parietal arc. The Oase 2 configuration is otherwise known solely from the Shanidar Neandertals among earlier Late Pleistocene humans.

Mladeč 3, 5 and 6 and Muierii 1 have prominent occipital buns, with strong sagittal convexity and clear sagittal concavity along the lambdoid suture (Fig. 24.1), indicative of late, additional growth along the lambdoid suture. Occipital buns are the product of differential infantile brain versus neurocranial growth rates (Trinkaus and LeMay 1982), and the relevant homology is therefore the development of a prominent bun. The details of its shape are less important, since those are influenced, as in all neurocranial growth, by the forms of adjoining structures. This homology is supported by the metrical analysis of Late Pleistocene occipital bones (Gunz and Harvati 2007), in which late archaic and early modern humans exhibit a continuum in the development of an occipital bun. Even though occipital buns are known in other samples (Pleistocene and recent), they are absent in the MPMH sample and abundant in the Neandertals, especially in Europe.

The otherwise modern Cioclovina 1 cranium exhibits a horizontal oval suprainiac fossa, a modest nuchal torus restricted to the median half of the occipital, and the absence of an external occipital protuberance (Soficaru et al. 2007; Fig. 24.2). The only contrasts with the stereotypical Neandertal pattern (Santa Luca 1978) are the modest depth of the fossa, the presence of the superior nuchal line on, rather than below, the nuchal torus, and the close proximity on the midline of the semispinalis capitis fossae. If one sets the nuchal line position and semispinalis fossae proximity aside as separate (and modern) features, the suprainiac morphology is distinct from that of MPMH and similar to that of the Neandertals. Since it is the combination of the first three features, rather than necessarily one of them alone, which defines the Neandertal pattern (Hublin 1978), this pattern can only be considered as homologous to the Neandertal pattern. Although this suite of features is occasionally present in other samples (Trinkaus 2004), it is absent from the MPMH sample [Aduma 3 has the fossa but combines it with a pronounced external occipital protuberance and the absence of a nuchal torus (Haile-Selassie et al. 2004)] and present in 100% of the Neandertals.

Mladeč 1 and 5 exhibit large juxtamastoid eminences, a Middle Pleistocene archaic trait present in most Neandertals but absent from the MPMH. Mladeč 2 and Oase 2 have medium sized crests, such as are found in the remainder of the Neandertals and only one MPMH cranium, Qafzeh 3. Cioclovina 1 and Muierii 2 lack such crests.

The Oase 1 mandible exhibits an exceptionally wide ramus (Fig. 24.3), and the anterior position of the Oase 2 zygomatic bones implies a similarly wide mandibular ramus (Fig. 24.1). Such a wide ramus is absent in the MPMH and the Neandertals, and it is a more generally archaic feature present in Middle Pleistocene specimens such as Arago 2, KMN-BK 67, Mauer 1 and Tighenif 3 (Lebel and Trinkaus 2002). It is nonetheless present in the approximately contemporaneous Nazlet Khater 2 mandible (Crevecoeur 2008).

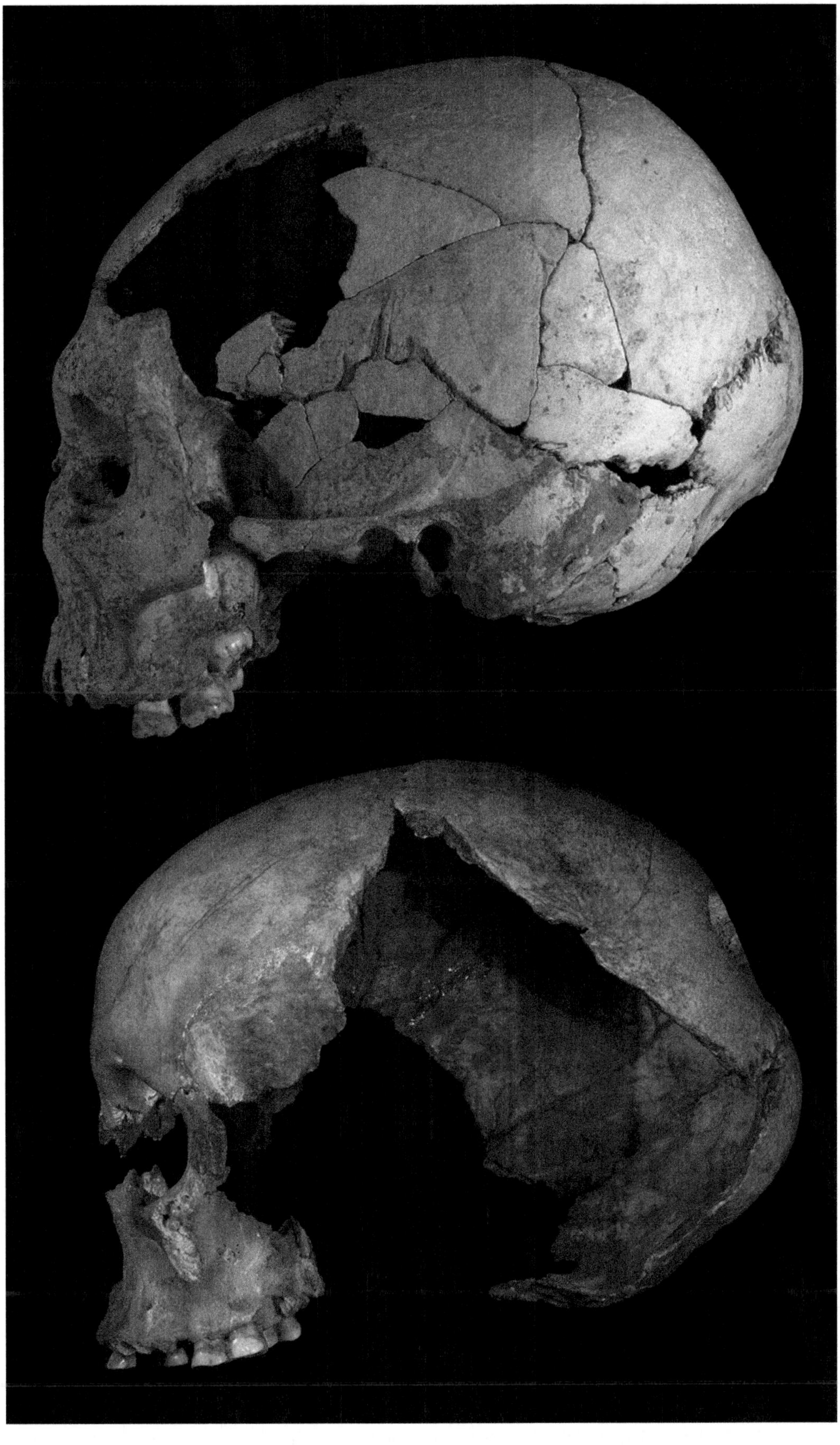

Fig. 24.1 The Oase 2 (*above*) and Muierii 1 (*below*) crania in *norma lateralis* left, not to the same scale. The long and low frontal profile of Oase 2 and the pronounced occipital bun of Muierii 1 are evident, as well as the anteriorly placed zygomatic region of Oase 2

Fig. 24.2 Posterior view of the Cioclovina 1 occipital region, showing the suprainiac fossa, medially evident nuchal torus, and absence of an external occipital protuberance

Mandibular ramus notch asymmetry is present in more than 70% of the Neandertals, absent in the MPMH, and absent in Oase 1. However, it is markedly present in the Muierii 1 mandible (Doboş et al. 2010; Fig. 24.3). In addition, the lingular bridging of the mandibular foramen, or horizontal oval pattern, is absent in the ancestral African modern humans and all earlier humans, but it is present in almost 50% of the Neandertals (Stefan and Trinkaus 1998a, b). Although it is absent from the Muierii 1 mandible, it is present, if unilaterally, on the Oase 1 mandible (Trinkaus et al. 2003; Doboş et al. 2010). In addition, a mandibular notch crest which is lateral relative to the mandibular condyle is plesiomorphic for the genus *Homo* and is found in all MPMH mandibles and Nazlet Khater 2. Among the Neandertals, about half of the mandibles have a medially positioned crest. Oase 2 has the lateral crest position, but Muierii 1 has a medially displaced one (Doboş et al. 2010).

Oase 1 and 2 and Mladeč 8 exhibit moderately large first molars and second molars near the upper limits of Late Pleistocene ranges of variation. However, the Oase 1 M_3s are the largest documented in the last 500,000 years of the genus *Homo*, and the M^3s of Oase 2 are the largest currently known for non-habiline members of the genus *Homo* (Trinkaus et al. 2003; Rougier et al. 2007) The Mladeč 8 third molars are absent, but its first two molars parallel the pattern in the Oase dentitions. These dentitions are generally archaic with respect to the other Late Pleistocene human dentitions.

Maxillary central incisor shoveling is ubiquitous among archaic *Homo*, including the Neandertals, but it is absent in the MPMH. Those early modern humans have the derived chisel-shaped incisor form. Early modern Europeans have variable maxillary incisor shoveling; it is absent from the Brassempouy sample (Henry-Gambier et al. 2004), but the series of maxillary central incisors from Les Rois exhibit the complete range of variation in marginal ridge and lingual tubercle development, from chisel-shaped to strongly shovel-shaped (Fig. 24.4).

The Earliest European Modern Humans: The Implications

Consequently, given the small sample size of the earliest modern Europeans, there is a surprising abundance of Neandertal and/or archaic *Homo* features, features that had been lost or were rare among the earliest African modern humans. These include a flat frontal profile, occipital buns, suprainiac morphology, juxtamastoid eminences, wide

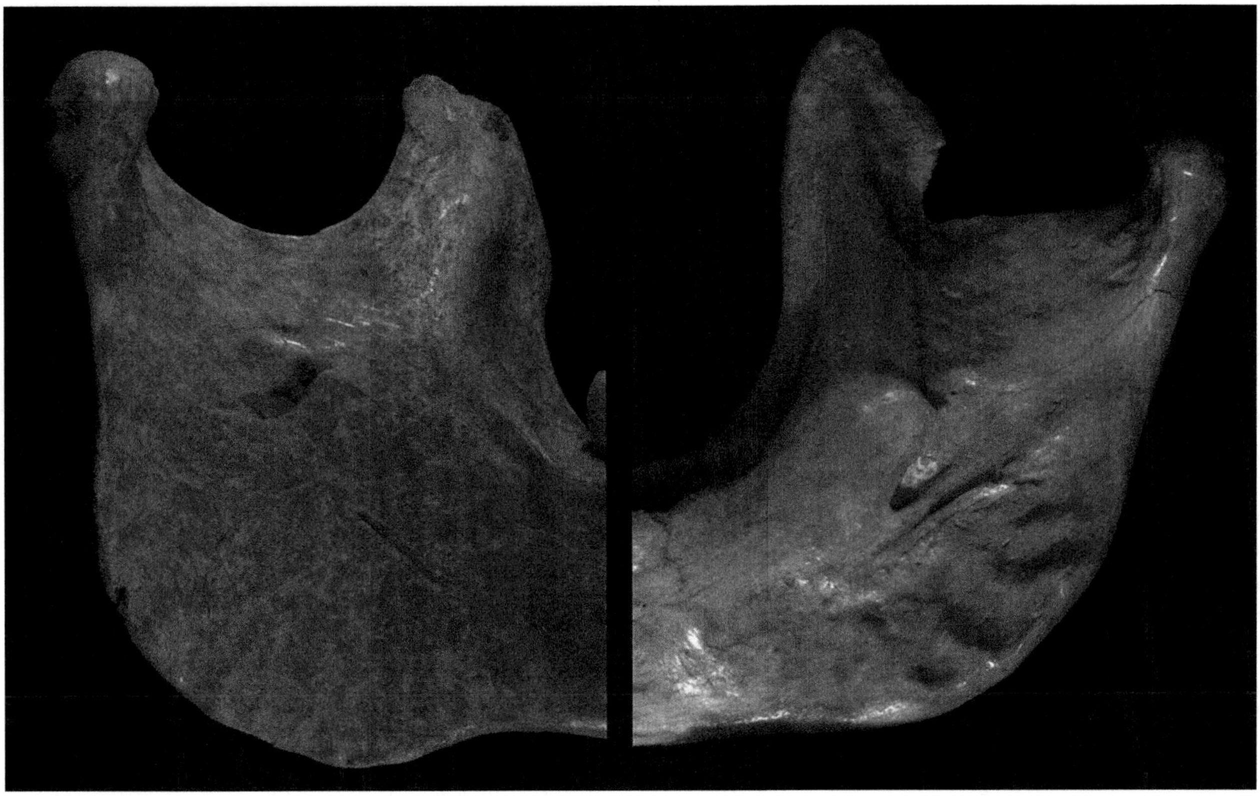

Fig. 24.3 Medial views of the Oase 1 (*left*) and Muierii 1 (*right*) mandibular rami, not to the same scale. Note the symmetrical (Oase 1) versus asymmetrical (Muierii 1) mandibular notches, the low (Oase 1) versus high (Muierii 1) coronoid processes, and the bridged (Oase 1) versus open (Muierii 1) mandibular foramina

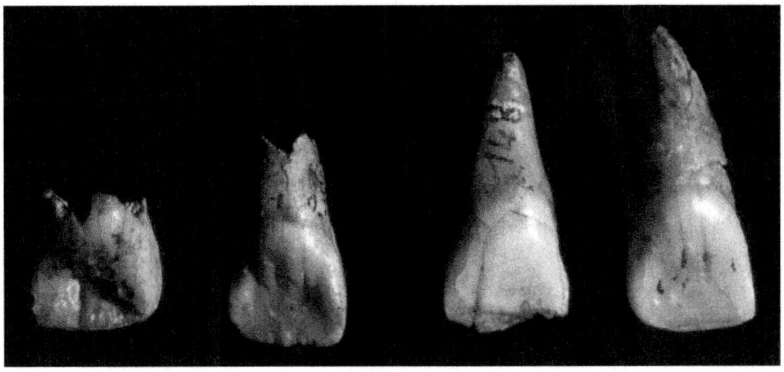

Fig. 24.4 Four isolated maxillary central incisors from Les Rois à Mouthiers, showing the range of variation in marginal ridge and lingual tubercle development (hence in shoveling) within one European early modern human sample

mandibular rami, mandibular foramen bridging, mandibular notch shape, notch crest to condyle position, large distal molars and maxillary incisor shoveling.

One could probably explain away one or two of these features as the products of random recombinations or reversals, in the context of purely MPMH ancestry. For example, a wide mandibular ramus could be considered part of the MPMH morphological suite, given the configuration of Nazlet Khater 2. Yet, wide rami are not known in the earlier and more probably ancestral MPMH, especially since Nazlet Khater 2 is no older than Oase 2. One could also argue that the presence of the Neandertal suprainiac pattern in the middle Middle Pleistocene Eyasi 1 cranium (Trinkaus 2004) or possibly in the Aduma 3 cranium (although there the fossa is associated with a prominent external occipital protuberance) (Haile-Selassie et al. 2004) means that its basis was still in the later MPMH sample; however, the full pattern is nonetheless absent from that terminal Middle/early Late Pleistocene sample and present in 100% of the Neandertals.

One could also argue that the MPMH sample is too small, especially for individual anatomical regions, to fully indicate its range of variation, and therefore its potential ancestry for these unusual traits in the European early modern humans. One can always make such an argument, that the absence of evidence is not the evidence of absence. However, this misses one of the key points – *any* phylogenetic assessment is not an issue of absolute certainty but an issue of *probability*. Given these unusual traits for an MPMH in the European early modern human sample, what is the probability that their ancestry was purely that of the MPMH, or alternatively the probability that they had no Neandertal ancestry?

Probabilistically, one can generously assign each of the ten features as having a 10% probability ($P=0.10$) of appearing in the early modern European sample by chance from a strictly MPMH ancestral population. Assuming that they are independent features, their cumulative probability of being present would be ten to the minus tenth ($P=10^{-10}$), or one in one ten-millionth. Even deleting half of the features as ambiguous would still keep the probability of finding these traits in the earliest European modern human sample vanishingly small, approximately one in one-hundred-thousand ($P=10^{-5}$).

Moreover, for each feature for which multiple specimens preserve the anatomical region, even within sites, there is considerable variability among these early modern Europeans. The sagittal frontal curvature range spans the combined MPMH and Neandertal (plus Gravettian) range of variation. Large occipital buns are present in four of seven crania. The suprainiac fossa pattern is present on one of the eight occipital bones. Molar megadontia is present in Mladeč 8 and Oase 1 and 2, but absent from Mladeč 1 and 2 and Muierii 1 and the Brassempouy, La Quina Aval and Les Rois dental samples. Juxtamastoid eminences vary from large to medium to absent. Mandibular ramus breadth contrasts between Oase 1 (and Oase 2) and Muierii 1 (plus Mladeč 1 and 2), and the two Romanian mandibles contrast in the three other ramal features. Incisor shoveling is variably present, even within the Les Rois sample. Furthermore, there is no consistent pattern from site-specific sample to sample as to which archaic and/or Neandertal features appear. Such variability and its random appearance is exactly what one would expect to be the product of two morphologically distinct populations blending over several millennia.

Given that the basic morphology pattern of these European early modern humans is that of recent humans, only a nontrivial degree of Neandertal admixture can adequately account for the combined presence of these archaic and/or Neandertal traits. Given that the distribution of the unusual characteristics is throughout the earliest modern European sample, this admixture was neither rare, nor unusual, nor geographically restricted. It *must have been* the dominant pattern.

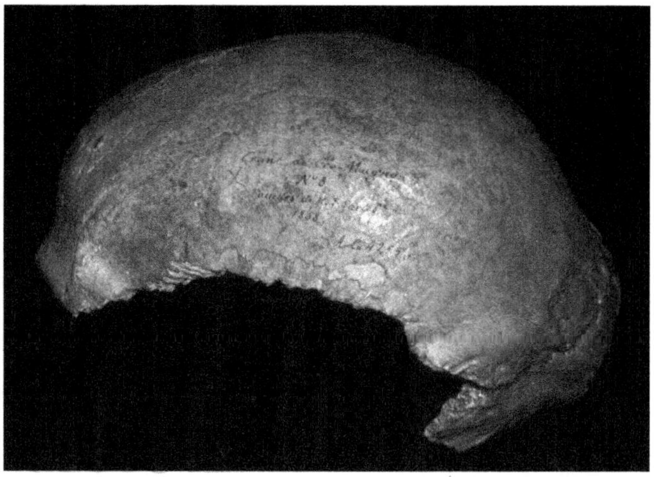

Fig. 24.5 The Cro-Magnon 3 cranium in *norma lateralis* left, illustrating the pronounced occipital bun in an otherwise morphologically modern cranium

The Subsequent European Modern Humans

Given this, one can also ask whether these "unusual" features persisted into the Mid Upper Paleolithic (the Gravettian *sensu lato*), the relatively abundant remains that date after ≈33,000 BP. Despite the greater time lapse since the Neandertals, except in at least Iberia, but in the context of multiple associated skeletons from burials and with better preservation, does the morphological mosaic persist?

Pronounced occipital buns are evident on Brno 2, Cro-Magnon 3, Dolní Věstonice 11, Pavlov 1 and Předmostí 2 and 7 (Fig. 24.5), along with smaller ones [hemi-buns (Smith 1984)] on several other Gravettian specimens. Several of the Gravettian immature mandibles, including Kostenki 3, Lagar Velho 1, Předmostí 2 and Sunghir 2, have retreating anterior symphyseal profiles relative to the alveolar plane, despite the presence of a projecting tuber symphyseos on all of their mandibles. Since the anterior symphyseal angle (infradentale-pogonion) changes little during development, it reflects the basic morphological pattern of these individuals despite their variably young ages.

Shovel-shaped maxillary incisors remain evident, if relatively uncommon. They are present to varying degrees of development on Dolní Věstonice 15, Lagar Velho 1 and Sunghir 2.

The derived Neandertal lineage dorsal sulcus pattern of the scapula axillary border (Trinkaus 2006b) is absent from the MPMH and from Muierii 1. It is present in three-quarters of the Neandertals, and it appears again in Barma Grande 2, Dolní Věstonice 14, Předmostí 14 and Sunghir 1, 2 and 3 (Fig. 24.6).

Most Gravettian clavicles are short similar to those of recent humans, but the Sunghir 1 skeleton has clavicles

Fig. 24.6 The Muierii 1 right scapula in lateral view (*left*) and the Sunghir 1 right scapula in dorsal view (*right*), not to the same scale. The Muierii 1 scapula has a bisulcate axillary border, whereas the Sunghir 1 scapula has a marked dorsal sulcus pattern. The Muierii 1 scapula also exhibits a relatively narrow glenoid fossa

which are as absolutely and relatively as long as those of the Neandertals (Kozlovskaya and Mednikova 2000; Fig. 24.7). This occurs in the context of the otherwise linear body proportions of Sunghir 1. Even though a pronounced muscular crest for the opponens pollicis muscle is occasionally present in early *Homo* and morphologically modern humans, they lack the flange-like morphology present on almost all Neandertals, young and old. The same flange is evident on the Sunghir 1 first metacarpals, especially on the left side.

Finally, all of the European early modern humans show some degree of tropical linear body proportions, including the Mladeč 27 femur as indicated by biomechanical modeling of its diaphyseal robusticity (Trinkaus 1981; Holliday 1997; Trinkaus et al. 2006a). Yet, Lagar Velho 1 has the low crural indices of the Neandertals (Ruff et al. 2002; Fig. 24.8), and Cro-Magnon 1 and 3, depending on how one sorts the mixed femora and tibiae, have crural indices that are either Neandertal-like or between those of the Neandertals and other Gravettian humans (Vallois and Billy 1965). Given the stability of such body proportions over extended periods of time, despite their ecogeographic variation across recent human populations (Trinkaus 1981; Holliday 1997), they can be used for shedding light on what are essentially populational processes. This is reinforced by the preservation of tropical crural indices in high latitude Gravettian skeletons, including Paviland 1 and the very cold climate Sunghir 1 and 2 (Kozlovskaya and Mednikova 2000; Holliday 2000).

Summary

From these considerations, several conclusions can be drawn about human population dynamics when Neandertals and modern humans met in Europe. The earliest modern humans in Europe experienced a gradual dispersal from east to west, across multiple millennia. They had a basically "modern" human morphological gestalt, with a dominance of clearly derived modern human characters. They also had an abundance of archaic characters, generally archaic and/or Neandertal characters. There was a persistence of such characters into the Gravettian, multiple millennia after the transition. One cannot explain these data by a process of total replacement, even if the "archaic" traits are not strictly "Neandertal" traits. They were already lost in the MPMH. Either there were multiple reversals, too many to be reasonable, or (more parsimoniously) admixture.

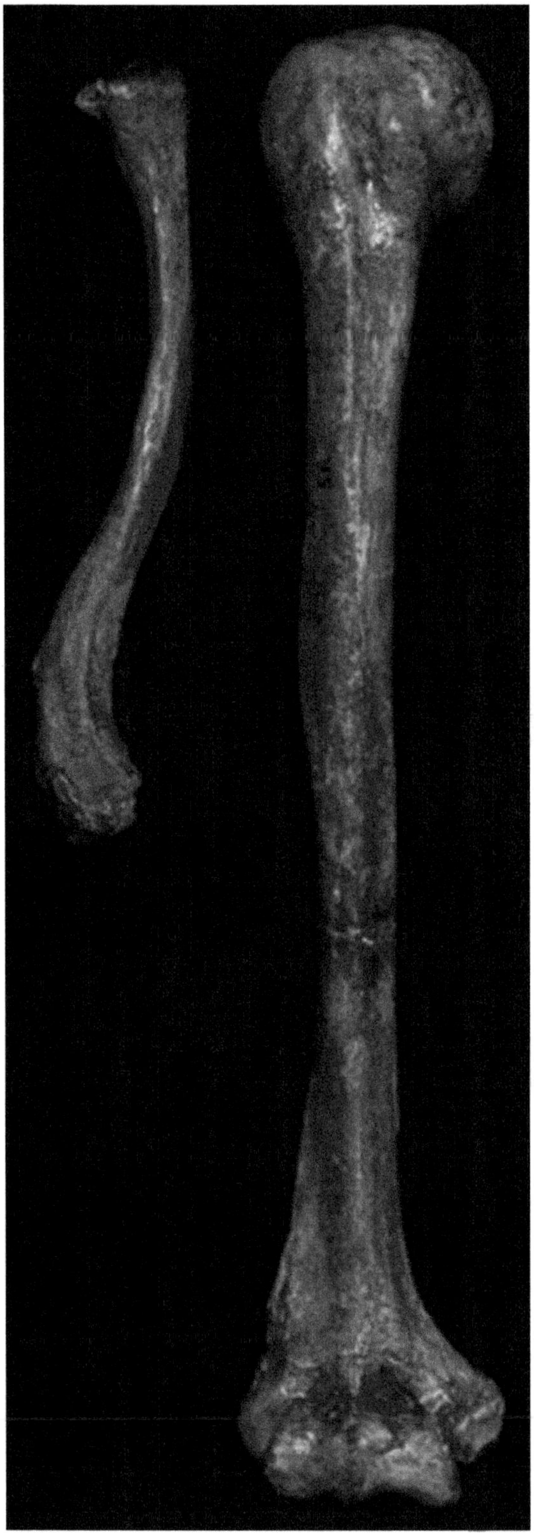

Fig. 24.7 The Sunghir 1 right clavicle and humerus, in dorsal and anterior views respectively, showing the marked elongation of its clavicle

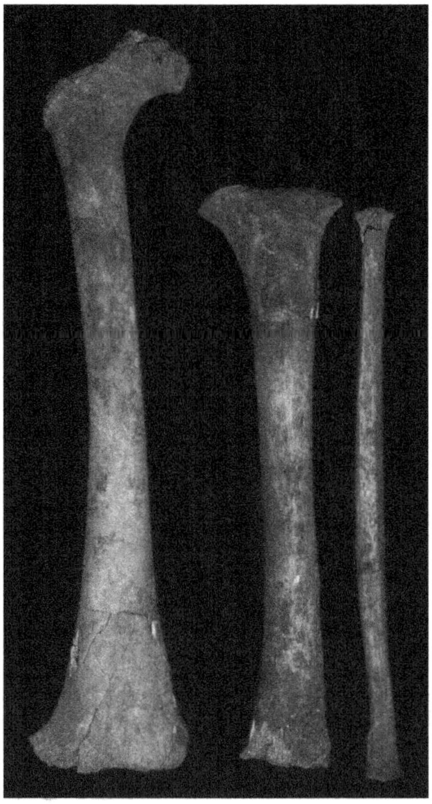

Fig. 24.8 The right femur, tibia and fibula of the juvenile Lagar Velho 1, shown without the epiphyses. Compared to the pattern seen in other Late Pleistocene juveniles, Lagar Velho 1 matches the pattern evident in the La Ferrassie and Dederiyeh Neandertals and contrasts with the relatively longer tibiae of the Gravettian Sunghir 2 skeleton

The behavioral implications of this conclusions are that there was a dynamic mosaic of populations on the landscape, and that the behavioral differences between the groups must have been subtle, certainly little more than the degree of variation known ethnohistorically across recent human populations. In other words, the indicated degree of assimilation implies that they saw each other as "people," as appropriate members of a social system and, by extension, appropriate partners and mates.

Human Paleobiology at the European Transition

Given this situation, can we move beyond sex and ancestry to consider the paleobiological indications of the mosaic? The archaeology suggests major behavioral changes, from the Mousterian to the later Aurignacian, with a mosaic through the Initial Upper Paleolithic and the earlier Aurignacian (Zilhão and d'Errico 2003; Conard 2006; Bar-Yosef and Zilhão 2006). Are these behavioral changes evident in the human remains, through functional anatomical aspects and in plastic or degenerative aspects? How "Neandertal" were the late Neandertals? Were they just like the earlier ones? How "modern" were the early modern humans? Were they just like the later ones? Given limited sample sizes, can we assess whether the changes were uniform, assuming that the few preserved pieces are representative?

The relevant samples consist of the before and after samples, and the two samples during the transition. The "before" sample is the Middle Paleolithic Neandertals, those >44,000 BP. The "after" sample is the Gravettian (or Mid Upper Paleolithic) modern humans, those between 33,000 and 24,000 BP. The "transitional" samples consist of the European Late Neandertals, those < 44,000 BP, and the European early modern humans, those >33,000 BP.

Late Neandertal Paleobiology

The late Neandertal sample consists of a scattering of remains from the sites of Arcy-Renne, Feldhofer, Lakonis, Oliveira, Saint-Césaire, Sima de las Palomas, Vindija and Zafarraya, of which the partial skeleton from Saint-Césaire becomes the most important given its relative completeness. Although clearly "Neandertal" in their specific or overall morphological patterns, some of these remains suggest a biological shift away from the pattern evident in earlier, Middle Paleolithic Neandertals.

Although the facial morphology of late Neandertals is indeed Neandertal, there are indications of overall facial gracilization. This was first proposed for the fragmentary remains from Vindija (Smith and Ranyard 1980; Wolpoff et al. 1981). It is evident in the Palomas 59 mandible. The Saint-Césaire 1 facial skeleton also exhibits this reduction, especially when it is placed in the context of its noticeably large, almost certainly male, postcranial remains (Trinkaus et al. 1999). Previous and persistent incorrect sexing of the skeleton as female, in ignorance of its postcranial dimensions, only emphasizes the facial gracilization evident in the skeleton.

In the upper limb, there is a shift towards the rounder and less anteroposteriorly strengthened humeri seen among Gravettian humans. The midshaft maximum to minimum diameter index of the Saint-Césaire left humerus falls at the bottom of the Neandertal range and well within the Gravettian range, indicating a shift in humeral loading regimes. Furthermore, the Saint-Césaire humerus has a prominent deltoid tuberosity (Fig. 24.9), a feature which contrasts with the modest deltoid tuberosities of Neandertals. Although variable, such deltoid tuberosities are more in evidence among the more recent modern humans.

In the femoral midshaft, it has long been noted that Neandertal femora, like archaic *Homo* femora in general, are subcircular in cross-section, lack a pilaster, and have variably developed medial buttresses (Twiesselmann 1961; Trinkaus 1976). A pilaster is a derived modern human trait which is frequently pronounced in early modern human femora (Trinkaus and Ruff 1999; Trinkaus 2006d). The Saint-Césaire femoral midshaft has the same, non-pilastric cross-sectional morphology as other Neandertals (Fig. 24.9). However, the distribution of bone within the shaft increases its anteroposterior bending strength, placing it biomechanically well within the Gravettian range of variation and separate from the Middle Paleolithic Neandertals. The Feldhofer 1 femora are biomechanically within the range of overlap between the earlier Neandertal and later Gravettian humans. Both of these changes take place within the context of the arctic body proportions characteristic of Neandertals in general (Trinkaus et al. 1999), and therefore they cannot be

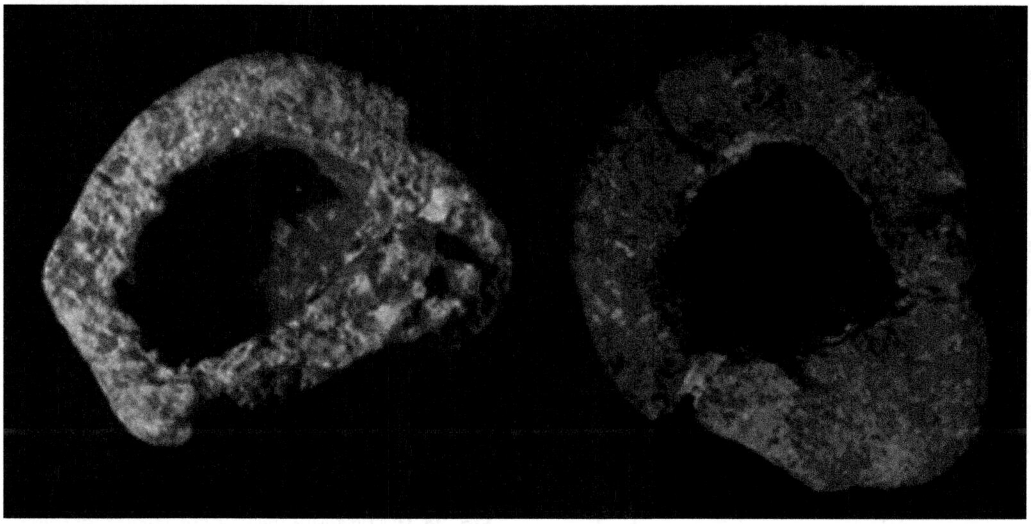

Fig. 24.9 Fossilization cross-sections of the Saint-Césaire 1 mid-proximal (65%) left humerus (*left*) and midshaft (50%) right femur (*right*), not to the same scale. Both views are looking proximally, and anterior is above. Note the pronounced deltoid ridge on the left humerus, and the pronounced posteromedial buttress on the femur which increased its anteroposterior rigidity

ascribed to changing femoral stress trajectories from contrasts in relative pelvic breadth (cf., Weaver 2003).

From these few comparisons, it is apparent that there were shifts in the behavioral loading regimes which these late Neandertals were placing on their limbs. The upper limb pattern had shifted, as is evident in the Saint-Césaire humerus, and the Feldhofer and especially Saint-Césaire femora indicate a shift in locomotor loading, possibly towards more long distance mobility (Ruff 2000).

Early Modern Human Paleobiology

Following on these late Neandertal paleobiological changes, there are aspects of the European early modern humans which are functionally less "modern" than their appellation implies. The sample includes remains from across Europe, but it is principally the Romanian and Czech sites which provide the most informative fossils.

Differential anterior versus posterior dental attrition largely separates Neandertals and Gravettian modern humans, with the Neandertals having more rapid anterior wear and the Gravettians having more even wear across the dental arcade. The Muierii 1 palate exhibits the even wear of Gravettian humans (Doboş et al. 2010). Yet, the Mladeč 8 canine, despite its absolutely and relatively large dimensions, has more advanced wear relative to its associated molars (Frayer et al. 2006).

Most Neandertal scapular glenoid fossae, especially on the right side, are relatively narrow, whereas the Gravettian ones are quite variable in their proportions but generally broader (Churchill and Trinkaus 1990; Churchill 1994; Trinkaus 2006c). This implies a significant increase in the loading of the glenohumeral articulation in the extremes of medial and lateral rotation, used particularly during the cocking phases of throwing (Churchill et al. 1996). The Muierii 1 scapula falls entirely with the Neandertals in this feature, slightly narrower than the Neandertal median value (Doboş et al. 2010; Fig. 24.6).

Despite a couple of low values, Neandertals generally have pronounced tuberosities for their pectoralis major muscles, a plastic feature reflecting thoracohumeral and scapulohumeral muscular hypertrophy. The two Mladeč humeri span most of the range for the Gravettian sample, with Mladeč 23 being close to the Neandertals and Mladeč 24 being relatively gracile even for a Gravettian human (Trinkaus et al. 2006a).

As noted already, Neandertal and Gravettian femoral midshafts contrast in shape and biomechanical properties. The only complete European early modern human femoral midshaft, the Mladeč 27 one, is completely modern and has a degree of anteroposterior reinforcement that is moderately high within the Gravettian distribution (Trinkaus et al. 2006a). However, the Mladeč 28 femur preserves only the proximal and mid-proximal shaft. At this mid-proximal level, most of the Gravettian femora remain well above the Neandertal range of variation in relative anteroposterior bone distribution. Feldhofer 1 is towards the top of the Neandertal distribution. Mladeč 27 is in the middle of the Gravettian distribution. But Mladeč 28, despite the presence of a small pilaster, has a bone distribution pattern that is completely Neandertal-like (Trinkaus et al. 2006a; Fig. 24.10).

And finally, Gravettian tali have a general reduction in the relative lengths of their talar trochleae, reflecting loading levels and patterns at the talocrural joint (Trinkaus 2006d). There is little overlap between them and the Neandertals. The Mladeč 30 talus, despite the inferred linear body proportions of the Mladeč sample based on the Mladeč 23 humerus and especially the Mladeč 27 femur (Trinkaus et al. 2006a), has the expanded talar trochlear length of the

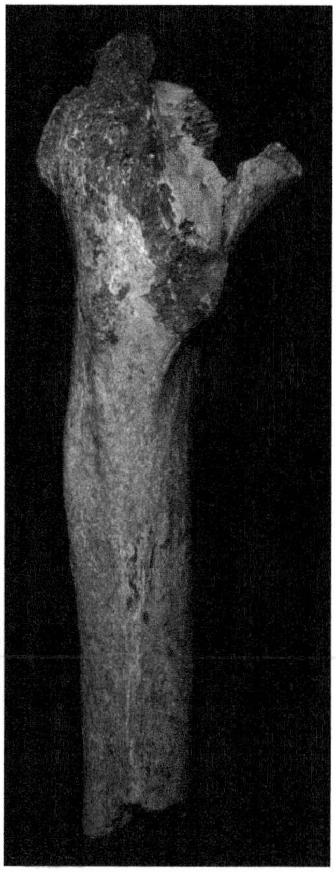

Fig. 24.10 Posterior view of the Mladeč 28 proximal femur, with its pronounced gluteal buttress, but small pilaster and relatively wide mid-proximal diaphysis

Neandertals. It is completely separate from the Mid Upper Paleolithic sample.

Paleobiological Issues

These paleobiological considerations of Late Neandertals and European early modern humans raise a series of issues. The issues concerning the late Neandertals are straight-forward. Their morphological gestalt is basically Neandertal – this has never been questioned. However, there is a reduction in facial robusticity. There is a shift in muscularity and loading patterns of the upper arm, related to changes in manipulation. There is a change in locomotor anatomy, related to anteroposterior loading and hence movement over the landscape. Probably correlated with these human biological changes are shifts in the archeological record, including shifts in lithic technology, expanded use of organic technology, and the use of ornaments reflecting more complicated social networks and socially modified behaviors.

The issues concerning early modern humans in Europe are more complicated. The first involves reproduction. Given the paleontological evidence for admixture between Neandertals and early modern humans, or assimilation of Neandertals into early modern human social groups, three things are evident. They were biologically close, despite the apparent morphological differences, most of which are anatomically deep and hence would have been invisible to them. They were behaviorally more than willing, and probably did not have a lot of mate choice anyway, given the undoubtedly low population densities. And they were socioculturally little different, and simply saw each other as people.

In terms of paleobiology, the early modern European fossils present considerable variability, with some specimens appearing functionally quite "archaic" or Neandertal-like, whereas others are fully "modern" or Gravettian-like. These involve differential canine wear, shoulder loading regimes with little extremes of rotation in at least one individual, and locomotor loading patterns with variable anteroposterior reinforcement.

There are also issues concerning the archaeological correlates of the early modern European human paleontological record. There are a number of major changes in human technology, evident in the lithic technology but especially apparent in the elaboration of organic technologies. These should affect the ways in which humans were manipulating their environment. However, where are the consistent arm changes? The Mladeč and Muierii remains provide a mosaic of shifts in upper limb anatomy. Moreover, Mladeč 8 and Muierii 1 exhibit contrasting patterns in relative anterior dental wear.

There is also growing evidence in the archaeological record for long distance movement of materials, especially marine mollusks hundreds of kilometers from the nearest maritime coastlines. But where are the consistent leg changes? They do not appear until the Mid Upper Paleolithic.

Summary

So where does this leave us? There was a gradual spread of modern humans, westward across Europe, after ≈41,000 BP, with variable admixture with late Neandertals. Late Neandertal biology and the associated archaeology indicate a clear behavioral shift to more "modern" patterns. This is evident in the Initial Upper Paleolithic and apparent in details of their biology. European early modern humans are associated with marked shifts in their cultural behavior, both technological and symbolic. However, there are variable shifts in biological reflections of that behavioral shift. It is apparent that simple models of an abrupt behavioral and phylogenetic transition for this period in Europe should be abandoned.

In sum, therefore, if we only contrast Middle Paleolithic Neandertals with Mid Upper Paleolithic early modern humans, the pattern is obvious. The contrasts are real. But what really happened in between?

We are not going to know the answers by continuing to contrast Middle Paleolithic Neandertals with the Mid Upper Paleolithic, or Gravettian, modern humans. We have to look at what derives from the period in between. The archaeological record is rich for this period, but its implications are too often buried in arguments regarding who learned what from whom. The human paleontological record is scanty, fragmentary, and widely dispersed. But ongoing work on directly dating specimens and looking at their biology from multiple biological perspectives is providing insights. The inferences provided here will be modified as work continues and the samples increase. But the complex mosaic that represents the fate of the Neandertals and the spread of modern humans in the northwestern Old World is only likely to become more interesting.

References

Arambourg, C. (1955). Sur l'attitude, en station verticale, des Néanderthaliens. *Comptes rendus de l'Académie des Sciences Paris, 240D*, 804–806.

Barker, G., Barton, H., Bird, M., Daly, P., Datan, I., Dykes, A., Farr, L., Gilbertson, D., Harrisson, B., Hunt, C., Higham, T., Kealhofer, L., Krigbaum, J., Lewis, H., McLaren, S., Paz, V., Pike, A., Piper, P., Pyatt, B., Rabett, R., Reynolds, T., Rose, J., Rushworth, G.,

Stephens, M., Stringer, C., Thompson, J., & Turney, C. (2007). The 'human revolution' in lowland tropical Southeast Asia: The antiquity and behavior of anatomically modern humans at Niah Cave (Sarawak, Borneo). *Journal of Human Evolution, 52*, 243–261.

Bar-Yosef, O., & Zilhão, J. (Eds.). (2006). *Towards a definition of the Aurignacian* (Trabalhos de Arqueologia, Vol. 45).

Berry, A. C., & Berry, R. J. (1967). Epigenetic variation in the human cranium. *Journal of Anatomy, 101*, 361–379.

Boule, M. (1908). L'homme fossile de La Chapelle-aux-Saints (Corrèze). *L'Anthropologie, 19*, 519–525.

Boule, M. (1911–1913). L'homme fossile de La Chapelle-aux-Saints. *Annales de Paléontologie, 6*, 111–172. 7, 21-56, 85-192; 8, 1–70.

Bräuer, G. (1982). Current problems and research on the origin of *Homo sapiens* in Africa. *Humanbiologia Budapestinensis, 9*, 69–78.

Brothwell, D. R. (1960). Upper Pleistocene human skull from Niah Cave, Sarawak. *Sarawak Museum Journal, 4*, 323–349.

Chilardi, S., Frayer, D. W., Gioia, P., Macchiarelli, R., & Mussi, M. (1996). Fontana Nuova di Ragusa (Sicily, Italy): Southernmost Aurignacian site in Europe. *Antiquity, 70*, 553–563.

Churchill, S. E. (1994). *Human upper body evolution in the Eurasian Later Pleistocene*. Ph.D.dissertation, University of New Mexico.

Churchill, S. E., & Trinkaus, E. (1990). Neandertal scapular glenoid morphology. *American Journal of Physical Anthropology, 83*, 147–160.

Churchill, S. E., Weaver, A. H., & Niewoehner, W. A. (1996). Late Pleistocene human technological and subsistence behavior: Functional interpretations of upper limb morphology. *Quaternaria Nova, 6*, 413–447.

Conard, N. J. (Ed.). (2006). *When Neanderthals and modern humans met*. Tübingen: Kerns Verlag.

Crevecoeur, I. (2008). *Étude Anthropologique du Squelette du Paléolithique Supérieur de Nazlet Khater 2 (Égypte)*. Leuven: Leuven University Press.

Crevecoeur, I., & Trinkaus, E. (2004). From the Nile to the Danube: A comparison of the Nazlet Khater 2 and Oase 1 early modern human mandibles. *Anthropologie (Brno), 42*, 229–239.

Cunningham, D. J. (1908). The evolution of the eyebrow region of the forehead, with special reference to the excessive supraorbital development in the Neanderthal race. *Transactions of the Royal Society of Edinburgh, 46*, 243–310.

Day, M., & Stringer, C. B. (1982). A reconsideration of the Omo Kibish remains and the *erectus-sapiens* transition. In H. de Lumley (Ed.), *L'Homo erectus et la Place de l'Homme de Tautavel parmi les Hominidés Fossiles* (pp. 814–846). Paris: CNRS.

Doboş, A., Soficaru, A., & Trinkaus, E. (2010). The Prehistory and Paleontology of the Peştera Muierii, Romania. *Études et Recherches Archéologiques de l'Université de Liège, 124*, 1–122.

Duarte, C., Maurício, J., Pettitt, P. B., Souto, P., Trinkaus, E., van der Plicht, H., & Zilhão, J. (1999). The early upper Paleolithic human skeleton from the Abrigo do Lagar Velho (Portugal) and modern human emergence in Iberia. *Proceedings of the National Academy of Sciences of the United States of America, 96*, 7604–7609.

Dujardin, V. (2003). Sondages à La Quina aval (Gardes-le-Pontaroux, Charente). *Antiquités Nationales, 33*, 21–26.

Ferembach, D. (1976). Les restes humains de la grotte de Dar-es-Soltane 2 (Maroc) Campagne 1975. *Bulletins et Mémoires de la Société d'Anthropologie de Paris Série XIII, 3*, 183–193.

Frayer, D. W., Jelínek, J., Oliva, M., & Wolpoff, M. H. (2006). Aurignacian male crania, jaws and teeth from the Mladeč Caves, Moravia, Czech Republic. In M. Teschler-Nicola (Ed.), *Early modern humans at the Moravian gate: The Mladeč Caves and their remains* (pp. 185–272). Vienna: Springer.

Gorjanović-Kramberger, D. (1906). *Der Diluviale Mensch von Krapina in Kroatien*. Wiesbaden: Kreidel's Verlag.

Grine, F. E., Bailey, R. M., Harvati, K., Nathan, R. P., Morris, A. G., Henderson, G. M., Ribot, I., & Pike, A. W. G. (2007). Late Pleistocene human skull from Hofmeyr, South Africa, and modern human origins. *Science, 315*, 226–229.

Gunz, P., & Harvati, K. (2007). The Neanderthal "chignon": Variation, integration, and homology. *Journal of Human Evolution, 52*, 262–274.

Haile-Selassie, Y., Asfaw, B., & White, T. D. (2004). Hominid cranial remains from Upper Pleistocene deposits at Aduma, Middle Awash, Ethiopia. *American Journal of Physical Anthropology, 123*, 1–10.

Henry-Gambier, D., Maureille, B., & White, R. (2004). Vestiges humains des niveaux de l'Aurignacien ancien du site de Brassempouy (Landes). *Bulletins et Mémoires de la Société d'Anthropologie de Paris, 16*, 49–87.

Holliday, T. W. (1997). Body proportions in late Pleistocene Europe and modern human origins. *Journal of Human Evolution, 32*, 423–447.

Holliday, T. W. (2000). The human remains from Paviland Cave: Body proportions of the Paviland 1 skeleton. In S. H. R. Aldhouse-Green (Ed.), *Paviland Cave and the 'Red Lady': A definitive report* (pp. 199–204). Bristol: Western Academic and Specialist Press Ltd.

Hublin, J. J. (1978). Quelques charactères apomorphes du crâne néandertalien et leur interprétation phylogénétique. *Comptes rendus de l'Académie des Sciences de Paris, 287D*, 923–926.

Hublin, J. J. (1983). Les origines de l'homme de type moderns en Europe. *Pour la Science, 64*, 62–71.

Hublin, J. J. (1998). Climat de l'Europe et origines des Néandertaliens. *Pour la Science, 245*, 52–59.

Hublin, J. J. (2000). Modern-nonmodern hominid interactions: A Mediterranean perspective. *Peabody Museum Papers, 8*, 157–182.

Klaatsch, H. (1901). Das Gliedmassenskelet des Neanderthalmenschen. *Anatomischer Anzeiger, 19*, 121–154.

Klaatsch, H., & Hauser, O. (1909). Homo mousteriensis Hauseri. Ein altdiluvialer Skelettfund in Departement Dordogne und seine Zugehörigkeit zum Neandertaltypus. *Archiv für Anthropologie, 7*, 287–297.

Kozlovskaya, M. V., & Mednikova, M. B. (2000). Catalogue of photos and tables on materials from Sunghirian graves 1 and 2. In T. I. Alexeev & N. O. Bader (Eds.), *Homo sungirensis. Upper Palaeolithic man: Ecological and evolutionary aspects of the investigation* (pp. 85–144). Moscow: Scientific World.

Lam, Y. M., Pearson, O. M., & Smith, C. M. (1996). Chin morphology and sexual dimorphism in the fossil hominid mandible sample from Klasies river mouth. *American Journal of Physical Anthropology, 100*, 545–557.

Lebel, S., & Trinkaus, E. (2002). Middle Pleistocene human remains from the Bau de l'Aubesier. *Journal of Human Evolution, 43*, 659–685.

Liu, W., Jin, C. Z., Zhang, Y. Q., Cai, Y. J., Xing, S., Wu, X. J., Cheng, H., Edwards, R. L., Pan, W. S., Qin, D. G., An, Z, S., Trinkaus, E., Wu, X. Z. (2010). Human remains from Zhirendong, south China, and modern human emergence in east Asia. *Proceedings of the National Academy of Sciences of the United States of America, 107*, 19201–19206.

Mallegni, F., & Segre-Naldini, E. (1992). A human maxilla (Fossellone 1) and scapula (Fossellone 2) recovered in the Pleistocene layers of the Fossellone Cave, Mt. Circeo, Italy. *Quaternaria Nova, 2*, 211–225.

McDougall, I., Brown, F. H., & Fleagle, J. G. (2005). Stratigraphic placement and age of modern humans from Kibish, Ethiopia. *Nature, 433*, 733–736.

Rougier, H., Milota, Ş., Rodrigo, R., Gherase, M., Sarcină, L., Moldovan, O., Zilhão, J. Constantin, S., Franciscus, R. G., Zollikofer, C. P. E., Ponce-du León, & M., Trinkaus, E. (2007). Peştera cu Oase 2 and the cranial morphology of early modern Europeans. *Proceedings of the National Academy of Sciences of the United States of America, 104*, 1164–1170.

Ruff, C. B. (2000). Biomechanical analysis of archaeological human skeletons. In M. A. Katzenberg & S. R. Saunders (Eds.), *Biological anthropology of the human skeleton* (pp. 71–102). New York: Wiley-Liss.

Ruff, C. B., Trinkaus, E., & Holliday, T. W. (2002). Body proportions and size. In J. Zilhão & E. Trinkaus (Eds.), *Portrait of the artist as a child. The Gravettian human skeleton from the Abrigo do Lagar*

Velho and its archeological context (Trabalhos de Arqueologia, Vol. 22, pp. 365–391).

Santa Luca, A. P. (1978). A re-examination of presumed Neandertal-like fossils. *Journal of Human Evolution, 7*, 619–636.

Schmitz, R. W. (Ed.). (2006). *Neanderthal 1856–2006*. Mainz: Verlag Philipp von Zabern.

Schwalbe, G. (1901). Der Neanderthalschädel. *Bonner Jahrbücher, 106*, 1–72.

Shang, H., & Trinkaus, E. (2010). *The Early Modern Human from Tianyuan Cave, China*. College Station: Texas A&M University Press.

Smith, F. H. (1984). Fossil hominids from the Upper Pleistocene of central Europe and the origin of modern Europeans. In F. H. Smith & F. Spencer (Eds.), *The origins of modern humans* (pp. 137–209). New York: Alan R. Liss.

Smith, F. H. (1993). Models and realities in modern human origins: The African fossil evidence. In M. J. Aitken, C. B. Stringer, & P. A. Mellars (Eds.), *The origin of modern humans and the impact of chronometric dating* (pp. 234–248). Princeton: Princeton University Press.

Smith, F. H., & Ranyard, G. C. (1980). Evolution of the supraorbital region in Upper Pleistocene fossil hominids from south-central Europe. *American Journal of Physical Anthropology, 53*, 589–610.

Soficaru, A., Petrea, C., Doboş, A., & Trinkaus, E. (2007). The human cranium from the Peştera Cioclovina Uscată, Romania: Context, age, taphonomy, morphology and paleopathology. *Current Anthropology, 48*, 611–619.

Stefan, V. H., & Trinkaus, E. (1998a). Discrete trait and dental morphometric affinities of the Tabun 2 mandible. *Journal of Human Evolution, 34*, 443–468.

Stefan, V. H., & Trinkaus, E. (1998b). La Quina 9 and Neandertal mandibular variability. *Bulletins et Mémoires de la Société d'Anthropologie de Paris ns, 10*, 293–324.

Straus, W. L., Jr., & Cave, A. J. E. (1957). Pathology and the posture of Neanderthal man. *The Quarterly Review of Biology, 32*, 348–363.

Thoma, A. (1984). Morphology and affinities of the Nazlet Khater man. *Journal of Human Evolution, 13*, 287–296.

Tillier, A. M. (1999). *Les Enfants Moustériens de Qafzeh. Interprétation Phylogénétique et Paléoauxologique*. Paris: CNRS Éditions.

Tobias, P. V. (1967). The hominid skeletal remains of Haua Fteah. In C. B. M. McBurney (Ed.), *The Haua Fteah (Cyrenaica) and the stone age of the South-East Mediterranean* (pp. 338–352). Cambridge: Cambridge University Press.

Toerien, M. J. (1957). Note on the cervical vertebrae of the La Chapelle man. *South African Journal of Science, 53*, 447–449.

Trinkaus, E. (1976). The evolution of the hominid femoral diaphysis during the Upper Pleistocene in Europe and the near East. *Zeitschrift für Morphologie und Anthropologie, 67*, 291–319.

Trinkaus, E. (1981). Neanderthal limb proportions and cold adaptation. In C. B. Stringer (Ed.), *Aspects of human evolution* (pp. 187–224). London: Taylor & Francis.

Trinkaus, E. (2004). Eyasi 1 and the suprainiac fossa. *American Journal of Physical Anthropology, 124*, 28–32.

Trinkaus, E. (2005). Early modern humans. *Annual Review of Anthropology, 34*, 207–230.

Trinkaus, E. (2006a). Modern human versus Neandertal evolutionary distinctiveness. *Current Anthropology, 47*, 597–620.

Trinkaus, E. (2006b). The Krapina scapulae. *Periodicum Biologorum, 108*, 341–351.

Trinkaus, E. (2006c). The upper limb remains. In E. Trinkaus & J. A. Svoboda (Eds.), *Early modern human evolution in Central Europe: The people of Dolní Věstonice and Pavlov* (pp. 327–372). New York: Oxford University Press.

Trinkaus, E. (2006d). The lower limb remains. In E. Trinkaus & J. A. Svoboda (Eds.), *Early modern human evolution in Central Europe: The people of Dolní Věstonice and Pavlov* (pp. 380–418). New York: Oxford University Press.

Trinkaus, E., & LeMay, M. (1982). Occipital bunning among later Pleistocene hominids. *American Journal of Physical Anthropology, 57*, 27–35.

Trinkaus, E., & Ruff, C. B. (1999). Diaphyseal cross-sectional geometry of near Eastern Middle Paleolithic humans: The femur. *Journal of Archaeological Science, 26*, 409–424.

Trinkaus, E., Churchill, S. E., Ruff, C. B., & Vandermeersch, B. (1999). Long bone shaft robusticity and body proportions of the Saint-Césaire 1 Châtelperronian Neandertal. *Journal of Archaeological Science, 26*, 753–773.

Trinkaus, E., Moldovan, O., Milota, Ş., Bîlgăr, A., Sarcina, L., Athreya, S., Bailey, S. E., Rodrigo, R., Gherase, M., Higham, T., Bronk Ramsey, C., & van der Plicht, J. (2003). An early modern human from the Peştera cu Oase, Romania. *Proceedings of the National Academy of Sciences of the United States of America, 100*, 11231–11236.

Trinkaus, E., Smith, F. H., Stockton, T. C., & Shackelford, L. L. (2006a). The human postcranial remains from Mladeč. In M. Teschler-Nicola (Ed.), *Early modern humans at the Moravian gate: The Mladeč Caves and their remains* (pp. 385–445). Vienna: Springer.

Trinkaus, E., Zilhão, J., Rougier, H., Rodrigo, R., Milota, S., Gherase, M., Sarcina, L., Moldovan, O., Băltean, I. C., Codrea, V., Bailey, S. E., Franciscus, R. G., de Ponce León, M., & Zollikofer, C. P. E. (2006b). The Peştera cu Oase and early modern humans in southeastern Europe. In N. J. Conard (Ed.), *When Neanderthals and modern humans met* (pp. 145–164). Tübingen: Kerns Verlag.

Twiesselmann, F. (1961). Le fémur néanderthalien de Fond-de-Forêt (Province de Liège). *Mémoires de l'Institut royal des Sciences naturelles de Belgique, 148*, 1–164.

Vallois, H. V., & Billy, G. (1965). Nouvelles recherches sur les hommes fossiles de l'Abri de Cro-Magnon. *L'Anthropologie, 69*, 47–74. 249–272.

Vallois, H. V., & Roche, J. (1958). La mandibule acheuléenne de Témara, Maroc. *Comptes rendus de l'Académie des Sciences Paris, 246*, 3113–3116.

Vandermeersch, B. (1981). *Les Hommes Fossiles de Qafzeh (Israel)*. Paris: CNRS.

Walker, M.J., Gibert, J., López, M.V., Lombardi, A.V., Pérez-Pérez, A., Zapata, J., Ortega, J., Higham, T., Pike, A., Schwenninger, J.L., Zilhão, J., & Trinkaus, E. (2008). Late Neandertals in southeastern Iberia: Sima de las Palomas del Cabezo Gordo, Murcia, Spain. *Proceedings of the National Academy of Sciences of the United States of America, 105*, 20631–20636.

Weaver, T. D. (2003). The shape of the Neandertal femur is primarily the consequence of a hyperpolar body form. *Proceedings of the National Academy of Sciences of the United States of America, 100*, 6926–6929.

White, T. D., Asfaw, B., DeGusta, D., Gilbert, H., Richards, G. D., Suwa, G., & Howell, F. C. (2003). Pleistocene *Homo sapiens* from Middle Awash, Ethiopia. *Nature, 423*, 742–747.

Wild, E. M., Teschler-Nicola, M., Kutschera, W., Steier, P., Trinkaus, E., & Wanek, W. (2005). First direct dating of early Upper Paleolithic human remains from Mladeč. *Nature, 435*, 332–335.

Wolpoff, M. H., Smith, F. H., Malez, M., Radovčić, J., & Rukavina, D. (1981). Upper Pleistocene human remains from Vindija Cave, Croatia, Yugoslavia. *American Journal of Physical Anthropology, 54*, 499–545.

Zilhão, J., & d'Errico, F. (Eds.). (2003). The chronology of the Aurignacian and of the Transitional Technocomplexes. Trabalhos de Arqueologia 33.

Chapter 25
Aliens from Outer Time? Why the "Human Revolution" Is Wrong, and Where Do We Go from Here?

João Zilhão

Abstract For the better part of the last quarter of a century, the "Human Revolution" paradigm both framed and inspired most research on modern human origins. It brought together genetic, archaeological and paleontological data to form a coherent narrative of recent human evolution positing that all present-day populations derived from a speciation event in East Africa that, some 150,000 years ago, generated a small founder group of anatomically, cognitively and behaviorally fully modern people. The rest would have been history: subsequent Out-of-Africa dispersal of these early African moderns, entailing the inevitable replacement, without admixture, of the less advanced, outcompeted species of Eurasian archaics, namely the Neandertals. Recent empirical developments have falsified the basic tenets of these views. The archaeology and paleontology of the time of contact now show that Neandertals and moderns featured similar levels of cultural achievement, that symbolic artifacts and personal ornaments had emerged in Neandertal Europe many millennia before the first in-dispersals of modern humans, and that significant admixture occurred as a result of such dispersals, as evidenced by the presence in post-contact populations of diagnostically Neandertal anatomical and cultural traits. The fossil DNA evidence is consistent with these results. Neandertals, therefore, can no longer be considered an evolutionary dead-end and productive explanations for their differentiation and eventual demise now must be sought in the realms of biogeography, demography and paleoethnography.

Keywords Neandertals • Modern Humans • Middle Paleolithic • Upper Paleolithic • Transitional Industries • Châtelperronian • Aurignacian • Radiocarbon dating

J. Zilhão (✉)
ICREA Research Professor, University of Barcelona, Departament de Prehistòria, Història Antiga i Arqueologia – Seminari d'Estudis i Recerques Prehistòriques, C/ Montalegre 6, 08001 Barcelona, Spain
e-mail: joao.zilhao@icrea.cat

Introduction

In a long-term perspective, the Middle-to-Upper Paleolithic transition (henceforth, simply the Transition) represents a watershed in human history. By the later Upper Paleolithic, around 25,000 years ago, all continents were occupied, all types of world ecosystems were exploited, and all aspects of ethnographically observed hunter-gatherer culture that can be documented archaeologically are indeed represented in one way or the other in the archaeological record. Such is unquestionably not the case in the Middle Paleolithic, at least not prior to 100,000 years ago. In this sense, there seems to be little reason to question the concept of an "Upper Paleolithic Revolution" (Gilman 1984), i.e., a "revolution" that, in fact, is a protracted process of technological improvement and demographic growth, combined in a feedback loop with developments towards more sophisticated modes of communication and social organization; put another way, a "revolution" that is a punctuated event only in a geological or evolutionary time scale.

In other formulations, however, this process is construed as a true punctuation in a historical time scale, one whose ultimate cause would lie in the emergence, late in human evolution, and in the framework of the speciation event that generated *Homo sapiens* (or "modern humans"), of the cognitive capabilities for ethnographically documented culture (or "modern behavior"): the "Human Revolution" (Mellars and Stringer 1989). Thirty years ago, there was valid and sufficient reason to see things that way, and the hypothesis has since dominated the field. Its attractiveness is easy to understand, especially in Europe, where the Transition broadly coincides with a major biological event (the disappearance of Neandertals), and where explaining this process as a simple byproduct of population replacement has the obvious advantage of simplicity. Simplicity, however, does not automatically equate with parsimony, much less with goodness-of-fit to the evidence.

It is my contention here that the Human Revolution paradigm, under which much interesting research was promoted, resulting in major advances in different fields (such as the genetics of fossils, or the archaeology of the late Middle

S. Condemi and G.-C. Weniger (eds.), *Continuity and Discontinuity in the Peopling of Europe: One Hundred Fifty Years of Neanderthal Study*,
Vertebrate Paleobiology and Paleoanthropology, DOI 10.1007/978-94-007-0492-3_25,

and early Late Pleistocene of Africa), has now become completely unproductive. Its effects on scientific practice are ever more deleterious, and have led to a trend where the explanation of modern behavior is reduced to the mechanical (if not simply tautological) reiteration that observing features of that behavior in the archaeological record proves that (a) modern humans do behave as modern humans, and (b) modern humans behave as modern humans because they have the capability to behave as modern humans…

This approach wastes the extraordinary potential of the Transition to explore issues of variation and change in a geographical, social and historical perspective, as illustrated by many recent regional studies (e.g., Richter 2000; Hopkinson 2004; van Peer 2004; Adler et al. 2006; Conard et al. 2006; Weniger 2011), as well as by comparative analyses of material culture that focus on understanding societies *per se*, not against an abstract standard of "modernity" (e.g., Vanhaeren and d'Errico 2006). That potential is also readily apparent in the realm of physical anthropology, where many examples show how functional approaches can productively elucidate issues of adaptation, life ways and culture at the time of the Transition (e.g., Aiello and Wheeler 2003; Trinkaus 2005a, b).

In order for more substantial progress along these lines to be possible, however, Paleoanthropology needs to shed the Human Revolution straightjacket. In the following, I will review the arguments that show the paradigm to be not only simple but also simply irreconcilable with current evidence, and to explore alternative ways of understanding the Upper Paleolithic, Neandertals and moderns that are consistent with the facts of the empirical record.

In the following, calendar dates derived from the oceanic or ice cap records, or obtained by Thermoluminescence (TL), Optically Stimulated Luminescence (OSL), Electromagnetic Spin Resonance (ESR) and Uranium-Thorium (U-Th) methods are given in years or thousands of years (ka) BP, and radiocarbon dates are expressed in years "^{14}C BP". For proper comparison, and since there is now a broad consensus between different calibration curves (Hughen et al. 2004; Shackleton et al. 2004; Fairbanks et al. 2005; Weninger et al. 2005), radiocarbon dates are associated with the corresponding calendar ages, expressed as "cal BP," and calculated with the CalPal software (Weninger and Jöris 2005).

The Human Revolution Paradigm

Until the 1980s, the consensus view of the Transition was that of an integrated process whereby, more or less simultaneously across the entire Old World, culturally Middle Paleolithic "paleoanthropian" populations evolved the biological features universally found among "neoanthropian" humans in tandem with the production of a significant number of behavioral innovations. The latter were seen as forming an integrated "Upper Paleolithic package", with blade-based technologies standing as a proxy for (a) the long-distance procurement of raw materials (b) the emergence of regional traditions of stone-tool making (c) the specialized hunting of a selective range of prey coupled with a broadening of the subsistence base to include birds and fish (d) an increase in the number of sites and in the density of archaeological levels, suggesting higher population numbers and larger co-resident groups, also manifested in the greater complexity apparent in the layout of residential sites (e) the manufacture of bone tools (f) the use of personal ornamentation, and (g) the production of sophisticated figurative art (Brézillon 1969; Mellars 1973; White 1982).

This view was to be challenged in the last quarter of the twentieth century, largely as a result of developments in human genetics, specifically, the phylogenetic implications derived from mtDNA patterns suggesting a single, recent African origin for all people alive today (Cann et al. 1987). This evidence eventually coalesced with three different lines of paleontological arguments pointing to a recent African ancestry for Europe's "neoanthropians": Firstly, the recognition that they had body proportions typical of tropical populations, in marked contrast with the continent's (Neandertal) paleoanthropians (Trinkaus 1981); secondly, the recognition that the "proto-Cro-Magnon" Skhul/Qafzeh people dated to the last interglacial, and were of broadly the same geological age suggested for the fully "neoanthropian" Omo-Kibish skulls (Vandermeersch 1981; Valladas et al. 1988); finally, the fact that a pattern of gradual emergence of the neoanthropian condition could be observed in the later Middle Pleistocene fossil record of Africa, whereas, in Europe, the emergence of neoanthropians was rather abrupt, suggesting they corresponded to an intrusive population (Bräuer 1984; Stringer et al. 1984).

Together, these developments carried the implication that the biological transition had occurred much earlier in Africa than in Europe, and that Europe's paleoanthropians represented a side-branch in the human evolutionary tree, a view that would be boosted by interpretations of the ancient mtDNA extracted from Neandertal fossils (Krings et al. 1997). Over the ten years separating the latter's work from that of Cann et al. (1987) no comparable change occurred, however, in the paradigmatic view that the cultural developments subsumed under the expression "Upper Paleolithic" were inextricably inter-woven with changes in brain structure and cognitive capabilities that were part and parcel of the overall process of skeletal neoanthropization. Thus, it was only logical to expect that, in the 1990s, the consensus view of the Transition would have evolved to one that attempted to reconcile the new genetic and paleontological evidence with traditional archaeological perceptions of the Middle and the Upper Paleolithic. But, in order to retain the logical consistency of notions of the Transition as a fully integrated biocultural phenomenon, the geographical scope of the process had to be restricted to Africa. Consequently, it was now proposed that the emergence of overall modernity – the Human Revolution – resulted from a speciation event among eastern African populations, the anatomically and culturally "modern" people generated by that event then spreading from there to the entire Old World, replacing along the way the autochthonous anatomically and culturally "archaic" coeval populations living in the rest of Africa and in Eurasia.

Where Europe was concerned, such a revision of the post-war view of the Transition carried the implication that, by definition, the continent's paleoanthropians could not have been involved in the process: Since they were not biologically modern, they couldn't possibly have been behaviorally modern either. Moreover, the first uncontroversial manifestations of figurative art in the Upper Paleolithic of the continent were found in the Aurignacian, seemingly in exclusive association with the skeletal remains of neoanthropians, leading to speculations that paleoanthropians lacked the cognitive requirements for symbolic thinking and might even have been deprived of fully human forms of language (Lieberman and Crelin 1971; Noble and Davidson 1996; Lieberman 2007). Thus, it was only natural to conclude that, in Europe, the story of the Middle to Upper Paleolithic transition was simply that of the extinction without descent of an "archaic" species (Neandertals), outcompeted by an expanding "modern" species ("humans") whose superiority ultimately derived from the enhanced cognitive capabilities, generated in the framework of the speciation event, of their "modern" brains (Stringer and Gamble 1993).

From the beginning, this solution to the reconciliation of the archaeological evidence with the largely genetically based "Out-of-Africa-with-complete-replacement" view of modern human origins faced two major problems. First, the fact that, for about 100,000 years, the archaeological record of African early modern humans showed little or no evidence of "Upper Paleolithic-ness". Second, the fact that, as suggested by the Saint-Césaire burial (Lévêque and Vandermeersch 1980) and the Grotte du Renne's human remains (Leroi-Gourhan 1958; Hublin et al. 1996), the Châtelperronian, widely recognized as an Upper Paleolithic entity representing the dawn of art (Leroi-Gourhan 1964; Bordes 1968), had been made by Neandertals. These two facts represented a direct challenge to the paradigmatic view of the Transition as a tightly integrated biocultural process. If that view was to be retained, satisfactory explanations had to be found for such major anomalies, and these two issues – the origins of "behavioral modernity" in Africa, and the significance of the Châtelperronian – have indeed been at the center of modern human origins research.

Where Africa is concerned, a first attempt at solving the problem was the suggestion that only after a mutation occurring some time around 50,000 years ago among later African moderns were language and symbolic thinking possible, and that such a mutation would have been the ultimate explanation for the quantum leap in culture and demography triggering the extraordinarily rapid expansion across Africa first, and Eurasia after, of the population that carried it (for a recent formulation, cf. Klein 2003). As pointed out by several authors (e.g., Lahr and Foley 1998; Shennan 2001), this solution is inconsistent with the genetic and archaeological data indicating that the Out-of-Africa spread of anatomically modern humans begun well before the temporal horizon postulated for the occurrence of the putative cognition-related mutation. As a result, an alternative view developed and eventually gained wide acceptance: that behavioral modernity was gradually acquired in the African lineage leading from the Kabwe-type populations to the later Middle Pleistocene anatomically moderns (McBrearty and Brooks 2000).

This view, however, implied that behavioral modernity had to be redefined in ways that made the notion compatible with the nature of the archaeological record of the African MSA (Middle Stone Age), in particular with the lack of figurative art until quite late in the sequence (despite McBrearty and Brooks' claims of an age in excess of 50,000 years ago for the painted slabs from Apollo 11 cave, the stratigraphic and dating context leaves no doubt that these slabs date to no more than ca. 28 ka ^{14}C BP, i.e., ca. 32.4 ka cal BP – Wendt 1974; Vogelsang 1998). Thus, emphasis was put on the fact that other realms of the archaeological record provided evidence for symbolism that, although indirect, was unequivocal, namely: (a) the geographical patterning observed in stylistical attributes of the lithic points of the MSA which, it was suggested, could be taken as a proxy for ethnicity (b) the use of pigments and (c) the occurrence of personal ornaments and abstract engravings in the Still Bay levels of Blombos cave, dated to ca. 75 ka BP. Modern behavior thus became equated with "fully symbolic *sapiens* behavior", recognizable archaeologically "when artifacts or features carry a clear symbolic message that is exosomatic" (Henshilwood and Marean 2003: 643–644).

Where Europe is concerned, maintaining the logical integrity of the model was achieved by suggesting that the association with the Châtelperronian of artifacts carrying such a "clear exosomatic symbolic message" – personal ornaments and decorated bone tools – was spurious. These arguments appeared under different formats, which can be grouped into two major families, one invoking natural, and the other cultural processes. It was suggested, for instance, that, at the Grotte du Renne, the key site documenting the use of personal ornaments by Châtelperronian Neandertals (Fig. 25.1), such ornaments were in fact intrusive from the overlying Aurignacian (e.g. Taborin 2002; White 2002). The alternative proposition was that those ornaments could represent (a) items scavenged by Neandertals from contemporary, abandoned Aurignacian modern human sites (b) evidence for

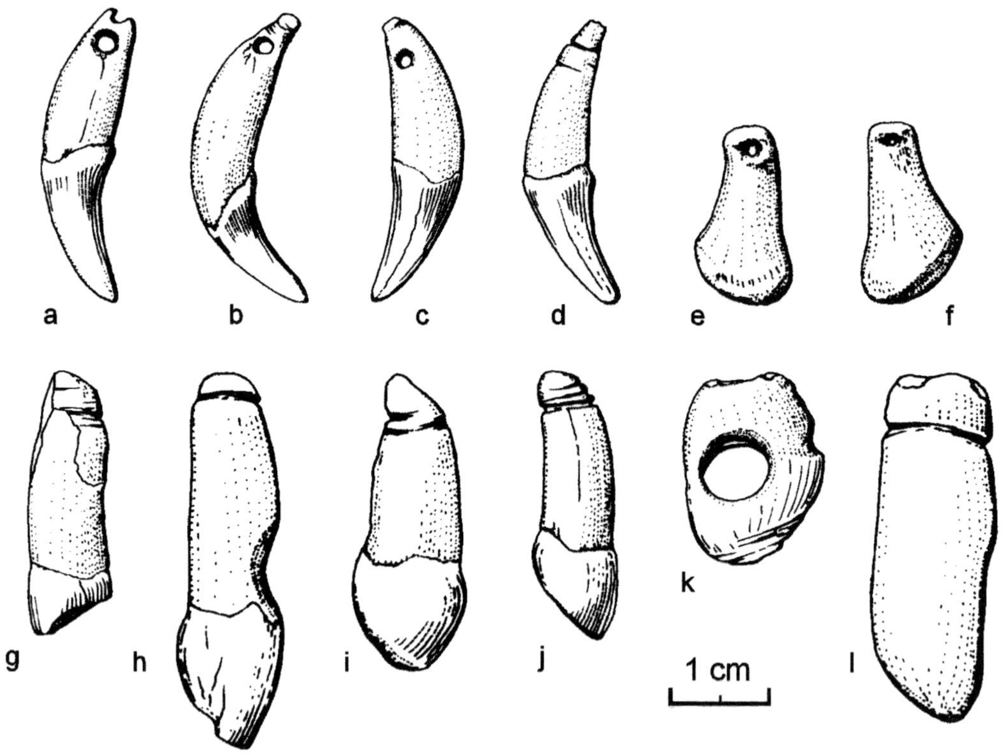

Fig. 25.1 Pierced and grooved pendants from the Châtelperronian levels of the Grotte du Renne (France): (**a**–**d**). fox canines; (**e**–**f**). reindeer phalanges; (**g**–**j**). bovid incisors; (**k**). red deer canine; (**l**). fossil belemnite (After Zilhão and d'Errico 1999b, modified)

trade between the two groups or (c) if at all manufactured indeed by the Neandertals themselves, "imitation without understanding" of the product of Aurignacian modern human symbolic crafts (Stringer and Gamble 1993; Hublin et al. 1996). Through their common denial of the Neandertals' capacity to independently develop symbolic material culture, these arguments effectively managed to reconcile the Châtelperronian evidence with the notion that cognitive and behavioral modernity were a species-specific attribute of the African species, *Homo sapiens*.

Paradigm Lost

Acculturation at the Grotte du Renne?

As shown in a series of papers published at the turn of the century (d'Errico et al. 1998, 2003; Zilhão and d'Errico 1999a, b, 2000, 2003a, b; Zilhão 2001; d' Errico 2003), neither the imitation nor the taphonomic or trade/scavenging solutions to the Châtelperronian problem were consistent with the empirical record. At the Grotte du Renne, most ornaments came from the lowermost Châtelperronian occupation (level X), on average separated by some 90 cm of mostly horizontal stratification from Aurignacian level VII, which, moreover, had yielded five times fewer ornaments than the underlying Châtelperronian sequence. This vertical distribution pattern (Fig. 25.2) refuted the hypothesis that the ornaments found in the Châtelperronian levels were intrusive, and *in situ* manufacture debris provided evidence that the associated bone tools (some of which decorated) had been produced at the site, not imported to it (Fig. 25.3). Consequently, the only alternative that could stand against the recognition that these levels represented an autochthonous, independent development of European Neandertals was that of acculturation, in any of its two flavors: close-contact acculturation, resulting in imitation (Mellars 1999), or long-distance acculturation, resulting in re-elaboration (Hublin 2000).

As pointed out by d'Errico et al. (1998), close-contact acculturation was contradicted by the obvious differences in choice of blank, manufacturing technique, and function, between the bone tools and ornaments from the Grotte du Renne's Châtelperronian and their putative Aurignacian sources of inspiration. For instance, while most Aurignacian tooth pendants are pierced, most Châtelperronian ones were made by carving a furrow around the tooth root so that a string could be tied around it for suspension; and, when piercing was used, the Châtelperronian approach was to puncture a hole through the root and then smooth and enlarge it, whereas the normal Aurignacian approach is first to thin

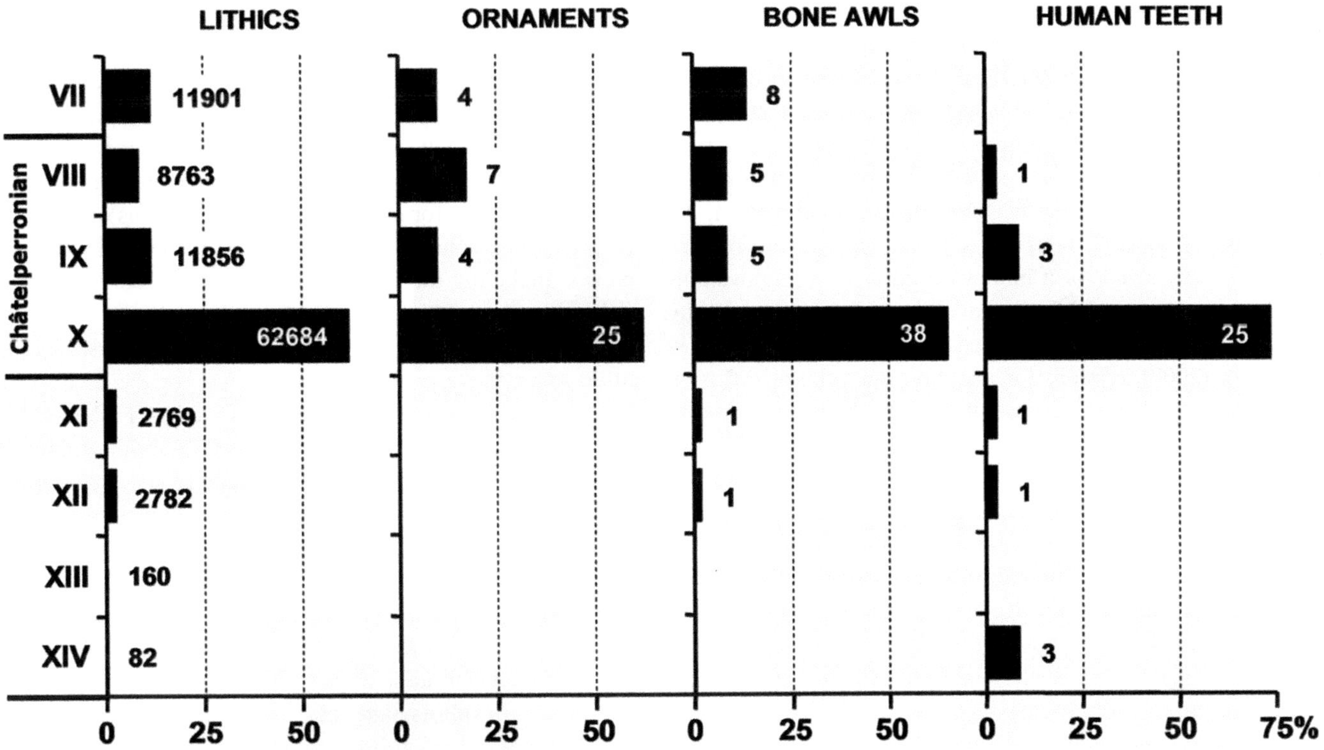

Fig. 25.2 Vertical distributions of key Grotte du Renne finds. Note that the two bone awls in levels Mousterian XI–XII are identical to those in overlying level X and probably correspond to Châtelperronian intrusions. Likewise, three ornaments in Châtelperronian level VIII are ivory beads identical to material in level VII that is typical for the Aurignacian elsewhere in the region, and, thus, probably correspond to intrusions (After Girard 1980; Connet 2002; Schmider 2002; d'Errico et al. 2003; Bailey and Hublin 2006; Zilhão 2007; Hublin, personal communication)

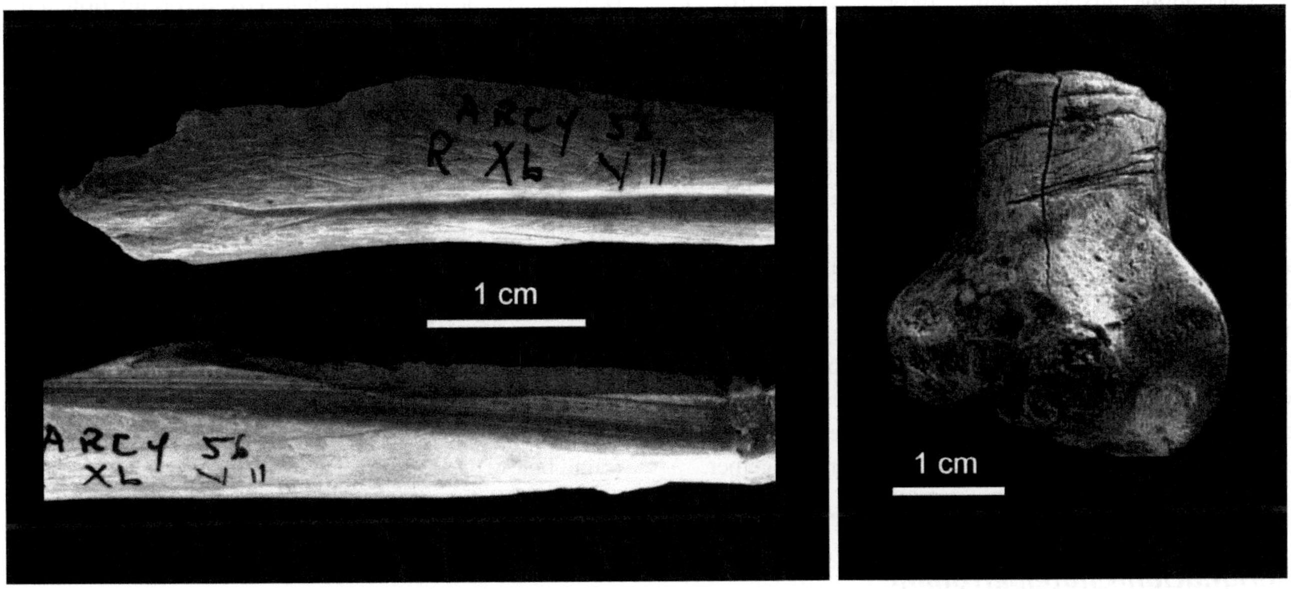

Fig. 25.3 Debris from bone tool manufacture in Châtelperronian level Xb from the Grotte du Renne. *Left*, close-up views of a fragment of long bone diaphysis showing traces of grooving with a flint point to produce a bone rod: above, striations produced by the flint point occasionally slipping out of the groove in the process; below, parallel internal striations produced by the repeated movement of the flint point. *Right*, distal ulna of a swan sawed with a lithic tool to produce a bone tube (After d'Errico et al 1998; Zilhão and d'Errico 2000, modified)

the root by scraping and only then drill the perforation. And while most Aurignacian bone tools are deer antler sagaie points of different types used as tips of composite projectiles, the Châtelperronian assemblage from the Grotte du Renne was almost exclusively composed of awls made out of thin, elongated bones (such as accessory metapodials of horse).

In the alternative view of long-distance acculturation, the elements of modern behavior seen in the Châtelperronian would simply stand for a bow-wave diffusion of innovations slightly in advance of the actual arrival of the anatomically modern innovators, as suggested by Mellars (1999), or, in Hublin's (2000) version, for adoption via cultural diffusion (without actual movements of people being necessarily involved). Hublin actually took the argument one step further to suggest that such would also have been the mechanism via which, from source areas inhabited by modern humans (Africa or the Near East), all other cultural innovations observed in the European record throughout the entire duration of the Middle Paleolithic, namely the practice of burial, had been brought into the continent. These propositions suffer from major empirical problems (such as, for instance, the fact that the earliest instance of deliberate burial currently known seems to be that of the Tabun Neandertal woman – Grün and Stringer 2000), but the main issue is that they are logically inconsistent to begin with.

In fact, in the context of the Human Revolution paradigm, long-distance acculturation models are faced with a double conundrum: (a) the putative speciation of Neandertals would have resulted from long-term geographical isolation, preventing gene flux but … allowing for complete cultural interconnectedness, and (b) since long-distance cultural diffusion implies the spread of concepts, not just objects, the recipient of the innovations must have had … those exact cognitive capabilities the lack of which the argument was designed to explain. Obviously, the simple fact of Neandertal-ness itself implies largely separate evolutionary trajectories between Europe and Africa. Therefore (a) if such cognitive capabilities existed in the Neandertal world, then independent invention, not long-distance acculturation, is the parsimonious explanation for burials or ornaments, and (b) if no such major isolation existed, and flux of both genes and memes between the two continents continued throughout and notwithstanding of the process of anatomically becoming either Neandertal or modern, then we are talking about populations of a single species interacting via the normal mechanisms documented by ecology, ethology, ethnography and history, and the rationale for thinking in terms of species-specific behaviors disappears altogether.

Aurignacian/Châtelperronian Interstratifications?

In order for close-contact acculturation to work, contemporaneity between the acculturator and the acculturated over neighboring or overlapping territories must be postulated. And, indeed, the scenarios put forward by Mellars (1989) and Allsworth-Jones (1990, cf. 2004 for a recent restatement of the argument) to explain the Transition in western and central Europe, respectively, relied heavily on the notion of a long-term contemporaneity between the Aurignacian (assumed to represent a proxy for the first European moderns) and the so-called "transitional" early Upper Paleolithic technocomplexes, including the Châtelperronian (assumed, conversely, to be an archaeological proxy for late Neandertals).

These transitional entities correspond to a diverse array of lithic assemblage-types that, in one way or the other, fit at least some aspects of the definition of the Upper Paleolithic. The Châtelperronian of the Franco-Cantabrian region is characterized by the production of blade blanks transformed in curve-backed Châtelperron points and knives; ornaments (found not only at the Grotte du Renne but also at Quinçay, Caune de Belvis, St.-Césaire and others), include pierced and grooved pendants made up of teeth, bones and fossils, as well as ivory discs and *Dentalium* tubes. The Uluzzian of Italy and Greece is a flake-based industry, with some production of non-Levallois blade blanks, characterized by its standardized backed microliths, mostly lunates (Fig. 25.4); the only securely associated ornaments are *Dentalium* tubes. In the Bachokirian of Bulgaria, the Upper Paleolithic aspect is due to the preponderance of endscrapers made on thick blade blanks, and the presence of pierced teeth and bone pendants. The Bohunician of Moravia and southern Poland features the production of morphologically Levallois points obtained by non-Levallois methods and used as blanks for retouched tool assemblages of characteristic Upper Paleolithic typology (Fig. 25.5). The Szeletian, defined by the production of finely crafted bifacial foliate points, seemingly follows the Bohunician in these regions, and is in all likelihood closely related to the Altmühlian with *blattspitzen* of southern Germany (Fig. 25.6); the unifacial blade-point industries (such as the Lincombian of England and Belgium, or the Jerzmanovician of eastern Germany and Poland), finally, are likely to represent the last stage, across the northern European plains, of this cultural tradition, to which probably belong the ivory discs recovered in poorly documented stratigraphic contexts at the Ilsenhöhle (Germany) and the Trou Magrite (Belgium) (Fig. 25.7).

When reliable stratigraphic information is available, such early manifestations of the European Upper Paleolithic are found below the Aurignacian. This pattern is naturally indicative of some form of linear chronological succession, but acceptance that such is the case has been hindered by the long-term impact of a mid-twentieth century controversy surrounding the putative existence in the classical region of southwestern France of two parallel "phyla" developing side by side for some 15,000 years – the Aurignacian and the "Perigordian". That controversy was eventually settled by the demonstration that (a) the Lower Perigordian and the

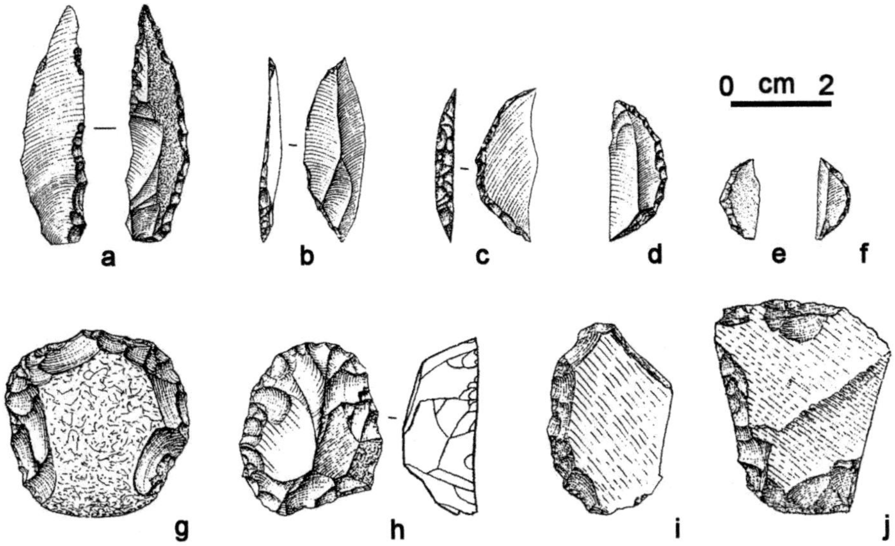

Fig. 25.4 Uluzzian artifacts from level EI-II of Cavallo cave, Italy: (**a**). microlithic backed point; (**b–f**). segments; (**g–h**). endscrapers; (**i–j**). sidescrapers (After Palma di Cesnola 1993, modified)

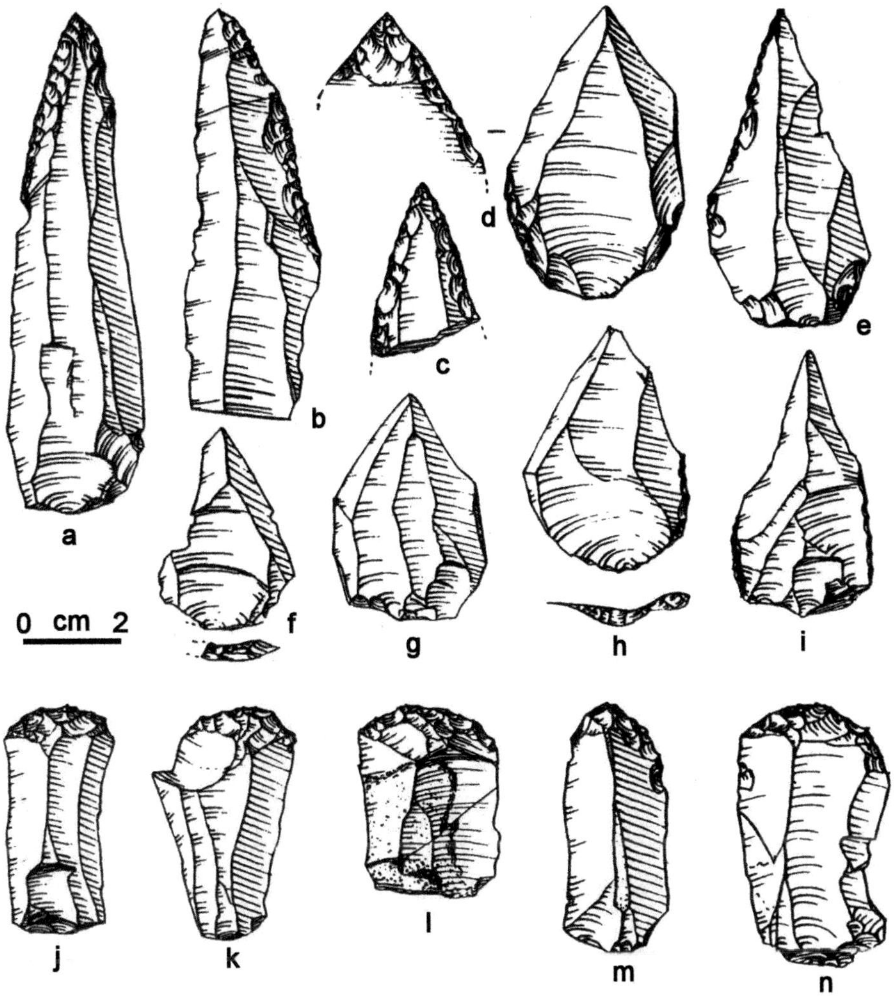

Fig. 25.5 Bohunician artifacts from level 5 of the open air site of Stránská skála III (Czech Republic): (**a–i**). points; (**j–n**). endscrapers (After Svoboda 1988, modified)

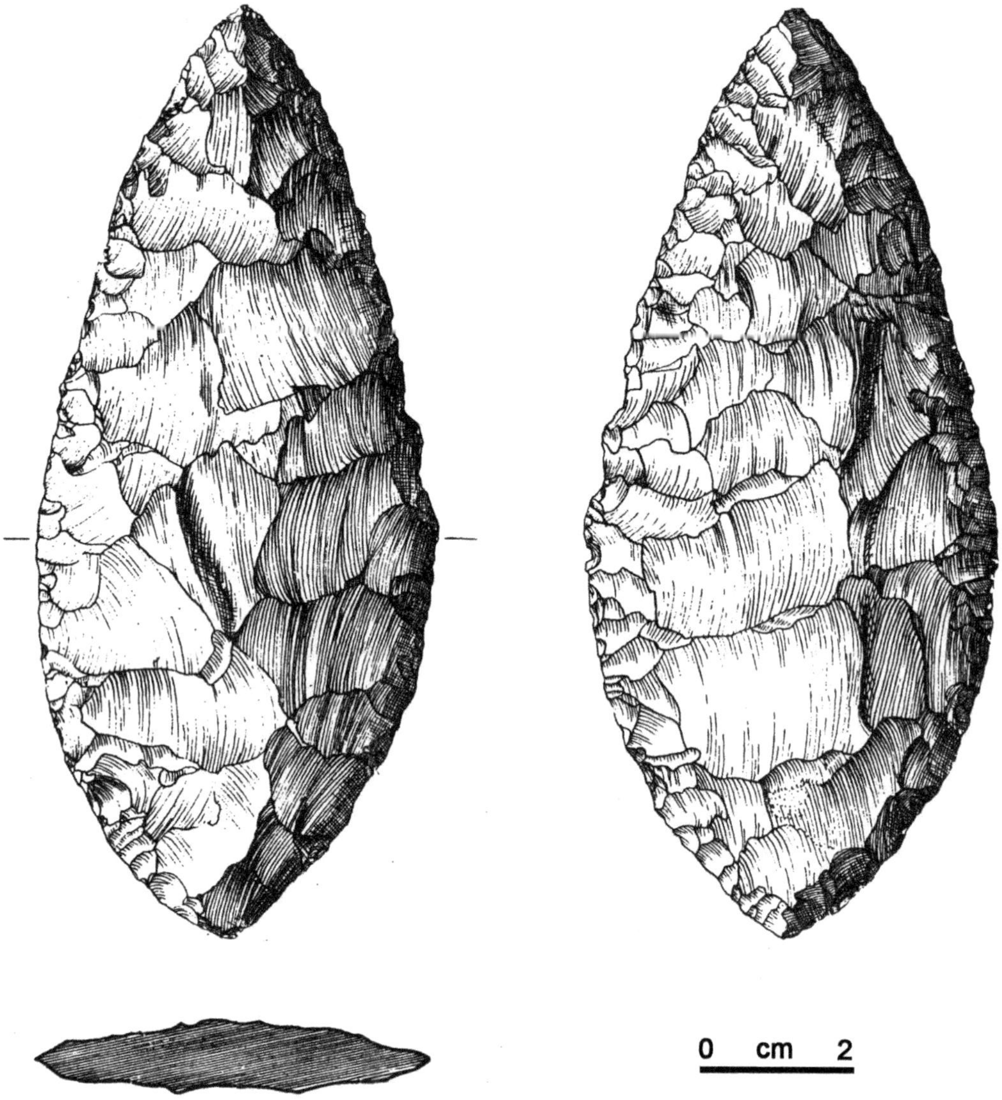

Fig. 25.6 Bifacial foliate point (*blattspitze*) from Horizon 2 of the Ilsenhöhle (Ranis), Germany (After Hülle 1977)

Fig. 25.7 Ornaments of the earliest Upper Paleolithic of central and eastern Europe: *left*, perforated fossil gastropod from level 2 of Willendorf II, Austria; *center*, spindle-shaped bone pendant from level 11 (Bachokirian) of the type-site; *right*, perforated ivory disc from horizon 2 of the Ilsenhöhle (Ranis), Germany (After Felgenhauer 1956–1959; Hülle 1977; Kozłowski 1982)

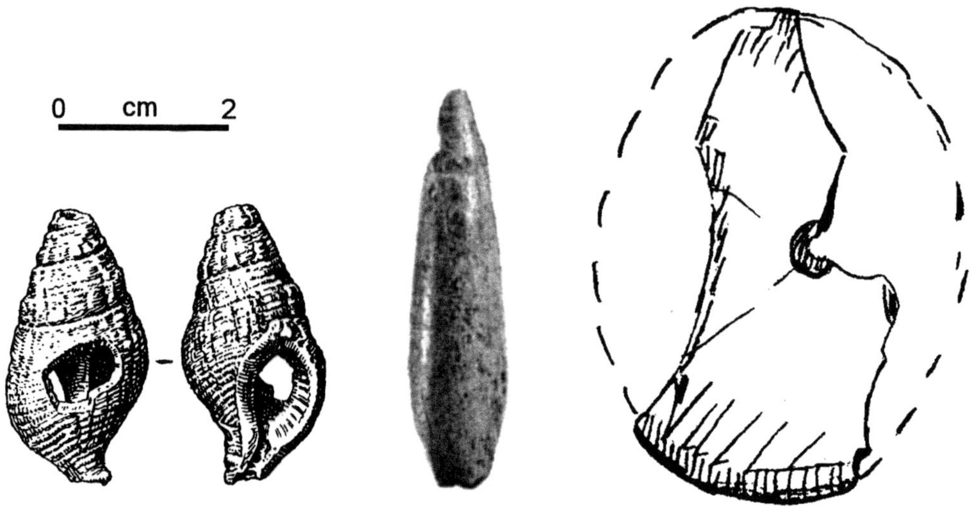

Upper Perigordian were separated by several millennia, no direct phyletic link or thread of technological continuity existing between the two (for which, accordingly, the alternative designations of, respectively, Châtelperronian and Gravettian, have since been retained) (Laville et al. 1980), and (b) the "Aurignacian V" capping the Upper Perigordian sequence at the site of Laugerie-Haute was in no way related to the true or "typical" Aurignacian (Sonneville-Bordes 1982) and in fact represents a transitional facies between the Gravettian and the Solutrean (Zilhão et al. 1999).

Where the earlier interval of the suggested Aurignacian/Perigordian contemporaneity was concerned, however, these studies did not question the validity of the interstratifications between Châtelperronian and Aurignacian reported by Bordes and Labrot (1967), Champagne and Espitalié (1981), and Bernaldo de Quirós (1982) for, respectively, Roc-de-Combe and Le Piage, in France, and El Pendo, in Spain. Thus, although originally claimed in a totally different theoretical context (in support of the notion that the Perigordian as a whole and the Aurignacian constituted two parallel cultural traditions of fully Upper Paleolithic neoanthropians), those interstratifications would eventually provide, three decades later, the empirical basis for the long-term contemporaneity between late Neandertals and early moderns required by the Human Revolution paradigm. At the same time, application of the new AMS radiocarbon dating technique produced results for El Castillo (Cabrera and Bischoff 1989) and L'Arbreda (Bischoff et al. 1989) that seemingly strengthened the interstratification evidence by indicating an Aurignacian presence in northern Spain prior to the emergence of the Châtelperronian in the Aquitaine basin (but see below). A search for the eastern European source of this precocious modern human spread logically ensued, and the putative Aurignacian affinities of the "transitional" levels from Bacho Kiro and Temnata, in Bulgaria, led to suggestions that the Bachokirian industry found therein was a good candidate for the role of ancestor (Kozłowski and Otte 2000; Otte and Kozłowski 2003).

Over the last decade, however, a considerably clearer and much simplified picture of the Transition has emerged. At El Pendo, the notion of interstratification cannot be retained anymore, as Montes and Sanguino (2001) and Montes et al. (2005) showed that the sequence is in secondary position, and that its different levels feature a mix of disparate Middle and Upper Paleolithic items throughout the entire thickness of the deposits. Likewise, at Roc-de-Combe and Le Piage, J.-G. Bordes (2002, 2003, 2006), following-up on reservations previously put forward by other workers (Demars 1990; 1996; d'Errico et al. 1998; Rigaud 2001), demonstrated beyond reasonable doubt the illusory nature of the interstratifications, in fact an artifact of both taphonomy and excavation error (Fig. 25.8). Gravina et al. (2005) and Mellars et al. (2007) have since resurrected the notion on the basis of the stratigraphy of Delporte's 1950s excavations at the Grotte

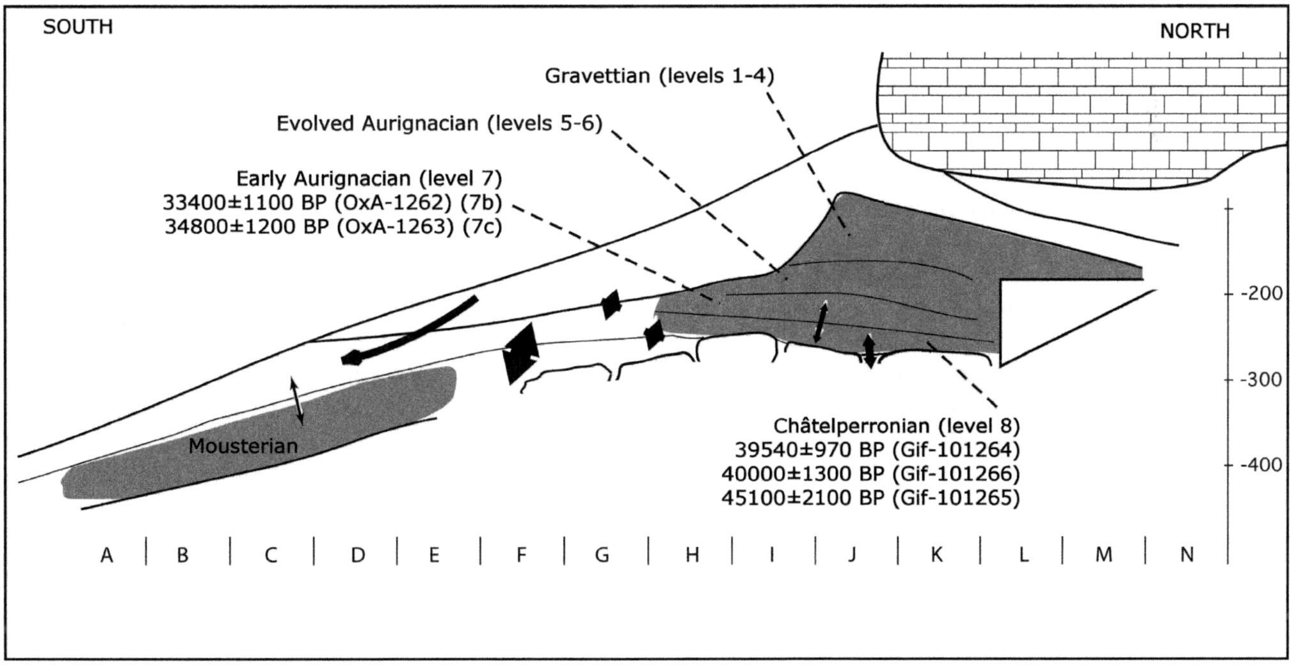

Fig. 25.8 Roc-de-Combe, synthetic sagittal projection showing J.-G. Bordes's reconstruction of the sequence, based on lithic taphonomy. The *arrows* indicate the direction and intensity of the post-depositional movements involved in the site formation process. Bordes and Labrot's "interstratification" was based on the disturbed area outside the *drip line*, while only the assemblages from the areas in *gray* are valid; these areas feature the chronostratigraphic succession normal for the region, and radiocarbon dating from securely provenanced samples yielded stratigraphically consistent results (After Bordes 2002, 2006, modified)

des Fées at Châtelperron, the type-site of the Châtelperronian. Their argument was contingent upon the notion that level B4 of this site was a "pure" Aurignacian lens interstratified between two "pure" blocks of Châtelperronian deposits. However, as shown by Zilhão et al. (2006, 2008), level B4 was made up for the most part of Châtelperronian material and the few Aurignacian pieces in levels B4–B5 (carinated "scrapers," endscrapers on blades with Aurignacian retouch, and bladelets with inverse or alternate retouch) at best represent isolated intrusions into otherwise *in situ* Châtelperronian deposits. Overlying levels B1–B3 were clearly disturbed, explaining both the fragmentary condition of the material and its mix of Mousterian, Châtelperronian, Aurignacian and even Solutrean items. In fact, bar a couple of very marginal, conceivably intact, non-interstratified remnants excavated in 1962, the evidence leaves no doubt that the entire sequence of "interstratified" deposits excavated by Delporte at the Grotte des Fées in 1951–1954 corresponds to backdirt from the mid-nineteenth-century excavations.

Radiometrically Late "Transitional" Industries?

Recent reviews of the relevant collections (Teyssandier 2003; Tsanova and Bordes 2003; Rigaud and Lucas 2006) fully confirmed previous reservations (Rigaud 2001; Zilhão and d'Errico 1999b; 2003b) concerning the purported Aurignacian affinities of the Bachokirian. In fact, its technological basis is related to the Levallois method, i.e., is of a Middle Paleolithic nature, at best comparable to the earlier phases of the IUP (Initial Upper Paleolithic) or Emiran of the Near East (Kuhn 2003), a view that Kozłowski (2004) has since accepted formally. This reassessment of the Bachokirian not as a precursor to the Aurignacian but as a transitional entity rooted in the preceding Middle Paleolithic is consistent with its chronology; as shown by critical analyses of the radiocarbon datasets (Zilhão and d'Errico 1999b, 2003a, b; Zilhão 2006a, b, 2007), multiple dates in the range of ca. 43–38 ka ^{14}C BP (46.5–42.7 ka cal BP) now exist for the transitional industries of Europe, from the Bachokirian in the East to the Châtelperronian in the West, with the results for level B5 of the Grotte des Fées being a significant addition to this growing corpus of data (Table 25.1; Fig. 25.9).

These conclusions, however, are not necessarily inconsistent with the acculturation argument if, as suggested initially by Leroyer and Leroi-Gourhan (1983) and most recently by Floss (2003), one hypothesizes that the most conspicuous "modern" aspects of the "transitional" technocomplexes only appeared in later, epigonical phases, post-dating the arrival of the Aurignacian. The results in the range of ca. 33–32 ka ^{14}C BP (38.2–36.9 ka cal BP) obtained for the Châtelperronian of the Grotte du Renne (Hedges et al. 1994; David et al. 2001), in particular, would provide substantial support for such a hypothesis, as would the dates in the same range obtained for the Szeletian/Altmühlian-to-Jerzmanovician sequence represented by the Ranis 2 and Ranis 3 find horizons of the Ilsenhöhle, in eastern Germany (Grünberg 2006); where the latter is concerned, this chronometric evidence would also be consistent with the time interval (ca. 38–30 ka ^{14}C BP; 42.7–33.4 ka cal BP) delimited by currently available dates for the Jerzmanovician of the type site, the Nietoperzowa cave, in Poland (Kozłowski and Kozłowski 1996). Given that, moreover, a significant proportion of the ornaments and decorated bone tools found in the transitional industries comes from only one of these sites, the Grotte du Renne, one might thus be legitimately led to speculate that such symbolic artifacts stand indeed for the close-contact acculturation of residual, post-35 ka ^{14}C BP (post-40.7 ka cal BP) Neandertal populations confined to refugial areas located beyond the range of the Aurignacian, whose presence in southern Germany and southern France is unquestionably attested since ca. 36.5 ka ^{14}C BP (ca. 42 ka cal BP).

As argued by Zilhão and d'Errico (2003a), based on the experiment conducted by Richter (2002) at the Sesselfelsgrotte, in Bavaria, these recent dates must be assessed bearing in mind the geochemical context of the samples' provenience. When the samples are of bone collected in exposed areas (exterior or close to the drip line) of cave and rockshelter sites, where the deposits underwent long-term post-depositional leaching, the evidence from the Sesselfelsgrotte conclusively shows that they yield much younger ages than those from interior areas of the same levels. The impact of this chemical processes in all likelihood explains the disparate results obtained for the Châtelperronian of the Grotte du Renne (for which there are also dates of ca. 45 and ca. 38 ka ^{14}C BP – 48.1 and 42.7 cal BP, respectively), as well as the marked rejuvenation of all results obtained for the key Châtelperronian-to-Aurignacian sequence of Labeko Koba, in Spain (Arrizabalaga and Altuna 2000; Arrizbalaga et al. 2003; Zilhão 2006c). The Ilsenhöhle samples come from deposits in similar situation (the site is a collapsed rockshelter, not a cave), and the ages obtained for the transitional levels, much younger than expected, probably simply represent yet another example of this problem.

Moreover, important methodological developments (Bronk Ramsey et al. 2004; Higham et al. 2006) have now confirmed that the rejuvenation of bone results often observed when comparing them with charcoal dates for the same contexts (Zilhão and d'Errico 1999b; Jöris et al. 2003) may indeed be an artifact of incomplete decontamination, a view now endorsed even by those who systematically used such rejuvenated results to support the main contentions of the Human Revolution paradigm (e.g., Mellars 2006). Neither the Ilsenhöhle nor the Grotte du Renne dates come from bone

Table 25.1 Selected radiocarbon dates for key sites documenting the archaeological and human paleontological context of the Transition in Europe and the Greater Mediterranean, calibrated with CalPal (Weninger and Jöris 2005). For each site, the latest result available was used for the pre-Aurignacian "Transitional" entities, while the earliest result available was used for the Early Ahmarian and the Protoaurignacian; the result for uppermost IUP level G of Üçağizli was also included, as it provides a *terminus post quem* for the Early Ahmarian. Following the criteria outlined in Zilhão and d'Errico (1999, 2003a), samples whose association with the dated context was questionable, or for which incomplete decontamination was apparent or probable, are not included

Site	Culture/Fossil	Level/Provenience	Sample type	Method	Lab number	Result BP	Age cal BP	cal BP 2σ range
Klisoura 1	Uluzzian	V, hearth 42	Burnt bone	AMS	GifA-99168	40,010±740	43,870±620	45,110–42,630
Bacho Kiro	Bachokirian	11-I	Charcoal	AMS	OxA-3183	37,650±1,450	42,640±960	44,560–40,720
Temnata	Bachokirian	4 B, TD-V [Interior]	Charcoal (mostly)	AMS	OxA-5173	36,900±1,300	41,760±1,150	44,060–39,460
Nietoperzowa	Jerzmanovician	6	Charcoal	Conventional	GrN-2181	38,500±1,240	43,170±790	44,750–41,590
Willendorf II	Upper Paleolithic	2	Charcoal	Conventional	GrN-11190	39,500/+1,500/−1200	43,670±890	45,450–41,890
Bohunice-Kejbaly	Bohunician	4a	Charcoal	Conventional	Q-1044	40,173±1,200	44,050±890	45,830–42,270
Stránská skála III (1982)	Bohunician	5, base of upper paleosol, locus III-1	Charcoal	Conventional	GrN-12297	38,200±1,100	43,020±730	44,480–41,560
Vedrovice V (1982-83)	Szeletian	4 [main concentration in 17/I]	Charcoal	Conventional	GrN-15514	37,600±800	42,600±510	43,620–41,580
Sesselfels	Micoquian/ Altmühlian	E3 (inside)	Burnt bone	Conventional	GrN-7153	37,100±1,000	42,290±600	43,490–41,090
Pin Hole Cave	Lincombian	East Passage, close to the roof	Tooth, fetal hyena	AMS	OxA-4754	37,800±1,600	42,700±1,070	44,840–40,560
Abri Dubalen (Brassempouy)	Châtelperronian	EBC2	Bone	AMS	GifA-101045	36,130±690	41,360±780	42,920–39,800
Châtelperron	Châtelperronian	B5	Bone	AMS	OxA-13622	39,150±600	43,430±580	44,590–42,270
Grotte du Renne	Châtelperronian	Xb1, Y10	Bone (horse)	AMS	OxA-8451/Ly-894	38,300±1,300	43,080±820	44,720–41,440
Caune de Belvis	Châtelperronian	[7]	Bone	AMS	AA-7390	35,425±1,140	40,640±1,130	42,900–38,380
Combe Saunière	Châtelperronian	X	Bone	AMS	OxA-6503 (tripeptide)	38,100±1,000	42,960±680	44,320–41,600
Grotte XVI	Châtelperronian	B	Bone	AMS	GifA-95581	35,000±1,200	40,310±1,210	42,730–37,890
La Quina, aval	Châtelperronian	4	Bone	AMS	OxA-10261/Ly-1367	35,950±450	41,440±580	42,600–40,280
Roc-de-Combe	Châtelperronian	8, K9	Bone	AMS	Gif-101264	39,540±970	43,630±700	45,030–42,230
Üçağizli	IUP (=Ksar 'Akil XXI)	G	Charcoal	AMS	AA37626	39,100±1,500	43,490±930	45,350–41,630
Kebara	Early Ahmarian	IIIBf (Q16d, 5.38 m, "hearth")	Charcoal	AMS	OxA-1567	35,600±1,600	40,520±1,470	43,460–37,580
Boker A	Early Ahmarian	1	Charcoal	Conventional	SMU-578	37,920±2,810	42,120±2,400	46,920–37,320
Krems-Hundsteig	Protoaurignacian	brown layer with hearths	Charcoal	Conventional	KN-654	35,500±2,000	40,150±1,920	43,990–36,310
Grotta di Fumane	Protoaurignacian	A2, near hearth S14	Charcoal	AMS	UtC-2048	36,500±600	41,960±330	42,620–41,300
Riparo Mochi	Protoaurignacian	G, Cut 59; east trench 1959	Charcoal	AMS	OxA-3591	35,700±850	40,910±990	42,890–38,930
Esquicho-Grapaou	Protoaurignacian	SLC 1b	Charcoal	Conventional	MC-2161	34,540±2,000	39,490±1,970	43,430–35,550
Isturitz	Protoaurignacian	4d, U27	Burnt bone	AMS	GifA-98232	36,510±610	41,960±340	42,640–41,280
Morin	Protoaurignacian	8	Charcoal	AMS	GifA-96263	36,590±770	41,950±450	42,850–41,050
Peştera cu Oase	Oase 1	Sala Mandibulei	Bone	AMS	OxA-11711/ GrA-6165	34,950/+990/−890	40,450±1,020	42,490–38,410
Kleine Feldhofer Grotte	Neandertal 2	Dump site	Bone	AMS	ETH-19660	39,240±670	43,470±590	44,650–42,290

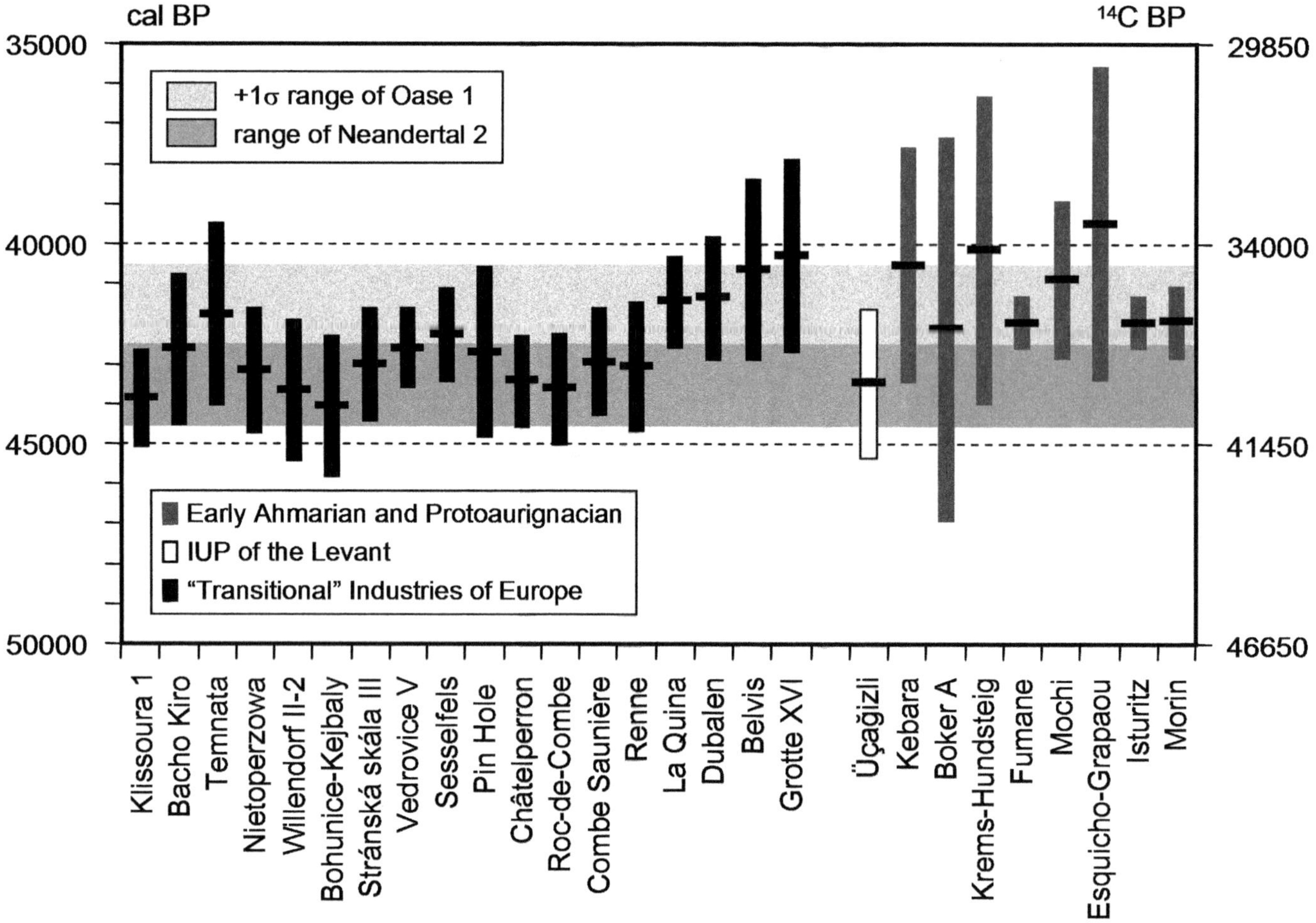

Fig. 25.9 Two-sigma plot of the dates in Table 25.1. In Europe, transitional industries fall in the Neandertals' time range, the tail of younger results illustrating the impact of incomplete decontamination on bone samples, not long-term contemporaneity with the Aurignacian (at Grotte XVI, for instance, two other results for the same level place it firmly beyond 38 ka ^{14}C BP, i.e., 43.5 ka cal BP). Assuming the Oase 1 fossil and the Protoaurignacian as proxies, the first modern human dispersals into Europe took place ca. 42 ka cal BP, and post-dated by several millennia the emergence of ornaments and decorated bone tools in Europe, from the Greek Uluzzian in the East to the French Châtelperronian in the West

samples pre-treated with the new ultrafiltration technique, which, given these sites' geochemical framework, renders almost inevitable the conclusion that they should be treated as minimum ages only.

The widespread impact of sampling issues in the artificial rejuvenation of the "transitional" industries is nowhere more apparent than in Italy. As pointed out by Mussi (2001), all the dates available for the Italian Uluzzian are conventional results obtained on bone samples in the 1960s and early 1970s. Given the technical limitations of the time, these were minimum ages only, and were indeed reported as such by the dating labs: >31 ka ^{14}C BP (R-352; >36.1 cal BP) for level EI-II of Cavallo, in the middle of the site's ca. 75 cm thick Uluzzian sequence, and >34 ka ^{14}C BP (F-106; >40.2 cal BP) for level *rpi* of Castelcivita, also towards the middle of a ca. 1 m thick sequence. That the Uluzzian emerged at about the same time as the Châtelperronian is now confirmed by the ca. 40 ka ^{14}C BP (43.9 ka cal BP) obtained for level V of Klisoura 1 cave, in Greece (Koumouzelis et al. 2001a, b), and the CI (Campanian Ignimbrite) tephras, a region-wide chronostratigraphic marker, provide an unambiguous *terminus ante quem* for the end of the technocomplex. These tephras formed in the framework of the caldera explosion of the Phlegraean Fields, in central Italy, whose age is now well-established, on the basis of 36 high-precision single-crystal ^{40}A/^{39}A measurements, at 39,280 ± 110 ka BP (de Vivo et al. 2001), or between ca. 34 and ca. 33 ka ^{14}C BP in the radiocarbon time scale. Fall-outs from the explosion accumulated through most of southeastern Europe, as far away as the Russian-Ukrainian plain (Fedele et al. 2003), and, in all known Italian cave site occurrences, the Uluzzian is found below CI deposits (and further separated from these by Protoaurignacian levels at Castelcivita, and by Uluzzian levels with intrusive Aurignacian material at Cavallo).

In France, the archaeological stratigraphy of the Grotte du Renne likewise provides a *terminus ante quem* for the site's Châtelperronian occupation. As argued by Bon and Bodu (2002), the lithic technology represented in the site's Aurignacian level VII displays strong affinities with the earliest Aurignacian of southern France (see below), and Julien et al. (2002) identify as split-based points two bone tool fragments found therein. The implication is that the ornament-rich level X, at the base of the Châtelperronian sequence, instead of being a late, epigonical, Aurignacian-influenced cultural manifestation, must predate significantly the arrival of the Aurignacian in western Europe.

A Radiometrically Early Aurignacian?

Where the Aurignacian is concerned, the critique of the chronometric dataset of the Transition showed that nowhere did it predate ca. 36.5 ka ^{14}C BP (ca. 42 ka cal BP), with consistent (and as yet unrefuted) explanations, based on issues of taphonomy and assemblage definition, having been offered for the few apparent anomalies (Zilhão and d'Errico 1999b, 2003a, b; Zilhão 2006c). In northern Spain, for instance, the results for the Mousterian-to-Aurignacian sequence in levels I-to-H of L'Arbreda cover the 41–34 ka ^{14}C BP interval (44.6–40.2 ka cal BP) but feature several instances of stratigraphic inversion, possibly as a result of the presence of residual charcoal material in the interior area of the shelter that was sampled for dating, and of the fact that a sloping stratigraphy was excavated in arbitrary horizontal spits; as a consequence, it is impossible to establish a one-to-one correlation between any of the available results and the different technocomplexes represented in the artifact collection by diagnostic lithic and bone tools. A related situation exists at El Castillo, where level 18 corresponds (a) in the interior area (early twentieth-century excavations), to a 1 m-thick palimpsest of both Mousterian and Aurignacian occupations, the former documented by dates in excess of 47 ka ^{14}C BP (50.6 ka cal BP) for its faunal component, and (b) in the exterior area (1980s excavations), to the conflation in a single, chimerical "Transitional Aurignacian" entity of a mix of material from different occupations contained in thin levels compressed under heavy rock fall, with available radiocarbon results nonetheless retaining some stratigraphic coherence, those for sublevel 18B overlapping with the ca. 36.5 ka ^{14}C BP (42 ka cal BP) pan-European horizon for the earliest Aurignacian, and those for sublevel 18C overlapping with the dates for the underlying Mousterian level 20 (Fig. 25.10).

The situation in central Europe is no different. The association with the ensemble of split-based bone points from Istállóskő cave, in Hungary, of the bone fragments conventionally dated to ca. 44.3 ka ^{14}C BP (47.5 ka cal BP) has long been questioned (e.g., Svoboda et al. 1996), and is now all but abandoned, given the much younger, normal results of ca. 33 ka ^{14}C BP (38.2 ka cal BP) obtained by Adams and Ringer (2004) for better-provenanced samples. Where southern Germany is concerned, the earliest Aurignacian is that contained in horizon AH-III of the Geissenklösterle, whose chronology is now accepted by all parties (cf. Teyssandier et al. 2006) to fall in the range of ca. 35–33 ka ^{14}C BP (40.7 ka cal BP). Two earlier results, in the range of ca. 39–38 ka ^{14}C BP (43.3–42.7 ka cal BP), exist for level 3 of the Austrian open air site of Willendorf II, whose lithic assemblage is clearly Aurignacian (Teyssandier 2003; Nigst 2006); however, there is another result of ca. 34 ka ^{14}C BP (40.2 cal BP), and the level was found in a geomorphological situation conducive to the presence of residual material in the soliflucted charcoal lenses whence the dated samples come (Haesaerts and Teyssandier 2003; Haesaerts et al. 2004). Finally, at Keilberg-Kirche (Uthmeier 1996, 2004), an *in situ* fireplace excavated in 1987 yielded charcoal of coniferous and deciduous trees (willow or seabuckthorn, suggesting an interstadial landscape) dated to ca. 38 ka ^{14}C BP (42.7 ka cal BP), and, four years later, a lithic assemblage of Aurignacian affinities was recovered in the salvage excavation of the surrounding area. However, these artifacts were contained in displaced loess lenses whose original deposition must have taken place under stadial, not interstadial conditions, and the presence of nosed endscrapers and carinated burins (Fig. 25.11) indicates a late, not an early Aurignacian. Given that the combined sedimentological and botanical evidence makes it clear that the site contained remains from at least two distinct human occupation episodes, the parsimonious interpretation of the hearth's date is that it relates to the few late Mousterian or Altmühlian lithics present in the assemblage (sidescrapers and bifacially flaked items), not to the more abundant but, typologically, conspicuously late Aurignacian material.

Were Moderns, Not Neandertals, the Makers of the Châtelperronian?

In these circumstances, and although "Neandertal acculturation" models based on long-term contemporaneity with the Aurignacian still thrive in one guise or the other, only one really promising line of reasoning remains for supporters of the notion that Neandertals were devoid of fully *sapiens* cognition and behavior (as measured by exactly the same archaeological standards designed to gauge their emergence in the African lineage): the suggestion that the Châtelperronian and the other broadly coeval "transitional" industries of central and eastern Europe were in fact the work of moderns, not Neandertals. Where the Châtelperronian is concerned, this

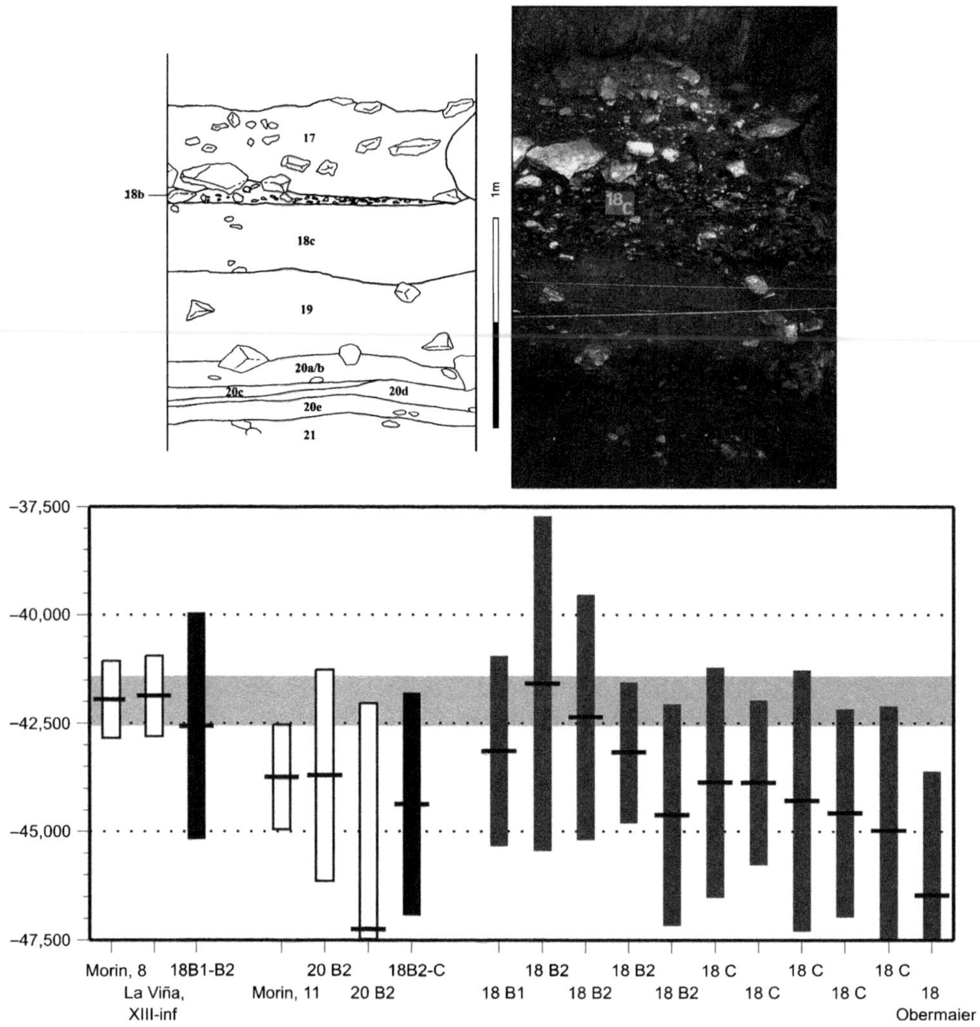

Fig. 25.10 El Castillo. *Above*: stratigraphy of the Transition levels. Note that sublevel 18C stratigraphically underlies 18B, as is also apparent in Cabrera and Bischoff's (1989) sample provenience schemes (After Maíllo et al. 2004) *Below*: calibrated radiocarbon dates for level 18 (*dark gray bars*), compared with the averages (*black bars*) for the two apparent clusters (younger, four B1–B2 dates; older, six B2–C dates) and the reliable results (*white bars*) for the latest Mousterian and the earliest Upper Paleolithic of the Cantabrian strip (After Zilhão 2006c)

was of course the long-held view of François Bordes (1968, 1972); one that led him to suggest that the Saint-Césaire Neandertal burial was that of a victim of the maker of the associated Châtelperronian artifacts (Bordes 1981).

Arguing along similar lines, Bar-Yosef (2006) recently suggested that the human remains found in the Châtelperronian levels of the Grotte du Renne, all of which, when diagnostic, are of Neandertal affinities (Leroi-Gourhan 1958; Hublin et al. 1996; Bailey and Hublin 2006), could well have been derived from the underlying Mousterian, or bear witness to the fact that the excavation techniques used at the site mixed the original stratigraphy. This notion, however, is as inconsistent with the overall horizontal and vertical find distribution patterns as the suggestion that the ornaments from the Châtelperronian levels are intrusive from the overlying Aurignacian. In fact, the inventory of the loose teeth (which, given their small size, would be the most affected by post-depositional displacement processes) shows that they belong to a number of different individuals, both adult and juvenile, distributed from the top (level VIII) to the bottom (sublevel Xc) of the Châtelperronian sequence, with a major concentration in sublevel Xb. This pattern does not change if we exclude from the counts those teeth (totaling 15) that were recovered in rows 3–7 of the grid, where the marked inclination of the levels could have originated stratigraphic complications indeed; the distribution of the remaining 14 (1 in VIII, 1 in IX, 11 in Xb, and 1 in Xc) is identical to that of the entire sample of 29. Moreover, the two teeth (a permanent lower incisor and a permanent upper premolar) from levels VIII and IX, separated from Mousterian level XI by the >50 cm thick level X, are clearly, metrically and morphologically, in the Neandertal, not the modern human range.

Thus, the vertical distribution pattern of the Neandertal teeth across the Grotte du Renne sequence is identical to that of the ornaments and bone tools, with the concentration in sublevel Xb matching that of the lithic materials upon which the industrial diagnosis of that level as Châtelperronian is based (Connet 2002) (cf. Fig. 25.2). Consequently, as also pointed out by Hublin and Bailey (2006), much as the fact that 25 out of the 40 ornaments come from level X and only 4

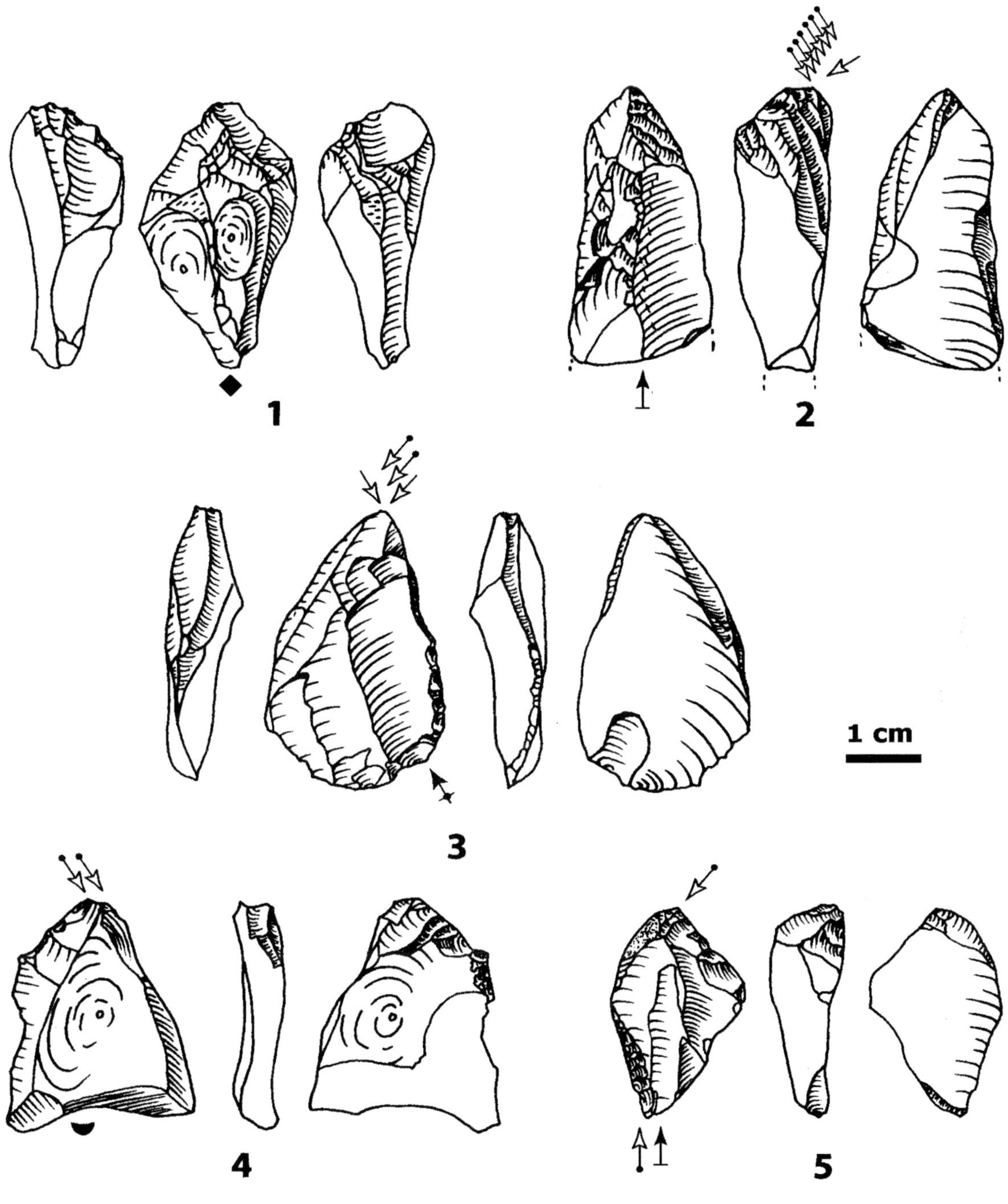

Fig. 25.11 Burins (*1–4*. carinated; *5*. on truncation) from Keilberg-Kirche (After Uthmeier 2004)

from Aurignacian level VII makes it impossible for the former to be intrusive from the latter, one can hardly conceive how the 29 human teeth in the Châtelperronian levels could all come from the Mousterian, which, over a sequence of similar thickness, yielded no more than five specimens (only one of which was from level XI, that immediately underlying the base of the Châtelperronian sequence; Hublin, personal communication). What kind of natural agency would move 83% of the Neandertal teeth putatively contained in the Mousterian levels up into the overlying Châtelperronian, leaving only 17% behind, without a similar displacement being equally conspicuous in other realms of material culture and without, conversely, large proportions of Châtelperronian objects, namely lithics, having been displaced downwards into the Mousterian?

The horizontal distribution patterns for level X are consistent with these stratigraphic observations. In fact, groups of teeth thought to come from the same individual (on the basis of both age-at-death and morphology) are tightly spatially clustered, not scattered, as one would expect if the original

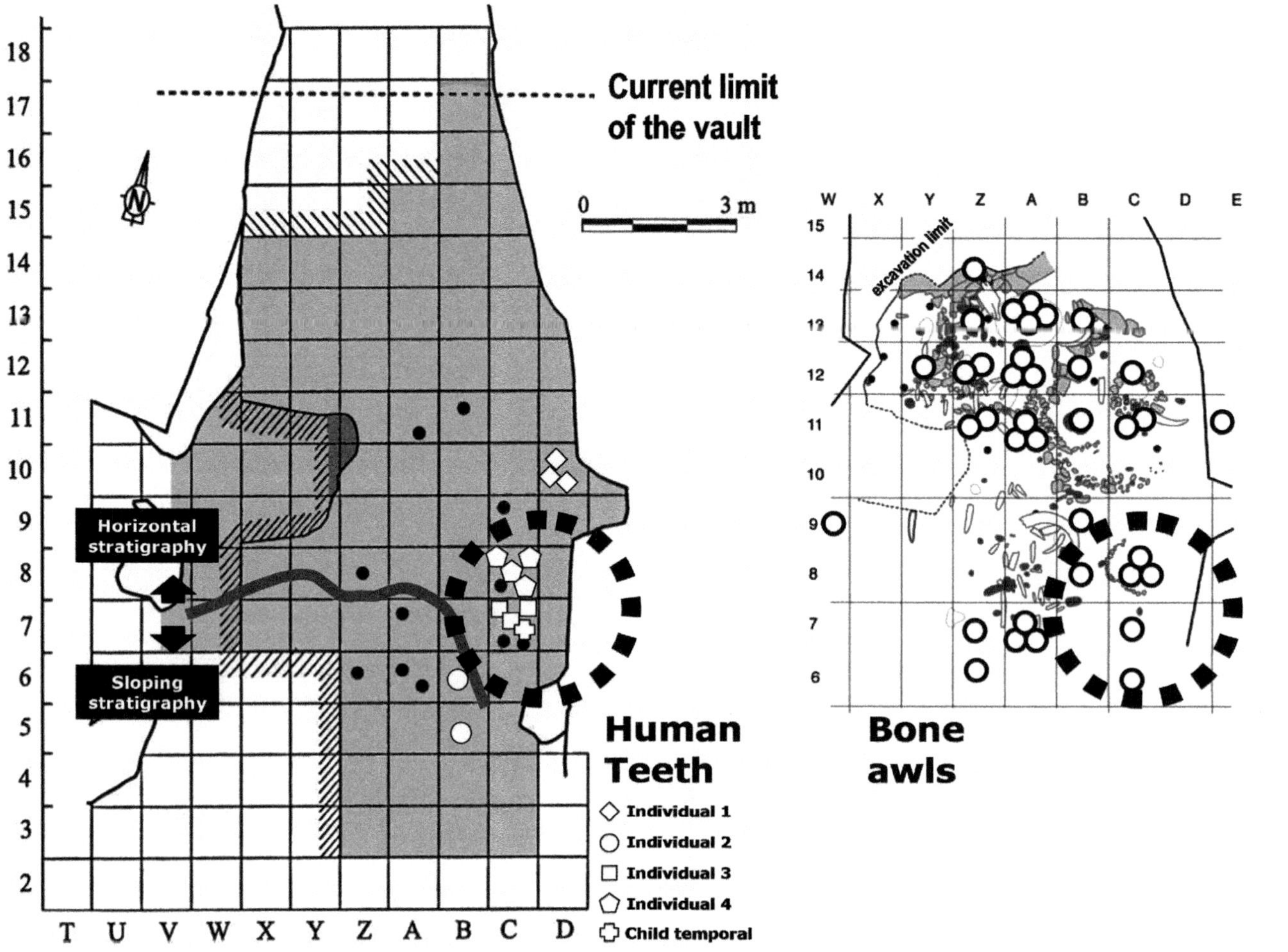

Fig. 25.12 Horizontal distribution of the human teeth and the bone awls in level X of the Grotte du Renne. Note how the material from a single individual clusters together (the child temporal may belong to individual 3) and how most finds come from an area of horizontal stratification that coincides with a major concentration of typical Châtelperronian bone awls (After d'Errico et al. 2003; Bailey and Hublin 2006)

position of the finds had been significantly disturbed (Fig. 25.12). The three teeth from individual 1 were found in square D10, sublevel Xb1; the three from individual 2 in adjacent squares B5 and B6, sublevel Xb2; the three from individual 3 in square C7, sublevel Xb2 (as was the temporal of an individual of the same age, 1 year old); and the four from individual 4 in square C8, sublevel Xb2. Only two of these twelve teeth come from the talus area with an inclined stratigraphy, the other ten coming from an area of level X where a significant concentration of typical Châtelperronian bone awls was also recovered (d'Errico et al. 2003).

The unquestionable association of the Grotte du Renne's Châtelperronian with diagnostic Neandertal remains also sheds significant light on the fact that the morphology of the two deciduous teeth found in level E of Grotta del Cavallo (Italy), at the base of the site's Uluzzian sequence, is indicative of Neandertal affinities. They are similar to Neandertal teeth in size, cusp morphology and taurodontism, the latter, in particular, being often present in Neandertal deciduous molars but having never been observed in early modern human juveniles (Churchill and Smith 2000). No one questions that the Uluzzian is a local industrial development rooted in the preceding Mousterian (Palma di Cesnola 1993; Mussi 2001), and that the roots of the Châtelperronian are to be found in the preceding MTA is widely accepted since François Bordes' review of the Transition in France. Bordes reconciled the link between the Châtelperronian and the MTA with his perception of the Upper Paleolithic as related solely to *Homo sapiens* people by postulating that the latter had made the MTA too. Today, given the fossil associations discussed above, the cultural continuity of the Châtelperronian and the Uluzzian with the preceding Mousterian (a continuity which, where the French sequence is concerned, was unanimously recognized until the Saint-Césaire discovery; e.g. Mellars 1973) must be extended to the biological realm, and carries the implication that the Transition in western Europe, having been an indigenous process, must have resulted from developments intrinsic to the Neandertal world.

Who Made the "Transitional" Industries of Eastern and Central Europe?

The resurrection of the notion that the Châtelperronian would have been made by moderns, not Neandertals, is a logical extension of the proposition that the true technological and typological roots of the pre-Aurignacian early Upper Paleolithic entities of eastern and central Europe lie not in the preceding regional Middle Paleolithic traditions of stone tool making but in the IUP of the Near East, with the Bohunician of Moravia and southern Poland (as well as the geographically intermediate Bachokirian) being the most clear case in point (cf. Svoboda and Bar-Yosef 2003, and the different papers therein). This hypothesis rests on three arguments: (1) that no Levallois blade production is observed in Moravia prior to the emergence of the Bohunician (2) that the latter's reduction strategy is identical to that documented in basal levels 1 and 2 of the Israeli site of Boker Tachtit, and (3) that the Bohunician must correspond, therefore, to an intrusion into the local sequence of a technology diffusing from the Levant, whence modern humans are supposed to have spread into Europe.

The first problem with these arguments lies in the implied assumption that the technological transition observed at Boker Tachtit is a unique event and, therefore, that, if an assemblage is found to fit into one of the stages of that transition, then its authorship must be credited to the carriers of such a cultural "mutation". Secondly, these arguments fail to consider that the diffusion of technologies can occur without migration being involved, and that the apparently intrusive nature of the Bohunician may simply be an artifact of the gap of at least ten millennia that currently exists in the Moravian sequence between the latest Micoquian of Kůlna cave and the earliest Bohunician. In fact, the sites of Piekary IIa and Księcia Józefa, in southern Poland, document the local development of volumetric, Upper Paleolithic methods of blade debitage out of Levallois flake-based technologies between ca. 50 and ca. 40 ka BP (Sitlivy et al. 1999a, b, 2004; Valladas et al. 2003), i.e., during the time period of the Moravian gap (Fig. 25.13). Parsimony dictates that there is no need to look into the Middle East for the source of the Bohunician when a better alternative, chronologically and geographically closer, is available.

Given the lack of human remains, the authorship of the Bohunician must remain an open issue, but the evidence for cultural continuity with the regional Middle Paleolithic is strongly in favor of the consensus view that the earliest Upper Paleolithic of Moravia relates to Neandertals, not moderns. The same reasoning pertains where the Bachokirian of Bulgaria is concerned. Contra Glén and Kaczanowski (1982), who suggested that the human teeth from Bacho Kiro had Neandertal affinities, Churchill and Smith (2000) subsequently concluded that aspects of size, shape and crown morphology aligned this material more with modern humans than with the Neandertals; this conclusion, however, applied only to the material found in the Aurignacian levels of the site, the single human remain from the Bachokirian ones being a taxonomically undiagnostic left mandibular fragment with deciduous first molar.

Further east, based on the chronostratigraphic evidence from Kostenki, Anikovich et al. (2007) argue for a pre-Aurignacian modern human spread into the Russian and Ukrainian plains. However, no evidence of anatomical modernity exists for the makers of the assemblages in question: those recovered in the lowermost cultural levels of loci 14 and 17 of the site, below CI tephras and in association with dates in the range of 36.5–35 ka ^{14}C BP (42–40.7 ka cal BP). The only human remains in these levels are two isolated, worn down, non-diagnostic teeth, so the argument is entirely supported by the fallacious premise that Upper Paleolithic archaeology, in any form, is an exclusive of modern humans. It is further suggested that these cultural levels stand for "an intrusion of modern humans onto the central East European Plain several 1,000 years before their spread across western and eastern Europe". Even under the assumption that modern humans made the earliest Upper Paleolithic of Kostenki, this claim is unsupported: the radiocarbon dates show contemporaneity, not precedence, with the earliest Aurignacian of the Danube basin, the Italian peninsula, and the Franco-Cantabrian region, and, calibrated, completely overlap with the OSL dates for the same levels. On the other hand, the assemblages are generalized, prismatic blade-based Upper Paleolithic tool-kits lacking any of the diagnostic tool-types of the Aurignacian and may well relate, in fact, to a northeastward expansion of the indigenous Polish developments seen at Piekary IIa and Księcia Józefa, i.e., to Neandertals, not moderns.

The dates for the pre-tephra levels of Kostenki also overlap with the time range of the Altmühlian and the Szeletian, as well as with that of the "Eastern Szeletian" of Buran-Kaya III, level C, in the Crimea; the latter, significantly, displays close cultural affinities with other Kostenki assemblages from below the tephra, such as the Streletskayan of locus 12, which is characterized by the production of fine, bifacially flaked foliate points (Monigal 2006) As discussed above, these assemblage-types are likely to be related to Neandertal, not modern human activity. Thus, the real situation at Kostenki seems to be that of a first Upper Paleolithic occupation of the area coeval with, but culturally different from, the earliest Aurignacian of European regions to the west, followed by several millennia of abandonment and a later Aurignacian resettlement ca. 33–32 ka ^{14}C BP (38.2–36.9 ka cal BP). The biological affinities of such an earliest Upper Paleolithic phase of occupation remain unknown, could conceivably lie with the Neandertals, and it is in any case very unlikely that moderns, if at all involved, were responsible for these cultural manifestations to the complete exclusion of Neandertals.

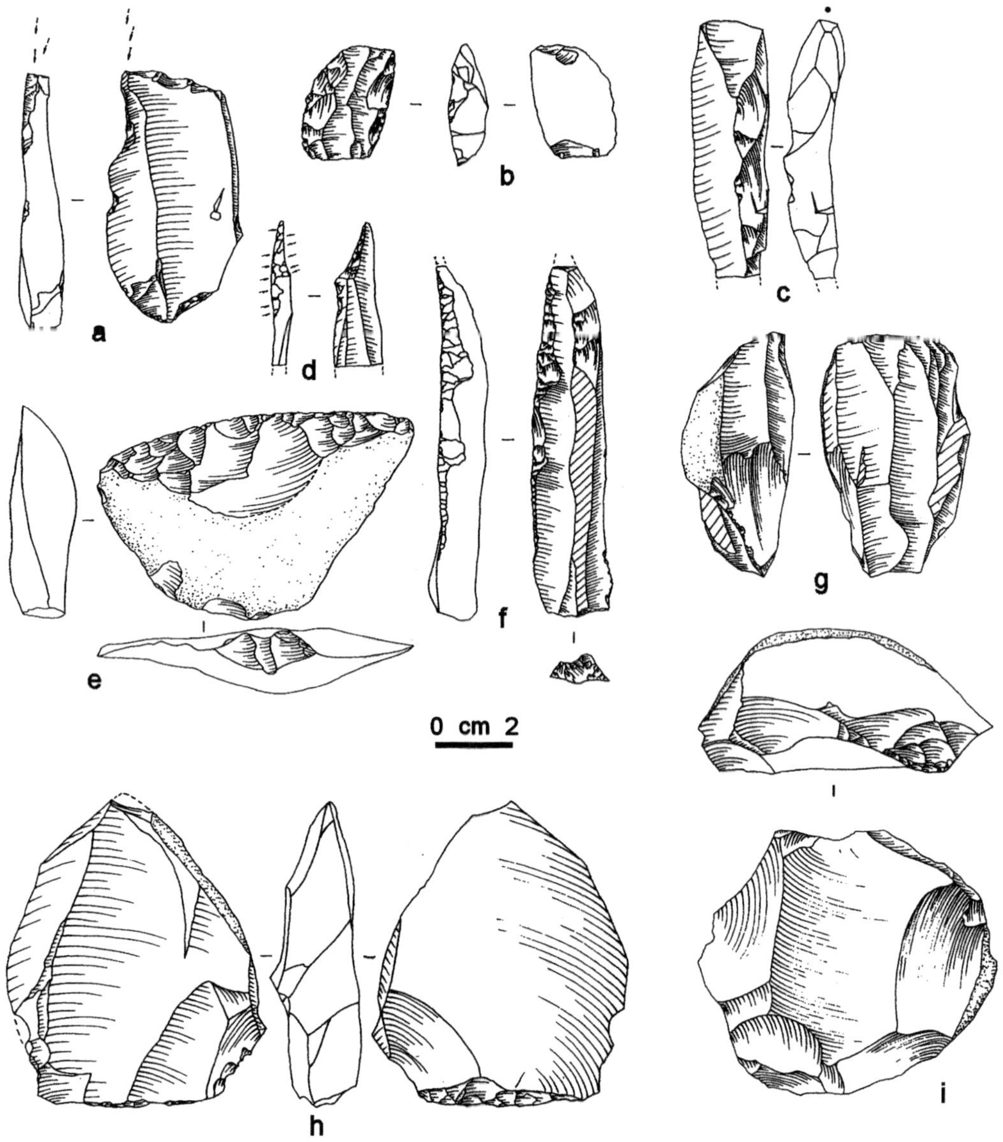

Fig. 25.13 Lithics from the "transitional" assemblage in level 7b of the open air, workshop site of Piekary IIa (Poland), TL-dated to ca. 53 ka cal BP: (**a**). burin; (**b**). truncated-facetted scraper; (**c**). crested blade; (**d**, **f**). backed blades; (**e**). transversal sidescraper; (**g**). bidirectional, prismatic blade core; (**h**). Levallois blank; (**i**). Levallois core; (After Valladas et al. 2003)

Paradigm Found

The Protoaurignacian

The interest promoted by the Human Revolution paradigm on the emergence of the Aurignacian led not only to major advances in the investigation of its radiometric chronology and putative interstratification with the Châtelperronian but also to a string of lithic technology studies that greatly improved our knowledge of the techno complex *per se* (Tixier and Reduron 1991; Le Brun-Ricalens 1993; Chiotti 1999; Bon 2000, 2002; Bon and Bodu 2002; Bordes 2000, 2002, 2006; Lucas 2000; Teyssandier 2000, 2003). These studies eventually confirmed traditional typology-based views (cf. de Sonneville-Bordes 1960) of the systematics of the Typical Aurignacian (i.e., with exclusion of the much later "Aurignacian V") (Fig. 25.14). As a result, there is now a wide consensus that the so-called Protoaurignacian, originally defined by G. Laplace and Italian authors (Palma di Cesnola 1993), for which the term "Fumanian" is some times also used (e.g., Bar-Yosef 2006; Mellars 2006) and generally considered to correspond to a cultural/geographic, Mediterranean "facies" of the classical Aurignacian, is instead a chronological "phase". Results from the re-excavation of the key cave site of Isturitz (Normand and Turq 2005),

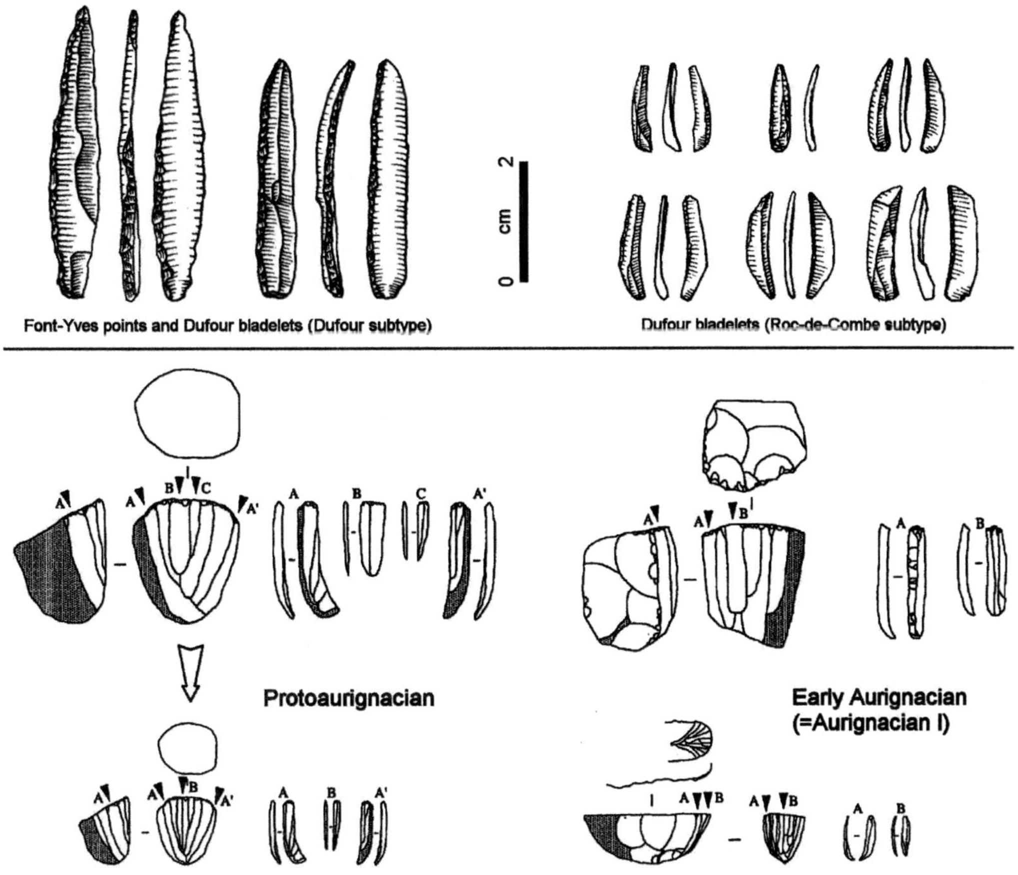

PHASE	BLADE/ BLADELET PRODUCTION	BLADELET CORE TYPE	DUFOUR BLADELET SUBTYPES	BONE TOOL TYPES
Late (Aurignacian III/IV)	**Dissociated**	**Burin**	• **Middle-sized, straight, elongated, Dufour subtype**	**Non-diagnostic**
Evolved (Aurignacian II)		**Nosed**	• **Short, twisted, Roc-de-Combe subtype**	**Lozengic**
Early (Aurignacian I)		**Carinated**	• **Middle-sized, curved, Dufour subtype** • **Mostly, blanks unretouched**	**Split-based**
Protoaurignacian	**Sequential**	**Prismatic**	• **Long, straight, Dufour subtype** • **Font-Yves points**	**Non-diagnostic**

Fig. 25.14 *Top*: comparison between the lithic technology and the typical bladelet tools of the Protoaurignacian and the Aurignacian I (Early Aurignacian) (After Bon 2000, 2002; Demars and Laurent 1989, modified) *Bottom*: basic technological features of the Aurignacian succession

in good accord with the revised stratigraphy of Le Piage (Bordes 2002, 2006), show that, in France as well as in Italy and Spain, this Protoaurignacian stratigraphically precedes the classical Early Aurignacian or Aurignacian I with split-based bone points.

The defining feature of the Protoaurignacian is the production of large amounts of bladelets extracted from unidirectional prismatic cores in the framework of a single, continuous reduction sequence (first for blade blanks, then for bladelets), which are trimmed in a very high proportion into either Font-Yves points or the longer, slender Dufour bladelets of the Dufour subtype (Demars and Laurent 1989). The Early Aurignacian, in turn, features a lithic production system characterized by (a) the production of large blades (extracted from single platform, crest-shaped, prismatic cores, and used as blanks for endscrapers and knives) and (b) the re-use of broken or exhausted tools (set up as thick, carinated "scrapers") and of the debris from prismatic core preparation and renewal (thick, often cortical flakes and blades) for the extraction of straight or curved bladelet blanks, which, although rarely retouched, also bear, when such is the case, a marginal, inverse, or alternate retouch of the so-called Dufour type. This Early Aurignacian is followed by an Evolved Aurignacian or Aurignacian II where thick "burins" (carinated or busked) and thick-nosed "scrapers" tend to replace carinated "scrapers" as the preferred type of bladelet core, generating characteristic small, twisted blanks, retouched into the particular Roc-de-Combe subtype of Dufour bladelets (Demars and Laurent 1989). Other types of points made of ivory, bone or deer antler emerge in this later facies, all with massive bases, mostly featuring flat or oval cross-sections and a broadly lozengic outline – the Mladeč (Lautsch) points.

The industrial affinities of all those Aurignacian occurrences which, after adequate taphonomic critique, were found to date to ca. 36.5–35 ka ^{14}C BP (42–40.7 ka cal BP), lie with this initial, Protoaurignacian phase, whereas all reliable dates for Aurignacian I and Aurignacian II assemblages post-date ca. 35 ka ^{14}C BP (Zilhão and d'Errico 2003a, 2003b; Teyssandier et al. 2006). Conversely, as discussed above, all recently obtained dates for the Châtelperronian (Roc-de-Combe, Grotte des Fées at Châtelperron) place its emergence well before 36.5 ka ^{14}C BP (42 ka cal BP), with other results (for Dubalen, Caune de Belvis and La Quina Aval, all from non-ultrafiltrated bone samples and, hence, possibly minimum ages only) overlapping this chronological horizon (cf. Fig. 25.9). Where the Uluzzian is concerned, its pre-CI age in the Italian sites and the dating evidence from Klisoura 1 confirm contemporaneity with the Châtelperronian. Thus, the Protoaurignacian occupies a chronometrically intermediate position consistent with the stratigraphic evidence: in western Europe, and considering only occurrences where the industrial diagnosis is fully unambiguous, Châtelperronian or Uluzzian occupations underlie Protoaurignacian ones at the sites of Labeko Koba, Le Piage and Castelcivita; and Protoaurignacian levels are in turn securely documented below levels with Aurignacian I split-based bone points at the sites of Morin, Labeko Koba, L'Arbreda, Isturitz, Le Piage, Mocchi and Fumane.

No sites in central and eastern European provide continuous sedimentary records spanning the entire MP-UP transition and, in particular, the two known Protoaurignacian occurrences – Tincova (western Romania) and Krems-Hundsteig (Lower Austria) – are open air loess sites with no underlying Middle or earlier Upper Paleolithic levels. The conventional date of ca. 35 ka ^{14}C BP (40.7 ka cal BP) for Krems-Hundsteig (Hahn 1977), however, is consistent with the notion that the Protoaurignacian corresponds to a pan-European cultural and chronological horizon, and preliminary reports on the assemblage from level VII of Kozarnika cave seem to establish the presence of the technocomplex in Bulgaria as well (Tsanova 2006). When reliable dates are available, all known occurrences of the Bohunician and the Szeletian (for instance at the open air Moravian sites of Bohunice, Stránská skála and Vedrovice V – Svoboda et al. 1996; Svoboda and Bar-Yosef 2003) are earlier. Conversely, dates from samples in reliable association with Aurignacian I or II assemblages, or obtained via the direct dating of either split-based or Mladeč bone points, are later (e.g., Adams and Ringer 2004; Svoboda and Bar-Yosef 2003; Rabeder and Pohar 2004), and identical to those obtained in Germany, Belgium, France and Spain. This evidence indicates that the industrial succession "transitional"-Protoaurignacian-Aurignacian I-Aurignacian II is valid across the entire continent, from Romania, Bulgaria and Greece, in the East, to the Franco-Cantabrian region, in the West.

Further to the north, in southern England, Belgium, central-northern Germany and Poland (as well as in the Russian-Ukrainian plain, as exemplified by the situation at Kostenki), it remains conceivable, however, that Neandertal-associated cultures survived at least into the time range of the Protoaurignacian and, possibly, of the Aurignacian I, both of which remain unknown in those areas. If the rejuvenation that almost certainly characterizes the relatively recent dates for the Jerzmanovician levels of the Nietoperzowa and the Ilsenhöhle is of no more than a few millennia, these two sites would provide further support for the possibility. That Neandertals survived in at least one part of Europe throughout the time interval covered by the Protoaurignacian and the Aurignacian I is any case now well established for Iberian regions south of the Ebro basin (Zilhão 1993; Walker et al. 2004a, b; cf. Zilhão 2006c for a recent review), and, as previously hypothesized (Zilhão 2001: 64–65, 2007: 15–18, 24), the Lincombian and the Jerzmanovician may well represent a similar phenomenon beyond the northern range of the Protoaurignacian.

The Nassarius *People*

The above review makes it clear that the Protoaurignacian falls into a well-defined temporal horizon and, wherever it is found, corresponds to a major chronostratigraphic discontinuity. Significantly, this break coincides with a similar phenomenon in the fossil record. Throughout the entire geographic range of the Protoaurignacian and the Aurignacian I, no diagnostic Neandertal remains have been directly dated to that temporal horizon; the ca. 29–28 ka ^{14}C BP (34.2 ka cal BP) results for two specimens from level G1 of Vindija were vastly underestimated, as shown by subsequent redating of the samples, using the ultrafiltration procedure, to >32.4 ka ^{14}C BP (>37.8 ka cal BP) (Higham et al. 2006; Zilhão 2009a). Conversely, it is at this time that evidence for the presence of anatomically modern people first appears in this part of the world: the human remains from the Peştera cu Oase, in Romania, one of which, the Oase 1 mandible, has been directly dated to ca. 35 ka ^{14}C BP (40.7 ka cal BP) (Trinkaus et al. 2003; Rougier et al. 2007; Zilhão et al. 2007; Trinkaus 2011).

In this context, it makes sense to construe the Protoaurignacian "intrusion" as a cultural manifestation of the dispersal of modern humans into Europe, a point upon which substantial agreement now seems to have been reached between authors of different persuasions (e.g., Mellars 2006; Zilhão 2006a, b, 2007). Two additional lines of reasoning support this conclusion. First, the fact that in both basic lithic technology (continuous reduction of single platform prismatic cores for the sequential extraction of slender blade and bladelet blanks – Fig. 25.15) and index fossils (the Font-Yves or El-Wad point, and the Dufour bladelets of the Dufour subtype), the Protoaurignacian is virtually indistinguishable from the Early Ahmarian of the Levant (Belfer-Cohen and Goring-Morris 2003; Zilhão 2006a, b, 2007), with which it is broadly coeval (cf. Fig. 25.9); and, whereas the Protoaurignacian is in clear technological discontinuity with the preceding "transitional" industries of Europe, the Early Ahmarian of the Levant seems to be in clear continuity with the preceding IUP/Emiran (Jones et al. 1983; Monigal 2003). The only evidence concerning the makers of this Levantine phylum is "Egbert," a juvenile modern human

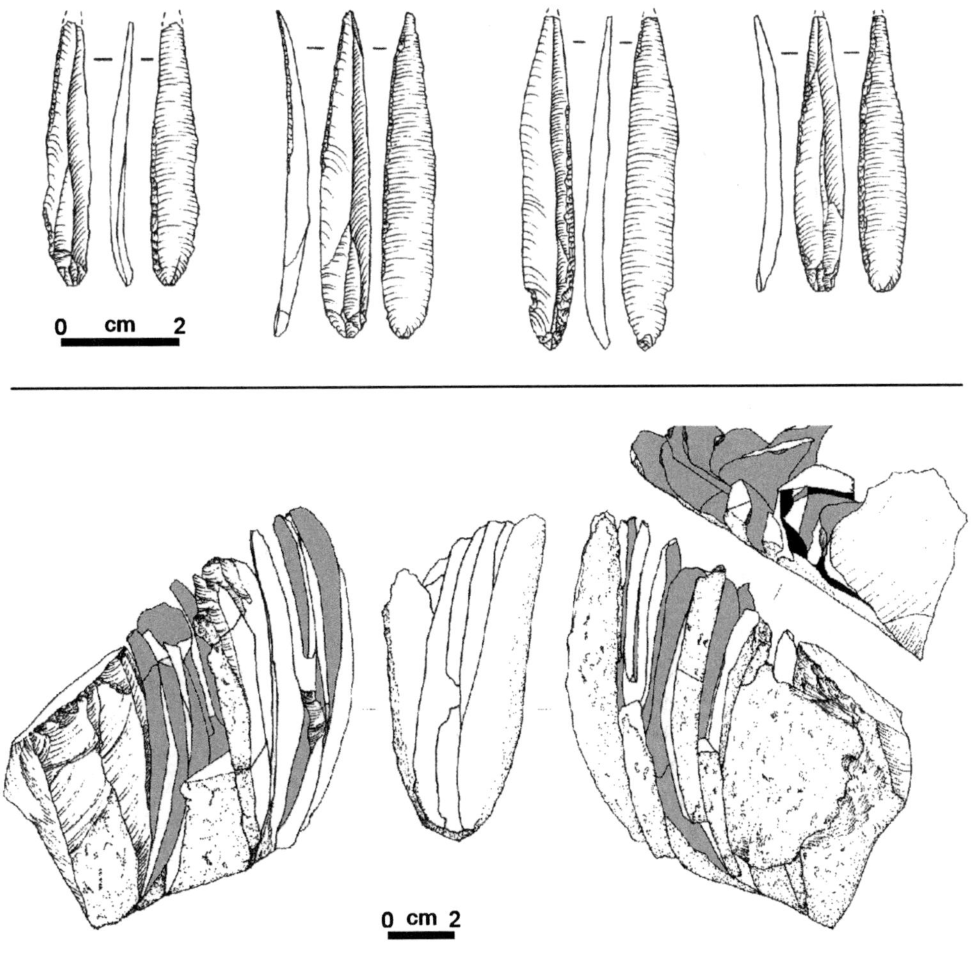

Fig. 25.15 Stone tools from the early Ahmarian (Boker A): *top*, el-Wad points; *bottom*, refitted core (ventral surfaces shaded *grey*) (After Jones et al. 1983; Monigal 2003, modified)

skeleton from the Early Ahmarian of Ksar 'Akil (Bergman and Stringer 1989); although only a cast of the skull survives, making direct dating impossible, the fact that the skeleton was found 1 m below the top of the Early Ahmarian sequence makes it unlikely, as pointed out by Mellars (2004), that this was an intrusive burial.

Since the diffusion of cultural innovations can occur without migration, it remains conceivable, however, that, in the westernmost parts of the Protoaurignacian range, farthest from the presumed source, the carriers of this technology were at some time in the process still biologically "pure" Neandertals. Bearing in mind such a possibility, linking the Protoaurignacian with modern human dispersal remains appropriate because such putative associations would still signal the penetration of moderns, if not into the immediate territories of those Neandertals, at least into territories that, despite the distance, were included in their web of exchange connections. Moreover, in most kinds of archaeological sites, occupation levels do not have the resolution to discriminate "Protoaurignacian Neandertals" from "palimpsest including skeletal remains of the last, immediately-preceding-the-Protoaurignacian Neandertals as well as cultural remains of the first, immediately-following-the-Neandertals Protoaurignacian moderns", and the resolution of chronometric methods in this time range is even poorer. To which must be added the fact that, if biological admixture is accepted as part of the equation (see below), diagnosing fragmentary human remains from the time of contact as either "modern" or "Neandertal" may not be possible (Bailey and Weaver 2006). So, the hypothesis that Protoaurignacian "memes" arrived in parts of Europe ahead of the actual dispersal of modern human "genes" may never be susceptible to empirical testing; unless and until the point is made beyond reasonable doubt by some sort of "smoking gun" (which would have to be a closed find, for instance, a Neandertal burial with characteristic Protoaurignacian offerings), the hypothesis of "exclusively cultural" diffusion, although conceivable in theory, is indistinguishable, for all practical purposes, from that of "cultural and demic" diffusion.

A second line of evidence supporting a link between the Protoaurignacian and a dispersal of modern humans into Europe concerns ornament types (Fig. 25.16; for a detailed review, cf. Zilhão 2007). Small, marine shell beads of diverse taxonomy (*Cyclope*, *Cyprea*, *Hinia*, *Homalopoma*, *Littorina*, *Nassarius (=Arcularia)*, *Trivia*, etc.) but similar basket-shaped morphology overwhelmingly dominate Protoaurignacian ornament assemblages (Taborin 1993; Kuhn and Stiner 1998; Broglio et al. 2002; Vanhaeren 2002; Broglio and Gurioli 2004); the same applies, although with a more restricted range of species, to the IUP and the Early Ahmarian of the Levant, where almost all beads are made of *Nassarius (=Arcularia) gibbosula* or, in low percentages, *Columbella rustica* (Kuhn et al. 2001). Given that ornament traditions are known to be long-lasting, much more so than stone tool making ones (Stiner 1999; Vanhaeren and d'Errico 2006), this preponderance of *Nassarius* and *Nassarius*-like

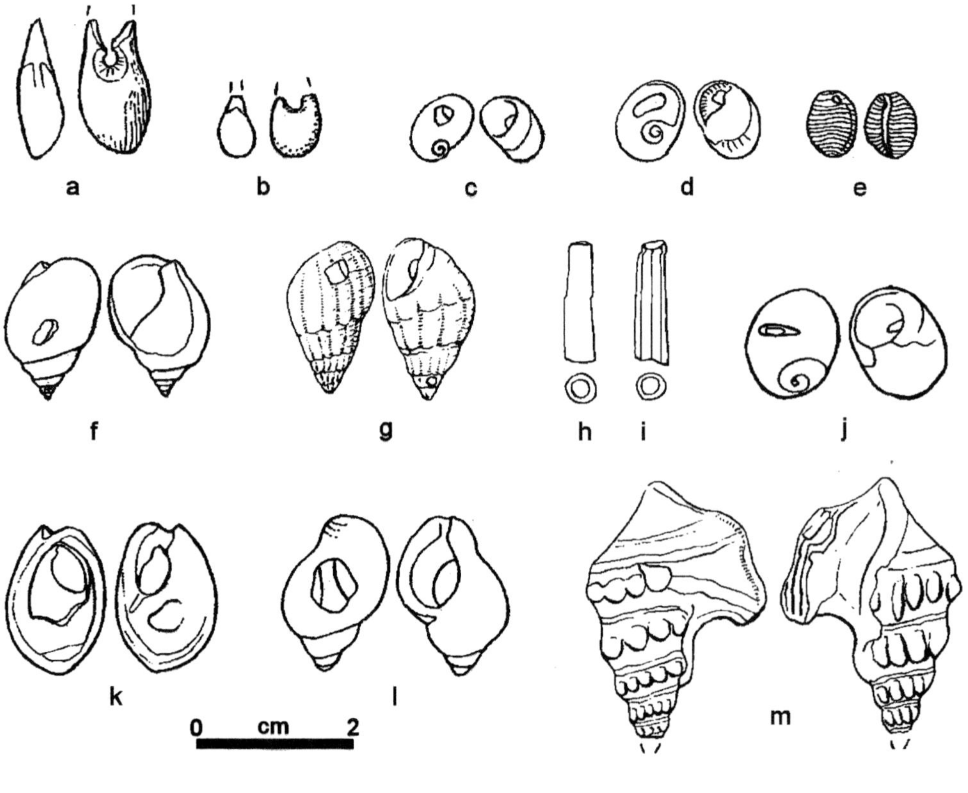

Fig. 25.16 Protoaurignacian beads from the Rothschild rockshelter (France): (**a**). pierced red deer canine; (**b**). steatite bead; (**c**). *Theodoxus fluviatilis*; (**d**). *Cyclope neritea*; (**e**). *Trivia europaea*; (**f**). *Sphaeronassa mutabilis*; (**g**). *Hinia reticulata*; (**h–i**). *Dentalium*; (**j**). *Littorina obtusata*; (**k**). *Nassarius (=Arcularia) gibbosula*; (**l**). *Nucella lapillus*; (**m**). *Aporrhaïs pespelecani* (After Barge 1983)

shells in the IUP, the Early Ahmarian and the Protoaurignacian cannot but evoke a cultural link between these populations and the Middle Stone Age of Africa, where the same genus was exclusively used for the manufacture of the oldest ornaments currently known: those from the Still Bay level of Blombos cave (Henshilwood et al. 2004; d'Errico et al. 2005), well dated to ca. 75 ka BP; those from the Mousterian of Skhul (Israel) and the Aterian of Oued Djebbana (Algeria), claimed to be of an even earlier, OIS-5 age (Vanhaeren et al. 2006); and those from the Middle Paleolithic (possibly Aterian) levels of the Grotte des Pigeons (Taforalt, Morocco), securely dated to ca. 82 ka BP (Bouzouggar et al. 2007). Anatomically, at least some of these people belonged to the early modern human stock that spread into Eurasia after 50 ka BP and, therefore, culturally, it seems justified to designate them collectively, despite and beyond the differences in lithic archaeology, as the "*Nassarius* people".

The similarity in lithic technology, lithic typology and items of personal ornamentation is consistent with the scant chronometric evidence available, supports construing the Protoaurignacian and the Early Ahmarian as two sides of the same cultural coin, and provides the keystone for a Mediterranean-wide chronostratigraphic correlation scheme and derived culture geography of the Transition (Figs. 25.17 and 25.18). This evidence leaves little doubt that the IUP of the Levant and the Châtelperronian, Uluzzian and equivalent technocomplexes of Europe are broadly contemporary. In this context, it is striking that the ultimately African tradition of basket-shaped, marine shell beads of the Levant could hardly be more distinct from that seen in the coeval transitional industries of Europe, characterized by pendants made of animal bones and teeth (completely unknown anywhere in Africa or the Levant before the time of the Aurignacian I of Europe), and featuring elongated mollusk fossils and tubes of *Dentalium* as the only instances of the use of shell blanks. Since such a sharp contrast (Fig. 25.19) cannot be attributed to raw-material availability, it must reflect instead the independent emergence, and possibly the functional distinction as well (d'Errico and Vanhaeren 2007), of the European tradition; and, obviously, it also represents an unsurpassable obstacle to "imitation" models of the emergence of ornaments among Neandertals (how can you imitate something that does not exist in the first place?). Why, where and when did Neandertals get the idea is open to discussion, but it most certainly could not have come from the *Nassarius* folk.

Contact and Admixture

A few ornaments typical of the European tradition (namely, pierced red deer canines) have also been found in Protoaurignacian contexts (for instance at Fumane, Mocchi or Isturitz), and are especially abundant in the Aurignacian I (where, as in the Châtelperronian from the Grotte du Renne, fox canines feature prominently). The parsimonious explanation for these elements of discontinuity between Europe's early modern human cultures and the African/Levantine tradition of personal ornamentation is that they were acquired from the indigenous, Neandertal societies where such novel elements had originally emerged.

Such a process of cultural blending is consistent with, and lends further support to the notion that significant admixture between in-dispersing modern humans and local anatomically archaic populations occurred all over Eurasia at the time of contact. In the fossil record, the case for such a model has been summarized by Trinkaus (2005c, 2006, 2011), and rests on the recognition that all known Aurignacian (as well as some early Gravettian) European modern humans feature a diverse mix of cranial and post-cranial features widely recognized as either diagnostic of the Neandertals or as generically primitive traits (in the cladistic sense) that had already been lost in the lineage of the *Nassarius* people prior to their Out-of-Africa dispersal.

Harvati et al. (2007) claim to have disproved this interpretation on the basis of a geometric morphometric analysis of the Romanian skull from Cioclovina, directly dated to ca. 28.5 ka ^{14}C BP (ca. 33 kacal BP) (Soficaru et al. 2007). From the fact that their study failed to find any of the conditions to be expected in "hybrids (whether F1 or later generation) between well-defined fossil human taxa, and, therefore, also between Neandertals and early modern humans," they feel confident in rejecting "the hypothesis that Cioclovina represents a hybrid individual between early modern humans and Neandertals" (Harvati et al. 2007: 734, 740). However, as pointed out by Soficaru and Trinkaus (personal communication), Harvati et al.'s study is methodologically flawed, as issues of admixture can only be resolved by looking at discretely inherited elements, which sort and resort in a Mendelian manner over generations. Since geometric morphometrics does not detect the presence/absence of such discrete features as have been used to diagnose Neandertal ancestry (e.g., the lingular bridging of the left mandibular foramen in the Oase 1 mandible), Harvati et al. in fact achieve no more than re-stating something that no one questions: that the fossils from Cioclovina and broadly coeval early modern European localities (Mladeč, Oase, Muierii, etc.) are all basically modern humans.

More importantly, Harvati et al.'s argument entirely misses the point. Hybrid individuals between "well-defined" Neandertal and modern human populations could only have existed at the time of contact, which, in Romania, was some ten millennia before Cioclovina. The expectations against which Harvati et al. assess the issue are derived from empirical studies concerning initial generations of hybrids among extant Primates. They could conceivably represent a valid standard if Cioclovina dated to ca. 42 kacal BP, but it doesn't: the issue

	CHÂTELPERRONIAN	PROTOAURIGNACIAN	AURIGNACIAN I
Cueva Morin	"Level 10"	9-8	7-6
Labeko Koba	IX base	VII	VI-V
L'Arbreda	H/I interface	H lower	H upper
Isturitz		4d	4b1-b2
Le Piage	brecciated remnants below K	K	GI
Esquicho-Grapaou		SLC1b	BR1
Mocchi		G	F
Fumane		A3-A1	D6-D3
	ULUZZIAN		
Castelcivita	10	9-6	
	EMIRAN (IUP)	EARLY AHMARIAN	
Ksar 'Akil	XXV-XXI	XX-XIV (plus XIII-IX?)	VIII-VII
Kebara		IV-III	II-I

Fig. 25.17 Correlation of early Upper Paleolithic stratified sequences in Europe and the near East (After Zilhão 2007)

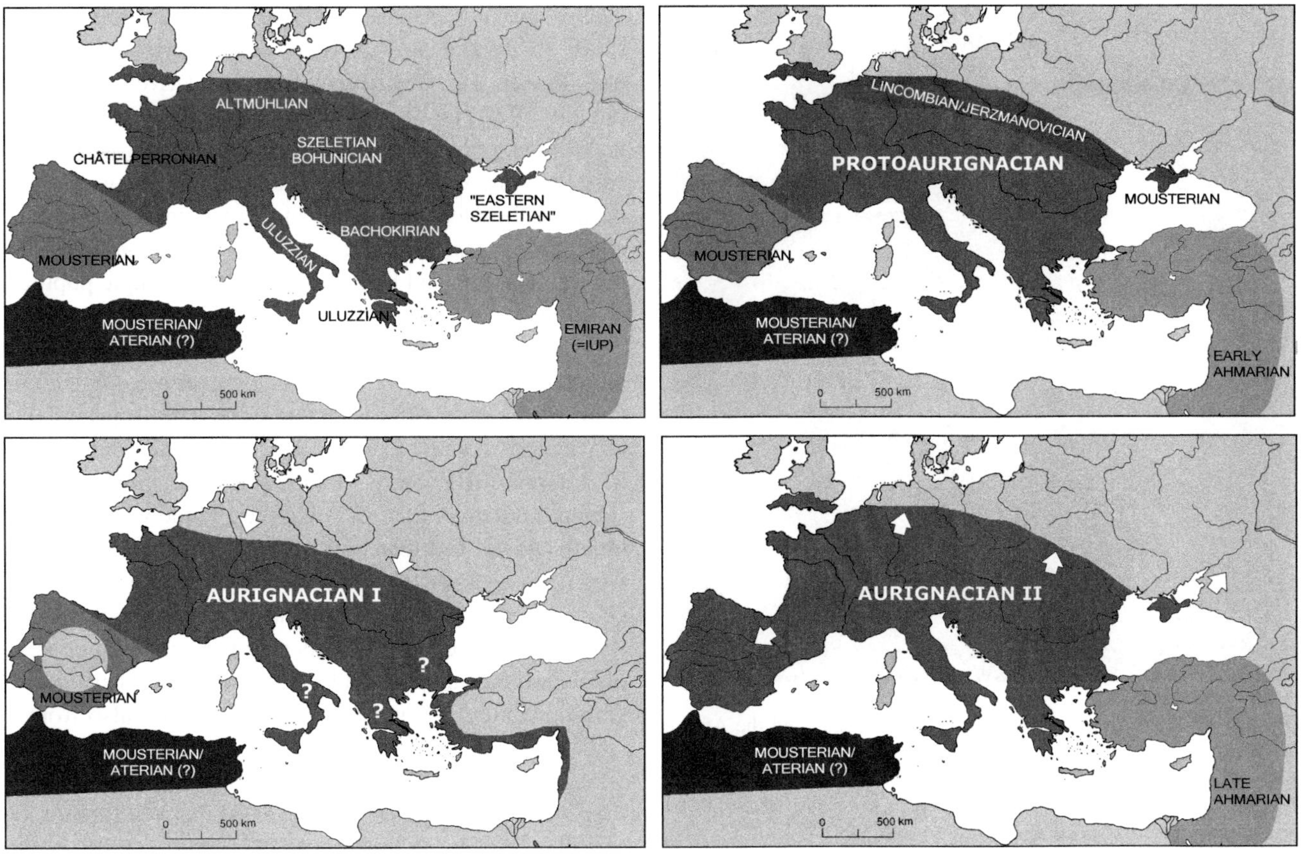

Fig. 25.18 Culture geography of the greater Mediterranean area during the transition. *Top*: before (*left*) and after (*right*) ca. 36.5 ka ^{14}C BP (42 ka cal BP). *Bottom left*: during the ca. 34.5–32.5 ka ^{14}C BP (37.8–35.1 ka cal BP) interval, at the time of Heinrich Event 4 and the Phlegraean fields caldera explosion. *Bottom right*: after 32.5 ^{14}C BP (35.1 ka cal BP) (After Zilhão 2001, 2006b, modified)

here is about the F500, not the F1, F2 or F3 generations! In Cioclovina's time, in fact, one of the two parent populations that had contributed to the ancestral admixture process (Romanian Neandertals) was long gone. And evolution, either by drift or selection, would have had sufficient time to erase the kinds of anomalies (greater or smaller size than either of the parent populations, supernumerary teeth, and extra cranial sutures) common in hybrid individuals and that Harvati et al. (wrongly) expected to be present in Cioclovina.

Thus, in claiming to refute a mixed ancestry for the population represented by the Cioclovina skull, Harvati et al. (2007) simply reproduce the error previously made by Tattersall and Schwartz (1999) in their comment to the announcement of the Lagar Velho child discovery (Duarte et al. 1999). The error consists in misinterpreting the evidence for admixture reported from Cioclovina and Lagar Velho as if such reports implied that these were hybrid individuals themselves. In fact, the real interpretation put forward by Duarte et al. (1999), Trinkaus (2005c, 2006, 2011) and Soficaru et al. (2007) is that both were "normal" representatives of "stable" populations, and that the Neandertal or archaic features in these and other coeval early modern Europeans could only be explained by gene flux between Neandertals and modern humans at the time of contact, many millennia before either Cioclovina or Lagar Velho were born: in short, the argument is about evolutionary, not immediate ancestry.

In this regard, the human paleontological evidence is fully consistent with the findings of Holliday's (2006) study of species interfertility versus time of divergence, which showed that the minimum amount of time for complete reproductive separation to occur among the many lineages of mammals for which fossil or molecular data are available is 1.4 million years, the estimated time of divergence between *Gazella thomsoni* and *G. rufifrons*. For hominids, where the interval between generations is at least four times longer, this result translates into five to six million years of divergence, ten times more than is estimated for the Neandertal/modern split from both the fossil and the molecular records (e.g., Green et al. 2006). As pointed out by Trinkaus (2011), the implication of these comparisons is that, regardless of personal or philosophical preferences concerning the appropriate taxonomic status for Neandertals, admixture (or the Assimilation Model; Smith 1984) should be considered the null hypothesis of modern human origins.

Fig. 25.19 Characteristic examples of the European and the African pre-Aurignacian traditions of personal ornamentation. *Left*: perforated wolf canine from the Châtelperronian levels of Quinçay, France (After Zilhão and d'Errico 1999b). *Right*: *Nassarius gibbosula* from the IUP of Üçağizli and *Nassarius kraussianus* from the MSA of Blombos (After Kuhn et al. 2001; d'Errico et al. 2005)

Recent developments in the genetics of both extant and fossil humans provide additional support for admixture. The phylogeographic analysis of the DNA of extant humans (including mtDNA and the Y chromosome, as well as non-recombining *loci* of the X chromosome and other parts of the genome – Templeton 2002, 2005) showed that the Out-of-Africa expansion of early modern humans could not have been a replacement event. If so, it would have wiped out the signatures of previous expansion events of the same kind, including that of *Homo erectus*, which is not the case, as such signatures are still extant in every gene region expected to contain information about events or processes substantially older than 150,000 years. And, using similar techniques, Evans et al. (2006) have also shown that, in one particular gene involved in the control of brain size during development (*microcephalin*), the adaptive allele, which occurs in 70% of today's humans, introgressed from an archaic lineage, most probably the Neandertals.

Despite unsolved problems relating to contamination issues (Wall and Kim 2007), preliminary results from the Neandertal genome project not only are consistent with this evidence, they also promise, if confirmed, to provide the definitive refutation that Neandertals and moderns were distinct at the biospecies level. Green et al. (2006) looked at sites where humans carry a single nucleotide polymorphism (SNP) to investigate how often the Neandertal had the ancestral (that is, the chimpanzee) or the derived (that is, the human) allele, and found out that the sequenced individual had the derived allele in 30% of the SNPs for which the comparison was possible; given the estimated time of divergence between the Neandertal and modern human lineages, they concluded that such a high percentage implies gene flux from modern humans and is incompatible with simple population split models.

This conclusion is of great significance because the sequenced Neandertal individual (specimen Vi-80/33.16) comes from level G3 of Vindija cave. The Uppsala laboratory directly dated it to 38,310 ± 2,130 ^{14}C BP (Ua-19009; ca. 42.8 ka cal BP) (Serre et al. 2004) but, given the large standard deviation, the fact that the sample was not ultrafiltrated, and the fact that similar small, non-ultrafiltrated samples from Neandertal remains recovered in overlying level G1 were shown to be rejuvenated (Higham et al. 2006; Zilhão 2008a; see above), this result should be treated simply as a minimum age and interpreted in light of its stratigraphic context. Neandertal specimen Vi-75, also from G3, yielded an infinite result of >42 ka ^{14}C BP (Ua-13873; >45.5 ka cal BP) (Krings et al. 2000), whereas the only date for cave bears from overlying level G1 that Wild et al. (2001) consider reliable is 46,800/+2,300/–1,800 ^{14}C BP (VERA-1428; ca. 50.3 ka cal BP). This dating and stratigraphic evidence leaves no doubt, therefore, that the Vi-80/33.16 Neandertal lived in central Europe many millennia prior to the time of contact; yet, the relevant sections of its nuclear genome show gene flux from modern humans! If these preliminary finds are confirmed, the implications will be inescapable: although geographical isolation led to morphological differentiation during the late Middle and early Late Pleistocene, gene flux between Europe and Africa was never completely interrupted throughout and, therefore, Neandertals and moderns must be construed as different populations of a single species, thus bringing structural ruin to the foundations of the Human Revolution edifice.

Towards a Paleoethnographic-Historical Approach

Since the notion of absolute biological barriers to interbreeding can now be left to rest in peace, no obstacles remain to look at contact and admixture at the time of the Transition with the proper approach, that of paleoethnography, as advocated by Zilhão and Trinkaus (2002) and Stringer (2006). As Stringer rightly puts it (p. 195), "the fundamental question is surely whether, when these groups encountered each other, they would have primarily seen each other as essentially 'same' or 'other' – potential friend, enemy, prey even?" Stringer's view is that "after an evolutionary separation much longer than any between living peoples, they would not only have looked very different in their bodies and faces, but probably also in their skin color, eyes, hair, body hair, gestures, body language and communication. To which could be added whatever clothing or body embellishments were in use as practicalities or social signals amongst the groups of the time. So even if the two species were reproductively compatible, the two populations may hardly ever have wanted to mate if they didn't like the look of each other. And if offspring did result, they may not have been favored as mates by the next generations of either parent group".

Such a scenario of mutual avoidance is conceivable, and it certainly cannot be excluded from the range of interaction types that might have pertained in the numerous and diverse encounter situations that must have occurred as groups of modern humans dispersed into Neandertal territory. The question, however, is whether it is reasonable to sustain that mutual avoidance was the general rule, and the answer must be no, for a number of reasons. First and foremost, there is the argument that "the proof of the pudding is in the eating". No matter what our expectations might be, the fact is that we do see in both the archaeological and the paleontological records sufficient evidence that certain aspects of early European modern human biology and culture reflect inheritance from the Neandertals; moreover, we see that such evidence is geographically widespread, from the Oase fossils in the East to the Lagar Velho child skeleton in the West, and from the *Dentalium* tubes in the Protoaurignacian of Mediterranean France to the pierced fox canines of the Aurignacian I of central and northern France. And, where the genetic record is concerned, although the *microcephalin* evidence is consistent with rare, even singular, episodes of interbreeding, the percentage of derived SNPs in the Vi-80/33.16 Neandertal is not.

Secondly, the aspects of biology to which Stringer refers (skin color, eyes, hair, body hair, etc.), which can be subsumed under the concept of "racial appearance", do not constitute significant obstacles to positive interaction between very distinct groups. As shown by psychological evidence (cf. Kurzban et al. 2001), and because of the correspondences that are established between allegiance and appearance, "race" functions in encounter situations much in the same way as dress, dialect or ethnic badges: a readily observable, rather arbitrary feature that "acquires social significance and cognitive efficacy when it validly cues patterns of alliance." In the experiment reported by Kurzban et al., subjects that had experienced a lifetime in which ethnicity was a good predictor of social alliances and coalitional affiliations were submitted to an alternative social world in which race was irrelevant to the prevailing system of alliance. The observed result was that of a dramatic decrease in the extent to which the "other" was categorized by "race" after less than four minutes of exposure to that new environment.

When modeling Neandertal-modern human encounters, "race" should thus be construed simply as yet another element of cultural difference. And, although time of separation from a common ancestor is of relevance when assessing the amount of "objective" biological separation because it carries implications to the outcome of interbreeding that are independent of either the perception or the volition of the potential participants in such encounters (such as whether offspring will be viable), that is not the case when assessing the impact of cultural contrasts on those participants' willingness to promote or avoid contact. The extent to which those contrasts happen to be large or small depends on the direction and pace of the historical trajectories that come face to face in such situations, not on for how long the participants' genetic lineages had evolved in isolation. For instance, there can be little doubt that, culturally, the contrast between people of the *Nassarius* (modern) and *Dentalium* (Neandertal) "races" at the time of contact in, say, Greece, must have been several orders of magnitude smaller than the contrast between the English colony of Sydney and the local Aboriginal people. Still, even in the framework of the most culturally and technologically contrasted encounter situations conceivable in human evolution (Australian hunter-gatherers versus European industrial age colonists), interaction ensued, at both the biological and the cultural level.

Therefore, the "rule of the pudding" dictates that the correct expectation in proper paleoethnographic terms for Neandertal-modern human encounters should be interaction, not mutual avoidance. This is all the more so since, in the sparsely populated world of the Transition, mutual avoidance and interaction must be framed in terms of **necessity**, not **preference**. As discussed elsewhere in more detail for the Iberian case (Zilhão and Trinkaus 2002), mutual avoidance in a scenario of equivalent cultural, economic and technological capabilities leads to a situation where, under exogamy, and given hunter-gatherer population densities, alliance and reproductive networks become stretched beyond the limits of the practical; in such a scenario, interaction is mandatory (Fig. 25.20). In order to work, therefore, mutual avoidance demands the postulate that in-dispersing modern humans held a significant adaptive advantage over local Neandertals. Conceivably, that advantage would have allowed them to pursue a strategy of initial occupation of only the empty spaces left in between the territories of the Neandertals, with population growth in those modern enclaves then generating a process of gradual expansion, leading to the marginalization and reduction of Neandertal niches, implying the break-up of their social and reproductive networks and, hence, ultimately, extinction.

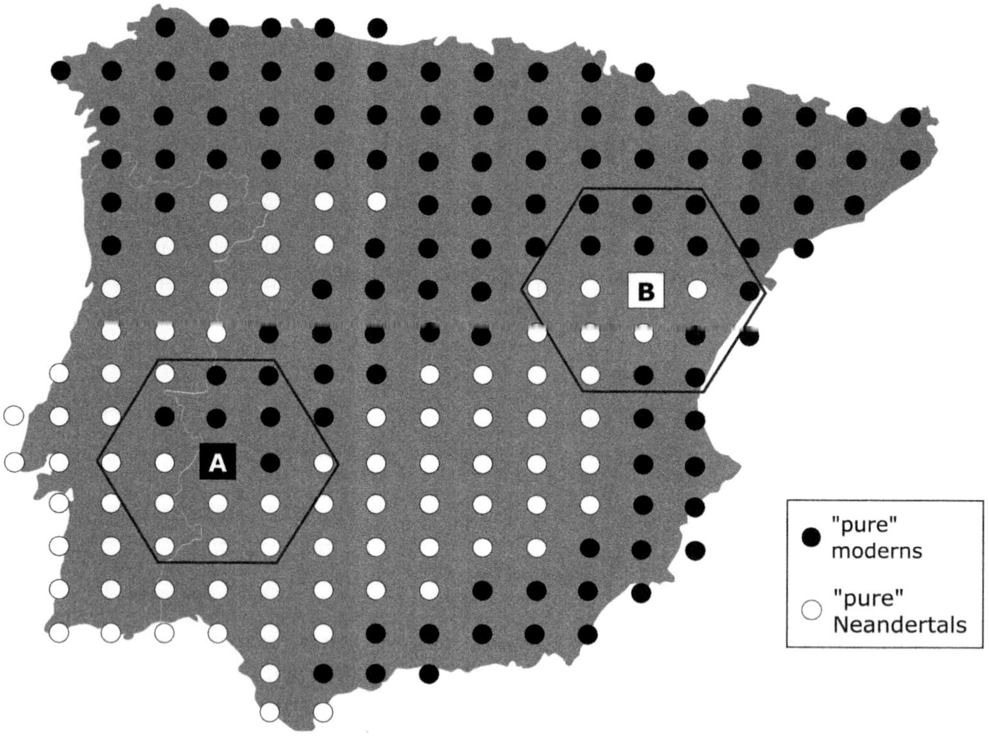

Fig. 25.20 Why mutual avoidance does not work. In this simple model for the Iberian case (After Zilhão and Trinkaus 2002), assumptions are a population density of 0.01/km² and an area available for occupation of ca. 500,000 km²; i.e., a total population of 5,000 people, with each circle representing a band of 25 individuals. If interbreeding did not take place, the mating networks for people at the head of the early modern humans' wedge of expansion would have to stretch for some 500 km. Based on ethnographic evidence, conditions to find mates are, in general, that one must be available in a universe of ca. 400 people, defined in such a way that residential moves >200 km are not required, i.e., a universe such as that defined by the *hexagons* around bands A and B, which, in the modeled situation, implies admixture

The difficulty for such mutual avoidance scenarios (e.g. Currat and Excoffier 2004) is that the archaeological record fails to provide any demonstration of the required adaptive advantage of early moderns (for a discussion, cf. Zilhão 2006a). The latest attempt to find "the" adaptive explanation for Neandertal extinction suggests that it resided in that, contrary to African moderns, European Neandertals would have lacked any form of division of labor, leading to a situation of marked imbalance in economic and reproductive productivity at the time of contact (Kuhn and Stiner 2006, following on Horan et al. 2005). Given that Neandertals inhabited cold-temperate or subarctic environments, the crux of Kuhn and Stiner's argument is that nowhere in the Neandertal record do we see any evidence that, as in all known ethnographic hunter-gatherer societies living in similar environments, "females [had] taken on the role of technology specialists". In particular, they invoke the absence of "the types of artifacts commonly used to make tailored, weather-resistant clothing and well-insulated artificial shelters – bone needles and awls, for example"; they claim, rightly, that such items do not appear until the early Upper Paleolithic, but they fail to consider that those early Upper Paleolithic contexts where the evidence does appear for the first time, namely the Châtelperronian, are Neandertal-associated. They further fail to consider that functional analysis and experimental replication showed that the collection of awls from the Grotte du Renne had been subjected to an intensive use – a minimum of 20,000 perforations on 2.5 mm thick leather, with many, given their fineness, having probably been used on lesser resistant materials, such as furs, bird hides or intestines (d'Errico et al. 2003). One can hardly think of what such intensive use might have been for if not the making of tailored clothes; thus, if, in subarctic environments, such tasks are primarily female ones, then the earliest real evidence for the existence of an institutionalized sexual division of labor is in fact found among Neandertal, not modern human societies.

Ironically, it may well be the case that these technologically "advanced" Neandertal societies living in the northernmost fringe of the Pleistocene human range may well have been those that least contributed to the gene pool of subsequent early and mid-Upper Paleolithic Europeans, as they probably were the only ones that really went extinct without descent (biologically, although not culturally, as the ornament types they had developed survived their demise). It has been suggested (Zilhão 2006a, b, 2009b; cf. Jacobi 2007, for a concurring view of the northern European process) that the isolated survival and delayed assimilation of Iberian Neandertals probably related not just to the specific biogeographical features of Iberia at this particular time but also to broader demographic patterns: soon after the first penetration of the continent by modern human groups, the impact of Heinrich Event 4 in regions to the north of the Pyrenees, aggravated by the effects of the Phlegraean Fields caldera explosion in central and eastern Europe, must have been responsible for a demographic crash that must also go a long way into explaining why Iberian Neandertals were "left alone." In the northern European plains, however, this major environmental crisis must have had the opposite effect on aboriginal Neandertal populations. As indicated by the fact that no evidence of human settlement exists north of the Rhine *limes* during the time of the Aurignacian I, which broadly coincides with that major environmental crisis, extinction is the most likely outcome for the Lincombian/Jerzmanovician groups, with resettlement of the area they previously occupied taking place only in Aurignacian II times, and by modern human groups descended from the mixed populations generated in the framework of the Protoaurignacian/Early Ahmarian expansion (Fig. 25.18). Significantly, the disappearance of the "Ebro frontier" also takes place in such Evolved Aurignacian times, suggesting that demographic recovery and indeed a marked increase in population numbers had occurred at this time in the core areas of the continent, triggering a north- and southward expansion of early modern European groups.

The Neandertal: Fossil Human or Fossil Mirror?

In the realm of biology, the Human Revolution paradigm was confronted from the beginning by supporters of the Multiregional Hypothesis, which, in its strict, original definition, was its exact opposite; at the time of the Transition in Europe, where one saw total discontinuity, the other argued for total continuity. Likewise, in the realm of culture, there always have been voices supporting the proposition that not only the transitional industries but also the Aurignacian had been made, at least in part, by the Neandertals. In sum, that we were dealing with a gradual, integrated process with no significant disruptions of regional biological and culture-historical trajectories (e.g., Straus 2003; Cabrera et al. 2006). The evidence indicates that this view is unrealistic, and that, as summarized in Fig. 25.21, "independent invention *cum* migration, interaction, admixture and cultural blending" is the true alternative to the different variants of "Neandertal acculturation followed by extinction without descent."

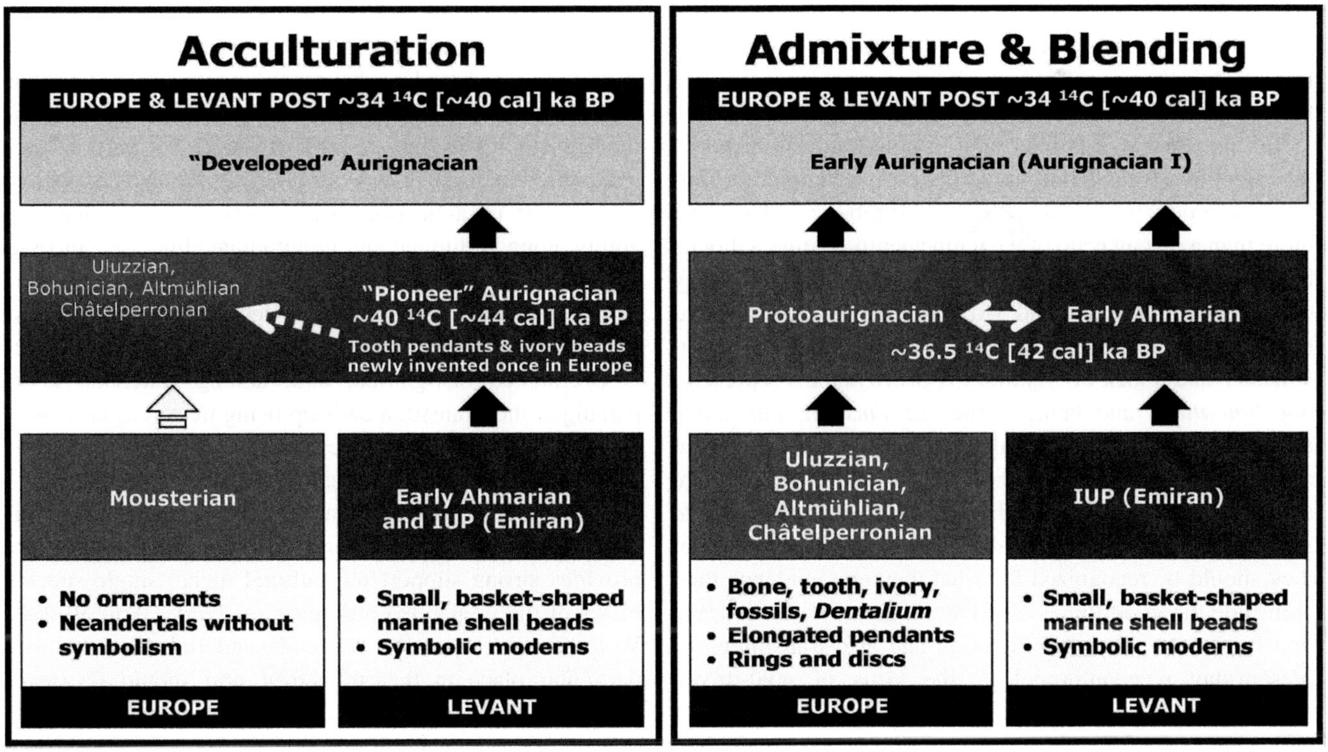

Fig. 25.21 Modeling the transition in Europe. The Human revolution requires acculturation. Admixture is consistent with the evidence for the blending of African and European ornament traditions apparent in the Protoaurignacian and the Aurignacian I

The body of data produced by archaeological research over the last decade comforts the biological admixture and cultural blending view of the process, and it seems fair to say that, since the mid-1990s, not a single piece of evidence surfaced that could provide support for the notion of the Neandertals' cognition-based cultural or adaptive inferiority. Yet, the notion thrives, underlying the pervasive use of double standards in the assessment of the evidence (Roebroeks and Corbey 2001; Speth 2004), and to an extent that must appear as absolutely extraordinary from the perspective of any other scientific discipline. One of the most fascinating examples is that concerning figurative art. The fact that it remains to be documented among Neandertals is often taken as the definitive proof of their handicaps. Yet, the earliest such art currently known is that represented by the sculpted figurines of the Evolved Aurignacian of southern Germany and the paintings of Chauvet cave, which date to no earlier than ca. 32 ka ^{14}C BP (36.9 kacal BP), i.e., some five millennia after the *Nassarius* folk came into Europe. Put another way, figurative art is as absent among Neandertals as it is among Europe's earliest modern humans, and its emergence in the Aurignacian II is best interpreted as a further indication of the population growth, and ensuing intensified levels of social interaction, inferred above on the basis of the marked expansion of the Aurignacian territory that takes place at this time (for a review, cf. Zilhão 2007). However, because, under the Human Revolution, taxonomic assignments must carry biological implications, the absence of figurative art prior to 42 kacal BP is often attributed, in the European record, to a cognitive handicap, while social or economic factors (if not that old favorite "it's there, we just haven't found it yet") are invoked to explain the African record.

In this context, it is important to note that statements to the effect that Neandertals could at best have copied or imitated the products of modern humans' symbolic crafts are no more than a distant echo of what nineteenth-century scholars had to say about Cro-Magnon art. As Roebroeks and Corbey (2001) point out, Piette and Mortillet, for instance, referred to that art as reflecting an "*esprit léger*", an absence of "*symbolisme*" and a lack of "*réflexion et prévoyance*", representing "*imitation*" and being borne "*de l'instant, non d'une réflexion esthétique*". Ironically, today, these notions, bred out of the progressivist views of French positivism, flourish, applied to Neandertals instead of Cro-Magnons, in the *sancta sanctorum* of Anglo-Saxon academia. It is about time that they should be recognized for what they are: evidence that, alongside the good things – evolution, natural selection, the first human fossils – the Victorian age also transmitted to present-day paleoanthropology the virus of goal-driven progress.

In the days of King's creation of *Homo neanderthalensis*, human fossils were used as supporting ancillary evidence in mainstream ethnological views of the racial ladder, to which they added time depth. Today, ranking human "races" in terms of cognition is no longer scientifically acceptable but, in western culture, the philosophical or religious need to place "us" at the top of the ladder of life (or of creation) still prevails, and explains the continued search for images of what "we" *are not* (or *not anymore*) that, by contrast, enhance the basics of what "we" *are* (cf. Sommer 2006): such is the place occupied by Neandertals in the context of the Human Revolution paradigm. Depending on different perceptions, going back to the eighteenth and nineteenth centuries, of the fundamental basis for the triumphant status of civilized society and industrial capitalism, so the tendency arose for Neanderthals to be represented as lacking in the corresponding behavioral feature. For instance, to give but a few examples, the Enlightenment emphasized the power of reason, Adam Smith stressed the importance of the division of labor, and David Ricardo expounded on the role of international trade and comparative advantage. Not surprisingly, explanations for the demise of the Neanderthals have correspondingly postulated competitive inferiority caused by their lacking in symbolic cognition, labor specialization by sex and age class, long-distance circulation of raw materials, or logistical organization of the subsistence base…

Of course, when the Neandertals (or at least some of them) are found to conform to the opposite of these models' expectations, then the argument is turned upside down. In a recent formulation (O'Connell 2006), for instance, it is the Neandertals' extreme focus on large mammal hunting that would have allowed modern humans, with greater behavioral flexibility and a broader subsistence base, to outcompete them. In truth, the archaeological record shows that, on the ground, the Transition was about people featuring not only different anatomies but also a diverse array of cultures and adaptations, ones whose intra-Neandertal and intra-modern human variability along latitudinal and longitudinal clines encompassed almost the entire range of settlement-subsistence strategies ethnographically documented. Because of the Human Revolution straightjacket, however, the evidence to that effect is overlooked or ignored, and, among adherents to the paradigm, the Transition ends up being treated as an abstract game played between two reified entities with little (if any) relation to actual empirical realities.

One hundred and fifty years after the Feldhofer cave discoveries, it is time to move on. The evidence reviewed above provides strong support to "cultural niche, single species" views of the human evolutionary trajectory (Wolpoff 2002; Wolpoff and Caspari 2011). The Neandertals have their own particular place in that trajectory, and should be treated accordingly, and in their own terms: as past, not "other".

Acknowledgements I thank the organizers for their invitation to contribute to the Congress, and the Humboldt Foundation for the support that made my participation possible.

References

Adams, B., & Ringer, Á. (2004). New C^{14} dates for the Hungarian Early Upper Paleolithic. *Current Anthropology, 45*(4), 541–551.

Adler, D. S., Bar-OZ, G., Belfer-Cohen, A., & Bar-Yosef, O. (2006). Ahead of the game. Middle and Upper Paleolithic hunting behaviors in the Southern Caucasus. *Current Anthropology, 47*(1), 89–118.

Aiello, L., & Wheeler, P. (2003). Neandertal thermoregulation and the glacial climate. In T. H. van Andel & W. Davies (Eds.), *Neanderthals and modern humans in the European landscape during the last glaciation* (pp. 147–166). Cambridge: McDonald Institute for Archaeological Research.

Allsworth-Jones, P. (1990). The Szeletian and the stratigraphic succession in Central Europe and adjacent areas: Main trends, recent results, and problems for resolution. In P. Mellars (Ed.), *The emergence of modern humans: An archaeological perspective* (pp. 160–243). Edinburgh: Edinburgh University Press.

Allsworth-Jones, P. (2004). The Szeletian revisited. *Anthropologie, 42*(3), 281–296.

Anikovich, M. V., Sinitsyn, A. A., Hoffecker, J. F., Holliday, V. T., Popov, V. V., Lisitsyn, S. N., Forman, S. L., Levkovskaya, G. M., Pospelova, G. A., Kuz'mina, I. E., Burova, N. D., Goldberg, P., Macphail, R. I., Giaccio, B., & Praslov, N. D. (2007). Early Upper Paleolithic in Eastern Europe and implications for the dispersal of modern humans. *Nature, 315*, 223–226.

Arrizabalaga, A., & Altuna, J. (Eds.). (2000). *Labeko Koba (Arrasate, País Vasco): Hienas y humanos en los albores del Paleolítico superior* (Munibe Antropología-Arkeologia, Vol. 52). San Sebastián-Donostia: Sociedad de Ciencias Aranzadi.

Arrizbalaga, A., Altuna, J., Areso, P., Elorza, M., García, M., Iriarte, M. J., Mariezkurrena, K., Mujika, J., Pemán, E., TarriÑo, A., Uriz, A., Viera, L., & Straus, L. G. (2003). The initial Upper Paleolithic in Northern Iberia: New evidence from Labeko Koba. *Current Anthropology, 44*(3), 413–421.

Bailey, S. E., & Hublin, J.-J. (2006). Dental remains from the Grotte du Renne at Arcy-sur-Cure (Yonne). *Journal of Human Evolution, 50*, 485–508.

Bailey, S. E., & Weaver, T. D. (2006). Who made the early Upper Paleolithic? The wisdom of the teeth. Puerto Rico: Annual Meeting of Paleoanthropology Society. Abstracts (p. 91).

Barge, H. (1983). Essai sur les parures du Paléolithique supérieur dans le sud de la France. La faune malacologique aurignacienne de l'abri Rothschild (Cabrières, Hérault). *Bulletin du Musée d'Anthropologie Préhistorique de Monaco, 27*, 69–83.

Bar-Yosef, O. (2006). Neanderthals and modern humans: A different interpretation. In N. J. Conard (Ed.), *When Neanderthals and modern humans met* (pp. 467–482). Tübingen: Kerns Verlag.

Belfer-Cohen, A., & Goring-Morris, A. N. (2003). Current issues in Levantine Upper Palaeolithic research. In A. N. Goring-Morris & A. Belfer-Cohen (Eds.), *More than meets the eye. Studies on Upper Palaeolithic diversity in the near East* (pp. 1–12). Oxford: Oxbow Books.

Bergman, C. A., & Stringer, C. B. (1989). Fifty years after: Egbert, an early Upper Palaeolithic juvenile from Ksar Akil, Lebanon. *Paleorient, 15*(2), 99–111.

Bernaldo de Quirós, F. (1982). *Los inicios del Paleolítico Superior Cantabrico*. Madrid: Centro de Investigación y Museo de Altamira.

Bischoff, J. L., Soler, N., Maroto, J., & Julià, R. (1989). Abrupt Mousterian/Aurignacian boundary at c. 40 ka bp: Accelerator 14C dates from l'Arbreda Cave (Catalunya, Spain). *Journal of Archaeological Science, 16*, 563–576.

Bon, F. (2000). La question de l'unité technique et économique de l'Aurignacien:Reflexions sur la variabilité des industries lithiques à partir de l'analyse comparée de trois sites des Pyrénées françaises. La Tuto de Camalhot, Régismont-le-Haut et Brassempouy. Ph.D. dissertation, University of Paris I.

Bon, F. (2002). *L'Aurignacien entre Mer et Océan. Réflexion sur l'unité des phases anciennes de l'Aurignacien dans le sud de la France* (Vol. XXIX). Paris: Mémoire de la Société Préhistorique Française.

Bon, F., & Bodu, P. (2002). Analyse technologique du débitage aurignacien. In B. Schmider (Ed.), *L'Aurignacien de la grotte du Renne. Les fouilles d'André Leroi-Gourhan à Arcy-sur-Cure (Yonne)* (Gallia Préhistoire supplément, Vol. XXXIV, pp. 115–133). Paris: Centre National de la Recherche Scientifique.

Bordes, F. (1968). *Le Paléolithique dans le monde*. Paris: Hachette.

Bordes, F. (1972). Du Paléolithique moyen au Paléolithique supérieur: continuité ou discontinuité? In F. Bordes (Ed.), *The origin of Homo sapiens* (pp. 211–217). Paris: UNESCO.

Bordes, F. (1981). Un néandertalien encombrant. *La Recherche, 12*, 643–645.

Bordes, J.-G. (2000). La séquence aurignacienne de Caminade revisitée: l'apport des raccords d'intérêt stratigraphique. *Paléo, 12*, 387–408.

Bordes, J.-G. (2002). Les interstratifications Châtelperronien/Aurignacien du Roc de Combe et du Piage (Lot, France). Analyse taphonomique des industries lithiques: Conséquences archéologiques. Ph.D. dissertation, University of Bordeaux I.

Bordes, J.-G. (2003). Lithic taphonomy of the Châtelperronian/Aurignacian interstratifications in Roc de Combe and Le Piage (Lot, France). In J. Zilhão & F. D'Errico (Eds.), *The Chronology of the Aurignacian and of the transitional technocomplexes. Dating, stratigraphies, cultural implications* (Trabalhos de Arqueologia, Vol. 33, pp. 223–244). Lisboa: Instituto Português de Arqueologia.

Bordes, J.-G. (2006). News from the West: A reevaluation of the classical Aurignacian sequence of the Périgord. In O. Bar-Yosef & J. Zilhão (Eds.), *Towards a definition of the Aurignacian* (Trabalhos de Arqueologia, Vol. 45, pp. 147–171). Lisboa: American School of Prehistoric Research/Instituto Português de Arqueologia.

Bordes, F., & Labrot, J. (1967). Stratigraphie de la grotte de Roc-de-Combe (Lot) et ses implications. *Bulletin de la Société Préhistorique Française. Études et Travaux, 64*, 15–28.

Bouzouggar, A., Barton, N., Vanhaeren, M., D'Errico, F., Collcutt, S., Higham, T., Hodge, E., Parfitt, S., Rhodes, E., Schwenninger, J.-L., Stringer, C., Turner, E., Ward, S., Moutmir, A., & Stambouli, A. (2007). 82,000-year-old shell beads from North Africa and implications for the origins of modern human behavior. *Proceedings of the National Academy of Sciences of the United States of America, 104*(24), 9964–9969.

Bräuer, G. (1984). A craniological approach to the origin of anatomically modern *Homo sapiens* in Africa and implications for the appearance of modern Europeans. In F. Smith & F. Spencer (Eds.), *The origins of modern humans: A world survey of the fossil evidence* (pp. 327–410). New York: Alan R. Liss.

Brézillon, M. (1969). *Dictionnaire de la Préhistoire*. Paris: Larousse.

Broglio, A., & Gurioli, F. (2004). The symbolic behaviour of the first modern humans: The Fumane cave evidence (Venetian pre-Alps). In M. Otte (Ed.), *La Spiritualité. Actes du colloque de la commission 8 de l'UISPP* (Paléolithique supérieur, pp. 97–102). Liège, 10–12 décembre 2003. Liège: Études et Recherches Archéologiques de l'Université de Liège 106.

Broglio, A., Bertola, S., Destefani, M., & Marini, D. (2002). L'Aurignaziano della Grotta di Fumane. In A. Aspes (Ed.), *Preistoria Veronese. Contributi e aggiornamenti* (Memorie del Museo Civico di Storia Naturale di Verona, Vol. II^{a} serie, pp. 29–36). Verona: Sezione Scienze dell'Uomo n° 5.

Bronk Ramsey, C., Higham, T., Bowles, A., & Hedges, R. (2004). Improvements to the pretreatment of bone at Oxford. *Radiocarbon, 46*, 155–163.

Cabrera, V., & Bischoff, J. L. (1989). Accelerator ^{14}C dates for Early Upper Paleolithic (Basal Aurignacian) at El Castillo Cave (Spain). *Journal of Archaeological Science, 16*, 577–584.

Cabrera, V., Maíllo, J. M., Pike-Tay, A., Garralda, M. D., & Bernaldo de Quiros, F. (2006). A Cantabrian perspective on late Neandertals. In

N. J. Conard (Ed.), *When Neanderthals and modern humans met* (pp. 441–465). Tübingen: Kerns Verlag.

Cann, R. L., Stoneking, M., & Wilson, A. C. (1987). Mitochondrial DNA and human evolution. *Nature, 325*, 31–36.

Champagne, F., & Espitalié, R. (1981). *Le Piage, site préhistorique du Lot*. Paris: Société Préhistorique Française.

Chiotti, L. (1999). Les industries lithiques des niveaux aurignaciens de l'Abri Pataud, Les Eyzies-de-Tayac (Dordogne): étude technologique et typologique. Ph.D. dissertation, Institut de Paléontologie Humaine, Paris.

Churchill, S. E., & Smith, F. H. (2000). Makers of the early Aurignacian of Europe. *Yearbook of Physical Anthropology, 43*, 61–115.

Conard, N. J., Bolus, M., Goldberg, P., & Münzel, S. C. (2006). The last Neanderthals and first modern humans in the Swabian Jura. In N. J. Conard (Ed.), *When Neanderthals and modern humans met* (pp. 305–341). Tübingen: Kerns Verlag.

Connet, N. (2002). Le Châtelperronien : Refléxions sur l'unité et l'identité techno-économique de l'industrie lithique. L'apport de l'analyse diachronique des industries lithiques des couches Châtelperroniennes de la grotte du Renne à Arcy-sur-Cure (Yonne). Ph.D. dissertation, University of Lille I.

Currat, M., & Excoffier, L. (2004). Modern humans did not admix with Neanderthals during their range expansion into Europe. *PLoS Biology, 2*(12), 2264–2274.

D'Errico, F. (2003). The invisible frontier. A multiple species model for the origin of behavioral modernity. *Evolutionary Anthropology, 12*, 188–202.

D'Errico, F., & Vanhaeren, M. (2007). Evolution or revolution? New evidence for the origin of symbolic behavior in Africa and Europe. In P. M. Mellars, O. Bar-Yosef, C. Stringer, & K. V. Boyle (Eds.), *Rethinking the human revolution* (pp. 275–286). Cambridge: McDonald Institute for Archaeological Research.

D'Errico, F., Zilhão, J., Baffier, D., Julien, M., & Pelegrin, J. (1998). Neanderthal acculturation in Western Europe? A critical review of the evidence and its interpretation. *Current Anthropology, 39*(Supplement), 1–44.

D'Errico, F., Julien, M., Liolios, D., Vanhaeren, M., & Baffier, D. (2003). Many awls in our argument. Bone tool manufacture and use in the Châtelperronian and Aurignacian levels of the Grotte du Renne at Arcy-sur-Cure. In J. Zilhão & F. D'Errico (Eds.), *The chronology of the Aurignacian and of the transitional technocomplexes. Dating, stratigraphies, cultural implications* (Trabalhos de Arqueologia, Vol. 33, pp. 247–270). Lisboa: Instituto Português de Arqueologia.

D'Errico, F., Henshilwood, Ch, Vanhaeren, M., & Van Niekerk, K. (2005). Nassarius kraussianus shell beads from Blombos Cave: Evidence for symbolic behaviour in the Middle Stone Age. *Journal of Human Evolution, 48*, 3–24.

David, F., Connet, N., Girard, M., Lhomme, V., Miskovsky, J.-C., & Roblin-Jouve, A. (2001). Le Châtelperronien de la grotte du Renne à Arcy-sur-Cure (Yonne). Données sédimentologiques et chronostratigraphiques. *Bulletin de la Société Préhistorique Française, 98*, 207–230.

de Sonneville-Bordes, D. (1960). *Le Paléolithique Supérieur en Périgord*. Bordeaux: Delmas.

De Vivo, B., Rolandi, G., Gans, P. B., Calvert, A., Bohrson, W. A., Spera, F. J., & Belkin, H. E. (2001). New constraints on the pyroclastic eruptive history of the Campanian volcanic plain (Italy). *Mineralogy and Petrology, 73*, 47–65.

Demars, P.-Y. (1990). Les interstratifications entre Aurignacien et Châtelperronien à Roc-de-Combe et au Piage (Lot). In C. Farizy (Ed.), *Paléolithique moyen récent et Paléolithique supérieur ancien en Europe* (Mémoires du Musée de Préhistoire de l'Ile de France, Vol. 3, pp. 235–239). Nemours: Musée de Préhistoire d'Ile-de-France.

Demars, P.-Y. (1996). La place du Piage et de Roc de Combe (Lot) dans la transition du Paléolithique moyen au Paléolithique supérieur. *Bulletin Préhistorique du Sud-Ouest. Nouvelles Etudes, 3*, 11–35.

Demars, P.-Y., & Laurent, P. (1989). *Types d'outils lithiques du Paléolithique supérieur en Europe*. Paris: Centre National de la Recherche Scientifique.

Duarte, C., Maurício, J., Pettitt, P. B., Souto, P., Trinkaus, E., Van Der Plicht, H., & Zilhão, J. (1999). The early Upper Paleolithic human skeleton from the Abrigo do Lagar Velho (Portugal) and modern human emergence in Iberia. *Proceedings of the National Academy of Sciences of the United States of America, 96*, 7604–7609.

Evans, P. D., Mekel-Bobrov, N., Vallender, E. J., Hudson, R. R., & Lahn, B. T. (2006). Evidence that the adaptive allele of the brain size gene microcephalin introgressed into Homo sapiens from an archaic Homo lineage. *Proceedings of the National Academy of Sciences of the United States of America, 103*(48), 18178–18183.

Fairbanks, R. G., Mortlock, R. A., Chiu, T-Ch, Cao, L., Kaplan, A., Guilderson, Th P, Fairbanks, T. W., Bloom, A. L., Grootes, P. M., & Nadeau, M.-J. (2005). Radiocarbon calibration curve spanning 0 to 50,000 years BP based on paired 230Th/234U/238U and 14C dates on pristine corals. *Quaternary Science Reviews, 24*, 1781–1796.

Fedele, F. G., Giaccio, B., Isaia, R., & Orsi, G. (2003). The Campanian ignimbrite eruption, Heinrich Event 4, and Palaeolithic change in Europe: A high-resolution investigation. In A. Robock & C. Oppenheimer (Eds.), *Volcanism and the Earth's atmosphere* (Geophysical Monograph, Vol. 139, pp. 301–325). Washington, DC: American Geophysical Union.

Felgenhauer, F. (1956–1959). *Willendorf in der Wachau. Monographie der Paläolith-Fundstellen I-VII*. 3 vols., with contributions by F. Brandtner, A. Papp, E. Thenius, E. Zirkl.. Wien: R. M. Rohrer (Mitteilungen der Prähistorischen Kommission der Österreichischen Akademie der Wissenschaften, VIII–IX).

Floss, H. (2003). Did they meet or not? Observations on Châtelperronian and Aurignacian settlement patterns in eastern France. In J. Zilhão & F. D'Errico (Eds.), *The chronology of the Aurignacian and of the transitional technocomplexes. Dating, stratigraphies, cultural implications* (Trabalhos de Arqueologia, Vol. 33, pp. 273–287). Lisboa: Instituto Português de Arqueologia.

Gilman, A. (1984). Explaining the Upper Palaeolithic revolution. In M. Spriggs (Ed.), *Marxist perspectives in archaeology* (pp. 115–126). Cambridge: Cambridge University Press.

Girard, C. (1980). Les industries moustériennes de la grotte du Renne à Arcy-sur-Cure. *Gallia Préhistoire, 23*(1), 1–36.

Glén, E., & Kaczanowski, K. (1982). Human remains. In J. Kozłowski (Ed.), *Excavation in the Bacho Kiro cave (Bulgaria). Final report* (pp. 75–79). Warsaw: Polish Scientific Publishers.

Gravina, B., Mellars, P., & Bronk Ramsey, C. (2005). Radiocarbon dating of interstratified Neanderthal and early modern human occupations at the Chatelperronian type-site. *Nature, 438*, 51–56.

Green, R. E., Krause, J., Ptak, S. E., Briggs, A. W., Ronan, M. T., Simons, J. F., Du, L., Egholm, M., Rothberg, J. M., Paunovic, M., & Pääbo, S. (2006). Analysis of one million base pairs of Neanderthal DNA. *Nature, 444*, 330–336.

Grün, R., & Stringer, C. (2000). Tabun revisited: Revised ESR chronology and new ESR and U-series analyses of dental material from Tabun C1. *Journal of Human Evolution, 39*, 601–612.

Grünberg, J. (2006). New AMS dates for Palaeolithic and Mesolithic camp sites and single finds in Saxony-Anhalt and Thuringia (Germany). *Proceedings of the Prehistoric Society, 72*, 95–112.

Haesaerts, P., & Teyssandier, N. (2003). The Early Upper Paleolithic occupation of Willendorf II (Lower Austria): A contribution to the chronostratigraphic and cultural context of the beginning of the Upper Paleolithic in central Europe. In J. Zilhão & F. D'Errico (Eds.), *The chronology of the Aurignacian and of the transitional technocomplexes. Dating, stratigraphies, cultural implications*

(Trabalhos de Arqueologia, Vol. 33, pp. 133–151). Lisboa: Instituto Português de Arqueologia.

Haesaerts, P., Damblon, F., Sinitsyn, A., & Van der Plicht, J. (2004). Kostienki 14 (Voronezh, Central Russia): New Data on Stratigraphy and Radiocarbon Chronology. In *Actes du XIVème Congrès UISPP, Section 6, Le Paléolithique Supérieur*. (BAR International Series, Vol. 1240, pp. 169–180).

Hahn, J. (1977). *Aurignacien, das ältere Jungpaläolithikum in Mittel- und Osteuropa*. Köln/Wien: Böhlau-Verlag.

Harvati, K., Gunz, P., & Grigorescu, D. (2007). Cioclovina (Romania): Affinities of an early modern European. *Journal of Human Evolution, 53*, 732–746.

Hedges, R. E., Housley, R. A., Bronk Ramsey, C., & Van Klinken, G. J. (1994). Radiocarbon dates from the Oxford AMS system: Datelist 18. *Archaeometry, 36*, 337–374.

Henshilwood, C., & Marean, C. (2003). The origin of modern human behavior. Critique of the models and their test implications. *Current Anthropology, 44*(5), 627–651.

Henshilwood, C., D'Errico, F., Vanhaeren, M., Van Niekerk, K., & Jacobs, Z. (2004). Middle Stone Age shell beads from South Africa. *Science, 304*, 404.

Higham, T. F. G., Jacobi, R. M., & Bronk Ramsey, C. (2006). AMS radiocarbon dating of ancient bone using ultrafiltration. *Radiocarbon, 48*(2), 179–195.

Holliday, T. W. (2006). Neanderthals and modern humans: An example of a mammalian syngameon? In K. Harvati & T. Harrison (Eds.), *Neanderthals revisited: New approaches and perspectives* (pp. 289–306). New York: Springer.

Hopkinson, T. (2004). Leaf Points, landscapes and environment change in the European late middle Palaeolithic. In N. J. Conard (Ed.), *Settelement dynamics of the Middle Paleolithic and middle stone age* (Vol. II, pp. 227–258). Tübingen: Kerns.

Horan, R. D., Bulte, E., & Shogren, J. F. (2005). How trade saved humanity from biological exclusion: An economic theory of Neanderthal extinction. *Journal of Economic Behavior and Organization, 58*, 1–29.

Hublin, J.-J. (2000). Modern-nonmodern hominid interactions: A Mediterranean perspective. In O. Bar-Yosef & D. Pilbeam (Eds.), *The geography of Neandertals and modern humans in Europe and the greater Mediterranean* (Peabody Museum Bulletin, Vol. 8, pp. 157–182). Cambridge: Peabody Museum, Harvard University.

Hublin, J.-J., & Bailey, S. E. (2006). Revisiting the last Neandertals. In N. J. Conard (Ed.), *When Neanderthals and modern humans met* (pp. 105–128). Tübingen: Kerns Verlag.

Hublin, J.-J., Spoor, F., Braun, M., Zonneveld, F., & Condemi, S. (1996). A late Neanderthal associated with Upper Palaeolithic artefacts. *Nature, 381*, 224–226.

Hughen, K., Lehman, S., Southon, J., Overpeck, J., Marchal, O., Herring, C., & Turnbull, J. (2004). 14C Activity and global carbon cycle changes over the past 50, 000 years. *Science, 303*, 202–207.

Hülle, W. M. (1977). *Die Ilsenhöhle unter Burg Ranis/Thüringen. Eine paläolitische Jägerstation*. Stuttgart: Gustav Fischer.

Jacobi, R. (2007). A collection of early Upper Palaeolithic artefacts from beedings, near Pulborough, West Sussex, and the context of similar finds from the British Isles. *Proceedings of the Prehistoric Society, 73*, 229–325.

Jones, M., Marks, A. E., & Kaufman, D. (1983). Boker. The artifacts. In A. E. Marks (Ed.), *Prehistory and paleoenvironments in the central Negev, Israel* (The Avdat/Aqev Area, Part 3, Vol. III, pp. 283–329). Dallas: Southern Methodist University.

Jöris, O., Álvarez, E., & Weninger, B. (2003). Radiocarbon evidence of the middle to Upper Paleolithic transition in Southwestern Europe. *Trabajos de Prehistoria, 60*(2), 15–38.

Julien, M., Baffier, D., & Liolios, D. (2002). L'industrie osseuse, In B. Schmider (Ed.), *L'Aurignacien de la grotte du Renne. Les fouilles d'André Leroi-Gourhan á Arcy-sur-Cure (Yonne)* (Gallia Préhistoire supplément, Vol. XXXIV, pp. 217–250). Paris: Centre National de la Recherche Scientifique.

Klein, R. G. (2003). Whither the Neanderthals? *Science, 299*, 1525–1527.

Koumouzelis, M., Ginter, B., Kozłowski, J. K., Pawlikowski, M., Bar-Yosef, O., Albert, R. M., Litynska-Zajac, M., Stworzewicz, E., Wojtal, P., Lipecki, G., Tomek, T., Bochenski, Z. M., & Pazdur, A. (2001a). The early Upper Palaeolithic in Greece: The excavations in Klisoura Cave. *Journal of Archaeological Science, 28*, 515–539.

Koumouzelis, M., Kozłowski, J. K., Escutenaire, C., Sitlivy, V., Sobczyk, K., Valladas, H., Tisnerat-Laborde, N., Wojtal, P., & Ginter, B. (2001b). La fin du Paléolithique moyen et le début du Paléolithique supérieur en Grèce: la séquence de la Grotte 1 de Klisoura. *L'Anthropologie, 105*, 469–504.

Kozłowski, J. (Ed.). (1982). *Excavation in the Bacho Kiro cave (Bulgaria). Final report*. Warsaw: Polish Scientific Publishers.

Kozłowski, J. (2004). Early Upper Paleolithic Levallois-derived industries in the Balkans and in the Middle Danube Basin. *Anthropologie, 42*(3), 263–280.

Kozłowski, J. K., & Kozłowski, S. K. (1996). *Le Paléolithique en Pologne*. Grenoble: Jérôme Millon.

Kozłowski, J. K., & Otte, M. (2000). La formation de l'Aurignacien en Europe. *L'Anthropologie, 104*, 3–15.

Krings, M., Stone, A., Schmitz, R. W., Krainitzki, H., Stoneking, M., & Pääbo, S. (1997). Neandertal DNA sequences and the origin of modern humans. *Cell, 90*, 19–30.

Krings, M., Capelli, C., Tschentscher, F., Geisert, H., Meyer, S., von Haeseler, A., Grossschmidt, K., Possnert, G., Paunovic, M., & Pääbo, S. (2000). A view of Neandertal genetic diversity. *Nature Genetics, 26*, 144–146.

Kuhn, S. L. (2003). In what sense is the Levantine initial Upper Paleolithic a "transitional" industry? In J. Zilhão & F. D'Errico (Eds.), *The chronology of the Aurignacian and of the transitional technocomplexes. Dating, stratigraphies, cultural implications* (Trabalhos de Arqueologia, Vol. 33, pp. 61–69). Lisboa: Instituto Português de Arqueologia.

Kuhn, S., & Stiner, M. (1998). The earliest Aurignacian of Riparo Mochi (Liguria, Italy). *Current Anthropology, 39*(Supplement), 175–189.

Kuhn, S., & Stiner, M. (2006). What's a Mother to Do? The Division of Labor among Neandertals and Modern Humans in Eurasia. *Current Anthropology, 47*(6), 953–980.

Kuhn, S. L., Stiner, M. C., Reese, D. S., & Gülec, E. (2001). Ornaments of the earliest Upper Paleolithic: New insights from the Levant. *Proceedings of the National Academy of Sciences of the United States of America, 98*(13), 7641–7646.

Kurzban, R., Tooby, J., & Cosmides, L. (2001). Can race be erased? Coalitional computation and social categorization. *Proceedings of the National Academy of Sciences of the United States of America, 98*(26), 15387–15392.

Lahr, M. M., & Foley, R. (1998). Towards a theory of Modern Human Origins: Geography, Demography, and Diversity in recent human evolution. *Yearbook of Physical Anthropology, 41*, 37–176.

LaVille, H., Rigaud, J-Ph, & Sackett, J. (1980). *Rock shelters of the Périgord*. New York: Academic.

Le Brun-Ricalens, F. (1993). Réflexions préliminaires sur le comportement litho-technologique et l'occupation du territoire du pays des Serres à l'Aurignacien. *Paléo, 5*, 127–153.

Leroi-Gourhan, A. (1958). Etude des restes humains fossiles provenant des Grottes d'Arcy-sur-Cure. *Annales de Paléontologie, 44*, 87–148.

Leroi-Gourhan, A. (1964). *Les religions de la Préhistoire*. Paris: Presses Universitaires de France.

Leroyer, C., & Leroi-Gourhan, A. (1983). Problèmes de chronologie : le Castelperronien et l'Aurignacien. *Bulletin de la Société Préhistorique Française* (Vol. 80, p. 41–44). Paris.

Lévêque, F., & Vandermeersch, B. (1980). Découverte de restes humains dans un niveau castelperronien à Saint-Césaire (Charente-Maritime). *Comptes rendus de l'Académie des Sciences de Paris, 291D*, 187–189.

Lieberman, P. (2007). The evolution of human speech. Its anatomical and neural bases. *Current Anthropology, 48*(1), 39–66.

Lieberman, P., & Crelin, E. S. (1971). On the speech of the Neanderthal man. *Linguistic Inquiry, 2*(2), 203–222.

Lucas, G. (2000). Les industries lithiques du Flageolet 1 (Dordogne) : Approche économique, technologique et analyse spatiale. Ph.D. dissertation, University of Bordeaux I.

Maíllo, J. M., Cabrera, V., & Bernaldo de Quirós, F. (2004). Le débitage lamellaire dans le Moustérien final de Cantabrie (Espagne): le cas de El Castillo et Cueva Morin. *L'anthropologie, 108*, 367–393.

Mcbrearty, S., & Brooks, A. (2000). The revolution that wasn't: A new interpretation of the origin of modern human behavior. *Journal of Human Evolution, 39*, 453–563.

Mellars, P. A. (1973). The character of the middle-upper paleolithic transition in south-west France. In C. Renfrew (Ed.), *The Explanation of cultural change. Models in prehistory* (pp. 255–276). London: Duckworth.

Mellars, P. A. (1989). Major issues in the emergence of modern humans. *Current Anthropology, 30*(3), 349–385.

Mellars, P. A. (1999). The Neanderthal problem continued. *Current Anthropology, 40*(3), 341–350.

Mellars, P. A. (2004). Neanderthals and the modern human colonization of Europe. *Nature, 432*, 461–465.

Mellars, P. A. (2006). A new radiocarbon revolution and the dispersal of modern humans in Eurasia. *Nature, 439*, 931–935.

Mellars, P. A., & Stringer, C. B. (Eds.). (1989). *The human revolution*. Edinburgh: Edinburgh University Press.

Mellars, P. A., Gravina, B., & Bronk Ramsey, Ch. (2007). Confirmation of Neanderthal/modern human interstratification at the Chatelperronian type-site. *Proceedings of the National Academy of Sciences of the United States of America, 104*(9), 3657–3662.

Monigal, K. (2003). Technology, economy and mobility at the beginning of the Levantine Upper Palaeolithic. In A. N. Goring-Morris & A. Belfer-Cohen (Eds.), *More than meets the eye. Studies on Upper Palaeolithic diversity in the near East* (pp. 118–133). Oxford: Oxbow Books.

Monigal, K. (2006). Transit lounge of Eastern Europe: Multicultural Crimea during the late Middle Paleolithic and early Upper Paleolithic. In N. J. Conard (Ed.), *When Neanderthals and modern humans met* (pp. 189–211). Tübingen: Kerns Verlag.

Montes, R., & Sanguino, J. (Eds.), (2001). *La Cueva del Pendo. Actuaciones Arqueológicas 1994–2000*. Santander: Ayuntamiento de Camargo/Gobierno de Cantabria/Parlamento de Cantabria.

Montes, R., Sanguino, J., Martín, P., Gómez, A. J., & Morcillo, C. (2005). La secuencia estratigráfica de la cueva de El Pendo (Escobedo de Camargo, Cantabria): Problemas geoarqueológicos de un referente cronocultural. In M. Santonja, A. Pérez-Gonzalez, & M. Machado (Eds.), *Geoarqueología y patrimonio en la Península Ibérica y el entorno mediterráneo* (pp. 127–138). Almazán (Soria): ADEMA.

Mussi, M. (2001). *Earliest Italy. An overview of the Italian Paleolithic and Mesolithic*. New York: Kluwer Academic.

Nigst, P. R. (2006). The first modern humans in the middle Danube Area? New evidence from Willendorf II (Eastern Austria). In N. J. Conard (Ed.), *When Neanderthals and modern humans met* (pp. 269–304). Tübingen: Kerns Verlag.

Noble, W., & Davidson, I. (1996). *Human evolution, language and mind*. Cambridge: Cambridge University Press.

Normand, Ch., & Turq, A. (2005). L'Aurignacien de la grotte d'Isturitz (France): la production lamellaire dans la séquence de la salle Saint-Martin. In F. Le Brun-Ricalens, J.-G. Bordes, & F. Bon (Eds.), *Productions lamellaires attribuées à l'Aurignacien. Chaînes opératoires et perspectives technoculturelles* (ArchéoLogiques, Vol. 1, pp. 375–394). Luxembourg: Musée National d'Histoire et d'Art.

O'Connell, J. F. (2006). How did modern humans displace Neanderthals? Insights from hunter-gatherer ethnography and archaeology. In N. J. Conard (Ed.), *When Neanderthals and modern humans met* (pp. 43–64). Tübingen: Kerns Verlag.

Otte, M., & Kozłowski, J. (2003). Constitution of the aurignacian through Eurasia. In J. Zilhão & F. D'Errico (Eds.), *The chronology of the Aurignacian and of the transitional technocomplexes. Dating, stratigraphies, cultural implications* (Trabalhos de Arqueologia, Vol. 33, pp. 19–27). Lisboa: Instituto Português de Arqueologia.

Palma Di Cesnola, A. (1993). *Il Paleolitico superiore in Italia*. Firenze: Garlatti e Razzai.

Rabeder, G., & Pohar, V. (2004). Stratigraphy and chronology of the cave sediments from Potočka zijalka (Slovenia). In M. Pacher, V. Pohar, & G. Rabeder (Eds.), *Potočka Zijalka. Palaeontological and archaeological results of the campaigns 1997–2000* (Mitteilungen der Kommission für Quartärforschung der Österreichischen Akademie der Wissenschaften, Vol. 13, pp. 235–246). Vienna: Verlag der Österreichischen Akademie der Wissenschaften.

Richter, J. (2000). Social memory among late Neanderthals. In J. Örschiedt & G. Weniger (Eds.), *Neanderthals and modern humans - Discussing the transition. Central and eastern Europe from 50.000 - 30.000 B. P* (pp. 123–132). Mettmann: Neandertal Museum.

Richter, J. (2002). Die ^{14}C daten aus der Sesselfelsgrotte und die Zeitstellung des Micoquien/M.M.O. *Germania, 80*(1), 1–22.

Rigaud, J-Ph. (2001). À propos de la contemporanéité du Castelperronien et de l'Aurignacien ancien dans le nord-est de l'Aquitaine: une révision des données et ses implications. In J. Zilhão, Th Aubry, & A. F. Carvalho (Eds.), *Les premiers hommes modernes de la Péninsule Ibérique* (Trabalhos de Arqueologia, Vol. 17, pp. 61–68). Lisboa: Instituto Português de Arqueologia.

Rigaud, J-Ph, & Lucas, G. (2006). The first aurignacian technocomplexes in Europe: A revision of the Bachokirian. In O. Bar-Yosef & J. Zilhão (Eds.), *Towards a definition of the Aurignacian* (Trabalhos de Arqueologia, Vol. 45, pp. 277–284). Lisboa: American School of Prehistoric Research/Instituto Português de Arqueologia.

Roebroeks, W., & Corbey, R. (2001). Biases and double standards in palaeoanthropology. In R. Corbey & W. Roebroeks (Eds.), *Studying human origins. Disciplinary history and epistemology* (Archeological Studies, Vol. 6, pp. 67–76). Amsterdam: University Press.

Rougier, H., Milota, Ş., Rodrigo, R., Gherase, M., Sarcina, L., Moldovan, O., Zilhão, J., Constantin, S., Franciscus, R. G., Zollikofer, C. P. E., Ponce de León, M., & Trinkaus, E. (2007). Peştera cu Oase 2 and the cranial morphology of early modern Europeans. *Proceedings of the National Academy of Sciences of the United States of America, 104*(4), 1165–1170.

Schmider, B. (2002). *L'Aurignacien de la grotte du Renne. Les fouilles d'André Leroi-Gourhan à Arcy-sur-Cure (Yonne)* (Gallia Préhistoire supplément, Vol. XXXIV). Paris: Centre National de la Recherche Scientifique.

Serre, D., Langaney, A., Chech, M., Teschler-Nicola, M., Paunovic, M., Mennecier, Ph, Hofreiter, M., Possnert, G., & Pääbo, S. (2004). No evidence of Neandertal mtDNA contribution to early modern humans. *PLoS Biology, 2*(3), 313–317.

Shackleton, N. J., Fairbanks, R. G., Chiu, T., & Parrenin, F. (2004). Absolute calibration of the Greenland time scale: Implications for Antarctic time scales and for Δ^{14}C. *Quaternary Science Reviews, 23*, 1513–1522.

Shennan, S. (2001). Demography and cultural innovation: A model and its implications for the emergence of modern human culture. *Cambridge Archaeological Journal, 11*(1), 5–16.

Sitlivy, V., Sobczyk, K., Kalicki, T., Escutenaire, C., Zieba, A., & Kaczor, K. (1999a). The new Paleolithic Site of Księcia Józefa

(Cracow, Poland) with blade and flake reduction. *Préhistoire Européene, 15*, 87–111.
Sitlivy, V., Sobczyk, K., Morawski, W., Zieba, A., & Escutenaire, C. (1999b). Piekary IIa Paleolithic industries: Preliminary results of a new multidisciplinary investigations. *Préhistoire Européene, 15*, 45–64.
Sitlivy, V., Sobczyk, K., Escutenaire, C., Kalicki, T., Kolenyk, A., Kozłowski, J., Mercier, N., Tisnerat-Laborde, N., Valladas, H., & Zieba, A. (2004). Late middle Paleolithic complexes of Cracow region, Poland. *In Actes du XIVème Congrès UISPP, Section, Le Pateolithique Supérieur*. (BAR International Series, Vol. 1240, pp. 305–317) Oxford.
Smith, F. (1984). Fossil hominids from the Upper Pleistocene of central Europe and the origin of modern Europeans. In F. Smith & F. Spencer (Eds.), *The origins of modern humans: A world survey of the fossil evidence* (pp. 137–209). New York: Alan R. Liss.
Soficaru, A., Petrea, C., Doboş, A., & Trinkaus, E. (2007). The Human Cranium from the Peştera Cioclovina Uscată, Romania. *Current Anthropology, 48*(4), 611–619.
Sommer, M. (2006). Mirror, mirror on the wall: Neanderthal as image and 'Distortion' in early 20th-Century French science and press. *Social Studies of Science, 36*(2), 207–240.
Sonneville-Bordes, D. (1982). L'évolution des industries aurignaciennes. In *Aurignacien et Gravettien en Europe, II, Cracovie – Nitra – 1980* (pp. 339–360). Liège: Études et Recherches Archéologiques de l'Université de Liège 13.
Speth, J. (2004). News flash: Negative evidence convicts Neanderthals of gross mental incompetence. *World Archaeology, 36*(4), 519–526.
Stiner, M. (1999). Palaeolithic mollusc exploitation at Riparo Mocchi (Balzi Rossi, Italy): Food and ornaments from the Aurignacian through the Epigravettian. *Antiquity, 73*, 735–754.
Straus, L. G. (2003). "The Aurignacian"? Some thoughts. In J. Zilhão & F. D'Errico (Eds.), *The chronology of the Aurignacian and of the transitional technocomplexes. Dating, stratigraphies, cultural implications* (Trabalhos de Arqueologia, Vol. 33, pp. 11–17). Lisboa: Instituto Português de Arqueologia.
Stringer, C. (2006). *Homo Britannicus. The incredible story of human life in Britain*. London: Allen Lane.
Stringer, C., & Gamble, C. (1993). In *Search of the Neanderthals*. London: Thames and Hudson.
Stringer, C., Hublin, J.-J., & Vandermeersch, V. (1984). The origin of anatomically modern humans in western Europe. In F. Smith & F. Spencer (Eds.), *The origins of modern humans. A world survey of the fossil evidence* (pp. 51–135). New York: Alan R. Liss.
Svoboda, J. (1988). Early Upper Paleolithic industries in Moravia: A review of recent evidence. In J. Kozłowski (Ed.), *L'Homme de Néandertal. Vol. 8. La mutation* (pp. 169–192). Liège: Études et Recherches Archéologiques de l'Université de Liège 35.
Svoboda, J., & Bar-Yosef, O. (2003). *Stránská skála. Origins of the Upper Paleolithic in the Brno Basin, Moravia, Czech Republic*. Cambridge, MA: Harvard University Press.
Svoboda, J., Ložek, V., & Vlček, E. (1996). *Hunters between East and West*. New York: Plenum.
Taborin, Y. (1993). *La parure en coquillage au Paléolithique*. Paris: Centre National de la Recherche Scientifique.
Taborin, Y. (2002). Les objets de parure et les curiosa. In B. Schmider (Ed.), *L'Aurignacien de la grotte du Renne. Les fouilles d'André Leroi-Gourhan à Arcy-sur-Cure (Yonne)* (Gallia Préhistoire supplément, Vol. XXXIV, pp. 251–256). Paris: Centre National de la Recherche Scientifique.
Tattersall, I., & Schwartz, J. H. (1999). Hominids and hybrids: The place of Neanderthals in human evolution. *Proceedings of the National Academy of Sciences of the United States of America, 96*, 7117–7119.
Templeton, A. (2002). Out of Africa again and again. *Nature, 416*, 45–50.
Templeton, A. (2005). Haplotype trees and modern human origins. *Yearbook of Physical Anthropology, 48*, 33–59.
Teyssandier, N. (2000). L'industrie aurignacienne du secteur II de Barbas (Creysse, Dordogne): Analyse technique et implications archéologiques. *Ateliers, 20*, 29–59.
Teyssandier, N. (2003). Les débuts de l'Aurignacien en Europe. Discussion à partir des sites de Geissenklösterle, Willendorf II, Krems-Hundssteig et Bacho Kiro. Ph.D. dissertation, University of Paris X.
Teyssandier, N., Bolus, M., & Conard, N. J. (2006). The early aurignacian in central Europe and its place in a European perspective. In O. Bar-Yosef & J. Zilhão (Eds.), *Towards a definition of the Aurignacian* (Trabalhos de Arqueologia, Vol. 45, pp. 241–256). Lisboa: American School of Prehistoric Research/Instituto Português de Arqueologia.
Tixier, J., & Reduron, M. (1991). Et passez au pays des silex: Rapportez-nous des lames! In *25 Ans d'Études technologiques en Préhistoire* (pp. 235–243). Juan-les-Pins: Éditions APDCA.
Trinkaus, E. (1981). Neanderthal limb proportions and cold adaptation. In C. B. Stringer (Ed.), *Aspects of human evolution* (pp. 187–224). London: Taylor and Francis.
Trinkaus, E. (2005a). Anatomical evidence for the antiquity of human footwear use. *Journal of Archaeological Science, 32*, 1515–1526.
Trinkaus, E. (2005b). The Adiposity Paradox in the middle Danubian Gravettian. *Anthropologie, 43*(2), 101–109.
Trinkaus, E. (2005c). Early modern humans. *Annual Reviews of Anthropology, 34*, 207–230.
Trinkaus, E. (2006). Modern human versus Neandertal evolutionary distinctiveness. *Current Anthropology, 47*(4), 597–620.
Trinkaus, E. (2011). Late Neandertals and early modern humans in Europe: Population dynamics and paleobiology. In G. Weniger & S. Condemi (Eds.), *Continuity and discontinuity in the peopling of Europe* (pp. 315–330). Dordrecht: Springer.
Trinkaus, E., Moldovan, O., Milota, Ş., Bîlgar, A., Sarcina, L., Athreya, S., Bailey, S. E., Rodrigo, R., Mircea, G., Higham, Th, Bronk Ramsey, C. H., & Plicht, J. v. d. (2003). An early modern human from the Peştera cu Oase, Romania. *Proceedings of the National Academy of Sciences of the United States of America, 100*, 11231–11236.
Tsanova, Ts. (2006). Entre l'Europe et le Proche-Orient, le début de la production lamellaire dans les Balkans de l'Est: l'exemple du niveau VII de la grotte Kozarnika en Bulgarie. In L. Oosterbeek, J. Raposo (Eds.), XV Congress of the International Union for Prehistoric and Protohistoric Sciences. Book of Abstracts. UISPP (pp. 544–545) Lisbon.
Tsanova, Ts., & Bordes, J.-G. (2003). Contribution au débat sur l'origine de l'Aurignacien: principaux résultats d'une étude technologique de l'industrie lithique de la couche 11 de Bacho Kiro. In Ts. Tsonev. E. Montagnari Kokelj (Eds.), *The Humanized Mineral World: Towards social and symbolic evaluation of prehistoric technologies in South Eastern Europe*. (Proceedings of the ESF workshop, Sofia - 3–6 September 2003, pp. 41–50). Liège: Études et Recherches Archéologiques de l'Université de Liège 103.
Uthmeier, Th. (1996). Ein bemerkenswert frühes Inventar des Aurignacien von der Freilandfundstelle "Keilberg-Kirche" bei Regensburg. *Archäologisches Korrespondenzblatt, 26*, 233–248.
Uthmeier, Th. (2004). *Micoquien Aurignacien und Gravettien in Bayern. Eine regionale Studie zum Übergang vom Mittel- zum Jungpaläolithikum*. Bonn: Verlag Rudolf Habelt.
Valladas, H., Reyss, J. L., Joron, J. L., Valladas, G., Bar-Yosef, O., & Vandermeersch, B. (1988). Thermoluminescence dating of Mousterian "Proto-Cro-Magnon" remains from Israel. *Nature, 331*, 614–616.
Valladas, H., Mercier, N., Escutenaire, C., Kalicki, T., Kozłowski, J., Sitlivy, V., Sobczyk, K., Zieba, A., & Van Vliet-Lanoe, B. (2003). The late Middle Paleolithic blade technologies and the transition to the Upper Paleolithic in southern Poland: TL dating contribution. *Eurasian Prehistory, 1*(1), 57–82.

Van Peer, Ph. (2004). Did Middle Stone Age moderns of Sub-Saharan African descent trigger an Upper Paleolithic revolution in the Lower Nile Valley? *Anthropologie, 42*(3), 215–225.

Vandermeersch, B. (1981). *Les Hommes fossiles de Qafzeh (Israël)*. Paris: Centre National de la Recherche Scientifique.

Vanhaeren, M. (2002). Les fonctions de la parure au Paléolithique supérieur: de l'individu à l'unité culturelle. Ph.D. dissertation, University of Bordeaux I.

Vanhaeren, M., & D'Errico, F. (2006). Aurignacian ethno-linguistic geography of Europe revealed by personal ornaments. *Journal of Archaeological Science, 33*, 1105–1128.

Vanhaeren, M., D'Errico, F., Stringer, C., James, S. L., Todd, J. A., & Mienis, H. K. (2006). Middle Paleolithic shell Beads in Israel and Algeria. *Science, 312*, 1785–1787.

Vogelsang, R. (1998). *Middle-stone-age-Fundstellen in Südwest-Namibia, Köln* (Africa Praehistorica, Vol. 11). Cologne, Germany: Heinrich-Barth-Institut.

Walker, M. J., Gibert, J., Eastham, A., Rodríguez, T., Carrión, J., Yll, E. I., Legaz, A., López, A., López, M., & Romero, G. (2004a). Neanderthals and their landscapes: Middle Palaeolithic land use in the Segura Basin and adjacent areas of southeastern Spain. In N. J. Conard (Ed.), *Settelement dynamics of the Middle Paleolithic and middle stone age* (Vol. II, pp. 461–511). Tübingen: Kerns.

Walker, M. J., Gibert, J., Rodríguez, T., López, M., Legaz, A., & López, A. (2004b). Two Neanderthal Man sites from Murcia (S.E. Spain): Sima de las Palomas del Cabezo Gordo and Cueva Negra del Estrecho del Rio Quípar. In *Actes du XIVème Congrès UISPP, Section 4, Premiers Hommes et Paléolithique Inférieur* (BAR International Series, Vol. 1272, 167–180).

Wall, J. D., & Kim, S. K. (2007). Inconsistencies in Neanderthal genomic DNA sequences. *PLoS Genetics, 3*(10), e175.

Wendt, W. E. (1974). "Art mobilier" aus der Apollo 11-Grotte in Südwest-Afrika. Die ältesten datierten Kunstwerke Afrikas. *Acta praehistorica et archaeologica, 5*, 1–42.

Weniger, G.-C. (2011). 150 years of Neanderthal research – A hopeless situation but not serious. In G. Weniger & S. Condemi (Eds.), *Continuity and discontinuity in the peopling of Europe* (pp. 379–382). Dordrecht: Springer.

Weninger, B., & Jöris, O. (2005). The Cologne radiocarbon calibration & paleoclimate research package (www.calpal.de).

Weninger, B., Danzeglocke, U., & Jöris, O. (2005). Comparison of dating results achieved using different radiocarbon-age Calibration curves and data. www.calpal.de 22 October 2005.

White, R. (1982). Rethinking the Middle/Upper Paleolithic transition. *Current Anthropology, 23*(2), 169–192.

White, R. (2002). Observations technologiques sur les objets de parure. In B. Schmider (Ed.), *L'Aurignacien de la grotte du Renne. Les fouilles d'André Leroi-Gourhan à Arcy-sur-Cure (Yonne)* (Gallia Préhistoire supplément, Vol. XXXIV, pp. 257–266). Paris: Centre National de la Recherche Scientifique.

Wild, E. M., Paunovic, M., Rabeder, G., Stefan, I., & Steier, P. (2001). Age determination of fossil bones from the Vindija Neanderthal site in Croatia. *Radiocarbon, 43*, 1021–1028.

Wolpoff, M. (2002). *Human paleontology*. Ann Arbor: University of Michigan.

Wolpoff, M. H., & Caspari, R. (2011). Neandertals and the roots of human recency. In G. Weniger & S. Condemi (Eds.), *Continuity and discontinuity in the peopling of Europe* (pp. 367–378). Dordrecht: Springer.

Zilhão, J. (1993). Le passage du Paléolithique moyen au Paléolithique supérieur dans le Portugal. In V. Cabrera (Ed.), *El Origen del Hombre Moderno en el Suroeste de Europa* (pp. 127–145). Madrid: Universidad Nacional de Educación a Distancia.

Zilhão, J. (2001). *Anatomically archaic, behaviorally modern: The last Neanderthals and their destiny*. Amsterdam: Stichting Nederlands Museum voor Anthropologie en Praehistoriae.

Zilhão, J. (2006a). Genes. Fossils and culture. An overview of the evidence for Neandertal-Modern human interaction and admixture. *Proceedings of the Prehistoric Society, 72*, 1–20.

Zilhão, J. (2006b). Neandertals and moderns mixed, and it matters. *Evolutionary Anthropology, 15*, 183–195.

Zilhão, J. (2006c). Chronostratigraphy of the Middle-to-Upper Paleolithic transition in the Iberian Peninsula. *Pyrenae, 37*, 7–84.

Zilhão, J. (2007). The emergence of ornaments and art: An archeological perspective on the origins of behavioral "modernity". *Journal of Archaeological Research, 15*(1), 1–54.

Zilhão, J. (2009a). Szeletian and early, not Aurignacian or late: A review of the chronology and cultural associations of the Vindija G1 Neandertals. In M. Camps, & P. Chauhan (Eds.), *A sourcebook of Paleolithic transitions. Methods, theories, and interpretations*. (pp. 407–426) New York: Springer.

Zilhão, J. (2009b). The Ebro frontier revisited. In M. Camps & C. Szmidt (Eds.), *The Mediterranean from 50,000 to 25,000 BP: Turning points and new directions* (pp. 293–311). Oxford: Oxbow Books.

Zilhão, J., & D'Errico, F. (1999a). Reply. In The Neanderthal problem continued. *Current Anthropology, 40*(3), 355–364.

Zilhão, J., & D'Errico, F. (1999b). The chronology and taphonomy of the earliest Aurignacian and its implications for the understanding of Neanderthal extinction. *Journal of World Prehistory, 13*(1), 1–68.

Zilhão, J., & D'Errico, F. (2000). La nouvelle "bataille aurignacienne" Une révision critique de la chronologie du Châtelperronien et de l'Aurignacien ancien. *L'Anthropologie, 104*(1), 17–50.

Zilhão, J., & D'Errico, F. (2003a). An Aurignacian "Garden of Eden" in southern Germany? An alternative interpretation of the Geissenklösterle and a critique of the Kulturpumpe model. *Paleo, 15*, 69–86.

Zilhão, J., & D'Errico, F. (2003b). The chronology of the aurignacian and transitional technocomplexes. Where do we stand? In J. Zilhão & F. D'Errico (Eds.), *The chronology of the Aurignacian and of the transitional technocomplexes. Dating, stratigraphies, cultural implications* (Trabalhos de Arqueologia, Vol. 33, pp. 313–348). Lisboa: Instituto Português de Arqueologia.

Zilhão, J., & Trinkaus, E. (2002). Historical implications. In J. Zilhão & E. Trinkaus (Eds.), *Portrait of the artist as a child. The Gravettian human skeleton from the Abrigo do Lagar Velho and its archeological context* (Trabalhos de Arqueologia, Vol. 22, pp. 542–558). Lisboa: Instituto Português de Arqueologia.

Zilhão, J., Aubry, Th., & Almeida, F. (1999). Un modèle technologique pour le passage du Gravettien au Solutréen dans le Sud-Ouest de l'Europe. In: D. Sacchi (Ed.), *Les faciès leptolithiques du nord-ouest méditerranéen: milieux naturels et culturels*. XXIVe Congrès Préhistorique de France. (26–30 Septembre 1994, Carcassonne, pp. 165–183) Carcassonne.

Zilhão, J., D'Errico, F., Bordes, J.-G., Lenoble, A., Texier, J.-P., & Rigaud, J-Ph. (2006). Analysis of Aurignacian interstratification at the Châtelperronian-type site and implications for the behavioral modernity of Neandertals. *Proceedings of the National Academy of Sciences of the United States of America, 103*(33), 12643–12648.

Zilhão, J., Trinkaus, E., Constantin, S., Milota, S., Gherase, M., Sacrina, L., Danciu, A., Rougier, H., Quilès, J., & Rodrigo, R. (2007). The Peştera cu Oase people, Europe's earliest modern humans. In P. M. Mellars, O. Bar-Yosef, C. Stringer, & K. V. Boyle (Eds.), *Rethinking the human revolution* (pp. 249–262). Cambridge: McDonald Institute for Archaeological Research.

Zilhão, J., D'Errico, F., Bordes, J.-G., Lenoble, A., Texier, J.-P., & Rigaud, J-Ph. (2008). Grotte des Fées (Châtelperron): History of research, stratigraphy, dating, and archaeology of the Châtelperronian type-site. *Paleoanthropology, 2008*, 1–42.

Chapter 26
Neandertals and the Roots of Human Recency

Milford H. Wolpoff and Rachel Caspari

That Neanderthals are thought of in terms of a "problem" or a "question" is remarkably similar to the way in which Germans thought about Jews prior to World War II. In both instances, the objects of such treatment were cast in the role of a collective "other" whose differences have been assumed to indicate the extent of their failure to qualify for fully human status.

C. Loring Brace[1]

Perhaps this is true, but there is another point; the Neandertal "other" has been the way we humans define ourselves.

Abstract The concept of modernity, or "humanness," has been difficult, if not impossible, to define. This has not prevented discussions of its appearance and evolution. In a 2003 essay the historian of science, Robert Proctor, suggested three intellectual transitions that have given rise to current understandings that "humanness" was attained recently. Two of the three transitions represent changes in the way phyletic diversity in the hominid record – the number of human species and genera that are recognized – is viewed. In this paper we explore the effect of these two transitions on our understandings of Neandertal humanity. We find that if these transitions lead to a conclusion that modernity is a phylogenetic attribute of humans, "humanness" must actually be old rather than recent and must apply to Neandertals. We propose that in contrast to the three areas explored by Proctor, a focus on major post-Neandertal demographic shifts and concomitant cultural and genetic changes presents a different intellectual foundation for understanding modernity.

Keywords Neandertal • Humanness • Modernity

M.H. Wolpoff (✉)
Department of Anthropology, University of Michigan, Ann Arbor, MI 48109-1092, USA
e-mail: wolpoff@umich.edu

R. Caspari
Department of Sociology, Anthropology and Social Work, Central Michigan University, Mt. Pleasant, MI 48859, USA
e-mail: caspa1r@cmich.edu

Introduction

Neandertals have been pivotal in any discussion of whether humanity is a recent phenomenon because they have always been important in how we define ourselves. In 2003 Robert Proctor, a historian of science, published an essay on the intellectual roots of a "human recency" hypothesis that *Homo sapiens* became "human" recently. Neandertals play a key role in this hypothesis.

Proctor isolates three realms of intellectual inquiry that have undergone transitions or "crises" that have led to what he sees as the current "consensus" of human recency:

1. Rethinking the Acheulean in archaeology
2. Recognition of phyletic diversity in human paleontology, attributed to the demise of the single species hypothesis which is described as a "dogmatic assumption" deriving from the "liberal, anti-racialist climate of the post-Auschwitz era"
3. Recognition of "biological recency" (i.e., recent African origins [RAO], or the Eve theory) in molecular anthropology

This implication is that these three transitions are revolutions that have thrown off the earlier blinders of bias, and refuted contentions that humanness is ancient. The second and third of these transitions are interrelated since both involve thinking about phyletic diversity, especially the issue of how many contemporary hominid species can be recognized. In this paper, we explore the effect of these two transitions on our understandings of Neandertal humanity and the meaning of

[1] From Brace (1964), this is the opening quotation of the Proctor (2003) paper.

S. Condemi and G.-C. Weniger (eds.), *Continuity and Discontinuity in the Peopling of Europe: One Hundred Fifty Years of Neanderthal Study*, Vertebrate Paleobiology and Paleoanthropology, DOI 10.1007/978-94-007-0492-3_26,

modernity itself, and suggest a better approach to understanding what it means to be a "modern human."

In the discussion of human recency, "humanness" can be seen as a proxy for "modernity" and both terms are equally difficult to define or even describe. Proctor takes the view that humanness has been a quality "granted" to fossils by scientists or by scientific consensus. "Granted" is a term that implies a subjective designation, outside the realm of empirical inquiry.[2] He implies that historical or political preconceptions biased the views of scientists that (mistakenly) "granted" humanity to older human remains, but that current views of human recency are less subjective, based on lines of inquiry stemming from "harder" data.

But are current views of human recency less subjective? Certainly the question of the humanness of prehistoric people is central, and one that we seek to address scientifically, but can the question be examined empirically, and if so, how? Scientific inquiry into the question of human recency is a central aspect of modern human origin studies,[3] but modernity itself has been difficult to define. The behavioral and the biological, represented by lithic and taxonomic typologies, are often conflated, and the focus is less often on the humans who are supposed to be modern than on Neandertals – the "other" whose absence of modernity helps define what modernity is.

If we accept that these three crises or transitions achieved the recognition that humanity is recent, they must in some way define humanity-. The first transition reflects a new consensus that tool use alone does not define humanity; as Proctor puts it, the Acheulean is no longer thought to reflect language or culture. The recognition that technology and "humaneness" are not linked is also underscored by increasing understanding of the behavioral complexity of non-human primates and other animals. Thus the bar is raised for evidence of "culture" or "humanity," necessitating clear evidence of symbolic activities, such as art, body ornamentation and stylized tools.

The second and third transitions are the ones we focus on here; both deal with the recognition of species and the concept that humaness is phyletic. The second transition involves the demise of the single species hypothesis.[4] This opened the door to new ways of thinking about humanness: if there are multiple ways of being a hominid, being a hominid doesn't necessarily mean being human. It further suggests there may also be multiple ways of being human – that humanity may apply to different species. This second transition is primarily an epistemological shift which Proctor attributes to sociopolitical factors: he ties the failure to recognize phyletic diversity to the liberal thinking of the post-Auschwitz years, a bias only overcome (in his view) with overwhelming new fossil evidence; in particular, the convincing evidence[5] that ER 3733 was penecontemporary with australopithecines.

The third transition represents the development of a new overwhelming form of evidence for "biological recency," meaning a young age for the existing human species. Proctor sees this transition as the consequence of revelations from molecular anthropology.

How do these views of human recency help us define humanity? If we accept that these transitions underlie contemporary understandings of modernity, what happens when we apply these factors to our thinking about Neandertals? We evaluate these transitions relative to the Neandertals in two ways. First, we examine the impact of the molecular transition on our understanding of Neandertal phylogenetics. Second, we address the epistemological issue Proctor raises by exploring whether the argument against Neandertal phyletic distinction is an artifact of liberal thinking.

The Phyletic Distinction of Modern Humans: Background

For most of the earlier half of the last century, it was quite rare to address the issue of a unique, recent origin for humanness or modernity. The origin of recent humanity is itself recent as a particular question of interest. Current thinking on the subject began with a seminal paper by Protsch (1975), summarizing his dissertation. Bräuer, Stringer, and others who later came to "champion Out-of-Africa" (Proctor 2003:224 footnote 27) intellectually followed from Protsch's work and accepted his conclusions. Protsch argued that people resembling recent humans were found in Africa earlier than anywhere else, and therefore "modern humans" must have originated there. While not all of Protsch's research has held up over the years, this paper has been highly influential; many of the dates he proposed for early modern Africans withstood the test of time and with discoveries such as Herto (White et al. 2003), Protsch's argument is commonly accepted.

[2] Proctor (2003:213) links humanness to scientists ideas about humanness through the unusual contention that humanness is "granted" to prehistoric forms by the paleoanthropologists who study them. To wit: "even older hominids were sometimes granted humanity," or "humanness is often not even granted to *Homo habilis*." This is an inaccurate description of how science works.

[3] "Humans," that is, in contrast to "hominids" or "hominines."

[4] Interestingly, he refers to this as a "demise" and not as a "disproof," thereby denying a scientific description to refutation, the most basic scientific process.

[5] Earlier claims of distinct australopithecine taxa in South Africa lacked this clarity because the purported taxa were not contemporary and could (indeed, may) have represented the same lineage at different times.

However, a single origin did not necessarily mean a phyletic origin. In fact, before the late1980s the origin of modernity was not considered phyletic at all. Neither Protsch, nor Bräuer (1978, 1984) who followed him, contended that early humans of modern form in Africa implied *unique* African origins. There was a distinct difference between how the origin of humanity was treated and consideration of the origin of *Homo sapiens* – they were not at all the same.

Models for the phyletic origin of *H. sapiens* took two forms. First, some assumed there was but a single human lineage, and that it branched off from other evolving hominid lineages quite early (reviewed by Hawks and Wolpoff 2003 and references therein). This idea of a *unique evolutionary origin* for *a human species* (*H. sapiens*) is an old one in paleoanthropology, and there is a strong intellectual thread from Haeckel (1883) and his theory of human origins in Lemuria to Howells (1942) and then to Tattersall (1995). The central contention in this thread is that recent humans are the latest populations of a species evolving in parallel with other hominid species, now extinct. However, historically, this explanation did not invoke human recency since humanity was often considered an attribute of the entire lineage.

A second phyletic model, sometimes incorporated into the first, was that modernity evolved within a broad group of competing races of the same species, evolving independently so that some became modern earlier than others and were thereby more successful (Haeckel 1883; Keith 1948; Coon 1963). This polygenic perspective differs from the first model, in that the taxa are sub-specific rather than specific; however, this may be a distinction without a difference, since gene-flow was considered to have minimal effects on the lineages. Because it dealt with race, this model had important social implications (Wolpoff and Caspari 1997). While the second model addressed the attainment of modernity through interracial competition, in neither case was *modernity itself* considered a consequence of recent phyletic divergence. These two models reappear in the discussion of polygenism, below.

A new contention about recency brought together African origins and the idea that modernity was a phyletic attribute: the Eve theory postulated an *African origin* for a *recently evolved modern human species.* The most important support for this synthesis was Proctor's third (molecular) transition: the interpretation of mitochondrial DNA that suggested the ancestors of recent humans first appeared in Africa, and replaced other populations as a new species (Stoneking and Cann 1989). The subsequent model of replacement without mixture in explanation of the process of recent populational origins was quickly accepted by some paleoanthropologists, beginning with Stringer and Andrews (1988).

The most significant implication of this synthesis is that modernity (or the recency of humanness) has a phyletic definition—modern humans are a new hominid species. A logical and necessary consequence of the branching model that associates modernity with an evolutionary lineage is that Neandertals are not human (Klein 1999; Tattersall 2002). Moreover, accepting that modernity is phyletic, human paleontologists have been driven to find an anatomical definition of this lineage and its humanness (Day and Stringer 1991; Lieberman 1995), as surely they must be able to if the theory is valid. Such definitions are invariably in contrast to Neandertals, who then *by definition* are not human. The most successful definitions, however, have not been anatomical (Wolpoff 1986) but genetic, based on the recovery of ancient mitochondrial DNA.

The Impact of the Molecular Transition on Understanding Neandertal Phylogenetics

In a section entitled "Molecular Anthropology" Proctor (2003:24) writes: "the idea of modern humans' developing slowly and separately in different parts of the world over a period of about a million years is today known as 'multiregionalism'." This is not multiregionalism. The theory he describes is known as polygenism today, and indeed has been for at least 100 years, and Proctor scored little better in his rendering of the role of molecular anthropology. "So far," he asserts (p. 225) "it seems pretty clear that the Out-of-Africanists are winning the field. Multiregionalists have no technical wonder comparable to sequencing … the extraction and sequencing of DNA from Neandertal bones [suggest] a last common ancestor with humans ca. 500,000 years ago." This is also inaccurate; mitochondrial histories are not populational histories as recent work underscores (Greene et al. 2010). But there is a far more important point. Sequencing is not the domain of either camp. It provides another source of data which has helped resolve many of the complexities of Neandertal/modern human relationships. Ironically, as it has turned out, sequencing supports multiregionalism by showing intermixture.

There is now significant information addressing the issue of mixture between the Neandertal deme and the so-called "modern human" deme that encountered Neandertals, which is the most direct evidence possible for addressing Neandertal phylogenetics. Sources of this information include:

1. Post-Neandertal anatomical variation in Europe
2. Neandertal nuclear genetic variation
3. Neandertal nuclear genes in living humans

All three address the issue of gene flow between Neandertals and other human populations, and combined they clearly show that there was gene flow into and out of Neandertal populations, who were therefore among the ancestors of later humans.

Firstly, Neandertal features.[6] This evidence of Neandertal features in later populations has been in the literature for some time, and continues to accumulate (Duarte et al. 1999; Frayer 1992, 1997; Frayer et al. 2006; Rougier et al. 2007; Soficaru et al. 2006; Trinkaus et al. 2003; Wolpoff and Caspari 1996; Wolpoff et al. 2001). It is not reviewed here, but is most definitely not what Proctor expected.

Secondly, knowledge of Neandertal nuclear genetic variation is, so far, based on the analysis of only a few specimens. Greene and colleagues (2010) show at least a 4–5% Neandertal contribution to some later populations. These are genes promoted by selection and reflect a significant degree of interbreeding. The *FOXP2* gene (Krause et al. 2007) in two El Sidrón (Spain) specimens was surprising to workers who assumed minimal Neandertal language ability. A melanocortin-1 receptor allele associated with depigmentation was found in Monte Lessini (Italy) and El Sidrón (Spain) specimens (Lalueza-Fox et al. 2007), although this allele differs from the surviving European allele whose origin seems no more than 10 kyr (Hawks et al. 2007). Analysis of the nuclear DNA from the Vi 33.16[7] individual shows evidence of genetic input from the other demes in the human population into the Neandertal population. The Vindija Neandertal differences from extant humans are quite similar to differences within humans.

Green et al. (2006) note that the mean divergence time estimate for Neandertal and contemporary human alleles is 516 kyr, while the mean divergence time between the same alleles among contemporary humans is 459 kyr. The 95% confidence intervals of these two estimates overlap (Table 26.1) and the authors conclude (p. 344): "Neanderthal genetic difference to humans must therefore be interpreted within the context of human diversity." These researchers argue that because some of the common human single nucleotide polymorphisms found in the Neandertal genome they analyzed have a more recent origin than this time span, a population split cannot account for their presence in the Neandertal. Therefore, they reason, the alleles must have entered the Neandertal gene pool more recently than the estimated divergence dates – that is, because of interbreeding.

Table 26.1 Divergence estimates of the Vindija Neandertal

	Neandertal and current alleles	Current human alleles
Mean divergence time	516 kyr	459 kyr
95% confidence interval	465–569 kyr	419–498 kyr

Divergence estimates of the Vindija Neandertal from modern alleles, and the same set of modern alleles from each other, from Green et al. (2006)

The third source, studies of nuclear variation in current populations, provides further information because these studies show that Neandertal alleles, and the alleles of other archaic populations, regularly entered the modern gene pool (Eswaran et al. 2005) and were under positive selection there (Hawks et al. 2008). Plagnol and Wall (2006) conservatively estimate about 5% nuclear gene ancestry from archaic humans and this was confirmed by Greene et al. (2010).

There is clearly no issue about whether there was *enough* gene flow for an effect in later populations because consequences are influenced by the magnitude of selection and human population expansion. Hawks and Cochrane (2006) have provided minimal estimates of what would be required to account for the observed introgressions[8] (see below) of genes under selection, and more regular interchanges between the populations would have the same effects. The question of whether there were regular interchanges has also been addressed archaeologically (D'Errico 2003; D'Errico et al. 2003; Zilhão 2001; Zilhão et al. 2006) because there is significant evidence of contact and interchange in cultural remains. But it cannot be emphasized too strongly that the evolutionary importance of genetic interchanges is related to selection acting on the genes that are interchanged and to human population growth.

A number of Neandertal descendent genes are related to brain size and function. The most compelling evidence is for one of the haplogroups of the *microcephalin* gene (Evans et al. 2005, 2006). All of the *microcephalin* haplogroups have coalescence times of close to 2 myr. Evans and colleagues claim there has been an adaptive introgression into the human population for one of the *microcephalin* haplogroups, where the gene moves from one population to another and has a selective advantage. In this particular case the Neandertals are the most likely source population because today the D haplogroup is rare in Africa, but common in Asia, and especially Europe. If the D haplogroup increased under selection a long time ago in Africa, it would be most common there. Non-D haplogroups are common in Africa (and elsewhere), but there is almost no evidence of recombination with the D haplogroups. This means that the D haplogroups attained high frequency because of selection recently when they entered the population ancestral to recent humans. There is a very similar case for the MAPT haplotype (Hardy et al. 2005), that has been linked to a possible role in Alzheimer's and Parkinson's disease.

[6] Features that are either uniquely Neandertal or very common in the Neandertal sample.

[7] This was incorrectly identified as Vi 80, but "80" references the year of discovery, not the specimen identification.

[8] Introgression refers to gene flow from one population to another, when there is evolution of genes under selection in one population that are later introduced to another population where they are spread by selection (Evans et al. 2006; Hawks and Cochran 2006).

What these have in common is that they are key alleles that entered the human population, out of archaic human varieties including Neandertals. These and other alleles increased in frequency over time under selection (Hawks et al. 2008), unlike the fate of some of the anatomical features that Neandertals contributed to the human gene pool (Frayer 1997). But the origin of these alleles is far older than the time when they first entered the human gene pool. This brings us to the second point; these (and other) beneficial genes must have been ubiquitous in Neandertal populations because they were under selection, and the fact that they did not find their way into other human populations earlier than they did addresses ancient population structure. These ancient populations, part of a world with many times fewer people than now alive, were far more isolated from each other than populations are today and may have formed subspecies (Wolpoff 2009).

So, the impact of the molecular transition as it can be understood today turns out to be quite the opposite of what Proctor expected. Now that there is direct evidence from Neandertal nuclear DNA, we recognize that Neandertals are not isolated and separately evolving. They are connected to humanity through the sharing of genes that, because they are under selection, are key genes in the evolutionary process.

Is the Argument Against Neandertal Phyletic Distinction an Artifact of Liberal Thinking?

The second phyletic model discussed above, the model of racial competition to account for the evolution of modernity, was very influential politically through the first half of the twentieth century. Proctor addresses its backlash in his argument that the holocaust and its consequences had an important effect on the interpretation of human evolution. Certainly, this model of racial competition had jarring social impact (Wolpoff and Caspari 1997). Ernst Haeckel's work provides a strong historical example. Haeckel supported the notion of common descent from a single ape ancestor for humans[9] and did not derive races from different ape species as several of his contemporaries did. According to him different human species each evolved from Pithecanthropi (different species of ape-men) living in different regions. Each attained human status; their human attributes were independently acquired through competition between them. [The ability of these species to interbreed was no impediment; like many other evolutionists of his time he accepted the notion that cross-species fertility was common in other animal species.] In the later decades of his life, Haeckel had a nationalist social agenda. His monism provided an interpretation of Darwinism within the Romantic framework of Naturphilosophie, romanticizing links between the human spirit, the land, and nature. Haeckel believed in a transcendental racial unity of the Deutsche Volk, a common spirit that bound them to the fatherland, and through Darwinism he found the mechanism explaining their natural racial superiority. Contrasting with the way that social Darwinism was used in Britain to justify *laissez-faire* capitalism by showing that individual competition was the natural way, Haeckel applied Darwinism through his theory of the competition between racial groups, explaining why the extermination and exploitation of other racial groups were the inevitable and desirable consequences of natural selection. Haeckel's contribution to biological education in Germany included the conviction that natural selection would result in European superiority as other races would be out-competed (exterminated). Nazism, for the most part developing after his death, was applied biological anthropology.[10]

Proctor believes that the subsequent impact of applied anthropology in Nazi Germany was to demote the significance of diversity, even to make it politically incorrect. For instance, "racial diversity became unfashionable after the revelation of the crimes of the Nazis" (2003:221). And in the fossil record, Proctor (2003:224) describes "a struggle over how to grant early hominids (e.g. the Neanderthals) dignity....the Multiregionalists basically maintain the UNESCO line that to deny them a close biological link to the present is to exclude them from the Family of Man."

Proctor agrees with Tattersall that there was a direct relation between this emerging perspective and the issue of humanness for Neandertals, and the link establishing it was the New Synthesis.

> Tattersall has suggested that the emphasis on population thinking during the peak prestige years of the New Synthesis also helped foster the idea that "no amount of variation" was too great to be contained within a single species (Tattersall 1995:116). This "big tent" recognition of human genetic diversity seems to have retarded the acceptance of new hominid lineages and to have made it difficult to believe that some lineages perished without issue (Proctor 2003:223).

This is an unusual interpretation of how paleoanthropologists practice science. Phenetic issues about the amount of variation that can be accepted within a fossil species have been addressed through analogies with ecologically and/or

[9] The lineage from apes to humans passed through a node that would later be allocated to the Neandertals, Haeckel named that place in his phylogeny "*Homo stupidus*."

[10] See *Race and Human Evolution* (Wolpoff and Caspari 1997).

phylogenetically similar species. The cladistic view is that species are defined by their beginnings and ends and the uniqueness of their evolutionary histories, and not by the amount of variation they contain (Wiley 1981). For the most part, Proctor's interpretation can be explained by the fact that he equates the New Synthesis with gradualism, by which he means an explanation of evolution in which all change is slow and accumulative, macroevolution is microevolution continuing for a very long time, and all speciation is anagenesis. For Proctor, like Tattersall before him, gradualism is a form of linear progressionism and its collapse was a key part of the transitions or "crises" he describes.

None of this is correct, describing neither gradualism nor the New Synthesis. The major role of populational thinking in the New Synthesis (Huxley 1942; Mayr and Provine 1980) was to connect the evolutionary mechanisms elaborated by population geneticists with the observations of paleontologists who focused on the origins of higher taxa, and include Mayr's attempts to explain the origin of evolutionary novelties (1963). It was Mayr (1954) who first proposed a model of genetic reorganization permitting the rapid emergence of evolutionary novelties at the time of speciation, and who understood that speciation was thereby a primary source of new variation. In the New Synthesis branching evolution is an engine of fundamental genetic change through the multiplication of new taxa and the replacement of old ones. A reluctance to recognize new lineages or to ignore the extinction of old ones could not logically follow from this revolution in evolutionary thinking.

The fact is that the New Synthesis did not demote the significance of diversity; quite the opposite. Nor can such a demotion[11] be attributed to political liberalism intruding into science. Proctor asserts (p.224):

"The most common fear seems to have been that by allowing multiple lineages of humans one would open the door to racism by excluding one or the other lineage from the ancestral sequence leading to modern humans."

After WWII political liberalism challenged the race concept and made it unpopular; Proctor believes this then affected interpretations of the fossil record. But how do we get from modern races to fossil species? If there were multiple hominid species and one of them led to living humans, the other species would hardly be demoted. They would be a group of sister species that could just be described as different from each other, and this is indeed the way some paleoanthropologists write about Neandertals (Laitman et al. 1992; Rak 1993); not as more primitive than humans, but as different from humans. What recent paleontologist has been more liberal in his political thinking than Gould, particularly over the issue of race (1981, 2002), and yet he supported the interpretation of multiple hominid lineages throughout his life.

[11] "the radical liberalism of the1950s and 60s," as Proctor (p. 216) puts it.

Let us be clear. We recognize the importance of sociopolitical influences on science, particularly the science of human origins (Wolpoff and Caspari 1997). But despite the importance of social influences, they do not necessarily dictate scientific results and empirical data have been used to refute social preconceptions. The one thing Proctor did not address is the possibility races were rejected for a scientific reason; the possibility that "race" was rejected as an organizing principal for living human variation because the evidence of comparative analysis and genetics showed it to be an incorrect interpretation since there are no human subspecies. This becomes intertwined with species issues in the fossil record precisely because it is unclear whether Neandertals should be regarded as a human species or a human race. Similarly, Proctor did not address the theoretical or empirical bases for incorporating Neandertals (and other hominids) into the human lineage. We think it is possible that Proctor did not focus on *scientific* explanations for the rejection of Neandertal phylogenetic distinction because he starts with the assumption that there can be no empirical basis for this position because a branching model validly defines modernity.

The Implications of Introgression[8]

The introgression of Neandertal genes into the human genome, and vice-versa, addresses a key contention in Proctor's work discussed above, the contention that liberal thinking about race colors the interpretation of fossils. In his logical progression, the holocaust was the ultimate consequence of racial views, so the post- holocaust liberal view was that races don't exist. And if human variation cannot be described as racial, it followed that all phyletic categories should be reduced, and hominid species, too, disappeared with the single species hypothesis.[12] Therefore, liberalism is blamed for putting off recognition of diversity.[13]

However, we contend the race concept was refuted because of the pattern of human genetic variation, not the amount, and the Single Species Hypothesis, based on the competitive exclusion principle, derived from ecological models, not human social factors. In both cases, it is the *pattern* of variation that is important for understanding relationship – the absence of population structure in the case of race, or the absence of adaptive shifts in the case of species. The models of human unity and the Single Species Hypothesis

[12] Proctor missed the even more massive taxonomic reductions that came with the revision of the hominoid primates by Simons and Pilbeam (1965).

[13] Proctor is not alone in these assertions about liberalism and science, but we think that in the history of science, just as in science itself, a common view, even a majority view, is not necessarily a correct view.

were very different from each other, with fundamentally different theoretical foundations. However, when the logic of "no races" is applied to the fossil record, it gives us a valuable instrument for recognizing differences.

Proctor's logic is that this reasoning about race leads to a lower level of taxonomy for Neandertals, a liberal interpretation extended to the fossil record. But when this reasoning about race is applied to the fossil record, quite the opposite is true – Neandertals are shown to be a subspecies, unlike human "races". This is underscored by the recognition of significant introgression between archaic populations (Hawks et al. 2008), which indicates that interbreeding between Neandertals and modern humans took place, but that those populations were more subdivided than populations of today. The recent introgression of genes that evolved in Neandertals under selection at a much earlier time shows that Neandertal populations were significantly isolated from other human populations (as well they may have been from each other (Reich et al. 2010)). Under selection, these genes would have been ubiquitous in Neandertal populations; yet, they took considerable time to enter the human gene pool, implying lack of widespread contact. That fact, combined with the magnitude of differences between Neandertals and their penecontemporaries, is more than sufficient to show they were a human subspecies. Templeton (1998) notes that in the zoological literature the meaning of "race" is "subspecies." And so it is not a liberal view that pushes scientists to consider Neandertals within the human species, but an analysis of patterns of variation. One of the most interesting consequences of the recent recognition of introgression is the first clear genetic demonstration that humans in the past, unlike humans today, formed distinct races (Wolpoff 2009).

We also question whether the interpretations of human races today are derivative of liberalism; despite overwhelming evidence to the contrary, the race concept persists (Caspari 2003, 2010). Current human variation does not attain the subspecies level. Populations are neither different enough nor separated enough for a subspecies interpretation of their variation to be valid. The ratio of within group to between group variance is very high in humans, genetic evidence of gene flow extends deep into the past, and there is no treeness[14] for human groups (Templeton 1998). In these assertions Templeton's conclusions are similar to virtually all anthropologists.

So Can We Accept Neandertals as Human?

If we regard humanness as a phylogenetic attribute, there is a simple answer to this question. If Neandertals belong to the same species as their penecontemporaries such as Omo, Herto, or Qafzeh (all widely regarded as modern human), then they too are modern humans. In considering the question of what can fit under a rubric of humanness Proctor (p. 229) writes: "we may choose to include *erectus* or to exclude them [in humanity]. This is a separate question from whether they could have bred with 'us'." But is this actually a separate question? We now have convincing evidence that Neandertals and other human populations exchanged genes. We know that Neandertal genes dispersed to other populations *under selection* and led to significant adaptive changes (Evans et al. 2005, 2006; Hawks et al. 2008; Greene et al. 2010). These observations reinforce the anatomical observations of mixed ancestry for post-Neandertal European populations (Trinkaus et al. 2003; Wolpoff et al. 2001) and inexorably lead to the conclusion that for both phylogenetic and biological reasons, since one intermingled population is not more human than the other, Neandertals are human.

Discussion

We have addressed Neandertals and the origin of humanness using criteria recently invoked by Proctor because they address key issues. Two of Proctor's three roots of human recency involve historical transitions in thinking about human phylogeny. These are the ones we have dealt with here, because they impinge on the way we interpret Neandertal humanness. One of the transitions was epistemological; Proctor argued that with the rejection of the single species hypothesis, it was possible to recognize more species in the hominid lineage, including, ultimately, a modern human species. Proctor considers the failure to recognize multiple species in the past the result of post-war liberal bias, expressed in the New Synthesis. The other transition was the advent of molecular anthropology, which, Proctor contends, provided evidence of a recent modern human species. Therefore, using these criteria, recognition of human recency rests on recognition of a modern human species.

But is a reluctance to taxonomize variation a consequence of liberalism or does it reflect biological reality? The question is quite different as addressed to variation below and above the level of *Homo sapiens*. Below the species level, taxonomizing variation has lead to the polygenic view that human geographic "races" have been essentially isolated from each other, and that their evolutionary histories are competitive but otherwise mutually independent. Since Haeckel, polygenism continued to be held by central, often influential, figures within paleoanthropology such as Hooton (1946), Keith (1948), and Coon (1963) up to the middle of the twentieth century (Bowler 1986; Wolpoff and Caspari 1997). Polygenic

[14] In valid subspecies, the relationship of one subspecies should be equal to all subspecies descendent from an earlier branch.

interpretations of human geographic diversity continued to reappear long after separate primate origins for human "races" had been thoroughly falsified (Brace 1981). Polygenism, a cornerstone of racism, was always poorly supported scientifically; it was *scientifically* rejected with the understanding of intra specific evolutionary processes within our species. If liberalism has played any role in the science, it has been in its promotion of the implications of the results of evolutionary research, and in firmly burying the corrosive theory of polygenism.

It is only recently that a focus on recognizing and understanding intra specific evolutionary processes became important, and this was a consequence of the rise of genetics in the first half of the twentieth century. It was the development of the current understanding of genetics, not liberal bias, which underlies the New Synthesis' focus on populations; moreover, the mechanisms of speciation and phyletic diversity were also major components of the synthesis. The New Synthesis was not primarily a product of post-war political sentiment, and did not minimize phyletic diversity. Rather it brought *process* to the understanding of both intra and inter-specific evolutionary variation. The rejection of the race concept, while championed by post-war liberals, was a product of the emerging recognition of limited genetic diversity within the human species, which had as much to do with the *pattern* of genetic variation as with the *amount*. The low between-group genetic variation in humans indicated that there are no human subspecies today.

But were there sub specific clades in the past, with more isolation and greater variation than human populations show today? What are the demographic consequences of the drastically smaller population sizes before the Neolithic, and the significantly shorter human life span before the Upper Paleolithic? These questions arise anew because of the ways genetics has developed in this twenty-first century. The molecular transition has provided important data involving modern human origin. As we discuss below, genetic evidence suggests more population subdivision in the past, but it does not support the phyletic recency of humanity. Multiple factors, including natural selection, affect genetic divergence times and genetic divergences are not population divergences. Mean genetic divergence time estimated for at least one Neandertal and current humans is only slightly older than the mean divergence times for living humans from each other (both are about 500 kyr). For living humans, we know that these mean genetic divergence times do not reflect population divergence (we have no human subspecies). But for Neandertals, genetic divergence is assumed to represent just that, since we enter into their analysis with the *assumption* of phyletic separation. Indeed, the divergence times used to interpret Neandertal nuclear DNA are derived from the fossil record.[15] The molecular transition's contribution to thinking about human recency has thereby been circular; its support for the phyletic distinction of Neandertal and modern human species starts with the assumption of phyletic distinction.

But recent genetic analysis moving beyond phylogeny holds interesting suggestions for population structure in the past. Focus on alleles under selection in recent humans, coupled with increased understanding of the Neandertal nuclear genome suggests the possibility that there were past human subspecies. The molecular data indicate introgression between two demes, members of the same species because they exchange genes, but demes more distinct from each other than any human populations today. This is what subspecies would be like, if we still had them. It is the thinking of the New Synthesis that provides evidence of past human subspecies.

Thus it is ironic that if we use Proctor's transitions which define humanity phyletically, then humanity is both more variable in the past, more ancient than usually thought, and includes the same Neandertal populations that serve as "the other" in classic definitions of modern humans. Only if the "Eve" or RAO theory were valid could we hope to find an unambiguous understanding of modernity (or humanness), and only then could we expect the biological and behavioral aspects of modernity to be linked. But this is not the case and modernity (or humanness) remains difficult to identify on anatomical grounds, and can only be examined empirically if we can agree on criteria. Because these criteria differ widely, there is little consensus about the appearance of humanity or modernity. If phyletic criteria are used, we argue above that modernity is old, and must apply to Neandertals as well. If behavioral criteria are used, modernity may mean something quite different and may appear much more recently. Interestingly, the molecular transition, while not providing support for human phyletic recency, does suggest a third meaning of modernity which is much more recent than the period traditionally focussed on by paleoanthropologists. Fueled by post-Neandertal demographic changes, genetic modernity may be a product of the last 10,000 years.

The behavioral changes in the last 30–40,000 years of human evolution, the period after Neandertals, are dramatic and have a demographic basis. Genetic, archaeological and paleontological data all point to population expansions within the last 40,000 years, likely fueled by an increase in adult survivorship Caspari and Lee (2004, 2006). The potential demographic consequences of increased adult survivorship are significant. Not only does increased survivorship imply greater lifetime fertility for individuals, the investment of older individuals in their children's families may provide

[15] Green and colleagues (2006: 335) use a split time "inferred from the fossil record," citing Hublin. This both assumes that there *was* a populational split, and that it can be estimated from fossils.

a selective advantage promoting further population increase. Population expansion may have provided social pressures that led to extensive trade networks, increased mobility, and more complex systems of cooperation and competition between groups, resulting in increased personal ornamentation and other material expressions of individual and group identity. Moreover, increased adult survivorship allows for the increased importance of transgenerational relationships that are critical to the development and survival of social groups with large amounts of complex information to transmit (Caspari and Lee 2004). Population growth and expansion increases the number of mutations and thereby the number of changes due to selection, resulting in significant genetic changes in the human species in our recent evolutionary history (Hawks et al. 2007).

In this context, we propose that Neandertals are modern human, in the sense that their penecontemporaries such as Omo, Herto, and Qafzeh are modern humans. But behaviorally and genetically Neandertals are not "just like us." Hawks et al. (2007:20756–20757) write that "to the extent that adaptive alleles continue to reflect demographic growth, the Neolithic and later periods would have experienced a rate of adaptive evolution >100 times higher than characterized most of human evolution." This caveat looms over any consideration of the recency of modernity.

Conclusion

We suggest that the three intellectual roots of human recency that Proctor outlines lead to aspects of "humanness" that are not recent at all, and that humanness can and should describe Neandertals no more or less than it describes other penecontemporary archaic human populations. But the true root of human recency is behavioral and lies in the demographic changes that result from a new pattern of life history. These demographic shifts are truly unique to recent humans and underlie the population expansions, related behavioral innovations, and increased rate of genetic change associated with "modern" humans and their continuing evolution.

Acknowledgements We are grateful to Dr. Silvana Condemi and Wighart von Koenigswald for our invitation to attend the "150 Years of Neanderthal Discovery" conference in Bonn and to contribute to this volume. We thank the two anonymous reviews for their help.

References

Bowler, P. J. (1986). *Theories of human evolution: A century of debate, 1844–1944*. Baltimore: John Hopkins University Press.

Brace, C. L. (1964). The fate of the "Classic" Neanderthals: A consideration of hominid catastrophism. *Current Anthropology, 5*, 3–43. 7, 204–214.

Brace, C. L. (1981). Tales of the phylogenetic woods: The evolution and significance of evolutionary trees. *American Journal of Physical Anthropology, 56*(4), 411–429.

Bräuer, G. (1978). The morphological differentiation of anatomically modern man in Africa, with special regard to recent finds from East Africa. *Zeitschrift für Morphologieund Anthropologie, 69*(3), 266–292.

Bräuer, G. (1984). The "Afro-European sapiens hypothesis" and hominid evolution in East Asia during the late middle and upper Pleistocene. In P. Andrews & J. L. Franzen (Eds.), *The early evolution of man, with special emphasis on Southeast Asia and Africa* (Courier Forschungs institut Senckenberg, Vol. 69, pp. 145–165).

Caspari, R. (2003). From types to populations: A century of race, physical anthropology and the American anthropological association. *American Anthropologist, 105*(1), 63–74.

Caspari, R. (2010). Deconstructing race: race, racial thinking and geographic variation. In: *Companion to Biological Anthropology*. C. Larsen (Ed.) Wiley-Liss, pp. 104–122.

Caspari, R., & Lee, S.-H. (2004). Older age becomes common late in human evolution. *Proceedings of the National Academy of Sciences of the United States of America, 101*, 10895–10900.

Caspari, R., & Lee, S.-H. (2006). Is human longevity a consequence of cultural change or modern biology? *American Journal of Physical Anthropology, 129*(4), 512–517.

Coon, C. S. (1963). *The origin of races*. New York: Knopf.

D'Errico, F. (2003). The invisible frontier: A multiple species model for the origin of behavioral modernity. *Evolutionary Anthropology, 12*, 188–202.

D'Errico, F., Julien, M., Liolios, D., VanHaren, M., & Baffier, D. (2003). Many awls in our argument. Bone tool manufacture and use in the Châtelperronian and Aurignacian levels of the Grotte de Renne at Arcy-sur-Cure. In J. Zilhão,. & F. D'Errico (Eds.), *The Chronology of the Aurignacian and of the Transitional Technocomplexes.Dating, Stratigraphies, Cultural Implications*. Proceedings of Symposium 6.1 of the XIVth Congress of the UISP, Instituto Portugês de Arqueologia, Lisbon (pp. 247–270).

Day, M. H., & Stringer, C. B. (1991). Les restes crâniens d'Omo-Kibish et leur classification a l'intérieur de Genre Homo. *L'Anthropologie, 95*(2/3), 573–594.

Deeb, S. S., Jørgensen, A. L., Battisi, L., Iwasaki, L., & Motulsky, A. G. (1994). Sequence divergence of the red and green visual pigments in the great apes and man. *Proceedings of the National Academy of Sciences of the United States of America, 91*, 7262–7266.

Duarte, C., Maurício, J., Pettitt, P. B., Souto, P., Trinkaus, E., van der Plicht, H., & Zilhão, J. (1999). The early upper Paleolithic human skeleton from the Abrigo do Lagar Velho (Portugal) and modern human emergence in Iberia. *Proceedings of the National Academy of Sciences of the United States of America, 96*, 7604–7609.

Eswaran, V., Harpending, H., & Rogers, A. R. (2005). Genomics refutes an exclusively African origin of humans. *Journal of Human Evolution, 49*(1), 1–18.

Evans, P. D., Gilbert, S. L., Mekel-Bobrov, N., Vallender, E. J., Anderson, J. R., Vaez-Azizi, L. M., Tishkoff, S. A., Hudson, R. R., & Lahn, B. T. (2005). *Microcephalin*, a gene regulating brain size, continues to evolve adaptively in humans. *Science, 309*, 1717–1720.

Evans, P. D., Mekel-Bobrov, N., Vallender, E. J., Hudson, R. R., & Lahn, B. T. (2006). Evidence that the adaptive allele of the brain size gene *microcephalin* introgressed into *Homo sapiens* from an archaic *Homo* lineage. *Proceedings of the National Academy of Sciences of the United States of America, 103*, 18178–18183.

Frayer, D. W. (1992). Evolution at the European edge: Neanderthal and the upper Paleolithic relationships. *Préhistoire Europèene/European Prehistory, 2*, 9–69.

Frayer, D. W. (1997). Perspectives on Neanderthals as ancestors. In G. A. Clark & C. M. Willermet (Eds.), *Conceptual issues in modern human origins research* (pp. 220–235). New York: Aldine de Gruyter.

Frayer, D. W., Jelínek, J., Oliva, M., & Wolpoff, M. H. (2006). Aurignacian male crania, jaws, and teeth from the Mladeč Caves, Moravia, Czech Republic. In M. Teschler-Nicola (Ed.), *Early modern humans at the Moravian gate: The Mladeč caves and their remains* (pp. 185–272). Wien: Springer.

Gould, S. J. (1981). *The mis measure of man*. New York: Norton.

Gould, S. J. (2002). *The structure of evolutionary theory*. Cambridge: Belknap Press of Harvard University Press.

Green, R. E., Krause, J., Ptak, S. E., Briggs, A. W., Ronan, M. T., Simons, J. F., Du, L., Egholm, M., Rothberg, J. M., Paunovic, M., & Pääbo, S. (2006). Analysis of one million base pairs of Neanderthal DNA. *Nature, 444*, 330–336.

Green, R. E., Krause, J., Briggs, A. W., Maricic, T., Stenzel, U., Kircher, M., Patterson, N., Li Heng, Zhai Weiwei, M. Fritz, M. H-Y., Hansen, N. F., Durand, E. Y., Malaspinas, A-S., Jensen, J. D., Marques-Bonet, T., Alkan, C., Prüfer, K., Meyer, M., Burbano, H. A., Good, J. M., Schultz, R., Aximu-Petri, A., Butthof, A., Höber, B., Höffner, B., Siegemund, M., Weihmann, A., Nusbaum, C., Lander, E. S., Russ, C., Novod, Affourtit, J., Egholm, M., Verna, C., Rudan, P., Brajkovic, D., Kucan, Ž., Gušic, I., Doronichev, V. B., Golovanova, L. V. Lalueza-Fox, C., de la Rasilla, M, Fortea, J., Rosas, A., Schmitz, R. W., Johnson, P. L. F., Eichler, E. E., Falush, D., Birney, E., Mullikin, J. C., Slatkin, M., Nielsen, R., Kelso, J., Lachmann, M., Reich, D., & Pääbo, S. (2010). A draft sequence of the Neandertal genome. *Science, 328*, 710–722.

Haeckel, E. (1883). *The history of creation, or the development of the earth and its inhabitants by natural causes. A popular exposition of the doctrine of evolution in general, and that of Darwin, Goethe, and Lamark in particular*. New York: Appleton.

Harding, R. M., Healy, E., Ray, A. J., Ellis, N. S., Flanagan, N., Todd, C., Dixon, C., Sajantila, A., Jackson, I. J., Birch-Machin, M. A., & Rees, J. L. (2000). Evidence for variable selective pressures at MC1R. *American Journal of Human Genetics, 66*(4), 1351–1361.

Hardy, J., Pittman, A., Myers, A., Gwinn-Hardy, K., Fung, H. C., de Silva, R., Hutton, M., & Duckworth, J. (2005). Evidence suggesting that *Homo neanderthalensis* contributed the H2 MAPT haplotype to *Homo sapiens*. *Biochemical Society Transactions, 33*, 582–585.

Hawks, J., & Cochran, G. (2006). Dynamics of adaptive introgression from archaic to modern humans. *PaleoAnthropology, 2006*, 101–115.

Hawks, J., & Wolpoff, M. H. (2003). Sixty years of modern human origins in the American anthropological association. *American Anthropologist, 105*(1), 87–98.

Hawks, J., Wang, E. T., Cochran, G. M., Harpending, H. C., & Moyzis, R. K. (2007). Recent acceleration of human adaptive evolution. *Proceedings of the National Academy of Sciences of the United States of America, 104*(52), 20753–20758.

Hawks, J., Cochran, G., Harpending, H. C., & Lahn, B. T. (2008). A genetic legacy from archaic *Homo*. *Trends in Genetics, 24*(1), 19–23.

Hooton, E. A. (1946). *Up from the Ape* (revisedth ed.). New York: MacMillan.

Howells, W. W. (1942). Fossil man and the origin of races. *American Anthropologist, 44*, 182–193.

Huxley, J. S. (1942). *Evolution, the modern synthesis*. London: Allen.

Keith, A. (1948). *A new theory of human evolution*. New York: Philosophical Library.

Klein, R. G. (1999). *The human career: Human biological and cultural origins* (2nd ed.). Chicago: University of Chicago Press.

Krause, J., Lalueza-Fox, C., Orlando, L., Enard, W., Green, R. E., Burbano, H. A., Hublin, J.-J., Bertranpetit, J., Hänni, C., Fortea, J., de la Rasilla, M., Rosas, A., & Pääbo, S. (2007). The derived Fox P2 variant of modern humans was shared with Neandertals. *Current Biology, 17*, 1–5.

Laitman, J. T., Reidenberg, J. S., & Gannon, P. J. (1992). Fossil skulls and hominid vocal tracts: New approaches to charting the evolution of human speech. In J. Wind, B. Chiarelli, B. Bichakjian, & A. Nocentini (Eds.), *Language origin: A multidisciplinary approach* (pp. 395–407). Dordrecht: Kluwer.

Lalueza-Fox, C., Römpler, H., Caramelli, D., Stäubert, C., Catalano, G., Hughes, D., Rohland, N., Pilli, E., Longo, L., Condemi, S., de la Rasilla, M., Fortea, J., Rosas, A., Stoneking, M., Schöneberg, T., Bertranpetit, J., & Hofreiter, M. (2007). A melanocortin 1 receptor allele suggests varying pigmentation among Neanderthals. *Science, 318*, 1453–1455.

Lieberman, D. E. (1995). Testing hypotheses about recent human evolution from skulls. Integrating morphology, function, development, and phylogeny. *Current Anthropology, 36*(2), 159–197.

Mayr, E. (1954). Changes of genetic environment and evolution. In J. S. Huxley, A. C. Hardy, & E. B. Ford (Eds.), *Evolution as a process* (pp. 188–213). London: Allen & Unwin.

Mayr, E. (1963). *Animal species and evolution*. Cambridge: Belknap Press of Harvard University Press.

Mayr, E., & Provine, W. B. (Eds.). (1980). *The evolutionary synthesis: Perspectives on the unification of biology*. Cambridge: Belknap Press of Harvard University Press.

Morral, N., Nunes, V., Casals, T., Chill6n, M., Gim6nez, J., Bertranpetit, J., & Estivill, X. (1993). Micro sattelite haplotypes for cystic fibrosis: Mutation frameworks and evolutionary tracers. *Human Molecular Genetics, 2*(7), 1015–1022.

Plagnol, V., & Wall, J. D. (2006). Possible ancestral structure in human populations. *PLoS Genetics, 2*(7), e105.

Proctor, R. N. (2003). Three roots of human recency. *Current Anthropology, 44*(2), 213–239.

Protsch, R. (1975). The absolute dating of upper Pleistocene sub-Saharan fossil hominids and their place in human evolution. *Journal of Human Evolution, 4*, 297–322.

Rak, Y. (1993). Morphological variation in *Homo neanderthalensis* and *Homo sapiens* in the Levant: A biogeographic model. In W. H. Kimbel & L. B. Martin (Eds.), *Species, species concepts, and primate evolution* (pp. 523–536). New York: Plenum.

Reich, D., Green, R. E., Kircher, M., Krause, J., Patterson, N., Durand, E. Y., Viola, B., Briggs, A. W., Stenzel, U., Johnson, P. F. L., Maricic, T., Good, J. M., Marques-Bonet, T., Alkan, C., Fu Qiaomei, Mallick, S., Li Heng, Meyer, M., Eichler, E. E., Stoneking, M., Richards, M., Talamo, S., Shunkov, M. V., Derevianko, A. P., Hublin, J-J., Kelso, J., Slatkin, M. & Pääbo, S. (2010). Genetic history of an archaic hominin group from Denisova Cave in Siberia. *Nature, 468*, 1053–1060.

Rougier, H., Milota, S., Rodrigo, R., Gherase, M., Sarcina, L., Moldovan, O., Zilhão, J., Constantin, S., Franciscus, R. G., Zollikofer, C. P. E., Ponce de León, M., & Trinkaus, E. (2007). Petera cu Oase 2 and the cranial morphology of early modern Europeans. *Proceedings of the National Academy of Sciences of the United States of America, 104*(4), 1165–1170.

Simons, E. L., & Pilbeam, D. (1965). Preliminary revision of the Dryopithecinae (Pongidae, Anthropoidea). *Folia Primatologia, 3*, 81–152.

Soficaru, A., Dobos, S., & Trinkaus, E. (2006). Early modern humans from the Petera Muierii, Baia de Fier, Romania. *Proceedings of the National Academy of Sciences of the United States of America, 103*, 17196–17201.

Stoneking, M., & Cann, R. L. (1989). African origins of human mitochondrial DNA. In P. Mellars & C. B. Stringer (Eds.), *The human revolution: Behavioural and biological perspectives on the origins of modern humans* (pp. 17–30). Edinburgh: Edinburgh University Press.

Stringer, C. B., & Andrews, P. (1988). Genetic and fossil evidence for the origin of modern humans. *Science, 239*, 1263–68.

Tattersall, I. (1995). *The fossil trail: How we know what we think we know about human evolution*. New York: Oxford University Press.

Tattersall, I. (2002). *The Monkey in the mirror. Essays on the science of what makes us human*. New York: Oxford University Press.

Templeton, A. R. (1998). Human races: A genetic and evolutionary perspective. *American Anthropologist, 100*(3), 632–650.

Trinkaus, E., Milota, Ş., Rodrigo, R., Mircea, G., & Moldovan, O. (2003). Early modern human cranial remains from the Peştera cuOase, Romania. *Journal of Human Evolution, 45*, 245–253.

White, T. D., Asfaw, B., Degusta, D., Gilbert, H., Richards, G. D., Suwa, G., & Howell, F. C. (2003). Pleistocene *Homo sapiens* from Middle Awash, Ethiopia. *Nature, 423*, 742–747.

Wiley, E. O. (1981). *Phylogenetics. The theory and practice of phylogenetic systematics*. New York: Wiley.

Wolpoff, M. H. (1986).Describing anatomically modern Homo sapiens: a distinction without a definable difference. In V. V. Novotný & A. Mizerová (Eds.), *Fossil Man. New Facts, New Ideas*. Papers in Honor of Jan Jelínek's Life Anniversary. Anthropos (Brno) (23: 41–53).

Wolpoff, M. H. (2009). How Neandertals inform human variation. *American Journal of Physical Anthropology, 139*(1), 91–102.

Wolpoff, M. H., & Caspari, R. (1996). An unparalleled parallelism. *Anthropologie (Brno), 34*(3), 215–223.

Wolpoff, M. H., & Caspari, R. (1997). *Race and human evolution*. New York: Simon & Schuster.

Wolpoff, M. H., Hawks, J. D., Frayer, D. W., & Hunley, K. (2001). Modern human ancestry at the peripheries: A test of the replacement theory. *Science, 291*, 293–297.

Zilhão, J. (2001). Neandertal/modern human interaction in Europe. In P. Thacker & M. Hays (Eds.), *Questioning the answers: Resolving fundamental problems of the early upper Paleolithic* (British archaeological reports international series, Vol. 1005, pp. 13–19).

Zilhão, J., d'Errico, F., Bordes, J.-G., Lenoble, A., Texier, J.-G., & Rigaud, J.-P. (2006). Analysis of Aurignacian inter stratification at the Châtelperronian-type site and implications for the behavioral modernity of Neandertals. *Proceedings of the National Academy of Sciences of the United States of America, 103*(33), 12643–12648.

Chapter 27
Epilogue: 150 Years of Neanderthal Research – A Hopeless Situation but Not Serious

Gerd-Christian Weniger

Since the Western world first became aware of Neanderthals, this Pleistocene human has been a regular focus of both public and specialist interest. In fact, we know far more about Neanderthals than we do about any other extinct human. Further, over the past 150 years no other palaeospecies has provided such a constant source of discussion and fierce debate among palaeoanthropologists (human paleontologists) and archaeologists.

Whereas at the time of the Neanderthal's discovery there was still no scientific discipline engaged explicitly in the study of the history of humankind in the Pleistocene, today at least three different specialist fields are significantly involved in research: archaeology, palaeoanthropology and palaeogenetics. In the first instance all three are reliant for their investigations upon direct evidence from these Ice Age people. However, the size of samples available to scientists of the three disciplines is subject to extreme variation. Whilst archaeologists have at their disposal more than 1,000 Middle Palaeolithic sites with several thousand archaeological levels, as well as some hundred thousand artefacts, palaeoanthropologists only have access to around 300 Neanderthal individuals represented by about 3,000 bone fragments; finally, palaeogeneticists must rely on less than 20 samples of mtDNA and even less of nuclear DNA from Neanderthal fossils. Based on their empirical data each of these disciplines has developed its own models of Neanderthal history.

Notwithstanding, all three disciplines are intrinsically linked, albeit that each draws upon quite different methods and is influenced by different research cultures and world views. For this reason, even after some 150 years of research, many results and interpretations from these different scientific branches are not particularly congruent. There is still no generally accepted theory concerning the rise of the Neanderthal as a Pleistocene form of *Homo* and its disappearance. Evidently, none of the three disciplines has so far been able to develop an authoritative theory on the basis of its own data alone. This difficult situation has resulted in the renunciation by the research community of holistic approaches to research, and in its place we are witnessing a demand for highly specialized, high-resolution laboratory procedures. Seen critically, one might describe this trend as an 'escape to manual processing'. It is driven by the hope that more precise laboratory analysis will provide presumed "hard evidence" and help avoid the predicament of "biohistorical" vagueness. Furthermore, behind this reaction lies the hope of being released from the responsibility of devising a plausible model for something that in the meantime has accumulated into a gigantic mountain of tangled data from the spheres of archaeology, palaeoanthropology, palaeogenetics, geology and climatology. Be this as it may, the only chance we might have to solve the Neanderthal problem is to develop models that try to integrate all available data.

Cardinal findings from archaeology, palaeoanthropology and palaeogenetics considered as largely confirmed are:

- From a palaeogenetic viewpoint there is so far no indication for the substantial exchange of genetic material between Neanderthals and anatomically modern humans.
- From a palaeoanthropological viewpoint there is no indication for a substantial fusion of both of these morphological forms.
- From an archaeological viewpoint there is no indication for significant cultural differences between Neanderthals and anatomically modern humans.

Although a multitude of different opinions still exists with respect to these core messages, a significant majority of the research community from each of the respective disciplines is in support of these conclusions. Taken seriously and arranged in a plausible way in a line of argument, these messages provoke the mandatory question as to whether Neanderthals and anatomically modern humans ever actually met in Europe.

The dispersal of radiocarbon dates between 45,000 BP and 25,000 BP shows a distinctive period of overlap between the Middle Palaeolithic and the subsequent Aurignacian, and

G.-C. Weniger (✉)
Neanderthal Museum, Talstrasse 300, D-40822
Mettmann, Germany
e-mail: weniger@neanderthal.de

S. Condemi and G.-C. Weniger (eds.), *Continuity and Discontinuity in the Peopling of Europe: One Hundred Fifty Years of Neanderthal Study*, Vertebrate Paleobiology and Paleoanthropology, DOI 10.1007/978-94-007-0492-3_27,

even into the Gravettian. However, the 'historical' probability of this pattern must be contested; a large number of studies which have focused on the radiocarbon chronology have shown that dating in this particular time period is marred by some considerable inconsistencies. For this reason it is perhaps advisable to return to the simple, and at the same time reliable, stratigraphical method of geology. So far in Europe there is no known interstratification of Middle Palaeolithic and Châtelperronian or Aurignacian assemblages, nor is there evidence of interstratification of Châtelperronian and subsequent Aurignacian complexes. Indeed, local stratigraphies indicate in contrast to radiocarbon chronology a clear chronological succession of these aforementioned cultural complexes.

Parallel to analyses of the relative-chronology, it is essential to carefully consider the climatic events which impacted Europe in the critical period between 45 and 30 ka. The long, precise and well correlated stratigraphic sequences from marine sediments and ice cores form the backbone of modern palaeoclimatological research. These proxies provide a high resolution record of climate trends during the last Ice Age in Europe. They range in chronological accuracy from centuries to just a few decades, and as such they are characterized by a much higher resolution than the available archaeological data. The data from these cores disclose that the period between 45 and 25 ka was distinguished by exceptionally acute climatic fluctuation. The last warm event associated with conditions that were similar to those currently prevailing in Europe occurred ca. 45 ka. Subsequently, there ensued a series of very short, rapid climate changes, the so called Dansgaard-Oeschger events.

The majority of the climate oscillations to have occurred in this period still cannot be adequately synchronized with climate data from terrestrial geological and palaeobotanical proxies. However, they show that north of the Pyrenees, the Alps and the Carpathian Mountains, hunter-gatherer populations of this period were subjected to continual climate change. Therefore, we can assume that within the space of just a few generations the northerly settlement boundary was regularly pushed back as far as the Mediterranean region. Climate driven migration of megafauna is also very well attested in Europe for the last Ice Age. The high resolution climate proxies suggest the development of a pronounced yo-yo effect with dramatic consequences for human populations. Cold phases would have been characterized by the abandonment of territories, increased mobility, reduction in population density, retreat into the Mediterranean region, and in severe cases the extinction of a percentage of the population. Warm phases would have made possible a northward re-dispersal of populations. Therefore, there is growing evidence that prevailing climate conditions would not have permitted settlement continuity in Europe up to 52° north latitude between 45 and 25 ka.

In addition to the standard climate oscillations of this period, on six occasions there ensued exceptional climate anomalies now known as Heinrich events which depart from the standard pattern of Dansgaard-Oeschger oscillations. The most important characteristic of Heinrich events is extreme aridity in the south. In the Mediterranean region semi-deserts developed, a phenomenon which is not only attested on the Iberian Peninsula but also in the Eastern Mediterranean; water levels at Lake Lisan – the Pleistocene Dead Sea – dropped by more than 50 m during Heinrich event 4. In fact, Heinrich event 4 shows a positive temporal correlation with the transition from Middle Palaeolithic to the Aurignacian at around 40 ka. On the basis of what is currently known about the Heinrich events, there is good reason to assume that these severe climate anomalies could have breached the standard cold phase migration patterns of hunter-gatherer populations. The extreme aridity to have prevailed in the retreat areas of the Mediterranean at these times could have resulted in the total collapse of human populations. Heinrich events were short in duration, lasting a maximum of 200 years. Typically, they ended abruptly and were followed by very swift climate amelioraton within just a few decades. Consequently, it is likely that these reversed conditions, which would have encouraged the development of extensive grasslands, would have quickly attracted migratory ungulate populations, and therefore humans also, from West Asia. These groups would have migrated into a largely deserted Europe, but a Europe which would have been very attractive due to its high ungulate biomass. These "pull factors" might even explain the tremendous speed of Aurignacian expansion.

Thus, the hypotheses of population dynamics described here unite the apparently incompatible core messages from archaeology, palaeoanthropology and palaeogenetics. Against the background of this demographic scenario the likelihood is rather slight that Neanderthals and anatomically modern humans would have encountered one another in Europe. However, if it ever did come to contacts, it is most likely that these would have occurred in the eastern territories of Neanderthal dispersal; in these areas, the morphological differences between the two forms of *Homo* are at their lowest. Further substantiation for this overall hypothesis is provided by the chronological position of Heinrich event 3 at around 30 ka. Seeing as the transition from Aurignacian to Gravettian occurred at this time, this Heinrich event might also have triggered similar demographic processes resulting in cultural change. As is the case with the aforementioned Middle Palaeolithic and Aurignacian complexes, stratigraphic sequences from archaeological sites fail to yield interstratification of Aurignacian and Gravettian assemblages.

The Heinrich events have the potential to work as climate specific catalysts, and as such, they have had a decisive influence upon settlement processes of Palaeolithic hunter-gatherer groups in Europe. Indeed, should this 'Repeated Replacement Model' be corroborated by further data in the future, then it might be concluded that the extinction of the Neanderthals was an altogether unspectacular incident – just another of numerous 'natural' exchange processes of European Ice Age populations. We have to keep in mind that human adaptation to cold climate conditions was a still ongoing process in European hunter-gatherer communities throughout the whole Late Pleistocene. Even during the LGM around 20 ka, the cultural ability of modern humans was not sufficiently developed to resist the harsh environmental conditions prevailing in Central Europe. Humans were forced to abandon the area for at least 1,000 years or even longer. Human cultural adaptation to arctic environment developed only later, in the early Holocene. Finally, the most spectacular thing about Neanderthal disappearance might actually lie in the seemingly unspectacular nature of the processes involved.

Index

S. Condemi and G.-C. Weniger (eds.), *Continuity and Discontinuity in the Peopling of Europe: One Hundred Fifty Years of Neanderthal Study*, Vertebrate Paleobiology and Paleoanthropology, DOI 10.1007/978-94-007-0492-3, © Springer Science+Business Media B.V. 2011

E

F

G

H

I

J

K

L

M

N

O

P

Q

R

9789400704916